AF841859

Dear Bob,

With this book I would like to thank you very much for the support of the IAPWS project "New Industrial Formulation IAPWS-IF97" by your part in the evaluation process.

Thank you, too, for your encouraging words particularly when we ran into difficult problems during this project.

Wolfgang Wagner

Bochum, August 4, 1998

Wolfgang Wagner • Alfred Kruse

Properties of Water and Steam
Zustandsgrößen von Wasser und Wasserdampf

Springer
Berlin
Heidelberg
New York
Barcelona
Budapest
Hongkong
London
Mailand
Paris
Santa Clara
Singapur
Tokio

Wolfgang Wagner • Alfred Kruse

Properties of Water and Steam

The Industrial Standard IAPWS-IF97
for the Thermodynamic Properties
and Supplementary Equations for Other Properties

Tables based on these Equations

Zustandsgrößen von Wasser und Wasserdampf

Der Industrie-Standard IAPWS-IF97
für die thermodynamischen Zustandsgrößen
und ergänzende Gleichungen für andere Eigenschaften

Tafeln auf der Grundlage dieser Gleichungen

Springer

Professor Dr.-Ing. Wolfgang Wagner
Dr.-Ing. Alfred Kruse

Lehrstuhl für Thermodynamik
Fakultät für Maschinenbau
Ruhr-Universität Bochum
D-44780 Bochum

ISBN 3-540-64339-7 Springer-Verlag Berlin Heidelberg New York

Cataloging-in-Publishing Data applied for

Die deutsche Bibliothek - CIP-Einheitsaufnahme

Wagner, Wolfgang:
Properties of water and steam: the industrial standard IAPWS-IF97 for the thermodynamic properties and supplementary equations for other properties ; tables based on these equations = Zustandsgrößen von Wasser und Wasserdampf / Wolfgang Wagner. Unter Mitarb. von Alfred Kruse. - Berlin ; Heidelberg ; New York ; Barcelona ; Budapest ; Hongkong ; London ; Mailand ; Paris ; Santa Clara ; Singapur ; Tokio:
Springer, 1998
ISBN 3-540-64339-7

Satz: Reproduktionsfertige Vorlage der Autoren
Umschlaggestaltung: de´blik, Berlin

SPIN: 10676269 60/3020 - 5 4 3 2 1 0 - Gedruckt auf säurefreiem Papier

Preface

In 1997, the International Association for the Properties of Water and Steam (IAPWS) adopted a new formulation for the thermodynamic properties of water and steam for industrial use. This formulation is called "IAPWS Industrial Formulation 1997 for the Thermodynamic Properties of Water and Steam" and "IAPWS Industrial Formulation 1997" or "IAPWS-IF97" for short. The new formulation IAPWS-IF97 replaces the previous industrial formulation, IFC-67, that has formed the basis for power-plant calculations and other industrial applications since the late 1960's. IAPWS-IF97 improves significantly both the accuracy and the speed of the calculation of thermodynamic properties. The differences from IFC-67 will require many users, particularly boiler and turbine manufacturers but also power-station companies and corresponding engineering offices, to modify design and application codes. In addition to these applications, IAPWS-IF97 is also of importance for energy-engineering applications in chemical industry and in other branches of industry. Therefore, this book presents the individual equations of IAPWS-IF97 for calculating the thermodynamic properties of water and steam for industrial use.

The IAPWS Industrial Formulation 1997 was developed in an international research project. This development was coordinated by the IAPWS Task Group "New Industrial Formulation" chaired by one of the authors of this book (W. W.). The final form of IAPWS-IF97 is based on contributions and equations by

J. R. Cooper	R. Mareš	Y. Takaishi
A. Dittmann	K. Oguchi	I. Tanishita
J. Kijima	H. Sato	J. Trübenbach
H.-J. Kretzschmar	I. Stöcker	W. Wagner
A. Kruse	O. Šifner	Th. Willkommen.

Besides these "developers" many other IAPWS colleagues, particularly the members of the two working groups "Industrial Calculations" and "Thermophysical Properties of Water and Steam", contributed to the entire success of this comprehensive project; we appreciate their contribution very much. We are especially grateful to the chairmen of these two working groups, B. Rukes and J. R. Cooper. In addition, we would like to thank the members of the IAPWS Task Group "New Industrial Formulation - Evaluation" for testing IAPWS-IF97 regarding the fulfilment of requirements and checking the influence on real power-cycle calculations; concerning these important pieces of work we are particularly grateful to the chairman of this task group, K. Miyagawa, and his colleagues R. Spencer, R. B. McClintock, and P. Bradley for their exceptional efforts.

In addition to IAPWS-IF97, the industrial standard for the thermodynamic properties of water and steam, the most recent equations for the transport properties dynamic viscosity and thermal conductivity are also presented. Moreover, equations for the surface tension, static dielectric constant, and refractive index are given.

The text of this book is bilingual. Part A contains the description of the above mentioned equations for the thermophysical properties in English and Part B the corresponding description in German. Comprehensive tables of the most important thermophysical properties of water and steam are given in Part C in both languages.

The values in the tables of Part C were exclusively calculated from the corresponding equations summarized in Part A and Part B, respectively. These tables, which are mainly based on the new industrial formulation IAPWS-IF97, replace the tables "Properties of Water and Steam in SI-Units" prepared by E. Schmidt and edited by U. Grigull (Springer-Verlag Berlin Heidelberg New York, R. Oldenbourg München, Fourth, Enlarged Printing, 1989) which are based on the previous industrial formulation IFC-67.

We wish to express our warmest thanks to Mr. C. Bonsen for his help in handling the computer programs for calculating the transport properties and for producing all the tables. We would also like to thank Mrs. A.-M. Sieg for typing the text of the manuscript. We are particularly grateful to the Deutsche Forschungsgemeinschaft for their financial support of that part of the development of IAPWS-IF97 which was carried out at the Ruhr-University Bochum.

Bochum, February 1998

W. Wagner
A. Kruse

Vorwort

Die „International Association for the Properties of Water and Steam" (IAPWS) hat 1997 eine neue Industrie-Formulation für die Berechnung der thermodynamischen Eigenschaften von Wasser und Wasserdampf verabschiedet. Der Name der neuen Formulation lautet „IAPWS Industrial Formulation 1997 for the Thermodynamic Properties of Water and Steam" mit den Kurzbezeichnungen „IAPWS Industrial Formulation 1997" bzw. „IAPWS-IF97". Die neue Formulation IAPWS-IF97 ersetzt die bisherige Industrie-Formulation IFC-67, die seit 1967 als Grundlage für die Berechnung von Kraftwerksprozessen sowie für weitere industrielle Anwendungen diente. Gegenüber der IFC-67 werden Genauigkeit und Rechengeschwindigkeit durch die neue Formulation deutlich erhöht. Aufgrund der signifikanten Unterschiede zwischen den beiden Formulationen werden viele Anwender, insbesondere Kessel- und Turbinenhersteller, aber auch Kraftwerksbetreiber und entsprechende Ingenieurbüros, ihre Auslegungs- und Anwendungsprogramme modifizieren müssen. Darüber hinaus ist die Kenntnis der IAPWS-IF97 auch für energietechnische Anwendungen in der chemischen Industrie und in anderen Industriezweigen von Bedeutung. Vor diesem Hintergrund werden in diesem Buch die einzelnen Gleichungen der IAPWS-IF97 für die industrielle Berechnung der thermodynamischen Eigenschaften von Wasser und Wasserdampf beschrieben.

Die IAPWS Industrie-Formulation 1997 wurde im Rahmen eines internationalen Forschungsprojektes entwickelt. Koordiniert wurden diese Arbeiten von der für diesen Zweck gegründeten IAPWS-Arbeitsgruppe „New Industrial Formulation", die von einem der Autoren dieses Buches (W. W.) geleitet wurde. Die endgültige Form der IAPWS-IF97 beruht auf Beiträgen und Gleichungen von

J. R. Cooper	R. Mareš	Y. Takaishi
A. Dittmann	K. Oguchi	I. Tanishita
J. Kijima	H. Sato	J. Trübenbach
H.-J. Kretzschmar	I. Stöcker	W. Wagner
A. Kruse	O. Šifner	Th. Willkommen.

Neben diesen „Entwicklern" haben auch viele andere Kollegen der IAPWS, insbesondere die Mitglieder der beiden Arbeitsgruppen „Industrial Calculations" und „Thermophysical Properties of Water and Steam", zum Gesamterfolg dieses umfassenden Projektes beigetragen, wofür wir uns ganz herzlich bedanken. Besonders dankbar sind wir den Vorsitzenden dieser beiden Arbeitsgruppen, B. Rukes und J. R. Cooper. Darüber hinaus bedanken wir uns bei den Mitgliedern der IAPWS-Arbeitsgruppe „New Industrial Formulation - Evaluation", die speziell für die Prüfung der IAPWS-IF97 bezüglich der Erfüllung aller Anforderungen und für die Untersuchung des Einflusses der neuen Formulation auf die Berechnung realer Kraftwerksprozesse gebildet worden war. Für diese wichtigen Arbeiten danken wir insbesondere dem Vorsitzenden dieser Arbeitsgruppe, K. Miyagawa, sowie seinen Kollegen R. Spencer, R. B. McClintock und P. Bradley.

Zusätzlich zur IAPWS-IF97, dem neuen Industrie-Standard für die thermodynamischen Eigenschaften von Wasser und Wasserdampf, geben wir auch die aktuellen Gleichungen für die Transportgrößen dynamische Viskosität und Wärmeleitfähigkeit sowie für die Größen Oberflächenspannung, statische Dielektrizitätskonstante und Brechungsindex an.

Der Text des Buches ist zweisprachig. In Teil A erfolgt die Beschreibung der erwähnten Gleichungen für die thermophysikalischen Eigenschaften in englisch, und der Teil B enthält die entsprechende Beschreibung in deutsch. Die umfassenden Tafeln der wichtigsten thermophysikalischen Eigenschaften von Wasser und Wasserdampf sind im zweisprachigen Teil C zusammengefaßt.

Die Werte in den Tafeln von Teil C wurden ausschließlich aus den entsprechenden Gleichungen berechnet, die in Teil A bzw. Teil B aufgeführt sind. Diese Tafeln, die hauptsächlich auf dem neuen Industrie-Standard IAPWS-IF97 beruhen, ersetzen das Tafelwerk „Zustandsgrößen von Wasser und Wasserdampf in SI-Einheiten" bearbeitet von E. Schmidt, herausgegeben von U. Grigull (Springer-Verlag Berlin Heidelberg New York, R. Oldenbourg München, 4. Auflage, 1989), das auf dem bisherigen Industrie-Standard IFC-67 beruht.

Wir bedanken uns bei Herrn C. Bonsen für seine Hilfe beim Erstellen der Computer-Programme für die Berechnung der Transportgrößen und die Erstellung der Tafeln. Frau A.-M. Sieg sind wir dankbar für das Schreiben des druckfertigen Manuskriptes. Unser besonderer Dank gilt der Deutschen Forschungsgemeinschaft für die finanzielle Unterstützung des Teils der Entwicklungsarbeiten für die IAPWS-IF97, der an der Ruhr-Universität Bochum durchgeführt worden ist.

Bochum, im Februar 1998

W. Wagner
A. Kruse

Contents
Inhaltsverzeichnis

Nomenclature

Quantities

a	Coefficient
c_p	Specific isobaric heat capacity
c_v	Specific isochoric heat capacity
f	Specific Helmholtz free energy, $f = u - Ts$
g	Specific Gibbs free energy, $g = h - Ts$; Harris-Alder g-factor
h	Specific enthalpy
Δh_v	Specific enthalpy of evaporation, $\Delta h_v = h'' - h'$
I	Exponent
i	Serial number; Exponent
J	Exponent
j	Serial number; Exponent
k	Boltzmann's constant
M	Molar mass
N_A	Avogadro's number
n	Coefficient; refractive index
Pr	Prandtl number, $Pr = \eta c_p \lambda^{-1}$
p	Pressure
R	Specific gas constant
R_m	Molar gas constant
s	Specific entropy
Δs_v	Specific entropy of evaporation, $\Delta s_v = s'' - s'$
T	Thermodynamic temperature[1]
t	Celsius temperature, t [°C] = T [K] – 273.15 K
u	Specific internal energy
v	Specific volume
w	Speed of sound
α	Mean molecular polarizability of the isolated water molecule
β	Transformed pressure, Eq. (3.22a)
γ	Dimensionless Gibbs free energy, $\gamma = g/(RT)$
Δ	Difference in any quantity
δ	Reduced density, $\delta = \rho/\rho^*$
ε	Relative static dielectric constant (relative static permittivity)
ε_o	Permittivity of vacuum (electric constant)

[1] All temperature values given in this book are temperatures according to the Internationa Temperature Scale of 1990 (ITS-90)

η	Reduced enthalpy, $\eta = h/h^*$; dynamic viscosity
θ	Reduced temperature, $\theta = T/T^*$
ϑ	Transformed temperature, Eq. (3.22b)
κ	Isentropic exponent, $\kappa = -vp^{-1}(\partial p/\partial v)_s$
Λ	Reduced thermal conductivity, $\Lambda = \lambda/\lambda^*$; reduced wavelength of light, $\Lambda = \lambda/\lambda^*$
λ	Thermal conductivity; wavelength of light
μ	Dipole moment of the isolated water molecule
ν	Kinematic viscosity, $\nu = \eta\rho^{-1}$
π	Reduced pressure, $\pi = p/p^*$
ρ	Mass density
σ	Reduced entropy, $\sigma = s/s^*$; surface tension
τ	Inverse reduced temperature, $\tau = T^*/T$
ϕ	Dimensionless Helmholtz free energy, $\phi = f/(RT)$
Ψ	Reduced dynamic viscosity, $\Psi = \eta/\eta^*$

Superscripts

o	Ideal-gas part; ideal gas
r	Residual part
*	Reducing quantity
′	Saturated liquid state
″	Saturated vapour state

Subscripts

b	Normal boiling point
c	Critical point
max	Maximum value
s	Saturation state
t	Triple point

Formelzeichen

Größen

a	Koeffizient
c_p	spezifische isobare Wärmekapazität
c_v	spezifische isochore Wärmekapazität
f	spezifische freie Energie (Helmholtz-Funktion), $f = u - Ts$
g	spezifische freie Enthalpie (Gibbs-Funktion), $g = h - Ts$; Harris-Alder g-Faktor
h	spezifische Enthalpie
Δh_v	spezifische Verdampfungsenthalpie, $\Delta h_v = h'' - h'$
I	Exponent
i	Laufindex ; Exponent
J	Exponent
j	Laufindex ; Exponent
k	Boltzmann-Konstante
M	Molmasse
N_A	Avogadro-Konstante
n	Koeffizient ; Brechungsindex
Pr	Prandtl-Zahl, $Pr = \eta c_p \lambda^{-1}$
p	Druck
R	spezifische Gaskonstante
R_m	molare (universelle) Gaskonstante
s	spezifische Entropie
Δs_v	spezifische Verdampfungsentropie, $\Delta s_v = s'' - s'$
T	thermodynamische Temperatur[1]
t	Celsius-Temperatur, t [°C] = T [K] – 273,15 K
u	spezifische innere Energie
v	spezifisches Volumen
w	Schallgeschwindigkeit
α	mittlere molekulare Polarisierbarkeit eines einzelnen Wassermoleküls
β	transformierter Druck, Gl. (3.22a)
γ	dimensionslose freie Enthalpie, $\gamma = g/(RT)$
Δ	Differenz einer Größe
δ	reduzierte Dichte, $\delta = \rho / \rho^*$
ε	relative statische Dielektrizitätskonstante (relative statische Permittivität)
ε_0	elektrische Feldkonstante (Permittivität des Vakuums)

[1] Alle in diesem Buch angegebenen Temperaturwerte sind Temperaturen nach der Internationalen Temperaturskala von 1990 (ITS-90).

η	reduzierte Enthalpie, $\eta = h/h^*$; dynamische Viskosität
θ	reduzierte Temperatur, $\theta = T/T^*$
ϑ	transformierte Temperatur, Gl. (3.22b)
κ	Isentropenexponent, $\kappa = -vp^{-1}(\partial p/\partial v)_s$
Λ	reduzierte Wärmeleitfähigkeit, $\Lambda = \lambda/\lambda^*$; reduzierte Wellenlänge des Lichtes, $\Lambda = \lambda/\lambda^*$
λ	Wärmeleitfähigkeit; Wellenlänge des Lichtes
μ	Dipolmoment eines einzelnen Wassermoleküls
ν	kinematische Viskosität, $\nu = \eta\rho^{-1}$
π	reduzierter Druck, $\pi = p/p^*$
ρ	Dichte
σ	reduzierte Entropie, $\sigma = s/s^*$; Oberflächenspannung
τ	inverse reduzierte Temperatur, $\tau = T^*/T$
ϕ	dimensionslose freie Energie, $\phi = f/(RT)$
Ψ	reduzierte dynamische Viskosität, $\Psi = \eta/\eta^*$

Hochgestellte Indizes

o	Anteil des idealen Gases; ideales Gas
r	residueller Anteil
*	Reduktionsgröße
′	siedende Flüssigkeit
″	gesättigter Dampf

Tiefgestellte Indizes

b	normaler Siedepunkt
c	kritischer Punkt
max	Maximalwert
s	Sättigungszustand (Siedezustand)
t	Tripelpunkt

Part A

Equations for the Calculation of the Thermophysical Properties of Water and Steam

1 Introduction

Part A of this book presents the current internationally agreed equations for industrial calculations of the most relevant thermophysical properties of water and steam.

All equations for calculating *thermodynamic* properties are covered by the new industrial standard "IAPWS Industrial Formulation 1997 for the Thermodynamic Properties of Water and Steam" (IAPWS-IF97). This new formulation was adopted by the International Association for the Properties of Water and Steam (IAPWS) at its annual meeting in Erlangen, Germany, in 1997. The new formulation, IAPWS-IF97, replaces the previous industrial standard IFC-67 [1]. The new industrial standard IAPWS-IF97 is comprehensively described in Chap. 3.

In addition to the equations for the thermodynamic properties of water and steam, Chap. 4 summarizes current equations for the *transport* properties dynamic viscosity and thermal conductivity and also presents correlation equations for the surface tension, static dielectric constant, and refractive index.

The tables of the most important properties of water and steam given in Part C of this book were exclusively calculated from the corresponding equations of Chaps. 3 and 4 in Part A.

2 Reference Constants

This chapter summarizes all reference constants needed for evaluating the equations given in Chaps. 3 and 4.

The specific gas constant of ordinary water,

$$R = 0.461\,526\ \mathrm{kJ\ kg^{-1}\ K^{-1}}, \tag{2.1}$$

results from the recommended values of the molar gas constant [2],

$$R_\mathrm{m} = 8.314\,51\ \mathrm{J\ mol^{-1}\ K^{-1}}, \tag{2.2}$$

and from the molar mass of ordinary water,

$$M = 18.015\,257\ \mathrm{g\ mol^{-1}}. \tag{2.3}$$

The value of the molar mass of ordinary water results from the molar mass of hydrogen, $M_\mathrm{H} = 1.007\,975\,97\ \mathrm{g\ mol^{-1}}$ (based on the molar mass of the isotopes ^{1}H and ^{2}H given in [3] and the isentopic concentration corresponding to the molar fraction of ^{1}H equal to 0.999 85 and of ^{2}H equal to 0.000 15 [4]), and the molar mass of oxygen, $M_\mathrm{O} = 15.999\,3047\ \mathrm{g\ mol^{-1}}$ (based on the molar mass of the isotopes ^{16}O, ^{17}O, and ^{18}O given in [3] and the isotopic concentrations corresponding to the molar fractions of ^{16}O, ^{17}O, and ^{18}O equal to 0.997 62, 0.000 38, and 0.002, respectively, considered to be characteristic for all natural occurrences of oxygen [4]).

The values of the critical parameters

$$T_\mathrm{c} = 647.096\ \mathrm{K}, \tag{2.4}$$

$$p_\mathrm{c} = 22.064\ \mathrm{MPa}, \tag{2.5}$$

$$\rho_\mathrm{c} = 322\ \mathrm{kg\ m^{-3}} \tag{2.6}$$

are from the corresponding IAPWS release [5]. The triple-point temperature is

$$T_\mathrm{t} = 273.16\ \mathrm{K} \tag{2.7}$$

according to the International Temperature Scale of 1990 (ITS-90) [6] and the triple-point pressure

$$p_\mathrm{t} = 611.657\ \mathrm{Pa} \tag{2.8}$$

was determined by Guildner et al. [7]. According to the scientific standard of the thermodynamic properties of ordinary water, the IAPWS-95 formulation [8], the temperature of the normal boiling point (at a pressure of 0.101 325 MPa (1 atm)) amounts to

$$T_\mathrm{b} = 373.1243\ \mathrm{K}. \tag{2.9}$$

3 IAPWS Industrial Formulation 1997 for the Thermodynamic Properties of Water and Steam

In the 1960's an industrial formulation for the thermodynamic properties of water and steam was developed called "The 1967 IFC Formulation for Industrial Use" (IFC-67) [1]. Since 1967 IFC-67 has been formally recognized to calculate thermodynamic properties of water and steam for any official use such as performance guarantee calculations of power cycles. In addition to this, IFC-67 has been used for innumerable other industrial applications. However, during the last few years a number of weaknesses of IFC-67 have appeared. This fact and the progress that has been achieved in mathematical methods to develop accurate equations of state led to the development of a new industrial formulation in an international research project initiated and coordinated by the International Association for the Properties of Water and Steam (IAPWS).

At the IAPWS meeting in Erlangen, Germany, in 1997, this formulation was adopted as the new international industrial standard for the thermodynamic properties of water and steam. The name of the new industrial standard is "IAPWS Industrial Formulation 1997 for the Thermodynamic Properties of Water and Steam" and "IAPWS Industrial Formulation 1997" or "IAPWS-IF97" for short. The new formulation IAPWS-IF97 replaces the previous industrial standard IFC-67 [1]. In comparison with IFC-67, IAPWS-IF97 improves significantly both the accuracy and the speed of the calculation of thermodynamic properties.

The project to develop a new industrial formulation was initiated and continuously supervised by the IAPWS Working Group "Industrial Calculations" of which many members are representatives of international power-cycle companies. This working group, chaired by B. Rukes (Germany), is the successor to the IAPWS "Subcommittee on Industrial Calculations". The official starting point for this comprehensive project was given at the IAPWS meeting in Buenos Aires in 1990 when the Task Group "New Industrial Formulation" was set up. This task group, which was chaired by one of the authors of this book (W. W.), consisted of 12 members from 7 countries and was directly responsible for the development of IAPWS-IF97. The Task Group "New Industrial Formulation - Evaluation", chaired by K. Miyagawa (Japan), accomplished very important work by testing the formulation regarding the fulfilment of requirements and checking the influence on real power-cycle calculations.

This chapter presents the entire information about the individual equations of IAPWS-IF97 which is necessary for calculating the thermodynamic properties of water and steam. Detailed information about the development of the single equations and their accuracy, the achieved consistency along the region boundaries, and details of the computation speed in comparison with the previous industrial standard IFC-67 can be found in the international publication on IAPWS-IF97 [9].

3.1 Characteristic Features of IAPWS-IF97

This section gives general information about the structure of the new formulation including the entire range of validity and makes some general statements about the quality of IAPWS-IF97 concerning accuracy, consistency along the region boundaries, and computation speed compared with the previous industrial standard IFC-67.

3.1.1 Structure of IAPWS-IF97

The IAPWS Industrial Formulation 1997 consists of a set of equations for different regions which cover the following range of validity:

$$273.15\ \mathrm{K} \le T \le 1073.15\ \mathrm{K} \qquad p \le 100\ \mathrm{MPa}$$
$$1073.15\ \mathrm{K} < T \le 2273.15\ \mathrm{K} \qquad p \le 10\ \mathrm{MPa}\,.$$

Figure 3.1 shows the 5 regions into which the entire range of validity of IAPWS-IF97 is divided. The boundaries of the regions can be directly taken from Fig. 3.1 except for the boundary between regions 2 and 3; this boundary is defined by the so-called B23-equation given in Sec. 3.3.1. Both regions 1 and 2 are individually covered by a fundamental equation for the specific Gibbs free energy $g(p,T)$, region 3 by a fundamental equation for the specific Helmholtz free energy $f(\rho,T)$, and the saturation curve, corresponding to region 4, by a saturation-pressure equation $p_s(T)$. The high-temperature region 5 is also covered by a $g(p,T)$ equation. These 5 equations, shown in rectangular boxes in Fig. 3.1, form the so-called *basic equations*.

The industrial standard IAPWS-IF97 has been coupled to the scientific standard, the "IAPWS Formulation 1995 for the Thermodynamic Properties of Ordinary Water Substance for General and Scientific Use" [8], hereafter abbreviated to IAPWS-95. This coupling was achieved by fitting the basic equations of regions 1 to 3 and 5 to values of the specific

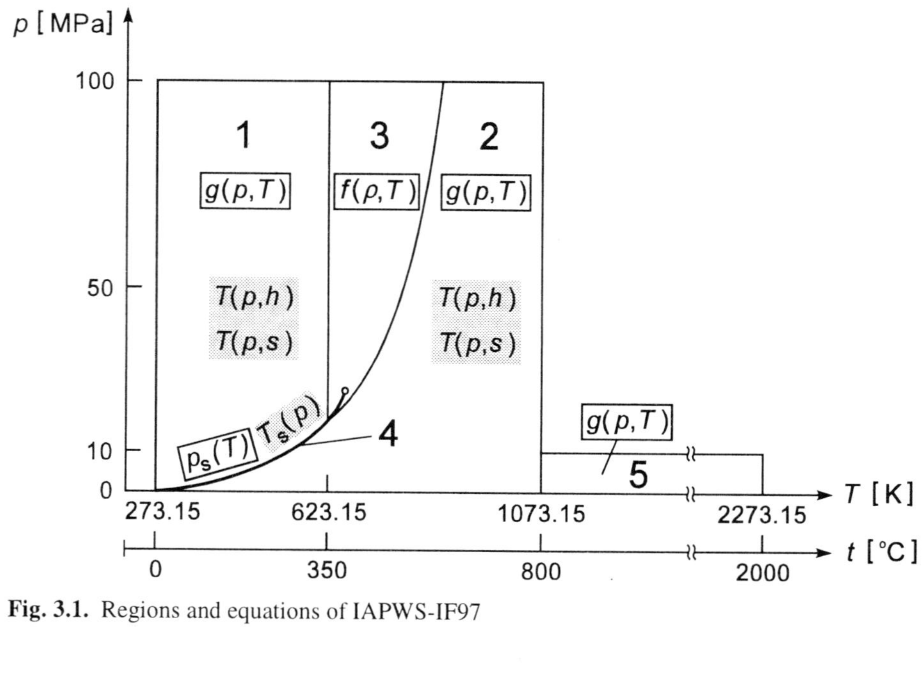

Fig. 3.1. Regions and equations of IAPWS-IF97

volume v, specific enthalpy h, specific isobaric heat capacity c_p, and speed of sound w calculated from IAPWS-95 [8]. Accordingly, the basic equation of region 4, the saturation-pressure equation, was fitted to values of the saturation pressure p_s from IAPWS-95.

In addition to these basic equations, so-called *backward equations* are provided for regions 1, 2, and 4 in form of $T(p,h)$ and $T(p,s)$ for regions 1 and 2, and $T_s(p)$ for region 4. These backward equations, marked in grey in Fig. 3.1, were developed in such a way that they are numerically very consistent with the corresponding basic equation. Thus, properties as functions of p,h and of p,s for regions 1 and 2, and of p for region 4 can be calculated without any iteration. Consequently, properties such as $T(p,h)$, $h(p,s)$, and $h'(p)$ can be calculated directly from the backward equation or by combining with the corresponding basic equation, for example, $h(p,s)$ via the relation $h\big(p,T(p,s)\big)$. As a result of this special concept for the development of the new industrial standard IAPWS-IF97, the most important properties can be calculated extremely quickly, see next section.

3.1.2 Quality of IAPWS-IF97

The achieved overall quality of the industrial formulation IAPWS-IF97 is characterized by the following general results regarding the 3 criterions accuracy, consistency along the region boundaries, and computation speed.

The accuracy of IAPWS-IF97 is illustrated by the fact that for its entire range of validity only 0.2 % of the calculated v values, 6 % of the c_p values, 2 % of the w values, and no p_s values are outside the uncertainty of the corresponding IAPWS-95 values [3]. When carrying out the same test with the previous industrial standard IFC-67, then between 47 % (for v) and 80 % (for p_s) of the IFC-67 values are outside the uncertainty of the corresponding IAPWS-95 values. Based on all comparisons made [9] it can be concluded that IAPWS-IF97 is more than one order of magnitude more accurate than IFC-67. The estimated uncertainties of IAPWS-IF97 regarding the properties v, c_p, w, and p_s in the entire range of validity are given in Sec. 3.7. Besides the representation of the properties for the stable homogeneous regions and at saturation, the corresponding IAPWS-IF97 equations also yield reasonable values for both the metastable superheated-liquid region and the metastable subcooled-vapour region close to the saturated liquid line and the saturated vapour line, respectively.

Compared with IFC-67, an additional important jump in quality was achieved by the fact that IAPWS-IF97 meets clearly the requirements regarding the consistencies along the region boundaries, see Fig. 3.1. IAPWS-IF97 remains clearly within the permitted inconsistencies according to the so-called Prague values [10] which have the following values:

a) Single phase

Specific volume:	± 0.05 %
Enthalpy:	± 0.2 kJ kg^{-1}
Heat capacity:	± 1 %
Entropy:	± 0.0002 kJ kg^{-1} K^{-1}
Gibbs free energy:	± 0.2 kJ kg^{-1}
Speed of sound:	± 1 %

b) Saturation

Saturation pressure:	± 0.05 %
Saturation temperature:	± 0.02 %
Gibbs free energy:	± 0.2 kJ kg^{-1}.

The region boundary for which the fulfilment of the consistency requirement is most difficult is the boundary between regions 2 and 3 and here particularly with respect to the isobaric heat capacity. For IAPWS-IF97, the maximum inconsistency in c_p at this boundary amounts to 0.35 % whereas the corresponding IFC-67 inconsistency is greater than 6 %.

The third and probably the greatest advantage is the very large improvement in the computation speed compared with IFC-67. For the most important regions 1, 2, and 4 altogether, where the computing time is particularly relevant, the computation-speed factor IAPWS-IF97 / IFC-67 amounts to 5.1. This value was determined by taking into account the frequencies of use of the most relevant property functions in these regions based on a survey of the international power-cycle companies and related industries. This means that for these important regions IAPWS-IF97 is more than 5 times faster than IFC-67 provided the individual equations are properly programmed. For region 3, IAPWS-IF97 is 3.6 times faster than IFC-67 and for region 5 even 12.2 times faster.

Details about accuracy, consistency along the region boundaries, and computation speed will be given in the comprehensive article on IAPWS-IF97 [9].

3.2 Equations for Region 1

This section contains all details relevant for using the basic equation and the backward equations of region 1 of IAPWS-IF97. Uncertainty estimates of the most relevant properties, calculated from IAPWS-IF97, can be found in Sec. 3.7.

3.2.1 Basic Equation

The basic equation for this region is a fundamental equation for the specific Gibbs free energy g. This equation is expressed in dimensionless form, $\gamma = g/(RT)$, and reads

$$\frac{g(p,T)}{RT} = \gamma(\pi,\tau) = \sum_{i=1}^{34} n_i (7.1-\pi)^{I_i} (\tau - 1.222)^{J_i}, \tag{3.1}$$

where $\pi = p/p^*$ and $\tau = T^*/T$ with $p^* = 16.53$ MPa and $T^* = 1386$ K; R is given by Eq. (2.1). The coefficients n_i and exponents I_i and J_i of Eq. (3.1) are listed in Table 3.1.

All thermodynamic properties can be derived from Eq. (3.1) by using the appropriate combinations of the dimensionless Gibbs free energy γ and its derivatives. Relations between the relevant thermodynamic properties and γ and its derivatives are summarized in Table 3.2. All required derivatives of the dimensionless Gibbs free energy γ, Eq. (3.1), are explicitly given in Table 3.3.

Since the 5th International Conference on the Properties of Steam in London in 1956, the specific internal energy and the specific entropy of the saturated liquid at the triple point have been set equal to zero:

$$u'_t = 0 \quad ; \quad s'_t = 0 \; . \tag{3.2}$$

In order to meet this condition at the temperature and pressure of the triple point, see Eqs. (2.7) and (2.8), the coefficients n_3 and n_4 in Eq. (3.1) have been adjusted accordingly. As a consequence, Eq. (3.1) yields for the specific enthalpy of the saturated liquid at the triple point

$$h'_t = 0.611\,783 \text{ J kg}^{-1} \; . \tag{3.3}$$

Table 3.1. Coefficients and exponents of Eq. (3.1)

i	I_i	J_i	n_i	i	I_i	J_i	n_i
1	0	−2	0.146 329 712 131 67	18	2	3	−0.441 418 453 308 46 × 10^{-5}
2	0	−1	−0.845 481 871 691 14	19	2	17	−0.726 949 962 975 94 × 10^{-15}
3	0	0	−0.375 636 036 720 40 × 10^{1}	20	3	−4	−0.316 796 448 450 54 × 10^{-4}
4	0	1	0.338 551 691 683 85 × 10^{1}	21	3	0	−0.282 707 979 853 12 × 10^{-5}
5	0	2	−0.957 919 633 878 72	22	3	6	−0.852 051 281 201 03 × 10^{-9}
6	0	3	0.157 720 385 132 28	23	4	−5	−0.224 252 819 080 00 × 10^{-5}
7	0	4	−0.166 164 171 995 01 × 10^{-1}	24	4	−2	−0.651 712 228 956 01 × 10^{-6}
8	0	5	0.812 146 299 835 68 × 10^{-3}	25	4	10	−0.143 417 299 379 24 × 10^{-12}
9	1	−9	0.283 190 801 238 04 × 10^{-3}	26	5	−8	−0.405 169 968 601 17 × 10^{-6}
10	1	−7	−0.607 063 015 658 74 × 10^{-3}	27	8	−11	−0.127 343 017 416 41 × 10^{-8}
11	1	−1	−0.189 900 682 184 19 × 10^{-1}	28	8	−6	−0.174 248 712 306 34 × 10^{-9}
12	1	0	−0.325 297 487 705 05 × 10^{-1}	29	21	−29	−0.687 621 312 955 31 × 10^{-18}
13	1	1	−0.218 417 171 754 14 × 10^{-1}	30	23	−31	0.144 783 078 285 21 × 10^{-19}
14	1	3	−0.528 383 579 699 30 × 10^{-4}	31	29	−38	0.263 357 816 627 95 × 10^{-22}
15	2	−3	−0.471 843 210 732 67 × 10^{-3}	32	30	−39	−0.119 476 226 400 71 × 10^{-22}
16	2	0	−0.300 017 807 930 26 × 10^{-3}	33	31	−40	0.182 280 945 814 04 × 10^{-23}
17	2	1	0.476 613 939 069 87 × 10^{-4}	34	32	−41	−0.935 370 872 924 58 × 10^{-25}

Table 3.2. Relations of thermodynamic properties to the dimensionless Gibbs free energy γ and its derivatives when using Eq. (3.1)

Property	Relation
Specific volume $v = (\partial g/\partial p)_T$	$v(\pi,\tau)\dfrac{p}{RT} = \pi\gamma_\pi$
Specific internal energy $u = g - T(\partial g/\partial T)_p - p(\partial g/\partial p)_T$	$\dfrac{u(\pi,\tau)}{RT} = \tau\gamma_\tau - \pi\gamma_\pi$
Specific entropy $s = -(\partial g/\partial T)_p$	$\dfrac{s(\pi,\tau)}{R} = \tau\gamma_\tau - \gamma$
Specific enthalpy $h = g - T(\partial g/\partial T)_p$	$\dfrac{h(\pi,\tau)}{RT} = \tau\gamma_\tau$
Specific isobaric heat capacity $c_p = (\partial h/\partial T)_p$	$\dfrac{c_p(\pi,\tau)}{R} = -\tau^2\gamma_{\tau\tau}$
Specific isochoric heat capacity $c_v = (\partial u/\partial T)_v$	$\dfrac{c_v(\pi,\tau)}{R} = -\tau^2\gamma_{\tau\tau} + \dfrac{(\gamma_\pi - \tau\gamma_{\pi\tau})^2}{\gamma_{\pi\pi}}$
Speed of sound $w = v\left[-(\partial p/\partial v)_s\right]^{0.5}$	$\dfrac{w^2(\pi,\tau)}{RT} = \dfrac{\gamma_\pi^2}{\dfrac{(\gamma_\pi - \tau\gamma_{\pi\tau})^2}{\tau^2\gamma_{\tau\tau}} - \gamma_{\pi\pi}}$

$$\gamma_\pi = \left(\frac{\partial\gamma}{\partial\pi}\right)_\tau,\quad \gamma_{\pi\pi} = \left(\frac{\partial^2\gamma}{\partial\pi^2}\right)_\tau,\quad \gamma_\tau = \left(\frac{\partial\gamma}{\partial\tau}\right)_\pi,\quad \gamma_{\tau\tau} = \left(\frac{\partial^2\gamma}{\partial\tau^2}\right)_\pi,\quad \gamma_{\pi\tau} = \left(\frac{\partial^2\gamma}{\partial\pi\,\partial\tau}\right)$$

Table 3.3. The dimensionless Gibbs free energy γ, Eq. (3.1), and its derivatives

$\gamma = \sum_{i=1}^{34} n_i (7.1-\pi)^{I_i} (\tau - 1.222)^{J_i}$	$\gamma_\tau = \sum_{i=1}^{34} n_i (7.1-\pi)^{I_i} J_i (\tau - 1.222)^{J_i - 1}$
$\gamma_\pi = \sum_{i=1}^{34} -n_i I_i (7.1-\pi)^{I_i - 1} (\tau - 1.222)^{J_i}$	$\gamma_{\tau\tau} = \sum_{i=1}^{34} n_i (7.1-\pi)^{I_i} J_i (J_i - 1)(\tau - 1.222)^{J_i - 2}$
$\gamma_{\pi\pi} = \sum_{i=1}^{34} n_i I_i (I_i - 1)(7.1-\pi)^{I_i - 2} (\tau - 1.222)^{J_i}$	$\gamma_{\pi\tau} = \sum_{i=1}^{34} -n_i I_i (7.1-\pi)^{I_i - 1} J_i (\tau - 1.222)^{J_i - 1}$

$$\gamma_\pi = \left(\frac{\partial \gamma}{\partial \pi}\right)_\tau, \quad \gamma_{\pi\pi} = \left(\frac{\partial^2 \gamma}{\partial \pi^2}\right)_\tau, \quad \gamma_\tau = \left(\frac{\partial \gamma}{\partial \tau}\right)_\pi, \quad \gamma_{\tau\tau} = \left(\frac{\partial^2 \gamma}{\partial \tau^2}\right)_\pi, \quad \gamma_{\pi\tau} = \left(\frac{\partial^2 \gamma}{\partial \pi \partial \tau}\right)$$

Equation (3.1) covers region 1 of IAPWS-IF97 defined by the following range of temperature and pressure, see Fig. 3.1:

$$273.15 \text{ K} \le T \le 623.15 \text{ K} \qquad p_s(T) \le p \le 100 \text{ MPa} .$$

In addition to the properties in the stable single-phase liquid region, Eq. (3.1) also yields reasonable values in the metastable superheated-liquid region close to the saturated liquid line. For temperatures between 273.15 K and 273.16 K at pressures below the melting pressure [11] (metastable subcooled liquid) all values are calculated by extrapolation from Eqs. (3.1) and (3.23).

To assist the user in computer-program verification of Eq. (3.1), Table 3.4 contains test values of the most relevant properties.

Table 3.4. Thermodynamic property values calculated from Eq. (3.1) for selected temperatures and pressures

Property	$T = 300$ K $p = 3$ MPa	$T = 300$ K $p = 80$ MPa	$T = 500$ K $p = 3$ MPa
v [m^3 kg^{-1}]	$0.100\,215\,168 \times 10^{-2}$	$0.971\,180\,894 \times 10^{-3}$	$0.120\,241\,800 \times 10^{-2}$
h [kJ kg^{-1}]	$0.115\,331\,273 \times 10^{3}$	$0.184\,142\,828 \times 10^{3}$	$0.975\,542\,239 \times 10^{3}$
u [kJ kg^{-1}]	$0.112\,324\,818 \times 10^{3}$	$0.106\,448\,356 \times 10^{3}$	$0.971\,934\,985 \times 10^{3}$
s [kJ kg^{-1} K^{-1}]	$0.392\,294\,792$	$0.368\,563\,852$	$0.258\,041\,912 \times 10^{1}$
c_p [kJ kg^{-1} K^{-1}]	$0.417\,301\,218 \times 10^{1}$	$0.401\,008\,987 \times 10^{1}$	$0.465\,580\,682 \times 10^{1}$
w [m s^{-1}]	$0.150\,773\,921 \times 10^{4}$	$0.163\,469\,054 \times 10^{4}$	$0.124\,071\,337 \times 10^{4}$

3.2.2 Backward Equations

For a sufficiently accurate calculation of properties as function of p,h or of p,s without any iteration (see Sec. 3.1.1), the two backward equations require extremely good numerical consistency with the basic equation. The exact requirements for these numerical consistencies were obtained from comprehensive test calculations for several characteristic power cycles. Based on these investigations, the numerical consistency requirement between the basic equation, Eq. (3.1), and the two backward equations, $T(p,h)$ and $T(p,s)$, corresponding to

Eqs. (3.4) and (3.5), respectively, was set to ± 25 mK. This means, for example, that the temperature value directly determined from the backward equation $T(p,h)$, Eq. (3.4), for given values of p and h must agree within ± 25 mK with the temperature value determined by iteration from the basic equation, Eq. (3.1), for the same p and h. This iteration from the basic equation is carried out via the function $h(p,T)$ of which the relation with the dimensionless Gibbs free energy is given in Table 3.2.

3.2.2.1 The Backward Equation $T(p, h)$

The backward equation $T(p,h)$ for region 1 has the following dimensionless form:

$$\frac{T(p,h)}{T^*} = \theta(\pi,\eta) = \sum_{i=1}^{20} n_i \pi^{I_i} (\eta+1)^{J_i} \;, \tag{3.4}$$

where $\theta = T/T^*$, $\pi = p/p^*$, and $\eta = h/h^*$ with $T^* = 1\ \mathrm{K}$, $p^* = 1\ \mathrm{MPa}$, and $h^* = 2500\ \mathrm{kJ\ kg^{-1}}$. The coefficients n_i and exponents I_i and J_i of Eq. (3.4) are listed in Table 3.5.

Table 3.5. Coefficients and exponents of Eq. (3.4)

i	I_i	J_i	n_i	i	I_i	J_i	n_i
1	0	0	$-0.238\,724\,899\,245\,21 \times 10^{3}$	11	1	4	$-0.659\,647\,494\,236\,38 \times 10^{1}$
2	0	1	$0.404\,211\,886\,379\,45 \times 10^{3}$	12	1	10	$0.939\,654\,008\,783\,63 \times 10^{-2}$
3	0	2	$0.113\,497\,468\,817\,18 \times 10^{3}$	13	1	32	$0.115\,736\,475\,053\,40 \times 10^{-6}$
4	0	6	$-0.584\,576\,160\,480\,39 \times 10^{1}$	14	2	10	$-0.258\,586\,412\,820\,73 \times 10^{-4}$
5	0	22	$-0.152\,854\,824\,131\,40 \times 10^{-3}$	15	2	32	$-0.406\,443\,630\,847\,99 \times 10^{-8}$
6	0	32	$-0.108\,667\,076\,953\,77 \times 10^{-5}$	16	3	10	$0.664\,561\,861\,916\,35 \times 10^{-7}$
7	1	0	$-0.133\,917\,448\,726\,02 \times 10^{2}$	17	3	32	$0.806\,707\,341\,030\,27 \times 10^{-10}$
8	1	1	$0.432\,110\,391\,835\,59 \times 10^{2}$	18	4	32	$-0.934\,777\,712\,139\,47 \times 10^{-12}$
9	1	2	$-0.540\,100\,671\,705\,06 \times 10^{2}$	19	5	32	$0.582\,654\,420\,206\,01 \times 10^{-14}$
10	1	3	$0.305\,358\,922\,039\,16 \times 10^{2}$	20	6	32	$-0.150\,201\,859\,535\,03 \times 10^{-16}$

Equation (3.4) covers the same range of validity as the basic equation, Eq. (3.1), except for the metastable superheated-liquid region, where Eq. (3.4) is not valid. The actual numerical inconsistencies between Eqs. (3.4) and (3.1) amount maximally to 23.6 mK and on average (root-mean-square value) to 13.4 mK.

To assist the user in computer-program verification of Eq. (3.4), Table 3.6 contains the corresponding test values.

Table 3.6. Temperature values calculated from Eq. (3.4) for selected pressures and enthalpies

p [MPa]	h [kJ kg^{-1}]	T [K]
3	500	$0.391\,798\,509 \times 10^{3}$
80	500	$0.378\,108\,626 \times 10^{3}$
80	1500	$0.611\,041\,229 \times 10^{3}$

3.2.2.2
The Backward Equation T(p, s)

The backward equation $T(p,s)$ for region 1 has the following dimensionless form:

$$\frac{T(p,s)}{T^*} = \theta(\pi,\sigma) = \sum_{i=1}^{20} n_i\, \pi^{I_i} (\sigma + 2)^{J_i}\,, \qquad (3.5)$$

where $\theta = T/T^*$, $\pi = p/p^*$, and $\sigma = s/s^*$ with $T^* = 1\ \mathrm{K}$, $p^* = 1\ \mathrm{MPa}$, and $s^* = 1\ \mathrm{kJ\ kg^{-1}\ K^{-1}}$. The coefficients n_i and exponents I_i and J_i of Eq. (3.5) are listed in Table 3.7.

Equation (3.5) covers the same range of validity as the basic equation, Eq. (3.1), except for the metastable superheated-liquid region, where Eq. (3.5) is not valid. The actual numerical inconsistencies between Eqs. (3.5) and (3.1) amount maximally to 21.8 mK and on average (root-mean-square value) to 5.8 mK.

To assist the user in computer-program verification of Eq. (3.5), Table 3.8 contains the corresponding test values.

Table 3.7. Coefficients and exponents of Eq. (3.5)

i	I_i	J_i	n_i	i	I_i	J_i	n_i
1	0	0	$0.174\,782\,680\,583\,07 \times 10^{3}$	11	1	12	$0.356\,721\,106\,073\,66 \times 10^{-9}$
2	0	1	$0.348\,069\,308\,928\,73 \times 10^{2}$	12	1	31	$0.173\,324\,969\,948\,95 \times 10^{-23}$
3	0	2	$0.652\,925\,849\,784\,55 \times 10^{1}$	13	2	0	$0.566\,089\,006\,548\,37 \times 10^{-3}$
4	0	3	$0.330\,399\,817\,754\,89$	14	2	1	$-0.326\,354\,831\,397\,17 \times 10^{-3}$
5	0	11	$-0.192\,813\,829\,231\,96 \times 10^{-6}$	15	2	2	$0.447\,782\,866\,906\,32 \times 10^{-4}$
6	0	31	$-0.249\,091\,972\,445\,73 \times 10^{-22}$	16	2	9	$-0.513\,221\,569\,085\,07 \times 10^{-9}$
7	1	0	$-0.261\,076\,364\,893\,32$	17	2	31	$-0.425\,226\,570\,422\,07 \times 10^{-25}$
8	1	1	$0.225\,929\,659\,815\,86$	18	3	10	$0.264\,004\,413\,606\,89 \times 10^{-12}$
9	1	2	$-0.642\,564\,633\,952\,26 \times 10^{-1}$	19	3	32	$0.781\,246\,004\,597\,23 \times 10^{-28}$
10	1	3	$0.788\,762\,892\,705\,26 \times 10^{-2}$	20	4	32	$-0.307\,321\,999\,036\,68 \times 10^{-30}$

Table 3.8. Temperature values calculated from Eq. (3.5) for selected pressures and entropies

p [MPa]	s [kJ kg^{-1} K^{-1}]	T [K]
3	0.5	$0.307\,842\,258 \times 10^{3}$
80	0.5	$0.309\,979\,785 \times 10^{3}$
80	3	$0.565\,899\,909 \times 10^{3}$

3.3
Equations for Region 2

This section contains all details relevant for using the basic equation and the backward equations of region 2 of IAPWS-IF97. The auxiliary equation for defining the boundary between regions 2 and 3 is given in Sec. 3.3.1. Uncertainty estimates of the most relevant properties, calculated from IAPWS-IF97, can be found in Sec. 3.7.

3.3.1
Auxiliary Equation for the Boundary between Regions 2 and 3

The boundary between regions 2 and 3 (see Fig. 3.1) is defined by the following simple quadratic pressure-temperature relation, the B23-equation

$$\pi = n_1 + n_2\theta + n_3\theta^2 \ , \tag{3.6}$$

where $\pi = p/p^*$ and $\theta = T/T^*$ with $p^* = 1\,\text{MPa}$ and $T^* = 1\,\text{K}$. The coefficients n_1 to n_3 of Eq. (3.6) are listed in Table 3.9. Equation (3.6) describes roughly an isentropic line; the entropy values along this boundary line are between $s = 5.047$ kJ kg^{-1} K^{-1} and $s = 5.261$ kJ kg^{-1} K^{-1}.

Alternatively Eq. (3.6) can be expressed explicitly in temperature as

$$\theta = n_4 + \left[(\pi - n_5)/n_3\right]^{0.5} \tag{3.7}$$

with θ and π as defined for Eq. (3.6) and the coefficients n_3 to n_5 listed in Table 3.9. Equations (3.6) and (3.7) cover the range from 623.15 K at 16.5292 MPa up to 863.15 K at 100 MPa.

Table 3.9. Coefficients of Eqs. (3.6) and (3.7)

i	n_i	i	n_i
1	$0.348\,051\,856\,289\,69 \times 10^{3}$	4	$0.572\,544\,598\,627\,46 \times 10^{3}$
2	$-0.116\,718\,598\,799\,75 \times 10^{1}$	5	$0.139\,188\,397\,788\,70 \times 10^{2}$
3	$0.101\,929\,700\,393\,26 \times 10^{-2}$		

For computer-program verification, Eqs. (3.6) and (3.7) must meet the following T-p point: $T = 0.623\,150\,000 \times 10^3$ K, $p = 0.165\,291\,643 \times 10^2$ MPa.

3.3.2
Basic Equation

The basic equation for this region is a fundamental equation for the specific Gibbs free energy g. This equation is expressed in dimensionless form, $\gamma = g/(RT)$, and is separated into 2 parts, an ideal-gas part γ^{o} and a residual part γ^{r}, so that

$$\frac{g(p,T)}{RT} = \gamma(\pi,\tau) = \gamma^{\text{o}}(\pi,\tau) + \gamma^{\text{r}}(\pi,\tau) \ , \tag{3.8}$$

where $\pi = p/p^*$ and $\tau = T^*/T$ with R given by Eq. (2.1).

The equation for the ideal-gas part γ^{o} of the dimensionless Gibbs free energy reads

$$\gamma^{\text{o}} = \ln\pi + \sum_{i=1}^{9} n_i^{\text{o}} \tau^{J_i^{\text{o}}} \ , \tag{3.9}$$

where $\pi = p/p^*$ and $\tau = T^*/T$ with $p^* = 1$ MPa and $T^* = 540$ K. The coefficients n_1^{o} and n_2^{o} were adjusted in such a way that the values for the specific internal energy and specific entropy, calculated from Eq. (3.8), relate to Eq. (3.2). Table 3.10 contains the coefficients n_i^{o} and exponents J_i^{o} of Eq. (3.9).

Table 3.10. Coefficients and exponents of Eq. (3.9)[a]

i	J_i^o	n_i^o	i	J_i^o	n_i^o
1[a]	0	$-0.969\,276\,865\,002\,17 \times 10^{1}$	6	-2	$0.142\,408\,191\,714\,44 \times 10^{1}$
2[a]	1	$0.100\,866\,559\,680\,18 \times 10^{2}$	7	-1	$-0.438\,395\,113\,194\,50 \times 10^{1}$
3	-5	$-0.560\,879\,112\,830\,20 \times 10^{-2}$	8	2	$-0.284\,086\,324\,607\,72$
4	-4	$0.714\,527\,380\,814\,55 \times 10^{-1}$	9	3	$0.212\,684\,637\,533\,07 \times 10^{-1}$
5	-3	$-0.407\,104\,982\,239\,28$			

[a] If Eq. (3.9) is incorporated into Eq. (3.11), instead of the values for n_1^o and n_2^o given above, the following values for these two coefficients must be used: $n_1^o = -0.969\,372\,683\,930\,49 \times 10^1$, $n_2^o = 0.100\,872\,759\,700\,06 \times 10^2$.

The form of the residual part γ^{r} of the dimensionless Gibbs free energy is as follows:

$$\gamma^{\mathrm{r}} = \sum_{i=1}^{43} n_i \, \pi^{I_i} (\tau - 0.5)^{J_i} \; , \tag{3.10}$$

where $\pi = p/p^*$ and $\tau = T^*/T$ with $p^* = 1$ MPa and $T^* = 540$ K. The coefficients n_i and exponents I_i and J_i of Eq. (3.10) are listed in Table 3.11.

All thermodynamic properties can be derived from Eq. (3.8) by using the appropriate combinations of the ideal-gas part γ^{o}, Eq. (3.9), and the residual part γ^{r}, Eq. (3.10), of the dimensionless Gibbs free energy and their derivatives. Relations between the relevant thermodynamic properties and γ^{o} and γ^{r} and their derivatives are summarized in Table 3.12. All required derivatives of the ideal-gas part and of the residual part of the dimensionless Gibbs free energy are explicitly given in Table 3.13 and Table 3.14, respectively.

Table 3.11. Coefficients and exponents of Eq. (3.10)

i	I_i	J_i	n_i	i	I_i	J_i	n_i
1	1	0	$-0.177\,317\,424\,732\,13 \times 10^{-2}$	23	7	0	$-0.590\,595\,643\,242\,70 \times 10^{-17}$
2	1	1	$-0.178\,348\,622\,923\,58 \times 10^{-1}$	24	7	11	$-0.126\,218\,088\,991\,01 \times 10^{-5}$
3	1	2	$-0.459\,960\,136\,963\,65 \times 10^{-1}$	25	7	25	$-0.389\,468\,424\,357\,39 \times 10^{-1}$
4	1	3	$-0.575\,812\,590\,834\,32 \times 10^{-1}$	26	8	8	$0.112\,562\,113\,604\,59 \times 10^{-10}$
5	1	6	$-0.503\,252\,787\,279\,30 \times 10^{-1}$	27	8	36	$-0.823\,113\,408\,979\,98 \times 10^{1}$
6	2	1	$-0.330\,326\,416\,702\,03 \times 10^{-4}$	28	9	13	$0.198\,097\,128\,020\,88 \times 10^{-7}$
7	2	2	$-0.189\,489\,875\,163\,15 \times 10^{-3}$	29	10	4	$0.104\,069\,652\,101\,74 \times 10^{-18}$
8	2	4	$-0.393\,927\,772\,433\,55 \times 10^{-2}$	30	10	10	$-0.102\,347\,470\,959\,29 \times 10^{-12}$
9	2	7	$-0.437\,972\,956\,505\,73 \times 10^{-1}$	31	10	14	$-0.100\,181\,793\,795\,11 \times 10^{-8}$
10	2	36	$-0.266\,745\,479\,140\,87 \times 10^{-4}$	32	16	29	$-0.808\,829\,086\,469\,85 \times 10^{-10}$
11	3	0	$0.204\,817\,376\,923\,09 \times 10^{-7}$	33	16	50	$0.106\,930\,318\,794\,09$
12	3	1	$0.438\,706\,672\,844\,35 \times 10^{-6}$	34	18	57	$-0.336\,622\,505\,741\,71$
13	3	3	$-0.322\,776\,772\,385\,70 \times 10^{-4}$	35	20	20	$0.891\,858\,453\,554\,21 \times 10^{-24}$
14	3	6	$-0.150\,339\,245\,421\,48 \times 10^{-2}$	36	20	35	$0.306\,293\,168\,762\,32 \times 10^{-12}$
15	3	35	$-0.406\,682\,535\,626\,49 \times 10^{-1}$	37	20	48	$-0.420\,024\,676\,982\,08 \times 10^{-5}$
16	4	1	$-0.788\,473\,095\,593\,67 \times 10^{-9}$	38	21	21	$-0.590\,560\,296\,856\,39 \times 10^{-25}$
17	4	2	$0.127\,907\,178\,522\,85 \times 10^{-7}$	39	22	53	$0.378\,269\,476\,134\,57 \times 10^{-5}$
18	4	3	$0.482\,253\,727\,185\,07 \times 10^{-6}$	40	23	39	$-0.127\,686\,089\,346\,81 \times 10^{-14}$
19	5	7	$0.229\,220\,763\,376\,61 \times 10^{-5}$	41	24	26	$0.730\,876\,105\,950\,61 \times 10^{-28}$
20	6	3	$-0.167\,147\,664\,510\,61 \times 10^{-10}$	42	24	40	$0.554\,147\,153\,507\,78 \times 10^{-16}$
21	6	16	$-0.211\,714\,723\,213\,55 \times 10^{-2}$	43	24	58	$-0.943\,697\,072\,412\,10 \times 10^{-6}$
22	6	35	$-0.238\,957\,419\,341\,04 \times 10^{2}$				

Table 3.12. Relations of thermodynamic properties to the ideal-gas part γ^{o} and the residual part γ^{r} of the dimensionless Gibbs free energy and their derivatives when using Eq. (3.8) or Eq. (3.11)

Property	Relation
Specific volume $v = (\partial g/\partial p)_T$	$v(\pi,\tau)\dfrac{p}{RT} = \pi\left(\gamma_\pi^{\mathrm{o}} + \gamma_\pi^{\mathrm{r}}\right)$
Specific internal energy $u = g - T(\partial g/\partial T)_p - p(\partial g/\partial p)_T$	$\dfrac{u(\pi,\tau)}{RT} = \tau\left(\gamma_\tau^{\mathrm{o}} + \gamma_\tau^{\mathrm{r}}\right) - \pi\left(\gamma_\pi^{\mathrm{o}} + \gamma_\pi^{\mathrm{r}}\right)$
Specific entropy $s = -(\partial g/\partial T)_p$	$\dfrac{s(\pi,\tau)}{R} = \tau\left(\gamma_\tau^{\mathrm{o}} + \gamma_\tau^{\mathrm{r}}\right) - \left(\gamma^{\mathrm{o}} + \gamma^{\mathrm{r}}\right)$
Specific enthalpy $h = g - T(\partial g/\partial T)_p$	$\dfrac{h(\pi,\tau)}{RT} = \tau\left(\gamma_\tau^{\mathrm{o}} + \gamma_\tau^{\mathrm{r}}\right)$
Specific isobaric heat capacity $c_p = (\partial h/\partial T)_p$	$\dfrac{c_p(\pi,\tau)}{R} = -\tau^2\left(\gamma_{\tau\tau}^{\mathrm{o}} + \gamma_{\tau\tau}^{\mathrm{r}}\right)$
Specific isochoric heat capacity $c_v = (\partial u/\partial T)_v$	$\dfrac{c_v(\pi,\tau)}{R} = -\tau^2\left(\gamma_{\tau\tau}^{\mathrm{o}} + \gamma_{\tau\tau}^{\mathrm{r}}\right) - \dfrac{\left(1 + \pi\gamma_\pi^{\mathrm{r}} - \tau\pi\gamma_{\pi\tau}^{\mathrm{r}}\right)^2}{1 - \pi^2\gamma_{\pi\pi}^{\mathrm{r}}}$
Speed of sound $w = v\left[-(\partial p/\partial v)_s\right]^{0.5}$	$\dfrac{w^2(\pi,\tau)}{RT} = \dfrac{1 + 2\pi\gamma_\pi^{\mathrm{r}} + \pi^2\gamma_\pi^{\mathrm{r}\,2}}{\left(1 - \pi^2\gamma_{\pi\pi}^{\mathrm{r}}\right) + \dfrac{\left(1 + \pi\gamma_\pi^{\mathrm{r}} - \tau\pi\gamma_{\pi\tau}^{\mathrm{r}}\right)^2}{\tau^2\left(\gamma_{\tau\tau}^{\mathrm{o}} + \gamma_{\tau\tau}^{\mathrm{r}}\right)}}$

$$\gamma_\pi^{\mathrm{r}} = \left(\frac{\partial\gamma^{\mathrm{r}}}{\partial\pi}\right)_\tau,\ \gamma_{\pi\pi}^{\mathrm{r}} = \left(\frac{\partial^2\gamma^{\mathrm{r}}}{\partial\pi^2}\right)_\tau,\ \gamma_\tau^{\mathrm{r}} = \left(\frac{\partial\gamma^{\mathrm{r}}}{\partial\tau}\right)_\pi,\ \gamma_{\tau\tau}^{\mathrm{r}} = \left(\frac{\partial^2\gamma^{\mathrm{r}}}{\partial\tau^2}\right)_\pi,\ \gamma_{\pi\tau}^{\mathrm{r}} = \left(\frac{\partial^2\gamma^{\mathrm{r}}}{\partial\pi\,\partial\tau}\right),\ \gamma_\tau^{\mathrm{o}} = \left(\frac{\partial\gamma^{\mathrm{o}}}{\partial\tau}\right)_\pi,\ \gamma_{\tau\tau}^{\mathrm{o}} = \left(\frac{\partial^2\gamma^{\mathrm{o}}}{\partial\tau^2}\right)_\pi$$

Table 3.13. The ideal-gas part γ^{o} of the dimensionless Gibbs free energy, Eq. (3.9), and its derivatives

$\gamma^{\mathrm{o}} = \ln\pi + \sum\limits_{i=1}^{9} n_i^{\mathrm{o}}\tau^{J_i^{\mathrm{o}}}$	$\gamma_\tau^{\mathrm{o}} = \sum\limits_{i=1}^{9} n_i^{\mathrm{o}} J_i^{\mathrm{o}}\tau^{J_i^{\mathrm{o}}-1}$
$\gamma_\pi^{\mathrm{o}} = \pi^{-1}$	$\gamma_{\tau\tau}^{\mathrm{o}} = \sum\limits_{i=1}^{9} n_i^{\mathrm{o}} J_i^{\mathrm{o}}\left(J_i^{\mathrm{o}} - 1\right)\tau^{J_i^{\mathrm{o}}-2}$
$\gamma_{\pi\pi}^{\mathrm{o}} = -\pi^{-2}$	$\gamma_{\pi\tau}^{\mathrm{o}} = 0$

$$\gamma_\pi^{\mathrm{o}} = \left(\frac{\partial\gamma^{\mathrm{o}}}{\partial\pi}\right)_\tau,\ \gamma_{\pi\pi}^{\mathrm{o}} = \left(\frac{\partial^2\gamma^{\mathrm{o}}}{\partial\pi^2}\right)_\tau,\ \gamma_\tau^{\mathrm{o}} = \left(\frac{\partial\gamma^{\mathrm{o}}}{\partial\tau}\right)_\pi,\ \gamma_{\tau\tau}^{\mathrm{o}} = \left(\frac{\partial^2\gamma^{\mathrm{o}}}{\partial\tau^2}\right)_\pi,\ \gamma_{\pi\tau}^{\mathrm{o}} = \left(\frac{\partial^2\gamma^{\mathrm{o}}}{\partial\pi\,\partial\tau}\right)$$

Equation (3.8) covers region 2 of IAPWS-IF97 defined by the following range of temperature and pressure, see Fig. 3.1:

$$
\begin{aligned}
273.15\ \mathrm{K} &\le T \le 623.15\ \mathrm{K} & 0 &< p \le p_{\mathrm{s}}(T)_{\mathrm{Eq.(3.2)}} \\
623.15\ \mathrm{K} &< T \le 863.15\ \mathrm{K} & 0 &< p \le p(T)_{\mathrm{Eq.(3.6)}} \\
863.15\ \mathrm{K} &< T \le 1073.15\ \mathrm{K} & 0 &< p \le 100\ \mathrm{MPa}
\end{aligned}
$$

In addition to the properties in the stable single-phase vapour region, Eq. (3.8) also yields reasonable values in the metastable-vapour region *for pressures above 10 MPa*. Equation (3.8) is not valid in the metastable-vapour region at pressures $p \leq 10$ MPa; for this part of the metastable-vapour region see Sec. 3.3.3. For temperatures between 273.15 K and 273.16 K at pressures above the sublimation pressure [11] (metastable subcooled vapour) all values are calculated by extrapolation from Eqs. (3.8) and (3.23).

To assist the user in computer-program verification of Eq. (3.8), Table 3.15 contains test values of the most relevant properties.

Table 3.14. The residual part γ^{r} of the dimensionless Gibbs free energy, Eq. (3.10), and its derivatives

$$\gamma^{\mathrm{r}} = \sum_{i=1}^{43} n_i\, \pi^{I_i} (\tau - 0.5)^{J_i} \qquad \gamma^{\mathrm{r}}_{\tau} = \sum_{i=1}^{43} n_i\, \pi^{I_i} J_i (\tau - 0.5)^{J_i - 1}$$

$$\gamma^{\mathrm{r}}_{\pi} = \sum_{i=1}^{43} n_i\, I_i\, \pi^{I_i - 1} (\tau - 0.5)^{J_i} \qquad \gamma^{\mathrm{r}}_{\tau\tau} = \sum_{i=1}^{43} n_i\, \pi^{I_i} J_i (J_i - 1)(\tau - 0.5)^{J_i - 2}$$

$$\gamma^{\mathrm{r}}_{\pi\pi} = \sum_{i=1}^{43} n_i\, I_i (I_i - 1) \pi^{I_i - 2} (\tau - 0.5)^{J_i} \qquad \gamma^{\mathrm{r}}_{\pi\tau} = \sum_{i=1}^{43} n_i\, I_i\, \pi^{I_i - 1} J_i (\tau - 0.5)^{J_i - 1}$$

$$\gamma^{\mathrm{r}}_{\pi} = \left(\frac{\partial \gamma^{\mathrm{r}}}{\partial \pi}\right)_{\tau},\ \gamma^{\mathrm{r}}_{\pi\pi} = \left(\frac{\partial^2 \gamma^{\mathrm{r}}}{\partial \pi^2}\right)_{\tau},\ \gamma^{\mathrm{r}}_{\tau} = \left(\frac{\partial \gamma^{\mathrm{r}}}{\partial \tau}\right)_{\pi},\ \gamma^{\mathrm{r}}_{\tau\tau} = \left(\frac{\partial^2 \gamma^{\mathrm{r}}}{\partial \tau^2}\right)_{\pi},\ \gamma^{\mathrm{r}}_{\pi\tau} = \left(\frac{\partial^2 \gamma^{\mathrm{r}}}{\partial \pi \partial \tau}\right)$$

Table 3.15. Thermodynamic property values calculated from Eq.(3.8) for selected temperatures and pressures

Property	$T = 300$ K $p = 0.0035$ MPa	$T = 700$ K $p = 0.0035$ MPa	$T = 700$ K $p = 30$ MPa
v [m^3 kg^{-1}]	$0.394\,913\,866 \times 10^{2}$	$0.923\,015\,898 \times 10^{2}$	$0.542\,946\,619 \times 10^{-2}$
h [kJ kg^{-1}]	$0.254\,991\,145 \times 10^{4}$	$0.333\,568\,375 \times 10^{4}$	$0.263\,149\,474 \times 10^{4}$
u [kJ kg^{-1}]	$0.241\,169\,160 \times 10^{4}$	$0.301\,262\,819 \times 10^{4}$	$0.246\,861\,076 \times 10^{4}$
s [kJ kg^{-1} K^{-1}]	$0.852\,238\,967 \times 10^{1}$	$0.101\,749\,996 \times 10^{2}$	$0.517\,540\,298 \times 10^{1}$
c_p [kJ kg^{-1} K^{-1}]	$0.191\,300\,162 \times 10^{1}$	$0.208\,141\,274 \times 10^{1}$	$0.103\,505\,092 \times 10^{2}$
w [m s^{-1}]	$0.427\,920\,172 \times 10^{3}$	$0.644\,289\,068 \times 10^{3}$	$0.480\,386\,523 \times 10^{3}$

3.3.3
Supplementary Equation for the Metastable-Vapour Region

As for the basic equation, Eq. (3.8), the supplementary equation for a part of the metastable-vapour region is given in the dimensionless form of the specific Gibbs free energy, $\gamma = g/(RT)$, consisting of an ideal-gas part γ^{o} and a residual part γ^{r}, so that

$$\frac{g(p,T)}{RT} = \gamma(\pi,\tau) = \gamma^{\mathrm{o}}(\pi,\tau) + \gamma^{\mathrm{r}}(\pi,\tau)\ , \tag{3.11}$$

where $\pi = p/p^*$ and $\tau = T^*/T$ with R given by Eq. (2.1).

The equation for the ideal-gas part γ^{o} is identical with Eq. (3.9) except for the values of the two coefficients n_1^{o} and n_2^{o}, see Table 3.10. For the use of Eq. (3.9) as part of Eq. (3.11) the coefficients n_1^{o} and n_2^{o} were slightly readjusted to meet the high consistency requirement between Eqs. (3.11) and (3.8) regarding the properties h and s along the saturated vapour line, see below.

The equation for the residual part γ^{r} reads

$$\gamma^{\mathrm{r}} = \sum_{i=1}^{13} n_i \, \pi^{I_i} (\tau - 0.5)^{J_i} \; , \tag{3.12}$$

where $\pi = p/p^*$ and $\tau = T^*/T$ with $p^* = 1$ MPa and $T^* = 540$ K. The coefficients n_i and exponents I_i and J_i of Eq. (3.12) are listed in Table 3.16. In the metastable-vapour region there are no experimental data to which an equation can be fitted. Thus, Eq. (3.11) is only based on input values extrapolated from the stable single-phase region 2. These extrapolations were not performed with IAPWS-95 but with a special low-density gas equation [8].

Table 3.16. Coefficients and exponents of Eq. (3.12)

i	I_i	J_i	n_i
1	1	0	$-0.733\,622\,601\,865\,06 \times 10^{-2}$
2	1	2	$-0.882\,238\,319\,431\,46 \times 10^{-1}$
3	1	5	$-0.723\,345\,552\,132\,45 \times 10^{-1}$
4	1	11	$-0.408\,131\,785\,344\,55 \times 10^{-2}$
5	2	1	$0.200\,978\,033\,802\,07 \times 10^{-2}$
6	2	7	$-0.530\,459\,218\,986\,42 \times 10^{-1}$
7	2	16	$-0.761\,904\,090\,869\,70 \times 10^{-2}$
8	3	4	$-0.634\,980\,376\,573\,13 \times 10^{-2}$
9	3	16	$-0.860\,430\,930\,285\,88 \times 10^{-1}$
10	4	7	$0.753\,215\,815\,227\,70 \times 10^{-2}$
11	4	10	$-0.792\,383\,754\,461\,39 \times 10^{-2}$
12	5	9	$-0.228\,881\,607\,784\,47 \times 10^{-3}$
13	5	10	$-0.264\,565\,014\,828\,10 \times 10^{-2}$

All thermodynamic properties can be derived from Eq. (3.11) by using the appropriate combinations of the ideal-gas part γ^{o}, Eq. (3.9), and the residual part γ^{r}, Eq. (3.12), of the dimensionless Gibbs free energy and their derivatives. Relations between the relevant thermodynamic properties and γ^{o} and γ^{r} and their derivatives are summarized in Table 3.12. All required derivatives of the ideal-gas part and of the residual part of the dimensionless Gibbs free energy are explicitly given in Table 3.13 and Table 3.17, respectively.

Equation (3.11) is valid in the metastable-vapour region from the saturated vapour line to the 5 % equilibrium moisture line (determined from the equilibrium h' and h'' values) at pressures from the triple-point pressure, see Eq. (2.8), up to 10 MPa. The consistency of Eq. (3.11) with the basic equation, Eq. (3.8), along the saturated vapour line is characterized by the following maximum inconsistencies regarding the properties v, h, c_p, s, g, and w:

$$|\Delta v|_{\max} = 0.014\ \% \qquad |\Delta s|_{\max} = 0.082\ \mathrm{J\,kg^{-1}\,K^{-1}}$$
$$|\Delta h|_{\max} = 0.043\ \mathrm{kJ\,kg^{-1}} \qquad |\Delta g|_{\max} = 0.023\ \mathrm{kJ\,kg^{-1}}$$
$$|\Delta c_p|_{\max} = 0.78\ \% \qquad |\Delta w|_{\max} = 0.051\ \% \,.$$

These maximum inconsistencies are clearly smaller than the consistency requirements along the region boundaries corresponding to the so-called Prague values [10], which are given in

Sec. 3.1.2. Along the 10 MPa isobar in the metastable-vapour region, the transition between Eq. (3.11) and Eq. (3.8) is not smooth but for practical calculations unimportant.

To assist the user in computer-program verification of Eq. (3.11), Table 3.18 contains test values of the most relevant properties.

Table 3.17. The residual part γ^{r} of the dimensionless Gibbs free energy, Eq. (3.12), and its derivatives

$$\gamma^{\mathrm{r}} = \sum_{i=1}^{13} n_i\, \pi^{I_i} (\tau - 0.5)^{J_i} \qquad \gamma^{\mathrm{r}}_{\tau} = \sum_{i=1}^{13} n_i\, \pi^{I_i} J_i (\tau - 0.5)^{J_i - 1}$$

$$\gamma^{\mathrm{r}}_{\pi} = \sum_{i=1}^{13} n_i\, I_i\, \pi^{I_i - 1} (\tau - 0.5)^{J_i} \qquad \gamma^{\mathrm{r}}_{\tau\tau} = \sum_{i=1}^{13} n_i\, \pi^{I_i} J_i (J_i - 1)(\tau - 0.5)^{J_i - 2}$$

$$\gamma^{\mathrm{r}}_{\pi\pi} = \sum_{i=1}^{13} n_i\, I_i (I_i - 1) \pi^{I_i - 2} (\tau - 0.5)^{J_i} \qquad \gamma^{\mathrm{r}}_{\pi\tau} = \sum_{i=1}^{13} n_i\, I_i\, \pi^{I_i - 1} J_i (\tau - 0.5)^{J_i - 1}$$

$$\gamma^{\mathrm{r}}_{\pi} = \left(\frac{\partial \gamma^{\mathrm{r}}}{\partial \pi}\right)_{\tau},\ \gamma^{\mathrm{r}}_{\pi\pi} = \left(\frac{\partial^2 \gamma^{\mathrm{r}}}{\partial \pi^2}\right)_{\tau},\ \gamma^{\mathrm{r}}_{\tau} = \left(\frac{\partial \gamma^{\mathrm{r}}}{\partial \tau}\right)_{\pi},\ \gamma^{\mathrm{r}}_{\tau\tau} = \left(\frac{\partial^2 \gamma^{\mathrm{r}}}{\partial \tau^2}\right)_{\pi},\ \gamma^{\mathrm{r}}_{\pi\tau} = \left(\frac{\partial^2 \gamma^{\mathrm{r}}}{\partial \pi \partial \tau}\right)$$

Table 3.18. Thermodynamic property values calculated from Eq. (3.11) for selected values of temperature and pressure

Property	$T = 450$ K $p = 1$ MPa	$T = 440$ K $p = 1$ MPa	$T = 450$ K $p = 1.5$ MPa
v [m^3 kg^{-1}]	0.192 516 540	0.186 212 297	0.121 685 206
h [kJ kg^{-1}]	0.276 881 115 × 10^4	0.274 015 123 × 10^4	0.272 134 539 × 10^4
u [kJ kg^{-1}]	0.257 629 461 × 10^4	0.255 393 894 × 10^4	0.253 881 758 × 10^4
s [kJ kg^{-1} K^{-1}]	0.656 660 377 × 10^1	0.650 218 759 × 10^1	0.629 170 440 × 10^1
c_p [kJ kg^{-1} K^{-1}]	0.276 349 265 × 10^1	0.298 166 443 × 10^1	0.362 795 578 × 10^1
w [m s^{-1}]	0.498 408 101 × 10^3	0.489 363 295 × 10^3	0.481 941 819 × 10^3

3.3.4 Backward Equations

For a sufficiently accurate calculation of properties as function of p, h or of p, s without any iteration (see Sec. 3.1.1), the two backward equations require extremely good numerical consistency with the basic equation. The exact requirements for these numerical consistencies were obtained from comprehensive test calculations for several characteristic power cycles. Based on these investigations, the numerical inconsistency requirement between the basic equation, Eq. (3.8), and the corresponding backward equations, $T(p,h)$ and $T(p,s)$, was set to ± 10 mK and ± 25 mK, respectively, depending on which part of region 2 is considered, see Tables 3.23 and 3.28. This means, for example, that the temperature value directly determined from the backward equation $T(p,h)$ for subregion 2a, Eq. (3.15), for given values of p and h must agree within ± 10 mK with the temperature value determined by iteration from the basic equation, Eq. (3.8), for the same p and h. This iteration from the basic equation is carried out via the function $h(p,T)$ of which the relation with the dimensionless Gibbs free energy is given in Table 3.12.

Due to the very high demand regarding the numerical consistency between the basic equation, Eq. (3.8), and the backward equations, region 2 is covered by *three* $T(p,h)$ and *three* $T(p,s)$ equations. Figure 3.2 shows how region 2 is divided into the 3 subregions for the backward equations. The boundary between subregions 2a and 2b is the isobar $p = 4$ MPa; the boundary between subregions 2b and 2c corresponds to the entropy line $s = 5.85$ kJ kg^{-1} K^{-1}.

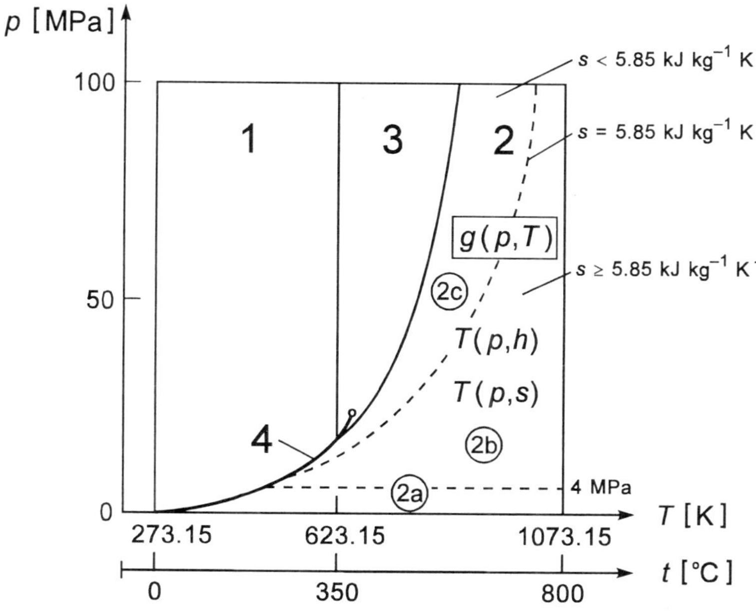

Fig. 3.2. Division of region 2 of IAPWS-IF97 into the 3 subregions 2a, 2b, and 2c for the backward equations $T(p,h)$ and $T(p,s)$

In order to know whether the $T(p,h)$ equation for subregion 2b or for subregion 2c has to be used for given values of p and h, a special correlation equation for the boundary between subregions 2b and 2c (which approximates $s = 5.85$ kJ kg^{-1} K^{-1}) is needed, see Fig. 3.2. This boundary equation, called B2bc-equation, is a simple quadratic pressure-enthalpy relation which reads

$$\pi = n_1 + n_2\eta + n_3\eta^2 \ , \tag{3.13}$$

where $\pi = p/p^*$ and $\eta = h/h^*$ with $p^* = 1$ MPa and $h^* = 1$ kJ kg^{-1}. The coefficients n_1 to n_3 of Eq. (3.13) are listed in Table 3.19. Based on its simple form, Eq. (3.13) does not describe exactly the isentropic line $s = 5.85$ kJ kg^{-1} K^{-1}; the entropy values corresponding to this p-h relation are between $s = 5.81$ kJ kg^{-1} K^{-1} and $s = 5.85$ kJ kg^{-1} K^{-1}. The enthalpy-explicit form of Eq. (3.13) is as follows:

$$\eta = n_4 + \left[(\pi - n_5)/n_3\right]^{0.5} \tag{3.14}$$

with η and π according to Eq. (3.13) and the coefficients n_3 to n_5 listed in Table 3.19. Equations (3.13) and (3.14) give the boundary line between subregions 2b and 2c from the saturation state at $T = 554.485$ K and $p_s = 6.54670$ MPa to $T = 1019.32$ K and $p = 100$ MPa.

For the backward equations $T(p,s)$ the boundary between subregions 2b and 2c is, based on the value $s = 5.85$ kJ kg^{-1} K^{-1} along this boundary, automatically defined for given values of p and s.

Table 3.19. Coefficients of Eqs. (3.13) and (3.14)

i	n_i	i	n_i
1	$0.905\,842\,785\,147\,23 \times 10^{3}$	4	$0.265\,265\,719\,084\,28 \times 10^{4}$
2	$-0.679\,557\,863\,992\,41$	5	$0.452\,575\,789\,059\,48 \times 10^{1}$
3	$0.128\,090\,027\,301\,36 \times 10^{-3}$		

For computer-program verification, Eqs. (3.13) and (3.14) must meet the following p-h point: $p = 0.100\,000\,000 \times 10^{3}$ MPa , $h = 0.351\,600\,432\,3 \times 10^{4}$ kJ kg^{-1}.

3.3.4.1 The Backward Equations T(p, h) for Subregions 2a, 2b, and 2c

The backward equation $T(p,h)$ for **subregion 2a** in its dimensionless form reads

$$\frac{T_{2a}(p,h)}{T^*} = \theta_{2a}(\pi,\eta) = \sum_{i=1}^{34} n_i\, \pi^{I_i} (\eta - 2.1)^{J_i} \ , \qquad (3.15)$$

where $\theta = T/T^*$, $\pi = p/p^*$, and $\eta = h/h^*$ with $T^* = 1$ K , $p^* = 1$ MPa, and $h^* = 2000$ kJ kg^{-1}. The coefficients n_i and exponents I_i and J_i of Eq. (3.15) are listed in Table 3.20.

The backward equation $T(p,h)$ for **subregion 2b** in its dimensionless form reads

$$\frac{T_{2b}(p,h)}{T^*} = \theta_{2b}(\pi,\eta) = \sum_{i=1}^{38} n_i (\pi - 2)^{I_i} (\eta - 2.6)^{J_i} \ , \qquad (3.16)$$

where $\theta = T/T^*$, $\pi = p/p^*$, and $\eta = h/h^*$ with $T^* = 1$ K , $p^* = 1$ MPa, and $h^* = 2000$ kJ kg^{-1}. The coefficients n_i and exponents I_i and J_i of Eq. (3.16) are listed in Table 3.21.

The backward equation $T(p,h)$ for **subregion 2c** in its dimensionless form reads

$$\frac{T_{2c}(p,h)}{T^*} = \theta_{2c}(\pi,\eta) = \sum_{i=1}^{23} n_i (\pi + 25)^{I_i} (\eta - 1.8)^{J_i} \ , \qquad (3.17)$$

where $\theta = T/T^*$, $\pi = p/p^*$, and $\eta = h/h^*$ with $T^* = 1$ K , $p^* = 1$ MPa, and $h^* = 2000$ kJ kg^{-1}. The coefficients n_i and exponents I_i and J_i of Eq. (3.17) are listed in Table 3.22.

Table 3.20. Coefficients and exponents of the backward equation $T(p,h)$ for subregion 2a, Eq. (3.15)

i	I_i	J_i	n_i	i	I_i	J_i	n_i
1	0	0	$0.108\,989\,523\,182\,88 \times 10^{4}$	18	2	7	$0.116\,708\,730\,771\,07 \times 10^{2}$
2	0	1	$0.849\,516\,544\,955\,35 \times 10^{3}$	19	2	36	$0.128\,127\,984\,040\,46 \times 10^{9}$
3	0	2	$-0.107\,817\,480\,918\,26 \times 10^{3}$	20	2	38	$-0.985\,549\,096\,232\,76 \times 10^{9}$
4	0	3	$0.331\,536\,548\,012\,63 \times 10^{2}$	21	2	40	$0.282\,245\,469\,730\,02 \times 10^{10}$
5	0	7	$-0.742\,320\,167\,902\,48 \times 10^{1}$	22	2	42	$-0.359\,489\,714\,107\,03 \times 10^{10}$
6	0	20	$0.117\,650\,487\,243\,56 \times 10^{2}$	23	2	44	$0.172\,273\,499\,131\,97 \times 10^{10}$
7	1	0	$0.184\,457\,493\,557\,90 \times 10^{1}$	24	3	24	$-0.135\,513\,342\,407\,75 \times 10^{5}$
8	1	1	$-0.417\,927\,005\,496\,24 \times 10^{1}$	25	3	44	$0.128\,487\,346\,646\,50 \times 10^{8}$
9	1	2	$0.624\,781\,969\,358\,12 \times 10^{1}$	26	4	12	$0.138\,657\,242\,832\,26 \times 10^{1}$
10	1	3	$-0.173\,445\,631\,081\,14 \times 10^{2}$	27	4	32	$0.235\,988\,325\,565\,14 \times 10^{6}$
11	1	7	$-0.200\,581\,768\,620\,96 \times 10^{3}$	28	4	44	$-0.131\,052\,365\,450\,54 \times 10^{8}$
12	1	9	$0.271\,960\,654\,737\,96 \times 10^{3}$	29	5	32	$0.739\,998\,354\,747\,66 \times 10^{4}$
13	1	11	$-0.455\,113\,182\,858\,18 \times 10^{3}$	30	5	36	$-0.551\,966\,970\,300\,60 \times 10^{6}$
14	1	18	$0.309\,196\,886\,047\,55 \times 10^{4}$	31	5	42	$0.371\,540\,859\,962\,33 \times 10^{7}$
15	1	44	$0.252\,266\,403\,578\,72 \times 10^{6}$	32	6	34	$0.191\,277\,292\,396\,60 \times 10^{5}$
16	2	0	$-0.617\,074\,228\,683\,39 \times 10^{-2}$	33	6	44	$-0.415\,351\,648\,356\,34 \times 10^{6}$
17	2	2	$-0.310\,780\,466\,295\,83$	34	7	28	$-0.624\,598\,551\,925\,07 \times 10^{2}$

Table 3.21. Coefficients and exponents of the backward equation $T(p,h)$ for subregion 2b, Eq (3.16)

i	I_i	J_i	n_i	i	I_i	J_i	n_i
1	0	0	$0.148\,950\,410\,795\,16 \times 10^{4}$	20	2	40	$0.712\,803\,519\,595\,51 \times 10^{-4}$
2	0	1	$0.743\,077\,983\,140\,34 \times 10^{3}$	21	3	1	$0.110\,328\,317\,899\,99 \times 10^{-3}$
3	0	2	$-0.977\,083\,187\,978\,37 \times 10^{2}$	22	3	2	$0.189\,552\,483\,879\,02 \times 10^{-3}$
4	0	12	$0.247\,424\,647\,056\,74 \times 10^{1}$	23	3	12	$0.308\,915\,411\,605\,37 \times 10^{-2}$
5	0	18	$-0.632\,813\,200\,160\,26$	24	3	24	$0.135\,555\,045\,549\,49 \times 10^{-2}$
6	0	24	$0.113\,859\,521\,296\,58 \times 10^{1}$	25	4	2	$0.286\,402\,374\,774\,56 \times 10^{-6}$
7	0	28	$-0.478\,118\,636\,486\,25$	26	4	12	$-0.107\,798\,573\,575\,12 \times 10^{-4}$
8	0	40	$0.852\,081\,234\,315\,44 \times 10^{-2}$	27	4	18	$-0.764\,627\,124\,548\,14 \times 10^{-4}$
9	1	0	$0.937\,471\,473\,779\,32$	28	4	24	$0.140\,523\,928\,183\,16 \times 10^{-4}$
10	1	2	$0.335\,931\,186\,049\,16 \times 10^{1}$	29	4	28	$-0.310\,838\,143\,314\,34 \times 10^{-4}$
11	1	6	$0.338\,093\,556\,014\,54 \times 10^{1}$	30	4	40	$-0.103\,027\,382\,121\,03 \times 10^{-5}$
12	1	12	$0.168\,445\,396\,719\,04$	31	5	18	$0.282\,172\,816\,350\,40 \times 10^{-6}$
13	1	18	$0.738\,757\,452\,366\,95$	32	5	24	$0.127\,049\,022\,719\,45 \times 10^{-5}$
14	1	24	$-0.471\,287\,374\,361\,86$	33	5	40	$0.738\,033\,534\,682\,92 \times 10^{-7}$
15	1	28	$0.150\,202\,731\,397\,07$	34	6	28	$-0.110\,301\,392\,389\,09 \times 10^{-7}$
16	1	40	$-0.217\,641\,142\,197\,50 \times 10^{-2}$	35	7	2	$-0.814\,563\,652\,078\,33 \times 10^{-13}$
17	2	2	$-0.218\,107\,553\,247\,61 \times 10^{-1}$	36	7	28	$-0.251\,805\,456\,829\,62 \times 10^{-10}$
18	2	8	$-0.108\,297\,844\,036\,77$	37	9	1	$-0.175\,652\,339\,694\,07 \times 10^{-17}$
19	2	18	$-0.463\,333\,246\,358\,12 \times 10^{-1}$	38	9	40	$0.869\,341\,563\,441\,63 \times 10^{-14}$

Table 3.22. Coefficients and exponents of the backward equation $T(p,h)$ for subregion 2c, Eq. (3.17)

i	I_i	J_i	n_i	i	I_i	J_i	n_i
1	−7	0	$-0.323\,683\,985\,552\,42 \times 10^{13}$	13	1	4	$0.379\,660\,012\,724\,86 \times 10^{1}$
2	−7	4	$0.732\,633\,509\,021\,81 \times 10^{13}$	14	1	8	$-0.108\,429\,848\,800\,77 \times 10^{2}$
3	−6	0	$0.358\,250\,899\,454\,47 \times 10^{12}$	15	2	4	$-0.453\,641\,726\,766\,60 \times 10^{-1}$
4	−6	2	$-0.583\,401\,318\,515\,90 \times 10^{12}$	16	6	0	$0.145\,591\,156\,586\,98 \times 10^{-12}$
5	−5	0	$-0.107\,830\,682\,174\,70 \times 10^{11}$	17	6	1	$0.112\,615\,974\,072\,30 \times 10^{-11}$
6	−5	2	$0.208\,255\,445\,631\,71 \times 10^{11}$	18	6	4	$-0.178\,049\,822\,406\,86 \times 10^{-10}$
7	−2	0	$0.610\,747\,835\,645\,16 \times 10^{6}$	19	6	10	$0.123\,245\,796\,908\,32 \times 10^{-6}$
8	−2	1	$0.859\,777\,225\,355\,80 \times 10^{6}$	20	6	12	$-0.116\,069\,211\,309\,84 \times 10^{-5}$
9	−1	0	$-0.257\,457\,236\,041\,70 \times 10^{5}$	21	6	16	$0.278\,463\,670\,885\,54 \times 10^{-4}$
10	−1	2	$0.310\,810\,884\,227\,14 \times 10^{5}$	22	6	20	$-0.592\,700\,384\,741\,76 \times 10^{-3}$
11	0	0	$0.120\,823\,158\,659\,36 \times 10^{4}$	23	6	22	$0.129\,185\,829\,918\,78 \times 10^{-2}$
12	0	1	$0.482\,197\,551\,092\,55 \times 10^{3}$				

Equations (3.15), (3.16), and (3.17) are only valid in the respective subregion 2a, 2b, and 2c which do not include the metastable-vapour region. The boundaries between these subregions are defined at the beginning of Sec. 3.3.4; the lowest pressure for which Eq. (3.15) is valid amounts to 611.153 Pa corresponding to the sublimation pressure [11] at 273.15 K. The tolerated and actual numerical inconsistencies between Eqs. (3.15) to (3.17) and Eq. (3.8), given as maximum and root-mean-square (RMS) temperature differences, are listed in Table 3.23.

Table 3.23. Tolerated and actual temperature differences between Eqs. (3.15) to (3.17) and Eq. (3.8)

Subregion	Equation	Temperature differences [mK]		
		Tolerated	Maximum	RMS
2a	3.15	10	9.3	2.9
2b	3.16	10	9.6	3.9
2c	3.17	25	23.7	10.4

To assist the user in computer-program verification of Eqs. (3.15) to (3.17), Table 3.24 contains the corresponding test values.

Table 3.24. Temperature values calculated from Eqs. (3.15) to (3.17) for selected pressures and enthalpies

Equation	p [MPa]	h [kJ kg^{-1}]	T [K]
3.15	0.001	3000	$0.534\,433\,241 \times 10^{3}$
	3	3000	$0.575\,373\,370 \times 10^{3}$
	3	4000	$0.101\,077\,577 \times 10^{4}$
3.16	5	3500	$0.801\,299\,102 \times 10^{3}$
	5	4000	$0.101\,531\,583 \times 10^{4}$
	25	3500	$0.875\,279\,054 \times 10^{3}$
3.17	40	2700	$0.743\,056\,411 \times 10^{3}$
	60	2700	$0.791\,137\,067 \times 10^{3}$
	60	3200	$0.882\,756\,860 \times 10^{3}$

3.3.4.2
The Backward Equations *T*(*p*, *s*) for Subregions 2a, 2b, and 2c

The backward equation $T(p,s)$ for **subregion 2a** in its dimensionless form reads

$$\frac{T_{2a}(p,s)}{T^*} = \theta_{2a}(\pi,\sigma) = \sum_{i=1}^{46} n_i \pi^{I_i} (\sigma - 2)^{J_i} , \tag{3.18}$$

where $\theta = T/T^*$, $\pi = p/p^*$, and $\sigma = s/s^*$ with $T^* = 1\ \mathrm{K}$, $p^* = 1\ \mathrm{MPa}$, and $s^* = 2\ \mathrm{kJ\ kg^{-1}\ K^{-1}}$. The coefficients n_i and exponents I_i and J_i of Eq. (3.18) are listed in Table 3.25.

The backward equation $T(p,s)$ for **subregion 2b** in its dimensionless form reads

$$\frac{T_{2b}(p,s)}{T^*} = \theta_{2b}(\pi,\sigma) = \sum_{i=1}^{44} n_i \pi^{I_i} (10 - \sigma)^{J_i} , \tag{3.19}$$

where $\theta = T/T^*$, $\pi = p/p^*$, and $\sigma = s/s^*$ with $T^* = 1\ \mathrm{K}$, $p^* = 1\ \mathrm{MPa}$, and $s^* = 0.7853\ \mathrm{kJ\ kg^{-1}\ K^{-1}}$. The coefficients n_i and exponents I_i and J_i of Eq. (3.19) are listed in Table 3.26.

The backward equation $T(p,s)$ for **subregion 2c** in its dimensionless form reads

$$\frac{T_{2c}(p,s)}{T^*} = \theta_{2c}(\pi,\sigma) = \sum_{i=1}^{30} n_i \pi^{I_i} (2 - \sigma)^{J_i} , \tag{3.20}$$

where $\theta = T/T^*$, $\pi = p/p^*$, and $\sigma = s/s^*$ with $T^* = 1\ \mathrm{K}$, $p^* = 1\ \mathrm{MPa}$, and $s^* = 2.9251\ \mathrm{kJ\ kg^{-1}\ K^{-1}}$. The coefficients n_i and exponents I_i and J_i of Eq. (3.20) are listed in Table 3.27.

Table 3.25. Coefficients and exponents of the backward equation $T(p,s)$ for subregion 2a, Eq. (3.18)

i	I_i	J_i	n_i	i	I_i	J_i	n_i
1	−1.5	−24	$-0.392\,359\,838\,619\,84 \times 10^{6}$	24	−0.25	−11	$-0.597\,806\,388\,727\,18 \times 10^{4}$
2	−1.5	−23	$0.515\,265\,738\,272\,70 \times 10^{6}$	25	−0.25	−6	$-0.704\,014\,639\,268\,62 \times 10^{3}$
3	−1.5	−19	$0.404\,824\,431\,610\,48 \times 10^{5}$	26	0.25	1	$0.338\,367\,841\,075\,53 \times 10^{3}$
4	−1.5	−13	$-0.321\,937\,909\,239\,02 \times 10^{3}$	27	0.25	4	$0.208\,627\,866\,351\,87 \times 10^{2}$
5	−1.5	−11	$0.969\,614\,242\,186\,94 \times 10^{2}$	28	0.25	8	$0.338\,341\,726\,561\,96 \times 10^{-1}$
6	−1.5	−10	$-0.228\,678\,463\,717\,73 \times 10^{2}$	29	0.25	11	$-0.431\,244\,284\,148\,93 \times 10^{-4}$
7	−1.25	−19	$-0.449\,429\,141\,243\,57 \times 10^{6}$	30	0.5	0	$0.166\,537\,913\,564\,12 \times 10^{3}$
8	−1.25	−15	$-0.501\,183\,360\,201\,66 \times 10^{4}$	31	0.5	1	$-0.139\,862\,920\,558\,98 \times 10^{3}$
9	−1.25	−6	$0.356\,844\,635\,600\,15$	32	0.5	5	$-0.788\,495\,479\,998\,72$
10	−1.0	−26	$0.442\,353\,358\,481\,90 \times 10^{5}$	33	0.5	6	$0.721\,324\,117\,538\,72 \times 10^{-1}$
11	−1.0	−21	$-0.136\,733\,888\,117\,08 \times 10^{5}$	34	0.5	10	$-0.597\,548\,393\,982\,83 \times 10^{-2}$
12	−1.0	−17	$0.421\,632\,602\,078\,64 \times 10^{6}$	35	0.5	14	$-0.121\,413\,589\,539\,04 \times 10^{-4}$
13	−1.0	−16	$0.225\,169\,258\,374\,75 \times 10^{5}$	36	0.5	16	$0.232\,270\,967\,338\,71 \times 10^{-6}$
14	−1.0	−9	$0.474\,421\,448\,656\,46 \times 10^{3}$	37	0.75	0	$-0.105\,384\,635\,661\,94 \times 10^{2}$
15	−1.0	−8	$-0.149\,311\,307\,976\,47 \times 10^{3}$	38	0.75	4	$0.207\,189\,254\,965\,02 \times 10^{1}$
16	−0.75	−15	$-0.197\,811\,263\,204\,52 \times 10^{6}$	39	0.75	9	$-0.721\,931\,552\,604\,27 \times 10^{-1}$
17	−0.75	−14	$-0.235\,543\,994\,707\,60 \times 10^{5}$	40	0.75	17	$0.207\,498\,870\,811\,20 \times 10^{-6}$
18	−0.5	−26	$-0.190\,706\,163\,020\,76 \times 10^{5}$	41	1.0	7	$-0.183\,406\,579\,113\,79 \times 10^{-1}$
19	−0.5	−13	$0.553\,756\,698\,831\,64 \times 10^{5}$	42	1.0	18	$0.290\,362\,723\,486\,96 \times 10^{-6}$
20	−0.5	−9	$0.382\,936\,914\,373\,63 \times 10^{4}$	43	1.25	3	$0.210\,375\,278\,936\,19$
21	−0.5	−7	$-0.603\,918\,605\,805\,67 \times 10^{3}$	44	1.25	15	$0.256\,812\,397\,299\,99 \times 10^{-3}$
22	−0.25	−27	$0.193\,631\,026\,203\,31 \times 10^{4}$	45	1.5	5	$-0.127\,990\,029\,337\,81 \times 10^{-1}$
23	−0.25	−25	$0.426\,606\,436\,986\,10 \times 10^{4}$	46	1.5	18	$-0.821\,981\,026\,520\,18 \times 10^{-5}$

Table 3.26. Coefficients and exponents of the backward equation $T(p,s)$ for subregion 2b, Eq. (3.19)

i	I_i	J_i	n_i	i	I_i	J_i	n_i
1	−6	0	$0.316\,876\,650\,834\,97 \times 10^{6}$	23	0	2	$0.417\,273\,471\,596\,10 \times 10^{2}$
2	−6	11	$0.208\,641\,758\,818\,58 \times 10^{2}$	24	0	4	$0.219\,325\,494\,345\,32 \times 10^{1}$
3	−5	0	$-0.398\,593\,998\,035\,99 \times 10^{6}$	25	0	5	$-0.103\,200\,500\,090\,77 \times 10^{1}$
4	−5	11	$-0.218\,160\,585\,188\,77 \times 10^{2}$	26	0	6	$0.358\,829\,435\,167\,03$
5	−4	0	$0.223\,697\,851\,942\,42 \times 10^{6}$	27	0	9	$0.525\,114\,537\,260\,66 \times 10^{-2}$
6	−4	1	$-0.278\,417\,034\,458\,17 \times 10^{4}$	28	1	0	$0.128\,389\,164\,507\,05 \times 10^{2}$
7	−4	11	$0.992\,074\,360\,714\,80 \times 10^{1}$	29	1	1	$-0.286\,424\,372\,193\,81 \times 10^{1}$
8	−3	0	$-0.751\,975\,122\,991\,57 \times 10^{5}$	30	1	2	$0.569\,126\,836\,648\,55$
9	−3	1	$0.297\,086\,059\,511\,58 \times 10^{4}$	31	1	3	$-0.999\,629\,545\,849\,31 \times 10^{-1}$
10	−3	11	$-0.344\,068\,785\,485\,26 \times 10^{1}$	32	1	7	$-0.326\,320\,377\,784\,59 \times 10^{-2}$
11	−3	12	$0.388\,155\,642\,491\,15$	33	1	8	$0.233\,209\,225\,767\,23 \times 10^{-3}$
12	−2	0	$0.175\,112\,950\,857\,50 \times 10^{5}$	34	2	0	$-0.153\,348\,098\,574\,50$
13	−2	1	$-0.142\,371\,128\,544\,49 \times 10^{4}$	35	2	1	$0.290\,722\,882\,399\,02 \times 10^{-1}$
14	−2	6	$0.109\,438\,033\,641\,67 \times 10^{1}$	36	2	5	$0.375\,347\,027\,411\,67 \times 10^{-3}$
15	−2	10	$0.899\,716\,193\,084\,95$	37	3	0	$0.172\,966\,917\,024\,11 \times 10^{-2}$
16	−1	0	$-0.337\,597\,400\,989\,58 \times 10^{4}$	38	3	1	$-0.385\,560\,508\,445\,04 \times 10^{-3}$
17	−1	1	$0.471\,628\,858\,183\,55 \times 10^{3}$	39	3	3	$-0.350\,177\,122\,926\,08 \times 10^{-4}$
18	−1	5	$-0.191\,882\,419\,936\,79 \times 10^{1}$	40	4	0	$-0.145\,663\,936\,314\,92 \times 10^{-4}$
19	−1	8	$0.410\,785\,804\,921\,96$	41	4	1	$0.564\,208\,572\,672\,69 \times 10^{-5}$
20	−1	9	$-0.334\,653\,781\,720\,97$	42	5	0	$0.412\,861\,500\,746\,05 \times 10^{-7}$
21	0	0	$0.138\,700\,347\,775\,05 \times 10^{4}$	43	5	1	$-0.206\,846\,711\,188\,24 \times 10^{-7}$
22	0	1	$-0.406\,633\,261\,958\,38 \times 10^{3}$	44	5	2	$0.164\,093\,936\,747\,25 \times 10^{-8}$

Table 3.27. Coefficients and exponents of the backward equation $T(p,s)$ for subregion 2c, Eq. (3.20)

i	I_i	J_i	n_i	i	I_i	J_i	n_i
1	−2	0	$0.909\,685\,010\,053\,65 \times 10^{3}$	16	3	1	$-0.145\,970\,082\,847\,53 \times 10^{-1}$
2	−2	1	$0.240\,456\,670\,884\,20 \times 10^{4}$	17	3	5	$0.566\,311\,756\,310\,27 \times 10^{-2}$
3	−1	0	$-0.591\,623\,263\,871\,30 \times 10^{3}$	18	4	0	$-0.761\,558\,645\,845\,77 \times 10^{-4}$
4	0	0	$0.541\,454\,041\,280\,74 \times 10^{3}$	19	4	1	$0.224\,403\,429\,193\,32 \times 10^{-3}$
5	0	1	$-0.270\,983\,084\,111\,92 \times 10^{3}$	20	4	4	$-0.125\,610\,950\,134\,13 \times 10^{-4}$
6	0	2	$0.979\,765\,250\,979\,26 \times 10^{3}$	21	5	0	$0.633\,231\,326\,609\,34 \times 10^{-6}$
7	0	3	$-0.469\,667\,729\,594\,35 \times 10^{3}$	22	5	1	$-0.205\,419\,896\,753\,75 \times 10^{-5}$
8	1	0	$0.143\,992\,746\,047\,23 \times 10^{2}$	23	5	2	$0.364\,053\,703\,900\,82 \times 10^{-7}$
9	1	1	$-0.191\,042\,042\,304\,29 \times 10^{2}$	24	6	0	$-0.297\,598\,977\,892\,15 \times 10^{-8}$
10	1	3	$0.532\,991\,671\,119\,71 \times 10^{1}$	25	6	1	$0.101\,366\,185\,297\,63 \times 10^{-7}$
11	1	4	$-0.212\,529\,753\,759\,34 \times 10^{2}$	26	7	0	$0.599\,257\,196\,923\,51 \times 10^{-11}$
12	2	0	$-0.311\,473\,344\,137\,60$	27	7	1	$-0.206\,778\,701\,051\,64 \times 10^{-10}$
13	2	1	$0.603\,348\,408\,946\,23$	28	7	3	$-0.208\,742\,781\,818\,86 \times 10^{-10}$
14	2	2	$-0.427\,648\,397\,025\,09 \times 10^{-1}$	29	7	4	$0.101\,621\,668\,250\,89 \times 10^{-9}$
15	3	0	$0.581\,855\,972\,552\,59 \times 10^{-2}$	30	7	5	$-0.164\,298\,282\,813\,47 \times 10^{-9}$

Equations (3.18), (3.19), and (3.20) are only valid in the respective subregion 2a, 2b, and 2c which do not include the metastable-vapour region. The boundaries between these subregions are defined at the beginning of Sec. 3.3.4; the lowest pressure for which Eq. (3.18) is valid amounts to 611.153 Pa corresponding to the sublimation pressure [11] at 273.15 K. The tolerated and actual numerical inconsistencies between Eqs. (3.18) to (3.20) and Eq. (3.8), given as maximum and root-mean-square (RMS) temperature differences, are listed in Table 3.28.

To assist the user in computer-program verification of Eqs. (3.18) to (3.20), Table 3.29 contains the corresponding test values.

Table 3.28. Tolerated and actual temperature differences between Eqs. (3.18) to (3.20) and Eq. (3.8)

Subregion	Equation	Temperature differences [mK]		
		Tolerated	Maximum	RMS
2a	3.18	10	8.8	1.2
2b	3.19	10	6.5	2.8
2c	3.20	25	19.0	8.3

Table 3.29. Temperature values calculated from Eqs. (3.18) to (3.20) for selected pressures and entropies

Equation	p [MPa]	s [kJ kg^{-1}K^{-1}]	T [K]
	0.1	7.5	$0.399\,517\,097 \times 10^3$
3.18	0.1	8	$0.514\,127\,081 \times 10^3$
	2.5	8	$0.103\,984\,917 \times 10^4$
	8	6	$0.600\,484\,040 \times 10^3$
3.19	8	7.5	$0.106\,495\,556 \times 10^4$
	90	6	$0.103\,801\,126 \times 10^4$
	20	5.75	$0.697\,992\,849 \times 10^3$
3.20	80	5.25	$0.854\,011\,484 \times 10^3$
	80	5.75	$0.949\,017\,998 \times 10^3$

3.4 Basic Equation for Region 3

This section contains all details relevant for using the basic equation of region 3 of IAPWS-IF97. The auxiliary equation for defining the boundary between regions 2 and 3 is given in Sec. 3.3.1 Uncertainty estimates of the most relevant properties, calculated from IAPWS-IF97, can be found in Sec. 3.7.

The basic equation for this region is a fundamental equation for the specific Helmholtz free energy f. This equation is expressed in dimensionless form, $\phi = f/(RT)$, and reads

$$\frac{f(\rho,T)}{RT} = \phi(\delta,\tau) = n_1 \ln\delta + \sum_{i=2}^{40} n_i \delta^{I_i} \tau^{J_i}, \tag{3.21}$$

where $\delta = \rho/\rho^*$, $\tau = T^*/T$ with $\rho^* = \rho_c$, $T^* = T_c$ and R, T_c, and ρ_c given by Eqs. (2.1), (2.4), and (2.6). The coefficients n_i and exponents I_i and J_i of Eq. (3.21) are listed in Table 3.30.

Table 3.30. Coefficients and exponents of Eq. (3.21)

i	I_i	J_i	n_i	i	I_i	J_i	n_i
1	0	0	$0.106\,580\,700\,285\,13 \times 10^{1}$	21	3	4	$-0.201\,899\,150\,235\,70 \times 10^{1}$
2	0	0	$-0.157\,328\,452\,902\,39 \times 10^{2}$	22	3	16	$-0.821\,476\,371\,739\,63 \times 10^{-2}$
3	0	1	$0.209\,443\,969\,743\,07 \times 10^{2}$	23	3	26	$-0.475\,960\,357\,349\,23$
4	0	2	$-0.768\,677\,078\,787\,16 \times 10^{1}$	24	4	0	$0.439\,840\,744\,735\,00 \times 10^{-1}$
5	0	7	$0.261\,859\,477\,879\,54 \times 10^{1}$	25	4	2	$-0.444\,764\,354\,287\,39$
6	0	10	$-0.280\,807\,811\,486\,20 \times 10^{1}$	26	4	4	$0.905\,720\,707\,197\,33$
7	0	12	$0.120\,533\,696\,965\,17 \times 10^{1}$	27	4	26	$0.705\,224\,500\,879\,67$
8	0	23	$-0.845\,668\,128\,125\,02 \times 10^{-2}$	28	5	1	$0.107\,705\,126\,263\,32$
9	1	2	$-0.126\,543\,154\,777\,14 \times 10^{1}$	29	5	3	$-0.329\,136\,232\,589\,54$
10	1	6	$-0.115\,244\,078\,066\,81 \times 10^{1}$	30	5	26	$-0.508\,710\,620\,411\,58$
11	1	15	$0.885\,210\,439\,843\,18$	31	6	0	$-0.221\,754\,008\,730\,96 \times 10^{-1}$
12	1	17	$-0.642\,077\,651\,816\,07$	32	6	2	$0.942\,607\,516\,650\,92 \times 10^{-1}$
13	2	0	$0.384\,934\,601\,866\,71$	33	6	26	$0.164\,362\,784\,479\,61$
14	2	2	$-0.852\,147\,088\,242\,06$	34	7	2	$-0.135\,033\,722\,413\,48 \times 10^{-1}$
15	2	6	$0.489\,722\,815\,418\,77 \times 10^{1}$	35	8	26	$-0.148\,343\,453\,524\,72 \times 10^{-1}$
16	2	7	$-0.305\,026\,172\,569\,65 \times 10^{1}$	36	9	2	$0.579\,229\,536\,280\,84 \times 10^{-3}$
17	2	22	$0.394\,205\,368\,791\,54 \times 10^{-1}$	37	9	26	$0.323\,089\,047\,037\,11 \times 10^{-2}$
18	2	26	$0.125\,584\,084\,243\,08$	38	10	0	$0.809\,648\,029\,962\,15 \times 10^{-4}$
19	3	0	$-0.279\,993\,296\,987\,10$	39	10	1	$-0.165\,576\,797\,950\,37 \times 10^{-3}$
20	3	2	$0.138\,997\,995\,694\,60 \times 10^{1}$	40	11	26	$-0.449\,238\,990\,618\,15 \times 10^{-4}$

In addition to representing the thermodynamic properties in the single-phase region, along the saturation line for temperatures from 623.15 K to T_c, Eq. (3.21) meets the phase-equilibrium condition for the coexisting liquid and vapour phase (equality of specific Gibbs free energy for both phases taken into account by the Maxwell criterion, see Table 3.31). Moreover, Eq. (3.21) reproduces exactly the critical parameters according to Eqs. (2.4) to (2.6) and yields zero for the first two pressure derivatives with respect to density at the critical point.

All thermodynamic properties can be derived from Eq. (3.21) by using the appropriate combinations of the dimensionless Helmholtz free energy ϕ and its derivatives. Relations between the relevant thermodynamic properties and ϕ and its derivatives are summarized in Table 3.31. All required derivatives of the dimensionless Helmholtz free energy are explicitly given in Table 3.32.

Equation (3.21) covers region 3 of IAPWS-IF97 defined by the following range of temperature and pressure, see Fig. 3.1:

$$623.15\ \mathrm{K} \le T \le T(p)_{\mathrm{Eq.(3.7)}} \qquad p(T)_{\mathrm{Eq.(3.6)}} \le p \le 100\ \mathrm{MPa}\,.$$

In addition to the properties in the stable single-phase region defined above, Eq. (3.21) also yields reasonable values in the metastable regions (superheated liquid and subcooled steam) close to the saturated liquid and saturated vapour line.

To assist the user in computer-program verification of Eq. (3.21), Table 3.33 contains test values of the most relevant properties.

Table 3.31. Relations of thermodynamic properties to the dimensionless Helmholtz free energy ϕ and its derivatives when using Eq. (3.21)

Property	Relation
Pressure $p=\rho^2(\partial f/\partial \rho)_T$	$\frac{p(\delta,\tau)}{\rho RT}=\delta\phi_\delta$
Specific internal energy $u=f-T(\partial f/\partial T)_\rho$	$\frac{u(\delta,\tau)}{RT}=\tau\phi_\tau$
Specific entropy $s=-(\partial f/\partial T)_\rho$	$\frac{s(\delta,\tau)}{R}=\tau\phi_\tau-\phi$
Specific enthalpy $h=f-T(\partial f/\partial T)_\rho+\rho(\partial f/\partial \rho)_T$	$\frac{h(\delta,\tau)}{RT}=\tau\phi_\tau+\delta\phi_\delta$
Specific isochoric heat capacity $c_v=(\partial u/\partial T)_\rho$	$\frac{c_v(\delta,\tau)}{R}=-\tau^2\phi_{\tau\tau}$
Specific isobaric heat capacity $c_p=(\partial h/\partial T)_p$	$\frac{c_p(\delta,\tau)}{R}=-\tau^2\phi_{\tau\tau}+\frac{(\delta\phi_\delta-\delta\tau\phi_{\delta\tau})^2}{2\delta\phi_\delta+\delta^2\phi_{\delta\delta}}$
Speed of sound $w=\left[(\partial p/\partial \rho)_s\right]^{0.5}$	$\frac{w^2(\delta,\tau)}{RT}=2\delta\phi_\delta+\delta^2\phi_{\delta\delta}-\frac{(\delta\phi_\delta-\delta\tau\phi_{\delta\tau})^2}{\tau^2\phi_{\tau\tau}}$
Phase-equilibrium condition (Maxwell criterion)	$\frac{p_s}{RT\rho'}=\delta'\phi_\delta(\delta',\tau)$; $\frac{p_s}{RT\rho''}=\delta''\phi_\delta(\delta'',\tau)$
	$\frac{p_s}{RT}\left(\frac{1}{\rho''}-\frac{1}{\rho'}\right)=\phi(\delta',\tau)-\phi(\delta'',\tau)$

$$\phi_\delta=\left(\frac{\partial\phi}{\partial\delta}\right)_\tau,\ \phi_{\delta\delta}=\left(\frac{\partial^2\phi}{\partial\delta^2}\right)_\tau,\ \phi_\tau=\left(\frac{\partial\phi}{\partial\tau}\right)_\delta,\ \phi_{\tau\tau}=\left(\frac{\partial^2\phi}{\partial\tau^2}\right)_\delta,\ \phi_{\delta\tau}=\left(\frac{\partial^2\phi}{\partial\delta\partial\tau}\right)$$

Table 3.32. The dimensionless Helmholtz free energy ϕ, Eq. (3.21), and its derivatives

$$\phi=n_1\ln\delta+\sum_{i=2}^{40}n_i\,\delta^{I_i}\,\tau^{J_i}$$

$$\phi_\tau=\sum_{i=2}^{40}n_i\,\delta^{I_i}\,J_i\,\tau^{J_i-1}$$

$$\phi_\delta=n_1\delta^{-1}+\sum_{i=2}^{40}n_i\,I_i\,\delta^{I_i-1}\,\tau^{J_i}$$

$$\phi_{\tau\tau}=\sum_{i=2}^{40}n_i\,\delta^{I_i}\,J_i(J_i-1)\,\tau^{J_i-2}$$

$$\phi_{\delta\delta}=-n_1\delta^{-2}+\sum_{i=2}^{40}n_i\,I_i(I_i-1)\,\delta^{I_i-2}\,\tau^{J_i}$$

$$\phi_{\delta\tau}=\sum_{i=2}^{40}n_i\,I_i\,\delta^{I_i-1}\,J_i\,\tau^{J_i-1}$$

$$\phi_\delta=\left(\frac{\partial\phi}{\partial\delta}\right)_\tau,\ \phi_{\delta\delta}=\left(\frac{\partial^2\phi}{\partial\delta^2}\right)_\tau,\ \phi_\tau=\left(\frac{\partial\phi}{\partial\tau}\right)_\delta,\ \phi_{\tau\tau}=\left(\frac{\partial^2\phi}{\partial\tau^2}\right)_\delta,\ \phi_{\delta\tau}=\left(\frac{\partial^2\phi}{\partial\delta\partial\tau}\right)$$

Table 3.33. Thermodynamic property values calculated from Eq. (3.21) for selected temperatures and densities

Property	$T = 650$ K, $\rho = 500$ kg m^{-3}	$T = 650$ K, $\rho = 200$ kg m^{-3}	$T = 750$ K, $\rho = 500$ kg m^{-3}
p [MPa]	$0.255\,837\,018 \times 10^{2}$	$0.222\,930\,643 \times 10^{2}$	$0.783\,095\,639 \times 10^{2}$
h [kJ kg^{-1}]	$0.186\,343\,019 \times 10^{4}$	$0.237\,512\,401 \times 10^{4}$	$0.225\,868\,845 \times 10^{4}$
u [kJ kg^{-1}]	$0.181\,226\,279 \times 10^{4}$	$0.226\,365\,868 \times 10^{4}$	$0.210\,206\,932 \times 10^{4}$
s [kJ kg^{-1} K^{-1}]	$0.405\,427\,273 \times 10^{1}$	$0.485\,438\,792 \times 10^{1}$	$0.446\,971\,906 \times 10^{1}$
c_p [kJ kg^{-1} K^{-1}]	$0.138\,935\,717 \times 10^{2}$	$0.446\,579\,342 \times 10^{2}$	$0.634\,165\,359 \times 10^{1}$
w [m s^{-1}]	$0.502\,005\,554 \times 10^{3}$	$0.383\,444\,594 \times 10^{3}$	$0.760\,696\,041 \times 10^{3}$

3.5 Equations for Region 4

This section contains all details relevant for using the equations for region 4 of IAPWS-IF97 (saturation line). Uncertainty estimates of the saturation pressure, calculated from IAPWS-IF97, can be found in Sec. 3.7.

The equation for describing the saturation line is an implicit quadratic equation which can be directly solved with regard to both saturation pressure p_s and saturation temperature T_s. This equation reads

$$\beta^2\vartheta^2 + n_1\beta^2\vartheta + n_2\beta^2 + n_3\beta\vartheta^2 + n_4\beta\vartheta + n_5\beta + n_6\vartheta^2 + n_7\vartheta + n_8 = 0\,, \tag{3.22}$$

where

$$\beta = \left(p_\mathrm{s} / p^*\right)^{0.25} \tag{3.22a}$$

and

$$\vartheta = \frac{T_\mathrm{s}}{T^*} + \frac{n_9}{\left(T_\mathrm{s} / T^*\right) - n_{10}} \tag{3.22b}$$

with $p^* = 1$ MPa and $T^* = 1$ K; for the coefficients n_1 to n_{10} see Table 3.34.

3.5.1 Saturation-Pressure Equation (Basic Equation)

The solution of Eq. (3.22) with regard to saturation pressure is as follows:

$$\frac{p_\mathrm{s}}{p^*} = \left[\frac{2C}{-B + \left(B^2 - 4AC\right)^{0.5}}\right]^4, \tag{3.23}$$

where $p^* = 1$ MPa and

$$A = \vartheta^2 + n_1\vartheta + n_2$$
$$B = n_3\vartheta^2 + n_4\vartheta + n_5$$
$$C = n_6\vartheta^2 + n_7\vartheta + n_8$$

with ϑ according to Eq. (3.22b). The coefficients n_i of Eq. (3.23) are listed in Table 3.34.

Table 3.34. Coefficients of Eqs. (3.22) to (3.24)

i	n_i	i	n_i
1	$0.116\,705\,214\,527\,67 \times 10^{4}$	6	$0.149\,151\,086\,135\,30 \times 10^{2}$
2	$-0.724\,213\,167\,032\,06 \times 10^{6}$	7	$-0.482\,326\,573\,615\,91 \times 10^{4}$
3	$-0.170\,738\,469\,400\,92 \times 10^{2}$	8	$0.405\,113\,405\,420\,57 \times 10^{6}$
4	$0.120\,208\,247\,024\,70 \times 10^{5}$	9	$-0.238\,555\,575\,678\,49$
5	$-0.323\,255\,503\,223\,33 \times 10^{7}$	10	$0.650\,175\,348\,447\,98 \times 10^{3}$

Equations (3.22) to (3.24) reproduce exactly the p-T values at the triple point according to Eqs. (2.7) and (2.8), at the normal boiling point according to Eq. (2.9) and at the critical point according to Eqs. (2.4) and (2.5).

Equation (3.23) is valid along the entire vapour-liquid saturation line from the triple-point temperature T_t to the critical temperature T_c and can be simply extrapolated to 273.15 K so that it covers the temperature range

$$273.15 \text{ K} \leq T \leq 647.096 \text{ K}.$$

To assist the user in computer-program verification of Eq. (3.23), Table 3.35 contains corresponding test values.

Table 3.35. Saturation-pressure values calculated from Eq. (3.23) for selected temperatures

T [K]	p_s [MPa]
300	$0.353\,658\,941 \times 10^{-2}$
500	$0.263\,889\,776 \times 10^{1}$
600	$0.123\,443\,146 \times 10^{2}$

3.5.2 Saturation-Temperature Equation (Backward Equation)

The saturation-temperature solution of Eq. (3.22) reads

$$\frac{T_s}{T^*} = \frac{n_{10} + D - \left[(n_{10}+D)^2 - 4(n_9 + n_{10}D)\right]^{0.5}}{2}, \tag{3.24}$$

where $T^* = 1$ K and

$$D = \frac{2G}{-F - \left(F^2 - 4EG\right)^{0.5}}$$

with

$$E = \beta^2 + n_3\beta + n_6$$
$$F = n_1\beta^2 + n_4\beta + n_7$$
$$G = n_2\beta^2 + n_5\beta + n_8$$

and β according to Eq. (3.22a). The coefficients n_i of Eq. (3.24) are listed in Table 3.34.

Equation (3.24) has the same range of validity as Eq. (3.23), which means that it covers the vapour-liquid saturation line according to the pressure range

$$611.213\ \mathrm{Pa} \le p \le 22.064\ \mathrm{MPa}\,.$$

The value of 611.213 Pa corresponds to the pressure when Eq. (3.24) is extrapolated to 273.15 K. Since the saturation-pressure equation, Eq. (3.23), and the saturation-temperature equation, Eq. (3.24), have been derived from the same implicit equation, Eq. (3.22), for describing the saturation line, both Eq. (3.23) and Eq. (3.24) are *completely* consistent with each other.

To assist the user in computer-program verification of Eq. (3.24), Table 3.36 contains corrresponding test values.

Table 3.36. Saturation-temperature values calculated from Eq. (3.24) for selected pressures

p [MPa]	T_s [K]
0.1	$0.372\,755\,919 \times 10^3$
1	$0.453\,035\,632 \times 10^3$
10	$0.584\,149\,488 \times 10^3$

3.6 Basic Equation for Region 5

This section contains all details relevant for using the equations for region 5 of IAPWS-IF97. Uncertainty estimates of the most relevant properties can be found in Sec. 3.7.

The basic equation for this high-temperature region is a fundamental equation for the specific Gibbs free energy g. This equation is expressed in dimensionless form, $\gamma = g/(RT)$, and is separated into 2 parts, an ideal-gas part γ^{o} and a residual part γ^{r}, so that

$$\frac{g(p,T)}{RT} = \gamma(\pi,\tau) = \gamma^{\mathrm{o}}(\pi,\tau) + \gamma^{\mathrm{r}}(\pi,\tau)\ , \tag{3.25}$$

where $\pi = p/p^*$ and $\tau = T^*/T$ with R given by Eq. (2.1).

The equation for the ideal-gas part γ^{o} of the dimensionless Gibbs free energy reads

$$\gamma^{\mathrm{o}} = \ln\pi + \sum_{i=1}^{6} n_i^{\mathrm{o}}\,\tau^{J_i^{\mathrm{o}}}\ , \tag{3.26}$$

where $\pi = p/p^*$ and $\tau = T^*/T$ with $p^* = 1$ MPa and $T^* = 1000$ K. The coefficients n_1^{o} and n_2^{o} were adjusted in such a way that the values for the specific internal energy and specific entropy, calculated from Eq. (3.25), relate to Eq. (3.2). Table 3.37 contains the coefficients n_i^{o} and exponents J_i^{o} of Eq. (3.26).

Table 3.37. Coefficients and exponents of Eq. (3.26)

i	J_i^{o}	n_i^{o}	i	J_i^{o}	n_i^{o}
1	0	$-0.131\,799\,836\,742\,01 \times 10^2$	4	-2	$0.369\,015\,349\,803\,33$
2	1	$0.685\,408\,416\,344\,34 \times 10^1$	5	-1	$-0.311\,613\,182\,139\,25 \times 10^1$
3	-3	$-0.248\,051\,489\,334\,66 \times 10^{-1}$	6	2	$-0.329\,616\,265\,389\,17$

The form of the residual part γ^{r} of the dimensionless Gibbs free energy is as follows:

$$\gamma^{\mathrm{r}} = \sum_{i=1}^{5} n_i \, \pi^{I_i} \, \tau^{J_i} \, , \tag{3.27}$$

where $\pi = p/p^*$ and $\tau = T^*/T$ with $p^* = 1$ MPa and $T^* = 1000$ K. The coefficients n_i and exponents I_i and J_i of Eq. (3.27) are listed in Table 3.38.

All thermodynamic properties can be derived from Eq. (3.25) by using the appropriate combinations of the ideal-gas part γ^{o}, Eq. (3.26), and the residual part γ^{r}, Eq. (3.27), of the dimensionless Gibbs free energy and their derivatives. Relations between the relevant thermodynamic properties and γ^{o} and γ^{r} and their derivatives are summarized in Table 3.39. All required derivatives of the ideal-gas part and of the residual part of the dimensionless Gibbs free energy are explicitly given in Table 3.40 and Table 3.41, respectively.

Table 3.38. Coefficients and exponents of Eq. (3.27)

i	I_i	J_i	n_i
1	1	0	$-0.125\,631\,835\,895\,92 \times 10^{-3}$
2	1	1	$0.217\,746\,787\,145\,71 \times 10^{-2}$
3	1	3	$-0.459\,428\,208\,999\,10 \times 10^{-2}$
4	2	9	$-0.397\,248\,283\,595\,69 \times 10^{-5}$
5	3	3	$0.129\,192\,282\,897\,84 \times 10^{-6}$

Table 3.39. Relations of thermodynamic properties to the ideal-gas part γ^{o} and the residual part γ^{r} of the dimensionless Gibbs free energy and their derivatives when using Eq. (3.25)

Property	Relation
Specific volume $v = (\partial g/\partial p)_T$	$v(\pi,\tau)\dfrac{p}{RT} = \pi\left(\gamma_\pi^{\mathrm{o}} + \gamma_\pi^{\mathrm{r}}\right)$
Specific internal energy $u = g - T(\partial g/\partial T)_p - p(\partial g/\partial p)_T$	$\dfrac{u(\pi,\tau)}{RT} = \tau\left(\gamma_\tau^{\mathrm{o}} + \gamma_\tau^{\mathrm{r}}\right) - \pi\left(\gamma_\pi^{\mathrm{o}} + \gamma_\pi^{\mathrm{r}}\right)$
Specific entropy $s = -(\partial g/\partial T)_p$	$\dfrac{s(\pi,\tau)}{R} = \tau\left(\gamma_\tau^{\mathrm{o}} + \gamma_\tau^{\mathrm{r}}\right) - \left(\gamma^{\mathrm{o}} + \gamma^{\mathrm{r}}\right)$
Specific enthalpy $h = g - T(\partial g/\partial T)_p$	$\dfrac{h(\pi,\tau)}{RT} = \tau\left(\gamma_\tau^{\mathrm{o}} + \gamma_\tau^{\mathrm{r}}\right)$
Specific isobaric heat capacity $c_p = (\partial h/\partial T)_p$	$\dfrac{c_p(\pi,\tau)}{R} = -\tau^2\left(\gamma_{\tau\tau}^{\mathrm{o}} + \gamma_{\tau\tau}^{\mathrm{r}}\right)$
Specific isochoric heat capacity $c_v = (\partial u/\partial T)_v$	$\dfrac{c_v(\pi,\tau)}{R} = -\tau^2\left(\gamma_{\tau\tau}^{\mathrm{o}} + \gamma_{\tau\tau}^{\mathrm{r}}\right) - \dfrac{(1+\pi\gamma_\pi^{\mathrm{r}} - \tau\pi\gamma_{\pi\tau}^{\mathrm{r}})^2}{1-\pi^2\gamma_{\pi\pi}^{\mathrm{r}}}$
Speed of sound $w = v\left[-(\partial p/\partial v)_s\right]^{0.5}$	$\dfrac{w^2(\pi,\tau)}{RT} = \dfrac{1+2\pi\gamma_\pi^{\mathrm{r}} + \pi^2\gamma_\pi^{\mathrm{r}\,2}}{\left(1-\pi^2\gamma_{\pi\pi}^{\mathrm{r}}\right) + \dfrac{\left(1+\pi\gamma_\pi^{\mathrm{r}} - \tau\pi\gamma_{\pi\tau}^{\mathrm{r}}\right)^2}{\tau^2\left(\gamma_{\tau\tau}^{\mathrm{o}} + \gamma_{\tau\tau}^{\mathrm{r}}\right)}}$

$\gamma_\pi^{\mathrm{r}} = \left(\dfrac{\partial\gamma^{\mathrm{r}}}{\partial\pi}\right)_\tau$, $\gamma_{\pi\pi}^{\mathrm{r}} = \left(\dfrac{\partial^2\gamma^{\mathrm{r}}}{\partial\pi^2}\right)_\tau$, $\gamma_\tau^{\mathrm{r}} = \left(\dfrac{\partial\gamma^{\mathrm{r}}}{\partial\tau}\right)_\pi$, $\gamma_{\tau\tau}^{\mathrm{r}} = \left(\dfrac{\partial^2\gamma^{\mathrm{r}}}{\partial\tau^2}\right)_\pi$, $\gamma_{\pi\tau}^{\mathrm{r}} = \left(\dfrac{\partial^2\gamma^{\mathrm{r}}}{\partial\pi\partial\tau}\right)$, $\gamma_\tau^{\mathrm{o}} = \left(\dfrac{\partial\gamma^{\mathrm{o}}}{\partial\tau}\right)_\pi$, $\gamma_{\tau\tau}^{\mathrm{o}} = \left(\dfrac{\partial^2\gamma^{\mathrm{o}}}{\partial\tau^2}\right)_\pi$

Table 3.40. The ideal-gas part γ^{o} of the dimensionless Gibbs free energy, Eq. (3.26), and its derivatives

$\gamma^{\mathrm{o}} = \ln\pi + \sum_{i=1}^{6} n_i^{\mathrm{o}} \tau^{J_i^{\mathrm{o}}}$	$\gamma_\tau^{\mathrm{o}} = \sum_{i=1}^{6} n_i^{\mathrm{o}} J_i^{\mathrm{o}} \tau^{J_i^{\mathrm{o}}-1}$
$\gamma_\pi^{\mathrm{o}} = \pi^{-1}$	$\gamma_{\tau\tau}^{\mathrm{o}} = \sum_{i=1}^{6} n_i^{\mathrm{o}} J_i^{\mathrm{o}} \left(J_i^{\mathrm{o}}-1\right) \tau^{J_i^{\mathrm{o}}-2}$
$\gamma_{\pi\pi}^{\mathrm{o}} = -\pi^{-2}$	$\gamma_{\pi\tau}^{\mathrm{o}} = 0$

$\gamma_\pi^{\mathrm{o}} = \left(\frac{\partial\gamma^{\mathrm{o}}}{\partial\pi}\right)_\tau,\ \gamma_{\pi\pi}^{\mathrm{o}} = \left(\frac{\partial^2\gamma^{\mathrm{o}}}{\partial\pi^2}\right)_\tau,\ \gamma_\tau^{\mathrm{o}} = \left(\frac{\partial\gamma^{\mathrm{o}}}{\partial\tau}\right)_\pi,\ \gamma_{\tau\tau}^{\mathrm{o}} = \left(\frac{\partial^2\gamma^{\mathrm{o}}}{\partial\tau^2}\right)_\pi,\ \gamma_{\pi\tau}^{\mathrm{o}} = \left(\frac{\partial^2\gamma^{\mathrm{o}}}{\partial\pi\,\partial\tau}\right)$

Table 3.41. The residual part γ^{r} of the dimensionless Gibbs free energy, Eq. (3.27), and its derivatives

$\gamma^{\mathrm{r}} = \sum_{i=1}^{5} n_i \pi^{I_i} \tau^{J_i}$	$\gamma_\tau^{\mathrm{r}} = \sum_{i=1}^{5} n_i \pi^{I_i} J_i \tau^{J_i-1}$
$\gamma_\pi^{\mathrm{r}} = \sum_{i=1}^{5} n_i I_i \pi^{I_i-1} \tau^{J_i}$	$\gamma_{\tau\tau}^{\mathrm{r}} = \sum_{i=1}^{5} n_i \pi^{I_i} J_i (J_i-1) \tau^{J_i-2}$
$\gamma_{\pi\pi}^{\mathrm{r}} = \sum_{i=1}^{5} n_i I_i (I_i-1) \pi^{I_i-2} \tau^{J_i}$	$\gamma_{\pi\tau}^{\mathrm{r}} = \sum_{i=1}^{5} n_i I_i \pi^{I_i-1} J_i \tau^{J_i-1}$

$\gamma_\pi^{\mathrm{r}} = \left(\frac{\partial\gamma^{\mathrm{r}}}{\partial\pi}\right)_\tau,\ \gamma_{\pi\pi}^{\mathrm{r}} = \left(\frac{\partial^2\gamma^{\mathrm{r}}}{\partial\pi^2}\right)_\tau,\ \gamma_\tau^{\mathrm{r}} = \left(\frac{\partial\gamma^{\mathrm{r}}}{\partial\tau}\right)_\pi,\ \gamma_{\tau\tau}^{\mathrm{r}} = \left(\frac{\partial^2\gamma^{\mathrm{r}}}{\partial\tau^2}\right)_\pi,\ \gamma_{\pi\tau}^{\mathrm{r}} = \left(\frac{\partial^2\gamma^{\mathrm{r}}}{\partial\pi\,\partial\tau}\right)$

Equation (3.25) covers region 5 of IAPWS-IF97 defined by the following temperature and pressure range, see Fig. 3.1:

$$1073.15\ \mathrm{K} \le T \le 2273.15\ \mathrm{K} \qquad 0 < p \le 10\ \mathrm{MPa}\,.$$

In this range Eq. (3.25) is only valid for pure undissociated water, any dissociation will have to be taken into account separately.

To assist the user in computer-program verification of Eq. (3.25), Table 3.42 contains test values of the most relevant properties.

Table 3.42. Thermodynamic property values calculated from Eq.(3.25) for selected temperatures and pressures

Property	T = 1500 K p = 0.5 MPa	T = 1500 K p = 8 MPa	T = 2000 K p = 8 MPa
v [m^3 kg^{-1}]	0.138 455 354 × 10^1	0.865 156 616 × 10^{-1}	0.115 743 146
h [kJ kg^{-1}]	0.521 976 332 × 10^4	0.520 609 634 × 10^4	0.658 380 291 × 10^4
u [kJ kg^{-1}]	0.452 748 654 × 10^4	0.451 397 105 × 10^4	0.565 785 774 × 10^4
s [kJ kg^{-1} K^{-1}]	0.965 408 431 × 10^1	0.836 546 724 × 10^1	0.915 671 044 × 10^1
c_p [kJ kg^{-1} K^{-1}]	0.261 610 228 × 10^1	0.264 453 866 × 10^1	0.285 306 750 × 10^1
w [m s^{-1}]	0.917 071 933 × 10^3	0.919 708 859 × 10^3	0.105 435 806 × 10^4

3.7 Uncertainties of IAPWS-IF97

In this section uncertainties are given for the properties specific volume, specific isobaric heat capacity, speed of sound, and saturation pressure when calculated from the corresponding equations of IAPWS-IF97. Uncertainties cannot be derived for the specific enthalpy, since there is no reasonable basis for their estimation; the values of the specific enthalpy depend on the selection of the zero point. For enthalpy differences, and here only these are of interest, one can conclude that the uncertainty of *isobaric* enthalpy differences is always smaller than the uncertainty of the isobaric heat capacity.

The estimated uncertainties for the above mentioned properties result from 2 contributions:

- Uncertainty of the scientific standard for the thermodynamic properties of water and steam, the IAPWS-95 formulation, from which the values of the properties were calculated to fit the equations of the industrial formulation IAPWS-IF97. The uncertainty of IAPWS-95 is mainly based on the estimated uncertainties of the selected experimental data of the properties [8], which had been used for the development of the scientific standard.
- Deviations of IAPWS-IF97 from IAPWS-95 regarding the properties considered.

The uncertainties of the properties v, c_p, and w calculated from IAPWS-IF97 for the single-phase region are indicated in Figs. 3.3 to 3.5 as tolerance values. As used here "tolerance" means the range of possible values as judged by IAPWS, and no statistical significance can be attached to it. With regard to the uncertainty for the specific isobaric heat capacity and the speed of sound, see Figs. 3.4 and 3.5, it should be noted that the uncertainties for these properties increase drastically when approaching the critical point. The statement "no definitive uncertainty estimates possible" for temperatures above 1273 K is based on the fact that this range is beyond the range of validity of IAPWS-95 and the corresponding input values for IAPWS-IF97 were extrapolated from IAPWS-95. From various tests of IAPWS-95 [8] it is expected that these extrapolations yield reasonable values.

The estimated uncertainties of the saturation pressure calculated from the IAPWS-IF97 are given in Fig. 3.6.

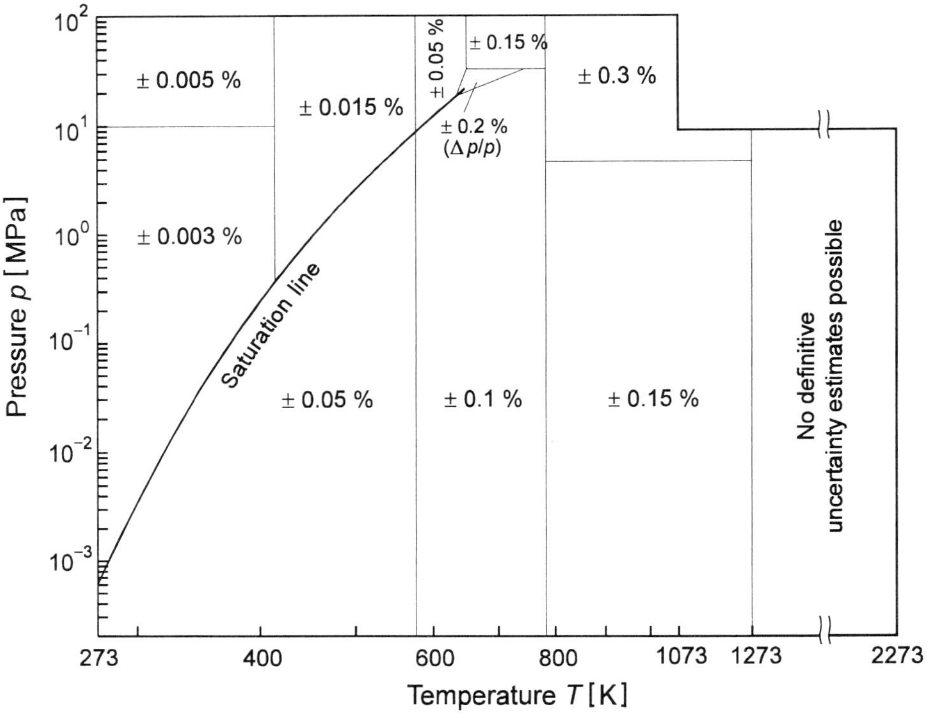

Fig. 3.3. Uncertainties in specific volume, $\Delta v/v$, estimated for the corresponding equations of IAPWS-IF97. In the enlarged critical region (triangle), the uncertainty is given as percentage uncertainty in pressure, $\Delta p/p$. This region is bordered by the two isochores 0.0019 $m^3\,kg^{-1}$ and 0.0069 $m^3\,kg^{-1}$ and by the 30 MPa isobar. The positions of the lines separating the uncertainty regions are approximate

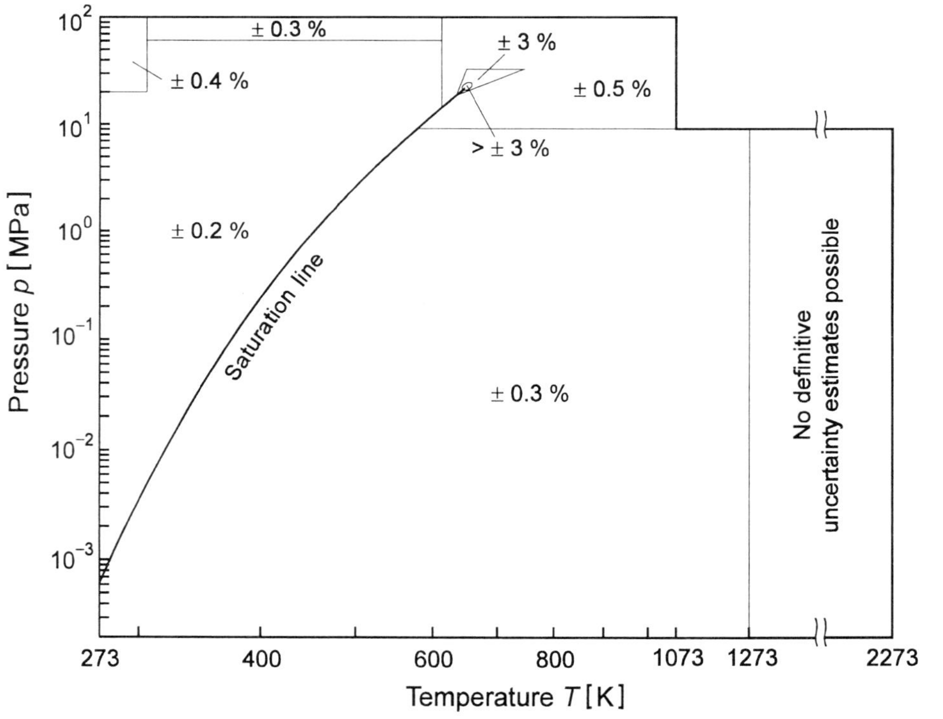

Fig. 3.4. Uncertainties in specific isobaric heat capacity, $\Delta c_p/c_p$, estimated for the corresponding equations of IAPWS-IF97. For the definition of the triangle around the critical point, see Fig. 3.3. The positions of the lines separating the uncertainty regions are approximate

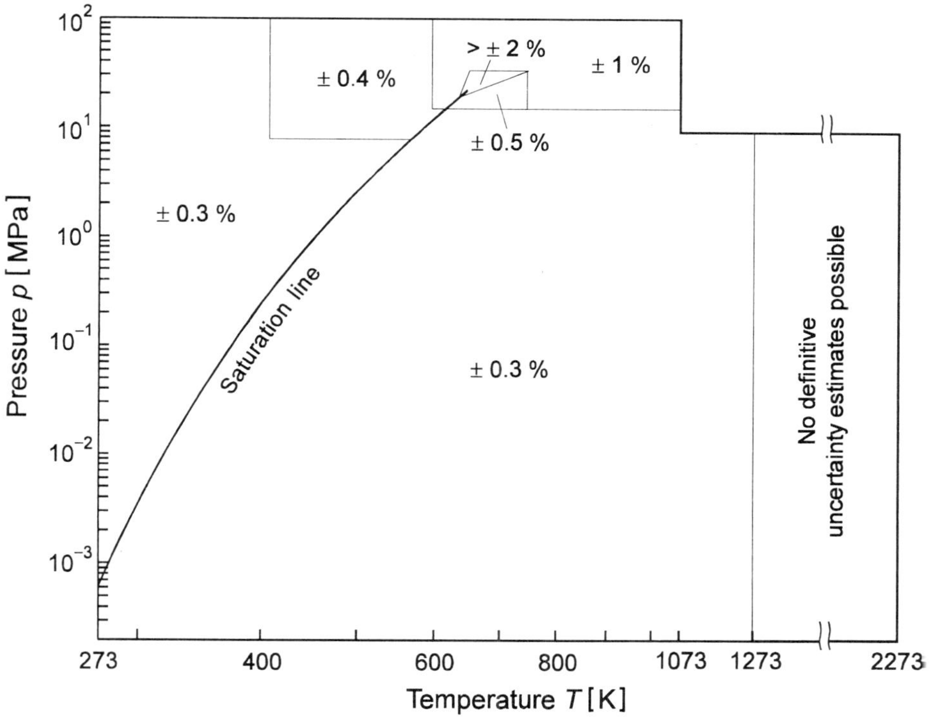

Fig. 3.5. Uncertainties in speed of sound, $\Delta w/w$, estimated for the corresponding equations of IAPWS-IF97. For the definition of the triangle around the critical point, see Fig. 3.3. The positions of the lines separating the uncertainty regions are approximate

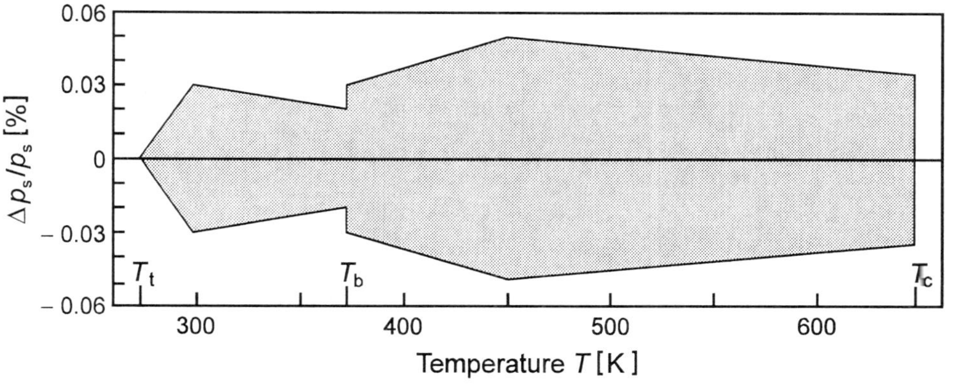

Fig. 3.6. Uncertainties in saturation pressure, $\Delta p_s/p_s$, estimated for the saturation-pressure equation, Eq. (3.23)

4 Equations for Transport Properties and Other Properties

Besides the development of international standard equations for the thermodynamic properties of water and steam IAPWS has also initiated and coordinated the development of equations for transport properties and for other properties. The latest IAPWS equations for the transport properties dynamic viscosity and thermal conductivity are presented in Sects. 4.1 and 4.2 and the IAPWS equations for the properties surface tension, static dielectric constant, and refractive index are described in Sects. 4.3 to 4.5. Except for the surface tension, all these equations contain the density as one of their input variables. If these properties are to be calculated for given values of pressure and temperature, then the density must always be determined first. In such cases, the density value should be calculated from the industrial formulation IAPWS-IF97 [9] described in Chap. 3. This procedure was used to calculate the values of the transport properties and other properties given in the corresponding tables in Part C of this book.

4.1 Dynamic Viscosity

The equation for the dynamic viscosity η of water and steam was originally formulated by Aleksandrov et al. [12] and is based on a set of experimental viscosity data selected by a special committee of IAPS (now IAPWS). This equation had formed the IAPWS equation for the dynamic viscosity from 1975 to 1997 [13–15]. In 1997, a slightly modified equation was adopted as the current viscosity equation of IAPWS [16] which is consistent with the temperature scale ITS-90 [6] and, regarding density, with the IAPWS-95 formulation [8], the current scientific standard for the thermodynamic properties of water and steam.

The reduced dynamic viscosity is represented by the equation

$$\Psi(\delta,\tau) = \Psi_0(\tau) \cdot \Psi_1(\delta,\tau) \tag{4.1}$$

where $\Psi = \eta/\eta^*$, $\delta = \rho/\rho^*$, and $\tau = T^*/T$ with $\eta^* = 55.071 \times 10^{-6}$ Pa s, $\rho^* = 317.763$ kg m^{-3}, and $T^* = 647.226$ K. The first term of the product in Eq. (4.1) covers the viscosity of steam in the ideal-gas limit and has the form

$$\Psi_0(\tau) = \left[\tau^{0.5} \sum_{i=0}^{3} n_i^{\mathrm{o}} \tau^i \right]^{-1}, \tag{4.2}$$

where $\tau = T^*/T$ with $T^* = 647.226$ K; the coefficients n_i^{o} are listed in Table 4.1. The equation for the second multiplicative factor in Eq. (4.1) reads

$$\Psi_1(\delta,\tau) = \exp\left[\delta \sum_{i=1}^{19} n_i (\delta - 1)^{I_i} (\tau - 1)^{J_i} \right], \tag{4.3}$$

where $\delta = \rho/\rho^*$ and $\tau = T^*/T$ with $\rho^* = 317.763$ kg m^{-3} and $T^* = 647.226$ K. The coefficients n_i and the exponents I_i and J_i are listed in Table 4.2.

Table 4.1. Coefficients of Eq. (4.2)

i	n_i^{o}	i	n_i^{o}
0	$0.100\,000 \times 10^{1}$	2	$0.579\,829$
1	$0.978\,197$	3	$-0.202\,354$

Table 4.2. Coefficients and exponents of Eq. (4.3)

i	I_i	J_i	n_i	i	I_i	J_i	n_i
1	0	0	$0.513\,204\,7$	11	2	2	$-0.126\,318\,4 \times 10^{1}$
2	0	1	$0.320\,565\,6$	12	3	0	$0.177\,806\,4$
3	0	4	$-0.778\,256\,7$	13	3	1	$0.460\,504\,0$
4	0	5	$0.188\,544\,7$	14	3	2	$0.234\,037\,9$
5	1	0	$0.215\,177\,8$	15	3	3	$-0.492\,417\,9$
6	1	1	$0.731\,788\,3$	16	4	0	$-0.417\,661\,0 \times 10^{-1}$
7	1	2	$0.124\,104\,4 \times 10^{1}$	17	4	3	$0.160\,043\,5$
8	1	3	$0.147\,678\,3 \times 10^{1}$	18	5	1	$-0.157\,838\,6 \times 10^{-1}$
9	2	0	$-0.281\,810\,7$	19	6	3	$-0.362\,948\,1 \times 10^{-2}$
10	2	1	$-0.107\,078\,6 \times 10^{1}$				

IAPWS endorses the validity of Eq. (4.1) for the dynamic viscosity in the following range of temperatures and pressures:

$$
\begin{aligned}
273.15\ \text{K} \le\ & T \le\ 423.15\ \text{K} & p \le 500\ \text{MPa}\\
423.15\ \text{K} <\ & T \le\ 873.15\ \text{K} & p \le 350\ \text{MPa}\\
873.15\ \text{K} <\ & T \le 1173.15\ \text{K} & p \le 300\ \text{MPa}\,.
\end{aligned}
$$

The estimated uncertainty in dynamic viscosity calculated from Eq. (4.1) is given in the IAPWS release [16].

If the dynamic viscosity is to be calculated from Eq. (4.1) for given values of *pressure* and temperature, then the input quantity reduced density δ has to be calculated first. For this density calculation, the use of the IAPWS-95 formulation [8] is recommended in the IAPWS release [16]. However, for industrial use, we recommend IAPWS-IF97 (see Chap. 3) for calculating the input density for Eq. (4.1). For this application, the difference between IAPWS-IF97 and IAPWS-95 is negligible compared with the estimated uncertainty of Eq. (4.1). The use of IAPWS-IF97 has clear advantages for industrial calculations and avoids the implementation of a second formulation for thermodynamic properties. Accordingly, the viscosity values listed in the corresponding tables in Part C of this book were calculated in the aforementioned way.

To assist the user in computer-program verification of Eq. (4.1), Table 4.3 contains corresponding test values.

Table 4.3. Dynamic viscosity values calculated from Eq. (4.1) for selected temperatures and pressures

Property	$T = 273.15$ K $p = 0.1$ MPa	$T = 773.15$ K $p = 0.1$ MPa	$T = 773.15$ K $p = 100$ MPa
ρ [kg m^{-3}] (IAPWS-IF97)	$0.999\,843\,633 \times 10^{3}$	$0.280\,459\,027$	$0.528\,195\,286 \times 10^{3}$
η [μPa s]	$0.179\,153\,095 \times 10^{4}$	$0.285\,746\,586 \times 10^{2}$	$0.660\,620\,604 \times 10^{2}$

4.2 Thermal Conductivity

For the calculation of the thermal conductivity λ, two different equations are currently recommended by IAPWS [17]; these equations have essentially the same data basis [18]. The so-called "Equation for Industrial Use" [19] is, regarding density, connected with IFC-67 [1], the predecessor of IAPWS-IF97. The other thermal conductivity equation, the "Equation for Scientific Use" [19], is, regarding density, connected with IAPWS-95 [8]. On account of the enormous differences between IFC-67 and IAPWS-95 we have chosen the "Equation for Scientific Use" [17] for calculating the thermal conductivity.

The equation for the thermal conductivity reads

$$\Lambda(\delta,\tau)=\Lambda_0(\tau)\cdot\Lambda_1(\delta,\tau)+\Lambda_2(\delta,\tau)\,, \tag{4.4}$$

where $\Lambda = \lambda/\lambda^*$, $\delta=\rho/\rho^*$, and $\tau=T^*/T$ with $\lambda^*=0.4945\ \mathrm{W\,m^{-1}\,K^{-1}}$, $\rho^*=317.763\ \mathrm{kg\,m^{-3}}$, and $T^*=647.226$ K. The factor $\Lambda_0(\tau)$ represents the thermal conductivity of steam in the ideal-gas limit and has the form

$$\Lambda_0(\tau)=\left[\tau^{0.5}\sum_{i=0}^{3} n_i^{\mathrm{o}}\,\tau^{i}\right]^{-1}, \tag{4.5}$$

where $\tau=T^*/T$ with $T^*=647.226$ K; the coefficients n_i^{o} are listed in Table 4.4. The factor Λ_1 is as follows:

$$\Lambda_1(\delta,\tau)=\exp\left[\delta\sum_{i=0}^{4}\sum_{j=0}^{5} n_{ij}(\tau-1)^i(\delta-1)^j\right], \tag{4.6}$$

where $\delta=\rho/\rho^*$ and $\tau=T^*/T$ with $\rho^*=317.763\ \mathrm{kg\,m^{-3}}$ and $T^*=647.226$ K. The coefficients n_{ij} are listed in Table 4.5. The additive term $\Lambda_2(\delta,\tau)$ of Eq. (4.4), which accounts for an enhancement of the thermal conductivity in the critical region, is necessary for the representation of the thermal conductivity within the given uncertainties [17] even outside the limited critical region and should be used in any case. The equation for the term Λ_2 has the form

$$\Lambda_2(\delta,\tau)=\frac{0.0013848}{\Psi(\delta,\tau)_{\mathrm{Eq.(4.1)}}}(\tau\,\delta)^{-2}\left(\frac{\partial\pi}{\partial(\tau^{-1})}\right)_{\delta}^{2}\left[\delta\left(\frac{\partial\delta}{\partial\pi}\right)_{\tau}\right]^{0.4678}\delta^{0.5}$$

$$\cdot\exp\left[-18.66\left(\tau^{-1}-1\right)^2-(\delta-1)^4\right], \tag{4.7}$$

where $\delta=\rho/\rho^*$, $\tau=T^*/T$, and $\pi = p/p^*$ with $\rho^*=317.763\ \mathrm{kg\,m^{-3}}$, $T^*=647.226$ K, and $p^*=22.115$ MPa. The reduced dynamic viscosity Ψ is calculated from Eq. (4.1) and the partial derivatives $(\partial\pi/\partial(\tau^{-1}))_\delta$ and $(\partial\delta/\partial\pi)_\tau$ should be calculated from the corresponding equation of IAPWS-IF97 (see Chap. 3).

Table 4.4. Coefficients of Eq. (4.5)

i	n_i^{o}	i	n_i^{o}
0	$0.100\,000\,0 \times 10^1$	2	$0.259\,909\,6 \times 10^1$
1	$0.697\,826\,7 \times 10^1$	3	$-0.998\,254\,0$

Table 4.5. Coefficients of Eq. (4.6)

j	n_{0j}	n_{1j}	n_{2j}	n_{3j}	n_{4j}
0	$0.132\,930\,46 \times 10^1$	$0.170\,183\,63 \times 10^1$	$0.522\,461\,58 \times 10^1$	$0.871\,276\,75 \times 10^1$	$-0.185\,259\,99 \times 10^1$
1	$-0.404\,524\,37$	$-0.221\,568\,45 \times 10^1$	$-0.101\,241\,11 \times 10^2$	$-0.950\,006\,11 \times 10^1$	$0.934\,046\,90$
2	$0.244\,094\,90$	$0.165\,110\,57 \times 10^1$	$0.498\,746\,87 \times 10^1$	$0.437\,866\,06 \times 10^1$	0
3	$0.186\,607\,51 \times 10^{-1}$	$-0.767\,360\,02$	$-0.272\,976\,94$	$-0.917\,837\,82$	0
4	$-0.129\,610\,68$	$0.372\,833\,44$	$-0.430\,833\,93$	0	0
5	$0.448\,099\,53 \times 10^{-1}$	$-0.112\,031\,60$	$0.133\,338\,49$	0	0

IAPWS endorses the validity of Eq. (4.4) for the thermal conductivity in the following range of temperatures and pressures:

$$
\begin{aligned}
273.15\ \mathrm{K} &\le T \le 398.15\ \mathrm{K} & p &\le 400\ \mathrm{MPa} \\
398.15\ \mathrm{K} &< T \le 523.15\ \mathrm{K} & p &\le 200\ \mathrm{MPa} \\
523.15\ \mathrm{K} &< T \le 673.15\ \mathrm{K} & p &\le 150\ \mathrm{MPa} \\
673.15\ \mathrm{K} &< T \le 1073.15\ \mathrm{K} & p &\le 100\ \mathrm{MPa}\,.
\end{aligned}
$$

The estimated uncertainty in thermal conductivity calculated from Eq. (4.4) is given in the IAPWS release [17].

If the thermal conductivity is to be calculated from Eq. (4.4) for given values of *pressure* and temperature, then the input quantity reduced density δ has to be calculated first. For this density calculation, the use of the IAPWS-95 formulation [8] is recommended in the IAPWS release [17]. However, for industrial use, we recommend IAPWS-IF97 (see Chap. 3) for calculating the input density for Eq. (4.4) and the partial derivatives of the thermal properties needed in Eq. (4.7). For this application, the difference between IAPWS-IF97 and IAPWS-95 is negligible compared with the estimated uncertainty of Eq. (4.4). The use of IAPWS-IF97 has clear advantages for industrial calculations and avoids the implementation of a second formulation for thermodynamic properties. Accordingly, the values of the thermal conductivity listed in the corresponding tables in Part C of this book were calculated in the aforementioned way.

To assist the user in computer-program verification of Eq. (4.4), Table 4.6 contains corresponding test values.

Table 4.6. Thermal conductivity values calculated from Eq. (4.4) for selected temperatures and pressures

Property	$T = 273.15$ K $p = 0.1$ MPa	$T = 773.15$ K $p = 0.1$ MPa	$T = 773.15$ K $p = 100$ MPa
ρ [kg m^{-3}] (IAPWS-IF97)	$0.999\,843\,633 \times 10^3$	$0.280\,459\,027$	$0.528\,195\,286 \times 10^3$
λ [mW K^{-1}]	$0.561\,075\,931 \times 10^3$	$0.669\,788\,623 \times 10^2$	$0.394\,700\,444 \times 10^3$

4.3 Surface Tension

In 1975, the experimental results of the surface tension σ of the interface between the liquid phase and vapour phase of water were critically examined by IAPWS. As a result, IAPWS recommended table values of surface tension, which were adjusted to ITS-90 [6] in 1994 [20]. An equation for the surface tension as function of temperature was fitted to these table values [20].

The equation for the surface tension has the following form:

$$\frac{\sigma}{\sigma^*} = 235.8\,(1-\theta)^{1.256}\left[1-0.625\,(1-\theta)\right], \tag{4.8}$$

where $\theta = T/T^*$ with $\sigma^* = 1\ \mathrm{mN\ m^{-1}}$ and $T^* = T_c$. This equation is valid between T_t and T_c given by Eqs. (2.8) and (2.4). The estimated uncertainty in surface tension calculated from Eq. (4.8) is given in the IAPWS release [20].

To assist the user in computer-program verification of Eq (4.8), Table 4.7 contains corresponding test values.

Table 4.7. Surface tension values calculated from Eq. (4.8) for selected temperatures

T [K]	σ [mN m^{-1}]
300	$0.716\,859\,625 \times 10^2$
500	$0.314\,719\,761 \times 10^2$
600	$0.837\,561\,087 \times 10^1$

4.4 Static Dielectric Constant

A collection of experimental data available for the relative static dielectric constant (relative static permittivity) ε was published in 1995 [21]. Based on a selected set of these data, a new equation for the static dielectric constant was developed [22]. This equation, which has its physical basis in the so-called g-factor proposed by Harris and Alder [23], was approved by IAPWS in 1997 [24].

The equation for the relative static dielectric constant reads

$$\varepsilon = \frac{1+A+5B+\left(9+2A+18B+A^2+10AB+9B^2\right)^{0.5}}{4-4B}, \tag{4.9}$$

where A and B are given by

$$A = \frac{10^3 N_A\, \mu^2 \rho g}{M \varepsilon_o k T} \tag{4.9a}$$

and

$$B = \frac{10^3 N_A\, \alpha \rho}{3 M \varepsilon_o}\ . \tag{4.9b}$$

The values of the constants k, N_A, α, ε_o, and μ used in Eqs. (4.9a) and (4.9b) are given in

Table 4.8; for the molar mass M see Eq. (2.3). The correlation equation for the Harris-Alder g-factor in Eq. (4.9a) reads

$$g = 1 + \sum_{i=1}^{11} n_i\,\delta^{I_i}\,\tau^{J_i} + n_{12}\,\delta\left(\frac{T}{228\,\mathrm{K}} - 1\right)^{-1.2}, \tag{4.10}$$

where $\delta = \rho/\rho^*$, $\tau = T^*/T$ with $\rho^* = \rho_c$ and $T^* = T_c$, and ρ_c and T_c given by Eqs. (2.6) and (2.4). Table 4.9 contains the coefficients n_i and exponents I_i and J_i of Eq. (4.10).

Table 4.8. Constants used in Eqs. (4.9a) and (4.9b)

Quantity	Value
Boltzmann's constant k	$1.380\,658 \times 10^{-23}$ J K^{-1}
Avogadro's number N_A	$6.022\,136\,7 \times 10^{23}$ mol^{-1}
Mean molecular polarizability α	1.636×10^{-40} C^2 J^{-1} m^2
Permittivity of vacuum ε_o	$8.854\,187\,817 \times 10^{-12}$ C^2 J^{-1} m^{-1}
Molecular dipole moment μ	6.138×10^{-30} C m

Table 4.9. Coefficients and exponents of Eq. (4.10)

i	I_i	J_i	n_i	i	I_i	J_i	n_i
1	1	0.25	0.978 224 486 826	7	4	2	$0.949\,327\,488\,264 \times 10^{-1}$
2	1	1	− 0.957 771 379 375	8	5	2	$-0.980\,469\,816\,509 \times 10^{-2}$
3	1	2.5	0.237 511 794 148	9	6	5	$0.165\,167\,634\,970 \times 10^{-4}$
4	2	1.5	0.714 692 244 396	10	7	0.5	$0.937\,359\,795\,772 \times 10^{-4}$
5	3	1.5	− 0.298 217 036 956	11	10	10	$-0.123\,179\,218\,720 \times 10^{-9}$
6	3	2.5	− 0.108 863 472 196	12			$0.196\,096\,504\,426 \times 10^{-2}$

Equation (4.9) covers the following range of temperatures and pressures [22, 24]:

$$273.15\ \mathrm{K} \le T \le 323.15\ \mathrm{K} \qquad 0 < p \le p_{\max}$$
$$323.15\ \mathrm{K} < T \le 873.15\ \mathrm{K} \qquad 0 < p \le 600\ \mathrm{MPa}\,.$$

The value of $p_{\max}$ corresponds to the lower of the ice VI melting pressure [11] or 1000 MPa. Equation (4.9) is also valid for the metastable subcooled liquid at atmospheric pressure (0.101325 MPa) for temperatures from 238.15 K to 273.15 K. Furthermore, Eq. (4.9) extrapolates smoothly up to at least 1200 K and 1200 MPa. The estimated uncertainty in the static dielectric constant calculated from Eq. (4.9) is given in reference [22] and in the IAPWS release [24].

If the static dielectric constant is to be calculated from Eq. (4.9) for given values of *pressure* and temperature, then the input quantity reduced density δ has to be calculated first. For this density calculation, the use of IAPWS-95 formulation [8] is recommended in the IAPWS release [24]. However, for industrial use, we recommend IAPWS-IF97 (see Chap. 3) for calculating the input density for Eq. (4.9). For this application, the difference between IAPWS-IF97 and IAPWS-95 is negligible compared with the estimated uncertainty of Eq. (4.9). The use of IAPWS-IF97 has clear advantages for industrial calculations and avoids the implementation of a second formulation for thermodynamic properties. Accordingly, the values of the static dielectric constant listed in the corresponding tables in Part C of this book were calculated in the aforementioned way.

To assist the user in computer-program verification of Eq. (4.9), Table 4.10 contains corresponding test values.

Table 4.10. Static dielectric constant values calculated from Eq. (4.9) for selected temperatures and pressures

Property	$T = 300$ K $p = 10$ MPa	$T = 870$ K $p = 10$ MPa	$T = 870$ K $p = 100$ MPa
ρ [kg m^{-3}] (IAPWS-IF97)	$0.100\,094\,934 \times 10^{4}$	$0.261\,720\,329 \times 10^{2}$	0.378 [illegible]31 821 × 10^{3}
ε	$0.781\,122\,784 \times 10^{2}$	$0.112\,721\,058 \times 10^{1}$	$0.498\,289\,143 \times 10^{1}$

4.5 Refractive Index

Based on a comprehensive collection of experimental data of the refractive index n of water and steam [25] an equation for the Lorentz-Lorenz function dependent on density, temperature, and wavelength was developed in 1990 [26, 27]. This equation was refitted to the experimental data after converting them to the temperature scale ITS-90 [6] and, regarding density, to the values from IAPWS-95 [8], the current scientific standard of IAPWS for the thermodynamic properties. The refitted equation was approved by IAPWS in 1997 [28].

This equation for the refractive index has the following form:

$$n = \left(\frac{2A+1}{1-A}\right)^{0.5}, \tag{4.11}$$

where the quantity A is given by

$$A = \delta\left(a_0 + a_1\delta + a_2\theta + a_3\Lambda^2\theta + a_4\Lambda^{-2} + \frac{a_5}{\Lambda^2 - \Lambda_{\mathrm{UV}}^2} + \frac{a_6}{\Lambda^2 - \Lambda_{\mathrm{IR}}^2} + a_7\delta^2\right) \tag{4.11a}$$

with $\delta = \rho/\rho^*$, $\theta = T/T^*$, $\Lambda = \lambda/\lambda^*$ and $\rho^* = 1000$ kg m^{-3}, $T^* = 273.15$ K, $\lambda^* = 0.589$ μm. The values of the reduced effective infrared resonance Λ_{IR} and the reduced effective ultraviolet resonance Λ_{UV} are given by $\Lambda_{\mathrm{IR}} = 5.432\,937$ and $\Lambda_{\mathrm{UV}} = 0.229\,202$. The coefficients a_i of Eq. (4.11a) are listed in Table 4.11.

Table 4.11. Coefficients of Eq. (4.11a)

i	a_i	i	a_i
0	$0.244\,257\,733$	4	$0.158\,920\,570 \times 10^{-2}$
1	$0.974\,634\,476 \times 10^{-2}$	5	$0.245\,934\,259 \times 10^{-2}$
2	$-0.373\,234\,996 \times 10^{-2}$	6	$0.900\,704\,920$
3	$0.268\,678\,472 \times 10^{-3}$	7	$-0.166\,626\,219 \times 10^{-1}$

IAPWS endorses Eq. (4.11) in the following validity range regarding temperature, density, and wavelength:

$$\begin{aligned} 261.15\ \mathrm{K} &\le T \le 773.15\ \mathrm{K} \\ 0 &< \rho \le 1060\ \mathrm{kg\ m^{-3}} \\ 0.2\ \mu\mathrm{m} &\le \lambda \le 1.1\ \mu\mathrm{m}\,. \end{aligned}$$

Extrapolation of the equation to longer wavelengths seems to be reasonable in liquid water at wavelengths up to 1.9 μm. The estimated uncertainty in the refractive index calculated from Eq. (4.11) is given in the IAPWS release [28].

If the refractive index is to be calculated from Eq. (4.11) for given values of *pressure* and temperature, then the input quantity reduced density δ has to be calculated first. For this density calculation, the use of IAPWS-95 formulation [8] is recommended in the IAPWS release [28]. However, for industrial use, we recommend IAPWS-IF97 (see Chap. 3) for calculating the input density for Eq. (4.11). For this application, the difference between IAPWS-IF97 and IAPWS-95 is negligible compared with the estimated uncertainty of Eq. (4.11). The use of IAPWS-IF97 has clear advantages for industrial calculations and avoids the implementation of a second formulation for thermodynamic properties. Accordingly, values of the refractive index listed in the corresponding tables in Part C of this book were calculated in the aforementioned way.

To assist the user in computer-program verification of Eq. (4.11), Table 4.12 contains corresponding test values.

Table 4.12. Refractive index values calculated from Eq. (4.11) for selected temperatures, pressures, and wavelengths

Property	T = 273.15 K p = 0.1 MPa	T = 773.15 K p = 0.1 MPa	T = 773.15 K p = 100 MPa
ρ [kg m^{-3}] (IAPWS-IF97)	$0.999\,843\,633 \times 10^3$	0.280 459 027	$0.528\,195\,286 \times 10^3$
		$\lambda = 0.2265\ \mu$m	
n	$0.139\,452\,744 \times 10^1$	$0.100\,010\,083 \times 10^1$	$0.119\,828\,0\,82 \times 10^1$
		$\lambda = 0.589\ \mu$m	
n	$0.133\,434\,461 \times 10^1$	$0.100\,008\,711 \times 10^1$	$0.117\,020\,512 \times 10^1$

References

[1] International Formulation Committee of the 6th International Conference on the Properties of Steam (1967) The 1967 IFC Formulation for Industrial Use. Verein Deutscher Ingenieure, Düsseldorf

[2] Cohen, E.R., Taylor, B.N. (1986) The 1986 Adjustment of the Fundamental Physical Constants. CODATA Bulletin No. 63, Committee on Data for Science and Technology, Int. Council of Scientific Unions, Pergamon Press, Oxford

[3] Audi, G.,Wapstra, A.H. (1993) The 1993 atomic mass evaluation, (I) Atomic mass table. Nuclear Physics A 565: 1–65

[4] IUPAC (1991) Commission on Atomic Weights and Isotopic Abundances, Subcommittee for Isotopic Abundance Measurements. Isotopic compositions of the elements 1989. Pure and Appl. Chem. 63: 991–1002

[5] IAPWS (1995) Release on The Values of Temperature, Pressure and Density of Ordinary and Heavy Water Substances at Their Respective Critical Points. In: White, Jr., H.J., Sengers, J.V., Neumann, D.B., Bellows, J.C. (eds) Physical Chemistry of Aqueous Systems: Meeting the Needs of Industry. Proceedings of the 12th International Conference on the Properties of Water and Steam. Begell House, New York, pp. A101–A102

[6] Preston-Thomas, H. (1990) The International Temperature Scale of 1990 (ITS-90). Metrologia 27: 3–10

[7] Guildner, L.A., Johnson, D.P., Jones, F.E. (1976) Vapor Pressure of Water at Its Triple Point. J. Res. Natl. Bur. Stand. 80A: 505–521

[8] Pruß, A., Wagner, W. (1998) New International Formulation for the Thermodynamic Properties of Ordinary Water Substance for General and Scientific Use. To be submitted to J. Phys. Chem. Ref. Data. See also: IAPWS (1996) Release on the IAPWS Formulation 1995 for the Thermodynamic Properties of Ordinary Water Substance for General and Scientific Use. IAPWS Secretariat[1]

[9] Wagner, W., Cooper, J.R., Dittmann, A., Kijima, J., Kretzschmar, H.-J., Kruse, A., Mareš, R., Oguchi, K., Sato, H., Stöcker, I., Šifner, O., Takaishi, Y., Tanishita, I., Trübenbach, J., and Willkommen, Th. (1998) The IAPWS Industrial Formulation 1997 for the Thermodynamic Properties of Water and Steam. Submitted to ASME Journal of Engineering for Gas Turbines and Power

[10] Minutes of the meetings of the International Formulation Committee of ICPS in Prague (1965)

[11] IAPWS (1994) Release on the Pressure along the Melting and the Sublimation Curves of Ordinary Water Substance. In: Wagner, W., Saul, A., Pruß, A., J. Phys. Chem. Ref. Data 23: 515–527. Also in: White, Jr., H.J., Sengers, J.V., Neumann, D.B., Bellows, J.C. (eds) Physical Chemistry of Aqueous Systems: Meeting the Needs of Industry. Proceedings of the 12th International Conference on the Properties of Water and Steam. Begell House, New York, pp. A9–A12

[12] Aleksandrov, A.A., Ivanov, A.I., Matveev, A.B. (1975) The Dynamic Viscosity of Water and Steam within a Wide Region of Temperatures and Pressures. Teploenergetika 22: 59–64 (English translation: Therm. Eng. 22: 77–83)

[13] Scheffler, K., Rosner, N., Straub, J., Grigull, U. (1978) Der neue internationale Standard der dynamischen Viskosität von Wasser und Wasserdampf. Brennstoff - Wärme - Kraft 30: 73-78

[14] Nagashima, A. (1977) Viscosity of Water Substance – New International Formulation and Its Background. J. Phys. Chem. Ref. Data 6: 1133–1166

[15] Nagashima, A. (1983) Erratum: Viscosity of Water Substance – New International Formulation and Its Background. J. Phys. Chem. Ref. Data 12: 403

[16] IAPWS (1997) Revised Release on the IAPS Formulation 1985 for the Viscosity of Ordinary Water Substance. IAPWS Secretariat[1]

[17] IAPWS (1997) Revised Release on the IAPS Formulation 1985 for the Thermal Conductivity of Ordinary Water Substance. IAPWS Secretariat[1]

[18] Scheffler, K., Rosner, N., Reimann, M. (1977) Available Input of the Thermal Conductivity of Water Substance. Institut A für Thermodynamik, Technische Universität, München

[19] Sengers, J.V., Watson, J.T.R., Basu, R.S., Kamgar-Parsi, B., Hendricks, R.C. (1984) Representative Equations for the Thermal Conductivity of Water Substance. J. Phys. Chem. Ref. Data 13: 893–933

[20] IAPWS (1994) Release on Surface Tension of Ordinary Water Substance. IAPWS Secretariat[1]

[21] Fernández, D.P., Mulev, Y., Goodwin, A.R.H., Levelt Sengers, J.M.H. (1995) A Database for the Static Dielectric Constant of Water and Steam. J. Phys. Chem. Ref. Data 24: 33–69

[1] The release is available from the Executive Secretary of IAPWS, Dr. R.B. Dooley, EPRI, 3412 Hillview Avenue, Palo Alto, California 94304, USA.

[22] Fernández, D.P., Goodwin, A.R.H., Lemmon, E.W., Levelt Sengers, J.M.H., Williams, R.C. (1997) A Formulation for the Static Permittivity of Water and Steam at Temperatures from 238 K to 873 K at Pressures up to 1200 MPa, Including Derivatives and Debye-Hückel Coefficients. J. Phys. Chem. Ref. Data 26: 1125–1166

[23] Harris, F.E., Alder, B.J. (1953) Dielectric Polarization in Polar Substances. J. Chem. Phys. 21: 1031–1038

[24] IAPWS (1997) Release on the Static Dielectric Constant of Ordinary Water Substance for Temperatures from 238 K to 873 K and Pressures up to 1000 MPa. IAPWS Secretariat[2]

[25] Thormälen, I., Straub, J., Grigull, U. (1985) Refractive Index of Water and Its Dependence on Wavelength, Temperature, and Density. J. Phys. Chem. Ref. Data 14: 933–945

[26] Schiebener, P., Straub, J., Levelt Sengers, J.M.H., Gallagher, J.S. (1990) Refractive Index of Water and Steam as Function of Wavelength, Temperature and Density. J. Phys. Chem. Ref. Data 19: 677–717

[27] Schiebener, P., Straub, J., Levelt Sengers, J.M.H., Gallagher, J.S. (1990) Erratum: Refractive Index of Water and Steam as Function of Wavelength, Temperature and Density. J. Phys. Chem. Ref. Data 19: 1617

[28] IAPWS (1997) Release on the Refractive Index of Ordinary Water Substance as a Function of Wavelength, Temperature and Pressure. IAPWS Secretariat[2]

[2] The release is available from the Executive Secretary of IAPWS, Dr. R.B. Dooley, EPRI, 3412 Hillview Avenue, Palo Alto, California 94304, USA.

Teil B

Gleichungen für die Berechnung der thermophysikalischen Eigenschaften von Wasser und Wasserdampf

1 Einleitung

Teil B dieses Buches enthält die neuen, international vereinbarten Gleichungen für die industrielle Berechnung einer großen Anzahl thermophysikalischer Eigenschaften von Wasser und Wasserdampf.

Die Gleichungen für die Ermittlung der *thermodynamischen* Eigenschaften sind im neuen Industrie-Standard „IAPWS Industrial Formulation 1997 for the Thermodynamic Properties of Water and Steam“ (IAPWS-IF97) enthalten. Diese neue Formulation IAPWS-IF97 wurde von der „International Association for the Properties of Water and Steam“ auf ihrem Jahrestreffen 1997 in Erlangen verabschiedet und ersetzt den bisherigen Industrie-Standard IFC-67. Der neue Industrie-Standard IAPWS-IF97 wird in Kap. 3 ausführlich beschrieben.

Zur Ergänzung der Gleichungen für die thermodynamischen Eigenschaften von Wasser und Wasserdampf enthält Kap. 4 die Gleichungen für die *Transportgrößen* dynamische Viskosität und Wärmeleitfähigkeit sowie Gleichungen für die Größen Oberflächenspannung, statische Dielektrizitätskonstante und Brechungsindex.

Die Tafeln der wichtigsten Eigenschaften von Wasser und Wasserdampf, die den Teil C dieses Buches bilden, wurden ausschließlich mit Hilfe der Gleichungen berechnet, die in Teil B zusammengestellt sind.

2 Referenzkonstanten

Dieses Kapitel faßt alle Referenzkonstanten zusammen, die zur Auswertung der Gleichungen in Kap. 3 und 4 benötigt werden.

Die spezifische Gaskonstante von Wasser,

$$R = 0{,}461\,526 \text{ kJ kg}^{-1}\text{ K}^{-1}, \tag{2.1}$$

ergibt sich aus dem international vereinbarten Wert für die molare Gaskonstante [2],

$$R_\text{m} = 8{,}314\,51 \text{ J mol}^{-1}\text{ K}^{-1}, \tag{2.2}$$

und der Molmasse von natürlichem Wasser,

$$M = 18{,}015\,257 \text{ g mol}^{-1}\,. \tag{2.3}$$

Dieser Wert der Molmasse folgt aus der Molmasse von Wasserstoff $M_\text{H} = 1{,}007\,975\,97 \text{ g mol}^{-1}$ (basierend auf der in [3] angegebenen Molmasse der Isotope ^{1}H und ^{2}H und der Isotopenkonzentration, charakterisiert durch die Molenbrüche von 0,999 85 für ^{1}H und von 0,000 15 für ^{2}H [4]), und der Molmasse von Sauerstoff, $M_\text{O} = 15{,}999\,3047 \text{ g mol}^{-1}$ (basierend auf den in [3] angegebenen Molmassen für die Isotope ^{16}O, ^{17}O und ^{18}O sowie der Isotopenkonzentration dieser drei Isotope, charakterisiert durch die Molenbrüche von 0,997 62, 0,000 38 und 0,002, entsprechend dem natürlichen Vorkommen von Sauerstoff [4]).

Die Werte der kritischen Parameter

$$T_\text{c} = 647{,}096 \text{ K}, \tag{2.4}$$

$$p_\text{c} = 22{,}064 \text{ MPa}, \tag{2.5}$$

$$\rho_\text{c} = 322 \text{ kg m}^{-3} \tag{2.6}$$

sind im entsprechenden IAPWS-Release angegeben [5]. Die Tripelpunktstemperatur beträgt

$$T_\text{t} = 273{,}16 \text{ K} \tag{2.7}$$

gemäß der Internationalen Temperaturskala von 1990 (ITS-90) [6], und der Tripelpunktsdruck wurde von Guildner et al. [7] zu

$$p_\text{t} = 611{,}657 \text{ Pa} \tag{2.8}$$

bestimmt. Entsprechend dem wissenschaftlichen Standard für die Berechnung der thermodynamischen Zustandsgrößen von natürlichem Wasser, der IAPWS-95 Formulation [8], hat die Temperatur des normalen Siedepunktes (bei 0,101 325 MPa (1 atm)) den Wert

$$T_\text{b} = 373{,}1243 \text{ K}\,. \tag{2.9}$$

3 Die IAPWS Industrie-Formulation 1997 für die thermodynamischen Eigenschaften von Wasser und Wasserdampf

In den 60er Jahren wurde eine Industrie-Formulation für die thermodynamischen Eigenschaften von Wasser und Wasserdampf entwickelt, die als „The 1967 IFC Formulation for Industrial Use“ (IFC-67) bezeichnet wird. Seit 1967 war die IFC-67 die international vereinbarte Grundlage bei vertragsrelevanten Berechnungen auf der Basis einschlägiger Normen, z. B. für Berechnungen im Rahmen von Garantiefragen bei Kraftwerksprozessen. Darüber hinaus wurde die IFC-67 für zahlreiche andere energietechnische Anwendungen eingesetzt. Im Laufe der Jahre stellten sich bei der IFC-67 jedoch eine Reihe von Unzulänglichkeiten heraus. Vor diesem Hintergrund und unter Berücksichtigung des großen Fortschritts in den numerischen Methoden zur Aufstellung hochgenauer Zustandsgleichungen wurde eine neue Industrie-Formulation entwickelt. Diese Arbeiten wurden von der „International Association for the Properties of Water and Steam“ (IAPWS) als internationales Forschungsprojekt initiiert und koordiniert.

Auf dem Jahrestreffen der IAPWS 1997 in Erlangen wurde diese Formulation als neuer internationaler Industrie-Standard für die thermodynamischen Eigenschaften von Wasser und Wasserdampf verabschiedet. Der Name des neuen Industrie-Standards lautet „IAPWS Industrial Formulation 1997 for the Thermodynamic Properties of Water and Steam“ mit den Kurzbezeichnungen „IAPWS Industrial Formulation 1997“ oder „IAPWS-IF97“. Die neue Formulation IAPWS-IF97 ersetzt den bisherigen Industrie-Standard IFC-67 [1]. Im Vergleich zur IFC-67 werden durch die IAPWS-97 die Genauigkeit und die Rechengeschwindigkeit bei der Berechnung der thermodynamischen Eigenschaften deutlich erhöht.

Das Projekt zur Entwicklung einer neuen Industrie-Formulation wurde von der IAPWS-Arbeitsgruppe „Industrial Calculations“, in der viele Mitglieder Vertreter der internationalen Kraftwerksindustrie sind, initiiert und kontinuierlich begleitet. Diese von B Rukes (Deutschland) geleitete Arbeitsgruppe ist die Nachfolgerin des IAPWS „Subcommittee on Industrial Calculations“. Der offizielle Beginn des umfassenden Projektes erfolgte auf dem Jahrestreffen der IAPWS 1990 in Buenos Aires mit der Gründung der IAPWS-Arbeitsgruppe „New Industrial Formulation“. Diese Arbeitsgruppe, die von einem der Autoren dieses Buches (W. W.) geleitet wurde und der 12 Mitglieder aus 7 Ländern angehörten, war direkt verantwortlich für die Entwicklung der IAPWS-IF97. Die IAPWS-Arbeitsgruppe „New Industrial Formulation-Evaluation“, die von K. Miyagawa (Japan) geleitet wurde, prüfte die neue Formulation bezüglich der Erfüllung der an die IAPWS-IF97 gestellten Anforderungen. Darüber hinaus führte sie umfassende Tests durch, um die Unterschiede zwischen der IAPWS-IF97 und der IFC-67 bei der Berechnung realer Kraftwerksprozesse zu ermitteln.

Dieses Kapitel enthält alle Informationen zu den einzelnen Gleichungen der IAPWS-IF97, die zur Berechnung der thermodynamischen Zustandsgrößen von Wasser und Wasserdampf mit der neuen Formulation notwendig sind. Weitergehende Informationen zur Entwicklung der Gleichungen, zu ihrer Genauigkeit, zur erreichten Konsistenz an den Bereichsgrenzen und zur Rechengeschwindigkeit im Vergleich zum bisherigen Industrie-Standard IFC-67 sind in der internationalen Veröffentlichung über die IAPWS-IF97 [9] zu finden.

3.1 Charakteristische Merkmale der IAPWS-IF97

Dieser Abschnitt faßt die allgemeinen Informationen zur Struktur der neuen Formulation einschließlich ihres Gültigkeitsbereiches zusammen und enthält generelle Aussagen zur Qualität der IAPWS-IF97 bezüglich Genauigkeit, Konsistenz an den Bereichsgrenzen und Rechengeschwindigkeit im Vergleich zum bisherigen Industrie-Standard IFC-67.

3.1.1 Struktur der IAPWS-IF97

Die IAPWS Industrie-Formulation 1997 besteht aus einem Satz von Gleichungen für verschiedene Bereiche, wobei insgesamt folgender Gültigkeitsbereich abgedeckt wird:

$$273{,}15\ \mathrm{K} \le T \le 1073{,}15\ \mathrm{K} \qquad p \le 100\ \mathrm{MPa}$$
$$1073{,}15\ \mathrm{K} < T \le 2273{,}15\ \mathrm{K} \qquad p \le 10\ \mathrm{MPa}\,.$$

Abbildung 3.1 zeigt die 5 Bereiche, in die der gesamte Gültigkeitsbereich der IAPWS-IF97 aufgeteilt ist. Die Grenzen zwischen diesen Bereichen lassen sich mit Ausnahme der Grenze zwischen den Bereichen 2 und 3 direkt aus Abb. 3.1 entnehmen. Diese Grenze wird durch die sog. B23-Gleichung definiert, die in Abschn. 3.3.1 angegeben ist. Für die Bereiche 1 und 2 existiert jeweils eine Fundamentalgleichung der spezifischen freien Enthalpie $g(p,T)$, für den Bereich 3 eine Fundamentalgleichung der spezifischen freien Energie $f(\rho,T)$, und die Sättigungslinie, die dem Bereich 4 entspricht, wird durch eine Dampfdruckgleichung $p_s(T)$ abgedeckt. Der Hochtemperaturbereich 5 wird ebenfalls durch eine Fundamentalgleichung der Form $g(p,T)$ beschrieben. Diese 5 Gleichungen, die in Abb. 3.1 eingerahmt sind, bilden die sog. *Basisgleichungen.*

Der Industrie-Standard IAPWS-IF97 wurde an den wissenschaftlichen Standard der IAPWS gekoppelt, der durch die „IAPWS Formulation 1995 for the Thermodynamic Properties of Ordinary Water Substance for General and Scientific Use“ [8] gegeben ist und im folgenden abge-

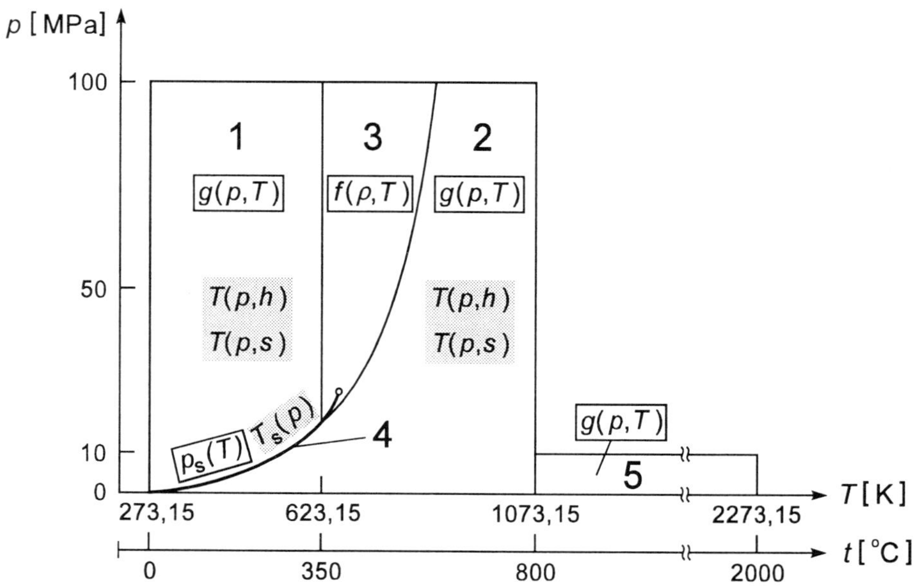

Abb. 3.1. Bereiche und Gleichungen der IAPWS-IF97

kürzt als IAPWS-95 bezeichnet wird. Diese Kopplung wurde dadurch erreicht, daß die Basisgleichungen der Bereiche 1 bis 3 und 5 an Werte der Zustandsgrößen spezifisches Volumen v, spezifische Enthalpie h, spezifische isobare Wärmekapazität c_p und Schallgeschwindigkeit w angepaßt wurden, die aus der IAPWS-95 [8] berechnet worden sind. Die Basisgleichung des Bereichs 4, die Dampfdruckgleichung, wurde an Werte des Dampfdruckes p_s angepaßt, die ebenfalls aus der IAPWS-95 ermittelt worden sind.

Zusätzlich zu den Basisgleichungen wurden für die Bereiche 1, 2 und 4 sog. *Rückwärtsgleichungen* entwickelt, nämlich Gleichungen der Form $T(p,h)$ und $T(p,s)$ für die Bereiche 1 und 2 sowie in der Form $T_s(p)$ für den Bereich 4. Bei der Entwicklung der Rückwärtsgleichungen, die in Abb. 3.1 grau hinterlegt sind, wurde sichergestellt, daß sie mit den entsprechenden Basisgleichungen sehr gut numerisch konsistent sind. Deshalb können mit Hilfe der Rückwärtsgleichungen für die Bereiche 1 und 2 Zustandsgrößen als Funktion von p,h und von p,s sowie für den Bereich 4 Sättigungsgrößen als Funktion von p ohne Iteration berechnet werden. Als Konsequenz lassen sich Funktionen, wie z. B. $T(p,h)$, $h(p,s)$ und $h'(p)$ direkt aus den Rückwärtsgleichungen berechnen oder in Kombination mit der entsprechenden Basisgleichung, z. B. $h(p,s)$ über die Beziehung $h(p,T(p,s))$. Als Ergebnis dieses speziellen Konzeptes für die Entwicklung des neuen Industrie-Standards IAPWS-IF97 ist die Berechnung der industriell wichtigsten Zustandsgrößen extrem schnell (s. nächster Abschnitt).

3.1.2
Qualität der IAPWS-IF97

Die erreichte Qualität der Industrie-Formulation IAPWS-IF97 läßt sich anhand der 3 Kriterien Genauigkeit, Konsistenz an den Bereichsgrenzen und Rechengeschwindigkeit charakterisieren, auf die im folgenden zusammenfassend eingegangen wird.

Die Genauigkeit der IAPWS-IF97 wird durch die Tatsache veranschaulicht, daß für den gesamten Gültigkeitsbereich nur 0,2 % der berechneten v-Werte, 6 % der c_p-Werte, 2 % der w-Werte und keine p_s-Werte außerhalb der Unsicherheiten der entsprechenden Werte der IAPWS-95 [8] liegen. Führt man die gleiche Untersuchung mit dem bisherigen Industrie-Standard IFC-67 durch, dann liegen zwischen 47 % (bei v) und 80 % (bei p_s) der mit der IFC-67 berechneten Werte außerhalb der Unsicherheiten der IAPWS-95. Auf der Basis aller durchgeführten Vergleiche [9] läßt sich die Schlußfolgerung ziehen, daß die IAPWS-IF97 um mehr als eine Größenordnung genauer ist als die IFC-67. Die für die IAPWS-IF97 abgeschätzten Unsicherheiten bezüglich der Zustandsgrößen v, c_p, w und p_s sind in Abschn. 3.7 zu finden. Neben der Wiedergabe der jeweiligen Zustandsgrößen in den stabilen homogenen Phasen und bei Sättigung liefern die Gleichungen der IAPWS-IF97 in der Nähe der entsprechenden Sättigungslinie auch sinnvolle Werte für die beiden metastabilen Bereiche überhitzte Flüssigkeit und unterkühlter Dampf.

Im Vergleich zur IFC-67 wurde ein weiterer wichtiger Qualitätssprung dadurch erreicht, daß die IAPWS-IF97 die Anforderungen bezüglich der Konsistenz an den Bereichsgrenzen (s. Abb. 3.1) sehr gut erfüllt. Die IAPWS-IF97 bleibt deutlich innerhalb der erlaubten Inkonsistenzen, die den im folgenden aufgeführten sog. Prag-Werten [10] entsprechen:

a) Homogene Phase

spezifisches Volumen:	$\pm$ 0,05 %
Enthalpie:	$\pm$ 0,2 kJ kg^{-1}
isobare Wärmekapazität:	$\pm$ 1 %
Entropie:	$\pm$ 0,0002 kJ kg^{-1} K^{-1}
freie Enthalpie:	$\pm$ 0,2 kJ kg^{-1}
Schallgeschwindigkeit:	$\pm$ 1 %

b) Sättigung

Dampfdruck:	$\pm$ 0,05 %
Siedetemperatur:	$\pm$ 0,02 %
freie Enthalpie:	$\pm$ 0,2 kJ kg^{-1}.

Am schwierigsten sind die Konsistenzanforderungen an der Grenze zwischen den Bereichen 2 und 3 zu erfüllen, und hier insbesondere für die isobare Wärmekapazität c_p. Bei der IAPWS-IF97 beträgt die maximale Inkonsistenz an dieser Bereichsgrenze in c_p nur 0,35 %, bei der IFC-67 jedoch mehr als 6 %.

Die dritte und wahrscheinlich wichtigste Qualitätssteigerung ist die sehr starke Erhöhung der Rechengeschwindigkeit im Vergleich zur IFC-67. Für die industriell wichtigsten Bereiche 1, 2 und 4 zusammen, in denen die Rechenzeit besonders relevant ist, beträgt der Rechengeschwindigkeitsverhältnis IAPWS-IF97/IFC-67 5,1. Dieses Ergebnis wurde unter Berücksichtigung der Aufrufhäufigkeiten der wichtigsten Zustandsfunktionen in diesen Bereichen ermittelt, deren Werte auf einer Umfrage in der internationalen Kraftwerksindustrie und bei verwandten Firmen beruhen. Für die wichtigsten Bereiche ist die IAPWS-IF97 also rund 5mal schneller als die IFC-67, vorausgesetzt, die einzelnen Gleichungen werden in geeigneter Weise programmiert. Für den Bereich 3 ist die IAPWS-IF97 3,6mal schneller als die IFC-67 und für den Bereich 5 sogar 12,2mal schneller.

Einzelheiten zu den Punkten Genauigkeit, Konsistenz an den Bereichsgrenzen und Rechengeschwindigkeit sind in dem internationalen Artikel über die IAPWS-IF97 zu finden [9].

3.2 Gleichungen für den Bereich 1

Dieser Abschnitt enthält alle Informationen, die für den Gebrauch der Basisgleichung und der Rückwärtsgleichungen des Bereichs 1 der IAPWS-IF97 benötigt werden. Die abgeschätzten Unsicherheiten der mit der IAPWS-IF97 berechneten Werte der wichtigsten Zustandsgrößen sind in Abschn. 3.7 zu finden.

3.2.1 Basisgleichung

Die Basisgleichung für diesen Bereich ist eine Fundamentalgleichung der spezifischen freien Enthalpie g. In dimensionsloser Form, $\gamma = g/(RT)$, lautet die Gleichung

$$\frac{g(p,T)}{RT} = \gamma(\pi,\tau) = \sum_{i=1}^{34} n_i (7{,}1-\pi)^{I_i} (\tau - 1{,}222)^{J_i} \tag{3.1}$$

mit $\pi = p/p^*$ und $\tau = T^*/T$ sowie $p^* = 16{,}53$ MPa und $T^* = 1386$ K; R ist durch Gl. (2.1) gegeben. Die Koeffizienten n_i und die Exponenten I_i and J_i der Gl. (3.1) sind in Tabelle 3.1 aufgeführt.

Aus Gl. (3.1) können alle thermodynamischen Zustandsgrößen durch geeignete Kopplung der dimensionslosen freien Enthalpie γ und ihrer Ableitungen ermittelt werden. Die Beziehungen zwischen den wichtigsten thermodynamischen Eigenschaften und γ sowie seinen Ableitungen sind in Tabelle 3.2 zusammengestellt. Alle erforderlichen Ableitungen der dimensionslosen freien Enthalpie γ, Gl. (3.1), sind in Tabelle 3.3 explizit aufgeführt.

Seit der 5. Internationalen Wasserdampfkonferenz 1956 in London werden die spezifische innere Energie und die spezifische Entropie der siedenden Flüssigkeit am Tripelpunkt zu Null gesetzt, also

$$u_\mathrm{t}' = 0 \quad ; \quad s_\mathrm{t}' = 0 \; . \tag{3.2}$$

Um diese Nullpunktfestlegung für die Temperatur und für den Druck am Tripelpunkt zu erfüllen, s. Gln. (2.7) und (2.8), wurden die Koeffizienten n_3 und n_4 in Gl. (3.1) entsprechend angepaßt. Als Konsequenz dieser Festlegung liefert Gl. (3.1) für die spezifische Enthalpie der siedenden Flüssigkeit am Tripelpunkt den Wert

$$h_\mathrm{t}' = 0{,}611\,783\ \mathrm{J\,kg^{-1}} \; . \tag{3.3}$$

Tabelle 3.1. Koeffizienten und Exponenten der Gl. (3.1)

i	I_i	J_i	n_i	i	I_i	J_i	n_i
1	0	−2	$0{,}146\,329\,712\,131\,67$	18	2	3	$-0{,}441\,418\,453\,308\,46 \cdot 10^{-5}$
2	0	−1	$-0{,}845\,481\,871\,691\,14$	19	2	17	$-0{,}726\,949\,962\,975\,94 \cdot 10^{-15}$
3	0	0	$-0{,}375\,636\,036\,720\,40 \cdot 10^{1}$	20	3	−4	$-0{,}316\,796\,448\,450\,54 \cdot 10^{-4}$
4	0	1	$0{,}338\,551\,691\,683\,85 \cdot 10^{1}$	21	3	0	$-0{,}282\,707\,979\,853\,12 \cdot 10^{-5}$
5	0	2	$-0{,}957\,919\,633\,878\,72$	22	3	6	$-0{,}852\,051\,281\,201\,03 \cdot 10^{-9}$
6	0	3	$0{,}157\,720\,385\,132\,28$	23	4	−5	$-0{,}224\,252\,819\,080\,00 \cdot 10^{-5}$
7	0	4	$-0{,}166\,164\,171\,995\,01 \cdot 10^{-1}$	24	4	−2	$-0{,}651\,712\,228\,956\,01 \cdot 10^{-6}$
8	0	5	$0{,}812\,146\,299\,835\,68 \cdot 10^{-3}$	25	4	10	$-0{,}143\,417\,299\,379\,24 \cdot 10^{-12}$
9	1	−9	$0{,}283\,190\,801\,238\,04 \cdot 10^{-3}$	26	5	−8	$-0{,}405\,169\,968\,601\,17 \cdot 10^{-6}$
10	1	−7	$-0{,}607\,063\,015\,658\,74 \cdot 10^{-3}$	27	8	−11	$-0{,}127\,343\,017\,416\,41 \cdot 10^{-8}$
11	1	−1	$-0{,}189\,900\,682\,184\,19 \cdot 10^{-1}$	28	8	−6	$-0{,}174\,248\,712\,306\,34 \cdot 10^{-9}$
12	1	0	$-0{,}325\,297\,487\,705\,05 \cdot 10^{-1}$	29	21	−29	$-0{,}687\,621\,312\,955\,31 \cdot 10^{-18}$
13	1	1	$-0{,}218\,417\,171\,754\,14 \cdot 10^{-1}$	30	23	−31	$0{,}144\,733\,078\,285\,21 \cdot 10^{-19}$
14	1	3	$-0{,}528\,383\,579\,699\,30 \cdot 10^{-4}$	31	29	−38	$0{,}263\,357\,816\,627\,95 \cdot 10^{-22}$
15	2	−3	$-0{,}471\,843\,210\,732\,67 \cdot 10^{-3}$	32	30	−39	$-0{,}119\,476\,226\,400\,71 \cdot 10^{-22}$
16	2	0	$-0{,}300\,017\,807\,930\,26 \cdot 10^{-3}$	33	31	−40	$0{,}182\,230\,945\,814\,04 \cdot 10^{-23}$
17	2	1	$0{,}476\,613\,939\,069\,87 \cdot 10^{-4}$	34	32	−41	$-0{,}935\,370\,872\,924\,58 \cdot 10^{-25}$

Tabelle 3.2. Beziehungen thermodynamischer Zustandsgrößen zur dimensionslosen freien Enthalpie γ und ihren Ableitungen auf der Basis von Gl. (3.1)

Zustandsgröße	Beziehung
spezifisches Volumen $v = (\partial g/\partial p)_T$	$v(\pi,\tau)\frac{p}{RT} = \pi\gamma_\pi$
spezifische innere Energie $u = g - T(\partial g/\partial T)_p - p(\partial g/\partial p)_T$	$\frac{u(\pi,\tau)}{RT} = \tau\gamma_\tau - \pi\gamma_\pi$
spezifische Entropie $s = -(\partial g/\partial T)_p$	$\frac{s(\pi,\tau)}{R} = \tau\gamma_\tau - \gamma$
spezifische Enthalpie $h = g - T(\partial g/\partial T)_p$	$\frac{h(\pi,\tau)}{RT} = \tau\gamma_\tau$
spezifische isobare Wärmekapazität $c_p = (\partial h/\partial T)_p$	$\frac{c_p(\pi,\tau)}{R} = -\tau^2\gamma_{\tau\tau}$
spezifische isochore Wärmekapazität $c_v = (\partial u/\partial T)_v$	$\frac{c_v(\pi,\tau)}{R} = -\tau^2\gamma_{\tau\tau} + \frac{(\gamma_\pi - \tau\gamma_{\pi\tau})^2}{\gamma_{\pi\pi}}$
Schallgeschwindigkeit $w = v[-(\partial p/\partial v)_s]^{0,5}$	$\frac{w^2(\pi,\tau)}{RT} = \frac{\gamma_\pi^2}{\frac{(\gamma_\pi - \tau\gamma_{\pi\tau})^2}{\tau^2\gamma_{\tau\tau}} - \gamma_{\pi\pi}}$

$$\gamma_\pi = \left(\frac{\partial\gamma}{\partial\pi}\right)_\tau, \quad \gamma_{\pi\pi} = \left(\frac{\partial^2\gamma}{\partial\pi^2}\right)_\tau, \quad \gamma_\tau = \left(\frac{\partial\gamma}{\partial\tau}\right)_\pi, \quad \gamma_{\tau\tau} = \left(\frac{\partial^2\gamma}{\partial\tau^2}\right)_\pi, \quad \gamma_{\pi\tau} = \left(\frac{\partial^2\gamma}{\partial\pi\partial\tau}\right)$$

Tabelle 3.3. Die dimensionslose freie Enthalpie γ, Gl. (3.1), und ihre Ableitungen

$\gamma = \sum_{i=1}^{34} n_i(7{,}1-\pi)^{I_i}(\tau-1{,}222)^{J_i}$	$\gamma_\tau = \sum_{i=1}^{34} n_i(7{,}1-\pi)^{I_i} J_i(\tau-1{,}222)^{J_i-1}$
$\gamma_\pi = \sum_{i=1}^{34} -n_i I_i(7{,}1-\pi)^{I_i-1}(\tau-1{,}222)^{J_i}$	$\gamma_{\tau\tau} = \sum_{i=1}^{34} n_i(7{,}1-\pi)^{I_i} J_i(J_i-1)(\tau-1{,}222)^{J_i-2}$
$\gamma_{\pi\pi} = \sum_{i=1}^{34} n_i I_i(I_i-1)(7{,}1-\pi)^{I_i-2}(\tau-1{,}222)^{J_i}$	$\gamma_{\pi\tau} = \sum_{i=1}^{34} -n_i I_i(7{,}1-\pi)^{I_i-1} J_i(\tau-1{,}222)^{J_i-1}$

$$\gamma_\pi = \left(\frac{\partial\gamma}{\partial\pi}\right)_\tau, \quad \gamma_{\pi\pi} = \left(\frac{\partial^2\gamma}{\partial\pi^2}\right)_\tau, \quad \gamma_\tau = \left(\frac{\partial\gamma}{\partial\tau}\right)_\pi, \quad \gamma_{\tau\tau} = \left(\frac{\partial^2\gamma}{\partial\tau^2}\right)_\pi, \quad \gamma_{\pi\tau} = \left(\frac{\partial^2\gamma}{\partial\pi\partial\tau}\right)$$

Gleichung (3.1) gilt für den Bereich 1 der IAPWS-IF97, der durch den folgenden Temperatur- und Druckbereich definiert ist, s. Abb. 3.1:

$$273{,}15\ \text{K} \le T \le 623{,}15\ \text{K} \qquad p_s(T) \le p \le 100\ \text{MPa}\,.$$

Neben den Zustandsgrößen des stabilen homogenen Flüssigkeitsgebietes liefert Gl. (3.1) auch sinnvolle Werte der Zustandsgrößen für die metastabile überhitzte Flüssigkeit in der Nähe der Siedelinie. Für Temperaturen zwischen 273,15 K und 273,16 K bei Drücken unterhalb des

Schmelzdruckes [11] (metastabile unterkühlte Flüssigkeit) werden die Werte aller Zustandsgrößen mit den Gln. (3.1) und (3.23) extrapoliert.

Zur Überprüfung von Computerprogrammen auf der Basis von Gl. (3.1) enthält Tabelle 3.4 entsprechende Testwerte für die wichtigsten Zustandsgrößen.

Tabelle 3.4. Aus Gl. (3.1) berechnete Werte thermodynamischer Zustandsgrößen für ausgewählte Temperaturen und Drücke

Zustandsgröße	$T = 300$ K $p = 3$ MPa	$T = 300$ K $p = 80$ MPa	$T = 500$ K $p = 3$ MPa
v [m^3 kg^{-1}]	$0{,}100\,215\,168 \cdot 10^{-2}$	$0{,}971\,180\,894 \cdot 10^{-3}$	$0{,}120\,241\,800 \cdot 10^{-2}$
h [kJ kg^{-1}]	$0{,}115\,331\,273 \cdot 10^{3}$	$0{,}184\,142\,828 \cdot 10^{3}$	$0{,}975\,542\,239 \cdot 10^{3}$
u [kJ kg^{-1}]	$0{,}112\,324\,818 \cdot 10^{3}$	$0{,}106\,448\,356 \cdot 10^{3}$	$0{,}971\,934\,985 \cdot 10^{3}$
s [kJ kg^{-1} K^{-1}]	$0{,}392\,294\,792$	$0{,}368\,563\,852$	$0{,}258\,041\,912 \cdot 10^{1}$
c_p [kJ kg^{-1} K^{-1}]	$0{,}417\,301\,218 \cdot 10^{1}$	$0{,}401\,008\,987 \cdot 10^{1}$	$0{,}465\,580\,682 \cdot 10^{1}$
w [m s^{-1}]	$0{,}150\,773\,921 \cdot 10^{4}$	$0{,}163\,469\,054 \cdot 10^{4}$	$0{,}124\,071\,337 \cdot 10^{4}$

3.2.2 Rückwärtsgleichungen

Um Zustandsgrößen als Funktion von p,h sowie von p,s ohne Iteration berechnen zu können (s. Abschn. 3.1.1), müssen die beiden Rückwärtsgleichungen sehr gut numerisch konsistent mit der Basisgleichung sein. Die genaue Anforderung bezüglich der numerischen Konsistenz wurde mit Hilfe umfassender Testrechnungen für eine Reihe charakteristischer Kraftwerksprozesse ermittelt. Auf der Basis dieser Untersuchungen wurde die numerische Konsistenzanforderung zwischen der Basisgleichung, Gl. (3.1), und den beiden Rückwärtsgleichungen $T(p,h)$ und $T(p,s)$, Gln. (3.4) und (3.5), zu $\pm$ 25 mK festgelegt. Das bedeutet zum Beispiel, daß die aus der Rückwärtsgleichung $T(p,h)$, Gl. (3.4), für vorgegebene Werte von p und h direkt berechnete Temperatur innerhalb von $\pm$ 25 mK mit dem Temperaturwert übereinstimmen muß, der für dieselben Werte von p und h aus der Basisgleichung, Gl. (3.1), iteriert werden kann. Diese Iteration aus der Basisgleichung erfolgt über die Funktion $h(p,T)$, deren Zusammenhang mit der dimensionslosen freien Enthalpie γ in Tabelle 3.2 angegeben ist.

3.2.2.1 *Die Rückwärtsgleichung T(p, h)*

Die Rückwärtsgleichung $T(p,h)$ für den Bereich 1 hat die folgende dimensionslose Form:

$$\frac{T(p,h)}{T^*} = \theta(\pi,\eta) = \sum_{i=1}^{20} n_i \pi^{I_i} (\eta+1)^{J_i} \tag{3.4}$$

mit $\theta = T/T^*$, $\pi = p/p^*$ und $\eta = h/h^*$ sowie $T^* = 1$ K, $p^* = 1$ MPa und $h^* = 2500$ kJ kg^{-1}. Die Koeffizienten n_i und die Exponenten I_i und J_i der Gl. (3.4) sind in Tabelle 3.5 aufgeführt.

Tabelle 3.5. Koeffizienten und Exponenten der Gl. (3.4)

i	I_i	J_i	n_i	i	I_i	J_i	n_i
1	0	0	$-0{,}238\,724\,899\,245\,21 \cdot 10^{3}$	11	1	4	$-0{,}659\,647\,494\,236\,38 \cdot 10^{1}$
2	0	1	$0{,}404\,211\,886\,379\,45 \cdot 10^{3}$	12	1	10	$0{,}939\,654\,008\,783\,63 \cdot 10^{-2}$
3	0	2	$0{,}113\,497\,468\,817\,18 \cdot 10^{3}$	13	1	32	$0{,}115\,736\,475\,053\,40 \cdot 10^{-6}$
4	0	6	$-0{,}584\,576\,160\,480\,39 \cdot 10^{1}$	14	2	10	$-0{,}258\,586\,412\,820\,73 \cdot 10^{-4}$
5	0	22	$-0{,}152\,854\,824\,131\,40 \cdot 10^{-3}$	15	2	32	$-0{,}406\,443\,630\,847\,99 \cdot 10^{-8}$
6	0	32	$-0{,}108\,667\,076\,953\,77 \cdot 10^{-5}$	16	3	10	$0{,}664\,561\,861\,916\,35 \cdot 10^{-7}$
7	1	0	$-0{,}133\,917\,448\,726\,02 \cdot 10^{2}$	17	3	32	$0{,}806\,707\,341\,030\,27 \cdot 10^{-10}$
8	1	1	$0{,}432\,110\,391\,835\,59 \cdot 10^{2}$	18	4	32	$-0{,}934\,777\,712\,139\,47 \cdot 10^{-12}$
9	1	2	$-0{,}540\,100\,671\,705\,06 \cdot 10^{2}$	19	5	32	$0{,}582\,654\,420\,206\,01 \cdot 10^{-14}$
10	1	3	$0{,}305\,358\,922\,039\,16 \cdot 10^{2}$	20	6	32	$-0{,}150\,201\,859\,535\,03 \cdot 10^{-16}$

Gleichung (3.4) hat denselben Gültigkeitsbereich wie die Basisgleichung, Gl. (3.1), mit Ausnahme der metastabilen überhitzten Flüssigkeit, für die Gl. (3.4) nicht gilt. Die tatsächlichen numerischen Inkonsistenzen zwischen den Gln. (3.4) und (3.1) betragen maximal 23,6 mK und im quadratischen Mittel 13,4 mK.

Zur Überprüfung von Computerprogrammen auf der Basis von Gl. (3.4) enthält Tabelle 3.6 entsprechende Testwerte.

Tabelle 3.6. Aus Gl. (3.4) berechnete Temperaturen für ausgewählte Werte von Druck und Enthalpie

p [MPa]	h [kJ kg^{-1}]	T [K]
3	500	$0{,}391\,798\,509 \cdot 10^{3}$
80	500	$0{,}378\,108\,626 \cdot 10^{3}$
80	1500	$0{,}611\,041\,229 \cdot 10^{3}$

3.2.2.2
Die Rückwärtsgleichung $T(p, s)$

Die Rückwärtsgleichung $T(p,s)$ für den Bereich 1 hat die folgende dimensionslose Form:

$$\frac{T(p,s)}{T^*} = \theta(\pi,\sigma) = \sum_{i=1}^{20} n_i\, \pi^{I_i} (\sigma + 2)^{J_i} \tag{3.5}$$

mit $\theta = T/T^*$, $\pi = p/p^*$ und $\sigma = s/s^*$ sowie $T^* = 1$ K, $p^* = 1$ MPa und $s^* = 1$ kJ kg^{-1} K^{-1}. Die Koeffizienten n_i und die Exponenten I_i und J_i der Gl. (3.5) sind in Tabelle 3.7 aufgeführt.

Gleichung (3.5) hat denselben Gültigkeitsbereich wie die Basisgleichung, Gl. (3.1), mit Ausnahme der metastabilen überhitzten Flüssigkeit, für die Gl. (3.5) nicht gilt. Die tatsächlichen numerischen Inkonsistenzen zwischen den Gln. (3.5) und (3.1) betragen maximal 21,8 mK und im quadratischen Mittel 5,8 mK.

Zur Überprüfung von Computerprogrammen auf der Basis von Gl. (3.5) enthält Tabelle 3.8 entsprechende Testwerte.

Tabelle 3.7. Koeffizienten und Exponenten der Gl. (3.5)

i	I_i	J_i	n_i	i	I_i	J_i	n_i
1	0	0	$0{,}174\,782\,680\,583\,07 \cdot 10^{3}$	11	1	12	$0{,}356\,721\,106\,073\,66 \cdot 10^{-9}$
2	0	1	$0{,}348\,069\,308\,928\,73 \cdot 10^{2}$	12	1	31	$0{,}173\,324\,969\,948\,95 \cdot 10^{-23}$
3	0	2	$0{,}652\,925\,849\,784\,55 \cdot 10^{1}$	13	2	0	$0{,}566\,089\,006\,548\,37 \cdot 10^{-3}$
4	0	3	$0{,}330\,399\,817\,754\,89$	14	2	1	$-0{,}326\,354\,831\,397\,17 \cdot 10^{-3}$
5	0	11	$-0{,}192\,813\,829\,231\,96 \cdot 10^{-6}$	15	2	2	$0{,}447\,782\,866\,906\,32 \cdot 10^{-4}$
6	0	31	$-0{,}249\,091\,972\,445\,73 \cdot 10^{-22}$	16	2	9	$-0{,}513\,221\,569\,085\,07 \cdot 10^{-9}$
7	1	0	$-0{,}261\,076\,364\,893\,32$	17	2	31	$-0{,}425\,226\,570\,422\,07 \cdot 10^{-25}$
8	1	1	$0{,}225\,929\,659\,815\,86$	18	3	10	$0{,}264\,004\,413\,606\,89 \cdot 10^{-12}$
9	1	2	$-0{,}642\,564\,633\,952\,26 \cdot 10^{-1}$	19	3	32	$0{,}781\,246\,004\,597\,23 \cdot 10^{-28}$
10	1	3	$0{,}788\,762\,892\,705\,26 \cdot 10^{-2}$	20	4	32	$-0{,}307\,321\,999\,036\,68 \cdot 10^{-30}$

Tabelle 3.8. Aus Gl. (3.5) berechnete Temperaturen für ausgewählte Werte von Druck und Entropie

p [MPa]	s [kJ kg^{-1} K^{-1}]	T [K]
3	0,5	$0{,}307\,842\,258 \cdot 10^{3}$
80	0,5	$0{,}309\,979\,785 \cdot 10^{3}$
80	3	$0{,}565\,899\,909 \cdot 10^{3}$

3.3 Gleichungen für den Bereich 2

Dieser Abschnitt enthält alle Informationen, die für den Gebrauch der Basisgleichung und der Rückwärtsgleichungen des Bereichs 2 der IAPWS-IF97 benötigt werden. Die Hilfsgleichung zur Definition der Grenze zwischen den Bereichen 2 und 3 ist in Abschn. 3.3.1 angegeben. Die abgeschätzten Unsicherheiten der mit der IAPWS-IF97 berechneten Werte der wichtigsten Zustandsgrößen sind in Abschn. 3.7 zu finden.

3.3.1 Hilfsgleichung für die Grenze zwischen den Bereichen 2 und 3

Die Grenze zwischen den Bereichen 2 und 3 (s. Abb. 3.1) ist durch die folgende, einfache quadratische Druck-Temperatur-Beziehung, die B23-Gleichung, definiert:

$$\pi = n_1 + n_2\theta + n_3\theta^2 \tag{3.6}$$

mit $\pi = p/p^*$ und $\theta = T/T^*$ sowie $p^* = 1\,\text{MPa}$ und $T^* = 1\,\text{K}$. Die Koeffizienten n_1 bis n_3 der Gl. (3.6) sind in Tabelle 3.9 aufgeführt. Gleichung (3.6) beschreibt grob eine isentrope Linie, wobei die Entropiewerte entlang dieser Grenzlinie zwischen $s = 5{,}047\,\text{kJ}\,\text{kg}^{-1}\,\text{K}^{-1}$ und $s = 5{,}261\,\text{kJ}\,\text{kg}^{-1}\,\text{K}^{-1}$ liegen.

Alternativ kann Gl. (3.6) auch explizit in der Temperatur ausgedrückt werden, nämlich

$$\theta = n_4 + \left[(\pi - n_5)/n_3\right]^{0,5} \tag{3.7}$$

mit θ und π entsprechend Gl. (3.6) sowie den Koeffizienten n_3 bis n_5, die in Tabelle 3.9 aufgeführt sind. Die Gln. (3.6) und (3.7) gelten für den Bereich von 623,15 K bei 16,5292 MPa bis 863,15 K bei 100 MPa.

Tabelle 3.9. Koeffizienten der Gln. (3.6) und (3.7)

i	n_i	i	n_i
1	$0{,}348\,051\,856\,289\,69 \cdot 10^{3}$	4	$0{,}572\,544\,598\,627\,46 \cdot 10^{3}$
2	$-0{,}116\,718\,598\,799\,75 \cdot 10^{1}$	5	$0{,}139\,188\,397\,788\,70 \cdot 10^{2}$
3	$0{,}101\,929\,700\,393\,26 \cdot 10^{-2}$		

Bei der Überprüfung von Computerprogrammen auf der Basis der Gln. (3.6) und (3.7) muß der folgende T,p-Punkt erfüllt werden: $T = 0{,}623\,150\,000 \cdot 10^3$ K , $p = 0{,}165\,291\,643 \cdot 10^2$ MPa.

3.3.2 Basisgleichung

Die Basisgleichung für diesen Bereich ist eine Fundamentalgleichung der freien Enthalpie g, die in dimensionsloser Form, $\gamma = g/(RT)$, verwendet wird, wobei g in den Anteil des idealen Gases γ^{o} und in den residuellen Anteil γ^{r} aufgeteilt ist. Es gilt also

$$\frac{g(p,T)}{RT} = \gamma(\pi,\tau) = \gamma^{\mathrm{o}}(\pi,\tau) + \gamma^{\mathrm{r}}(\pi,\tau) \qquad (3.8)$$

mit $\pi = p/p^*$ und $\tau = T^*/T$ sowie R gemäß Gl. (2.1).

Die Gleichung für den Anteil γ^{o} des idealen Gases der dimensionslosen freien Enthalpie lautet

$$\gamma^{\mathrm{o}} = \ln\pi + \sum_{i=1}^{9} n_i^{\mathrm{o}} \tau^{J_i^{\mathrm{o}}} \qquad (3.9)$$

mit $\pi = p/p^*$ und $\tau = T^*/T$ sowie $p^* = 1$ MPa und $T^* = 540$ K. Die Koeffizienten n_1^{o} und n_2^{o} wurden derart angepaßt, daß die aus Gl. (3.8) berechneten Werte für die spezifische innere Energie und die spezifische Entropie zu dem durch Gl. (3.2) definierten Bezugspunkt konsistent sind. Die Koeffizienten n_i^{o} und die Exponenten J_i^{o} der Gl. (3.9) sind in Tabelle 3.10 aufgeführt.

Der residuelle Anteil γ^{r} der dimensionslosen freien Enthalpie hat folgende Form:

$$\gamma^{\mathrm{r}} = \sum_{i=1}^{43} n_i \pi^{I_i} (\tau - 0{,}5)^{J_i} \qquad (3.10)$$

mit $\pi = p/p^*$ und $\tau = T^*/T$ sowie $p^* = 1$ MPa und $T^* = 540$ K. Die Koeffizienten n_i und die Exponenten I_i und J_i der Gl. (3.10) sind in Tabelle 3.11 aufgeführt.

Aus Gl. (3.8) können alle thermodynamischen Zustandsgrößen durch geeignete Kopplung von γ^{o}, Gl. (3.9), und γ^{r}, Gl. (3.10), sowie ihrer Ableitungen berechnet werden. Tabelle 3.12 enthält diese Beziehungen für die wichtigsten thermodynamischen Zustandsgrößen. Alle für die Auswertung dieser Berechnungsgleichungen notwendigen Ableitungen des Anteils des idealen Gases γ^{o} und des residuellen Anteils γ^{r} der dimensionslosen freien Enthalpie γ sind in den Tabellen 3.13 und 3.14 explizit angegeben.

Tabelle 3.10. Koeffizienten und Exponenten der Gl. (3.9)[a]

i	J_i^o	n_i^o	i	J_i^o	n_i^o
1[a]	0	$-0{,}969\,276\,865\,002\,17 \cdot 10^{1}$	6	-2	$0{,}142\,408\,191\,714\,44 \cdot 10^{1}$
2[a]	1	$0{,}100\,866\,559\,680\,18 \cdot 10^{2}$	7	-1	$-0{,}438\,395\,113\,194\,50 \cdot 10^{1}$
3	-5	$-0{,}560\,879\,112\,830\,20 \cdot 10^{-2}$	8	2	$-0{,}284\,086\,324\,607\,72$
4	-4	$0{,}714\,527\,380\,814\,55 \cdot 10^{-1}$	9	3	$0{,}212\,684\,637\,533\,07 \cdot 10^{-1}$
5	-3	$-0{,}407\,104\,982\,239\,28$			

[a] Falls Gl. (3.9) zur Berechnung von γ^o in Gl. (3.11) benutzt wird, müssen die oben angegebenen Werte für n_1^o und n_2^o durch die beiden folgenden Werte ersetzt werden: $n_1^o = -0{,}969\,372\,683\,930\,49 \cdot 10^{1}$, $n_2^o = 0{,}100\,872\,759\,700\,06 \cdot 10^{2}$.

Tabelle 3.11. Koeffizienten und Exponenten der Gl. (3.10)

i	I_i	J_i	n_i	i	I_i	J_i	n_i
1	1	0	$-0{,}177\,317\,424\,732\,13 \cdot 10^{-2}$	23	7	0	$-0{,}590\,595\,643\,242\,70 \cdot 10^{-17}$
2	1	1	$-0{,}178\,348\,622\,923\,58 \cdot 10^{-1}$	24	7	11	$-0{,}126\,218\,088\,991\,01 \cdot 10^{-5}$
3	1	2	$-0{,}459\,960\,136\,963\,65 \cdot 10^{-1}$	25	7	25	$-0{,}389\,468\,424\,357\,39 \cdot 10^{-1}$
4	1	3	$-0{,}575\,812\,590\,834\,32 \cdot 10^{-1}$	26	8	8	$0{,}112\,562\,113\,604\,59 \cdot 10^{-10}$
5	1	6	$-0{,}503\,252\,787\,279\,30 \cdot 10^{-1}$	27	8	36	$-0{,}823\,113\,408\,979\,98 \cdot 10^{1}$
6	2	1	$-0{,}330\,326\,416\,702\,03 \cdot 10^{-4}$	28	9	13	$0{,}198\,097\,128\,020\,88 \cdot 10^{-7}$
7	2	2	$-0{,}189\,489\,875\,163\,15 \cdot 10^{-3}$	29	10	4	$0{,}104\,069\,652\,101\,74 \cdot 10^{-18}$
8	2	4	$-0{,}393\,927\,772\,433\,55 \cdot 10^{-2}$	30	10	10	$-0{,}102\,347\,470\,959\,29 \cdot 10^{-12}$
9	2	7	$-0{,}437\,972\,956\,505\,73 \cdot 10^{-1}$	31	10	14	$-0{,}100\,181\,793\,795\,11 \cdot 10^{-8}$
10	2	36	$-0{,}266\,745\,479\,140\,87 \cdot 10^{-4}$	32	16	29	$-0{,}808\,829\,086\,469\,85 \cdot 10^{-10}$
11	3	0	$0{,}204\,817\,376\,923\,09 \cdot 10^{-7}$	33	16	50	$0{,}106\,930\,318\,794\,09$
12	3	1	$0{,}438\,706\,672\,844\,35 \cdot 10^{-6}$	34	18	57	$-0{,}336\,622\,505\,741\,71$
13	3	3	$-0{,}322\,776\,772\,385\,70 \cdot 10^{-4}$	35	20	20	$0{,}891\,858\,453\,554\,21 \cdot 10^{-24}$
14	3	6	$-0{,}150\,339\,245\,421\,48 \cdot 10^{-2}$	36	20	35	$0{,}306\,293\,168\,762\,32 \cdot 10^{-12}$
15	3	35	$-0{,}406\,682\,535\,626\,49 \cdot 10^{-1}$	37	20	48	$-0{,}420\,024\,676\,982\,08 \cdot 10^{-5}$
16	4	1	$-0{,}788\,473\,095\,593\,67 \cdot 10^{-9}$	38	21	21	$-0{,}590\,560\,296\,856\,39 \cdot 10^{-25}$
17	4	2	$0{,}127\,907\,178\,522\,85 \cdot 10^{-7}$	39	22	53	$0{,}378\,269\,476\,134\,57 \cdot 10^{-5}$
18	4	3	$0{,}482\,253\,727\,185\,07 \cdot 10^{-6}$	40	23	39	$-0{,}127\,686\,089\,346\,81 \cdot 10^{-14}$
19	5	7	$0{,}229\,220\,763\,376\,61 \cdot 10^{-5}$	41	24	26	$0{,}730\,876\,105\,950\,61 \cdot 10^{-28}$
20	6	3	$-0{,}167\,147\,664\,510\,61 \cdot 10^{-10}$	42	24	40	$0{,}554\,147\,153\,507\,78 \cdot 10^{-16}$
21	6	16	$-0{,}211\,714\,723\,213\,55 \cdot 10^{-2}$	43	24	58	$-0{,}943\,697\,072\,412\,10 \cdot 10^{-6}$
22	6	35	$-0{,}238\,957\,419\,341\,04 \cdot 10^{2}$				

Gleichung (3.8) gilt für den Bereich 2 der IAPWS-IF97, der durch folgenden Temperatur- und Druckbereich definiert ist, s. Abb. 3.1:

$$\begin{aligned} 273{,}15\ \mathrm{K} &\le T \le 623{,}15\ \mathrm{K} & 0 &< p \le p_s(T)_{\mathrm{Gl.(3.23)}} \\ 623{,}15\ \mathrm{K} &< T \le 863{,}15\ \mathrm{K} & 0 &< p \le p(T)_{\mathrm{Gl.(3.6)}} \\ 863{,}15\ \mathrm{K} &< T \le 1073{,}15\ \mathrm{K} & 0 &< p \le 100\ \mathrm{MPa}\,. \end{aligned}$$

Zusätzlich zu den Zustandsgrößen des stabilen homogenen Gasgebietes liefert Gl. (3.8) auch sinnvolle Werte für die Zustandsgrößen des metastabilen unterkühlten Dampfes in der Nähe der Taulinie *für Drücke oberhalb von 10 MPa*. Gleichung (3.8) gilt dagegen nicht für den metastabilen unterkühlten Dampf bei Drücken $p \le 10$ MPa; für diesen Bereich des metastabilen Dampfes s. Abschn. 3.3.3. Für Temperaturen zwischen 273,15 K und 273,16 K bei Drücken oberhalb des Sublimationsdruckes [11] (metastabiler unterkühlter Dampf) werden die Werte der Zustandsgrößen mit den Gln. (3.8) und (3.23) extrapoliert.

Zur Überprüfung von Computerprogrammen auf der Basis von Gl. (3.8) enthält Tabelle 3.15 entsprechende Testwerte für die wichtigsten Zustandsgrößen.

Tabelle 3.12. Beziehungen thermodynamischer Zustandsgrößen zum Anteil des idealen Gases γ^{o} und zum residuellen Anteil γ^{r} sowie ihren Ableitungen auf der Basis der dimensionslosen freien Enthalpie gemäß Gl. (3.8) oder Gl. (3.11)

Zustandsgröße	Beziehung
spezifisches Volumen $v = (\partial g/\partial p)_T$	$v(\pi,\tau)\dfrac{p}{RT} = \pi\left(\gamma_\pi^{\mathrm{o}} + \gamma_\pi^{\mathrm{r}}\right)$
spezifische innere Energie $u = g - T(\partial g/\partial T)_p - p(\partial g/\partial p)_T$	$\dfrac{u(\pi,\tau)}{RT} = \tau\left(\gamma_\tau^{\mathrm{o}} + \gamma_\tau^{\mathrm{r}}\right) - \pi\left(\gamma_\pi^{\mathrm{o}} + \gamma_\pi^{\mathrm{r}}\right)$
spezifische Entropie $s = -(\partial g/\partial T)_p$	$\dfrac{s(\pi,\tau)}{R} = \tau\left(\gamma_\tau^{\mathrm{o}} + \gamma_\tau^{\mathrm{r}}\right) - \left(\gamma^{\mathrm{o}} + \gamma^{\mathrm{r}}\right)$
spezifische Enthalpie $h = g - T(\partial g/\partial T)_p$	$\dfrac{h(\pi,\tau)}{RT} = \tau\left(\gamma_\tau^{\mathrm{o}} + \gamma_\tau^{\mathrm{r}}\right)$
spezifische isobare Wärmekapazität $c_p = (\partial h/\partial T)_p$	$\dfrac{c_p(\pi,\tau)}{R} = -\tau^2\left(\gamma_{\tau\tau}^{\mathrm{o}} + \gamma_{\tau\tau}^{\mathrm{r}}\right)$
spezifische isochore Wärmekapazität $c_v = (\partial u/\partial T)_v$	$\dfrac{c_v(\pi,\tau)}{R} = -\tau^2\left(\gamma_{\tau\tau}^{\mathrm{o}} + \gamma_{\tau\tau}^{\mathrm{r}}\right) - \dfrac{(1+\pi\gamma_\pi^{\mathrm{r}} - \tau\pi\gamma_{\pi\tau}^{\mathrm{r}})^2}{1-\pi^2\gamma_{\pi\pi}^{\mathrm{r}}}$
Schallgeschwindigkeit $w = v\left[-(\partial p/\partial v)_s\right]^{0,5}$	$\dfrac{w^2(\pi,\tau)}{RT} = \dfrac{1+2\pi\gamma_\pi^{\mathrm{r}} + \pi^2\gamma_\pi^{\mathrm{r}\,2}}{\left(1-\pi^2\gamma_{\pi\pi}^{\mathrm{r}}\right) + \dfrac{\left(1+\pi\gamma_\pi^{\mathrm{r}} - \tau\pi\gamma_{\pi\tau}^{\mathrm{r}}\right)^2}{\tau^2\left(\gamma_{\tau\tau}^{\mathrm{o}} + \gamma_{\tau\tau}^{\mathrm{r}}\right)}}$

$$\gamma_\pi^{\mathrm{r}} = \left(\frac{\partial\gamma^{\mathrm{r}}}{\partial\pi}\right)_\tau,\ \gamma_{\pi\pi}^{\mathrm{r}} = \left(\frac{\partial^2\gamma^{\mathrm{r}}}{\partial\pi^2}\right)_\tau,\ \gamma_\tau^{\mathrm{r}} = \left(\frac{\partial\gamma^{\mathrm{r}}}{\partial\tau}\right)_\pi,\ \gamma_{\tau\tau}^{\mathrm{r}} = \left(\frac{\partial^2\gamma^{\mathrm{r}}}{\partial\tau^2}\right)_\pi,\ \gamma_{\pi\tau}^{\mathrm{r}} = \left(\frac{\partial^2\gamma^{\mathrm{r}}}{\partial\pi\partial\tau}\right),\ \gamma_\tau^{\mathrm{o}} = \left(\frac{\partial\gamma^{\mathrm{o}}}{\partial\tau}\right)_\pi,\ \gamma_{\tau\tau}^{\mathrm{o}} = \left(\frac{\partial^2\gamma^{\mathrm{o}}}{\partial\tau^2}\right)_\pi$$

Tabelle 3.13. Der Anteil des idealen Gases γ^{o} der dimensionslosen freien Enthalpie, Gl. (3.9), und seine Ableitungen

$\gamma^{\mathrm{o}} = \ln\pi + \sum_{i=1}^{9} n_i^{\mathrm{o}}\tau^{J_i^{\mathrm{o}}}$	$\gamma_\tau^{\mathrm{o}} = \sum_{i=1}^{9} n_i^{\mathrm{o}} J_i^{\mathrm{o}}\tau^{J_i^{\mathrm{o}}-1}$
$\gamma_\pi^{\mathrm{o}} = \pi^{-1}$	$\gamma_{\tau\tau}^{\mathrm{o}} = \sum_{i=1}^{9} n_i^{\mathrm{o}} J_i^{\mathrm{o}}\left(J_i^{\mathrm{o}}-1\right)\tau^{J_i^{\mathrm{o}}-2}$
$\gamma_{\pi\pi}^{\mathrm{o}} = -\pi^{-2}$	$\gamma_{\pi\tau}^{\mathrm{o}} = 0$

$$\gamma_\pi^{\mathrm{o}} = \left(\frac{\partial\gamma^{\mathrm{o}}}{\partial\pi}\right)_\tau,\ \gamma_{\pi\pi}^{\mathrm{o}} = \left(\frac{\partial^2\gamma^{\mathrm{o}}}{\partial\pi^2}\right)_\tau,\ \gamma_\tau^{\mathrm{o}} = \left(\frac{\partial\gamma^{\mathrm{o}}}{\partial\tau}\right)_\pi,\ \gamma_{\tau\tau}^{\mathrm{o}} = \left(\frac{\partial^2\gamma^{\mathrm{o}}}{\partial\tau^2}\right)_\pi,\ \gamma_{\pi\tau}^{\mathrm{o}} = \left(\frac{\partial^2\gamma^{\mathrm{o}}}{\partial\pi\,\partial\tau}\right)$$

Tabelle 3.14. Der residuelle Anteil γ^{r} der dimensionslosen freien Enthalpie, Gl. (3.10), und seine Ableitungen

$\gamma^{\mathrm{r}} = \sum_{i=1}^{43} n_i \pi^{I_i} (\tau - 0{,}5)^{J_i}$	$\gamma^{\mathrm{r}}_{\tau} = \sum_{i=1}^{43} n_i \pi^{I_i} J_i (\tau - 0{,}5)^{J_i - 1}$
$\gamma^{\mathrm{r}}_{\pi} = \sum_{i=1}^{43} n_i I_i \pi^{I_i - 1} (\tau - 0{,}5)^{J_i}$	$\gamma^{\mathrm{r}}_{\tau\tau} = \sum_{i=1}^{43} n_i \pi^{I_i} J_i (J_i - 1)(\tau - 0{,}5)^{J_i - 2}$
$\gamma^{\mathrm{r}}_{\pi\pi} = \sum_{i=1}^{43} n_i I_i (I_i - 1) \pi^{I_i - 2} (\tau - 0{,}5)^{J_i}$	$\gamma^{\mathrm{r}}_{\pi\tau} = \sum_{i=1}^{43} n_i I_i \pi^{I_i - 1} J_i (\tau - 0{,}5)^{J_i - 1}$

$\gamma^{\mathrm{r}}_{\pi} = \left(\frac{\partial \gamma^{\mathrm{r}}}{\partial \pi}\right)_{\tau}, \; \gamma^{\mathrm{r}}_{\pi\pi} = \left(\frac{\partial^2 \gamma^{\mathrm{r}}}{\partial \pi^2}\right)_{\tau}, \; \gamma^{\mathrm{r}}_{\tau} = \left(\frac{\partial \gamma^{\mathrm{r}}}{\partial \tau}\right)_{\pi}, \; \gamma^{\mathrm{r}}_{\tau\tau} = \left(\frac{\partial^2 \gamma^{\mathrm{r}}}{\partial \tau^2}\right)_{\pi}, \; \gamma^{\mathrm{r}}_{\pi\tau} = \left(\frac{\partial^2 \gamma^{\mathrm{r}}}{\partial \pi \partial \tau}\right)$

Tabelle 3.15. Aus Gl. (3.8) berechnete Werte thermodynamischer Zustandsgrößen für ausgewählte Temperaturen und Drücke

Zustandsgröße	$T = 300$ K $p = 0{,}0035$ MPa	$T = 700$ K $p = 0{,}0035$ MPa	$T = 700$ K $p = 30$ MPa
v [m³ kg⁻¹]	$0{,}394\,913\,866 \cdot 10^{2}$	$0{,}923\,015\,898 \cdot 10^{2}$	$0{,}542\,946\,619 \cdot 10^{-2}$
h [kJ kg⁻¹]	$0{,}254\,991\,145 \cdot 10^{4}$	$0{,}333\,568\,375 \cdot 10^{4}$	$0{,}263\,149\,474 \cdot 10^{4}$
u [kJ kg⁻¹]	$0{,}241\,169\,160 \cdot 10^{4}$	$0{,}301\,262\,819 \cdot 10^{4}$	$0{,}246\,861\,076 \cdot 10^{4}$
s [kJ kg⁻¹ K⁻¹]	$0{,}852\,238\,967 \cdot 10^{1}$	$0{,}101\,749\,996 \cdot 10^{2}$	$0{,}517\,540\,298 \cdot 10^{1}$
c_p [kJ kg⁻¹ K⁻¹]	$0{,}191\,300\,162 \cdot 10^{1}$	$0{,}208\,141\,274 \cdot 10^{1}$	$0{,}103\,505\,092 \cdot 10^{2}$
w [m s⁻¹]	$0{,}427\,920\,172 \cdot 10^{3}$	$0{,}644\,289\,068 \cdot 10^{3}$	$0{,}480\,386\,523 \cdot 10^{3}$

3.3.3 Ergänzende Gleichung für das metastabile Dampfgebiet

Ebenso wie die Basisgleichung, Gl. (3.8), ist die ergänzende Gleichung für das metastabile Dampfgebiet als Fundamentalgleichung der dimensionslosen freien Enthalpie, $\gamma = g/(RT)$, formuliert. Dieser Ansatz besteht wiederum aus einem Anteil des idealen Gases γ^{o} und einem residuellen Anteil γ^{r} und lautet

$$\frac{g(p,T)}{RT} = \gamma(\pi,\tau) = \gamma^{\mathrm{o}}(\pi,\tau) + \gamma^{\mathrm{r}}(\pi,\tau) \tag{3.11}$$

mit $\pi = p/p^*$ und $\tau = T^*/T$ sowie R gemäß Gl. (2.1).

Die Gleichung für den Anteil des idealen Gases γ^{o} ist identisch mit Gl. (3.9) mit Ausnahme der Werte für die beiden Koeffizienten n_1^{o} und n_2^{o} (s. Tabelle 3.10). Für den Gebrauch von Gl. (3.9) als Teil von Gl. (3.11) wurden die Koeffizienten n_1^{o} und n_2^{o} geringfügig verändert, um die hohe Konsistenzanforderung zwischen den Gln. (3.11) und (3.8) bezüglich der Größen h und s entlang der Taulinie erfüllen zu können, s. weiter unten.

Die Gleichung für den residuellen Anteil γ^{r} lautet

$$\gamma^{\mathrm{r}} = \sum_{i=1}^{13} n_i \pi^{I_i} (\tau - 0{,}5)^{J_i} \tag{3.12}$$

mit $\pi = p/p^*$ und $\tau = T^*/T$ sowie $p^* = 1$ MPa und $T^* = 540$ K. Die Koeffizienten n_i und die

Exponenten I_i und J_i der Gl. (3.12) sind in Tabelle 3.16 aufgeführt. Da im metastabilen Dampfgebiet keine experimentellen Daten existieren, an die sich die Gleichung anpassen läßt, wurde Gl. (3.11) an Werte von Zustandsgrößen angepaßt, die aus dem stabilen homogenen Gasgebiet extrapoliert worden sind. Diese Extrapolationen wurden nicht mit der IAPWS-95 Formulation durchgeführt, sondern mit einer speziellen Gasgleichung für den Bereich niedriger Dichten [8].

Tabelle 3.16. Koeffizienten und Exponenten der Gl. (3.12)

i	I_i	J_i	n_i
1	1	0	$-0{,}733\,622\,601\,865\,06 \cdot 10^{-2}$
2	1	2	$-0{,}882\,238\,319\,431\,46 \cdot 10^{-1}$
3	1	5	$-0{,}723\,345\,552\,132\,45 \cdot 10^{-1}$
4	1	11	$-0{,}408\,131\,785\,344\,55 \cdot 10^{-2}$
5	2	1	$0{,}200\,978\,033\,802\,07 \cdot 10^{-2}$
6	2	7	$-0{,}530\,459\,218\,986\,42 \cdot 10^{-1}$
7	2	16	$-0{,}761\,904\,090\,869\,70 \cdot 10^{-2}$
8	3	4	$-0{,}634\,980\,376\,573\,13 \cdot 10^{-2}$
9	3	16	$-0{,}860\,430\,930\,285\,88 \cdot 10^{-1}$
10	4	7	$0{,}753\,215\,815\,227\,70 \cdot 10^{-2}$
11	4	10	$-0{,}792\,383\,754\,461\,39 \cdot 10^{-2}$
12	5	9	$-0{,}228\,881\,607\,784\,47 \cdot 10^{-3}$
13	5	10	$-0{,}264\,565\,014\,828\,10 \cdot 10^{-2}$

Aus Gl. (3.11) können alle thermodynamischen Zustandsgrößen durch geeignete Kopplung von γ^{o}, Gl. (3.9), und γ^{r}, Gl. (3.12), sowie ihren Ableitungen berechnet werden. Tabelle 3.12 enthält diese Beziehungen für die wichtigsten thermodynamischen Zustandsgrößen. Alle für die Auswertung dieser Berechnungsgleichungen notwendigen Ableitungen des Anteils des idealen Gases γ^{o} und des residuellen Anteiles γ^{r} der dimensionslosen freien Enthalpie γ sind in den Tabellen 3.13 und 3.17 explizit angegeben.

Gleichung (3.11) gilt für das metastabile Dampfgebiet von der Taulinie bis zu einer Linie konstanten Dampfgehaltes von 95 % (berechnet aus den Gleichgewichtswerten h' und h'') für Drücke vom Tripelpunktsdruck, s. Gl. (2.8), bis 10 MPa. Die numerische Konsistenz von Gl. (3.11) mit der Basisgleichung, Gl. (3.8), entlang der Taulinie ist durch folgende maximale Inkonsistenzen bezüglich der Zustandsgrößen v, h, c_p, s, g und w charakterisiert:

$$|\Delta v|_{\max} = 0{,}014\ \% \qquad |\Delta s|_{\max} = 0{,}082\ \mathrm{J\ kg^{-1}\ K^{-1}}$$
$$|\Delta h|_{\max} = 0{,}043\ \mathrm{kJ\ kg^{-1}} \qquad |\Delta g|_{\max} = 0{,}023\ \mathrm{kJ\ kg^{-1}}$$
$$|\Delta c_p|_{\max} = 0{,}78\ \% \qquad |\Delta w|_{\max} = 0{,}051\ \%\,.$$

Diese maximalen Inkonsistenzen sind deutlich kleiner als die Konsistenzanforderungen entlang der Bereichsgrenzen entsprechend den sog. Prag-Werten [10], die in Abschn. 3.1.2 angegeben sind. Entlang der Isobare $p = 10$ MPa im metastabilen Dampfgebiet ist der Übergang zwischen Gl. (3.11) und Gl. (3.8) nicht glatt, was jedoch für praktische Berechnungen keine Bedeutung hat.

Zur Überprüfung von Computerprogrammen auf der Basis von Gl. (3.11) enthält Tabelle 3.18 entsprechende Testwerte für die wichtigsten Zustandsgrößen.

Tabelle 3.17. Der residuelle Anteil γ^{r} der dimensionslosen freien Enthalpie, Gl. (3.12), und seine Ableitungen

$$\gamma^{\mathrm{r}} = \sum_{i=1}^{13} n_i \, \pi^{I_i} (\tau - 0{,}5)^{J_i} \qquad \gamma^{\mathrm{r}}_{\tau} = \sum_{i=1}^{13} n_i \, \pi^{I_i} J_i (\tau - 0{,}5)^{J_i - 1}$$

$$\gamma^{\mathrm{r}}_{\pi} = \sum_{i=1}^{13} n_i \, I_i \, \pi^{I_i - 1} (\tau - 0{,}5)^{J_i} \qquad \gamma^{\mathrm{r}}_{\tau\tau} = \sum_{i=1}^{13} n_i \, \pi^{I_i} J_i (J_i - 1)(\tau - 0{,}5)^{J_i - 2}$$

$$\gamma^{\mathrm{r}}_{\pi\pi} = \sum_{i=1}^{13} n_i \, I_i (I_i - 1) \pi^{I_i - 2} (\tau - 0{,}5)^{J_i} \qquad \gamma^{\mathrm{r}}_{\pi\tau} = \sum_{i=1}^{13} n_i \, I_i \, \pi^{I_i - 1} J_i (\tau - 0{,}5)^{J_i - 1}$$

$$\gamma^{\mathrm{r}}_{\pi} = \left(\frac{\partial \gamma^{\mathrm{r}}}{\partial \pi}\right)_{\tau}, \; \gamma^{\mathrm{r}}_{\pi\pi} = \left(\frac{\partial^2 \gamma^{\mathrm{r}}}{\partial \pi^2}\right)_{\tau}, \; \gamma^{\mathrm{r}}_{\tau} = \left(\frac{\partial \gamma^{\mathrm{r}}}{\partial \tau}\right)_{\pi}, \; \gamma^{\mathrm{r}}_{\tau\tau} = \left(\frac{\partial^2 \gamma^{\mathrm{r}}}{\partial \tau^2}\right)_{\pi}, \; \gamma^{\mathrm{r}}_{\pi\tau} = \left(\frac{\partial^2 \gamma^{\mathrm{r}}}{\partial \pi \, \partial \tau}\right)$$

Tabelle 3.18. Aus Gl. (3.11) berechnete Werte thermodynamischer Zustandsgrößen für ausgewählte Temperaturen und Drücke

Zustandsgröße	$T = 450$ K $p = 1$ MPa	$T = 440$ K $p = 1$ MPa	$T = 450$ K $p = 1{,}5$ MPa
v [m^3 kg^{-1}]	0,192 516 540	0,186 212 297	0,121 685 206
h [kJ kg^{-1}]	$0{,}276\,881\,115 \cdot 10^4$	$0{,}274\,015\,123 \cdot 10^4$	$0{,}272\,134\,539 \cdot 10^4$
u [kJ kg^{-1}]	$0{,}257\,629\,461 \cdot 10^4$	$0{,}255\,393\,894 \cdot 10^4$	$0{,}253\,881\,758 \cdot 10^4$
s [kJ kg^{-1} K^{-1}]	$0{,}656\,660\,377 \cdot 10^1$	$0{,}650\,218\,759 \cdot 10^1$	$0{,}629\,170\,440 \cdot 10^1$
c_p [kJ kg^{-1} K^{-1}]	$0{,}276\,349\,265 \cdot 10^1$	$0{,}298\,166\,443 \cdot 10^1$	$0{,}362\,795\,578 \cdot 10^1$
w [m s^{-1}]	$0{,}498\,408\,101 \cdot 10^3$	$0{,}489\,363\,295 \cdot 10^3$	$0{,}481\,941\,819 \cdot 10^3$

3.3.4 Rückwärtsgleichungen

Um Zustandsgrößen als Funktion von p, h sowie von p, s ohne Iteration ausreichend genau berechnen zu können (s. Abschn. 3.1.1), müssen die beiden Rückwärtsgleichungen mit der Basisgleichung sehr gut numerisch konsistent sein. Die genaue Anforderung für diese numerische Konsistenz wurde mit Hilfe umfassender Testrechnungen für eine Reihe charakteristischer Kraftwerksprozesse ermittelt. Auf der Basis dieser Untersuchungen wurde die numerische Konsistenzanforderung zwischen der Basisgleichung, Gl (3.8), und den beiden Rückwärtsgleichungen $T(p,h)$ und $T(p,s)$, abhängig davon, welches Gebiet im Bereich 2 betrachtet wird, zu $\pm$ 10 mK bzw. zu $\pm$ 25 mK festgesetzt, s. die Tabellen 3.23 und 3.28. Das bedeutet zum Beispiel, daß die aus der Rückwärtsgleichung $T(p,h)$, Gl. (3.15), für vorgegebene Werte von p und h direkt berechnete Temperatur innerhalb von $\pm$ 10 mK mit dem Temperaturwert übereinstimmen muß, der für dieselben Werte von p und h aus der Basisgleichung, Gl. (3.8), iteriert werden kann. Diese Iteration aus der Basisgleichung erfolgt über die Funktion $h(p,T)$, deren Zusammenhang mit der dimensionslosen freien Enthalpie γ in Tabelle 3.12 angegeben ist.

Aufgrund der sehr hohen Anforderungen bezüglich der numerischen Konsistenz zwischen der Basisgleichung, Gl. (3.8), und den Rückwärtsgleichungen sind für den Bereich 2 *drei* Gleichungen für $T(p, h)$ und *drei* Gleichungen für $T(p, s)$ notwendig, um diese Anforderungen im gesamten Bereich zu erfüllen. Abbildung 3.2 zeigt, in welcher Weise der Bereich 2 in die 3 Unterbereiche für die Rückwärtsgleichungen aufgeteilt ist. Die Grenze zwischen den Unterbereichen 2a und 2b ist durch die Isobare $p = 4$ MPa gegeben, und die Grenze zwischen den Unterbereichen 2b und 2c entspricht der Isentrope $s = 5{,}85$ kJ kg^{-1} K^{-1}.

Für die Entscheidung, ob für gegebene Werte von p und h die Gleichung $T(p,h)$ für den Unterbereich 2b oder 2c benutzt werden muß, wird eine entsprechende Korrelationsgleichung zur Beschreibung der Grenze zwischen den Unterbereichen 2b und 2c benötigt, die die Linie $s = 5{,}85$ kJ kg^{-1} K^{-1} annähert, s. Abb. 3.2. Diese als B2bc-Gleichung bezeichnete Grenzgleichung ist eine einfache Druck-Enthalpie-Beziehung und lautet

$$\pi = n_1 + n_2\eta + n_3\eta^2 \tag{3.13}$$

mit $\pi = p/p^*$ und $\eta = h/h^*$ sowie $p^* = 1$ MPa und $h^* = 1$ kJ kg^{-1}. Die Koeffizienten n_1 bis n_3 der Gl. (3.13) sind in Tabelle 3.19 aufgeführt. Aufgrund der sehr einfachen Form von Gl. (3.13) beschreibt diese nicht exakt die Isentrope $s = 5{,}85$ kJ kg^{-1} K^{-1}, sondern die zugehörigen Entropiewerte liegen zwischen $s = 5{,}81$ kJ kg^{-1} K^{-1} und $s = 5{,}85$ kJ kg^{-1} K^{-1}. Die in der Enthalpie explizite Form der Gl. (3.13) lautet

$$\eta = n_4 + \left[(\pi - n_5)/n_3\right]^{0,5} \tag{3.14}$$

mit π und η entsprechend Gl. (3.13) und den Koeffizienten n_3 bis n_5 in Tabelle 3.19. Die Gln. (3.13) und (3.14) beschreiben die Grenzlinie zwischen den Unterbereichen 2b und 2c vom Sättigungszustand bei $T = 554{,}485$ K und $p_s = 6{,}54670$ MPa bis $T = 1019{,}32$ K und $p = 100$ MPa.

Für die Rückwärtsgleichungen $T(p,s)$ ist die Grenze zwischen den Unterbereichen 2b und 2c, die der Isentrope $s = 5{,}85$ kJ kg^{-1} K^{-1} innerhalb der oben aufgeführten Werte für Temperatur und Druck entspricht, automatisch für vorgegebene Werte von p und s definiert.

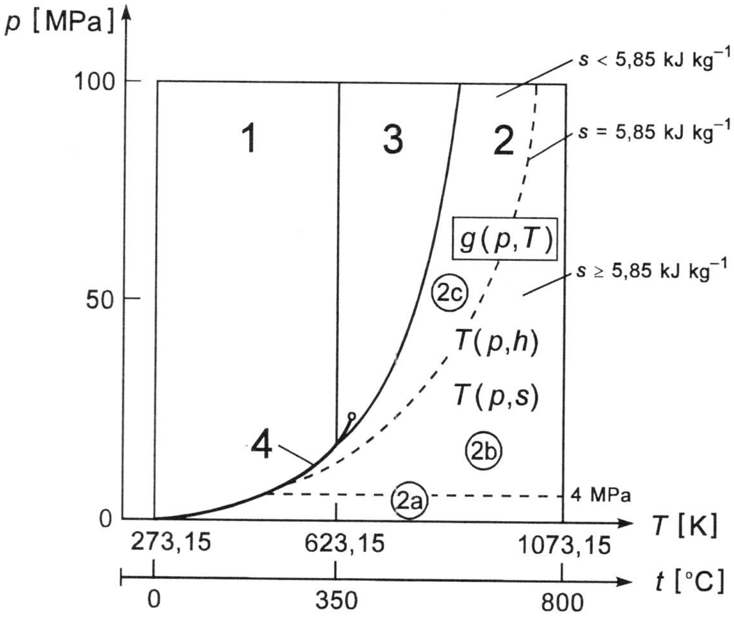

Abb. 3.2. Aufteilung des Bereiches 2 der IAPWS-IF97 in die 3 Unterbereiche 2a, 2b und 2c für die Rückwärtsgleichungen $T(p,h)$ und $T(p,s)$

Tabelle 3.19. Koeffizienten der Gln. (3.13) und (3.14)

i	n_i	i	n_i
1	$0{,}905\,842\,785\,147\,23 \cdot 10^{3}$	4	$0{,}265\,265\,719\,084\,28 \cdot 10^{4}$
2	$-\,0{,}679\,557\,863\,992\,41$	5	$0{,}452\,575\,789\,059\,48 \cdot 10^{1}$
3	$0{,}128\,090\,027\,301\,36 \cdot 10^{-3}$		

Bei der Überprüfung von Computerprogrammen auf der Basis der Gln. (3.13) und (3.14) muß folgender p,h-Punkt erfüllt werden: $p = 0{,}100\,000\,000 \cdot 10^3$ MPa, $h = 0{,}351\,600\,432\,3 \cdot 10^4$ kJ kg^{-1}.

3.3.4.1 Die Rückwärtsgleichungen $T(p, h)$ für die Unterbereiche 2a, 2b und 2c

Die Rückwärtsgleichung $T(p,h)$ für den **Unterbereich 2a** hat folgende dimensionslose Form:

$$\frac{T_{2\mathrm{a}}(p,h)}{T^*} = \theta_{2\mathrm{a}}(\pi,\eta) = \sum_{i=1}^{34} n_i\, \pi^{I_i} (\eta - 2{,}1)^{J_i} \tag{3.15}$$

mit $\theta = T/T^*$, $\pi = p/p^*$ und $\eta = h/h^*$ sowie $T^* = 1$ K, $p^* = 1$ MPa und $h^* = 2000$ kJ kg^{-1}. Die Koeffizienten n_i und die Exponenten I_i und J_i der Gl. (3.15) sind in Tabelle 3.20 aufgeführt.

Die Rückwärtsgleichung $T(p,h)$ für den **Unterbereich 2b** hat folgende dimensionslose Form:

$$\frac{T_{2\mathrm{b}}(p,h)}{T^*} = \theta_{2\mathrm{b}}(\pi,\eta) = \sum_{i=1}^{38} n_i (\pi - 2)^{I_i} (\eta - 2{,}6)^{J_i} \tag{3.16}$$

mit $\theta = T/T^*$, $\pi = p/p^*$ und $\eta = h/h^*$ sowie $T^* = 1$ K, $p^* = 1$ MPa und $h^* = 2000$ kJ kg^{-1}. Die Koeffizienten n_i und die Exponenten I_i und J_i der Gl. (3.16) sind in Tabelle 3.21 aufgeführt.

Die Rückwärtsgleichung $T(p,h)$ für den **Unterbereich 2c** hat folgende dimensionslose Form:

$$\frac{T_{2\mathrm{c}}(p,h)}{T^*} = \theta_{2\mathrm{c}}(\pi,\eta) = \sum_{i=1}^{23} n_i (\pi + 25)^{I_i} (\eta - 1{,}8)^{J_i} \tag{3.17}$$

mit $\theta = T/T^*$, $\pi = p/p^*$ und $\eta = h/h^*$ sowie $T^* = 1$ K, $p^* = 1$ MPa und $h^* = 2000$ kJ kg^{-1}. Die Koeffizienten n_i und die Exponenten I_i und J_i der Gl. (3.17) sind in Tabelle 3.22 aufgeführt.

Tabelle 3.20. Koeffizienten und Exponenten der Rückwärtsgleichung $T(p,h)$ für den Unterbereich 2a, Gl. (3.15)

i	I_i	J_i	n_i	i	I_i	J_i	n_i
1	0	0	0,108 989 523 182 88 · 10^{4}	18	2	7	0,116 708 730 771 07 · 10^{2}
2	0	1	0,849 516 544 955 35 · 10^{3}	19	2	36	0,128 127 984 040 46 · 10^{9}
3	0	2	– 0,107 817 480 918 26 · 10^{3}	20	2	38	– 0,985 549 096 232 76 · 10^{9}
4	0	3	0,331 536 548 012 63 · 10^{2}	21	2	40	0,282 245 469 730 02 · 10^{10}
5	0	7	– 0,742 320 167 902 48 · 10^{1}	22	2	42	– 0,359 489 714 107 03 · 10^{10}
6	0	20	0,117 650 487 243 56 · 10^{2}	23	2	44	0,172 273 499 131 97 · 10^{10}
7	1	0	0,184 457 493 557 90 · 10^{1}	24	3	24	– 0,135 513 342 407 75 · 10^{5}
8	1	1	– 0,417 927 005 496 24 · 10^{1}	25	3	44	0,128 487 346 646 50 · 10^{8}
9	1	2	0,624 781 969 358 12 · 10^{1}	26	4	12	0,138 657 242 832 26 · 10^{1}
10	1	3	– 0,173 445 631 081 14 · 10^{2}	27	4	32	0,235 988 325 565 14 · 10^{6}
11	1	7	– 0,200 581 768 620 96 · 10^{3}	28	4	44	– 0,131 052 365 450 54 · 10^{8}
12	1	9	0,271 960 654 737 96 · 10^{3}	29	5	32	0,739 998 354 747 66 · 10^{4}
13	1	11	– 0,455 113 182 858 18 · 10^{3}	30	5	36	– 0,551 966 970 300 60 · 10^{6}
14	1	18	0,309 196 886 047 55 · 10^{4}	31	5	42	0,371 540 859 962 33 · 10^{7}
15	1	44	0,252 266 403 578 72 · 10^{6}	32	6	34	0,191 277 292 396 60 · 10^{5}
16	2	0	– 0,617 074 228 683 39 · 10^{-2}	33	6	44	– 0,415 351 648 356 34 · 10^{6}
17	2	2	– 0,310 780 466 295 83	34	7	28	– 0,624 598 551 925 07 · 10^{2}

Tabelle 3.21. Koeffizienten und Exponenten der Rückwärtsgleichung $T(p,h)$ für den Unterbereich 2b, Gl. (3.16)

i	I_i	J_i	n_i	i	I_i	J_i	n_i
1	0	0	0,148 950 410 795 16 · 10^{4}	20	2	40	0,712 803 519 595 51 · 10^{-4}
2	0	1	0,743 077 983 140 34 · 10^{3}	21	3	1	0,110 328 317 899 99 · 10^{-3}
3	0	2	– 0,977 083 187 978 37 · 10^{2}	22	3	2	0,189 552 483 879 02 · 10^{-3}
4	0	12	0,247 424 647 056 74 · 10^{1}	23	3	12	0,308 915 411 605 37 · 10^{-2}
5	0	18	– 0,632 813 200 160 26	24	3	24	0,135 555 045 549 49 · 10^{-2}
6	0	24	0,113 859 521 296 58 · 10^{1}	25	4	2	0,286 402 374 774 56 · 10^{-6}
7	0	28	– 0,478 118 636 486 25	26	4	12	– 0,107 798 573 575 12 · 10^{-4}
8	0	40	0,852 081 234 315 44 · 10^{-2}	27	4	18	– 0,764 627 124 548 14 · 10^{-4}
9	1	0	0,937 471 473 779 32	28	4	24	0,140 523 928 183 16 · 10^{-4}
10	1	2	0,335 931 186 049 16 · 10^{1}	29	4	28	– 0,310 838 143 314 34 · 10^{-4}
11	1	6	0,338 093 556 014 54 · 10^{1}	30	4	40	– 0,103 027 382 121 03 · 10^{-5}
12	1	12	0,168 445 396 719 04	31	5	18	0,282 172 816 350 40 · 10^{-6}
13	1	18	0,738 757 452 366 95	32	5	24	0,127 049 022 719 45 · 10^{-5}
14	1	24	– 0,471 287 374 361 86	33	5	40	0,738 033 534 682 92 · 10^{-7}
15	1	28	0,150 202 731 397 07	34	6	28	– 0,110 301 392 389 09 · 10^{-7}
16	1	40	– 0,217 641 142 197 50 · 10^{-2}	35	7	2	– 0,814 563 652 078 33 · 10^{-13}
17	2	2	– 0,218 107 553 247 61 · 10^{-1}	36	7	28	– 0,251 805 456 829 62 · 10^{-10}
18	2	8	– 0,108 297 844 036 77	37	9	1	– 0,175 652 339 694 07 · 10^{-17}
19	2	18	– 0,463 333 246 358 12 · 10^{-1}	38	9	40	0,869 341 563 441 63 · 10^{-14}

Tabelle 3.22. Koeffizienten und Exponenten der Rückwärtsgleichung $T(p,h)$ für den Unterbereich 2c, Gl. (3.17)

i	I_i	J_i	n_i	i	I_i	J_i	n_i
1	−7	0	$-0{,}323\,683\,985\,552\,42 \cdot 10^{13}$	13	1	4	$0{,}379\,660\,012\,724\,86 \cdot 10^{1}$
2	−7	4	$0{,}732\,633\,509\,021\,81 \cdot 10^{13}$	14	1	8	$-0{,}108\,429\,848\,800\,77 \cdot 10^{2}$
3	−6	0	$0{,}358\,250\,899\,454\,47 \cdot 10^{12}$	15	2	4	$-0{,}453\,641\,726\,766\,60 \cdot 10^{-1}$
4	−6	2	$-0{,}583\,401\,318\,515\,90 \cdot 10^{12}$	16	6	0	$0{,}145\,591\,156\,586\,98 \cdot 10^{-12}$
5	−5	0	$-0{,}107\,830\,682\,174\,70 \cdot 10^{11}$	17	6	1	$0{,}112\,615\,974\,072\,30 \cdot 10^{-11}$
6	−5	2	$0{,}208\,255\,445\,631\,71 \cdot 10^{11}$	18	6	4	$-0{,}178\,049\,822\,406\,86 \cdot 10^{-10}$
7	−2	0	$0{,}610\,747\,835\,645\,16 \cdot 10^{6}$	19	6	10	$0{,}123\,245\,796\,908\,32 \cdot 10^{-6}$
8	−2	1	$0{,}859\,777\,225\,355\,80 \cdot 10^{6}$	20	6	12	$-0{,}116\,069\,211\,309\,84 \cdot 10^{-5}$
9	−1	0	$-0{,}257\,457\,236\,041\,70 \cdot 10^{5}$	21	6	16	$0{,}278\,463\,670\,885\,54 \cdot 10^{-4}$
10	−1	2	$0{,}310\,810\,884\,227\,14 \cdot 10^{5}$	22	6	20	$-0{,}592\,700\,384\,741\,76 \cdot 10^{-3}$
11	0	0	$0{,}120\,823\,158\,659\,36 \cdot 10^{4}$	23	6	22	$0{,}129\,185\,829\,918\,78 \cdot 10^{-2}$
12	0	1	$0{,}482\,197\,551\,092\,55 \cdot 10^{3}$				

Die Gln. (3.15), (3.16) und (3.17) sind nur im stabilen homogenen Gebiet des jeweiligen Unterbereichs 2a, 2b und 2c gültig; sie gelten also nicht für den metastabilen unterkühlten Dampf. Die Grenzen zwischen den Unterbereichen sind am Anfang von Abschn. 3.3.4 definiert (s. auch Abb. 3.2); der niedrigste Druck, für den Gl. (3.15) gültig ist, entspricht dem Sublimationsdruck bei 273,15 K, der 611,153 Pa beträgt [11]. Tabelle 3.23 zeigt die erlaubte und die tatsächliche numerische Inkonsistenz zwischen den Gln. (3.15) bis (3.17) und Gl. (3.8).

Tabelle 3.23. Erlaubte und tatsächliche Inkonsistenzen in der Temperatur zwischen den Gln. (3.15) bis (3.17) und Gl. (3.8)

Unterbereich	Gleichung	Temperaturdifferenz [mK]		
		erlaubt	maximal	quadratisches Mittel
2a	3.15	10	9,3	2,9
2b	3.16	10	9,6	3,9
2c	3.17	25	23,7	10,4

Zur Überprüfung von Computerprogrammen auf der Basis der Gln. (3.15) bis (3.17) enthält Tabelle 3.24 entsprechende Testwerte.

Tabelle 3.24. Aus den Gln. (3.15) bis (3.17) berechnete Temperaturen für ausgewählte Werte von Druck und Enthalpie

Gleichung	p [MPa]	h [kJ kg^{-1}]	T [K]
3.15	0,001	3000	$0{,}534\,433\,241 \cdot 10^{3}$
	3	3000	$0{,}575\,373\,370 \cdot 10^{3}$
	3	4000	$0{,}101\,077\,577 \cdot 10^{4}$
3.16	5	3500	$0{,}801\,299\,102 \cdot 10^{3}$
	5	4000	$0{,}101\,531\,583 \cdot 10^{4}$
	25	3500	$0{,}875\,279\,054 \cdot 10^{3}$
3.17	40	2700	$0{,}743\,056\,411 \cdot 10^{3}$
	60	2700	$0{,}791\,137\,067 \cdot 10^{3}$
	60	3200	$0{,}882\,756\,860 \cdot 10^{3}$

3.3.4.2
Die Rückwärtsgleichungen *T*(*p*, *s*) für die Unterbereiche 2a, 2b und 2c

Die Rückwärtsgleichung $T(p,s)$ für den **Unterbereich 2a** hat folgende dimensionslose Form:

$$\frac{T_{2a}(p,s)}{T^*} = \theta_{2a}(\pi,\sigma) = \sum_{i=1}^{46} n_i\, \pi^{I_i} (\sigma - 2)^{J_i} \tag{3.18}$$

mit $\theta = T/T^*$, $\pi = p/p^*$ und $\sigma = s/s^*$ sowie $T^* = 1$ K, $p^* = 1$ MPa und $s^* = 2$ kJ kg^{-1} K^{-1}. Die Koeffizienten n_i und die Exponenten I_i und J_i der Gl. (3.18) sind in Tabelle 3.25 aufgeführt.

Die Rückwärtsgleichung $T(p,s)$ für den **Unterbereich 2b** hat folgende dimensionslose Form:

$$\frac{T_{2b}(p,s)}{T^*} = \theta_{2b}(\pi,\sigma) = \sum_{i=1}^{44} n_i\, \pi^{I_i} (10 - \sigma)^{J_i} \tag{3.19}$$

mit $\theta = T/T^*$, $\pi = p/p^*$ und $\sigma = s/s^*$ sowie $T^* = 1$ K, $p^* = 1$ MPa und $s^* = 0{,}7853$ kJ kg^{-1} K^{-1}. Die Koeffizienten n_i und die Exponenten I_i und J_i der Gl. (3.19) sind in Tabelle 3.26 aufgeführt.

Die Rückwärtsgleichung $T(p,s)$ für den **Unterbereich 2c** hat folgende dimensionslose Form:

$$\frac{T_{2c}(p,s)}{T^*} = \theta_{2c}(\pi,\sigma) = \sum_{i=1}^{30} n_i\, \pi^{I_i} (2 - \sigma)^{J_i} \tag{3.20}$$

mit $\theta = T/T^*$, $\pi = p/p^*$ und $\sigma = s/s^*$ sowie $T^* = 1$ K, $p^* = 1$ MPa und $s^* = 2{,}9251$ kJ kg^{-1} K^{-1}. Die Koeffizienten n_i und die Exponenten I_i und J_i der Gl. (3.20) sind in Tabelle 3.27 aufgeführt.

Tabelle 3.25. Koeffizienten und Exponenten der Rückwärtsgleichung $T(p,s)$ für den Unterbereich 2a, Gl. (3.18)

i	I_i	J_i	n_i	i	I_i	J_i	n_i
1	−1,5	−24	$-0{,}392\,359\,838\,619\,84 \cdot 10^{6}$	24	−0,25	−11	$-0{,}597\,806\,388\,727\,18 \cdot 10^{4}$
2	−1,5	−23	$0{,}515\,265\,738\,272\,70 \cdot 10^{6}$	25	−0,25	−6	$-0{,}704\,014\,639\,268\,62 \cdot 10^{3}$
3	−1,5	−19	$0{,}404\,824\,431\,610\,48 \cdot 10^{5}$	26	0,25	1	$0{,}338\,367\,841\,075\,53 \cdot 10^{3}$
4	−1,5	−13	$-0{,}321\,937\,909\,239\,02 \cdot 10^{3}$	27	0,25	4	$0{,}208\,627\,866\,351\,87 \cdot 10^{2}$
5	−1,5	−11	$0{,}969\,614\,242\,186\,94 \cdot 10^{2}$	28	0,25	8	$0{,}338\,341\,726\,561\,96 \cdot 10^{-1}$
6	−1,5	−10	$-0{,}228\,678\,463\,717\,73 \cdot 10^{2}$	29	0,25	11	$-0{,}431\,244\,284\,148\,93 \cdot 10^{-4}$
7	−1,25	−19	$-0{,}449\,429\,141\,243\,57 \cdot 10^{6}$	30	0,5	0	$0{,}166\,537\,913\,564\,12 \cdot 10^{3}$
8	−1,25	−15	$-0{,}501\,183\,360\,201\,66 \cdot 10^{4}$	31	0,5	1	$-0{,}139\,862\,920\,558\,98 \cdot 10^{3}$
9	−1,25	−6	$0{,}356\,844\,635\,600\,15$	32	0,5	5	$-0{,}788\,495\,479\,998\,72$
10	−1,0	−26	$0{,}442\,353\,358\,481\,90 \cdot 10^{5}$	33	0,5	6	$0{,}721\,324\,117\,538\,72 \cdot 10^{-1}$
11	−1,0	−21	$-0{,}136\,733\,888\,117\,08 \cdot 10^{5}$	34	0,5	10	$-0{,}597\,548\,393\,982\,83 \cdot 10^{-2}$
12	−1,0	−17	$0{,}421\,632\,602\,078\,64 \cdot 10^{6}$	35	0,5	14	$-0{,}121\,413\,589\,539\,04 \cdot 10^{-4}$
13	−1,0	−16	$0{,}225\,169\,258\,374\,75 \cdot 10^{5}$	36	0,5	16	$0{,}232\,270\,967\,338\,71 \cdot 10^{-6}$
14	−1,0	−9	$0{,}474\,421\,448\,656\,46 \cdot 10^{3}$	37	0,75	0	$-0{,}105\,384\,635\,661\,94 \cdot 10^{2}$
15	−1,0	−8	$-0{,}149\,311\,307\,976\,47 \cdot 10^{3}$	38	0,75	4	$0{,}207\,189\,254\,965\,02 \cdot 10^{1}$
16	−0,75	−15	$-0{,}197\,811\,263\,204\,52 \cdot 10^{6}$	39	0,75	9	$-0{,}721\,931\,552\,604\,27 \cdot 10^{-1}$
17	−0,75	−14	$-0{,}235\,543\,994\,707\,60 \cdot 10^{5}$	40	0,75	17	$0{,}207\,498\,870\,811\,20 \cdot 10^{-6}$
18	−0,5	−26	$-0{,}190\,706\,163\,020\,76 \cdot 10^{5}$	41	1,0	7	$-0{,}183\,406\,579\,113\,79 \cdot 10^{-1}$
19	−0,5	−13	$0{,}553\,756\,698\,831\,64 \cdot 10^{5}$	42	1,0	18	$0{,}290\,362\,723\,486\,96 \cdot 10^{-6}$
20	−0,5	−9	$0{,}382\,936\,914\,373\,63 \cdot 10^{4}$	43	1,25	3	$0{,}210\,375\,278\,936\,19$
21	−0,5	−7	$-0{,}603\,918\,605\,805\,67 \cdot 10^{3}$	44	1,25	15	$0{,}256\,812\,397\,299\,99 \cdot 10^{-3}$
22	−0,25	−27	$0{,}193\,631\,026\,203\,31 \cdot 10^{4}$	45	1,5	5	$-0{,}127\,990\,029\,337\,81 \cdot 10^{-1}$
23	−0,25	−25	$0{,}426\,606\,436\,986\,10 \cdot 10^{4}$	46	1,5	18	$-0{,}821\,981\,026\,520\,18 \cdot 10^{-5}$

Tabelle 3.26. Koeffizienten und Exponenten der Rückwärtsgleichung $T(p,s)$ für den Unterbereich 2b, Gl. (3.19)

i	I_i	J_i	n_i	i	I_i	J_i	n_i
1	−6	0	$0{,}316\,876\,650\,834\,97 \cdot 10^{6}$	23	0	2	$0{,}417\,273\,471\,596\,10 \cdot 10^{2}$
2	−6	11	$0{,}208\,641\,758\,818\,58 \cdot 10^{2}$	24	0	4	$0{,}219\,325\,494\,345\,32 \cdot 10^{1}$
3	−5	0	$-0{,}398\,593\,998\,035\,99 \cdot 10^{6}$	25	0	5	$-0{,}103\,200\,500\,090\,77 \cdot 10^{1}$
4	−5	11	$-0{,}218\,160\,585\,188\,77 \cdot 10^{2}$	26	0	6	$0{,}358\,829\,435\,167\,03$
5	−4	0	$0{,}223\,697\,851\,942\,42 \cdot 10^{6}$	27	0	9	$0{,}525\,114\,537\,260\,66 \cdot 10^{-2}$
6	−4	1	$-0{,}278\,417\,034\,458\,17 \cdot 10^{4}$	28	1	0	$0{,}128\,389\,164\,507\,05 \cdot 10^{2}$
7	−4	11	$0{,}992\,074\,360\,714\,80 \cdot 10^{1}$	29	1	1	$-0{,}286\,424\,372\,193\,81 \cdot 10^{1}$
8	−3	0	$-0{,}751\,975\,122\,991\,57 \cdot 10^{5}$	30	1	2	$0{,}569\,126\,836\,648\,55$
9	−3	1	$0{,}297\,086\,059\,511\,58 \cdot 10^{4}$	31	1	3	$-0{,}999\,629\,545\,849\,31 \cdot 10^{-1}$
10	−3	11	$-0{,}344\,068\,785\,485\,26 \cdot 10^{1}$	32	1	7	$-0{,}326\,320\,377\,784\,59 \cdot 10^{-2}$
11	−3	12	$0{,}388\,155\,642\,491\,15$	33	1	8	$0{,}233\,209\,225\,767\,23 \cdot 10^{-3}$
12	−2	0	$0{,}175\,112\,950\,857\,50 \cdot 10^{5}$	34	2	0	$-0{,}153\,348\,098\,574\,50$
13	−2	1	$-0{,}142\,371\,128\,544\,49 \cdot 10^{4}$	35	2	1	$0{,}290\,722\,882\,399\,02 \cdot 10^{-1}$
14	−2	6	$0{,}109\,438\,033\,641\,67 \cdot 10^{1}$	36	2	5	$0{,}375\,347\,027\,411\,67 \cdot 10^{-3}$
15	−2	10	$0{,}899\,716\,193\,084\,95$	37	3	0	$0{,}172\,966\,917\,024\,11 \cdot 10^{-2}$
16	−1	0	$-0{,}337\,597\,400\,989\,58 \cdot 10^{4}$	38	3	1	$-0{,}385\,560\,508\,445\,04 \cdot 10^{-3}$
17	−1	1	$0{,}471\,628\,858\,183\,55 \cdot 10^{3}$	39	3	3	$-0{,}350\,177\,122\,926\,08 \cdot 10^{-4}$
18	−1	5	$-0{,}191\,882\,419\,936\,79 \cdot 10^{1}$	40	4	0	$-0{,}145\,663\,936\,314\,92 \cdot 10^{-4}$
19	−1	8	$0{,}410\,785\,804\,921\,96$	41	4	1	$0{,}564\,208\,572\,672\,69 \cdot 10^{-5}$
20	−1	9	$-0{,}334\,653\,781\,720\,97$	42	5	0	$0{,}412\,861\,500\,746\,05 \cdot 10^{-7}$
21	0	0	$0{,}138\,700\,347\,775\,05 \cdot 10^{4}$	43	5	1	$-0{,}206\,846\,711\,188\,24 \cdot 10^{-7}$
22	0	1	$-0{,}406\,633\,261\,958\,38 \cdot 10^{3}$	44	5	2	$0{,}164\,093\,936\,747\,25 \cdot 10^{-8}$

Tabelle 3.27. Koeffizienten und Exponenten der Rückwärtsgleichung $T(p,s)$ für den Unterbereich 2c, Gl. (3.20)

i	I_i	J_i	n_i	i	I_i	J_i	n_i
1	−2	0	$0{,}909\,685\,010\,053\,65 \cdot 10^{3}$	16	3	1	$-0{,}145\,970\,082\,847\,53 \cdot 10^{-1}$
2	−2	1	$0{,}240\,456\,670\,884\,20 \cdot 10^{4}$	17	3	5	$0{,}566\,311\,756\,310\,27 \cdot 10^{-2}$
3	−1	0	$-0{,}591\,623\,263\,871\,30 \cdot 10^{3}$	18	4	0	$-0{,}761\,558\,645\,845\,77 \cdot 10^{-4}$
4	0	0	$0{,}541\,454\,041\,280\,74 \cdot 10^{3}$	19	4	1	$0{,}224\,403\,429\,193\,32 \cdot 10^{-3}$
5	0	1	$-0{,}270\,983\,084\,111\,92 \cdot 10^{3}$	20	4	4	$-0{,}125\,610\,950\,134\,13 \cdot 10^{-4}$
6	0	2	$0{,}979\,765\,250\,979\,26 \cdot 10^{3}$	21	5	0	$0{,}633\,231\,326\,609\,34 \cdot 10^{-6}$
7	0	3	$-0{,}469\,667\,729\,594\,35 \cdot 10^{3}$	22	5	1	$-0{,}205\,419\,896\,753\,75 \cdot 10^{-5}$
8	1	0	$0{,}143\,992\,746\,047\,23 \cdot 10^{2}$	23	5	2	$0{,}364\,053\,703\,900\,82 \cdot 10^{-7}$
9	1	1	$-0{,}191\,042\,042\,304\,29 \cdot 10^{2}$	24	6	0	$-0{,}297\,598\,977\,892\,15 \cdot 10^{-8}$
10	1	3	$0{,}532\,991\,671\,119\,71 \cdot 10^{1}$	25	6	1	$0{,}101\,366\,185\,297\,63 \cdot 10^{-7}$
11	1	4	$-0{,}212\,529\,753\,759\,34 \cdot 10^{2}$	26	7	0	$0{,}599\,257\,196\,923\,51 \cdot 10^{-11}$
12	2	0	$-0{,}311\,473\,344\,137\,60$	27	7	1	$-0{,}206\,778\,701\,051\,64 \cdot 10^{-10}$
13	2	1	$0{,}603\,348\,408\,946\,23$	28	7	3	$-0{,}208\,742\,781\,818\,86 \cdot 10^{-10}$
14	2	2	$-0{,}427\,648\,397\,025\,09 \cdot 10^{-1}$	29	7	4	$0{,}101\,621\,668\,250\,89 \cdot 10^{-9}$
15	3	0	$0{,}581\,855\,972\,552\,59 \cdot 10^{-2}$	30	7	5	$-0{,}164\,298\,282\,813\,47 \cdot 10^{-9}$

Die Gln. (3.18), (3.19) und (3.20) sind nur im stabilen homogenen Gebiet des jeweiligen Unterbereichs 2a, 2b und 2c gültig; sie gelten also nicht für den metastabilen unterkühlten Dampf. Die Grenzen zwischen den Unterbereichen sind am Anfang von Abschn. 3.3.4 definiert (siehe auch Abb. 3.2); der niedrigste Druck, für den Gl. (3.18) gültig ist, entspricht dem Sublimationsdruck bei 273,15 K, der 611,153 Pa beträgt [11]. Tabelle 3.24 zeigt die erlaubten und die tatsächlichen numerischen Inkonsistenzen zwischen den Gln. (3.18) bis (3.20) und Gl. (3.8).

Zur Überprüfung von Computerprogrammen auf der Basis der Gln. (3.18) bis (3.20) enthält Tabelle 3.29 entsprechende Testwerte.

Tabelle 3.28. Erlaubte und tatsächliche Inkonsistenzen in der Temperatur zwischen den Gln. (3.18) bis (3.20) und Gl. (3.8)

Unterbereich	Gleichung	Temperaturdifferenz [mK]		
		erlaubt	maximal	quadratisches Mittel
2a	3.18	10	8,8	1,2
2b	3.19	10	6,5	2,8
2c	3.20	25	19,0	8,3

Tabelle 3.29. Aus den Gln. (3.18) bis (3.20) berechnete Temperaturen für ausgewählte Werte von Druck und Entropie

Gleichung	p [MPa]	s [kJ kg^{-1}K^{-1}]	T [K]
3.18	0,1	7,5	$0{,}399\,517\,097 \cdot 10^3$
	0,1	8	$0{,}514\,127\,081 \cdot 10^3$
	2,5	8	$0{,}103\,984\,917 \cdot 10^4$
3.19	8	6	$0{,}600\,484\,040 \cdot 10^3$
	8	7,5	$0{,}106\,495\,556 \cdot 10^4$
	90	6	$0{,}103\,801\,126 \cdot 10^4$
3.20	20	5,75	$0{,}697\,992\,849 \cdot 10^3$
	80	5,25	$0{,}854\,011\,484 \cdot 10^3$
	80	5,75	$0{,}949\,017\,998 \cdot 10^3$

3.4 Basisgleichung für den Bereich 3

Dieser Abschnitt enthält alle Informationen, die für den Gebrauch der Basisgleichung des Bereichs 3 der IAPWS-IF97 benötigt werden. Die Hilfsgleichung zur Definition der Grenze zwischen den Bereichen 2 und 3 ist in Abschn. 3.3.1 angegeben. Die abgeschätzten Unsicherheiten der mit der IAPWS-IF97 berechneten Werte der wichtigsten Zustandsgrößen sind in Abschn. 3.7 zu finden.

Die Basisgleichung für diesen Bereich ist eine Fundamentalgleichung für die spezifische freie Energie f in dimensionsloser Form, $\phi = f/(RT)$, und lautet

$$\frac{f(\rho,T)}{RT} = \phi(\delta,\tau) = n_1 \ln\delta + \sum_{i=2}^{40} n_i \delta^{I_i} \tau^{J_i} \tag{3.21}$$

mit $\delta = \rho/\rho^*$ und $\tau = T^*/T$ sowie $\rho^* = \rho_c$ und $T^* = T_c$, wobei R, T_c und ρ_c durch die Gln. (2.1), (2.4) und (2.6) gegeben sind. Die Koeffizienten n_i und die Exponenten I_i und J_i der Gl. (3.21) sind in Tabelle 3.30 aufgeführt.

Tabelle 3.30. Koeffizienten und Exponenten der Gl. (3.21)

i	I_i	J_i	n_i	i	I_i	J_i	n_i
1	0	0	$0{,}106\,580\,700\,285\,13 \cdot 10^{1}$	21	3	4	$-0{,}201\,899\,150\,235\,70 \cdot 10^{1}$
2	0	0	$-0{,}157\,328\,452\,902\,39 \cdot 10^{2}$	22	3	16	$-0{,}821\,476\,371\,739\,63 \cdot 10^{-2}$
3	0	1	$0{,}209\,443\,969\,743\,07 \cdot 10^{2}$	23	3	26	$-0{,}475\,960\,357\,349\,23$
4	0	2	$-0{,}768\,677\,078\,787\,16 \cdot 10^{1}$	24	4	0	$0{,}439\,840\,744\,735\,00 \cdot 10^{-1}$
5	0	7	$0{,}261\,859\,477\,879\,54 \cdot 10^{1}$	25	4	2	$-0{,}444\,764\,354\,287\,39$
6	0	10	$-0{,}280\,807\,811\,486\,20 \cdot 10^{1}$	26	4	4	$0{,}905\,720\,707\,197\,33$
7	0	12	$0{,}120\,533\,696\,965\,17 \cdot 10^{1}$	27	4	26	$0{,}705\,224\,500\,879\,67$
8	0	23	$-0{,}845\,668\,128\,125\,02 \cdot 10^{-2}$	28	5	1	$0{,}107\,705\,126\,263\,32$
9	1	2	$-0{,}126\,543\,154\,777\,14 \cdot 10^{1}$	29	5	3	$-0{,}329\,136\,232\,589\,54$
10	1	6	$-0{,}115\,244\,078\,066\,81 \cdot 10^{1}$	30	5	26	$-0{,}508\,710\,620\,411\,58$
11	1	15	$0{,}885\,210\,439\,843\,18$	31	6	0	$-0{,}221\,754\,008\,730\,96 \cdot 10^{-1}$
12	1	17	$-0{,}642\,077\,651\,816\,07$	32	6	2	$0{,}942\,607\,516\,650\,92 \cdot 10^{-1}$
13	2	0	$0{,}384\,934\,601\,866\,71$	33	6	26	$0{,}164\,362\,784\,479\,61$
14	2	2	$-0{,}852\,147\,088\,242\,06$	34	7	2	$-0{,}135\,033\,722\,413\,48 \cdot 10^{-1}$
15	2	6	$0{,}489\,722\,815\,418\,77 \cdot 10^{1}$	35	8	26	$-0{,}148\,343\,453\,524\,72 \cdot 10^{-1}$
16	2	7	$-0{,}305\,026\,172\,569\,65 \cdot 10^{1}$	36	9	2	$0{,}579\,229\,536\,280\,84 \cdot 10^{-3}$
17	2	22	$0{,}394\,205\,368\,791\,54 \cdot 10^{-1}$	37	9	26	$0{,}323\,089\,047\,037\,11 \cdot 10^{-2}$
18	2	26	$0{,}125\,584\,084\,243\,08$	38	10	0	$0{,}809\,648\,029\,962\,15 \cdot 10^{-4}$
19	3	0	$-0{,}279\,993\,296\,987\,10$	39	10	1	$-0{,}165\,576\,797\,950\,37 \cdot 10^{-3}$
20	3	2	$0{,}138\,997\,995\,694\,60 \cdot 10^{1}$	40	11	26	$-0{,}449\,238\,990\,618\,15 \cdot 10^{-4}$

Zusätzlich zur Wiedergabe der thermodynamischen Zustandsgrößen des homogenen Gebietes erfüllt Gl. (3.21) an der Sättigungslinie (Temperaturbereich $623{,}15\ \mathrm{K} \le T \le T_c$) auch die Phasengleichgewichtsbedingung für die koexistierende flüssige und dampfförmige Phase (Gleichheit der spezifischen freien Enthalpie in beiden Phasen, berücksichtigt durch das Maxwell-Kriterium, s. Tabelle 3.31). Darüber hinaus erfüllt Gl. (3.21) exakt die Werte der kritischen Parameter entsprechend den Gln. (2.4) bis (2.6) sowie am kritischen Punkt die Nebenbedingungen $(\partial p/\partial \rho)_T = 0$ und $(\partial^2 p/\partial \rho^2)_T = 0$.

Aus Gl. (3.21) können alle thermodynamischen Zustandsgrößen durch geeignete Kopplung der dimensionslosen freien Energie ϕ und ihrer Ableitungen ermittelt werden. Die Beziehungen zwischen den wichtigsten thermodynamischen Eigenschaften und ϕ sowie seinen Ableitungen sind in Tabelle 3.31 zusammengestellt. Alle erforderlichen Ableitungen der dimensionslosen freien Energie ϕ, Gl. (3.21), sind in Tabelle 3.32 explizit aufgeführt.

Gleichung (3.21) gilt für den Bereich 3 der IAPWS-IF97, der durch folgenden Temperatur- und Druckbereich definiert ist, s. Abb. 3.1:

$$623{,}15\ \mathrm{K} \le T \le T(p)_{\mathrm{Gl.(3.7)}} \qquad p(T)_{\mathrm{Gl.(3.6)}} \le p \le 100\ \mathrm{MPa}\,.$$

Neben den Zustandsgrößen des homogenen Gebietes des Bereichs 3 liefert Gl (3.21) auch sinnvolle Werte für die Zustandsgrößen der metastabilen überhitzten Flüssigkeit in der Nähe der Siedelinie und des metastabilen unterkühlten Dampfes in der Nähe der Taulinie.

Zur Überprüfung von Computerprogrammen auf der Basis von Gl. (3.21) enthält Tabelle 3.33 entsprechende Testwerte für die wichtigsten Zustandsgrößen.

Tabelle 3.31. Beziehungen thermodynamischer Zustandsgrößen zur dimensionslosen freien Energie ϕ und ihren Ableitungen auf der Basis von Gl. (3.21)

Zustandsgröße	Beziehung
Druck $p = \rho^2 (\partial f / \partial \rho)_T$	$\frac{p(\delta,\tau)}{\rho RT} = \delta \phi_\delta$
spezifische innere Energie $u = f - T(\partial f / \partial T)_\rho$	$\frac{u(\delta,\tau)}{RT} = \tau \phi_\tau$
spezifische Entropie $s = -(\partial f / \partial T)_\rho$	$\frac{s(\delta,\tau)}{R} = \tau \phi_\tau - \phi$
spezifische Enthalpie $h = f - T(\partial f / \partial T)_\rho + \rho(\partial f / \partial \rho)_T$	$\frac{h(\delta,\tau)}{RT} = \tau \phi_\tau + \delta \phi_\delta$
spezifische isochore Wärmekapazität $c_v = (\partial u / \partial T)_\rho$	$\frac{c_v(\delta,\tau)}{R} = -\tau^2 \phi_{\tau\tau}$
spezifische isobare Wärmekapazität $c_p = (\partial h / \partial T)_p$	$\frac{c_p(\delta,\tau)}{R} = -\tau^2 \phi_{\tau\tau} + \frac{(\delta \phi_\delta - \delta \tau \phi_{\delta\tau})^2}{2\delta \phi_\delta + \delta^2 \phi_{\delta\delta}}$
Schallgeschwindigkeit $w = [(\partial p / \partial \rho)_s]^{0,5}$	$\frac{w^2(\delta,\tau)}{RT} = 2\delta \phi_\delta + \delta^2 \phi_{\delta\delta} - \frac{(\delta \phi_\delta - \delta \tau \phi_{\delta\tau})^2}{\tau^2 \phi_{\tau\tau}}$
Phasengleichgewichtsbedingung (Maxwell-Kriterium)	$\frac{p_s}{RT\rho'} = \delta' \phi_\delta(\delta',\tau)$; $\frac{p_s}{RT\rho''} = \delta'' \phi_\delta(\delta'',\tau)$ $\frac{p_s}{RT}\left(\frac{1}{\rho''} - \frac{1}{\rho'}\right) = \phi(\delta',\tau) - \phi(\delta'',\tau)$

$\phi_\delta = \left(\frac{\partial \phi}{\partial \delta}\right)_\tau, \ \phi_{\delta\delta} = \left(\frac{\partial^2 \phi}{\partial \delta^2}\right)_\tau, \ \phi_\tau = \left(\frac{\partial \phi}{\partial \tau}\right)_\delta, \ \phi_{\tau\tau} = \left(\frac{\partial^2 \phi}{\partial \tau^2}\right)_\delta, \ \phi_{\delta\tau} = \left(\frac{\partial^2 \phi}{\partial \delta \partial \tau}\right)$

Tabelle 3.32. Die dimensionslose freie Energie ϕ, Gl. (3.21), und ihre Ableitungen

$\phi = n_1 \ln \delta + \sum_{i=2}^{40} n_i \, \delta^{I_i} \tau^{J_i}$	$\phi_\tau = \sum_{i=2}^{40} n_i \, \delta^{I_i} J_i \tau^{J_i - 1}$
$\phi_\delta = n_1 \delta^{-1} + \sum_{i=2}^{40} n_i \, I_i \delta^{I_i - 1} \tau^{J_i}$	$\phi_{\tau\tau} = \sum_{i=2}^{40} n_i \, \delta^{I_i} J_i (J_i - 1) \tau^{J_i - 2}$
$\phi_{\delta\delta} = -n_1 \delta^{-2} + \sum_{i=2}^{40} n_i \, I_i (I_i - 1) \delta^{I_i - 2} \tau^{J_i}$	$\phi_{\delta\tau} = \sum_{i=2}^{40} n_i \, I_i \delta^{I_i - 1} J_i \tau^{J_i - 1}$

$\phi_\delta = \left(\frac{\partial \phi}{\partial \delta}\right)_\tau, \ \phi_{\delta\delta} = \left(\frac{\partial^2 \phi}{\partial \delta^2}\right)_\tau, \ \phi_\tau = \left(\frac{\partial \phi}{\partial \tau}\right)_\delta, \ \phi_{\tau\tau} = \left(\frac{\partial^2 \phi}{\partial \tau^2}\right)_\delta, \ \phi_{\delta\tau} = \left(\frac{\partial^2 \phi}{\partial \delta \partial \tau}\right)$

Tabelle 3.33. Aus Gl. (3.21) berechnete Werte thermodynamischer Zustandsgrößen für ausgewählte Temperaturen und Dichten

Zustandsgröße	$T = 650$ K $\rho = 500$ kg m^{-3}	$T = 650$ K $\rho = 200$ kg m^{-3}	$T = 750$ K $\rho = 500$ kg m^{-3}
p [MPa]	$0{,}255\,837\,018 \cdot 10^{2}$	$0{,}222\,930\,643 \cdot 10^{2}$	$0{,}783\,095\,639 \cdot 10^{2}$
h [kJ kg^{-1}]	$0{,}186\,343\,019 \cdot 10^{4}$	$0{,}237\,512\,401 \cdot 10^{4}$	$0{,}225\,868\,845 \cdot 10^{4}$
u [kJ kg^{-1}]	$0{,}181\,226\,279 \cdot 10^{4}$	$0{,}226\,365\,868 \cdot 10^{4}$	$0{,}210\,206\,932 \cdot 10^{4}$
s [kJ kg^{-1} K^{-1}]	$0{,}405\,427\,273 \cdot 10^{1}$	$0{,}485\,438\,792 \cdot 10^{1}$	$0{,}446\,971\,906 \cdot 10^{1}$
c_p [kJ kg^{-1} K^{-1}]	$0{,}138\,935\,717 \cdot 10^{2}$	$0{,}446\,579\,342 \cdot 10^{2}$	$0{,}634\,165\,359 \cdot 10^{1}$
w [m s^{-1}]	$0{,}502\,005\,554 \cdot 10^{3}$	$0{,}383\,444\,594 \cdot 10^{3}$	$0{,}760\,696\,041 \cdot 10^{3}$

3.5 Gleichungen für den Bereich 4

Dieser Abschnitt enthält alle Informationen, die für den Gebrauch der Gleichungen des Bereichs 4 der IAPWS-IF97 (Sättigungslinie) benötigt werden. Die abgeschätzten Unsicherheiten der mit der IAPWS-IF97 berechneten Werte des Dampfdruckes sind in Abschn. 3.7 zu finden.

Die Gleichung zur Beschreibung der Sättigungslinie ist eine implizite quadratische Gleichung, die sowohl bezüglich des Dampfdruckes p_s als auch bezüglich der Siedetemperatur T_s direkt aufgelöst werden kann. Diese Gleichung hat folgende Form:

$$\beta^2\vartheta^2 + n_1\beta^2\vartheta + n_2\beta^2 + n_3\beta\vartheta^2 + n_4\beta\vartheta + n_5\beta + n_6\vartheta^2 + n_7\vartheta + n_8 = 0 \tag{3.22}$$

mit

$$\beta = \left(p_s / p^*\right)^{0,25} \tag{3.22a}$$

und

$$\vartheta = \frac{T_s}{T^*} + \frac{n_9}{\left(T_s / T^*\right) - n_{10}} \tag{3.22b}$$

sowie $p^* = 1$ MPa und $T^* = 1$ K; die Koeffizienten n_1 bis n_{10} sind in Tabelle 3.34 aufgeführt.

3.5.1 Dampfdruckgleichung (Basisgleichung)

Die Lösung der Gl. (3.22) bezüglich des Dampfdruckes lautet

$$\frac{p_s}{p^*} = \left[\frac{2C}{-B + \left(B^2 - 4AC\right)^{0,5}}\right]^4 \tag{3.23}$$

mit $p^* = 1$ MPa und

$$A = \vartheta^2 + n_1\vartheta + n_2$$

$$B = n_3\vartheta^2 + n_4\vartheta + n_5$$

$$C = n_6\vartheta^2 + n_7\vartheta + n_8\,,$$

wobei ϑ gemäß Gl. (3.22b) definiert ist. Die Koeffizienten n_i der Gl. (3.23) sind in Tabelle 3.34 aufgeführt.

Tabelle 3.34. Koeffizienten der Gln. (3.22) bis (3.24)

i	n_i	i	n_i
1	$0{,}116\,705\,214\,527\,67 \cdot 10^{4}$	6	$0{,}149\,151\,086\,135\,30 \cdot 10^{2}$
2	$-0{,}724\,213\,167\,032\,06 \cdot 10^{6}$	7	$-0{,}482\,326\,573\,615\,91 \cdot 10^{4}$
3	$-0{,}170\,738\,469\,400\,92 \cdot 10^{2}$	8	$0{,}405\,113\,405\,420\,57 \cdot 10^{6}$
4	$0{,}120\,208\,247\,024\,70 \cdot 10^{5}$	9	$-0{,}238\,555\,575\,678\,49$
5	$-0{,}323\,255\,503\,223\,33 \cdot 10^{7}$	10	$0{,}650\,175\,348\,447\,98 \cdot 10^{3}$

Die Gln. (3.22) bis (3.24) erfüllen exakt die p,T-Werte am Tripelpunkt entsprechend den Gln. (2.7) und (2.8), am normalen Siedepunkt gemäß Gl. (2.9) und am kritischen Punkt entsprechend den Gln. (2.4) und (2.5).

Die Dampfdruckgleichung, Gl. (3.23), gilt für die gesamte Sättigungslinie von der Tripelpunktstemperatur T_{t} bis zur kritischen Temperatur T_{c} und kann bis 273,15 K extrapoliert werden, so daß sie folgenden Temperaturbereich abdeckt:

$$273{,}15\ \mathrm{K} \leq T \leq 647{,}096\ \mathrm{K}\,.$$

Zur Überprüfung von Computerprogrammen auf der Basis von Gl. (3.23) enthält Tabelle 3.35 entsprechende Testwerte.

Tabelle 3.35. Aus Gl. (3.23) berechnete Dampfdrücke für ausgewählte Temperaturen

T [K]	p_{s} [MPa]
300	$0{,}353\,658\,941 \cdot 10^{-2}$
500	$0{,}263\,889\,776 \cdot 10^{1}$
600	$0{,}123\,443\,146 \cdot 10^{2}$

3.5.2
Siedetemperaturgleichung (Rückwärtsgleichung)

Die Lösung der Gl. (3.22) bezüglich der Siedetemperatur lautet:

$$\frac{T_{\mathrm{s}}}{T^*} = \frac{n_{10} + D - \left[(n_{10} + D)^2 - 4(n_9 + n_{10}D)\right]^{0,5}}{2} \tag{3.24}$$

mit $T^* = 1$ K und

$$D = \frac{2G}{-F - \left(F^2 - 4EG\right)^{0,5}}$$

sowie

$$E = \beta^2 + n_3\beta + n_6$$

$$F = n_1\beta^2 + n_4\beta + n_7$$

$$G = n_2\beta^2 + n_5\beta + n_8$$

und β gemäß Gl. (3.22a). Die Koeffizienten n_i der Gl. (3.24) sind in Tabelle 3.34 aufgeführt.

Gleichung (3.24) hat den gleichen Gültigkeitsbereich wie Gl. (3.23), d. h., sie deckt die gesamte Sättigungslinie Dampf-Flüssigkeit entsprechend folgendem Druckbereich ab:

$$611{,}213 \text{ Pa} \le p \le 22{,}064 \text{ MPa} .$$

Der Wert von 611,213 Pa entspricht dem Druck, wenn Gl. (3.24) bis zu 273,15 K extrapoliert wird. Weil die Dampfdruckgleichung, Gl. (3.23), und die Siedetemperaturgleichung, Gl. (3.24), derselben impliziten Gleichung für die Sättigungslinie, Gl. (3.22), entstammen, sind die Gln. (3.23) und (3.24) numerisch *exakt* konsistent zueinander.

Zur Überprüfung von Computerprogrammen auf der Basis von Gl. (3.24) enthält Tabelle 3.36 entsprechende Testwerte.

Tabelle 3.36. Aus Gl. (3.24) berechnete Siedetemperaturen für ausgewählte Drücke

p [MPa]	T_s [K]
0,1	$0{,}372\,755\,919 \cdot 10^3$
1	$0{,}453\,035\,632 \cdot 10^3$
10	$0{,}584\,149\,488 \cdot 10^3$

3.6 Basisgleichung für den Bereich 5

Dieser Abschnitt enthält alle Informationen, die für den Gebrauch der Gleichungen des Bereichs 5 der IAPWS-IF97 benötigt werden. Die abgeschätzten Unsicherheiten der mit der IAPWS-IF97 berechneten Werte der wichtigsten Zustandsgrößen sind in Abschn. 3.7 zu finden.

Die Basisgleichung für diesen Bereich ist eine Fundamentalgleichung der freien Enthalpie g in dimensionsloser Form, $\gamma = g/(RT)$, wobei g in den Anteil des idealen Gases γ^{o} und in den residuellen Anteil γ^{r} aufgeteilt wird. Es gilt also

$$\frac{g(p,T)}{RT} = \gamma(\pi,\tau) = \gamma^{\mathrm{o}}(\pi,\tau) + \gamma^{\mathrm{r}}(\pi,\tau) \tag{3.25}$$

mit $\pi = p/p^*$ und $\tau = T^*/T$ sowie R gemäß Gl. (2.1).

Die Gleichung für den Anteil γ^{o} des idealen Gases der dimensionslosen freien Enthalpie lautet

$$\gamma^{\mathrm{o}} = \ln\pi + \sum_{i=1}^{6} n_i^{\mathrm{o}} \tau^{J_i^{\mathrm{o}}} \tag{3.26}$$

mit $\pi = p/p^*$ und $\tau = T^*/T$ sowie $p^* = 1$ MPa und $T^* = 1000$ K. Die Koeffizienten n_1^{o} und n_2^{o} wurden derart angepaßt, daß die aus Gl. (3.25) berechneten Werte für die spezifische innere Energie und die spezifische Entropie zu dem durch Gl. (3.2) definierten Bezugspunkt konsistent sind. Tabelle 3.37 enthält die Koeffizienten n_i^{o} und die Exponenten J_i^{o} der Gl. (3.26).

Tabelle 3.37. Koeffizienten und Exponenten der Gl. (3.26)

i	J_i^o	n_i^o	i	J_i^o	n_i^o
1	0	$-0{,}131\,799\,836\,742\,01 \cdot 10^{2}$	4	-2	$0{,}369\,015\,349\,803\,33$
2	1	$0{,}685\,408\,416\,344\,34 \cdot 10^{1}$	5	-1	$-0{,}311\,613\,182\,139\,25 \cdot 10^{1}$
3	-3	$-0{,}248\,051\,489\,334\,66 \cdot 10^{-1}$	6	2	$-0{,}329\,616\,265\,389\,17$

Der residuelle Anteil γ^{r} der dimensionslosen freien Enthalpie hat folgende Form:

$$\gamma^{\mathrm{r}} = \sum_{i=1}^{5} n_i \, \pi^{I_i} \, \tau^{J_i} \tag{3.27}$$

mit $\pi = p/p^*$ und $\tau = T^*/T$ sowie $p^* = 1$ MPa und $T^* = 1000$ K. Die Koeffizienten n_i und die Exponenten I_i und J_i der Gl. (3.27) sind in Tabelle 3.38 aufgeführt.

Tabelle 3.38. Koeffizienten und Exponenten der Gl. (3.27)

i	I_i	J_i	n_i
1	1	0	$-0{,}125\,631\,835\,895\,92 \cdot 10^{-3}$
2	1	1	$0{,}217\,746\,787\,145\,71 \cdot 10^{-2}$
3	1	3	$-0{,}459\,428\,208\,999\,10 \cdot 10^{-2}$
4	2	9	$-0{,}397\,248\,283\,595\,69 \cdot 10^{-5}$
5	3	3	$0{,}129\,192\,282\,897\,84 \cdot 10^{-6}$

Aus Gl. (3.25) können alle thermodynamischen Zustandsgrößen durch geeignete Kopplung von γ^{o}, Gl. (3.26), und γ^{r}, Gl. (3.27), sowie ihren Ableitungen berechnet werden. Tabelle 3.39 enthält diese Beziehungen für die wichtigsten thermodynamischen Zustandsgrößen. Alle für die Auswertung dieser Berechnungsgleichungen notwendigen Ableitungen des Anteils des idealen Gases γ^{o} und des residuellen Anteils γ^{r} der dimensionslosen freien Enthalpie γ sind in den Tabellen 3.40 und 3.41 explizit angegeben.

Gleichung (3.25) gilt für den Bereich 5 der IAPWS-IF97, der durch folgenden Temperatur- und Druckbereich definiert ist, s. Abb. 3.1:

$$1073{,}15\ \mathrm{K} \le T \le 2273{,}15\ \mathrm{K} \qquad 0 < p \le 10\ \mathrm{MPa}\,.$$

Gleichung (3.25) ist in diesem Bereich nur für reines, nichtdissoziiertes Wasser gültig, d. h., mögliche Dissoziationen müssen zusätzlich berücksichtigt werden.

Zur Überprüfung von Computerprogrammen auf der Basis von Gl. (3.25) enthält Tabelle 3.42 entsprechende Testwerte für die wichtigsten Zustandsgrößen.

Tabelle 3.39. Beziehungen thermodynamischer Zustandsgrößen zum Anteil des idealen Gases γ^{o} und zum residuellen Anteil γ^{r} sowie ihren Ableitungen auf der Basis der dimensionslosen freien Enthalpie gemäß Gl. (3.25)

Zustandsgröße	Beziehung
spezifisches Volumen $v = (\partial g/\partial p)_T$	$v(\pi,\tau)\dfrac{p}{RT} = \pi\left(\gamma_\pi^{\mathrm{o}} + \gamma_\pi^{\mathrm{r}}\right)$
spezifische innere Energie $u = g - T(\partial g/\partial T)_p - p(\partial g/\partial p)_T$	$\dfrac{u(\pi,\tau)}{RT} = \tau\left(\gamma_\tau^{\mathrm{o}} + \gamma_\tau^{\mathrm{r}}\right) - \pi\left(\gamma_\pi^{\mathrm{o}} + \gamma_\pi^{\mathrm{r}}\right)$
spezifische Entropie $s = -(\partial g/\partial T)_p$	$\dfrac{s(\pi,\tau)}{R} = \tau\left(\gamma_\tau^{\mathrm{o}} + \gamma_\tau^{\mathrm{r}}\right) - \left(\gamma^{\mathrm{o}} + \gamma^{\mathrm{r}}\right)$
spezifische Enthalpie $h = g - T(\partial g/\partial T)_p$	$\dfrac{h(\pi,\tau)}{RT} = \tau\left(\gamma_\tau^{\mathrm{o}} + \gamma_\tau^{\mathrm{r}}\right)$
spezifische isobare Wärmekapazität $c_p = (\partial h/\partial T)_p$	$\dfrac{c_p(\pi,\tau)}{R} = -\tau^2\left(\gamma_{\tau\tau}^{\mathrm{o}} + \gamma_{\tau\tau}^{\mathrm{r}}\right)$
spezifische isochore Wärmekapazität $c_v = (\partial u/\partial T)_v$	$\dfrac{c_v(\pi,\tau)}{R} = -\tau^2\left(\gamma_{\tau\tau}^{\mathrm{o}} + \gamma_{\tau\tau}^{\mathrm{r}}\right) - \dfrac{(1+\pi\gamma_\pi^{\mathrm{r}} - \tau\pi\gamma_{\pi\tau}^{\mathrm{r}})^2}{1-\pi^2\gamma_{\pi\pi}^{\mathrm{r}}}$
Schallgeschwindigkeit $w = v\left[-(\partial p/\partial v)_s\right]^{0,5}$	$\dfrac{w^2(\pi,\tau)}{RT} = \dfrac{1+2\pi\gamma_\pi^{\mathrm{r}} + \pi^2\gamma_\pi^{\mathrm{r}\,2}}{\left(1-\pi^2\gamma_{\pi\pi}^{\mathrm{r}}\right) + \dfrac{\left(1+\pi\gamma_\pi^{\mathrm{r}} - \tau\pi\gamma_{\pi\tau}^{\mathrm{r}}\right)^2}{\tau^2\left(\gamma_{\tau\tau}^{\mathrm{o}} + \gamma_{\tau\tau}^{\mathrm{r}}\right)}}$

$$\gamma_\pi^{\mathrm{r}} = \left(\frac{\partial\gamma^{\mathrm{r}}}{\partial\pi}\right)_\tau,\ \gamma_{\pi\pi}^{\mathrm{r}} = \left(\frac{\partial^2\gamma^{\mathrm{r}}}{\partial\pi^2}\right)_\tau,\ \gamma_\tau^{\mathrm{r}} = \left(\frac{\partial\gamma^{\mathrm{r}}}{\partial\tau}\right)_\pi,\ \gamma_{\tau\tau}^{\mathrm{r}} = \left(\frac{\partial^2\gamma^{\mathrm{r}}}{\partial\tau^2}\right)_\pi,\ \gamma_{\pi\tau}^{\mathrm{r}} = \left(\frac{\partial^2\gamma^{\mathrm{r}}}{\partial\pi\partial\tau}\right),\ \gamma_\tau^{\mathrm{o}} = \left(\frac{\partial\gamma^{\mathrm{o}}}{\partial\tau}\right)_\pi,\ \gamma_{\tau\tau}^{\mathrm{o}} = \left(\frac{\partial^2\gamma^{\mathrm{o}}}{\partial\tau^2}\right)_\pi$$

Tabelle 3.40. Der Anteil des idealen Gases γ^{o} der dimensionslosen freien Enthalpie, Gl. (3.26), und seine Ableitungen

$\gamma^{\mathrm{o}} = \ln\pi + \sum\limits_{i=1}^{6} n_i^{\mathrm{o}}\tau^{J_i^{\mathrm{o}}}$	$\gamma_\tau^{\mathrm{o}} = \sum\limits_{i=1}^{6} n_i^{\mathrm{o}} J_i^{\mathrm{o}}\tau^{J_i^{\mathrm{o}}-1}$
$\gamma_\pi^{\mathrm{o}} = \pi^{-1}$	$\gamma_{\tau\tau}^{\mathrm{o}} = \sum\limits_{i=1}^{6} n_i^{\mathrm{o}} J_i^{\mathrm{o}}\left(J_i^{\mathrm{o}}-1\right)\tau^{J_i^{\mathrm{o}}-2}$
$\gamma_{\pi\pi}^{\mathrm{o}} = -\pi^{-2}$	$\gamma_{\pi\tau}^{\mathrm{o}} = 0$

$$\gamma_\pi^{\mathrm{o}} = \left(\frac{\partial\gamma^{\mathrm{o}}}{\partial\pi}\right)_\tau,\ \gamma_{\pi\pi}^{\mathrm{o}} = \left(\frac{\partial^2\gamma^{\mathrm{o}}}{\partial\pi^2}\right)_\tau,\ \gamma_\tau^{\mathrm{o}} = \left(\frac{\partial\gamma^{\mathrm{o}}}{\partial\tau}\right)_\pi,\ \gamma_{\tau\tau}^{\mathrm{o}} = \left(\frac{\partial^2\gamma^{\mathrm{o}}}{\partial\tau^2}\right)_\pi,\ \gamma_{\pi\tau}^{\mathrm{o}} = \left(\frac{\partial^2\gamma^{\mathrm{o}}}{\partial\pi\,\partial\tau}\right)$$

Tabelle 3.41. Der residuelle Anteil γ^{r} der dimensionslosen freien Enthalpie, Gl. (3.27), und seine Ableitungen

$$\gamma^{\mathrm{r}} = \sum_{i=1}^{5} n_i\, \pi^{I_i}\, \tau^{J_i} \qquad \gamma^{\mathrm{r}}_{\tau} = \sum_{i=1}^{5} n_i\, \pi^{I_i} J_i\, \tau^{J_i-1}$$

$$\gamma^{\mathrm{r}}_{\pi} = \sum_{i=1}^{5} n_i\, I_i\, \pi^{I_i-1}\, \tau^{J_i} \qquad \gamma^{\mathrm{r}}_{\tau\tau} = \sum_{i=1}^{5} n_i\, \pi^{I_i} J_i (J_i-1)\, \tau^{J_i-2}$$

$$\gamma^{\mathrm{r}}_{\pi\pi} = \sum_{i=1}^{5} n_i\, I_i (I_i-1) \pi^{I_i-2}\, \tau^{J_i} \qquad \gamma^{\mathrm{r}}_{\pi\tau} = \sum_{i=1}^{5} n_i\, I_i\, \pi^{I_i-1} J_i\, \tau^{J_i-1}$$

$$\gamma^{\mathrm{r}}_{\pi} = \left(\frac{\partial \gamma^{\mathrm{r}}}{\partial \pi}\right)_{\tau},\ \gamma^{\mathrm{r}}_{\pi\pi} = \left(\frac{\partial^2 \gamma^{\mathrm{r}}}{\partial \pi^2}\right)_{\tau},\ \gamma^{\mathrm{r}}_{\tau} = \left(\frac{\partial \gamma^{\mathrm{r}}}{\partial \tau}\right)_{\pi},\ \gamma^{\mathrm{r}}_{\tau\tau} = \left(\frac{\partial^2 \gamma^{\mathrm{r}}}{\partial \tau^2}\right)_{\pi},\ \gamma^{\mathrm{r}}_{\pi\tau} = \left(\frac{\partial^2 \gamma^{\mathrm{r}}}{\partial \pi \partial \tau}\right)$$

Tabelle 3.42. Aus Gl. (3.25) berechnete Werte thermodynamischer Zustandsgrößen für ausgewählte Temperaturen und Drücke

Zustandsgröße	T = 1500 K p = 0,5 MPa	T = 1500 K p = 8 MPa	T = 2000 K p = 8 MPa
v [m^3 kg^{-1}]	$0{,}138\,455\,354 \cdot 10^{1}$	$0{,}865\,156\,616 \cdot 10^{-1}$	$0{,}115\,743\,146$
h [kJ kg^{-1}]	$0{,}521\,976\,332 \cdot 10^{4}$	$0{,}520\,609\,634 \cdot 10^{4}$	$0{,}658\,380\,291 \cdot 10^{4}$
u [kJ kg^{-1}]	$0{,}452\,748\,654 \cdot 10^{4}$	$0{,}451\,397\,105 \cdot 10^{4}$	$0{,}565\,785\,774 \cdot 10^{4}$
s [kJ kg^{-1} K^{-1}]	$0{,}965\,408\,431 \cdot 10^{1}$	$0{,}836\,546\,724 \cdot 10^{1}$	$0{,}915\,671\,044 \cdot 10^{1}$
c_p [kJ kg^{-1} K^{-1}]	$0{,}261\,610\,228 \cdot 10^{1}$	$0{,}264\,453\,866 \cdot 10^{1}$	$0{,}285\,306\,750 \cdot 10^{1}$
w [m s^{-1}]	$0{,}917\,071\,933 \cdot 10^{3}$	$0{,}919\,708\,859 \cdot 10^{3}$	$0{,}105\,435\,806 \cdot 10^{4}$

3.7 Unsicherheiten der IAPWS-IF97

In diesem Abschnitt werden die Unsicherheiten für die aus der IAPWS-IF97 berechneten Werte der Zustandsgrößen spezifisches Volumen, spezifische isobare Wärmekapazität, Schallgeschwindigkeit und Dampfdruck angegeben. Für die spezifische Enthalpie können keine Unsicherheitsangaben gemacht werden, weil eine vernünftige Basis für deren Abschätzung fehlt; die Werte der spezifischen Enthalpie hängen von der Wahl des Nullpunktes ab. Von Interesse sind überhaupt nur Enthalpiedifferenzen, wobei die Unsicherheit von *isobaren* Enthalpiedifferenzen stets kleiner ist als die Unsicherheit der isobaren Wärmekapazität.

Die abgeschätzten Unsicherheiten für die oben genannten Zustandsgrößen setzen sich aus 2 Anteilen zusammen:

- Unsicherheit des wissenschaftlichen Standards für die thermodynamischen Eigenschaften von Wasser und Wasserdampf, der IAPWS-95 Formulation [8], aus der die Werte der Zustandsgrößen berechnet worden sind, an die die Gleichungen der Industrie-Formulation IAPWS-IF97 angepaßt wurden. Die Unsicherheit der IAPWS-95 beruht im wesentlichen auf den abgeschätzten Unsicherheiten der ausgewählten Meßwerte der entsprechenden Zustandsgrößen [8], die zur Entwicklung des wissenschaftlichen Standards benutzt worden sind.

- Abweichungen der IAPWS-IF97 von der IAPWS-95 bezüglich der betrachteten Zustandsgrößen.

Die Unsicherheiten der aus der IAPWS-IF97 für das homogene Zustandsgebiet berechneten Zustandsgrößen v, c_p und w sind in den Abb. 3.3 bis 3.5 als Toleranzwerte angegeben. Diese „Toleranzen" entsprechen dem Bereich der möglichen Unsicherheit, wie er von der IAPWS eingeschätzt wurde, wobei ihnen keine statistische Bedeutung beigemessen werden kann. Bezüglich der isobaren Wärmekapazität und der Schallgeschwindigkeit (s. Abb. 3.4 und 3.5) wird darauf hingewiesen, daß die Unsicherheit in den berechneten Werten dieser Zustandsgrößen bei Annäherung an den kritischen Punkt drastisch ansteigt. Die für den Temperaturbereich oberhalb von 1273 K gemachte Aussage „keine genaue Abschätzung der Unsicherheit möglich" basiert auf der Tatsache, daß dieser Bereich außerhalb des Gültigkeitsgebietes der IAPWS-95 liegt und die zur Anpassung der IAPWS-IF97 benutzten Werte der Zustandsgrößen mit der IAPWS-95 extrapoliert wurden. Auf der Grundlage verschiedener Testrechnungen mit der IAPWS-95 kann jedoch erwartet werden, daß diese Extrapolationen sinnvolle Werte liefern.

Die abgeschätzten Unsicherheiten der aus der IAPWS-IF97 berechneten Werte für den Dampfdruck sind aus Abb. 3.6 zu entnehmen.

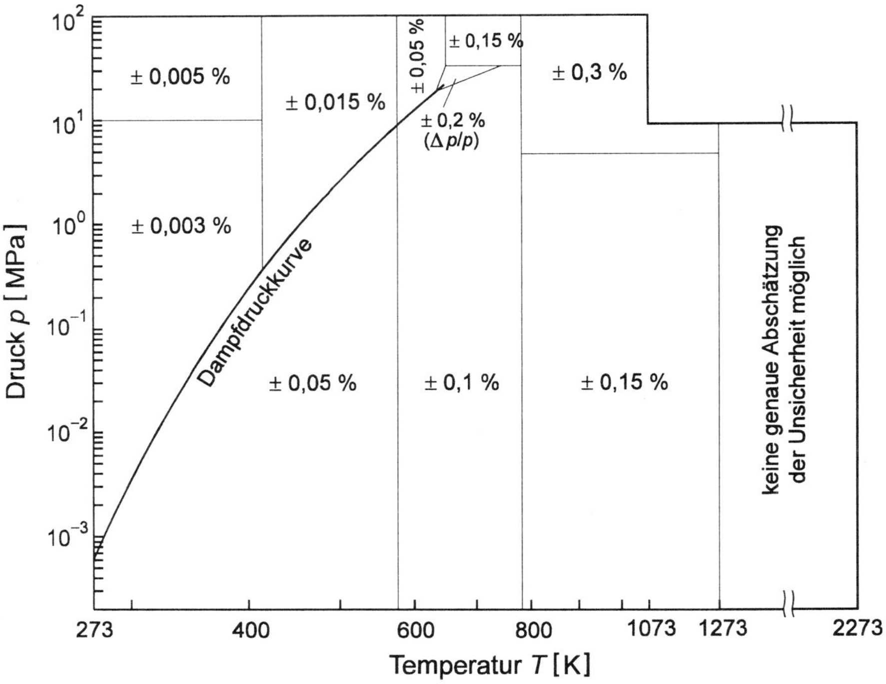

Abb. 3.3. Unsicherheiten im spezifischen Volumen, $\Delta v/v$, abgeschätzt für die entsprechenden Gleichungen der IAPWS-IF97. Für das erweiterte kritische Gebiet (Dreieck) ist die Unsicherheit als prozentuale Unsicherheit im Druck, $\Delta p/p$, angegeben. Dieser Bereich wird durch die beiden Isochoren $v = 0{,}0019\ \mathrm{m^3\,kg^{-1}}$ und $v = 0{,}0069\ \mathrm{m^3\,kg^{-1}}$ sowie durch die Isobare $p = 30$ MPa begrenzt. Die Grenzlinien zwischen den Gebieten mit unterschiedlicher Unsicherheit sind als ungefähre Angaben zu betrachten

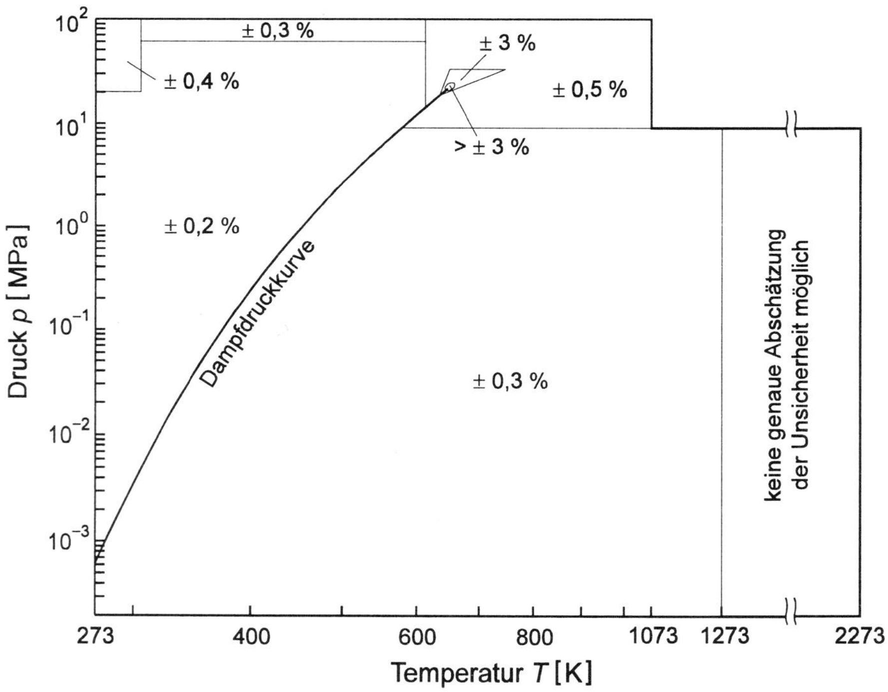

Abb. 3.4. Unsicherheiten in der spezifischen isobaren Wärmekapazität, $\Delta c_p/c_p$, abgeschätzt für die entsprechenden Gleichungen der IAPWS-IF97. Die Definition des dreieckigen Bereiches um den kritischen Punkt kann Abb. 3.3 entnommen werden. Die Grenzlinien zwischen den Gebieten mit unterschiedlicher Unsicherheit sind als ungefähre Angaben zu betrachten

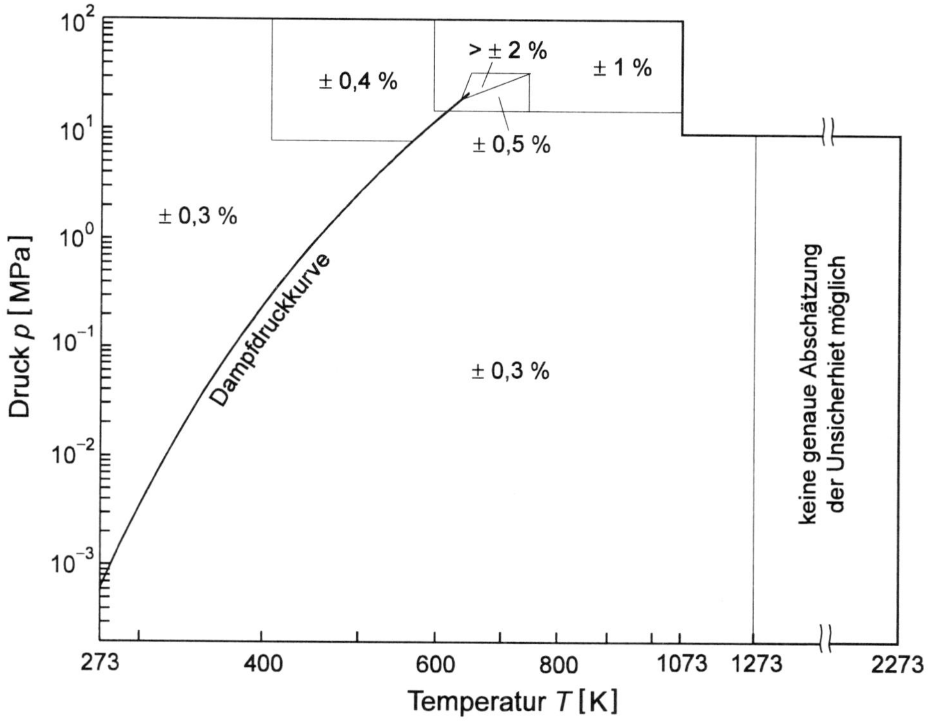

Abb. 3.5. Unsicherheiten in der Schallgeschwindigkeit, $\Delta w/w$, abgeschätzt für die entsprechenden Gleichungen der IAPWS-IF97. Die Definition des dreieckigen Bereiches um den kritischen Punkt kann Abb. 3.3 entnommen werden. Die Grenzlinien zwischen den Gebieten mit unterschiedlicher Unsicherheit sind als ungefähre Angaben zu betrachten

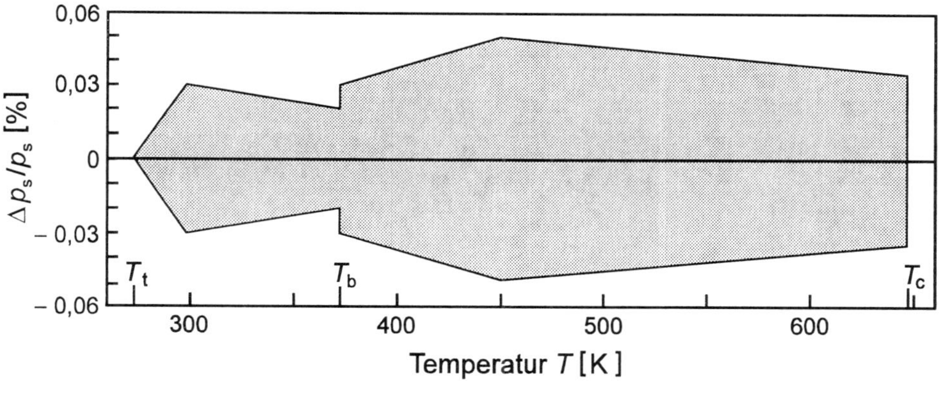

Abb. 3.6. Unsicherheiten im Dampfdruck, $\Delta p_s/p_s$, abgeschätzt für die Dampfdruckgleichung der IAPWS-IF97, Gl. (3.23)

4 Gleichungen für die Transportgrößen und für andere Eigenschaften

Neben der Entwicklung von internationalen Standardgleichungen für die thermodynamischen Zustandsgrößen von Wasser und Wasserdampf hat die IAPWS auch die Entwicklung von Gleichungen für die Transportgrößen und für andere Eigenschaften initiiert und koordiniert. Als Ergebnis dieser Arbeiten werden in den Abschn. 4.1 und 4.2 die neuesten IAPWS-Gleichungen für die Transportgrößen dynamische Viskosität und Wärmeleitfähigkeit und in den Abschn. 4.3 bis 4.5 die IAPWS-Gleichungen für die Eigenschaften Oberflächenspannung, statische Dielektrizitätskonstante und Brechungsindex angegeben. Mit Ausnahme der Oberflächenspannung enthalten die Gleichungen für alle anderen aufgezählten Zustandsgrößen die Dichte als eine der Eingangsvariablen. Wenn diese Zustandsgrößen - wie es industriell der Normalfall ist - für vorgegebene Werte von Druck und Temperatur berechnet werden sollen, muß immer zunächst die Dichte bestimmt werden. Diese Berechnungen der Dichte sollten mit der Industrie-Formulation IAPWS-IF97 [9] durchgeführt werden, die in Kap. 3 beschrieben ist. In dieser Weise wurden die Werte der Transportgrößen und der anderen Eigenschaften in den entsprechenden Tafeln in Teil C dieses Buches berechnet.

4.1 Dynamische Viskosität

Die Gleichung für die dynamische Viskosität η von Wasser und Wasserdampf wurde ursprünglich von Aleksandrov et al. [12] entwickelt und basiert auf einem Satz von Viskositätsmeßwerten, der von einer speziellen Arbeitsgruppe der IAPS (jetzt IAPWS) ausgewählt wurde. Diese Gleichung bildete von 1975 bis 1997 die IAPWS-Gleichung für die dynamische Viskosität [13–15]. Im Jahre 1997 wurde eine geringfügig modifizierte Gleichung als derzeit gültige Viskositätsgleichung der IAPWS verabschiedet [16]. Diese Gleichung ist konsistent zur Temperaturskala ITS-90 [6] und, bezüglich der Dichte, zum wissenschaftlichen Standard für die thermodynamischen Zustandsgrößen von Wasser und Wasserdampf, der IAPWS-95 Formulation [8].

Die Gleichung für die reduzierte dynamische Viskosität lautet

$$\Psi(\delta,\tau) = \Psi_0(\tau)\cdot\Psi_1(\delta,\tau) \tag{4.1}$$

mit $\Psi = \eta/\eta^*$, $\delta = \rho/\rho^*$ und $\tau = T^*/T$ sowie $\eta^* = 55{,}071\cdot 10^{-6}$ Pa s, $\rho^* = 317{,}763$ kg m^{-3} und $T^* = 647{,}226$ K. Der erste Term des Produktes in Gl. (4.1) beschreibt die Viskosität des Wasserdampfes im Grenzübergang zum idealen Gas und hat die Form

$$\Psi_0(\tau) = \left[\tau^{0,5}\sum_{i=0}^{3} n_i^{\mathrm{o}}\,\tau^{i}\right]^{-1} \tag{4.2}$$

mit $\tau = T^*/T$ und $T^* = 647{,}226$ K; die Koeffizienten n_i^{o} sind in Tabelle 4.1 aufgeführt. Die

Gleichung für den zweiten multiplikativen Faktor in Gl. (4.1) lautet

$$\Psi_1(\delta,\tau) = \exp\left[\delta \sum_{i=1}^{19} n_i (\delta - 1)^{I_i} (\tau - 1)^{J_i}\right] \tag{4.3}$$

mit $\delta = \rho/\rho^*$ und $\tau = T^*/T$ sowie $\rho^* = 317{,}763\ \mathrm{kg\ m^{-3}}$ und $T^* = 647{,}226\ \mathrm{K}$. Die Koeffizienten n_i und die Exponenten I_i und J_i sind in Tabelle 4.2 aufgeführt.

Tabelle 4.1. Koeffizienten der Gl. (4.2)

i	n_i^{o}	i	n_i^{o}
0	$0{,}100\,000 \cdot 10^{1}$	2	$0{,}579\,829$
1	$0{,}978\,197$	3	$-0{,}202\,354$

Tabelle 4.2. Koeffizienten und Exponenten der Gl. (4.3)

i	I_i	J_i	n_i	i	I_i	J_i	n_i
1	0	0	$0{,}513\,204\,7$	11	2	2	$-0{,}126\,318\,4 \cdot 10^{1}$
2	0	1	$0{,}320\,565\,6$	12	3	0	$0{,}177\,806\,4$
3	0	4	$-0{,}778\,256\,7$	13	3	1	$0{,}460\,504\,0$
4	0	5	$0{,}188\,544\,7$	14	3	2	$0{,}234\,037\,9$
5	1	0	$0{,}215\,177\,8$	15	3	3	$-0{,}492\,417\,9$
6	1	1	$0{,}731\,788\,3$	16	4	0	$-0{,}417\,661\,0 \cdot 10^{-1}$
7	1	2	$0{,}124\,104\,4 \cdot 10^{1}$	17	4	3	$0{,}160\,043\,5$
8	1	3	$0{,}147\,678\,3 \cdot 10^{1}$	18	5	1	$-0{,}157\,838\,6 \cdot 10^{-1}$
9	2	0	$-0{,}281\,810\,7$	19	6	3	$-0{,}362\,948\,1 \cdot 10^{-2}$
10	2	1	$-0{,}107\,078\,6 \cdot 10^{1}$				

Der Gültigkeitsbereich der Gl. (4.1) für die dynamische Viskosität wurde von der IAPWS für den folgenden Bereich von Temperatur und Druck festgelegt:

$$\begin{aligned} 273{,}15\ \mathrm{K} &\le T \le 423{,}15\ \mathrm{K} \qquad & p &\le 500\ \mathrm{MPa} \\ 423{,}15\ \mathrm{K} &< T \le 873{,}15\ \mathrm{K} \qquad & p &\le 350\ \mathrm{MPa} \\ 873{,}15\ \mathrm{K} &< T \le 1173{,}15\ \mathrm{K} \qquad & p &\le 300\ \mathrm{MPa}. \end{aligned}$$

Die abgeschätzten Unsicherheiten der mit Gl. (4.1) berechneten Werte für die dynamische Viskosität sind im IAPWS-Release [16] angegeben.

Wenn die dynamische Viskosität für gegebene Werte von *Druck* und Temperatur berechnet werden soll, muß zunächst für diese Werte die reduzierte Dichte δ als Eingangsgröße für Gl. (4.1) ermittelt werden. Im IAPWS-Release [16] wird bezüglich dieser Dichteberechnung auf die IAPWS-95 Formulation [8] hingewiesen. Für industrielle Anwendungen empfehlen wir jedoch, die IAPWS-IF97 (s. Kap. 3) zur Bestimmung der Eingangsdichte für Gl. (4.1) zu benutzen, da der Unterschied zwischen den Werten aus der IAPWS-IF97 und der IAPWS-95 im Vergleich zur abgeschätzten Unsicherheit der Gl. (4.1) vernachlässigbar ist. Die Verwendung der IAPWS-IF97 hat für industrielle Berechnungen gegenüber der IAPWS-95 jedoch deutliche Vorteile und vermeidet die Implementierung einer zweiten Formulation für die thermodynamischen Zustandsgrößen. Dementsprechend wurden die Viskositätswerte in den entsprechenden Tafeln in Teil C dieses Buches in der von uns empfohlenen Art berechnet.

Zur Überprüfung von Computerprogrammen auf der Basis von Gl. (4.1) enthält Tabelle 4.3 entsprechende Testwerte.

Tabelle 4.3. Aus Gl. (4.1) berechnete Werte der dynamischen Viskosität für ausgewählte Temperaturen und Drücke

Zustandsgröße	$T = 273{,}15$ K $p = 0{,}1$ MPa	$T = 773{,}15$ K $p = 0{,}1$ MPa	$T = 773{,}15$ K $p = 100$ MPa
ρ [kg m^{-3}] (IAPWS-IF97)	$0{,}999\,843\,633 \cdot 10^3$	$0{,}280\,459\,027$	$0{,}528\,195\,286 \cdot 10^3$
η [μPa s]	$0{,}179\,153\,095 \cdot 10^4$	$0{,}285\,746\,586 \cdot 10^2$	$0{,}660\,620\,604 \cdot 10^2$

4.2 Wärmeleitfähigkeit

Für die Berechnung der Wärmeleitfähigkeit λ werden von der IAPWS derzeit 2 unterschiedliche Gleichungen empfohlen [17], die im wesentlichen auf derselben Datenbasis [18] beruhen. Die sog. „Equation for Industrial Use“ [19] ist bezüglich der Dichte mit der IFC-67 [1], dem Vorgänger der jetzigen Industrie-Formulation IAPWS-IF97, verbunden. Die andere Wärmeleitfähigkeitsgleichung, die „Equation for Scientific Use“ [19], basiert bezüglich der Dichte auf der IAPWS-95 [8], dem wissenschaftlichen Standard für die Berechnung der thermodynamischen Zustandsgrößen von Wasser und Wasserdampf. Wegen der großen Differenzen zwischen der IFC-67 und der IAPWS-95 haben wir die „Equation for Scientific Use“ [17] für die Berechnung der Wärmeleitfähigkeit ausgewählt.

Die Gleichung für die Wärmeleitfähigkeit hat die Form

$$\Lambda(\delta,\tau) = \Lambda_0(\tau) \cdot \Lambda_1(\delta,\tau) + \Lambda_2(\delta,\tau) \tag{4.4}$$

mit $\Lambda = \lambda/\lambda^*$, $\delta = \rho/\rho^*$ und $\tau = T^*/T$ sowie $\lambda^* = 0{,}4945\ \mathrm{W\,m^{-1}\,K^{-1}}$, $\rho^* = 317{,}763\ \mathrm{kg\,m^{-3}}$ und $T^* = 647{,}226$ K. Der Faktor $\Lambda_0(\tau)$ erfaßt die Wärmeleitfähigkeit von Wasserdampf im Grenzübergang zum idealen Gas und lautet

$$\Lambda_0(\tau) = \left[\tau^{0,5} \sum_{i=0}^{3} n_i^{\mathrm{o}}\, \tau^i\right]^{-1} \tag{4.5}$$

mit $\tau = T^*/T$ und $T^* = 647{,}226$ K; die Koeffizienten n_i^{o} sind in Tabelle 4.4 aufgeführt. Der Faktor Λ_1 hat folgende Form:

$$\Lambda_1(\delta,\tau) = \exp\left[\delta \sum_{i=0}^{4} \sum_{j=0}^{5} n_{ij} (\tau - 1)^i (\delta - 1)^j\right] \tag{4.6}$$

mit $\delta = \rho/\rho^*$ und $\tau = T^*/T$ sowie $\rho^* = 317{,}763\ \mathrm{kg\,m^{-3}}$ und $T^* = 647{,}226$ K. Die Koeffizienten n_{ij} sind in Tabelle 4.5 aufgeführt. Der additive Term $\Lambda_2(\delta,\tau)$ der Gl. (4.4), der die Erhöhung der Wärmeleitfähigkeit im kritischen Gebiet beschreibt, ist für die Wiedergabe der Wärmeleitfähigkeit innerhalb der angegebenen Unsicherheiten [17] auch außerhalb des kritischen Gebietes notwendig und sollte in jedem Fall benutzt werden. Die Gleichung für den Term Λ_2 lautet

$$\Lambda_2(\delta,\tau)=\frac{0{,}0013848}{\Psi(\delta,\tau)_{\text{Gl.(4.1)}}}(\tau\,\delta)^{-2}\left(\frac{\partial\pi}{\partial(\tau^{-1})}\right)_{\delta}^{2}\left[\delta\left(\frac{\partial\delta}{\partial\pi}\right)_{\tau}\right]^{0{,}4678}\delta^{0{,}5}$$

$$\cdot\exp\left[-18{,}66\left(\tau^{-1}-1\right)^{2}-(\delta-1)^{4}\right] \qquad (4.7)$$

mit $\delta=\rho/\rho^*$, $\tau=T^*/T$ und $\pi = p/p^*$ sowie $\rho^* = 317{,}763\ \mathrm{kg\,m^{-3}}$, $T^* = 647{,}226$ K und $p^* = 22{,}115$ MPa. Die reduzierte dynamische Viskosität Ψ wird aus Gl. (4.1) berechnet, und die partiellen Ableitungen $(\partial\pi/\partial(\tau^{-1}))_\delta$ und $(\partial\delta/\partial\pi)_\tau$ sollten aus der entsprechenden Gleichung der IAPWS-IF97 (s. Kap. 3) ermittelt werden.

Tabelle 4.4. Koeffizienten der Gl. (4.5)

i	n_i^{o}	i	n_i^{o}
0	$0{,}100\,000\,0 \cdot 10^1$	2	$0{,}259\,909\,6 \cdot 10^1$
1	$0{,}697\,826\,7 \cdot 10^1$	3	$-0{,}998\,254\,0$

Tabelle 4.5. Koeffizienten der Gl. (4.6)

j	n_{0j}	n_{1j}	n_{2j}	n_{3j}	n_{4j}
0	$0{,}132\,930\,46 \cdot 10^1$	$0{,}170\,183\,63 \cdot 10^1$	$0{,}522\,461\,58 \cdot 10^1$	$0{,}871\,276\,75 \cdot 10^1$	$-0{,}185\,259\,99 \cdot 10^1$
1	$-0{,}404\,524\,37$	$-0{,}221\,568\,45 \cdot 10^1$	$-0{,}101\,241\,11 \cdot 10^2$	$-0{,}950\,006\,11 \cdot 10^1$	$0{,}934\,046\,90$
2	$0{,}244\,094\,90$	$0{,}165\,110\,57 \cdot 10^1$	$0{,}498\,746\,87 \cdot 10^1$	$0{,}437\,866\,06 \cdot 10^1$	0
3	$0{,}186\,607\,51 \cdot 10^{-1}$	$-0{,}767\,360\,02$	$-0{,}272\,976\,94$	$-0{,}917\,837\,82$	0
4	$-0{,}129\,610\,68$	$0{,}372\,833\,44$	$-0{,}430\,833\,93$	0	0
5	$0{,}448\,099\,53 \cdot 10^{-1}$	$-0{,}112\,031\,60$	$0{,}133\,338\,49$	0	0

Der Gültigkeitsbereich der Gl. (4.4) für die Wärmeleitfähigkeit wurde von der IAPWS für den folgenden Bereich von Temperatur und Druck festgesetzt:

$$\begin{array}{ll} 273{,}15\ \mathrm{K} \le T \le 398{,}15\ \mathrm{K} & p \le 400\ \mathrm{MPa} \\ 398{,}15\ \mathrm{K} < T \le 523{,}15\ \mathrm{K} & p \le 200\ \mathrm{MPa} \\ 523{,}15\ \mathrm{K} < T \le 673{,}15\ \mathrm{K} & p \le 150\ \mathrm{MPa} \\ 673{,}15\ \mathrm{K} < T \le 1073{,}15\ \mathrm{K} & p \le 100\ \mathrm{MPa}\,. \end{array}$$

Die abgeschätzten Unsicherheiten der mit Gl. (4.4) berechneten Werte für die Wärmeleitfähigkeit sind im IAPWS-Release [17] angegeben.

Wenn die Wärmeleitfähigkeit aus Gl. (4.4) für gegebene Werte von *Druck* und Temperatur berechnet werden soll, muß zunächst für diese Werte die reduzierte Dichte δ als Eingangsgröße für Gl. (4.4) ermittelt werden. Im IAPWS-Release [17] wird bezüglich dieser Dichteberechnung auf die IAPWS-95 Formulation [8] hingewiesen. Für industrielle Anwendungen empfehlen wir jedoch, die IAPWS-IF97 (s. Kap. 3) zur Bestimmung der Eingangsdichte für Gl. (4.4) und für die in Gl. (4.7) benötigten partiellen Ableitungen der thermischen Zustandsgrößen zu benutzen, da der Unterschied zwischen den Werten aus der IAPWS-IF97 und der IAPWS-95 im Vergleich zur abgeschätzten Unsicherheit der Gl. (4.4) vernachlässigbar ist. Die Verwendung der IAPWS-IF97 hat für industrielle Berechnungen gegenüber der IAPWS-95 jedoch deutliche

Vorteile und vermeidet die Implementierung einer zweiten Formulation für die thermodynamischen Zustandsgrößen. Dementsprechend wurden die Werte für die Wärmeleitfähigkeit in den entsprechenden Tafeln in Teil C dieses Buches in der von uns empfohlenen Art berechnet.

Zur Überprüfung von Computerprogrammen auf der Basis von Gl. (4.4) enthält Tabelle 4.6 entsprechende Testwerte.

Tabelle 4.6. Aus Gl. (4.4) berechnete Werte der Wärmeleitfähigkeit für ausgewählte Temperaturen und Drücke

Zustandsgröße	$T = 273{,}15$ K $p = 0{,}1$ MPa	$T = 773{,}15$ K $p = 0{,}1$ MPa	$T = 773{,}15$ K $p = 100$ MPa
ρ [kg m^{-3}] (IAPWS-IF97)	$0{,}999\,843\,633 \cdot 10^3$	$0{,}280\,459\,027$	$0{,}528\,195\,286 \cdot 10^3$
λ [mW K^{-1}]	$0{,}561\,075\,931 \cdot 10^3$	$0{,}669\,788\,623 \cdot 10^2$	$0{,}394\,700\,444 \cdot 10^3$

4.3 Oberflächenspannung

Die Meßwerte der Oberflächenspannung σ an der Grenzfläche zwischen der Flüssigkeits- und der Dampfphase des Wassers wurden 1975 von der IAPWS kritisch gesichtet. Als Ergebnis dieser Arbeiten hatte die IAPWS Tabellenwerte der Oberflächenspannung empfohlen, die 1994 mit der Temperaturskala ITS-90 [6] in Einklang gebracht wurden [20]. An diese Tabellenwerte wurde eine einfache Gleichung für die Oberflächenspannung in Abhängigkeit von der Temperatur angepaßt [20].

Die Gleichung für die Oberflächenspannung hat die folgende Form:

$$\frac{\sigma}{\sigma^*} = 235{,}8\;(1-\theta)^{1{,}256}\left[1 - 0{,}625\,(1-\theta)\right] \tag{4.8}$$

mit $\theta = T/T^*$ sowie $\sigma^* = 1$ mN m^{-1} und $T^* = T_c$. Gleichung (4.8) gilt auf der gesamten Sättigungslinie von T_t bis T_c, deren Werte durch die Gln. (2.8) und (2.4) gegeben sind. Die abgeschätzten Unsicherheiten der mit Gl. (4.8) berechneten Werte für die Oberflächenspannung sind im IAPWS-Release [20] angegeben.

Zur Überprüfung von Computerprogrammen auf der Basis von Gl. (4.8) enthält Tabelle 4.7 entsprechende Testwerte.

Tabelle 4.7. Aus Gl. (4.8) berechnete Werte der Oberflächenspannung für ausgewählte Temperaturen

T [K]	σ [mN m^{-1}]
300	$0{,}716\,859\,625 \cdot 10^2$
500	$0{,}314\,719\,761 \cdot 10^2$
600	$0{,}837\,561\,087 \cdot 10^1$

4.4
Statische Dielektrizitätskonstante

Im Jahre 1995 wurde eine Übersicht über die Meßwerte für die relative statische Dielektrizitätskonstante (relative statische Permittivität) ε veröffentlicht [21]. Auf der Basis eines ausgewählten Satzes dieser Daten wurde eine neue Gleichung für die statische Dielektrizitätskonstante entwickelt [22]. Diese Gleichung, deren physikalische Grundlage der von Harris und Alder [23] vorgeschlagene sog. g-Faktor ist, wurde 1997 von der IAPWS angenommen [24].

Die Gleichung für die relative statische Dielektrizitätskonstante hat folgende Form:

$$\varepsilon = \frac{1 + A + 5B + \left(9 + 2A + 18B + A^2 + 10AB + 9B^2\right)^{0,5}}{4 - 4B}, \tag{4.9}$$

wobei die Größen A und B durch die Beziehungen

$$A = \frac{10^3 N_\mathrm{A}\, \mu^2 \rho g}{M \varepsilon_\mathrm{o} k T} \tag{4.9a}$$

und

$$B = \frac{10^3 N_\mathrm{A} \alpha \rho}{3 M \varepsilon_\mathrm{o}} \tag{4.9b}$$

gegeben sind. Die Werte für die Konstanten k, N_A, α, ε_o und μ, die in den Gln. (4.9a) und (4.9b) benötigt werden, sind in Tabelle 4.8 zusammengestellt; für die Molmasse M wird der Wert aus Gl. (2.3) benutzt. Die Korrelationsgleichung für den Harris-Alder g-Faktor in Gl. (4.9a) lautet

$$g = 1 + \sum_{i=1}^{11} n_i\, \delta^{I_i} \tau^{J_i} + n_{12}\, \delta \left(\frac{T}{228\,\mathrm{K}} - 1 \right)^{-1,2} \tag{4.10}$$

mit $\delta = \rho/\rho^*$, $\tau = T^*/T$ sowie $\rho^* = \rho_\mathrm{c}$ und $T^* = T_\mathrm{c}$; ρ_c und T_c sind durch die Gln. (2.6) und (2.4) gegeben. Tabelle 4.9 enthält die Koeffizienten n_i und die Exponenten I_i und J_i der Gl. (4.10).

Tabelle 4.8. Werte der Konstanten in den Gln. (4.9a) und (4.9b)

Größe	Wert
Boltzmann Konstante k	$1{,}380\,658 \cdot 10^{-23}\ \mathrm{J\ K^{-1}}$
Avogadro-Zahl N_A	$6{,}022\,136\,7 \cdot 10^{23}\ \mathrm{mol^{-1}}$
Mittlere molare Polarisierbarkeit α	$1{,}636 \cdot 10^{-40}\ \mathrm{C^2\ J^{-1}\ m^2}$
Permittivität des Vakuums ε_o	$8{,}854\,187\,817 \cdot 10^{-12}\ \mathrm{C^2\ J^{-1}\ m^{-1}}$
Molekulares Dipolmoment μ	$6{,}138 \cdot 10^{-30}\ \mathrm{C\ m}$

Tabelle 4.9. Koeffizienten und Exponenten der Gl. (4.10)

i	I_i	J_i	n_i	i	I_i	J_i	n_i
1	1	0,25	0,978 224 486 826	7	4	2	$0{,}949\,327\,488\,264 \cdot 10^{-1}$
2	1	1	− 0,957 771 379 375	8	5	2	$-0{,}980\,469\,816\,509 \cdot 10^{-2}$
3	1	2,5	0,237 511 794 148	9	6	5	$0{,}165\,167\,634\,970 \cdot 10^{-4}$
4	2	1,5	0,714 692 244 396	10	7	0,5	$0{,}937\,359\,795\,772 \cdot 10^{-4}$
5	3	1,5	− 0,298 217 036 956	11	10	10	$-0{,}123\,179\,218\,720 \cdot 10^{-9}$
6	3	2,5	− 0,108 863 472 196	12			$0{,}196\,096\,504\,426 \cdot 10^{-2}$

Gleichung (4.9) gilt für den folgenden Temperatur- und Druckbereich [22, 24]:

$$273{,}15\ \mathrm{K} \le T \le 323{,}15\ \mathrm{K} \qquad 0 < p \le p_{\max}$$
$$323{,}15\ \mathrm{K} < T \le 873{,}15\ \mathrm{K} \qquad 0 < p \le 600\ \mathrm{MPa}\,.$$

Der Wert $p_{\max}$ entspricht dem Schmelzdruck von Eis VI [11] bzw. maximal 1000 MPa. Darüber hinaus gilt Gl. (4.9) für die metastabile unterkühlte Flüssigkeit bei Atmosphärendruck (0,101 325 MPa) für Temperaturen von 238,15 K bis 273,15 K. Für Extrapolationen kann Gl. (4.9) zumindest bis 1200 K und 1200 MPa benutzt werden. Die abgeschätzten Unsicherheiten der mit Gl. (4.9) berechneten Werte für die statische Dielektrizitätskonstante sind in der Veröffentlichung [22] und im IAPWS-Release [24] angegeben.

Wenn die statische Dielektrizitätskonstante aus Gl. (4.9) für gegebene Werte von *Druck* und Temperatur berechnet werden soll, muß zunächst für diese Werte die reduzierte Dichte δ als Eingangsgröße für Gl. (4.9) ermittelt werden. Im IAPWS-Release [24] wird bezüglich dieser Dichteberechnung auf die IAPWS-95 Formulation [8] hingewiesen. Für industrielle Anwendungen empfehlen wir jedoch, die IAPWS-IF97 (s. Kap. 3) zur Bestimmung der Eingangsdichte für Gl. (4.9) zu benutzen, da der Unterschied zwischen den Werten aus der IAPWS-IF97 und der IAPWS-95 im Vergleich zur abgeschätzten Unsicherheit der Gl. (4.9) vernachlässigbar ist. Die Verwendung der IAPWS-IF97 hat für industrielle Berechnungen gegenüber der IAPWS-95 deutliche Vorteile und vermeidet die Implementierung einer zweiten Formulation für die thermodynamischen Zustandsgrößen. Dementsprechend wurden die Werte der Dielektrizitätskonstanten in den entsprechenden Tafeln in Teil C dieses Buches in der von uns empfohlenen Art berechnet.

Zur Überprüfung von Computerprogrammen auf der Basis von Gl. (4.9) enthält Tabelle 4.10 entsprechende Testwerte.

Tabelle 4.10. Aus Gl. (4.9) berechnete Werte der statischen Dielektrizitätskonstanten für ausgewählte Temperaturen und Drücke

Zustandsgröße	$T = 300$ K $p = 10$ MPa	$T = 870$ K $p = 10$ MPa	$T = 870$ K $p = 100$ MPa
ρ [kg m^{-3}] (IAPWS-IF97)	$0{,}100\,094\,934 \cdot 10^{4}$	$0{,}261\,720\,329 \cdot 10^{2}$	$0{,}378\,031\,821 \cdot 10^{3}$
ε	$0{,}781\,122\,784 \cdot 10^{2}$	$0{,}112\,721\,058 \cdot 10^{1}$	$0{,}498\,289\,143 \cdot 10^{1}$

4.5 Brechungsindex

Auf der Grundlage einer umfassenden Sammlung experimenteller Daten des Brechungsindexes n von Wasser und Wasserdampf [25] wurde 1990 eine Gleichung für die Lorentz-Lorenz-Funktion in Abhängigkeit von Dichte, Temperatur und Wellenlänge entwickelt [26, 27]. Diese Gleichung wurde später noch einmal an die Meßwerte angepaßt, nachdem sie auf die Temperaturskala ITS-90 [6] und, bezüglich der Dichte, auf den wissenschaftlichen Standard der IAPWS für die thermodynamischen Zustandsgrößen, die IAPWS-95 Formulation, [8] umgerechnet worden waren. Die modifizierte Gleichung wurde 1997 von der IAPWS angenommen [28].

Die Gleichung für den Brechungsindex lautet

$$n = \left(\frac{2A+1}{1-A}\right)^{0,5} \tag{4.11}$$

mit der Größe A gemäß

$$A = \delta\left(a_0 + a_1\delta + a_2\theta + a_3\Lambda^2\theta + a_4\Lambda^{-2} + \frac{a_5}{\Lambda^2 - \Lambda_{\mathrm{UV}}^2} + \frac{a_6}{\Lambda^2 - \Lambda_{\mathrm{IR}}^2} + a_7\delta^2\right) \tag{4.11a}$$

und $\delta = \rho/\rho^*$, $\theta = T/T^*$, $\Lambda = \lambda/\lambda^*$ sowie $\rho^* = 1000\ \mathrm{kg\,m^{-3}}$, $T^* = 273{,}15\ \mathrm{K}$, $\lambda^* = 0{,}589\ \mu\mathrm{m}$. Die Werte der reduzierten effektiven Infrarot-Resonanz Λ_{IR} und der reduzierten effektiven Ultraviolett-Resonanz Λ_{UV} betragen $\Lambda_{\mathrm{IR}} = 5{,}432\,937$ und $\Lambda_{\mathrm{UV}} = 0{,}229\,202$. Die Koeffizienten a_i der Gl. (4.11a) sind in Tabelle 4.11 aufgeführt.

Tabelle 4.11. Koeffizienten der Gl. (4.11a)

i	a_i	i	a_i
0	0,244 257 733	4	$0{,}158\,920\,570 \cdot 10^{-2}$
1	$0{,}974\,634\,476 \cdot 10^{-2}$	5	$0{,}245\,934\,259 \cdot 10^{-2}$
2	$-0{,}373\,234\,996 \cdot 10^{-2}$	6	0,900 704 920
3	$0{,}268\,678\,472 \cdot 10^{-3}$	7	$-0{,}166\,626\,219 \cdot 10^{-1}$

Der Gültigkeitsbereich der Gl. (4.11) für den Brechungsindex wurde von der IAPWS für den folgenden Bereich von Temperatur, Dichte und Wellenlänge festgesetzt:

$$\begin{aligned} 261{,}15\ \mathrm{K} &\le T \le 773{,}15\ \mathrm{K} \\ 0 &< \rho \le 1060\ \mathrm{kg\,m^{-3}} \\ 0{,}2\ \mu\mathrm{m} &\le \lambda \le 1{,}1\ \mu\mathrm{m}\,. \end{aligned}$$

Für flüssiges Wasser kann Gl. (4.11) bis zu Wellenlängen von 1,9 μm sinnvoll extrapoliert werden. Die abgeschätzten Unsicherheiten der mit Gl. (4.11) berechneten Werte für den Brechungsindex sind im IAPWS-Release [28] angegeben.

Wenn der Brechungsindex aus Gl. (4.11) für gegebene Werte von *Druck* und Temperatur berechnet werden soll, muß zunächst für diese Werte die reduzierte Dichte δ als Eingangsgröße für Gl. (4.11) ermittelt werden. Im IAPWS-Release [28] wird bezüglich dieser Dichteberechnung auf die IAPWS-95 Formulation [8] hingewiesen. Für industrielle Anwendungen empfehlen wir jedoch, die IAPWS-IF97 (s. Kap. 3) zur Bestimmung der Eingangsdichte für Gl. (4.11) zu benutzen, da der Unterschied zwischen den Werten aus der IAPWS-IF97 und der IAPWS-95 im Vergleich zur abgeschätzten Unsicherheit der Gl. (4.11) vernachlässigbar ist. Die Verwendung der IAPWS-IF97 hat für industrielle Berechnungen gegenüber der IAPWS-95 deutliche Vorteile und vermeidet die Implementierung einer zweiten Formulation für die thermodynamischen Zustandsgrößen. Dementsprechend wurden die Werte des Brechungsindexes in den entsprechenden Tafeln in Teil C dieses Buches in der von uns empfohlenen Art berechnet.

Zur Überprüfung von Computerprogrammen auf der Basis von Gl. (4.11) enthält Tabelle 4.12 entsprechende Testwerte.

Tabelle 4.12. Aus Gl. (4.11) berechnete Werte des Brechungsindexes für ausgewählte Temperaturen, Drücke und Wellenlängen

Zustandsgröße	T = 273,15 K p = 0,1 MPa	T = 773,15 K p = 0,1 MPa	T = 773,15 K p = 100 MPa
ρ [kg m^{-3}] (IAPWS-IF97)	$0{,}999\,843\,633 \cdot 10^3$	0,280 459 027	$0{,}528\,195\,286 \cdot 10^3$
		$\lambda = 0{,}2265\ \mu$m	
n	$0{,}139\,452\,744 \cdot 10^1$	$0{,}100\,010\,083 \cdot 10^1$	$0{,}119\,828\,0\,82 \cdot 10^1$
		$\lambda = 0{,}589\ \mu$m	
n	$0{,}133\,434\,461 \cdot 10^1$	$0{,}100\,008\,711 \cdot 10^1$	$0{,}117\,020\,512 \cdot 10^1$

Literatur

[1] International Formulation Committee of the 6th International Conference on the Properties of Steam (1967) The 1967 IFC Formulation for Industrial Use. Verein Deutscher Ingenieure, Düsseldorf

[2] Cohen, E.R., Taylor, B.N. (1986) The 1986 Adjustment of the Fundamental Physical Constants. CODATA Bulletin No. 63, Committee on Data for Science and Technology, Int. Council of Scientific Unions, Pergamon Press, Oxford

[3] Audi, G.,Wapstra, A.H. (1993) The 1993 atomic mass evaluation, (I) Atomic mass table. Nuclear Physics A 565: 1–65

[4] IUPAC (1991) Commission on Atomic Weights and Isotopic Abundances, Subcommittee for Isotopic Abundance Measurements. Isotopic compositions of the elements 1989. Pure and Appl. Chem. 63: 991–1002

[5] IAPWS (1995) Release on The Values of Temperature, Pressure and Density of Ordinary and Heavy Water Substances at Their Respective Critical Points. In: White, Jr., H.J., Sengers, J.V., Neumann, D.B., Bellows, J.C. (eds) Physical Chemistry of Aqueous Systems: Meeting the Needs of Industry. Proceedings of the 12th International Conference on the Properties of Water and Steam. Begell House, New York, pp. A101–A102

[6] Preston-Thomas, H. (1990) The International Temperature Scale of 1990 (ITS-90). Metrologia 27: 3–10

[7] Guildner, L.A., Johnson, D.P., Jones, F.E. (1976) Vapor Pressure of Water at Its Triple Point. J. Res. Natl. Bur. Stand. 80A: 505–521

[8] Pruß, A., Wagner, W. (1998) New International Formulation for the Thermodynamic Properties of Ordinary Water Substance for General and Scientific Use. Wird zur Veröffentlichung eingereicht bei J. Phys. Chem. Ref. Data. See also: IAPWS (1996) Release on the IAPWS Formulation 1995 for the Thermodynamic Properties of Ordinary Water Substance for General and Scientific Use. IAPWS Secretariat[1]

[9] Wagner, W., Cooper, J.R., Dittmann, A., Kijima, J., Kretzschmar, H.-J., Kruse, A., Mareš, R., Oguchi, K., Sato, H., Stöcker, I., Šifner, O., Takaishi, Y., Tanishita, I., Trübenbach, J., and Willkommen, Th. (1998) The IAPWS Industrial Formulation 1997 for the Thermodynamic Properties of Water and Steam. Eingereicht bei ASME Journal of Engineering for Gas Turbines and Power

[10] Minutes of the meetings of the International Formulation Committee of ICPS in Prague (1965)

[11] IAPWS (1994) Release on the Pressure along the Melting and the Sublimation Curves of Ordinary Water Substance. In: Wagner, W., Saul, A., Pruß, A., J. Phys. Chem. Ref. Data 23: 515–527. Also in: White, Jr., H.J., Sengers, J.V., Neumann, D.B., Bellows, J.C. (eds) Physical Chemistry of Aqueous Systems: Meeting the Needs of Industry. Proceedings of the 12th International Conference on the Properties of Water and Steam. Begell House, New York, pp. A9–A12

[12] Aleksandrov, A.A., Ivanov, A.I., Matveev, A.B. (1975) The Dynamic Viscosity of Water and Steam within a Wide Region of Temperatures and Pressures. Teploenergetika 22: 59–64 (English translation: Therm. Eng. 22: 77–83)

[13] Scheffler, K., Rosner, N., Straub, J., Grigull, U. (1978) Der neue internationale Standard der dynamischen Viskosität von Wasser und Wasserdampf. Brennstoff - Wärme - Kraft 30: 73-78

[14] Nagashima, A. (1977) Viscosity of Water Substance – New International Formulation and Its Background. J. Phys. Chem. Ref. Data 6: 1133–1166

[15] Nagashima, A. (1983) Erratum: Viscosity of Water Substance – New International Formulation and Its Background. J. Phys. Chem. Ref. Data 12: 403

[16] IAPWS (1997) Revised Release on the IAPS Formulation 1985 for the Viscosity of Ordinary Water Substance. IAPWS Secretariat[1]

[17] IAPWS (1997) Revised Release on the IAPS Formulation 1985 for the Thermal Conductivity of Ordinary Water Substance. IAPWS Secretariat[1]

[18] Scheffler, K., Rosner, N., Reimann, M. (1977) Available Input of the Thermal Conductivity of Water Substance. Institut A für Thermodynamik, Technische Universität, München

[19] Sengers, J.V., Watson, J.T.R., Basu, R.S., Kamgar-Parsi, B., Hendricks, R.C. (1984) Representative Equations for the Thermal Conductivity of Water Substance. J. Phys. Chem. Ref. Data 13: 893–933

[20] IAPWS (1994) Release on Surface Tension of Ordinary Water Substance. IAPWS Secretariat[1]

[21] Fernández, D.P., Mulev, Y., Goodwin, A.R.H., Levelt Sengers, J.M.H. (1995) A Database for the Static Dielectric Constant of Water and Steam. J. Phys. Chem. Ref. Data 24: 33–69

---

[1] Das Release ist erhältlich vom Executive Secretary der IAPWS, Dr. R.B. Dooley, EPRI, 3412 Hillview Avenue, Palo Alto, California 94304, USA.

[22] Fernández, D.P., Goodwin, A.R.H., Lemmon, E.W., Levelt Sengers, J.M.H., Williams, R.C. (1997) A Formulation for the Static Permittivity of Water and Steam at Temperatures from 238 K to 873 K at Pressures up to 1200 MPa, Including Derivatives and Debye-Hückel Coefficients. J. Phys. Chem. Ref. Data 26: 1125–1166
[23] Harris, F.E., Alder, B.J. (1953) Dielectric Polarization in Polar Substances. J. Chem. Phys. 21: 1031–1038
[24] IAPWS (1997) Release on the Static Dielectric Constant of Ordinary Water Substance for Temperatures from 238 K to 873 K and Pressures up to 1000 MPa. IAPWS Secretariat[2]
[25] Thormälen, I., Straub, J., Grigull, U. (1985) Refractive Index of Water and Its Dependence on Wavelength, Temperature, and Density. J. Phys. Chem. Ref. Data 14: 933–945
[26] Schiebener, P., Straub, J., Levelt Sengers, J.M.H., Gallagher, J.S. (1990) Refractive Index of Water and Steam as Function of Wavelength, Temperature and Density. J. Phys. Chem. Ref. Data 19: 677–717
[27] Schiebener, P., Straub, J., Levelt Sengers, J.M.H., Gallagher, J.S. (1990) Erratum: Refractive Index of Water and Steam as Function of Wavelength, Temperature and Density. J. Phys. Chem. Ref. Data 19: 1617
[28] IAPWS (1997) Release on the Refractive Index of Ordinary Water Substance as a Function of Wavelength, Temperature and Pressure. IAPWS Secretariat[2]

[2] Das Release ist erhältlich vom Executive Secretary der IAPWS, Dr. R.B. Dooley, EPRI, 3412 Hillview Avenue, Palo Alto, California 94304, USA.

Part C / Teil C

Tables of the Properties of Water and Steam

Tafeln der Zustandsgrößen von Wasser und Wasserdampf

1 Saturation State (Temperature Table)
Sättigungszustand (Temperaturtafel)

The temperature table of the saturation states contains values for the saturated liquid ($'$) and saturated vapour ($''$) of the following thermophysical properties in the temperature range from 0 °C to the critical temperature $t_c = 373.946$ °C:

- Saturation pressure p_s
- Specific volume v
- Specific enthalpy h
- Specific enthalpy of evaporation Δh_v
- Specific entropy s
- Specific entropy of evaporation Δs_v
- Specific isobaric heat capacity c_p
- Speed of sound w
- Isentropic exponent κ
- Dynamic viscosity η
- Thermal conductivity λ.

For given temperatures, the saturation pressure p_s was calculated from the saturation-pressure equation of IAPWS-IF97, Eq. (3.23). For temperatures $t > 350$ °C (region 3 of IAPWS-IF97) the specific volumes of the saturated liquid and saturated vapour, v' and v'', were determined from Eq. (3.21) via the phase-equilibrium condition (see Table 3.31). If these values were calculated by iterations for given t and p_s, the results would be slightly different. With t and p_s (for region 3 of IAPWS-IF97 with t and v' or t and v'') all properties were determined from the corresponding equations in Chaps. 3 and 4. The *thermodynamic* properties were exclusively calculated from the industrial standard IAPWS-IF97.

Die Temperaturtafel der Sättigungszustände enthält für die siedende Flüssigkeit ($'$) und den gesättigten Dampf ($''$) Werte der folgenden thermophysikalischen Eigenschaften im Temperaturbereich von 0 °C bis zur kritischen Temperatur $t_c = 373{,}946$ °C:

- Dampfdruck p_s
- spezifisches Volumen v
- spezifische Enthalpie h
- spezifische Verdampfungsenthalpie Δh_v
- spezifische Entropie s
- spezifische Verdampfungsentropie Δs_v
- spezifische isobare Wärmekapazität c_p
- Schallgeschwindigkeit w
- Isentropenexponent κ
- dynamische Viskosität η
- Wärmeleitfähigkeit λ.

Der Dampfdruck p_s wurde für vorgegebene Temperaturen aus der Dampfdruckgleichung der IAPWS-IF97, Gl. (3.23), berechnet. Für Temperaturen $t > 350$ °C (Bereich 3 der IAPWS-IF97) wurden die spezifischen Volumina der siedenden Flüssigkeit und des gesättigten Dampfes, v' und v'', aus Gl. (3.21) über die Phasengleichgewichtsbedingung (s. Tabelle 3.31) bestimmt. Würde man diese Größen für vorgegebene Werte von t and p_s iterieren, erhielte man geringfügig andere Ergebnisse. Für t and p_s (im Bereich 3 der IAPWS-IF97 für t und v' bzw. t und v'') wurden alle Zustandsgrößen mit den Gleichungen in Kap. 3 und 4 ermittelt. Die *thermodynamischen* Zustandsgrößen wurden ausschließlich mit dem Industrie-Standard IAPWS-IF97 berechnet.

Table 1 Saturation State (Temperature Table)
Sättigungszustand (Temperaturtafel)

t [°C]	T [K]	p_s [MPa]	v' [$m^3\ kg^{-1}$]	v''	h'	h'' [$kJ\ kg^{-1}$]	Δh_v	s'	s'' [$kJ\ kg^{-1}\ K^{-1}$]	Δs_v
0	273.15	0.000611	0.001000	206.140	− 0.042	2500.89	2500.93	− 0.0002	9.1558	9.1559
0.01	273.16	0.000612	0.001000	205.997	0.001	2500.91	2500.91	0.0000	9.1555	9.1555
1	274.15	0.000657	0.001000	192.445	4.177	2502.73	2498.55	0.0153	9.1291	9.1138
2	275.15	0.000706	0.001000	179.764	8.392	2504.57	2496.17	0.0306	9.1027	9.0721
3	276.15	0.000758	0.001000	168.014	12.604	2506.40	2493.80	0.0459	9.0765	9.0306
4	277.15	0.000814	0.001000	157.121	16.813	2508.24	2491.42	0.0611	9.0506	8.9895
5	278.15	0.000873	0.001000	147.017	21.019	2510.07	2489.05	0.0763	9.0249	8.9486
6	279.15	0.000935	0.001000	137.638	25.224	2511.91	2486.68	0.0913	8.9994	8.9081
7	280.15	0.001002	0.001000	128.928	29.426	2513.74	2484.31	0.1064	8.9742	8.8678
8	281.15	0.001073	0.001000	120.834	33.626	2515.57	2481.94	0.1213	8.9492	8.8278
9	282.15	0.001148	0.001000	113.309	37.824	2517.40	2479.58	0.1362	8.9244	8.7882
10	283.15	0.001228	0.001000	106.309	42.021	2519.23	2477.21	0.1511	8.8998	8.7488
11	284.15	0.001313	0.001000	99.793	46.216	2521.06	2474.84	0.1659	8.8755	8.7096
12	285.15	0.001403	0.001001	93.724	50.410	2522.89	2472.48	0.1806	8.8514	8.6708
13	286.15	0.001498	0.001001	88.070	54.602	2524.71	2470.11	0.1953	8.8275	8.6322
14	287.15	0.001599	0.001001	82.798	58.794	2526.54	2467.75	0.2099	8.8038	8.5939
15	288.15	0.001706	0.001001	77.881	62.984	2528.36	2465.38	0.2245	8.7804	8.5559
16	289.15	0.001819	0.001001	73.291	67.173	2530.19	2463.01	0.2390	8.7571	8.5181
17	290.15	0.001938	0.001001	69.006	71.361	2532.01	2460.65	0.2534	8.7341	8.4806
18	291.15	0.002065	0.001001	65.003	75.548	2533.83	2458.28	0.2678	8.7112	8.4434
19	292.15	0.002198	0.001002	61.261	79.734	2535.65	2455.92	0.2822	8.6886	8.4064
20	293.15	0.002339	0.001002	57.761	83.920	2537.47	2453.55	0.2965	8.6661	8.3696
21	294.15	0.002488	0.001002	54.487	88.105	2539.29	2451.18	0.3108	8.6439	8.3331
22	295.15	0.002645	0.001002	51.422	92.289	2541.10	2448.81	0.3250	8.6218	8.2969
23	296.15	0.002811	0.001003	48.552	96.473	2542.92	2446.45	0.3391	8.6000	8.2609
24	297.15	0.002986	0.001003	45.863	100.656	2544.73	2444.08	0.3532	8.5783	8.2251
25	298.15	0.003170	0.001003	43.341	104.838	2546.54	2441.71	0.3673	8.5568	8.1895
26	299.15	0.003364	0.001003	40.977	109.021	2548.35	2439.33	0.3813	8.5355	8.1542
27	300.15	0.003568	0.001004	38.758	113.202	2550.16	2436.96	0.3952	8.5144	8.1192
28	301.15	0.003783	0.001004	36.675	117.384	2551.97	2434.59	0.4091	8.4934	8.0843
29	302.15	0.004009	0.001004	34.719	121.565	2553.78	2432.21	0.4230	8.4727	8.0497
30	303.15	0.004247	0.001004	32.882	125.745	2555.58	2429.84	0.4368	8.4521	8.0153
31	304.15	0.004497	0.001005	31.154	129.926	2557.39	2427.46	0.4506	8.4317	7.9812
32	305.15	0.004759	0.001005	29.529	134.106	2559.19	2425.08	0.4643	8.4115	7.9472
33	306.15	0.005035	0.001005	28.001	138.286	2560.99	2422.70	0.4780	8.3914	7.9135
34	307.15	0.005325	0.001006	26.562	142.465	2562.79	2420.32	0.4916	8.3715	7.8800
35	308.15	0.005629	0.001006	25.208	146.645	2564.58	2417.94	0.5052	8.3518	7.8467
36	309.15	0.005947	0.001006	23.932	150.824	2566.38	2415.56	0.5187	8.3323	7.8136
37	310.15	0.006282	0.001007	22.729	155.004	2568.17	2413.17	0.5322	8.3129	7.7807
38	311.15	0.006632	0.001007	21.595	159.183	2569.96	2410.78	0.5457	8.2936	7.7480
39	312.15	0.007000	0.001007	20.526	163.362	2571.75	2408.39	0.5591	8.2746	7.7155
40	313.15	0.007384	0.001008	19.517	167.541	2573.54	2406.00	0.5724	8.2557	7.6832
41	314.15	0.007787	0.001008	18.565	171.720	2575.33	2403.61	0.5858	8.2369	7.6512
42	315.15	0.008209	0.001009	17.665	175.899	2577.11	2401.21	0.5990	8.2183	7.6193
43	316.15	0.008650	0.001009	16.816	180.079	2578.89	2398.82	0.6123	8.1999	7.5876
44	317.15	0.009112	0.001009	16.013	184.258	2580.67	2396.42	0.6255	8.1816	7.5561

Table 1 Saturation State (Temperature Table) – Continuation
Sättigungszustand (Temperaturtafel) – Fortsetzung

t [°C]	c_p' [kJ kg^{-1} K^{-1}]	c_p''	w' [m s^{-1}]	w''	κ'	κ''	η' [10^{-6} kg m^{-1} s^{-1}]	η''	λ' [10^{-3} W m^{-1} K^{-1}]	λ''
0	4.2199	1.8882	1402.28	408.88	3216537	1.3269	1791.78	9.22	561.02	17.07
0.01	4.2199	1.8882	1402.33	408.89	3214432	1.3269	1791.16	9.22	561.04	17.07
1	4.2165	1.8889	1407.23	409.61	3013280	1.3268	1731.09	9.24	562.92	17.12
2	4.2134	1.8895	1412.07	410.34	2824038	1.3268	1673.61	9.26	564.82	17.18
3	4.2105	1.8902	1416.82	411.07	2647765	1.3267	1619.12	9.29	566.72	17.23
4	4.2078	1.8909	1421.48	411.80	2483499	1.3266	1567.41	9.31	568.62	17.28
5	4.2054	1.8917	1426.04	412.53	2330358	1.3266	1518.29	9.34	570.52	17.34
6	4.2031	1.8924	1430.50	413.25	2187526	1.3265	1471.59	9.36	572.42	17.39
7	4.2011	1.8932	1434.87	413.97	2054253	1.3264	1427.15	9.39	574.32	17.45
8	4.1992	1.8940	1439.14	414.69	1929848	1.3264	1384.83	9.41	576.22	17.51
9	4.1974	1.8949	1443.31	415.41	1813675	1.3263	1344.48	9.44	578.11	17.56
10	4.1958	1.8957	1447.40	416.13	1705144	1.3262	1305.98	9.46	580.00	17.62
11	4.1943	1.8966	1451.39	416.84	1603713	1.3262	1269.23	9.49	581.88	17.68
12	4.1930	1.8975	1455.28	417.55	1508880	1.3261	1234.11	9.51	583.75	17.74
13	4.1917	1.8985	1459.09	418.26	1420182	1.3260	1200.53	9.54	585.62	17.80
14	4.1905	1.8994	1462.81	418.97	1337189	1.3259	1168.39	9.57	587.48	17.86
15	4.1894	1.9004	1466.43	419.68	1259505	1.3258	1137.62	9.59	589.33	17.92
16	4.1884	1.9014	1469.97	420.38	1186763	1.3257	1108.13	9.62	591.17	17.98
17	4.1875	1.9025	1473.42	421.08	1118624	1.3257	1079.85	9.65	593.00	18.04
18	4.1866	1.9035	1476.79	421.79	1054772	1.3256	1052.72	9.67	594.82	18.10
19	4.1858	1.9046	1480.07	422.48	994916	1.3255	1026.67	9.70	596.62	18.16
20	4.1851	1.9057	1483.26	423.18	938785	1.3254	1001.65	9.73	598.42	18.23
21	4.1844	1.9069	1486.37	423.88	886129	1.3253	977.59	9.75	600.19	18.29
22	4.1838	1.9080	1489.41	424.57	836714	1.3252	954.46	9.78	601.96	18.35
23	4.1832	1.9092	1492.35	425.26	790326	1.3251	932.19	9.81	603.70	18.42
24	4.1827	1.9104	1495.22	425.95	746762	1.3250	910.76	9.84	605.44	18.48
25	4.1822	1.9116	1498.02	426.63	705837	1.3249	890.11	9.87	607.15	18.55
26	4.1817	1.9129	1500.73	427.32	667378	1.3248	870.20	9.90	608.85	18.62
27	4.1813	1.9141	1503.37	428.00	631223	1.3247	851.01	9.92	610.53	18.68
28	4.1809	1.9154	1505.93	428.68	597223	1.3246	832.49	9.95	612.19	18.75
29	4.1806	1.9167	1508.41	429.36	565239	1.3245	814.62	9.98	613.84	18.82
30	4.1803	1.9180	1510.83	430.04	535140	1.3244	797.35	10.01	615.46	18.89
31	4.1800	1.9194	1513.17	430.72	506807	1.3243	780.68	10.04	617.07	18.96
32	4.1798	1.9207	1515.44	431.39	480127	1.3242	764.56	10.07	618.65	19.02
33	4.1795	1.9221	1517.64	432.06	454995	1.3241	748.98	10.10	620.22	19.09
34	4.1794	1.9235	1519.77	432.73	431314	1.3240	733.90	10.13	621.76	19.16
35	4.1792	1.9249	1521.83	433.40	408992	1.3239	719.31	10.16	623.29	19.24
36	4.1791	1.9263	1523.83	434.07	387945	1.3238	705.19	10.19	624.79	19.31
37	4.1790	1.9278	1525.75	434.73	368094	1.3237	691.52	10.22	626.27	19.38
38	4.1789	1.9292	1527.62	435.40	349364	1.3235	678.26	10.25	627.74	19.45
39	4.1788	1.9307	1529.41	436.06	331687	1.3234	665.42	10.28	629.17	19.52
40	4.1788	1.9322	1531.15	436.72	314999	1.3233	652.97	10.31	630.59	19.60
41	4.1788	1.9337	1532.82	437.37	299238	1.3232	640.90	10.34	631.99	19.67
42	4.1788	1.9353	1534.43	438.03	284350	1.3231	629.18	10.37	633.36	19.75
43	4.1788	1.9368	1535.98	438.68	270281	1.3230	617.81	10.40	634.72	19.82
44	4.1789	1.9384	1537.47	439.33	256982	1.3229	606.78	10.43	636.05	19.90

Table 1 Saturation State (Temperature Table) – Continuation
Sättigungszustand (Temperaturtafel) – Fortsetzung

t [°C]	T [K]	p_s [MPa]	v' [m³ kg⁻¹]	v''	h'	h'' [kJ kg⁻¹]	Δh_v	s'	s'' [kJ kg⁻¹ K⁻¹]	Δs_v
45	318.15	0.009594	0.001010	15.253	188.437	2582.45	2394.02	0.6386	8.1634	7.5248
46	319.15	0.010099	0.001010	14.535	192.617	2584.23	2391.61	0.6517	8.1454	7.4937
47	320.15	0.010626	0.001011	13.856	196.796	2586.00	2389.21	0.6648	8.1276	7.4628
48	321.15	0.011176	0.001011	13.213	200.976	2587.77	2386.80	0.6778	8.1099	7.4320
49	322.15	0.011751	0.001012	12.604	205.156	2589.54	2384.39	0.6908	8.0923	7.4015
50	323.15	0.012351	0.001012	12.028	209.336	2591.31	2381.97	0.7038	8.0749	7.3711
51	324.15	0.012977	0.001013	11.482	213.517	2593.08	2379.56	0.7167	8.0576	7.3409
52	325.15	0.013631	0.001013	10.964	217.697	2594.84	2377.14	0.7296	8.0405	7.3109
53	326.15	0.014312	0.001014	10.473	221.878	2596.60	2374.72	0.7424	8.0235	7.2811
54	327.15	0.015022	0.001014	10.007	226.059	2598.35	2372.30	0.7552	8.0066	7.2514
55	328.15	0.015761	0.001015	9.565	230.241	2600.11	2369.87	0.7680	7.9899	7.2219
56	329.15	0.016532	0.001015	9.145	234.423	2601.86	2367.44	0.7807	7.9733	7.1926
57	330.15	0.017335	0.001016	8.747	238.605	2603.61	2365.01	0.7934	7.9568	7.1634
58	331.15	0.018171	0.001016	8.369	242.788	2605.36	2362.57	0.8060	7.9405	7.1344
59	332.15	0.019041	0.001017	8.009	246.971	2607.10	2360.13	0.8186	7.9243	7.1056
60	333.15	0.019946	0.001017	7.668	251.154	2608.85	2357.69	0.8312	7.9082	7.0770
61	334.15	0.020887	0.001018	7.343	255.338	2610.58	2355.25	0.8438	7.8922	7.0485
62	335.15	0.021866	0.001018	7.034	259.523	2612.32	2352.80	0.8563	7.8764	7.0201
63	336.15	0.022884	0.001019	6.740	263.708	2614.05	2350.35	0.8687	7.8607	6.9919
64	337.15	0.023942	0.001019	6.460	267.893	2615.78	2347.89	0.8811	7.8451	6.9639
65	338.15	0.025041	0.001020	6.194	272.079	2617.51	2345.43	0.8935	7.8296	6.9361
66	339.15	0.026183	0.001020	5.940	276.266	2619.23	2342.97	0.9059	7.8142	6.9083
67	340.15	0.027368	0.001021	5.699	280.453	2620.96	2340.50	0.9182	7.7990	6.8808
68	341.15	0.028599	0.001022	5.468	284.641	2622.67	2338.03	0.9305	7.7839	6.8534
69	342.15	0.029876	0.001022	5.249	288.829	2624.39	2335.56	0.9428	7.7689	6.8261
70	343.15	0.031201	0.001023	5.040	293.018	2626.10	2333.08	0.9550	7.7540	6.7990
71	344.15	0.032575	0.001023	4.840	297.208	2627.81	2330.60	0.9672	7.7392	6.7720
72	345.15	0.034000	0.001024	4.650	301.398	2629.51	2328.11	0.9793	7.7245	6.7452
73	346.15	0.035478	0.001025	4.468	305.589	2631.21	2325.62	0.9915	7.7100	6.7185
74	347.15	0.037009	0.001025	4.295	309.781	2632.91	2323.13	1.0035	7.6955	6.6920
75	348.15	0.038595	0.001026	4.129	313.974	2634.60	2320.63	1.0156	7.6812	6.6656
76	349.15	0.040239	0.001026	3.971	318.167	2636.29	2318.13	1.0276	7.6669	6.6393
77	350.15	0.041941	0.001027	3.820	322.361	2637.98	2315.62	1.0396	7.6528	6.6132
78	351.15	0.043703	0.001028	3.675	326.556	2639.66	2313.11	1.0516	7.6388	6.5872
79	352.15	0.045527	0.001028	3.537	330.752	2641.34	2310.59	1.0635	7.6248	6.5613
80	353.15	0.047415	0.001029	3.405	334.949	2643.01	2308.07	1.0754	7.6110	6.5356
81	354.15	0.049368	0.001030	3.279	339.146	2644.68	2305.54	1.0873	7.5973	6.5100
82	355.15	0.051387	0.001030	3.158	343.345	2646.35	2303.01	1.0991	7.5837	6.4846
83	356.15	0.053476	0.001031	3.043	347.544	2648.01	2300.47	1.1109	7.5701	6.4592
84	357.15	0.055636	0.001032	2.932	351.745	2649.67	2297.93	1.1227	7.5567	6.4340
85	358.15	0.057867	0.001032	2.826	355.946	2651.33	2295.38	1.1344	7.5434	6.4090
86	359.15	0.060174	0.001033	2.724	360.148	2652.98	2292.83	1.1461	7.5301	6.3840
87	360.15	0.062556	0.001034	2.627	364.352	2654.62	2290.27	1.1578	7.5170	6.3592
88	361.15	0.065017	0.001035	2.534	368.556	2656.26	2287.70	1.1694	7.5039	6.3345
89	362.15	0.067559	0.001035	2.445	372.762	2657.90	2285.14	1.1811	7.4909	6.3099

Table 1 Saturation State (Temperature Table) – Continuation
Sättigungszustand (Temperaturtafel) – Fortsetzung

t [°C]	c_p' [kJ kg^{-1} K^{-1}]	c_p''	w' [m s^{-1}]	w''	κ'	κ''	η' [10^{-6} kg m^{-1} s^{-1}]	η''	λ' [10^{-3} W m^{-1} K^{-1}]	λ''
45	4.1790	1.9400	1538.89	439.98	244408	1.3228	596.06	10.46	637.35	19.97
46	4.1791	1.9416	1540.26	440.63	232515	1.3227	585.65	10.49	638.64	20.05
47	4.1792	1.9432	1541.58	441.28	221263	1.3226	575.53	10.52	639.91	20.13
48	4.1794	1.9449	1542.83	441.92	210614	1.3224	565.70	10.55	641.15	20.21
49	4.1796	1.9466	1544.03	442.56	200533	1.3223	556.14	10.58	642.37	20.29
50	4.1798	1.9482	1545.18	443.20	190987	1.3222	546.84	10.62	643.57	20.36
51	4.1800	1.9500	1546.26	443.84	181945	1.3221	537.79	10.65	644.74	20.44
52	4.1802	1.9517	1547.30	444.48	173378	1.3220	528.99	10.68	645.90	20.52
53	4.1805	1.9534	1548.28	445.11	165258	1.3219	520.43	10.71	647.03	20.61
54	4.1808	1.9552	1549.21	445.74	157561	1.3218	512.09	10.74	648.14	20.69
55	4.1811	1.9570	1550.09	446.37	150261	1.3217	503.97	10.77	649.23	20.77
56	4.1814	1.9588	1550.91	447.00	143338	1.3216	496.06	10.81	650.30	20.85
57	4.1818	1.9607	1551.69	447.63	136768	1.3214	488.35	10.84	651.35	20.93
58	4.1821	1.9625	1552.42	448.25	130533	1.3213	480.84	10.87	652.37	21.02
59	4.1825	1.9644	1553.09	448.88	124614	1.3212	473.52	10.90	653.38	21.10
60	4.1829	1.9664	1553.72	449.50	118993	1.3211	466.38	10.93	654.36	21.19
61	4.1834	1.9683	1554.30	450.11	113655	1.3210	459.43	10.97	655.33	21.27
62	4.1838	1.9703	1554.83	450.73	108582	1.3209	452.64	11.00	656.27	21.36
63	4.1843	1.9723	1555.32	451.34	103762	1.3208	446.02	11.03	657.19	21.44
64	4.1848	1.9743	1555.76	451.95	99179	1.3206	439.56	11.06	658.10	21.53
65	4.1853	1.9764	1556.15	452.56	94822	1.3205	433.25	11.10	658.98	21.62
66	4.1859	1.9785	1556.50	453.17	90678	1.3204	427.09	11.13	659.84	21.71
67	4.1864	1.9806	1556.81	453.77	86736	1.3203	421.08	11.16	660.69	21.80
68	4.1870	1.9828	1557.07	454.38	82984	1.3202	415.22	11.19	661.51	21.89
69	4.1876	1.9850	1557.29	454.98	79414	1.3200	409.48	11.23	662.32	21.98
70	4.1882	1.9873	1557.46	455.57	76014	1.3199	403.88	11.26	663.11	22.07
71	4.1889	1.9895	1557.59	456.17	72777	1.3198	398.41	11.29	663.88	22.16
72	4.1896	1.9919	1557.68	456.76	69693	1.3197	393.06	11.33	664.63	22.25
73	4.1902	1.9942	1557.73	457.35	66755	1.3195	387.84	11.36	665.36	22.34
74	4.1910	1.9966	1557.74	457.94	63955	1.3194	382.73	11.39	666.08	22.44
75	4.1917	1.9990	1557.71	458.52	61286	1.3193	377.73	11.43	666.77	22.53
76	4.1924	2.0015	1557.64	459.11	58742	1.3191	372.85	11.46	667.45	22.63
77	4.1932	2.0041	1557.53	459.69	56315	1.3190	368.07	11.49	668.12	22.72
78	4.1940	2.0066	1557.38	460.26	54000	1.3189	363.39	11.53	668.76	22.82
79	4.1948	2.0092	1557.19	460.84	51791	1.3187	358.82	11.56	669.39	22.91
80	4.1956	2.0119	1556.97	461.41	49683	1.3186	354.34	11.59	670.00	23.01
81	4.1965	2.0146	1556.70	461.98	47671	1.3185	349.96	11.63	670.60	23.11
82	4.1974	2.0174	1556.40	462.55	45750	1.3183	345.67	11.66	671.18	23.21
83	4.1983	2.0202	1556.06	463.11	43915	1.3182	341.47	11.69	671.74	23.31
84	4.1992	2.0231	1555.69	463.67	42162	1.3180	337.36	11.73	672.29	23.41
85	4.2001	2.0260	1555.28	464.23	40488	1.3179	333.34	11.76	672.82	23.51
86	4.2011	2.0290	1554.84	464.79	38888	1.3177	329.39	11.79	673.34	23.61
87	4.2020	2.0321	1554.36	465.34	37358	1.3176	325.53	11.83	673.84	23.71
88	4.2030	2.0352	1553.84	465.89	35896	1.3174	321.75	11.86	674.33	23.81
89	4.2041	2.0383	1553.29	466.44	34497	1.3173	318.04	11.90	674.80	23.92

Table 1 Saturation State (Temperature Table) – Continuation
Sättigungszustand (Temperaturtafel) – Fortsetzung

t [°C]	T [K]	p_s [MPa]	v' [m³ kg⁻¹]	v''	h'	h'' [kJ kg⁻¹]	Δh_v	s'	s'' [kJ kg⁻¹ K⁻¹]	Δs_v
90	363.15	0.070182	0.001036	2.3591	376.97	2659.53	2282.56	1.1927	7.4781	6.2854
91	364.15	0.072890	0.001037	2.2771	381.18	2661.16	2279.98	1.2042	7.4653	6.2611
92	365.15	0.075685	0.001037	2.1983	385.38	2662.78	2277.39	1.2158	7.4526	6.2368
93	366.15	0.078568	0.001038	2.1228	389.59	2664.39	2274.80	1.2273	7.4400	6.2127
94	367.15	0.081542	0.001039	2.0502	393.81	2666.01	2272.20	1.2387	7.4275	6.1887
95	368.15	0.084609	0.001040	1.9806	398.02	2667.61	2269.60	1.2502	7.4150	6.1648
96	369.15	0.087771	0.001040	1.9138	402.23	2669.22	2266.98	1.2616	7.4027	6.1411
97	370.15	0.091031	0.001041	1.8497	406.45	2670.81	2264.37	1.2730	7.3904	6.1174
98	371.15	0.094390	0.001042	1.7880	410.66	2672.40	2261.74	1.2844	7.3782	6.0938
99	372.15	0.097852	0.001043	1.7288	414.88	2673.99	2259.11	1.2957	7.3661	6.0704
100	373.15	0.101418	0.001043	1.6719	419.10	2675.57	2256.47	1.3070	7.3541	6.0471
101	374.15	0.105091	0.001044	1.6171	423.32	2677.15	2253.83	1.3183	7.3421	6.0238
102	375.15	0.108873	0.001045	1.5645	427.54	2678.72	2251.18	1.3296	7.3303	6.0007
103	376.15	0.112768	0.001046	1.5140	431.76	2680.28	2248.52	1.3408	7.3185	5.9777
104	377.15	0.116776	0.001047	1.4653	435.99	2681.84	2245.85	1.3520	7.3068	5.9548
105	378.15	0.120902	0.001047	1.4185	440.21	2683.39	2243.18	1.3632	7.2951	5.9320
106	379.15	0.125147	0.001048	1.3734	444.44	2684.94	2240.50	1.3743	7.2836	5.9092
107	380.15	0.129514	0.001049	1.3301	448.67	2686.48	2237.81	1.3854	7.2721	5.8866
108	381.15	0.134007	0.001050	1.2883	452.90	2688.02	2235.12	1.3965	7.2607	5.8641
109	382.15	0.138626	0.001051	1.2481	457.13	2689.55	2232.41	1.4076	7.2493	5.8417
110	383.15	0.143376	0.001052	1.2094	461.36	2691.07	2229.70	1.4187	7.2380	5.8194
111	384.15	0.148259	0.001052	1.1721	465.60	2692.58	2226.99	1.4297	7.2268	5.7972
112	385.15	0.153277	0.001053	1.1362	469.83	2694.09	2224.26	1.4407	7.2157	5.7750
113	386.15	0.158435	0.001054	1.1015	474.07	2695.60	2221.53	1.4517	7.2047	5.7530
114	387.15	0.163734	0.001055	1.0681	478.31	2697.09	2218.78	1.4626	7.1937	5.7310
115	388.15	0.169177	0.001056	1.0359	482.55	2698.58	2216.03	1.4735	7.1827	5.7092
116	389.15	0.174768	0.001057	1.0049	486.80	2700.07	2213.27	1.4844	7.1719	5.6874
117	390.15	0.180509	0.001058	0.9750	491.04	2701.55	2210.51	1.4953	7.1611	5.6658
118	391.15	0.186404	0.001059	0.9461	495.29	2703.02	2207.73	1.5062	7.1504	5.6442
119	392.15	0.192455	0.001059	0.9182	499.53	2704.48	2204.94	1.5170	7.1397	5.6227
120	393.15	0.198665	0.001060	0.8913	503.78	2705.93	2202.15	1.5278	7.1291	5.6013
121	394.15	0.205039	0.001061	0.8653	508.04	2707.38	2199.35	1.5386	7.1186	5.5800
122	395.15	0.211578	0.001062	0.8403	512.29	2708.82	2196.53	1.5494	7.1081	5.5587
123	396.15	0.218287	0.001063	0.8161	516.55	2710.26	2193.71	1.5601	7.0977	5.5376
124	397.15	0.225168	0.001064	0.7927	520.80	2711.69	2190.88	1.5708	7.0873	5.5165
125	398.15	0.232224	0.001065	0.7701	525.06	2713.11	2188.04	1.5815	7.0770	5.4955
126	399.15	0.239460	0.001066	0.7483	529.32	2714.52	2185.19	1.5922	7.0668	5.4746
127	400.15	0.246878	0.001067	0.7272	533.59	2715.92	2182.34	1.6028	7.0566	5.4538
128	401.15	0.254481	0.001068	0.7068	537.85	2717.32	2179.47	1.6134	7.0465	5.4330
129	402.15	0.262274	0.001069	0.6871	542.12	2718.71	2176.59	1.6240	7.0364	5.4124
130	403.15	0.270260	0.001070	0.6681	546.39	2720.09	2173.70	1.6346	7.0264	5.3918
131	404.15	0.278441	0.001071	0.6497	550.66	2721.46	2170.80	1.6452	7.0165	5.3713
132	405.15	0.286823	0.001072	0.6318	554.93	2722.83	2167.89	1.6557	7.0066	5.3508
133	406.15	0.295407	0.001073	0.6146	559.21	2724.18	2164.97	1.6662	6.9967	5.3305
134	407.15	0.304199	0.001074	0.5979	563.49	2725.53	2162.04	1.6767	6.9869	5.3102

Table 1 Saturation State (Temperature Table) – Continuation
Sättigungszustand (Temperaturtafel) – Fortsetzung

t [°C]	c_p' [kJ kg^{-1} K^{-1}]	c_p''	w' [m s^{-1}]	w''	κ'	κ''	η' [10^{-6} kg m^{-1} s^{-1}]	η''	λ' [10^{-3} W m^{-1} K^{-1}]	λ''
90	4.2051	2.0415	1552.71	466.98	33160	1.3171	314.40	11.93	675.26	24.02
91	4.2062	2.0448	1552.09	467.52	31880	1.3169	310.84	11.96	675.71	24.12
92	4.2072	2.0482	1551.45	468.06	30656	1.3168	307.35	12.00	676.14	24.23
93	4.2083	2.0516	1550.76	468.60	29484	1.3166	303.93	12.03	676.55	24.34
94	4.2095	2.0551	1550.05	469.13	28362	1.3164	300.57	12.06	676.96	24.44
95	4.2106	2.0586	1549.30	469.66	27288	1.3163	297.28	12.10	677.35	24.55
96	4.2118	2.0623	1548.52	470.18	26259	1.3161	294.05	12.13	677.72	24.66
97	4.2130	2.0660	1547.71	470.71	25274	1.3159	290.89	12.17	678.09	24.77
98	4.2142	2.0697	1546.87	471.23	24330	1.3157	287.78	12.20	678.44	24.88
99	4.2154	2.0736	1546.00	471.74	23425	1.3155	284.74	12.24	678.77	24.99
100	4.2166	2.0775	1545.09	472.26	22559	1.3153	281.75	12.27	679.10	25.10
101	4.2179	2.0815	1544.16	472.77	21727	1.3152	278.81	12.30	679.41	25.21
102	4.2192	2.0856	1543.19	473.27	20930	1.3150	275.93	12.34	679.71	25.32
103	4.2205	2.0897	1542.20	473.78	20166	1.3148	273.10	12.37	680.00	25.43
104	4.2219	2.0940	1541.18	474.28	19433	1.3146	270.33	12.41	680.28	25.55
105	4.2232	2.0983	1540.13	474.77	18730	1.3144	267.61	12.44	680.54	25.66
106	4.2246	2.1027	1539.04	475.26	18055	1.3142	264.93	12.47	680.79	25.78
107	4.2260	2.1072	1537.93	475.75	17408	1.3139	262.30	12.51	681.03	25.89
108	4.2274	2.1118	1536.80	476.24	16786	1.3137	259.72	12.54	681.26	26.01
109	4.2289	2.1164	1535.63	476.72	16189	1.3135	257.19	12.58	681.48	26.13
110	4.2304	2.1212	1534.43	477.20	15616	1.3133	254.70	12.61	681.69	26.24
111	4.2319	2.1260	1533.21	477.68	15065	1.3131	252.25	12.65	681.89	26.36
112	4.2334	2.1310	1531.96	478.15	14536	1.3128	249.85	12.68	682.07	26.48
113	4.2349	2.1360	1530.69	478.62	14028	1.3126	247.49	12.72	682.25	26.60
114	4.2365	2.1411	1529.38	479.08	13540	1.3124	245.17	12.75	682.41	26.72
115	4.2381	2.1464	1528.05	479.54	13071	1.3121	242.89	12.78	682.57	26.85
116	4.2397	2.1517	1526.69	480.00	12620	1.3119	240.65	12.82	682.71	26.97
117	4.2413	2.1571	1525.31	480.45	12186	1.3117	238.44	12.85	682.85	27.09
118	4.2430	2.1626	1523.90	480.90	11769	1.3114	236.28	12.89	682.97	27.22
119	4.2447	2.1683	1522.47	481.35	11368	1.3112	234.15	12.92	683.08	27.34
120	4.2464	2.1740	1521.00	481.79	10982	1.3109	232.05	12.96	683.19	27.47
121	4.2481	2.1798	1519.52	482.23	10611	1.3106	229.99	12.99	683.28	27.59
122	4.2499	2.1858	1518.00	482.66	10253	1.3104	227.97	13.03	683.36	27.72
123	4.2516	2.1918	1516.47	483.09	9910	1.3101	225.97	13.06	683.44	27.85
124	4.2535	2.1979	1514.90	483.52	9579	1.3098	224.01	13.09	683.50	27.98
125	4.2553	2.2042	1513.32	483.94	9260	1.3096	222.08	13.13	683.56	28.11
126	4.2571	2.2105	1511.70	484.36	8953	1.3093	220.19	13.16	683.61	28.24
127	4.2590	2.2170	1510.07	484.78	8657	1.3090	218.32	13.20	683.64	28.37
128	4.2609	2.2236	1508.41	485.19	8373	1.3087	216.48	13.23	683.67	28.50
129	4.2629	2.2302	1506.72	485.60	8099	1.3084	214.67	13.27	683.69	28.63
130	4.2648	2.2370	1505.01	486.00	7834	1.3082	212.89	13.30	683.70	28.76
131	4.2668	2.2439	1503.28	486.40	7580	1.3079	211.14	13.34	683.70	28.90
132	4.2689	2.2509	1501.52	486.79	7334	1.3076	209.42	13.37	683.69	29.03
133	4.2709	2.2580	1499.74	487.19	7098	1.3073	207.72	13.40	683.67	29.17
134	4.2730	2.2653	1497.94	487.57	6870	1.3070	206.05	13.44	683.64	29.31

Table 1 Saturation State (Temperature Table) – Continuation
Sättigungszustand (Temperaturtafel) – Fortsetzung

t	T	p_s	v'	v''	h'	h''	Δh_v	s'	s''	Δs_v
[°C]	[K]	[MPa]	[m^3 kg^{-1}]			[kJ kg^{-1}]			[kJ kg^{-1} K^{-1}]	
135	408.15	0.313201	0.001075	0.5818	567.77	2726.87	2159.10	1.6872	6.9772	5.2900
136	409.15	0.322417	0.001076	0.5662	572.05	2728.20	2156.15	1.6977	6.9675	5.2698
137	410.15	0.331852	0.001077	0.5511	576.33	2729.53	2153.19	1.7081	6.9579	5.2498
138	411.15	0.341508	0.001078	0.5364	580.62	2730.84	2150.22	1.7185	6.9483	5.2298
139	412.15	0.351390	0.001079	0.5222	584.91	2732.15	2147.24	1.7289	6.9388	5.2098
140	413.15	0.361501	0.001080	0.5085	589.20	2733.44	2144.24	1.7393	6.9293	5.1900
141	414.15	0.371845	0.001081	0.4952	593.49	2734.73	2141.24	1.7496	6.9198	5.1702
142	415.15	0.382427	0.001082	0.4823	597.79	2736.01	2138.22	1.7600	6.9105	5.1505
143	416.15	0.393250	0.001083	0.4698	602.09	2737.28	2135.19	1.7703	6.9011	5.1308
144	417.15	0.404318	0.001084	0.4577	606.39	2738.54	2132.15	1.7806	6.8918	5.1112
145	418.15	0.415635	0.001085	0.4460	610.69	2739.80	2129.10	1.7909	6.8826	5.0917
146	419.15	0.427205	0.001086	0.4346	615.00	2741.04	2126.04	1.8011	6.8734	5.0723
147	420.15	0.439033	0.001087	0.4236	619.31	2742.27	2122.96	1.8114	6.8642	5.0529
148	421.15	0.451122	0.001088	0.4129	623.62	2743.50	2119.88	1.8216	6.8551	5.0335
149	422.15	0.463477	0.001089	0.4026	627.93	2744.71	2116.78	1.8318	6.8461	5.0143
150	423.15	0.476101	0.001091	0.3925	632.25	2745.92	2113.67	1.8420	6.8370	4.9951
151	424.15	0.489000	0.001092	0.3827	636.57	2747.12	2110.54	1.8521	6.8281	4.9759
152	425.15	0.502177	0.001093	0.3733	640.89	2748.30	2107.41	1.8623	6.8191	4.9569
153	426.15	0.515637	0.001094	0.3641	645.22	2749.48	2104.26	1.8724	6.8102	4.9378
154	427.15	0.529383	0.001095	0.3552	649.55	2750.64	2101.10	1.8825	6.8014	4.9189
155	428.15	0.543422	0.001096	0.3465	653.88	2751.80	2097.92	1.8926	6.7926	4.9000
156	429.15	0.557755	0.001097	0.3381	658.21	2752.95	2094.74	1.9027	6.7838	4.8811
157	430.15	0.572390	0.001098	0.3299	662.55	2754.08	2091.54	1.9127	6.7751	4.8623
158	431.15	0.587329	0.001100	0.3220	666.89	2755.21	2088.32	1.9228	6.7664	4.8436
159	432.15	0.602577	0.001101	0.3143	671.23	2756.33	2085.10	1.9328	6.7577	4.8249
160	433.15	0.618139	0.001102	0.3068	675.57	2757.43	2081.86	1.9428	6.7491	4.8063
161	434.15	0.634020	0.001103	0.2995	679.92	2758.53	2078.60	1.9528	6.7405	4.7878
162	435.15	0.650224	0.001104	0.2925	684.28	2759.61	2075.33	1.9627	6.7320	4.7693
163	436.15	0.666755	0.001106	0.2856	688.63	2760.68	2072.05	1.9727	6.7235	4.7508
164	437.15	0.683619	0.001107	0.2789	692.99	2761.75	2068.76	1.9826	6.7150	4.7324
165	438.15	0.700820	0.001108	0.2725	697.35	2762.80	2065.45	1.9926	6.7066	4.7140
166	439.15	0.718364	0.001109	0.2662	701.71	2763.84	2062.13	2.0025	6.6982	4.6957
167	440.15	0.736254	0.001110	0.2600	706.08	2764.87	2058.79	2.0123	6.6898	4.6775
168	441.15	0.754495	0.001112	0.2541	710.45	2765.89	2055.44	2.0222	6.6815	4.6593
169	442.15	0.773094	0.001113	0.2483	714.83	2766.90	2052.07	2.0321	6.6732	4.6411
170	443.15	0.792053	0.001114	0.2426	719.21	2767.89	2048.69	2.0419	6.6649	4.6230
171	444.15	0.811379	0.001116	0.2371	723.59	2768.88	2045.29	2.0518	6.6567	4.6050
172	445.15	0.831077	0.001117	0.2318	727.97	2769.85	2041.88	2.0616	6.6485	4.5870
173	446.15	0.851150	0.001118	0.2266	732.36	2770.82	2038.45	2.0714	6.6404	4.5690
174	447.15	0.871606	0.001119	0.2215	736.75	2771.77	2035.01	2.0811	6.6322	4.5511
175	448.15	0.892447	0.001121	0.2166	741.15	2772.70	2031.55	2.0909	6.6241	4.5332
176	449.15	0.913681	0.001122	0.2118	745.55	2773.63	2028.08	2.1007	6.6161	4.5154
177	450.15	0.935311	0.001123	0.2071	749.95	2774.55	2024.59	2.1104	6.6080	4.4976
178	451.15	0.957343	0.001125	0.2026	754.36	2775.45	2021.09	2.1201	6.6000	4.4799
179	452.15	0.979783	0.001126	0.1982	758.77	2776.34	2017.57	2.1298	6.5920	4.4622

Table 1 Saturation State (Temperature Table) – Continuation
Sättigungszustand (Temperaturtafel) – Fortsetzung

t [°C]	c_p' [kJ kg^{-1} K^{-1}]	c_p''	w' [m s^{-1}]	w''	κ'	κ''	η' [10^{-6} kg m^{-1} s^{-1}]	η''	λ' [10^{-3} W m^{-1} K^{-1}]	λ''
135	4.2751	2.2726	1496.11	487.96	6650	1.3067	204.40	13.47	683.61	29.44
136	4.2772	2.2800	1494.26	488.34	6438	1.3064	202.78	13.51	683.56	29.58
137	4.2794	2.2876	1492.39	488.71	6233	1.3060	201.19	13.54	683.51	29.72
138	4.2816	2.2953	1490.49	489.08	6036	1.3057	199.61	13.58	683.45	29.86
139	4.2838	2.3030	1488.57	489.45	5845	1.3054	198.07	13.61	683.38	30.00
140	4.2860	2.3109	1486.63	489.81	5661	1.3051	196.54	13.65	683.30	30.14
141	4.2883	2.3189	1484.66	490.17	5484	1.3048	195.04	13.68	683.21	30.28
142	4.2906	2.3270	1482.68	490.53	5313	1.3044	193.56	13.72	683.11	30.42
143	4.2930	2.3352	1480.67	490.88	5148	1.3041	192.10	13.75	683.01	30.57
144	4.2953	2.3436	1478.64	491.22	4988	1.3038	190.66	13.78	682.90	30.71
145	4.2978	2.3520	1476.58	491.57	4834	1.3035	189.24	13.82	682.77	30.86
146	4.3002	2.3606	1474.51	491.90	4685	1.3031	187.85	13.85	682.64	31.00
147	4.3027	2.3692	1472.41	492.24	4542	1.3028	186.47	13.89	682.51	31.15
148	4.3052	2.3780	1470.29	492.57	4403	1.3024	185.12	13.92	682.36	31.30
149	4.3077	2.3869	1468.15	492.89	4268	1.3021	183.78	13.96	682.20	31.45
150	4.3103	2.3959	1465.98	493.22	4139	1.3018	182.46	13.99	682.04	31.59
151	4.3129	2.4050	1463.80	493.53	4014	1.3014	181.16	14.03	681.87	31.74
152	4.3155	2.4142	1461.59	493.85	3892	1.3011	179.88	14.06	681.69	31.90
153	4.3182	2.4235	1459.36	494.16	3775	1.3007	178.62	14.10	681.50	32.05
154	4.3209	2.4329	1457.11	494.46	3662	1.3004	177.37	14.13	681.31	32.20
155	4.3236	2.4425	1454.84	494.76	3553	1.3000	176.14	14.16	681.10	32.35
156	4.3264	2.4521	1452.55	495.06	3447	1.2997	174.93	14.20	680.89	32.51
157	4.3292	2.4619	1450.24	495.35	3345	1.2993	173.73	14.23	680.67	32.66
158	4.3321	2.4717	1447.90	495.64	3246	1.2990	172.55	14.27	680.44	32.82
159	4.3350	2.4817	1445.55	495.93	3150	1.2986	171.39	14.30	680.20	32.97
160	4.3379	2.4918	1443.17	496.21	3057	1.2982	170.24	14.34	679.96	33.13
161	4.3409	2.5020	1440.77	496.48	2967	1.2979	169.11	14.37	679.70	33.29
162	4.3439	2.5123	1438.35	496.75	2881	1.2975	167.99	14.41	679.44	33.45
163	4.3470	2.5227	1435.91	497.02	2797	1.2972	166.89	14.44	679.17	33.61
164	4.3501	2.5332	1433.45	497.28	2715	1.2968	165.80	14.47	678.89	33.77
165	4.3532	2.5438	1430.97	497.54	2636	1.2964	164.72	14.51	678.61	33.93
166	4.3564	2.5545	1428.47	497.80	2560	1.2961	163.66	14.54	678.31	34.09
167	4.3596	2.5653	1425.95	498.05	2486	1.2957	162.61	14.58	678.01	34.25
168	4.3628	2.5763	1423.41	498.30	2415	1.2953	161.58	14.61	677.70	34.42
169	4.3662	2.5873	1420.85	498.54	2346	1.2950	160.56	14.65	677.38	34.58
170	4.3695	2.5985	1418.26	498.78	2279	1.2946	159.55	14.68	677.05	34.75
171	4.3729	2.6098	1415.66	499.01	2214	1.2942	158.56	14.72	676.72	34.91
172	4.3763	2.6212	1413.04	499.24	2151	1.2939	157.58	14.75	676.38	35.08
173	4.3798	2.6327	1410.39	499.46	2090	1.2935	156.61	14.78	676.02	35.25
174	4.3833	2.6443	1407.73	499.68	2031	1.2931	155.65	14.82	675.66	35.42
175	4.3869	2.6560	1405.04	499.90	1973	1.2928	154.70	14.85	675.30	35.59
176	4.3906	2.6679	1402.34	500.11	1918	1.2924	153.77	14.89	674.92	35.76
177	4.3942	2.6798	1399.61	500.32	1864	1.2920	152.84	14.92	674.53	35.93
178	4.3980	2.6919	1396.87	500.52	1812	1.2916	151.93	14.96	674.14	36.10
179	4.4017	2.7041	1394.10	500.72	1761	1.2913	151.03	14.99	673.74	36.27

Table 1 Saturation State (Temperature Table) – Continuation
Sättigungszustand (Temperaturtafel) – Fortsetzung

t [°C]	T [K]	p_s [MPa]	v' [m³ kg⁻¹]	v'' [m³ kg⁻¹]	h' [kJ kg⁻¹]	h'' [kJ kg⁻¹]	Δh_v [kJ kg⁻¹]	s' [kJ kg⁻¹ K⁻¹]	s'' [kJ kg⁻¹ K⁻¹]	Δs_v [kJ kg⁻¹ K⁻¹]
180	453.15	1.0026	0.001127	0.19386	763.19	2777.22	2014.03	2.1395	6.5841	4.4445
181	454.15	1.0259	0.001129	0.18966	767.61	2778.09	2010.48	2.1492	6.5761	4.4269
182	455.15	1.0496	0.001130	0.18557	772.03	2778.94	2006.91	2.1589	6.5682	4.4094
183	456.15	1.0737	0.001132	0.18159	776.46	2779.78	2003.32	2.1685	6.5604	4.3918
184	457.15	1.0983	0.001133	0.17770	780.89	2780.61	1999.72	2.1782	6.5525	4.3743
185	458.15	1.1233	0.001134	0.17392	785.32	2781.43	1996.10	2.1878	6.5447	4.3569
186	459.15	1.1487	0.001136	0.17023	789.76	2782.23	1992.47	2.1974	6.5369	4.3395
187	460.15	1.1746	0.001137	0.16663	794.21	2783.02	1988.81	2.2070	6.5292	4.3221
188	461.15	1.2009	0.001139	0.16313	798.66	2783.80	1985.14	2.2166	6.5214	4.3048
189	462.15	1.2277	0.001140	0.15971	803.11	2784.56	1981.45	2.2262	6.5137	4.2875
190	463.15	1.2550	0.001141	0.15638	807.57	2785.31	1977.74	2.2358	6.5060	4.2702
191	464.15	1.2828	0.001143	0.15313	812.03	2786.05	1974.02	2.2453	6.4983	4.2530
192	465.15	1.3110	0.001144	0.14996	816.49	2786.77	1970.28	2.2549	6.4907	4.2358
193	466.15	1.3397	0.001146	0.14686	820.96	2787.48	1966.52	2.2644	6.4831	4.2187
194	467.15	1.3689	0.001147	0.14385	825.44	2788.18	1962.74	2.2739	6.4755	4.2015
195	468.15	1.3986	0.001149	0.14091	829.92	2788.86	1958.94	2.2834	6.4679	4.1844
196	469.15	1.4288	0.001150	0.13803	834.40	2789.53	1955.12	2.2929	6.4603	4.1674
197	470.15	1.4595	0.001152	0.13523	838.89	2790.18	1951.29	2.3024	6.4528	4.1504
198	471.15	1.4907	0.001153	0.13250	843.39	2790.82	1947.44	2.3119	6.4453	4.1334
199	472.15	1.5224	0.001155	0.12983	847.89	2791.45	1943.56	2.3213	6.4378	4.1164
200	473.15	1.5547	0.001157	0.12722	852.39	2792.06	1939.67	2.3308	6.4303	4.0995
201	474.15	1.5875	0.001158	0.12468	856.90	2792.66	1935.76	2.3402	6.4228	4.0826
202	475.15	1.6208	0.001160	0.12220	861.42	2793.24	1931.82	2.3497	6.4154	4.0657
203	476.15	1.6546	0.001161	0.11977	865.94	2793.81	1927.87	2.3591	6.4080	4.0489
204	477.15	1.6891	0.001163	0.11740	870.46	2794.36	1923.90	2.3685	6.4006	4.0321
205	478.15	1.7240	0.001164	0.11509	874.99	2794.90	1919.90	2.3779	6.3932	4.0153
206	479.15	1.7596	0.001166	0.11283	879.53	2795.42	1915.89	2.3873	6.3858	3.9985
207	480.15	1.7956	0.001168	0.11062	884.07	2795.93	1911.86	2.3967	6.3785	3.9818
208	481.15	1.8323	0.001169	0.10847	888.62	2796.42	1907.80	2.4060	6.3711	3.9651
209	482.15	1.8696	0.001171	0.10636	893.17	2796.89	1903.72	2.4154	6.3638	3.9484
210	483.15	1.9074	0.001173	0.10430	897.73	2797.35	1899.62	2.4248	6.3565	3.9318
211	484.15	1.9458	0.001174	0.10229	902.29	2797.80	1895.50	2.4341	6.3492	3.9151
212	485.15	1.9848	0.001176	0.10032	906.86	2798.22	1891.36	2.4434	6.3420	3.8985
213	486.15	2.0245	0.001178	0.09840	911.44	2798.64	1887.20	2.4528	6.3347	3.8819
214	487.15	2.0647	0.001180	0.09652	916.02	2799.03	1883.01	2.4621	6.3275	3.8654
215	488.15	2.1055	0.001181	0.09469	920.61	2799.41	1878.80	2.4714	6.3202	3.8488
216	489.15	2.1470	0.001183	0.09289	925.20	2799.77	1874.57	2.4807	6.3130	3.8323
217	490.15	2.1891	0.001185	0.09114	929.80	2800.12	1870.31	2.4900	6.3058	3.8158
218	491.15	2.2319	0.001187	0.08942	934.41	2800.45	1866.04	2.4993	6.2986	3.7993
219	492.15	2.2753	0.001188	0.08774	939.02	2800.76	1861.73	2.5085	6.2914	3.7829
220	493.15	2.3193	0.001190	0.08610	943.64	2801.05	1857.41	2.5178	6.2842	3.7664
221	494.15	2.3640	0.001192	0.08449	948.27	2801.33	1853.06	2.5271	6.2771	3.7500
222	495.15	2.4093	0.001194	0.08292	952.90	2801.59	1848.69	2.5363	6.2699	3.7336
223	496.15	2.4553	0.001196	0.08139	957.54	2801.83	1844.29	2.5456	6.2628	3.7172
224	497.15	2.5020	0.001198	0.07988	962.19	2802.05	1839.87	2.5548	6.2557	3.7008

Table 1 Saturation State (Temperature Table) – Continuation
Sättigungszustand (Temperaturtafel) – Fortsetzung

t [°C]	c_p' [kJ kg^{-1} K^{-1}]	c_p''	w' [m s^{-1}]	w''	κ'	κ''	η' [10^{-6} kg m^{-1} s^{-1}]	η''	λ' [10^{-3} W m^{-1} K^{-1}]	λ''
180	4.4056	2.7164	1391.32	500.92	1712.53	1.2909	150.14	15.03	673.33	36.45
181	4.4094	2.7288	1388.51	501.11	1664.93	1.2905	149.26	15.06	672.91	36.62
182	4.4134	2.7414	1385.69	501.29	1618.77	1.2902	148.39	15.09	672.48	36.80
183	4.4173	2.7541	1382.85	501.47	1573.99	1.2898	147.53	15.13	672.05	36.97
184	4.4214	2.7669	1379.98	501.65	1530.55	1.2894	146.68	15.16	671.60	37.15
185	4.4255	2.7798	1377.10	501.82	1488.40	1.2891	145.84	15.20	671.15	37.33
186	4.4296	2.7928	1374.19	501.99	1447.51	1.2887	145.01	15.23	670.69	37.51
187	4.4338	2.8060	1371.27	502.15	1407.83	1.2883	144.19	15.27	670.22	37.69
188	4.4381	2.8193	1368.33	502.31	1369.32	1.2879	143.38	15.30	669.74	37.87
189	4.4424	2.8328	1365.36	502.46	1331.94	1.2876	142.58	15.34	669.26	38.05
190	4.4468	2.8464	1362.38	502.61	1295.67	1.2872	141.78	15.37	668.76	38.24
191	4.4512	2.8601	1359.38	502.76	1260.45	1.2868	141.00	15.40	668.26	38.42
192	4.4557	2.8739	1356.36	502.90	1226.26	1.2864	140.23	15.44	667.75	38.60
193	4.4603	2.8879	1353.32	503.03	1193.06	1.2861	139.46	15.47	667.22	38.79
194	4.4649	2.9021	1350.26	503.16	1160.83	1.2857	138.70	15.51	666.69	38.97
195	4.4696	2.9163	1347.17	503.29	1129.53	1.2853	137.95	15.54	666.16	39.16
196	4.4743	2.9308	1344.08	503.41	1099.14	1.2850	137.21	15.58	665.61	39.35
197	4.4792	2.9454	1340.96	503.52	1069.61	1.2846	136.48	15.61	665.05	39.54
198	4.4840	2.9601	1337.82	503.63	1040.93	1.2842	135.75	15.65	664.48	39.73
199	4.4890	2.9750	1334.66	503.74	1013.08	1.2838	135.03	15.68	663.91	39.92
200	4.4940	2.9900	1331.48	503.84	986.01	1.2835	134.32	15.71	663.33	40.11
201	4.4991	3.0052	1328.29	503.93	959.72	1.2831	133.62	15.75	662.73	40.31
202	4.5043	3.0206	1325.07	504.03	934.17	1.2827	132.92	15.78	662.13	40.50
203	4.5095	3.0361	1321.84	504.11	909.34	1.2823	132.23	15.82	661.52	40.69
204	4.5148	3.0518	1318.58	504.19	885.21	1.2820	131.55	15.85	660.90	40.89
205	4.5202	3.0677	1315.31	504.27	861.76	1.2816	130.88	15.89	660.27	41.09
206	4.5256	3.0837	1312.02	504.34	838.97	1.2812	130.21	15.92	659.63	41.28
207	4.5312	3.0999	1308.71	504.40	816.82	1.2808	129.55	15.96	658.98	41.48
208	4.5368	3.1163	1305.38	504.46	795.28	1.2804	128.89	15.99	658.32	41.68
209	4.5425	3.1329	1302.03	504.52	774.34	1.2801	128.25	16.03	657.65	41.88
210	4.5482	3.1496	1298.66	504.57	753.98	1.2797	127.60	16.06	656.98	42.09
211	4.5541	3.1666	1295.28	504.61	734.19	1.2793	126.97	16.10	656.29	42.29
212	4.5600	3.1837	1291.87	504.65	714.94	1.2789	126.34	16.13	655.59	42.49
213	4.5661	3.2010	1288.44	504.68	696.22	1.2785	125.72	16.17	654.89	42.70
214	4.5722	3.2186	1285.00	504.71	678.02	1.2782	125.10	16.20	654.17	42.91
215	4.5784	3.2363	1281.54	504.73	660.31	1.2778	124.49	16.24	653.44	43.11
216	4.5846	3.2542	1278.06	504.75	643.09	1.2774	123.88	16.27	652.71	43.32
217	4.5910	3.2723	1274.56	504.76	626.34	1.2770	123.28	16.31	651.96	43.53
218	4.5975	3.2907	1271.04	504.76	610.04	1.2766	122.69	16.34	651.20	43.74
219	4.6041	3.3092	1267.50	504.76	594.18	1.2763	122.10	16.38	650.44	43.96
220	4.6107	3.3280	1263.94	504.76	578.76	1.2759	121.52	16.41	649.66	44.17
221	4.6175	3.3470	1260.37	504.75	563.74	1.2755	120.94	16.45	648.87	44.38
222	4.6243	3.3662	1256.77	504.73	549.14	1.2751	120.37	16.48	648.07	44.60
223	4.6313	3.3856	1253.16	504.71	534.92	1.2747	119.80	16.52	647.27	44.82
224	4.6383	3.4053	1249.53	504.68	521.09	1.2743	119.24	16.55	646.45	45.03

Table 1 Saturation State (Temperature Table) – Continuation
Sättigungszustand (Temperaturtafel) – Fortsetzung

t [°C]	T [K]	p_s [MPa]	v' [m³ kg⁻¹]	v''	h'	h'' [kJ kg⁻¹]	Δh_v	s'	s'' [kJ kg⁻¹ K⁻¹]	Δs_v
225	498.15	2.5494	0.001199	0.07841	966.84	2802.26	1835.42	2.5641	6.2485	3.6845
226	499.15	2.5975	0.001201	0.07697	971.50	2802.45	1830.95	2.5733	6.2414	3.6681
227	500.15	2.6463	0.001203	0.07556	976.17	2802.61	1826.45	2.5825	6.2343	3.6518
228	501.15	2.6957	0.001205	0.07418	980.84	2802.76	1821.93	2.5917	6.2272	3.6355
229	502.15	2.7459	0.001207	0.07283	985.52	2802.90	1817.38	2.6009	6.2201	3.6192
230	503.15	2.7968	0.001209	0.07151	990.21	2803.01	1812.80	2.6102	6.2131	3.6029
231	504.15	2.8484	0.001211	0.07022	994.91	2803.10	1808.20	2.6194	6.2060	3.5866
232	505.15	2.9008	0.001213	0.06895	999.61	2803.18	1803.57	2.6285	6.1989	3.5704
233	506.15	2.9538	0.001215	0.06771	1004.32	2803.23	1798.91	2.6377	6.1918	3.5541
234	507.15	3.0077	0.001217	0.06649	1009.04	2803.27	1794.23	2.6469	6.1848	3.5379
235	508.15	3.0622	0.001219	0.06530	1013.77	2803.28	1789.52	2.6561	6.1777	3.5216
236	509.15	3.1176	0.001221	0.06414	1018.50	2803.28	1784.78	2.6653	6.1707	3.5054
237	510.15	3.1737	0.001223	0.06300	1023.24	2803.26	1780.01	2.6744	6.1636	3.4892
238	511.15	3.2306	0.001225	0.06188	1028.00	2803.21	1775.22	2.6836	6.1566	3.4730
239	512.15	3.2882	0.001227	0.06078	1032.76	2803.15	1770.39	2.6928	6.1496	3.4568
240	513.15	3.3467	0.001229	0.05971	1037.52	2803.06	1765.54	2.7019	6.1425	3.4406
241	514.15	3.4059	0.001232	0.05866	1042.30	2802.95	1760.65	2.7111	6.1355	3.4244
242	515.15	3.4659	0.001234	0.05763	1047.08	2802.82	1755.74	2.7203	6.1285	3.4082
243	516.15	3.5268	0.001236	0.05662	1051.88	2802.68	1750.80	2.7294	6.1214	3.3920
244	517.15	3.5884	0.001238	0.05563	1056.68	2802.50	1745.82	2.7385	6.1144	3.3759
245	518.15	3.6509	0.001240	0.05466	1061.49	2802.31	1740.82	2.7477	6.1074	3.3597
246	519.15	3.7142	0.001243	0.05371	1066.31	2802.10	1735.78	2.7568	6.1003	3.3435
247	520.15	3.7784	0.001245	0.05278	1071.14	2801.86	1730.72	2.7660	6.0933	3.3273
248	521.15	3.8434	0.001247	0.05186	1075.98	2801.60	1725.62	2.7751	6.0863	3.3112
249	522.15	3.9092	0.001249	0.05097	1080.83	2801.32	1720.49	2.7843	6.0793	3.2950
250	523.15	3.9759	0.001252	0.05009	1085.69	2801.01	1715.33	2.7934	6.0722	3.2788
251	524.15	4.0435	0.001254	0.04922	1090.55	2800.68	1710.13	2.8025	6.0652	3.2627
252	525.15	4.1120	0.001256	0.04838	1095.43	2800.33	1704.90	2.8117	6.0582	3.2465
253	526.15	4.1813	0.001259	0.04755	1100.32	2799.96	1699.64	2.8208	6.0511	3.2303
254	527.15	4.2515	0.001261	0.04674	1105.22	2799.56	1694.34	2.8299	6.0441	3.2141
255	528.15	4.3227	0.001264	0.04594	1110.13	2799.13	1689.01	2.8391	6.0370	3.1980
256	529.15	4.3947	0.001266	0.04516	1115.04	2798.69	1683.64	2.8482	6.0300	3.1818
257	530.15	4.4677	0.001269	0.04439	1119.97	2798.21	1678.24	2.8573	6.0229	3.1656
258	531.15	4.5415	0.001271	0.04364	1124.91	2797.71	1672.80	2.8664	6.0158	3.1494
259	532.15	4.6163	0.001274	0.04290	1129.86	2797.19	1667.33	2.8756	6.0088	3.1332
260	533.15	4.6921	0.001276	0.04218	1134.83	2796.64	1661.82	2.8847	6.0017	3.1170
261	534.15	4.7688	0.001279	0.04146	1139.80	2796.07	1656.27	2.8939	5.9946	3.1008
262	535.15	4.8464	0.001281	0.04077	1144.78	2795.47	1650.68	2.9030	5.9875	3.0845
263	536.15	4.9250	0.001284	0.04008	1149.78	2794.84	1645.06	2.9121	5.9804	3.0683
264	537.15	5.0046	0.001287	0.03941	1154.79	2794.19	1639.40	2.9213	5.9733	3.0520
265	538.15	5.0851	0.001289	0.03875	1159.81	2793.51	1633.70	2.9304	5.9662	3.0358
266	539.15	5.1667	0.001292	0.03810	1164.84	2792.80	1627.96	2.9396	5.9590	3.0195
267	540.15	5.2492	0.001295	0.03746	1169.88	2792.07	1622.18	2.9487	5.9519	3.0032
268	541.15	5.3327	0.001297	0.03684	1174.94	2791.30	1616.36	2.9579	5.9448	2.9869
269	542.15	5.4173	0.001300	0.03622	1180.01	2790.51	1610.50	2.9670	5.9376	2.9706

Table 1 Saturation State (Temperature Table) – Continuation
Sättigungszustand (Temperaturtafel) – Fortsetzung

t [°C]	c_p' [kJ kg^{-1} K^{-1}]	c_p''	w' [m s^{-1}]	w''	κ'	κ''	η' [10^{-6} kg m^{-1} s^{-1}]	η''	λ' [10^{-3} W m^{-1} K^{-1}]	λ''
225	4.6455	3.4252	1245.87	504.64	507.63	1.2739	118.68	16.59	645.62	45.25
226	4.6528	3.4453	1242.20	504.60	494.52	1.2735	118.13	16.62	644.78	45.48
227	4.6601	3.4657	1238.52	504.55	481.76	1.2732	117.58	16.66	643.93	45.70
228	4.6676	3.4863	1234.81	504.50	469.35	1.2728	117.04	16.69	643.07	45.92
229	4.6752	3.5072	1231.08	504.44	457.26	1.2724	116.50	16.73	642.19	46.15
230	4.6829	3.5283	1227.34	504.38	445.49	1.2720	115.96	16.76	641.31	46.37
231	4.6908	3.5497	1223.57	504.30	434.03	1.2716	115.43	16.80	640.41	46.60
232	4.6987	3.5713	1219.79	504.23	422.87	1.2712	114.91	16.84	639.51	46.83
233	4.7068	3.5933	1215.99	504.14	412.01	1.2708	114.39	16.87	638.59	47.06
234	4.7150	3.6154	1212.16	504.05	401.43	1.2704	113.87	16.91	637.66	47.30
235	4.7233	3.6379	1208.32	503.96	391.12	1.2700	113.36	16.94	636.72	47.53
236	4.7317	3.6606	1204.46	503.85	381.09	1.2696	112.85	16.98	635.77	47.77
237	4.7403	3.6837	1200.58	503.74	371.32	1.2692	112.34	17.02	634.81	48.00
238	4.7490	3.7070	1196.69	503.63	361.80	1.2688	111.84	17.05	633.83	48.24
239	4.7579	3.7306	1192.77	503.50	352.52	1.2684	111.34	17.09	632.85	48.48
240	4.7668	3.7545	1188.83	503.38	343.49	1.2680	110.85	17.13	631.85	48.73
241	4.7760	3.7788	1184.88	503.24	334.69	1.2676	110.36	17.16	630.84	48.97
242	4.7852	3.8033	1180.90	503.10	326.12	1.2672	109.88	17.20	629.81	49.22
243	4.7947	3.8282	1176.90	502.95	317.77	1.2668	109.39	17.23	628.78	49.47
244	4.8042	3.8534	1172.89	502.79	309.63	1.2664	108.91	17.27	627.73	49.72
245	4.8139	3.8789	1168.85	502.63	301.70	1.2660	108.44	17.31	626.67	49.97
246	4.8238	3.9048	1164.80	502.46	293.97	1.2656	107.97	17.35	625.60	50.22
247	4.8339	3.9310	1160.72	502.29	286.44	1.2652	107.50	17.38	624.52	50.48
248	4.8441	3.9576	1156.62	502.10	279.10	1.2648	107.03	17.42	623.42	50.74
249	4.8544	3.9846	1152.51	501.91	271.95	1.2644	106.57	17.46	622.31	51.00
250	4.8650	4.0119	1148.37	501.71	264.98	1.2640	106.11	17.49	621.19	51.26
251	4.8757	4.0396	1144.21	501.51	258.18	1.2636	105.66	17.53	620.05	51.53
252	4.8866	4.0677	1140.03	501.30	251.56	1.2632	105.20	17.57	618.90	51.80
253	4.8977	4.0962	1135.83	501.08	245.11	1.2628	104.75	17.61	617.74	52.07
254	4.9089	4.1251	1131.61	500.85	238.81	1.2624	104.31	17.65	616.56	52.34
255	4.9204	4.1545	1127.37	500.62	232.68	1.2620	103.86	17.68	615.38	52.61
256	4.9320	4.1843	1123.11	500.38	226.70	1.2616	103.42	17.72	614.17	52.89
257	4.9439	4.2145	1118.82	500.13	220.86	1.2612	102.98	17.76	612.96	53.17
258	4.9560	4.2452	1114.51	499.87	215.18	1.2608	102.54	17.80	611.73	53.46
259	4.9682	4.2764	1110.18	499.61	209.63	1.2604	102.11	17.84	610.49	53.74
260	4.9807	4.3080	1105.83	499.33	204.23	1.2600	101.68	17.88	609.23	54.03
261	4.9934	4.3402	1101.45	499.05	198.96	1.2596	101.25	17.92	607.96	54.32
262	5.0064	4.3728	1097.05	498.77	193.81	1.2592	100.82	17.96	606.68	54.62
263	5.0195	4.4060	1092.63	498.47	188.80	1.2587	100.40	17.99	605.38	54.92
264	5.0330	4.4398	1088.18	498.17	183.91	1.2583	99.98	18.03	604.07	55.22
265	5.0466	4.4741	1083.71	497.85	179.14	1.2579	99.56	18.07	602.75	55.53
266	5.0605	4.5090	1079.21	497.53	174.49	1.2575	99.14	18.11	601.41	55.84
267	5.0747	4.5445	1074.69	497.20	169.95	1.2571	98.73	18.15	600.05	56.15
268	5.0891	4.5806	1070.15	496.87	165.52	1.2567	98.32	18.20	598.68	56.47
269	5.1038	4.6173	1065.57	496.52	161.21	1.2563	97.90	18.24	597.30	56.79

Table 1 Saturation State (Temperature Table) – Continuation
Sättigungszustand (Temperaturtafel) – Fortsetzung

t [°C]	T [K]	p_s [MPa]	v' [m³ kg⁻¹]	v''	h'	h'' [kJ kg⁻¹]	Δh_v	s'	s'' [kJ kg⁻¹ K⁻¹]	Δs_v
270	543.15	5.5028	0.001303	0.03562	1185.09	2789.69	1604.60	2.9762	5.9304	2.9542
271	544.15	5.5894	0.001306	0.03503	1190.19	2788.84	1598.65	2.9853	5.9232	2.9379
272	545.15	5.6771	0.001309	0.03445	1195.30	2787.96	1592.66	2.9945	5.9160	2.9215
273	546.15	5.7657	0.001312	0.03388	1200.42	2787.05	1586.63	3.0037	5.9088	2.9051
274	547.15	5.8555	0.001315	0.03332	1205.55	2786.11	1580.56	3.0129	5.9016	2.8887
275	548.15	5.9463	0.001318	0.03277	1210.70	2785.14	1574.44	3.0221	5.8943	2.8723
276	549.15	6.0381	0.001321	0.03223	1215.87	2784.14	1568.28	3.0312	5.8871	2.8558
277	550.15	6.1311	0.001324	0.03169	1221.05	2783.11	1562.07	3.0404	5.8798	2.8393
278	551.15	6.2251	0.001327	0.03117	1226.24	2782.05	1555.81	3.0496	5.8725	2.8228
279	552.15	6.3202	0.001330	0.03066	1231.45	2780.95	1549.51	3.0589	5.8652	2.8063
280	553.15	6.4165	0.001333	0.03015	1236.67	2779.82	1543.15	3.0681	5.8578	2.7898
281	554.15	6.5138	0.001336	0.02966	1241.91	2778.66	1536.75	3.0773	5.8505	2.7732
282	555.15	6.6123	0.001339	0.02917	1247.16	2777.47	1530.30	3.0865	5.8431	2.7566
283	556.15	6.7119	0.001342	0.02869	1252.43	2776.24	1523.80	3.0958	5.8357	2.7399
284	557.15	6.8126	0.001346	0.02822	1257.72	2774.97	1517.25	3.1050	5.8283	2.7232
285	558.15	6.9145	0.001349	0.02776	1263.02	2773.67	1510.65	3.1143	5.8208	2.7065
286	559.15	7.0176	0.001352	0.02730	1268.34	2772.34	1504.00	3.1236	5.8134	2.6898
287	560.15	7.1218	0.001356	0.02686	1273.68	2770.97	1497.29	3.1329	5.8059	2.6730
288	561.15	7.2272	0.001359	0.02642	1279.03	2769.56	1490.53	3.1421	5.7984	2.6562
289	562.15	7.3338	0.001363	0.02598	1284.41	2768.12	1483.71	3.1515	5.7908	2.6393
290	563.15	7.4416	0.001366	0.02556	1289.80	2766.63	1476.84	3.1608	5.7832	2.6225
291	564.15	7.5506	0.001370	0.02514	1295.20	2765.11	1469.91	3.1701	5.7756	2.6055
292	565.15	7.6609	0.001373	0.02473	1300.63	2763.55	1462.92	3.1794	5.7680	2.5886
293	566.15	7.7723	0.001377	0.02432	1306.08	2761.95	1455.87	3.1888	5.7603	2.5715
294	567.15	7.8850	0.001381	0.02392	1311.54	2760.31	1448.76	3.1982	5.7526	2.5545
295	568.15	7.9990	0.001385	0.02353	1317.03	2758.63	1441.60	3.2076	5.7449	2.5374
296	569.15	8.1142	0.001388	0.02315	1322.54	2756.90	1434.37	3.2170	5.7372	2.5202
297	570.15	8.2306	0.001392	0.02277	1328.06	2755.14	1427.07	3.2264	5.7294	2.5030
298	571.15	8.3484	0.001396	0.02239	1333.61	2753.33	1419.72	3.2358	5.7215	2.4857
299	572.15	8.4674	0.001400	0.02202	1339.18	2751.47	1412.29	3.2453	5.7137	2.4684
300	573.15	8.5877	0.001404	0.02166	1344.77	2749.57	1404.80	3.2547	5.7058	2.4510
301	574.15	8.7093	0.001408	0.02131	1350.39	2747.63	1397.24	3.2642	5.6978	2.4336
302	575.15	8.8323	0.001412	0.02096	1356.02	2745.64	1389.62	3.2737	5.6898	2.4161
303	576.15	8.9566	0.001417	0.02061	1361.68	2743.60	1381.92	3.2833	5.6818	2.3985
304	577.15	9.0822	0.001421	0.02027	1367.37	2741.51	1374.15	3.2928	5.6737	2.3809
305	578.15	9.2092	0.001425	0.01994	1373.07	2739.38	1366.30	3.3024	5.6656	2.3632
306	579.15	9.3375	0.001430	0.01961	1378.81	2737.19	1358.38	3.3120	5.6575	2.3455
307	580.15	9.4672	0.001434	0.01928	1384.57	2734.95	1350.39	3.3216	5.6492	2.3277
308	581.15	9.5983	0.001439	0.01896	1390.35	2732.66	1342.31	3.3312	5.6410	2.3098
309	582.15	9.7308	0.001443	0.01865	1396.16	2730.32	1334.16	3.3409	5.6327	2.2918
310	583.15	9.8647	0.001448	0.01834	1402.00	2727.92	1325.92	3.3506	5.6243	2.2737
311	584.15	10.0001	0.001453	0.01803	1407.87	2725.47	1317.60	3.3603	5.6159	2.2556
312	585.15	10.1368	0.001457	0.01773	1413.77	2722.96	1309.20	3.3700	5.6074	2.2374
313	586.15	10.2750	0.001462	0.01744	1419.69	2720.39	1300.70	3.3798	5.5989	2.2191
314	587.15	10.4147	0.001467	0.01714	1425.65	2717.77	1292.12	3.3896	5.5903	2.2007

Table 1 Saturation State (Temperature Table) – Continuation
Sättigungszustand (Temperaturtafel) – Fortsetzung

t [°C]	c_p' [kJ kg^{-1} K^{-1}]	c_p''	w' [m s^{-1}]	w''	κ'	κ''	η' [10^{-6} kg m^{-1} s^{-1}]	η''	λ' [10^{-3} W m^{-1} K^{-1}]	λ''
270	5.1188	4.6547	1060.98	496.16	156.99	1.2559	97.50	18.28	595.91	57.11
271	5.1341	4.6928	1056.35	495.80	152.88	1.2554	97.09	18.32	594.49	57.44
272	5.1496	4.7316	1051.70	495.43	148.87	1.2550	96.68	18.36	593.07	57.77
273	5.1655	4.7711	1047.03	495.05	144.96	1.2546	96.28	18.40	591.63	58.11
274	5.1817	4.8114	1042.32	494.65	141.15	1.2542	95.88	18.44	590.18	58.45
275	5.1982	4.8524	1037.59	494.25	137.42	1.2538	95.48	18.48	588.71	58.80
276	5.2150	4.8942	1032.84	493.84	133.79	1.2533	95.08	18.53	587.22	59.15
277	5.2322	4.9369	1028.05	493.42	130.24	1.2529	94.69	18.57	585.73	59.51
278	5.2497	4.9804	1023.23	493.00	126.78	1.2525	94.29	18.61	584.22	59.87
279	5.2676	5.0248	1018.39	492.56	123.41	1.2521	93.90	18.66	582.69	60.24
280	5.2859	5.0701	1013.51	492.11	120.11	1.2516	93.51	18.70	581.15	60.61
281	5.3045	5.1163	1008.61	491.65	116.90	1.2512	93.12	18.74	579.59	60.99
282	5.3236	5.1636	1003.68	491.18	113.76	1.2508	92.73	18.79	578.02	61.38
283	5.3430	5.2118	998.71	490.70	110.70	1.2503	92.34	18.83	576.44	61.77
284	5.3629	5.2611	993.72	490.21	107.71	1.2499	91.95	18.88	574.84	62.17
285	5.3832	5.3114	988.70	489.71	104.79	1.2495	91.57	18.92	573.23	62.57
286	5.4040	5.3629	983.64	489.20	101.95	1.2490	91.18	18.97	571.61	62.98
287	5.4252	5.4155	978.56	488.67	99.17	1.2486	90.80	19.01	569.97	63.40
288	5.4469	5.4693	973.44	488.14	96.46	1.2481	90.42	19.06	568.31	63.83
289	5.4691	5.5243	968.30	487.59	93.81	1.2477	90.04	19.11	566.64	64.26
290	5.4918	5.5806	963.12	487.04	91.23	1.2472	89.66	19.15	564.96	64.71
291	5.5151	5.6381	957.91	486.47	88.71	1.2468	89.28	19.20	563.27	65.16
292	5.5389	5.6971	952.67	485.89	86.26	1.2463	88.90	19.25	561.56	65.62
293	5.5632	5.7574	947.40	485.30	83.86	1.2459	88.52	19.30	559.84	66.09
294	5.5882	5.8192	942.10	484.69	81.52	1.2454	88.15	19.35	558.10	66.56
295	5.6137	5.8825	936.78	484.08	79.23	1.2450	87.77	19.40	556.36	67.05
296	5.6399	5.9473	931.42	483.45	77.00	1.2445	87.39	19.45	554.59	67.55
297	5.6668	6.0137	926.03	482.81	74.83	1.2440	87.02	19.50	552.82	68.06
298	5.6943	6.0818	920.61	482.16	72.71	1.2436	86.64	19.55	551.04	68.58
299	5.7226	6.1516	915.16	481.49	70.64	1.2431	86.27	19.60	549.24	69.11
300	5.7515	6.2231	909.69	480.81	68.62	1.2427	85.90	19.65	547.43	69.65
301	5.7812	6.2965	904.19	480.12	66.66	1.2422	85.52	19.70	545.61	70.20
302	5.8118	6.3717	898.66	479.41	64.74	1.2417	85.15	19.76	543.78	70.77
303	5.8431	6.4489	893.10	478.69	62.86	1.2413	84.78	19.81	541.93	71.35
304	5.8753	6.5282	887.52	477.96	61.04	1.2408	84.40	19.87	540.08	71.94
305	5.9083	6.6096	881.91	477.21	59.26	1.2404	84.03	19.92	538.22	72.55
306	5.9423	6.6931	876.28	476.45	57.52	1.2399	83.66	19.98	536.34	73.18
307	5.9773	6.7790	870.62	475.68	55.83	1.2394	83.29	20.03	534.46	73.82
308	6.0132	6.8672	864.94	474.89	54.18	1.2390	82.91	20.09	532.56	74.47
309	6.0502	6.9580	859.23	474.09	52.57	1.2385	82.54	20.15	530.66	75.14
310	6.0883	7.0513	853.50	473.27	51.00	1.2381	82.17	20.21	528.75	75.83
311	6.1275	7.1473	847.74	472.44	49.47	1.2377	81.79	20.27	526.83	76.54
312	6.1679	7.2461	841.95	471.59	47.98	1.2373	81.42	20.33	524.90	77.27
313	6.2096	7.3479	836.14	470.73	46.53	1.2368	81.04	20.39	522.96	78.01
314	6.2525	7.4528	830.30	469.85	45.11	1.2364	80.67	20.45	521.02	78.78

Table 1 Saturation State (Temperature Table) – Continuation
Sättigungszustand (Temperaturtafel) – Fortsetzung

t [°C]	T [K]	p_s [MPa]	v' [m³ kg⁻¹]	v''	h'	h'' [kJ kg⁻¹]	Δh_v	s'	s'' [kJ kg⁻¹ K⁻¹]	Δs_v
315	588.15	10.556	0.001472	0.016856	1431.63	2715.08	1283.45	3.3994	5.5816	2.1822
316	589.15	10.698	0.001478	0.016572	1437.65	2712.33	1274.68	3.4093	5.5729	2.1636
317	590.15	10.842	0.001483	0.016292	1443.70	2709.51	1265.81	3.4192	5.5641	2.1449
318	591.15	10.988	0.001488	0.016016	1449.78	2706.63	1256.85	3.4291	5.5553	2.1261
319	592.15	11.135	0.001494	0.015744	1455.90	2703.68	1247.79	3.4391	5.5463	2.1072
320	593.15	11.284	0.001499	0.015476	1462.05	2700.67	1238.62	3.4491	5.5373	2.0882
321	594.15	11.434	0.001505	0.015211	1468.24	2697.58	1229.34	3.4592	5.5282	2.0691
322	595.15	11.586	0.001510	0.014951	1474.46	2694.42	1219.96	3.4692	5.5191	2.0498
323	596.15	11.739	0.001516	0.014693	1480.73	2691.19	1210.46	3.4794	5.5098	2.0305
324	597.15	11.894	0.001522	0.014439	1487.03	2687.87	1200.84	3.4895	5.5005	2.0110
325	598.15	12.051	0.001528	0.014189	1493.37	2684.48	1191.11	3.4997	5.4911	1.9913
326	599.15	12.209	0.001534	0.013941	1499.76	2681.01	1181.25	3.5100	5.4816	1.9715
327	600.15	12.368	0.001541	0.013697	1506.18	2677.45	1171.27	3.5203	5.4719	1.9516
328	601.15	12.530	0.001547	0.013457	1512.66	2673.81	1161.15	3.5307	5.4622	1.9316
329	602.15	12.693	0.001554	0.013219	1519.17	2670.08	1150.90	3.5411	5.4524	1.9113
330	603.15	12.858	0.001561	0.012984	1525.74	2666.25	1140.51	3.5516	5.4425	1.8909
331	604.15	13.024	0.001567	0.012752	1532.35	2662.33	1129.97	3.5621	5.4324	1.8703
332	605.15	13.192	0.001575	0.012523	1539.02	2658.30	1119.29	3.5727	5.4223	1.8496
333	606.15	13.362	0.001582	0.012297	1545.74	2654.18	1108.44	3.5833	5.4120	1.8287
334	607.15	13.533	0.001589	0.012073	1552.51	2649.95	1097.44	3.5940	5.4016	1.8075
335	608.15	13.707	0.001597	0.011852	1559.34	2645.60	1086.26	3.6048	5.3910	1.7862
336	609.15	13.882	0.001604	0.011634	1566.23	2641.14	1074.91	3.6157	5.3803	1.7646
337	610.15	14.059	0.001612	0.011418	1573.18	2636.57	1063.38	3.6266	5.3694	1.7428
338	611.15	14.237	0.001621	0.011204	1580.20	2631.86	1051.66	3.6376	5.3584	1.7208
339	612.15	14.418	0.001629	0.010993	1587.29	2627.03	1039.74	3.6488	5.3473	1.6985
340	613.15	14.600	0.001638	0.010784	1594.45	2622.07	1027.62	3.6599	5.3359	1.6760
341	614.15	14.784	0.001646	0.010577	1601.68	2616.96	1015.28	3.6712	5.3244	1.6531
342	615.15	14.970	0.001655	0.010372	1608.99	2611.71	1002.72	3.6826	5.3127	1.6300
343	616.15	15.158	0.001665	0.010170	1616.38	2606.30	989.92	3.6941	5.3008	1.6066
344	617.15	15.348	0.001675	0.009969	1623.86	2600.74	976.87	3.7058	5.2886	1.5829
345	618.15	15.540	0.001685	0.009770	1631.44	2595.01	963.57	3.7175	5.2763	1.5588
346	619.15	15.734	0.001695	0.009573	1639.11	2589.10	950.00	3.7294	5.2637	1.5344
347	620.15	15.930	0.001706	0.009377	1646.88	2583.02	936.14	3.7414	5.2509	1.5095
348	621.15	16.127	0.001717	0.009184	1654.75	2576.74	921.99	3.7535	5.2378	1.4843
349	622.15	16.327	0.001728	0.008992	1662.75	2570.27	907.53	3.7658	5.2245	1.4587
350	623.15	16.529	0.001740	0.008801	1670.86	2563.59	892.73	3.7783	5.2109	1.4326
351	624.15	16.733	0.001753	0.008612	1679.14	2556.72	877.57	3.7910	5.1970	1.4060
352	625.15	16.939	0.001765	0.008424	1687.54	2549.56	862.02	3.8039	5.1828	1.3789
353	626.15	17.147	0.001779	0.008237	1696.10	2542.15	846.05	3.8169	5.1681	1.3512
354	627.15	17.358	0.001793	0.008051	1704.81	2534.45	829.64	3.8302	5.1531	1.3229
355	628.15	17.570	0.001808	0.007866	1713.71	2526.46	812.75	3.8438	5.1377	1.2939
356	629.15	17.785	0.001823	0.007682	1722.80	2518.15	795.34	3.8577	5.1218	1.2642
357	630.15	18.002	0.001840	0.007497	1732.10	2509.48	777.38	3.8718	5.1054	1.2336
358	631.15	18.221	0.001857	0.007314	1741.64	2500.43	758.80	3.8863	5.0885	1.2022
359	632.15	18.443	0.001875	0.007130	1751.42	2490.96	739.54	3.9011	5.0710	1.1699

Table 1 Saturation State (Temperature Table) – Continuation
Sättigungszustand (Temperaturtafel) – Fortsetzung

t [°C]	c_p' [kJ kg^{-1} K^{-1}]	c_p''	w' [m s^{-1}]	w''	κ'	κ''	η' [10^{-6} kg m^{-1} s^{-1}]	η''	λ' [10^{-3} W m^{-1} K^{-1}]	λ''
315	6.2968	7.5610	824.43	468.96	43.7319	1.2360	80.29	20.51	519.07	79.57
316	6.3426	7.6726	818.53	468.05	42.3854	1.2356	79.92	20.58	517.11	80.38
317	6.3898	7.7879	812.60	467.12	41.0725	1.2353	79.54	20.64	515.15	81.22
318	6.4386	7.9069	806.63	466.18	39.7920	1.2349	79.16	20.71	513.18	82.08
319	6.4891	8.0301	800.63	465.23	38.5431	1.2346	78.79	20.78	511.21	82.97
320	6.5414	8.1575	794.58	464.25	37.3247	1.2342	78.41	20.85	509.23	83.88
321	6.5956	8.2894	788.48	463.26	36.1359	1.2339	78.03	20.92	507.24	84.83
322	6.6517	8.4262	782.34	462.25	34.9756	1.2336	77.64	20.99	505.25	85.80
323	6.7099	8.5682	776.14	461.23	33.8429	1.2333	77.26	21.06	503.25	86.80
324	6.7703	8.7156	769.88	460.18	32.7368	1.2331	76.88	21.13	501.25	87.84
325	6.8331	8.8689	763.55	459.12	31.6563	1.2328	76.49	21.21	499.25	88.92
326	6.8985	9.0284	757.15	458.04	30.6007	1.2326	76.10	21.28	497.24	90.03
327	6.9665	9.1946	750.67	456.94	29.5690	1.2324	75.71	21.36	495.22	91.18
328	7.0374	9.3679	744.11	455.82	28.5603	1.2323	75.32	21.44	493.21	92.37
329	7.1114	9.5489	737.45	454.68	27.5740	1.2321	74.93	21.52	491.18	93.60
330	7.1888	9.7381	730.70	453.52	26.6093	1.2320	74.54	21.61	489.15	94.88
331	7.2698	9.9361	723.85	452.33	25.6655	1.2320	74.14	21.69	487.12	96.21
332	7.3546	10.1435	716.88	451.13	24.7421	1.2319	73.74	21.78	485.07	97.59
333	7.4435	10.3612	709.81	449.90	23.8387	1.2319	73.34	21.87	483.03	99.03
334	7.5370	10.5897	702.62	448.65	22.9549	1.2319	72.93	21.96	480.97	100.52
335	7.6354	10.8300	695.31	447.37	22.0904	1.2320	72.52	22.05	478.91	102.07
336	7.7391	11.0829	687.88	446.07	21.2450	1.2321	72.11	22.15	476.84	103.69
337	7.8485	11.3495	680.33	444.74	20.4188	1.2322	71.70	22.24	474.76	105.38
338	7.9642	11.6306	672.67	443.39	19.6118	1.2324	71.28	22.34	472.68	107.14
339	8.0867	11.9274	664.90	442.00	18.8242	1.2327	70.86	22.45	470.59	108.97
340	8.2166	12.2412	657.03	440.59	18.0563	1.2329	70.43	22.55	468.48	110.89
341	8.3545	12.5732	649.08	439.15	17.3085	1.2333	70.00	22.66	466.37	112.90
342	8.5012	12.9247	641.05	437.67	16.5812	1.2337	69.57	22.78	464.26	115.00
343	8.6575	13.2973	632.96	436.17	15.8750	1.2341	69.13	22.89	462.13	117.19
344	8.8242	13.6925	624.84	434.63	15.1905	1.2346	68.68	23.01	460.00	119.49
345	9.0023	14.1120	616.71	433.05	14.5283	1.2352	68.23	23.13	457.87	121.90
346	9.1927	14.5576	608.60	431.44	13.8890	1.2359	67.77	23.26	455.73	124.43
347	9.3967	15.0312	600.53	429.80	13.2731	1.2366	67.31	23.39	453.59	127.08
348	9.6153	15.5349	592.54	428.11	12.6814	1.2375	66.84	23.53	451.46	129.86
349	9.8499	16.0709	584.65	426.39	12.1142	1.2384	66.36	23.67	449.33	132.78
350	10.1020	16.6415	576.91	424.63	11.5719	1.2395	65.87	23.82	447.21	135.85
351	10.4079	17.3581	567.48	422.65	10.9813	1.2395	65.38	23.97	445.07	139.29
352	10.7203	18.0835	558.74	420.82	10.4394	1.2410	64.87	24.13	442.92	142.85
353	11.0629	18.8772	549.88	418.90	9.9124	1.2423	64.36	24.30	440.76	146.63
354	11.4402	19.7494	540.88	416.89	9.4001	1.2436	63.83	24.48	438.61	150.66
355	11.8581	20.7121	531.75	414.80	8.9020	1.2449	63.29	24.66	436.45	154.96
356	12.3234	21.7804	522.47	412.61	8.4179	1.2462	62.74	24.85	434.31	159.58
357	12.8451	22.9727	513.03	410.32	7.9475	1.2474	62.17	25.05	432.18	164.54
358	13.4342	24.3122	503.42	407.92	7.4905	1.2487	61.58	25.26	430.08	169.90
359	14.1048	25.8285	493.64	405.41	7.0465	1.2500	60.98	25.49	428.01	175.70

Table 1 Saturation State (Temperature Table) – Continuation
Sättigungszustand (Temperaturtafel) – Fortsetzung

t [°C]	T [K]	p_s [MPa]	v' [m³ kg⁻¹]	v''	h' [kJ kg⁻¹]	h''	Δh_v	s' [kJ kg⁻¹ K⁻¹]	s''	Δs_v
360	633.15	18.666	0.001895	0.006946	1761.50	2481.03	719.54	3.9164	5.0528	1.1364
361	634.15	18.893	0.001915	0.006761	1771.89	2470.59	698.70	3.9321	5.0339	1.1018
362	635.15	19.121	0.001937	0.006575	1782.63	2459.56	676.93	3.9483	5.0141	1.0658
363	636.15	19.352	0.001961	0.006387	1793.79	2447.88	654.09	3.9652	4.9934	1.0282
364	637.15	19.586	0.001987	0.006198	1805.42	2435.44	630.02	3.9827	4.9715	0.9888
365	638.15	19.822	0.002016	0.006006	1817.61	2422.12	604.51	4.0011	4.9484	0.9473
366	639.15	20.061	0.002047	0.005809	1830.46	2407.75	577.29	4.0204	4.9237	0.9032
367	640.15	20.302	0.002082	0.005608	1844.12	2392.11	547.99	4.0410	4.8971	0.8560
368	641.15	20.546	0.002122	0.005399	1858.80	2374.89	516.09	4.0631	4.8681	0.8050
369	642.15	20.793	0.002168	0.005180	1874.81	2355.63	480.82	4.0873	4.8360	0.7488
370	643.15	21.043	0.002222	0.004947	1892.66	2333.60	440.94	4.1142	4.7998	0.6856
371	644.15	21.296	0.002290	0.004692	1913.26	2307.51	394.26	4.1453	4.7573	0.6121
372	645.15	21.553	0.002382	0.004398	1938.50	2274.60	336.10	4.1835	4.7045	0.5210
373	646.15	21.813	0.002526	0.004019	1974.01	2227.30	253.28	4.2375	4.6295	0.3920
373.946	647.096	22.064	0.003106		2087.55		0.00	4.4120		0.0000

Table 1 Saturation State (Temperature Table) – Continuation
Sättigungszustand (Temperaturtafel) – Fortsetzung

t [°C]	c_p' [kJ kg⁻¹ K⁻¹]	c_p''	w' [m s⁻¹]	w''	κ'	κ''	η' [10⁻⁶ kg m⁻¹ s⁻¹]	η''	λ' [10⁻³ W m⁻¹ K⁻¹]	λ''
360	14.8753	27.5595	483.68	402.78	6.6154	1.2513	60.36	25.73	426.00	182.01
361	15.7701	29.5550	473.52	400.01	6.1968	1.2527	59.71	25.98	424.08	188.93
362	16.8221	31.8817	463.15	397.10	5.7904	1.2543	59.04	26.26	422.27	196.55
363	18.0767	34.6308	452.56	394.02	5.3959	1.2560	58.34	26.55	420.63	205.01
364	19.5987	37.9307	441.72	390.76	5.0129	1.2579	57.60	26.87	419.22	214.49
365	21.4835	41.9679	430.62	387.29	4.6411	1.2600	56.83	27.21	418.16	225.25
366	23.8767	47.0237	419.23	383.57	4.2800	1.2625	56.00	27.59	417.59	237.62
367	27.0134	53.5439	407.52	379.54	3.9290	1.2653	55.12	28.02	417.77	252.11
368	31.2966	62.2786	395.45	375.15	3.5873	1.2688	54.16	28.50	419.09	269.51
369	37.4769	74.5952	382.96	370.29	3.2540	1.2730	53.10	29.05	422.28	291.09
370	47.1222	93.2737	369.98	364.80	2.9272	1.2783	51.92	29.70	428.72	319.17
371	64.1039	124.9504	356.40	358.39	2.6043	1.2855	50.54	30.50	441.51	358.55
372	101.0191	190.6467	342.05	350.48	2.2794	1.2959	48.83	31.57	469.31	421.89
373	230.6909	403.6750	326.65	339.51	1.9367	1.3147	46.43	33.23	550.51	559.58

2 Saturation State (Pressure Table)

Sättigungszustand (Drucktafel)

The pressure table of the saturation states contains values for the saturated liquid (′) and saturated vapour (″) of the following thermodynamic properties in the pressure range from the triple-point pressure $p_t = 611.657$ Pa to the critical pressure $p_c = 22.064$ MPa:

- Saturation temperature t_s
- Specific volume v
- Specific enthalpy h
- Specific enthalpy of evaporation Δh_v
- Specific entropy s
- Specific entropy of evaporation Δs_v.

For further properties please pay attention to Table 1. For pressures $p \leq 16.5292$ MPa the saturation temperature t_s was calculated from the saturation-temperature equation of IAPWS-IF97, Eq. (3.24). For pressures $p > 16.5292$ MPa (region 3 of IAPWS-IF97) the saturation temperature and the specific volumes of the saturated liquid and saturated vapour, v' and v'', were determined from Eq. (3.21) via the phase-equilibrium condition (see Table 3.31). If these volumes were calculated by iterations for given p and t_s, the results would be slightly different. With p and t_s (for region 3 of IAPWS-IF97 with t_s and v' or t_s and v'') all thermodynamic properties were determined from the corresponding equations of the industrial standard IAPWS-IF97 presented in Chap. 3.

Die Drucktafel der Sättigungszustände enthält für die siedende Flüssigkeit (′) und den gesättigten Dampf (″) Werte der folgenden thermodynamischen Eigenschaften für Drücke vom Tripelpunktsdruck $p_t = 611{,}657$ Pa bis zum kritischen Druck $p_c = 22{,}064$ MPa:

- Siedetemperatur t_s
- spezifisches Volumen v
- spezifische Enthalpie h
- spezifische Verdampfungsenthalpie Δh_v
- spezifische Entropie s
- spezifische Verdampfungsentropie Δs_v.

Weitere Zustandsgrößen sind in Tafel 1 aufgeführt. Die Siedetemperatur t_s wurde für vorgegebene Drücke $p \leq 16{,}5292$ MPa direkt aus der Siedetemperaturgleichung der IAPWS-IF97, Gl. (3.24), berechnet. Für Drücke $p > 16{,}5292$ MPa (Bereich 3 der IAPWS-IF97) wurden die Siedetemperatur und die spezifischen Volumina der siedenden Flüssigkeit und des gesättigten Dampfes, v' und v'', aus Gl. (3.21) mit Hilfe der Phasengleichgewichtsbedingung (s. Tabelle 3.31) bestimmt. Würde man die Volumina für vorgegebene Werte von p and t_s iterieren, erhielte man geringfügig andere Ergebnisse. Für p and t_s (im Bereich 3 der IAPWS-IF97 für t_s und v' bzw. t_s und v'') wurden alle thermodynamischen Zustandsgrößen mit den in Kap. 3 angegebenen Gleichungen des Industrie-Standards IAPWS-IF97 berechnet.

Table 2 Saturation State (Pressure Table) Sättigungszustand (Drucktafel)

p [MPa]	t_s [°C]	v' [m³ kg⁻¹]	v''	h'	h'' [kJ kg⁻¹]	Δh_v	s'	s'' [kJ kg⁻¹ K⁻¹]	Δs_v
0.000611657	0.01	0.001000	205.997	0.001	2500.91	2500.91	0.0000	9.1555	9.1555
0.0007	1.88	0.001000	181.223	7.890	2504.35	2496.46	0.0288	9.1058	9.0770
0.0008	3.76	0.001000	159.646	15.809	2507.80	2491.99	0.0575	9.0567	8.9992
0.0009	5.44	0.001000	142.763	22.888	2510.89	2488.00	0.0830	9.0135	8.9305
0.0010	6.97	0.001000	129.183	29.298	2513.68	2484.38	0.1059	8.9749	8.8690
0.0015	13.02	0.001001	87.962	54.685	2524.75	2470.06	0.1956	8.8270	8.6315
0.0020	17.50	0.001001	66.990	73.435	2532.91	2459.48	0.2606	8.7227	8.4621
0.0025	21.08	0.001002	54.242	88.430	2539.43	2451.00	0.3119	8.6422	8.3303
0.0030	24.08	0.001003	45.655	100.990	2544.88	2443.89	0.3543	8.5766	8.2222
0.0035	26.67	0.001003	39.468	111.836	2549.57	2437.74	0.3907	8.5213	8.1306
0.0040	28.96	0.001004	34.792	121.404	2553.71	2432.31	0.4224	8.4735	8.0510
0.0045	31.01	0.001005	31.132	129.981	2557.41	2427.43	0.4507	8.4314	7.9807
0.0050	32.88	0.001005	28.186	137.765	2560.77	2423.00	0.4763	8.3939	7.9177
0.0055	34.58	0.001006	25.763	144.901	2563.83	2418.93	0.4995	8.3600	7.8605
0.0060	36.16	0.001006	23.734	151.494	2566.67	2415.17	0.5209	8.3291	7.8083
0.0065	37.63	0.001007	22.010	157.627	2569.30	2411.67	0.5407	8.3008	7.7601
0.0070	39.00	0.001007	20.525	163.366	2571.76	2408.39	0.5591	8.2746	7.7155
0.0075	40.29	0.001008	19.234	168.760	2574.06	2405.30	0.5763	8.2502	7.6739
0.0080	41.51	0.001008	18.099	173.852	2576.24	2402.39	0.5925	8.2274	7.6349
0.0085	42.66	0.001009	17.095	178.676	2578.30	2399.62	0.6078	8.2060	7.5982
0.0090	43.76	0.001009	16.200	183.262	2580.25	2396.99	0.6223	8.1859	7.5636
0.0095	44.81	0.001010	15.396	187.634	2582.11	2394.48	0.6361	8.1669	7.5308
0.0100	45.81	0.001010	14.671	191.812	2583.89	2392.07	0.6492	8.1489	7.4997
0.0150	53.97	0.001014	10.020	225.935	2598.30	2372.37	0.7548	8.0071	7.2523
0.0200	60.06	0.001017	7.648	251.400	2608.95	2357.55	0.8320	7.9072	7.0753
0.0250	64.96	0.001020	6.203	271.925	2617.45	2345.52	0.8931	7.8302	6.9371
0.0300	69.10	0.001022	5.229	289.229	2624.55	2335.32	0.9439	7.7675	6.8235
0.0350	72.68	0.001024	4.525	304.251	2630.67	2326.42	0.9876	7.7146	6.7270
0.0400	75.86	0.001026	3.993	317.566	2636.05	2318.48	1.0259	7.6690	6.6431
0.0450	78.71	0.001028	3.576	329.554	2640.86	2311.31	1.0601	7.6288	6.5687
0.0500	81.32	0.001030	3.240	340.476	2645.21	2304.74	1.0910	7.5930	6.5020
0.0550	83.71	0.001032	2.964	350.523	2649.19	2298.67	1.1192	7.5606	6.4414
0.0600	85.93	0.001033	2.732	359.837	2652.85	2293.02	1.1452	7.5311	6.3859
0.0650	87.99	0.001035	2.535	368.527	2656.25	2287.72	1.1694	7.5040	6.3346
0.0700	89.93	0.001036	2.365	376.680	2659.42	2282.74	1.1919	7.4790	6.2871
0.0750	91.76	0.001037	2.217	384.365	2662.39	2278.02	1.2130	7.4557	6.2427
0.0800	93.49	0.001038	2.087	391.639	2665.18	2273.54	1.2328	7.4339	6.2011
0.0850	95.13	0.001040	1.972	398.547	2667.82	2269.27	1.2516	7.4135	6.1618
0.0900	96.69	0.001041	1.869	405.128	2670.31	2265.19	1.2694	7.3942	6.1248
0.0950	98.18	0.001042	1.777	411.415	2672.69	2261.27	1.2864	7.3760	6.0897
0.1000	99.61	0.001043	1.694	417.436	2674.95	2257.51	1.3026	7.3588	6.0562
0.1100	102.29	0.001045	1.550	428.775	2679.18	2250.40	1.3328	7.3268	5.9940
0.1200	104.78	0.001047	1.428	439.299	2683.06	2243.76	1.3608	7.2976	5.9369
0.1300	107.11	0.001049	1.325	449.132	2686.65	2237.52	1.3867	7.2708	5.8842
0.1400	109.29	0.001051	1.237	458.367	2689.99	2231.62	1.4109	7.2460	5.8352

Table 2 Saturation State (Pressure Table) – Continuation
Sättigungszustand (Drucktafel) – Fortsetzung

p	t_s	v'	v''	h'	h''	Δh_v	s'	s''	Δs_v
[MPa]	[°C]	[$m^3\ kg^{-1}$]			[$kJ\ kg^{-1}$]			[$kJ\ kg^{-1}\ K^{-1}$]	
0.15	111.35	0.001053	1.1594	467.08	2693.11	2226.03	1.4335	7.2229	5.7894
0.16	113.30	0.001054	1.0914	475.34	2696.04	2220.71	1.4549	7.2014	5.7464
0.17	115.15	0.001056	1.0312	483.18	2698.81	2215.62	1.4752	7.1811	5.7059
0.18	116.91	0.001058	0.9775	490.67	2701.42	2210.75	1.4944	7.1620	5.6677
0.19	118.60	0.001059	0.9293	497.82	2703.89	2206.07	1.5127	7.1440	5.6313
0.20	120.21	0.001061	0.8857	504.68	2706.24	2201.56	1.5301	7.1269	5.5968
0.21	121.76	0.001062	0.8462	511.27	2708.48	2197.21	1.5468	7.1106	5.5638
0.22	123.25	0.001063	0.8101	517.62	2710.62	2193.00	1.5628	7.0951	5.5323
0.23	124.69	0.001065	0.7771	523.73	2712.66	2188.93	1.5782	7.0802	5.5021
0.24	126.07	0.001066	0.7467	529.64	2714.62	2184.98	1.5930	7.0660	5.4731
0.25	127.41	0.001067	0.7187	535.35	2716.50	2181.15	1.6072	7.0524	5.4452
0.26	128.71	0.001068	0.6928	540.88	2718.31	2177.42	1.6210	7.0393	5.4183
0.27	129.97	0.001070	0.6687	546.25	2720.04	2173.79	1.6343	7.0267	5.3924
0.28	131.19	0.001071	0.6463	551.46	2721.72	2170.26	1.6472	7.0146	5.3674
0.29	132.37	0.001072	0.6254	556.53	2723.33	2166.81	1.6597	7.0029	5.3432
0.30	133.53	0.001073	0.6058	561.46	2724.89	2163.44	1.6718	6.9916	5.3198
0.31	134.65	0.001074	0.5874	566.26	2726.40	2160.14	1.6835	6.9806	5.2971
0.32	135.74	0.001075	0.5702	570.93	2727.86	2156.92	1.6950	6.9700	5.2751
0.33	136.81	0.001076	0.5540	575.50	2729.27	2153.77	1.7061	6.9597	5.2537
0.34	137.85	0.001078	0.5387	579.96	2730.64	2150.68	1.7169	6.9498	5.2329
0.35	138.86	0.001079	0.5242	584.31	2731.97	2147.65	1.7275	6.9401	5.2126
0.36	139.85	0.001080	0.5105	588.57	2733.25	2144.68	1.7378	6.9307	5.1929
0.37	140.82	0.001081	0.4975	592.74	2734.51	2141.77	1.7478	6.9215	5.1737
0.38	141.77	0.001082	0.4852	596.81	2735.72	2138.91	1.7576	6.9126	5.1550
0.39	142.70	0.001083	0.4735	600.81	2736.91	2136.10	1.7672	6.9039	5.1367
0.40	143.61	0.001084	0.4624	604.72	2738.06	2133.33	1.7766	6.8954	5.1188
0.41	144.50	0.001085	0.4518	608.56	2739.18	2130.62	1.7858	6.8872	5.1014
0.42	145.38	0.001085	0.4417	612.33	2740.27	2127.94	1.7948	6.8791	5.0843
0.43	146.24	0.001086	0.4320	616.03	2741.33	2125.31	1.8036	6.8712	5.0676
0.44	147.08	0.001087	0.4227	619.66	2742.37	2122.72	1.8122	6.8635	5.0513
0.45	147.91	0.001088	0.4139	623.22	2743.39	2120.16	1.8206	6.8560	5.0353
0.46	148.72	0.001089	0.4054	626.73	2744.38	2117.65	1.8289	6.8486	5.0197
0.47	149.52	0.001090	0.3973	630.18	2745.34	2115.16	1.8371	6.8414	5.0043
0.48	150.30	0.001091	0.3895	633.57	2746.28	2112.72	1.8450	6.8343	4.9892
0.49	151.08	0.001092	0.3820	636.90	2747.21	2110.30	1.8529	6.8274	4.9745
0.50	151.84	0.001093	0.3748	640.19	2748.11	2107.92	1.8606	6.8206	4.9600
0.52	153.32	0.001094	0.3612	646.60	2749.85	2103.25	1.8756	6.8074	4.9318
0.54	154.76	0.001096	0.3486	652.83	2751.52	2098.69	1.8901	6.7947	4.9045
0.56	156.15	0.001097	0.3368	658.88	2753.12	2094.24	1.9042	6.7824	4.8782
0.58	157.51	0.001099	0.3258	664.77	2754.66	2089.89	1.9179	6.7706	4.8528
0.60	158.83	0.001101	0.3156	670.50	2756.14	2085.64	1.9311	6.7592	4.8281
0.62	160.12	0.001102	0.3059	676.09	2757.56	2081.47	1.9440	6.7481	4.8041
0.64	161.37	0.001104	0.2969	681.54	2758.93	2077.39	1.9565	6.7374	4.7809
0.66	162.59	0.001105	0.2884	686.86	2760.25	2073.39	1.9686	6.7269	4.7583
0.68	163.79	0.001107	0.2804	692.06	2761.52	2069.46	1.9805	6.7168	4.7363

Table 2 Saturation State (Pressure Table) – Continuation
Sättigungszustand (Drucktafel) – Fortsetzung

p [MPa]	t_s [°C]	v'	v'' [m³ kg⁻¹]	h'	h'' [kJ kg⁻¹]	Δh_v	s'	s'' [kJ kg⁻¹ K⁻¹]	Δs_v
0.70	164.95	0.001108	0.272764	697.14	2762.75	2065.61	1.9921	6.7070	4.7149
0.72	166.09	0.001109	0.265582	702.12	2763.94	2061.82	2.0034	6.6974	4.6940
0.74	167.21	0.001111	0.258775	706.99	2765.08	2058.10	2.0144	6.6881	4.6737
0.76	168.30	0.001112	0.252314	711.76	2766.19	2054.43	2.0252	6.6790	4.6539
0.78	169.37	0.001113	0.246172	716.43	2767.26	2050.83	2.0357	6.6702	4.6345
0.80	170.41	0.001115	0.240328	721.02	2768.30	2047.28	2.0460	6.6615	4.6156
0.82	171.44	0.001116	0.234758	725.52	2769.31	2043.79	2.0561	6.6531	4.5970
0.84	172.45	0.001117	0.229445	729.93	2770.28	2040.35	2.0659	6.6449	4.5789
0.86	173.43	0.001119	0.224370	734.27	2771.23	2036.96	2.0756	6.6368	4.5612
0.88	174.41	0.001120	0.219518	738.53	2772.15	2033.61	2.0851	6.6289	4.5438
0.90	175.36	0.001121	0.214874	742.72	2773.04	2030.31	2.0944	6.6212	4.5268
0.92	176.29	0.001122	0.210425	746.85	2773.90	2027.06	2.1035	6.6137	4.5102
0.94	177.21	0.001124	0.206158	750.90	2774.74	2023.84	2.1125	6.6063	4.4938
0.96	178.12	0.001125	0.202064	754.89	2775.56	2020.67	2.1213	6.5991	4.4778
0.98	179.01	0.001126	0.198130	758.82	2776.35	2017.53	2.1299	6.5919	4.4620
1.00	179.89	0.001127	0.194349	762.68	2777.12	2014.44	2.1384	6.5850	4.4465
1.10	184.07	0.001133	0.177436	781.20	2780.67	1999.47	2.1789	6.5520	4.3731
1.20	187.96	0.001139	0.163250	798.50	2783.77	1985.27	2.2163	6.5217	4.3054
1.30	191.61	0.001144	0.151175	814.76	2786.49	1971.73	2.2512	6.4936	4.2425
1.40	195.05	0.001149	0.140768	830.13	2788.89	1958.76	2.2839	6.4675	4.1836
1.50	198.30	0.001154	0.131702	844.72	2791.01	1946.29	2.3147	6.4431	4.1284
1.60	201.38	0.001159	0.123732	858.61	2792.88	1934.27	2.3438	6.4200	4.0762
1.70	204.31	0.001163	0.116668	871.89	2794.53	1922.64	2.3715	6.3983	4.0268
1.80	207.12	0.001168	0.110362	884.61	2795.99	1911.37	2.3978	6.3776	3.9798
1.90	209.81	0.001172	0.104698	896.84	2797.26	1900.42	2.4229	6.3579	3.9350
2.00	212.38	0.001177	0.099581	908.62	2798.38	1889.76	2.4470	6.3392	3.8921
2.10	214.87	0.001181	0.094934	919.99	2799.36	1879.37	2.4701	6.3212	3.8511
2.20	217.26	0.001185	0.090695	930.98	2800.20	1869.22	2.4924	6.3040	3.8116
2.30	219.56	0.001189	0.086812	941.63	2800.92	1859.30	2.5138	6.2874	3.7736
2.40	221.80	0.001193	0.083242	951.95	2801.54	1849.58	2.5344	6.2714	3.7370
2.50	223.96	0.001197	0.079947	961.98	2802.04	1840.06	2.5544	6.2560	3.7015
2.60	226.05	0.001201	0.076897	971.74	2802.45	1830.71	2.5738	6.2411	3.6673
2.70	228.09	0.001205	0.074065	981.24	2802.78	1821.54	2.5925	6.2266	3.6341
2.80	230.06	0.001209	0.071428	990.50	2803.02	1812.51	2.6107	6.2126	3.6019
2.90	231.99	0.001213	0.068967	999.54	2803.18	1803.63	2.6284	6.1990	3.5706
3.00	233.86	0.001217	0.066664	1008.37	2803.26	1794.89	2.6456	6.1858	3.5402
3.10	235.68	0.001220	0.064504	1017.00	2803.28	1786.28	2.6624	6.1729	3.5105
3.20	237.46	0.001224	0.062475	1025.45	2803.24	1777.79	2.6787	6.1604	3.4817
3.30	239.20	0.001228	0.060564	1033.72	2803.13	1769.41	2.6946	6.1481	3.4535
3.40	240.90	0.001231	0.058761	1041.83	2802.96	1761.14	2.7102	6.1362	3.4260
3.50	242.56	0.001235	0.057058	1049.78	2802.74	1752.97	2.7254	6.1245	3.3991
3.60	244.19	0.001239	0.055446	1057.57	2802.47	1744.90	2.7403	6.1131	3.3728
3.70	245.78	0.001242	0.053918	1065.23	2802.15	1736.91	2.7548	6.1019	3.3471
3.80	247.33	0.001246	0.052468	1072.76	2801.78	1729.02	2.7690	6.0910	3.3219
3.90	248.86	0.001249	0.051089	1080.15	2801.36	1721.21	2.7830	6.0802	3.2973

Table 2 Saturation State (Pressure Table) – Continuation
Sättigungszustand (Drucktafel) – Fortsetzung

p [MPa]	t_s [°C]	v' [$m^3\ kg^{-1}$]	v''	h'	h'' [$kJ\ kg^{-1}$]	Δh_v	s'	s'' [$kJ\ kg^{-1}\ K^{-1}$]	Δs_v
4.0	250.36	0.001253	0.049777	1087.43	2800.90	1713.47	2.7967	6.0697	3.2731
4.1	251.83	0.001256	0.048526	1094.58	2800.39	1705.81	2.8101	6.0594	3.2493
4.2	253.27	0.001259	0.047333	1101.63	2799.85	1698.22	2.8232	6.0492	3.2260
4.3	254.68	0.001263	0.046193	1108.57	2799.27	1690.70	2.8362	6.0393	3.2031
4.4	256.07	0.001266	0.045103	1115.40	2798.65	1683.25	2.8488	6.0294	3.1806
4.5	257.44	0.001270	0.044059	1122.14	2798.00	1675.85	2.8613	6.0198	3.1585
4.6	258.78	0.001273	0.043060	1128.79	2797.31	1668.52	2.8736	6.0103	3.1367
4.7	260.10	0.001276	0.042101	1135.34	2796.59	1661.24	2.8857	6.0010	3.1153
4.8	261.40	0.001280	0.041181	1141.81	2795.83	1654.02	2.8975	5.9917	3.0942
4.9	262.68	0.001283	0.040296	1148.20	2795.04	1646.85	2.9092	5.9827	3.0734
5.0	263.94	0.001286	0.039446	1154.50	2794.23	1639.73	2.9207	5.9737	3.0530
5.1	265.18	0.001290	0.038628	1160.73	2793.38	1632.65	2.9321	5.9649	3.0328
5.2	266.41	0.001293	0.037840	1166.88	2792.51	1625.62	2.9433	5.9562	3.0129
5.3	267.61	0.001296	0.037081	1172.97	2791.60	1618.64	2.9543	5.9475	2.9933
5.4	268.80	0.001300	0.036349	1178.98	2790.67	1611.69	2.9652	5.9390	2.9739
5.5	269.97	0.001303	0.035642	1184.92	2789.72	1604.79	2.9759	5.9307	2.9548
5.6	271.12	0.001306	0.034960	1190.81	2788.74	1597.93	2.9865	5.9224	2.9359
5.7	272.26	0.001309	0.034300	1196.63	2787.73	1591.10	2.9969	5.9141	2.9173
5.8	273.38	0.001313	0.033663	1202.39	2786.70	1584.31	3.0072	5.9060	2.8988
5.9	274.49	0.001316	0.033046	1208.09	2785.64	1577.55	3.0174	5.8980	2.8806
6.0	275.59	0.001319	0.032449	1213.73	2784.56	1570.83	3.0274	5.8901	2.8626
6.1	276.67	0.001323	0.031870	1219.32	2783.46	1564.14	3.0374	5.8822	2.8448
6.2	277.73	0.001326	0.031310	1224.86	2782.33	1557.48	3.0472	5.8744	2.8272
6.3	278.79	0.001329	0.030766	1230.34	2781.19	1550.84	3.0569	5.8667	2.8098
6.4	279.83	0.001332	0.030239	1235.78	2780.02	1544.24	3.0665	5.8591	2.7926
6.5	280.86	0.001336	0.029728	1241.17	2778.83	1537.66	3.0760	5.8515	2.7755
6.6	281.88	0.001339	0.029231	1246.51	2777.62	1531.11	3.0854	5.8440	2.7586
6.7	282.88	0.001342	0.028748	1251.81	2776.39	1524.58	3.0947	5.8366	2.7419
6.8	283.88	0.001345	0.028279	1257.06	2775.13	1518.07	3.1039	5.8292	2.7253
6.9	284.86	0.001349	0.027823	1262.27	2773.86	1511.59	3.1130	5.8219	2.7089
7.0	285.83	0.001352	0.027380	1267.44	2772.57	1505.13	3.1220	5.8146	2.6926
7.1	286.79	0.001355	0.026948	1272.57	2771.26	1498.69	3.1309	5.8074	2.6765
7.2	287.74	0.001358	0.026528	1277.65	2769.93	1492.27	3.1398	5.8003	2.6605
7.3	288.68	0.001362	0.026119	1282.70	2768.58	1485.87	3.1485	5.7932	2.6447
7.4	289.62	0.001365	0.025720	1287.72	2767.21	1479.49	3.1572	5.7862	2.6290
7.5	290.54	0.001368	0.025331	1292.70	2765.82	1473.12	3.1658	5.7792	2.6134
7.6	291.45	0.001371	0.024952	1297.64	2764.41	1466.78	3.1743	5.7722	2.5979
7.7	292.35	0.001375	0.024583	1302.55	2762.99	1460.44	3.1827	5.7653	2.5826
7.8	293.25	0.001378	0.024223	1307.42	2761.55	1454.12	3.1911	5.7584	2.5673
7.9	294.13	0.001381	0.023871	1312.27	2760.09	1447.82	3.1994	5.7516	2.5522
8.0	295.01	0.001385	0.023528	1317.08	2758.61	1441.53	3.2077	5.7448	2.5372
8.1	295.88	0.001388	0.023192	1321.86	2757.12	1435.25	3.2158	5.7381	2.5223
8.2	296.74	0.001391	0.022865	1326.61	2755.60	1428.99	3.2239	5.7314	2.5075
8.3	297.59	0.001395	0.022545	1331.34	2754.07	1422.74	3.2320	5.7247	2.4928
8.4	298.44	0.001398	0.022232	1336.03	2752.52	1416.49	3.2399	5.7181	2.4782

Table 2 Saturation State (Pressure Table) – Continuation
Sättigungszustand (Drucktafel) – Fortsetzung

p [MPa]	t_s [°C]	v' [$m^3\ kg^{-1}$]	v'' [$m^3\ kg^{-1}$]	h' [$kJ\ kg^{-1}$]	h'' [$kJ\ kg^{-1}$]	Δh_v [$kJ\ kg^{-1}$]	s' [$kJ\ kg^{-1}\ K^{-1}$]	s'' [$kJ\ kg^{-1}\ K^{-1}$]	Δs_v [$kJ\ kg^{-1}\ K^{-1}$]
8.5	299.27	0.001401	0.021926	1340.70	2750.96	1410.26	3.2478	5.7115	2.4637
8.6	300.10	0.001405	0.021627	1345.34	2749.38	1404.04	3.2557	5.7050	2.4493
8.7	300.92	0.001408	0.021334	1349.96	2747.78	1397.82	3.2635	5.6984	2.4349
8.8	301.74	0.001411	0.021048	1354.54	2746.16	1391.62	3.2712	5.6919	2.4207
8.9	302.55	0.001415	0.020767	1359.11	2744.53	1385.42	3.2789	5.6855	2.4065
9.0	303.35	0.001418	0.020493	1363.65	2742.88	1379.23	3.2866	5.6790	2.3924
9.1	304.14	0.001422	0.020224	1368.17	2741.22	1373.05	3.2942	5.6726	2.3784
9.2	304.93	0.001425	0.019961	1372.66	2739.53	1366.87	3.3017	5.6662	2.3645
9.3	305.71	0.001428	0.019703	1377.14	2737.83	1360.70	3.3092	5.6598	2.3507
9.4	306.48	0.001432	0.019450	1381.59	2736.12	1354.53	3.3166	5.6535	2.3369
9.5	307.25	0.001435	0.019203	1386.02	2734.38	1348.37	3.3240	5.6472	2.3232
9.6	308.01	0.001439	0.018960	1390.43	2732.64	1342.21	3.3313	5.6409	2.3095
9.7	308.77	0.001442	0.018721	1394.81	2730.87	1336.06	3.3386	5.6346	2.2960
9.8	309.52	0.001446	0.018488	1399.18	2729.09	1329.90	3.3459	5.6283	2.2824
9.9	310.26	0.001449	0.018259	1403.54	2727.29	1323.75	3.3531	5.6221	2.2690
10.0	311.00	0.001453	0.018034	1407.87	2725.47	1317.61	3.3603	5.6159	2.2556
10.1	311.73	0.001456	0.017813	1412.18	2723.64	1311.46	3.3674	5.6097	2.2423
10.2	312.46	0.001460	0.017596	1416.48	2721.79	1305.31	3.3745	5.6035	2.2290
10.3	313.18	0.001463	0.017383	1420.76	2719.93	1299.17	3.3816	5.5973	2.2158
10.4	313.90	0.001467	0.017174	1425.02	2718.04	1293.02	3.3886	5.5912	2.2026
10.5	314.61	0.001470	0.016969	1429.27	2716.14	1286.88	3.3956	5.5850	2.1895
10.6	315.31	0.001474	0.016767	1433.50	2714.23	1280.73	3.4025	5.5789	2.1764
10.7	316.01	0.001478	0.016569	1437.72	2712.30	1274.58	3.4094	5.5728	2.1634
10.8	316.71	0.001481	0.016374	1441.92	2710.35	1268.43	3.4163	5.5667	2.1504
10.9	317.40	0.001485	0.016182	1446.11	2708.38	1262.27	3.4231	5.5606	2.1375
11.0	318.08	0.001489	0.015994	1450.28	2706.39	1256.12	3.4300	5.5545	2.1246
11.1	318.76	0.001492	0.015809	1454.44	2704.39	1249.96	3.4367	5.5485	2.1117
11.2	319.44	0.001496	0.015626	1458.58	2702.37	1243.79	3.4435	5.5424	2.0989
11.3	320.11	0.001500	0.015447	1462.72	2700.34	1237.62	3.4502	5.5363	2.0861
11.4	320.77	0.001503	0.015271	1466.84	2698.28	1231.45	3.4569	5.5303	2.0734
11.5	321.44	0.001507	0.015097	1470.95	2696.21	1225.26	3.4636	5.5243	2.0607
11.6	322.09	0.001511	0.014926	1475.05	2694.12	1219.08	3.4702	5.5182	2.0480
11.7	322.75	0.001515	0.014758	1479.13	2692.02	1212.88	3.4768	5.5122	2.0354
11.8	323.39	0.001519	0.014593	1483.21	2689.89	1206.68	3.4834	5.5062	2.0228
11.9	324.04	0.001522	0.014430	1487.27	2687.75	1200.47	3.4899	5.5001	2.0102
12.0	324.68	0.001526	0.014269	1491.33	2685.58	1194.26	3.4965	5.4941	1.9977
12.1	325.31	0.001530	0.014111	1495.37	2683.40	1188.03	3.5030	5.4881	1.9851
12.2	325.95	0.001534	0.013955	1499.41	2681.20	1181.79	3.5095	5.4821	1.9726
12.3	326.57	0.001538	0.013801	1503.43	2678.98	1175.55	3.5159	5.4761	1.9601
12.4	327.20	0.001542	0.013650	1507.45	2676.74	1169.29	3.5224	5.4700	1.9477
12.5	327.82	0.001546	0.013501	1511.46	2674.49	1163.02	3.5288	5.4640	1.9353
12.6	328.43	0.001550	0.013354	1515.47	2672.21	1156.74	3.5352	5.4580	1.9228
12.7	329.04	0.001554	0.013208	1519.46	2669.91	1150.45	3.5415	5.4520	1.9104
12.8	329.65	0.001558	0.013065	1523.45	2667.59	1144.14	3.5479	5.4459	1.8980
12.9	330.26	0.001562	0.012924	1527.43	2665.25	1137.82	3.5543	5.4399	1.8857

Table 2 Saturation State (Pressure Table) – Continuation
Sättigungszustand (Drucktafel) – Fortsetzung

p [MPa]	t_s [°C]	v' [$m^3 kg^{-1}$]	v''	h' [$kJ kg^{-1}$]	h''	Δh_v	s' [$kJ kg^{-1} K^{-1}$]	s''	Δs_v
13.0	330.86	0.001566	0.012785	1531.40	2662.89	1131.49	3.5606	5.4339	1.8733
13.1	331.45	0.001571	0.012648	1535.37	2660.51	1125.14	3.5669	5.4278	1.8610
13.2	332.05	0.001575	0.012512	1539.33	2658.11	1118.78	3.5732	5.4218	1.8486
13.3	332.64	0.001579	0.012379	1543.29	2655.69	1112.40	3.5794	5.4157	1.8363
13.4	333.22	0.001583	0.012247	1547.24	2653.24	1106.00	3.5857	5.4097	1.8240
13.5	333.81	0.001588	0.012116	1551.19	2650.77	1099.58	3.5920	5.4036	1.8116
13.6	334.39	0.001592	0.011988	1555.14	2648.28	1093.15	3.5982	5.3975	1.7993
13.7	334.96	0.001596	0.011861	1559.08	2645.77	1086.70	3.6044	5.3914	1.7870
13.8	335.53	0.001601	0.011735	1563.01	2643.24	1080.22	3.6106	5.3853	1.7747
13.9	336.10	0.001605	0.011611	1566.95	2640.68	1073.73	3.6168	5.3792	1.7624
14.0	336.67	0.001610	0.011489	1570.88	2638.09	1067.21	3.6230	5.3730	1.7500
14.1	337.23	0.001614	0.011368	1574.81	2635.49	1060.68	3.6292	5.3669	1.7377
14.2	337.79	0.001619	0.011248	1578.74	2632.85	1054.12	3.6353	5.3607	1.7254
14.3	338.35	0.001623	0.011130	1582.66	2630.20	1047.53	3.6415	5.3546	1.7131
14.4	338.90	0.001628	0.011014	1586.59	2627.51	1040.93	3.6477	5.3484	1.7007
14.5	339.45	0.001633	0.010898	1590.51	2624.81	1034.29	3.6538	5.3422	1.6884
14.6	340.00	0.001638	0.010784	1594.44	2622.07	1027.63	3.6599	5.3359	1.6760
14.7	340.54	0.001642	0.010671	1598.37	2619.31	1020.95	3.6661	5.3297	1.6636
14.8	341.08	0.001647	0.010560	1602.29	2616.52	1014.23	3.6722	5.3234	1.6512
14.9	341.62	0.001652	0.010449	1606.22	2613.71	1007.49	3.6783	5.3171	1.6388
15.0	342.16	0.001657	0.010340	1610.15	2610.86	1000.71	3.6844	5.3108	1.6264
15.1	342.69	0.001662	0.010232	1614.08	2607.99	993.91	3.6906	5.3045	1.6139
15.2	343.22	0.001667	0.010125	1618.02	2605.09	987.07	3.6967	5.2981	1.6014
15.3	343.75	0.001672	0.010019	1621.96	2602.16	980.20	3.7028	5.2917	1.5889
15.4	344.27	0.001677	0.009915	1625.90	2599.21	973.30	3.7089	5.2853	1.5764
15.5	344.79	0.001682	0.009811	1629.85	2596.22	966.37	3.7150	5.2789	1.5638
15.6	345.31	0.001688	0.009709	1633.80	2593.20	959.39	3.7212	5.2724	1.5513
15.7	345.83	0.001693	0.009607	1637.76	2590.15	952.39	3.7273	5.2659	1.5386
15.8	346.34	0.001699	0.009506	1641.72	2587.06	945.34	3.7334	5.2594	1.5260
15.9	346.85	0.001704	0.009407	1645.69	2583.95	938.26	3.7395	5.2529	1.5133
16.0	347.36	0.001710	0.009308	1649.67	2580.80	931.13	3.7457	5.2463	1.5006
16.1	347.86	0.001715	0.009210	1653.66	2577.62	923.97	3.7518	5.2397	1.4878
16.2	348.36	0.001721	0.009114	1657.65	2574.41	916.76	3.7580	5.2330	1.4750
16.3	348.86	0.001727	0.009018	1661.65	2571.16	909.51	3.7641	5.2263	1.4622
16.4	349.36	0.001732	0.008923	1665.66	2567.88	902.22	3.7703	5.2196	1.4493
16.5	349.86	0.001738	0.008828	1669.68	2564.57	894.88	3.7765	5.2129	1.4364
16.6	350.35	0.001744	0.008736	1673.74	2561.26	887.52	3.7827	5.2062	1.4235
16.7	350.84	0.001750	0.008643	1677.80	2557.87	880.07	3.7889	5.1993	1.4104
16.8	351.33	0.001757	0.008551	1681.86	2554.43	872.57	3.7952	5.1924	1.3973
16.9	351.81	0.001763	0.008460	1685.94	2550.96	865.02	3.8014	5.1855	1.3841
17.0	352.29	0.001769	0.008370	1690.04	2547.44	857.40	3.8077	5.1785	1.3709
17.1	352.77	0.001776	0.008280	1694.15	2543.88	849.73	3.8140	5.1715	1.3576
17.2	353.25	0.001782	0.008191	1698.27	2540.27	841.99	3.8203	5.1644	1.3442
17.3	353.73	0.001789	0.008102	1702.42	2536.61	834.19	3.8266	5.1573	1.3307
17.4	354.20	0.001796	0.008015	1706.59	2532.91	826.32	3.8330	5.1501	1.3172

Table 2 Saturation State (Pressure Table) – Continuation
Sättigungszustand (Drucktafel) – Fortsetzung

p [MPa]	t_s [°C]	v' [$m^3\ kg^{-1}$]	v''	h'	h'' [$kJ\ kg^{-1}$]	Δh_v	s'	s'' [$kJ\ kg^{-1}\ K^{-1}$]	Δs_v
17.5	354.67	0.001803	0.007927	1710.77	2529.16	818.39	3.8393	5.1429	1.3035
17.6	355.14	0.001810	0.007841	1714.98	2525.35	810.38	3.8457	5.1356	1.2898
17.7	355.61	0.001817	0.007755	1719.21	2521.50	802.29	3.8522	5.1282	1.2760
17.8	356.07	0.001824	0.007669	1723.46	2517.59	794.13	3.8587	5.1207	1.2621
17.9	356.54	0.001832	0.007584	1727.74	2513.62	785.88	3.8652	5.1132	1.2481
18.0	357.00	0.001839	0.007499	1732.04	2509.59	777.55	3.8717	5.1056	1.2339
18.1	357.45	0.001847	0.007415	1736.38	2505.51	769.13	3.8783	5.0980	1.2197
18.2	357.91	0.001855	0.007331	1740.74	2501.36	760.62	3.8849	5.0902	1.2053
18.3	358.36	0.001863	0.007248	1745.13	2497.14	752.01	3.8916	5.0824	1.1908
18.4	358.82	0.001872	0.007165	1749.56	2492.86	743.30	3.8983	5.0745	1.1762
18.5	359.27	0.001880	0.007082	1754.02	2488.50	734.48	3.9050	5.0664	1.1614
18.6	359.71	0.001889	0.007000	1758.52	2484.07	725.55	3.9119	5.0583	1.1465
18.7	360.16	0.001898	0.006918	1763.06	2479.56	716.50	3.9187	5.0501	1.1314
18.8	360.60	0.001907	0.006836	1767.64	2474.97	707.32	3.9257	5.0418	1.1161
18.9	361.04	0.001916	0.006755	1772.27	2470.29	698.02	3.9326	5.0333	1.1007
19.0	361.48	0.001925	0.006673	1776.94	2465.52	688.58	3.9397	5.0247	1.0850
19.1	361.92	0.001935	0.006592	1781.67	2460.66	678.99	3.9468	5.0160	1.0692
19.2	362.36	0.001945	0.006511	1786.45	2455.69	669.25	3.9541	5.0072	1.0531
19.3	362.79	0.001956	0.006430	1791.28	2450.62	659.34	3.9614	4.9982	1.0368
19.4	363.22	0.001966	0.006349	1796.18	2445.44	649.26	3.9688	4.9890	1.0203
19.5	363.65	0.001977	0.006267	1801.15	2440.14	638.99	3.9762	4.9797	1.0035
19.6	364.08	0.001989	0.006186	1806.18	2434.71	628.53	3.9838	4.9702	0.9864
19.7	364.50	0.002001	0.006105	1811.30	2429.14	617.85	3.9915	4.9605	0.9690
19.8	364.92	0.002013	0.006023	1816.50	2423.43	606.94	3.9994	4.9506	0.9512
19.9	365.34	0.002026	0.005941	1821.79	2417.57	595.78	4.0074	4.9405	0.9331
20.0	365.76	0.002039	0.005859	1827.17	2411.53	584.36	4.0155	4.9301	0.9146
20.1	366.18	0.002052	0.005776	1832.67	2405.32	572.65	4.0237	4.9195	0.8957
20.2	366.59	0.002067	0.005693	1838.28	2398.91	560.63	4.0322	4.9086	0.8764
20.3	367.01	0.002082	0.005609	1844.02	2392.29	548.26	4.0409	4.8973	0.8565
20.4	367.42	0.002097	0.005524	1849.91	2385.43	535.52	4.0497	4.8857	0.8360
20.5	367.82	0.002114	0.005438	1855.96	2378.31	522.35	4.0588	4.8738	0.8150
20.6	368.23	0.002131	0.005352	1862.19	2370.91	508.73	4.0682	4.8614	0.7932
20.7	368.63	0.002149	0.005264	1868.62	2363.20	494.58	4.0779	4.8486	0.7706
20.8	369.04	0.002169	0.005174	1875.28	2355.13	479.85	4.0879	4.8352	0.7472
20.9	369.44	0.002190	0.005082	1882.20	2346.65	464.45	4.0984	4.8212	0.7228
21.0	369.83	0.002212	0.004989	1889.43	2337.72	448.29	4.1093	4.8065	0.6972
21.1	370.23	0.002236	0.004893	1897.02	2328.25	431.23	4.1207	4.7910	0.6703
21.2	370.62	0.002262	0.004793	1905.03	2318.15	413.11	4.1329	4.7746	0.6417
21.3	371.01	0.002291	0.004690	1913.57	2307.30	393.72	4.1458	4.7570	0.6112
21.4	371.40	0.002324	0.004581	1922.77	2295.53	372.76	4.1597	4.7380	0.5783
21.5	371.79	0.002360	0.004466	1932.79	2282.59	349.80	4.1749	4.7172	0.5424
21.6	372.18	0.002403	0.004343	1943.94	2268.14	324.20	4.1918	4.6942	0.5024
21.7	372.56	0.002453	0.004207	1956.67	2251.59	294.91	4.2111	4.6679	0.4567
21.8	372.94	0.002517	0.004053	1971.85	2231.89	260.04	4.2342	4.6367	0.4025
21.9	373.33	0.002604	0.003868	1991.41	2206.80	215.39	4.2641	4.5973	0.3332
22.0	373.71	0.002750	0.003609	2021.91	2169.25	147.34	4.3109	4.5386	0.2278
22.064	373.95	0.003106		2087.55		0.00	4.4120		0.0000

3 Ideal-Gas State

Zustand des idealen Gases

This table contains values of the following thermodynamic properties in the ideal-gas state for temperatures from 0 °C to 2000 °C:

- Specific isobaric heat capacity c_p^{o}
- Specific enthalpy h^{o}
- Specific entropy s^{o} (at 611.657 Pa)
- Speed of sound w^{o}.

These thermodynamic properties were calculated from the corresponding equations of the industrial standard IAPWS-IF97 presented in Chap. 3. The values of the specific enthalpy and the specific entropy in the ideal-gas state relate to the reference state (zero points) defined by Eq. (3.2). The values of s^{o} were calculated for the pressure $p = p_{\mathrm{t}} = 611.657$ Pa, which is the pressure of the reference state. Entropy values for pressures $p \neq p_{\mathrm{t}}$ require the subtraction of the term $R \ln(p/p_{\mathrm{t}})$ from the tabulated values.

Diese Tafel enthält Werte der folgenden thermodynamischen Eigenschaften im Zustand des idealen Gases für Temperaturen von 0 °C bis 2000 °C:

- spezifische isobare Wärmekapazität c_p^{o}
- spezifische Enthalpie h^{o}
- spezifische Entropie s^{o} (bei 611,657 Pa)
- Schallgeschwindigkeit w^{o}.

Diese thermodynamischen Zustandsgrößen wurden mit den in Kap. 3 angegebenen Gleichungen des Industrie-Standards IAPWS-IF97 berechnet. Die Werte der spezifischen Enthalpie sowie der spezifischen Entropie im Zustand des idealen Gases sind auf den Referenzzustand (Nullpunkte) gemäß Gl. (3.2) bezogen. Die Werte von s^{o} wurden für den Druck $p = p_{\mathrm{t}} = 611{,}657$ Pa berechnet, der dem Druck des Referenzzustandes entspricht. Entropiewerte für Drücke $p \neq p_{\mathrm{t}}$ lassen sich aus den Tafelwerten durch Subtraktion des Ausdruckes $R \ln(p/p_{\mathrm{t}})$ berechnen.

Table 3 Ideal-Gas State
Zustand des idealen Gases

t [°C]	c_p^o [kJ kg^{-1} K^{-1}]	h^o [kJ kg^{-1}]	s^o [kJ kg^{-1} K^{-1}]	w^o [m s^{-1}]	t [°C]	c_p^o [kJ kg^{-1} K^{-1}]	h^o [kJ kg^{-1}]	s^o [kJ kg^{-1} K^{-1}]	w^o [m s^{-1}]
0	1.8589	2501.42	9.1571	409.52	400	2.0634	3280.09	10.8991	632.60
10	1.8611	2520.02	9.2240	416.86	410	2.0700	3300.76	10.9295	636.99
20	1.8634	2538.64	9.2886	424.08	420	2.0767	3321.49	10.9597	641.34
30	1.8658	2557.29	9.3512	431.16	430	2.0833	3342.29	10.9895	645.66
40	1.8685	2575.96	9.4118	438.11	440	2.0900	3363.16	11.0189	649.94
50	1.8714	2594.66	9.4705	444.93	450	2.0968	3384.09	11.0481	654.18
60	1.8745	2613.39	9.5276	451.64	460	2.1035	3405.09	11.0769	658.39
70	1.8780	2632.15	9.5831	458.23	470	2.1103	3426.16	11.1055	662.56
80	1.8816	2650.95	9.6371	464.71	480	2.1171	3447.30	11.1337	666.71
90	1.8856	2669.78	9.6897	471.09	490	2.1240	3468.51	11.1617	670.82
100	1.8897	2688.66	9.7410	477.36	500	2.1309	3489.78	11.1894	674.89
110	1.8941	2707.58	9.7910	483.53	510	2.1378	3511.12	11.2168	678.94
120	1.8986	2726.54	9.8399	489.61	520	2.1447	3532.54	11.2440	682.96
130	1.9034	2745.55	9.8876	495.60	530	2.1517	3554.02	11.2709	686.95
140	1.9082	2764.61	9.9343	501.51	540	2.1586	3575.57	11.2976	690.91
150	1.9133	2783.72	9.9800	507.33	550	2.1656	3597.19	11.3240	694.84
160	1.9184	2802.87	10.0248	513.07	560	2.1726	3618.88	11.3502	698.74
170	1.9237	2822.08	10.0686	518.73	570	2.1796	3640.64	11.3761	702.61
180	1.9291	2841.35	10.1116	524.32	580	2.1867	3662.47	11.4019	706.46
190	1.9346	2860.67	10.1538	529.84	590	2.1937	3684.38	11.4274	710.29
200	1.9402	2880.04	10.1951	535.29	600	2.2008	3706.35	11.4527	714.08
210	1.9458	2899.47	10.2358	540.67	610	2.2078	3728.39	11.4778	717.86
220	1.9516	2918.96	10.2757	545.99	620	2.2149	3750.51	11.5027	721.61
230	1.9573	2938.50	10.3149	551.24	630	2.2220	3772.69	11.5274	725.33
240	1.9632	2958.10	10.3535	556.43	640	2.2291	3794.95	11.5519	729.03
250	1.9691	2977.77	10.3915	561.57	650	2.2362	3817.27	11.5762	732.71
260	1.9751	2997.49	10.4288	566.65	660	2.2433	3839.67	11.6004	736.36
270	1.9811	3017.27	10.4656	571.68	670	2.2504	3862.14	11.6243	740.00
280	1.9872	3037.11	10.5018	576.65	680	2.2574	3884.68	11.6481	743.61
290	1.9933	3057.01	10.5374	581.56	690	2.2645	3907.29	11.6717	747.20
300	1.9995	3076.98	10.5726	586.43	700	2.2716	3929.97	11.6951	750.77
310	2.0057	3097.00	10.6072	591.25	710	2.2787	3952.72	11.7184	754.32
320	2.0120	3117.09	10.6413	596.02	720	2.2858	3975.54	11.7415	757.84
330	2.0183	3137.24	10.6750	600.75	730	2.2929	3998.43	11.7644	761.35
340	2.0246	3157.46	10.7083	605.43	740	2.2999	4021.40	11.7872	764.84
350	2.0310	3177.73	10.7411	610.06	750	2.3070	4044.43	11.8098	768.31
360	2.0374	3198.08	10.7735	614.65	760	2.3141	4067.54	11.8323	771.76
370	2.0438	3218.48	10.8054	619.20	770	2.3211	4090.71	11.8546	775.20
380	2.0503	3238.95	10.8370	623.71	780	2.3282	4113.96	11.8768	778.61
390	2.0569	3259.49	10.8682	628.17	790	2.3352	4137.28	11.8988	782.01

Table 3 Ideal-Gas State – Continuation
Zustand des idealen Gases – Fortsetzung

t [°C]	c_p^o [kJ kg^{-1} K^{-1}]	h^o [kJ kg^{-1}]	s^o [kJ kg^{-1} K^{-1}]	w^o [m s^{-1}]	t [°C]	c_p^o [kJ kg^{-1} K^{-1}]	h^o [kJ kg^{-1}]	s^o [kJ kg^{-1} K^{-1}]	w^o [m s^{-1}]
800	2.3423	4160.66	11.9207	785.38	1400	2.7041	5681.30	13.0403	964.95
825	2.3606	4219.49	11.9749	793.72	1425	2.7160	5749.05	13.0805	971.69
850	2.3774	4278.71	12.0282	802.02	1450	2.7276	5817.10	13.1203	978.39
875	2.3941	4338.36	12.0808	810.21	1475	2.7389	5885.43	13.1596	985.05
900	2.4109	4398.42	12.1325	818.31	1500	2.7501	5954.04	13.1986	991.66
925	2.4276	4458.90	12.1835	826.31	1525	2.7609	6022.93	13.2372	998.23
950	2.4442	4519.80	12.2338	834.22	1550	2.7715	6092.08	13.2754	1004.76
975	2.4606	4581.11	12.2834	842.05	1575	2.7819	6161.50	13.3132	1011.25
1000	2.4769	4642.83	12.3324	849.80	1600	2.7921	6231.18	13.3506	1017.70
1025	2.4930	4704.95	12.3807	857.46	1625	2.8020	6301.10	13.3877	1024.11
1050	2.5089	4767.47	12.4284	865.06	1650	2.8117	6371.28	13.4245	1030.48
1075	2.5245	4830.39	12.4755	872.58	1675	2.8212	6441.69	13.4608	1036.81
1100	2.5400	4893.70	12.5221	880.04	1700	2.8305	6512.33	13.4969	1043.11
1125	2.5551	4957.39	12.5680	887.43	1725	2.8397	6583.21	13.5326	1049.37
1150	2.5700	5021.45	12.6134	894.76	1750	2.8486	6654.32	13.5679	1055.59
1175	2.5847	5085.89	12.6583	902.02	1775	2.8574	6725.64	13.6030	1061.77
1200	2.5991	5150.69	12.7027	909.23	1800	2.8661	6797.19	13.6377	1067.92
1225	2.6132	5215.84	12.7465	916.38	1850	2.8829	6940.91	13.7062	1080.12
1250	2.6270	5281.34	12.7899	923.47	1900	2.8992	7085.47	13.7735	1092.18
1275	2.6406	5347.19	12.8328	930.51	1950	2.9152	7230.83	13.8396	1104.10
1300	2.6538	5413.37	12.8752	937.50	2000	2.9307	7376.98	13.9046	1115.89
1325	2.6668	5479.88	12.9171	944.43					
1350	2.6795	5546.71	12.9586	951.32					
1375	2.6919	5613.85	12.9997	958.16					

4 Single-Phase Region (Regions 1 to 3 of IAPWS-IF97)

Homogenes Zustandsgebiet (Bereiche 1 bis 3 der IAPWS-IF97)

This table contains values of the following thermophysical properties in the single-phase region for temperatures from 0 °C to 800 °C and for pressures from the triple-point pressure $p_t = 611.657$ Pa to 100 MPa (regions 1 to 3 of IAPWS-IF97):

- Specific volume v
- Specific enthalpy h
- Specific entropy s
- Specific isobaric heat capacity c_p
- Speed of sound w
- Isentropic exponent κ
- Dynamic viscosity η
- Thermal conductivity λ.

These properties were exclusively determined from the corresponding equations in Chaps. 3 and 4. The *thermodynamic* properties were calculated from the industrial standard IAPWS-IF97. For the calculation of the saturation properties please refer to the introduction of Table 1, p. 103.

Diese Tafel enthält Werte der folgenden thermophysikalischen Eigenschaften im homogenen Zustandsgebiet für Temperaturen von 0 °C bis 800 °C und für Drücke vom Tripelpunktsdruck $p_t = 611{,}657$ Pa bis 100 MPa (Bereiche 1 bis 3 der IAPWS-IF97):

- spezifisches Volumen v
- spezifische Enthalpie h
- spezifische Entropie s
- spezifische isobare Wärmekapazität c_p
- Schallgeschwindigkeit w
- Isentropenexponent κ
- dynamische Viskosität η
- Wärmeleitfähigkeit λ .

Diese Zustandsgrößen wurden ausschließlich mit den Gleichungen in Kap. 3 und 4 berechnet. Die *thermodynamischen* Zustandsgrößen wurden mit dem Industrie-Standard IAPWS-IF97 bestimmt. Bezüglich der Berechnung der Sättigungsgrößen gelten die in der Einführung zu Tafel 1, S. 103, gemachten Ausführungen.

Table 4 **Single-Phase Region** (Regions 1 to 3 of IAPWS-IF97) **Homogenes Zustandsgebiet** (Bereiche 1 bis 3 der IAPWS-IF97)

$p = p_t = 0.000611657$ MPa

t [°C]	v [m³ kg⁻¹]	h [kJ kg⁻¹]	s [kJ kg⁻¹ K⁻¹]	c_p [kJ kg⁻¹ K⁻¹]	w [m s⁻¹]	κ	$\eta \times 10^6$ [kg m⁻¹ s⁻¹]	$\lambda \times 10^3$ [W m⁻¹ K⁻¹]
0	0.001000	− 0.04	− 0.0002	4.2199	1402.28	3214201	1791.78	561.02
$t_s = 0.01$				Saturation				
Liquid	0.001000	0.00	0.0000	4.2199	1402.33	3214432	1791.16	561.04
Vapour	205.9975	2500.91	9.1555	1.8882	408.89	1.3269	9.22	17.07
5	209.7779	2510.30	9.1896	1.8764	412.86	1.3284	9.34	17.33
10	213.5616	2519.67	9.2229	1.8716	416.65	1.3289	9.46	17.60
15	217.3429	2529.02	9.2557	1.8698	420.33	1.3290	9.59	17.88
20	221.1227	2538.37	9.2878	1.8693	423.95	1.3289	9.73	18.17
25	224.9016	2547.71	9.3195	1.8695	427.53	1.3287	9.87	18.47
30	228.6797	2557.06	9.3505	1.8701	431.07	1.3285	10.02	18.79
35	232.4572	2566.42	9.3811	1.8708	434.57	1.3282	10.17	19.11
40	236.2342	2575.77	9.4113	1.8718	438.03	1.3279	10.32	19.43
45	240.0108	2585.13	9.4409	1.8728	441.47	1.3276	10.47	19.77
50	243.7870	2594.50	9.4701	1.8740	444.87	1.3272	10.63	20.11
60	251.3384	2613.25	9.5273	1.8767	451.59	1.3265	10.96	20.82
70	258.8887	2632.03	9.5828	1.8797	458.19	1.3258	11.29	21.56
80	266.4383	2650.85	9.6369	1.8831	464.67	1.3249	11.64	22.32
90	273.9871	2669.70	9.6895	1.8867	471.05	1.3240	11.99	23.10
100	281.5355	2688.58	9.7408	1.8907	477.33	1.3231	12.35	23.90
110	289.0834	2707.51	9.7909	1.8949	483.51	1.3221	12.72	24.72
120	296.6309	2726.48	9.8397	1.8993	489.59	1.3211	13.09	25.57
130	304.1782	2745.50	9.8875	1.9039	495.58	1.3201	13.47	26.43
140	311.7253	2764.56	9.9342	1.9087	501.49	1.3190	13.85	27.30
150	319.2721	2783.67	9.9799	1.9137	507.31	1.3179	14.23	28.20
160	326.8188	2802.84	10.0247	1.9188	513.05	1.3168	14.62	29.11
170	334.3654	2822.05	10.0685	1.9240	518.72	1.3156	15.01	30.03
180	341.9118	2841.32	10.1115	1.9294	524.31	1.3145	15.41	30.97
190	349.4581	2860.64	10.1537	1.9348	529.83	1.3133	15.81	31.93
200	357.0044	2880.01	10.1951	1.9404	535.28	1.3121	16.21	32.90
210	364.5506	2899.45	10.2357	1.9460	540.66	1.3109	16.61	33.88
220	372.0967	2918.93	10.2757	1.9517	545.98	1.3097	17.02	34.87
230	379.6428	2938.48	10.3149	1.9575	551.23	1.3085	17.42	35.88
240	387.1888	2958.08	10.3535	1.9633	556.43	1.3073	17.83	36.90
250	394.7348	2977.75	10.3914	1.9692	561.56	1.3061	18.24	37.93
260	402.2807	2997.47	10.4288	1.9752	566.64	1.3049	18.65	38.97
270	409.8266	3017.25	10.4655	1.9812	571.67	1.3037	19.06	40.03
280	417.3725	3037.09	10.5017	1.9873	576.64	1.3025	19.47	41.10
290	424.9184	3057.00	10.5374	1.9934	581.56	1.3013	19.89	42.17
300	432.4642	3076.96	10.5725	1.9996	586.43	1.3001	20.30	43.26
310	440.0100	3096.99	10.6072	2.0058	591.25	1.2989	20.72	44.36
320	447.5558	3117.08	10.6413	2.0120	596.02	1.2977	21.13	45.47
330	455.1015	3137.23	10.6750	2.0183	600.74	1.2965	21.55	46.58
340	462.6473	3157.44	10.7083	2.0247	605.42	1.2953	21.96	47.71

Table 4 Single-Phase Region (Regions 1 to 3 of IAPWS-IF97) – Continuation
Homogenes Zustandsgebiet (Bereiche 1 bis 3 der IAPWS-IF97) – Fortsetzung

$p = p_t = 0.000611657$ MPa								
t	v	h	s	c_p	w	κ	$\eta \times 10^6$	$\lambda \times 10^3$
[°C]	[m^3 kg^{-1}]	[kJ kg^{-1}]	[kJ kg^{-1} K^{-1}]	[kJ kg^{-1} K^{-1}]	[m s^{-1}]		[kg m^{-1} s^{-1}]	[W m^{-1} K^{-1}]
350	470.1930	3177.72	10.7411	2.0310	610.06	1.2941	22.38	48.85
360	477.7388	3198.07	10.7735	2.0374	614.65	1.2929	22.79	49.99
370	485.2845	3218.47	10.8054	2.0439	619.20	1.2917	23.21	51.15
380	492.8302	3238.94	10.8370	2.0504	623.70	1.2905	23.62	52.31
390	500.3758	3259.48	10.8682	2.0569	628.17	1.2893	24.04	53.49
400	507.9215	3280.08	10.8991	2.0635	632.60	1.2881	24.45	54.67
410	515.4672	3300.75	10.9295	2.0701	636.99	1.2869	24.87	55.86
420	523.0129	3321.48	10.9597	2.0767	641.34	1.2858	25.28	57.06
430	530.5585	3342.28	10.9895	2.0834	645.65	1.2846	25.70	58.26
440	538.1042	3363.15	11.0189	2.0901	649.93	1.2834	26.11	59.47
450	545.6498	3384.08	11.0481	2.0968	654.18	1.2822	26.52	60.70
460	553.1954	3405.09	11.0769	2.1036	658.39	1.2811	26.93	61.92
470	560.7411	3426.16	11.1055	2.1103	662.56	1.2799	27.34	63.16
480	568.2867	3447.29	11.1337	2.1172	666.70	1.2788	27.75	64.40
490	575.8323	3468.50	11.1617	2.1240	670.81	1.2776	28.16	65.65
500	583.3779	3489.77	11.1894	2.1309	674.89	1.2765	28.57	66.91
510	590.9235	3511.12	11.2168	2.1378	678.94	1.2753	28.98	68.17
520	598.4691	3532.53	11.2440	2.1447	682.96	1.2742	29.39	69.45
530	606.0147	3554.01	11.2709	2.1517	686.95	1.2731	29.80	70.72
540	613.5603	3575.56	11.2976	2.1586	690.90	1.2720	30.20	72.01
550	621.1059	3597.18	11.3240	2.1656	694.83	1.2708	30.61	73.30
560	628.6515	3618.88	11.3502	2.1726	698.74	1.2697	31.01	74.59
570	636.1970	3640.64	11.3761	2.1796	702.61	1.2686	31.41	75.89
580	643.7426	3662.47	11.4019	2.1867	706.46	1.2675	31.81	77.20
590	651.2882	3684.37	11.4274	2.1937	710.29	1.2664	32.21	78.52
600	658.8338	3706.34	11.4527	2.2008	714.08	1.2654	32.61	79.84
610	666.3793	3728.39	11.4778	2.2079	717.86	1.2643	33.01	81.16
620	673.9249	3750.50	11.5027	2.2149	721.60	1.2632	33.41	82.49
630	681.4705	3772.69	11.5274	2.2220	725.33	1.2622	33.81	83.83
640	689.0160	3794.94	11.5519	2.2291	729.03	1.2611	34.20	85.17
650	696.5616	3817.27	11.5762	2.2362	732.71	1.2601	34.60	86.52
660	704.1072	3839.66	11.6004	2.2433	736.36	1.2590	34.99	87.87
670	711.6527	3862.13	11.6243	2.2504	740.00	1.2580	35.38	89.23
680	719.1983	3884.67	11.6481	2.2575	743.61	1.2570	35.77	90.59
690	726.7438	3907.28	11.6717	2.2645	747.20	1.2560	36.16	91.96
700	734.2894	3929.96	11.6951	2.2716	750.77	1.2550	36.55	93.33
720	749.3805	3975.54	11.7415	2.2858	757.84	1.2530	37.32	96.08
740	764.4716	4021.39	11.7872	2.2999	764.84	1.2510	38.09	98.86
760	779.5626	4067.53	11.8323	2.3141	771.76	1.2491	38.86	101.65
780	794.6537	4113.96	11.8768	2.3282	778.61	1.2472	39.62	104.46
800	809.7448	4160.66	11.9207	2.3423	785.38	1.2454	40.37	107.29

Table 4 Single-Phase Region (Regions 1 to 3 of IAPWS-IF97) – Continuation
Homogenes Zustandsgebiet (Bereiche 1 bis 3 der IAPWS-IF97) – Fortsetzung

				p = 0.001 MPa				
t	v	h	s	c_p	w	κ	$\eta \times 10^6$	$\lambda \times 10^3$
[°C]	[$m^3\,kg^{-1}$]	[$kJ\,kg^{-1}$]	[$kJ\,kg^{-1}\,K^{-1}$]	[$kJ\,kg^{-1}\,K^{-1}$]	[$m\,s^{-1}$]		[$kg\,m^{-1}\,s^{-1}$]	[$W\,m^{-1}\,K^{-1}$]
0	0.001000	– 0.04	– 0.0002	4.2199	1402.28	1965990	1791.78	561.02
5	0.001000	21.02	0.0763	4.2054	1426.04	2033412	1518.29	570.52
t_s = 6.97				**Saturation**				
Liquid	0.001000	29.30	0.1059	4.2011	1434.73	2058166	1428.47	574.26
Vapour	129.1833	2513.68	8.9749	1.8932	413.95	1.3265	9.38	17.45
10	130.5895	2519.41	8.9953	1.8846	416.36	1.3275	9.46	17.61
15	132.9069	2528.81	9.0282	1.8774	420.17	1.3283	9.59	17.89
20	135.2222	2538.19	9.0604	1.8743	423.85	1.3285	9.73	18.18
25	137.5362	2547.55	9.0921	1.8732	427.45	1.3285	9.87	18.49
30	139.8493	2556.92	9.1233	1.8730	431.00	1.3283	10.02	18.80
35	142.1617	2566.28	9.1539	1.8733	434.51	1.3281	10.17	19.12
40	144.4736	2575.65	9.1841	1.8739	437.99	1.3278	10.32	19.44
45	146.7851	2585.02	9.2138	1.8747	441.43	1.3275	10.47	19.78
50	149.0961	2594.40	9.2430	1.8757	444.83	1.3272	10.63	20.12
60	153.7173	2613.17	9.3002	1.8780	451.56	1.3265	10.96	20.83
70	158.3374	2631.96	9.3558	1.8808	458.16	1.3257	11.29	21.56
80	162.9567	2650.79	9.4099	1.8840	464.65	1.3249	11.64	22.32
90	167.5753	2669.64	9.4625	1.8875	471.03	1.3240	11.99	23.10
100	172.1934	2688.54	9.5138	1.8913	477.31	1.3231	12.35	23.91
110	176.8111	2707.47	9.5639	1.8954	483.49	1.3221	12.72	24.73
120	181.4284	2726.44	9.6128	1.8997	489.58	1.3211	13.09	25.57
130	186.0454	2745.46	9.6606	1.9043	495.57	1.3201	13.47	26.43
140	190.6622	2764.53	9.7073	1.9090	501.48	1.3190	13.85	27.31
150	195.2789	2783.65	9.7530	1.9140	507.30	1.3179	14.23	28.20
160	199.8953	2802.81	9.7978	1.9190	513.04	1.3168	14.62	29.11
170	204.5116	2822.03	9.8416	1.9242	518.71	1.3156	15.01	30.03
180	209.1278	2841.30	9.8846	1.9295	524.30	1.3145	15.41	30.97
190	213.7439	2860.62	9.9268	1.9350	529.82	1.3133	15.81	31.93
200	218.3599	2880.00	9.9682	1.9405	535.27	1.3121	16.21	32.90
210	222.9759	2899.43	10.0088	1.9461	540.65	1.3109	16.61	33.88
220	227.5918	2918.92	10.0488	1.9518	545.97	1.3097	17.02	34.87
230	232.2076	2938.47	10.0880	1.9576	551.23	1.3085	17.42	35.88
240	236.8234	2958.07	10.1266	1.9634	556.42	1.3073	17.83	36.90
250	241.4391	2977.73	10.1645	1.9693	561.56	1.3061	18.24	37.93
260	246.0548	2997.46	10.2019	1.9753	566.64	1.3049	18.65	38.98
270	250.6705	3017.24	10.2386	1.9813	571.67	1.3037	19.06	40.03
280	255.2861	3037.08	10.2748	1.9874	576.64	1.3025	19.47	41.10
290	259.9017	3056.99	10.3105	1.9935	581.56	1.3013	19.89	42.17
300	264.5173	3076.95	10.3456	1.9996	586.43	1.3001	20.30	43.26
310	269.1328	3096.98	10.3803	2.0058	591.24	1.2989	20.72	44.36
320	273.7484	3117.07	10.4144	2.0121	596.02	1.2977	21.13	45.47
330	278.3639	3137.22	10.4481	2.0184	600.74	1.2965	21.55	46.58
340	282.9794	3157.44	10.4814	2.0247	605.42	1.2953	21.96	47.71

Table 4 Single-Phase Region (Regions 1 to 3 of IAPWS-IF97) – Continuation
Homogenes Zustandsgebiet (Bereiche 1 bis 3 der IAPWS-IF97) – Fortsetzung

p = 0.001 MPa

t [°C]	v [m^3 kg^{-1}]	h [kJ kg^{-1}]	s [kJ kg^{-1} K^{-1}]	c_p [kJ kg^{-1} K^{-1}]	w [m s^{-1}]	κ	$\eta \times 10^6$ [kg m^{-1} s^{-1}]	$\lambda \times 10^3$ [W m^{-1} K^{-1}]
350	287.5949	3177.72	10.5142	2.0311	610.05	1.2941	22.38	48.85
360	292.2104	3198.06	10.5466	2.0375	614.65	1.2929	22.79	50.00
370	296.8259	3218.47	10.5785	2.0439	619.20	1.2917	23.21	51.15
380	301.4413	3238.94	10.6101	2.0504	623.70	1.2905	23.62	52.31
390	306.0568	3259.47	10.6413	2.0569	628.17	1.2893	24.04	53.49
400	310.6722	3280.08	10.6722	2.0635	632.60	1.2881	24.45	54.67
410	315.2876	3300.74	10.7026	2.0701	636.99	1.2869	24.87	55.86
420	319.9030	3321.48	10.7328	2.0767	641.34	1.2858	25.28	57.06
430	324.5184	3342.28	10.7626	2.0834	645.65	1.2846	25.70	58.26
440	329.1338	3363.15	10.7920	2.0901	649.93	1.2834	26.11	59.47
450	333.7492	3384.08	10.8212	2.0968	654.18	1.2822	26.52	60.70
460	338.3646	3405.08	10.8500	2.1036	658.38	1.2811	26.93	61.92
470	342.9800	3426.15	10.8786	2.1104	662.56	1.2799	27.34	63.16
480	347.5954	3447.29	10.9068	2.1172	666.70	1.2788	27.75	64.40
490	352.2108	3468.50	10.9348	2.1240	670.81	1.2776	28.16	65.65
500	356.8261	3489.77	10.9625	2.1309	674.89	1.2765	28.57	66.91
510	361.4415	3511.11	10.9899	2.1378	678.94	1.2753	28.98	68.18
520	366.0568	3532.53	11.0171	2.1447	682.96	1.2742	29.39	69.45
530	370.6722	3554.01	11.0440	2.1517	686.94	1.2731	29.80	70.72
540	375.2875	3575.56	11.0707	2.1586	690.90	1.2720	30.20	72.01
550	379.9029	3597.18	11.0971	2.1656	694.83	1.2708	30.61	73.30
560	384.5182	3618.87	11.1233	2.1726	698.74	1.2697	31.01	74.59
570	389.1336	3640.63	11.1493	2.1797	702.61	1.2686	31.41	75.90
580	393.7489	3662.47	11.1750	2.1867	706.46	1.2675	31.81	77.20
590	398.3642	3684.37	11.2005	2.1937	710.29	1.2664	32.21	78.52
600	402.9796	3706.34	11.2258	2.2008	714.08	1.2654	32.61	79.84
610	407.5949	3728.38	11.2509	2.2079	717.86	1.2643	33.01	81.16
620	412.2102	3750.50	11.2758	2.2149	721.60	1.2632	33.41	82.49
630	416.8255	3772.68	11.3005	2.2220	725.33	1.2622	33.81	83.83
640	421.4409	3794.94	11.3250	2.2291	729.03	1.2611	34.20	85.17
650	426.0562	3817.26	11.3494	2.2362	732.71	1.2601	34.60	86.52
660	430.6715	3839.66	11.3735	2.2433	736.36	1.2590	34.99	87.87
670	435.2868	3862.13	11.3974	2.2504	740.00	1.2580	35.38	89.23
680	439.9021	3884.67	11.4212	2.2575	743.61	1.2570	35.77	90.59
690	444.5174	3907.28	11.4448	2.2645	747.20	1.2560	36.16	91.96
700	449.1327	3929.96	11.4682	2.2716	750.77	1.2550	36.55	93.33
720	458.3633	3975.53	11.5146	2.2858	757.84	1.2530	37.32	96.08
740	467.5939	4021.39	11.5603	2.2999	764.84	1.2510	38.09	98.86
760	476.8245	4067.53	11.6054	2.3141	771.76	1.2491	38.86	101.65
780	486.0551	4113.95	11.6499	2.3282	778.61	1.2472	39.62	104.46
800	495.2857	4160.66	11.6938	2.3423	785.38	1.2454	40.37	107.29

Table 4 Single-Phase Region (Regions 1 to 3 of IAPWS-IF97) – Continuation
Homogenes Zustandsgebiet (Bereiche 1 bis 3 der IAPWS-IF97) – Fortsetzung

				p = 0.002 MPa				
t [°C]	v [m³ kg⁻¹]	h [kJ kg⁻¹]	s [kJ kg⁻¹ K⁻¹]	c_p [kJ kg⁻¹ K⁻¹]	w [m s⁻¹]	κ	$\eta \times 10^6$ [kg m⁻¹ s⁻¹]	$\lambda \times 10^3$ [W m⁻¹ K⁻¹]
0	0.001000	– 0.04	– 0.0002	4.2199	1402.28	982998	1791.78	561.02
5	0.001000	21.02	0.0763	4.2054	1426.04	1016709	1518.29	570.52
10	0.001000	42.02	0.1511	4.1958	1447.40	1047117	1305.98	580.00
15	0.001001	62.98	0.2245	4.1894	1466.43	1074198	1137.62	589.33
t_s = 17.50				**Saturation**				
Liquid	0.001001	73.43	0.2606	4.1871	1475.10	1086483	1066.28	593.90
Vapour	66.9896	2532.91	8.7227	1.9030	421.43	1.3256	9.66	18.07
20	67.5721	2537.67	8.7390	1.8951	423.40	1.3265	9.73	18.22
25	68.7333	2547.12	8.7710	1.8863	427.17	1.3274	9.87	18.52
30	69.8930	2556.54	8.8023	1.8822	430.81	1.3277	10.01	18.82
35	71.0519	2565.94	8.8331	1.8805	434.36	1.3277	10.16	19.14
40	72.2101	2575.34	8.8634	1.8798	437.86	1.3275	10.32	19.47
45	73.3677	2584.74	8.8931	1.8798	441.31	1.3273	10.47	19.80
50	74.5250	2594.14	8.9224	1.8801	444.73	1.3270	10.63	20.14
60	76.8386	2612.95	8.9798	1.8815	451.47	1.3263	10.96	20.85
70	79.1511	2631.78	9.0354	1.8836	458.09	1.3256	11.29	21.58
80	81.4627	2650.62	9.0896	1.8863	464.59	1.3248	11.64	22.34
90	83.7737	2669.50	9.1423	1.8894	470.98	1.3239	11.99	23.12
100	86.0842	2688.41	9.1937	1.8929	477.27	1.3230	12.35	23.92
110	88.3942	2707.36	9.2438	1.8967	483.45	1.3221	12.72	24.74
120	90.7039	2726.35	9.2927	1.9009	489.54	1.3211	13.09	25.58
130	93.0133	2745.38	9.3405	1.9052	495.54	1.3200	13.47	26.44
140	95.3225	2764.45	9.3872	1.9098	501.45	1.3190	13.85	27.31
150	97.6315	2783.58	9.4330	1.9146	507.28	1.3179	14.23	28.21
160	99.9403	2802.75	9.4777	1.9196	513.02	1.3167	14.62	29.12
170	102.2490	2821.97	9.5216	1.9247	518.69	1.3156	15.01	30.04
180	104.5576	2841.24	9.5646	1.9300	524.28	1.3144	15.41	30.98
190	106.8660	2860.57	9.6068	1.9354	529.80	1.3133	15.81	31.93
200	109.1744	2879.95	9.6482	1.9409	535.25	1.3121	16.21	32.90
210	111.4827	2899.39	9.6889	1.9464	540.64	1.3109	16.61	33.88
220	113.7910	2918.88	9.7288	1.9521	545.96	1.3097	17.02	34.88
230	116.0992	2938.43	9.7680	1.9578	551.22	1.3085	17.42	35.88
240	118.4073	2958.04	9.8066	1.9636	556.41	1.3073	17.83	36.90
250	120.7154	2977.70	9.8446	1.9695	561.55	1.3061	18.24	37.93
260	123.0235	2997.43	9.8819	1.9755	566.63	1.3049	18.65	38.98
270	125.3315	3017.21	9.9187	1.9815	571.66	1.3037	19.06	40.03
280	127.6395	3037.06	9.9549	1.9875	576.63	1.3025	19.47	41.10
290	129.9475	3056.96	9.9906	1.9936	581.55	1.3013	19.89	42.18
300	132.2555	3076.93	10.0257	1.9997	586.42	1.3001	20.30	43.26
310	134.5634	3096.96	10.0603	2.0059	591.24	1.2989	20.72	44.36
320	136.8713	3117.05	10.0945	2.0122	596.01	1.2977	21.13	45.47
330	139.1792	3137.20	10.1282	2.0185	600.73	1.2965	21.55	46.59
340	141.4871	3157.42	10.1614	2.0248	605.41	1.2953	21.96	47.71

Table 4 Single-Phase Region (Regions 1 to 3 of IAPWS-IF97) – Continuation
Homogenes Zustandsgebiet (Bereiche 1 bis 3 der IAPWS-IF97) – Fortsetzung

$p = 0.002$ MPa								
t	v	h	s	c_p	w	κ	$\eta \times 10^6$	$\lambda \times 10^3$
[°C]	[$m^3\ kg^{-1}$]	[$kJ\ kg^{-1}$]	[$kJ\ kg^{-1}\ K^{-1}$]	[$kJ\ kg^{-1}\ K^{-1}$]	[$m\ s^{-1}$]		[$kg\ m^{-1}\ s^{-1}$]	[$W\ m^{-1}\ K^{-1}$]
350	143.7949	3177.70	10.1943	2.0311	610.05	1.2941	22.38	48.85
360	146.1028	3198.04	10.2266	2.0376	614.64	1.2929	22.79	50.00
370	148.4106	3218.45	10.2586	2.0440	619.19	1.2917	23.21	51.15
380	150.7185	3238.92	10.2902	2.0505	623.70	1.2905	23.62	52.32
390	153.0263	3259.46	10.3214	2.0570	628.17	1.2893	24.04	53.49
400	155.3341	3280.06	10.3522	2.0636	632.59	1.2881	24.45	54.67
410	157.6419	3300.73	10.3827	2.0701	636.98	1.2869	24.87	55.86
420	159.9497	3321.46	10.4129	2.0768	641.33	1.2857	25.28	57.06
430	162.2574	3342.26	10.4426	2.0834	645.65	1.2846	25.70	58.26
440	164.5652	3363.13	10.4721	2.0901	649.93	1.2834	26.11	59.48
450	166.8730	3384.07	10.5013	2.0969	654.17	1.2822	26.52	60.70
460	169.1807	3405.07	10.5301	2.1036	658.38	1.2811	26.93	61.93
470	171.4885	3426.14	10.5587	2.1104	662.56	1.2799	27.34	63.16
480	173.7962	3447.28	10.5869	2.1172	666.70	1.2788	27.75	64.40
490	176.1040	3468.48	10.6149	2.1241	670.81	1.2776	28.16	65.65
500	178.4117	3489.76	10.6426	2.1309	674.89	1.2765	28.57	66.91
510	180.7194	3511.10	10.6700	2.1378	678.94	1.2753	28.98	68.18
520	183.0272	3532.52	10.6972	2.1448	682.95	1.2742	29.39	69.45
530	185.3349	3554.00	10.7241	2.1517	686.94	1.2731	29.80	70.72
540	187.6426	3575.55	10.7508	2.1587	690.90	1.2720	30.20	72.01
550	189.9503	3597.17	10.7772	2.1657	694.83	1.2708	30.61	73.30
560	192.2580	3618.86	10.8034	2.1727	698.74	1.2697	31.01	74.59
570	194.5657	3640.63	10.8293	2.1797	702.61	1.2686	31.41	75.90
580	196.8734	3662.46	10.8551	2.1867	706.46	1.2675	31.81	77.20
590	199.1811	3684.36	10.8806	2.1938	710.28	1.2664	32.21	78.52
600	201.4888	3706.33	10.9059	2.2008	714.08	1.2654	32.61	79.84
610	203.7965	3728.38	10.9310	2.2079	717.85	1.2643	33.01	81.16
620	206.1042	3750.49	10.9559	2.2150	721.60	1.2632	33.41	82.49
630	208.4119	3772.68	10.9806	2.2220	725.33	1.2622	33.81	83.83
640	210.7196	3794.93	11.0051	2.2291	729.03	1.2611	34.20	85.17
650	213.0273	3817.26	11.0294	2.2362	732.71	1.2601	34.60	86.52
660	215.3350	3839.66	11.0536	2.2433	736.36	1.2590	34.99	87.87
670	217.6427	3862.12	11.0775	2.2504	739.99	1.2580	35.38	89.23
680	219.9504	3884.66	11.1013	2.2575	743.61	1.2570	35.77	90.59
690	222.2580	3907.27	11.1249	2.2646	747.20	1.2560	36.16	91.96
700	224.5657	3929.95	11.1483	2.2716	750.77	1.2550	36.55	93.33
720	229.1811	3975.53	11.1947	2.2858	757.84	1.2530	37.32	96.09
740	233.7964	4021.39	11.2404	2.2999	764.84	1.2510	38.09	98.86
760	238.4117	4067.53	11.2855	2.3141	771.76	1.2491	38.86	101.65
780	243.0271	4113.95	11.3300	2.3282	778.61	1.2473	39.62	104.46
800	247.6424	4160.65	11.3739	2.3423	785.38	1.2454	40.37	107.29

Table 4 Single-Phase Region (Regions 1 to 3 of IAPWS-IF97) – Continuation
Homogenes Zustandsgebiet (Bereiche 1 bis 3 der IAPWS-IF97) – Fortsetzung

				p = 0.005 MPa				
t [°C]	v [m³ kg⁻¹]	h [kJ kg⁻¹]	s [kJ kg⁻¹ K⁻¹]	c_p [kJ kg⁻¹ K⁻¹]	w [m s⁻¹]	κ	$\eta \times 10^6$ [kg m⁻¹ s⁻¹]	$\lambda \times 10^3$ [W m⁻¹ K⁻¹]
0	0.001000	– 0.04	– 0.0002	4.2199	1402.29	393202	1791.77	561.02
5	0.001000	21.02	0.0763	4.2054	1426.04	406686	1518.29	570.53
10	0.001000	42.02	0.1511	4.1958	1447.40	418850	1305.98	580.00
15	0.001001	62.99	0.2245	4.1894	1466.44	429682	1137.62	589.33
20	0.001002	83.92	0.2965	4.1851	1483.27	439207	1001.65	598.42
25	0.001003	104.84	0.3673	4.1822	1498.02	447467	890.10	607.15
30	0.001004	125.75	0.4368	4.1803	1510.83	454516	797.35	615.46
t_s = 32.88				**Saturation**				
Liquid	0.001005	137.77	0.4763	4.1796	1517.37	458044	750.89	620.02
Vapour	28.1863	2560.77	8.3939	1.9219	431.98	1.3241	10.09	19.09
35	28.3849	2564.84	8.4072	1.9143	433.63	1.3249	10.16	19.22
40	28.8514	2574.38	8.4379	1.9032	437.34	1.3259	10.31	19.54
45	29.3170	2583.88	8.4680	1.8976	440.91	1.3262	10.47	19.87
50	29.7822	2593.36	8.4976	1.8947	444.40	1.3262	10.63	20.21
60	30.7112	2612.29	8.5553	1.8924	451.21	1.3258	10.95	20.90
70	31.6392	2631.21	8.6112	1.8922	457.87	1.3252	11.29	21.63
80	32.5663	2650.14	8.6656	1.8933	464.40	1.3245	11.64	22.38
90	33.4927	2669.08	8.7185	1.8951	470.82	1.3237	11.99	23.15
100	34.4186	2688.05	8.7700	1.8977	477.12	1.3228	12.35	23.95
110	35.3441	2707.04	8.8202	1.9007	483.33	1.3219	12.71	24.77
120	36.2692	2726.06	8.8692	1.9042	489.43	1.3209	13.09	25.61
130	37.1941	2745.12	8.9171	1.9081	495.44	1.3199	13.46	26.46
140	38.1187	2764.22	8.9639	1.9122	501.36	1.3189	13.84	27.34
150	39.0431	2783.37	9.0097	1.9167	507.20	1.3178	14.23	28.23
160	39.9673	2802.56	9.0545	1.9214	512.95	1.3167	14.62	29.13
170	40.8914	2821.80	9.0984	1.9263	518.63	1.3155	15.01	30.06
180	41.8154	2841.08	9.1415	1.9313	524.22	1.3144	15.41	30.99
190	42.7393	2860.42	9.1837	1.9365	529.75	1.3132	15.81	31.95
200	43.6631	2879.82	9.2251	1.9419	535.21	1.3121	16.21	32.91
210	44.5868	2899.26	9.2658	1.9473	540.59	1.3109	16.61	33.89
220	45.5105	2918.76	9.3057	1.9529	545.92	1.3097	17.01	34.89
230	46.4341	2938.32	9.3450	1.9586	551.18	1.3085	17.42	35.89
240	47.3577	2957.93	9.3836	1.9643	556.38	1.3073	17.83	36.91
250	48.2812	2977.61	9.4216	1.9701	561.52	1.3061	18.24	37.94
260	49.2047	2997.34	9.4589	1.9760	566.60	1.3049	18.65	38.98
270	50.1282	3017.13	9.4957	1.9819	571.63	1.3037	19.06	40.04
280	51.0516	3036.98	9.5319	1.9879	576.60	1.3025	19.47	41.10
290	51.9750	3056.89	9.5676	1.9940	581.52	1.3013	19.89	42.18
300	52.8984	3076.86	9.6027	2.0001	586.39	1.3001	20.30	43.27
310	53.8217	3096.89	9.6374	2.0063	591.22	1.2989	20.72	44.36
320	54.7451	3116.98	9.6715	2.0125	595.99	1.2977	21.13	45.47
330	55.6684	3137.14	9.7052	2.0188	600.72	1.2965	21.55	46.59
340	56.5917	3157.36	9.7385	2.0251	605.40	1.2953	21.96	47.72

Table 4 Single-Phase Region (Regions 1 to 3 of IAPWS-IF97) – Continuation
Homogenes Zustandsgebiet (Bereiche 1 bis 3 der IAPWS-IF97) – Fortsetzung

p = 0.005 MPa								
t [°C]	v [m^3 kg^{-1}]	h [kJ kg^{-1}]	s [kJ kg^{-1} K^{-1}]	c_p [kJ kg^{-1} K^{-1}]	w [m s^{-1}]	κ	$\eta \times 10^6$ [kg m^{-1} s^{-1}]	$\lambda \times 10^3$ [W m^{-1} K^{-1}]
350	57.5150	3177.64	9.7713	2.0314	610.03	1.2941	22.38	48.85
360	58.4382	3197.99	9.8037	2.0378	614.63	1.2929	22.79	50.00
370	59.3615	3218.40	9.8357	2.0442	619.18	1.2917	23.21	51.15
380	60.2847	3238.87	9.8673	2.0507	623.68	1.2905	23.62	52.32
390	61.2080	3259.41	9.8985	2.0572	628.15	1.2893	24.04	53.49
400	62.1312	3280.01	9.9293	2.0637	632.58	1.2881	24.45	54.67
410	63.0544	3300.68	9.9598	2.0703	636.97	1.2869	24.87	55.86
420	63.9776	3321.42	9.9899	2.0769	641.32	1.2857	25.28	57.06
430	64.9008	3342.22	10.0197	2.0836	645.64	1.2846	25.70	58.26
440	65.8240	3363.09	10.0492	2.0903	649.92	1.2834	26.11	59.48
450	66.7472	3384.03	10.0783	2.0970	654.16	1.2822	26.52	60.70
460	67.6704	3405.03	10.1072	2.1038	658.37	1.2811	26.93	61.93
470	68.5936	3426.10	10.1357	2.1105	662.55	1.2799	27.34	63.16
480	69.5167	3447.24	10.1640	2.1173	666.69	1.2788	27.75	64.41
490	70.4399	3468.45	10.1920	2.1242	670.80	1.2776	28.16	65.66
500	71.3631	3489.73	10.2197	2.1311	674.88	1.2765	28.57	66.91
510	72.2862	3511.07	10.2471	2.1380	678.93	1.2753	28.98	68.18
520	73.2094	3532.49	10.2743	2.1449	682.95	1.2742	29.39	69.45
530	74.1325	3553.97	10.3012	2.1518	686.94	1.2731	29.80	70.73
540	75.0556	3575.52	10.3278	2.1588	690.90	1.2720	30.20	72.01
550	75.9788	3597.14	10.3543	2.1658	694.83	1.2708	30.61	73.30
560	76.9019	3618.84	10.3805	2.1728	698.73	1.2697	31.01	74.60
570	77.8250	3640.60	10.4064	2.1798	702.61	1.2686	31.41	75.90
580	78.7482	3662.43	10.4322	2.1868	706.46	1.2675	31.81	77.21
590	79.6713	3684.34	10.4577	2.1938	710.28	1.2664	32.21	78.52
600	80.5944	3706.31	10.4830	2.2009	714.08	1.2654	32.61	79.84
610	81.5175	3728.35	10.5081	2.2080	717.85	1.2643	33.01	81.16
620	82.4407	3750.47	10.5330	2.2150	721.60	1.2632	33.41	82.50
630	83.3638	3772.65	10.5577	2.2221	725.32	1.2622	33.81	83.83
640	84.2869	3794.91	10.5822	2.2292	729.02	1.2611	34.20	85.17
650	85.2100	3817.24	10.6065	2.2363	732.70	1.2601	34.60	86.52
660	86.1331	3839.64	10.6307	2.2434	736.36	1.2590	34.99	87.87
670	87.0562	3862.10	10.6546	2.2504	739.99	1.2580	35.38	89.23
680	87.9793	3884.64	10.6784	2.2575	743.60	1.2570	35.77	90.59
690	88.9024	3907.26	10.7020	2.2646	747.19	1.2560	36.16	91.96
700	89.8255	3929.94	10.7254	2.2717	750.76	1.2550	36.55	93.33
720	91.6717	3975.51	10.7718	2.2858	757.84	1.2530	37.32	96.09
740	93.5179	4021.37	10.8175	2.3000	764.84	1.2511	38.09	98.86
760	95.3641	4067.51	10.8626	2.3141	771.76	1.2491	38.86	101.65
780	97.2102	4113.94	10.9071	2.3282	778.61	1.2473	39.62	104.46
800	99.0564	4160.64	10.9510	2.3423	785.38	1.2454	40.37	107.29

Table 4 Single-Phase Region (Regions 1 to 3 of IAPWS-IF97) – Continuation
Homogenes Zustandsgebiet (Bereiche 1 bis 3 der IAPWS-IF97) – Fortsetzung

				p = 0.01 MPa				
t [°C]	v [m^3 kg^{-1}]	h [kJ kg^{-1}]	s [kJ kg^{-1} K^{-1}]	c_p [kJ kg^{-1} K^{-1}]	w [m s^{-1}]	κ	$\eta \times 10^6$ [kg m^{-1} s^{-1}]	$\lambda \times 10^3$ [W m^{-1} K^{-1}]
0	0.001000	−0.03	−0.0002	4.2199	1402.30	196603	1791.76	561.03
5	0.001000	21.03	0.0763	4.2053	1426.05	203346	1518.28	570.53
10	0.001000	42.03	0.1511	4.1958	1447.41	209428	1305.98	580.00
15	0.001001	62.99	0.2245	4.1894	1466.45	214844	1137.61	589.34
20	0.001002	83.93	0.2965	4.1851	1483.27	219606	1001.64	598.42
25	0.001003	104.84	0.3673	4.1822	1498.03	223736	890.10	607.15
30	0.001004	125.75	0.4368	4.1803	1510.84	227261	797.35	615.46
35	0.001006	146.65	0.5052	4.1792	1521.84	230209	719.31	623.29
40	0.001008	167.54	0.5724	4.1788	1531.15	232610	652.97	630.59
45	0.001010	188.44	0.6386	4.1790	1538.90	234495	596.06	637.36
t_s = 45.81				**Saturation**				
Liquid	0.001010	191.81	0.6492	4.1791	1540.01	234753	587.63	638.40
Vapour	14.6706	2583.89	8.1489	1.9413	440.51	1.3227	10.49	20.04
50	14.8674	2591.99	8.1741	1.9272	443.67	1.3240	10.62	20.31
60	15.3353	2611.18	8.2326	1.9124	450.74	1.3248	10.95	21.00
70	15.8018	2630.27	8.2891	1.9071	457.50	1.3246	11.28	21.71
80	16.2674	2649.33	8.3438	1.9051	464.09	1.3240	11.63	22.45
90	16.7323	2668.38	8.3970	1.9048	470.55	1.3233	11.98	23.22
100	17.1967	2687.43	8.4488	1.9057	476.89	1.3225	12.34	24.01
110	17.6607	2706.50	8.4992	1.9074	483.12	1.3216	12.71	24.82
120	18.1243	2725.58	8.5484	1.9098	489.25	1.3207	13.08	25.65
130	18.5876	2744.69	8.5964	1.9128	495.28	1.3197	13.46	26.50
140	19.0507	2763.84	8.6433	1.9163	501.22	1.3187	13.84	27.37
150	19.5136	2783.02	8.6892	1.9201	507.07	1.3176	14.23	28.26
160	19.9763	2802.24	8.7340	1.9243	512.83	1.3166	14.62	29.16
170	20.4389	2821.51	8.7780	1.9288	518.52	1.3154	15.01	30.08
180	20.9013	2840.82	8.8211	1.9336	524.13	1.3143	15.41	31.02
190	21.3637	2860.18	8.8634	1.9385	529.66	1.3132	15.80	31.97
200	21.8260	2879.59	8.9048	1.9436	535.13	1.3120	16.21	32.93
210	22.2882	2899.05	8.9455	1.9489	540.52	1.3108	16.61	33.91
220	22.7504	2918.57	8.9855	1.9543	545.85	1.3097	17.01	34.90
230	23.2124	2938.14	9.0248	1.9598	551.12	1.3085	17.42	35.91
240	23.6745	2957.76	9.0634	1.9654	556.32	1.3073	17.83	36.92
250	24.1365	2977.45	9.1014	1.9711	561.46	1.3061	18.24	37.95
260	24.5985	2997.19	9.1388	1.9769	566.55	1.3049	18.65	39.00
270	25.0604	3016.98	9.1756	1.9827	571.58	1.3037	19.06	40.05
280	25.5223	3036.84	9.2118	1.9887	576.56	1.3025	19.47	41.11
290	25.9842	3056.76	9.2475	1.9947	581.48	1.3013	19.89	42.19
300	26.4460	3076.73	9.2827	2.0007	586.36	1.3001	20.30	43.28
310	26.9078	3096.77	9.3173	2.0069	591.18	1.2989	20.72	44.37
320	27.3696	3116.87	9.3515	2.0130	595.96	1.2977	21.13	45.48
330	27.8314	3137.03	9.3852	2.0192	600.68	1.2965	21.55	46.60
340	28.2932	3157.26	9.4185	2.0255	605.37	1.2953	21.96	47.72

Table 4 Single-Phase Region (Regions 1 to 3 of IAPWS-IF97) – Continuation
Homogenes Zustandsgebiet (Bereiche 1 bis 3 der IAPWS-IF97) – Fortsetzung

$p = 0.01$ MPa

t [°C]	v [m^3 kg^{-1}]	h [kJ kg^{-1}]	s [kJ kg^{-1} K^{-1}]	c_p [kJ kg^{-1} K^{-1}]	w [m s^{-1}]	κ	$\eta\times 10^6$ [kg m^{-1} s^{-1}]	$\lambda\times 10^3$ [W m^{-1} K^{-1}]
350	28.7550	3177.54	9.4513	2.0318	610.00	1.2941	22.38	48.86
360	29.2167	3197.89	9.4837	2.0382	614.60	1.2929	22.79	50.01
370	29.6785	3218.31	9.5157	2.0446	619.15	1.2917	23.21	51.16
380	30.1402	3238.79	9.5473	2.0510	623.66	1.2905	23.62	52.32
390	30.6019	3259.33	9.5785	2.0575	628.13	1.2893	24.04	53.50
400	31.0636	3279.94	9.6093	2.0641	632.56	1.2881	24.45	54.68
410	31.5253	3300.61	9.6398	2.0706	636.95	1.2869	24.87	55.87
420	31.9870	3321.35	9.6699	2.0772	641.30	1.2857	25.28	57.06
430	32.4486	3342.15	9.6997	2.0839	645.62	1.2846	25.70	58.27
440	32.9103	3363.03	9.7292	2.0905	649.90	1.2834	26.11	59.48
450	33.3720	3383.96	9.7584	2.0972	654.15	1.2822	26.52	60.70
460	33.8336	3404.97	9.7872	2.1040	658.36	1.2811	26.93	61.93
470	34.2953	3426.04	9.8158	2.1107	662.53	1.2799	27.34	63.17
480	34.7569	3447.19	9.8440	2.1175	666.68	1.2788	27.75	64.41
490	35.2185	3468.40	9.8720	2.1244	670.79	1.2776	28.16	65.66
500	35.6802	3489.67	9.8997	2.1312	674.87	1.2765	28.57	66.92
510	36.1418	3511.02	9.9271	2.1381	678.92	1.2753	28.98	68.18
520	36.6034	3532.44	9.9543	2.1450	682.94	1.2742	29.39	69.45
530	37.0650	3553.92	9.9812	2.1520	686.92	1.2731	29.80	70.73
540	37.5267	3575.48	10.0079	2.1589	690.88	1.2720	30.20	72.01
550	37.9883	3597.10	10.0343	2.1659	694.82	1.2708	30.61	73.30
560	38.4499	3618.79	10.0605	2.1729	698.72	1.2697	31.01	74.60
570	38.9115	3640.56	10.0865	2.1799	702.60	1.2686	31.41	75.90
580	39.3731	3662.39	10.1122	2.1869	706.45	1.2675	31.81	77.21
590	39.8347	3684.30	10.1378	2.1940	710.27	1.2664	32.21	78.52
600	40.2963	3706.27	10.1631	2.2010	714.07	1.2654	32.61	79.84
610	40.7579	3728.32	10.1882	2.2081	717.84	1.2643	33.01	81.17
620	41.2195	3750.43	10.2131	2.2151	721.59	1.2632	33.41	82.50
630	41.6810	3772.62	10.2378	2.2222	725.32	1.2622	33.81	83.83
640	42.1426	3794.88	10.2623	2.2293	729.02	1.2611	34.20	85.18
650	42.6042	3817.20	10.2866	2.2364	732.70	1.2601	34.60	86.52
660	43.0658	3839.60	10.3107	2.2434	736.35	1.2590	34.99	87.88
670	43.5274	3862.07	10.3347	2.2505	739.99	1.2580	35.38	89.23
680	43.9890	3884.61	10.3585	2.2576	743.60	1.2570	35.77	90.59
690	44.4505	3907.23	10.3821	2.2647	747.19	1.2560	36.16	91.96
700	44.9121	3929.91	10.4055	2.2718	750.76	1.2550	36.55	93.33
720	45.8352	3975.48	10.4519	2.2859	757.84	1.2530	37.32	96.09
740	46.7584	4021.34	10.4976	2.3001	764.84	1.2511	38.09	98.86
760	47.6815	4067.49	10.5427	2.3142	771.76	1.2491	38.86	101.66
780	48.6046	4113.91	10.5872	2.3283	778.60	1.2473	39.62	104.47
800	49.5278	4160.62	10.6311	2.3424	785.38	1.2454	40.37	107.29

Table 4 Single-Phase Region (Regions 1 to 3 of IAPWS-IF97) – Continuation
Homogenes Zustandsgebiet (Bereiche 1 bis 3 der IAPWS-IF97) – Fortsetzung

				p = 0.025 MPa				
t [°C]	v [m^3 kg^{-1}]	h [kJ kg^{-1}]	s [kJ kg^{-1} K^{-1}]	c_p [kJ kg^{-1} K^{-1}]	w [m s^{-1}]	κ	$\eta\times10^6$ [kg m^{-1} s^{-1}]	$\lambda\times10^3$ [W m^{-1} K^{-1}]
0	0.001000	−0.02	−0.0002	4.2198	1402.32	78644	1791.72	561.03
5	0.001000	21.04	0.0763	4.2053	1426.07	81341	1518.25	570.54
10	0.001000	42.04	0.1511	4.1957	1447.43	83774	1305.96	580.01
15	0.001001	63.01	0.2245	4.1894	1466.47	85941	1137.60	589.34
20	0.001002	83.94	0.2965	4.1850	1483.30	87845	1001.64	598.43
25	0.001003	104.86	0.3673	4.1821	1498.05	89498	890.10	607.16
30	0.001004	125.76	0.4368	4.1802	1510.86	90907	797.35	615.47
35	0.001006	146.66	0.5052	4.1791	1521.86	92087	719.31	623.30
40	0.001008	167.56	0.5724	4.1787	1531.18	93047	652.97	630.60
45	0.001010	188.45	0.6386	4.1789	1538.92	93801	596.06	637.36
50	0.001012	209.35	0.7038	4.1797	1545.20	94360	546.84	643.57
60	0.001017	251.16	0.8312	4.1829	1553.73	94938	466.39	654.37
t_s = 64.96				**Saturation**				
Liquid	0.001020	271.93	0.8931	4.1853	1556.14	94978	433.48	658.95
Vapour	6.2034	2617.45	7.8302	1.9763	452.54	1.3205	11.10	21.62
70	6.2989	2627.36	7.8593	1.9598	456.22	1.3217	11.27	21.96
80	6.4878	2646.86	7.9153	1.9430	463.11	1.3223	11.62	22.67
90	6.6759	2666.24	7.9694	1.9348	469.73	1.3220	11.97	23.41
100	6.8634	2685.57	8.0219	1.9302	476.18	1.3215	12.33	24.18
110	7.0505	2704.86	8.0729	1.9278	482.50	1.3208	12.70	24.97
120	7.2372	2724.13	8.1226	1.9269	488.70	1.3200	13.07	25.79
130	7.4237	2743.40	8.1710	1.9272	494.79	1.3191	13.45	26.62
140	7.6099	2762.67	8.2182	1.9285	500.78	1.3182	13.83	27.48
150	7.7958	2781.97	8.2643	1.9306	506.68	1.3172	14.22	28.36
160	7.9817	2801.29	8.3095	1.9333	512.48	1.3162	14.61	29.25
170	8.1673	2820.64	8.3536	1.9366	518.20	1.3152	15.00	30.16
180	8.3529	2840.02	8.3969	1.9403	523.84	1.3141	15.40	31.09
190	8.5383	2859.45	8.4393	1.9444	529.40	1.3130	15.80	32.03
200	8.7237	2878.91	8.4809	1.9488	534.89	1.3118	16.20	32.99
210	8.9090	2898.42	8.5217	1.9535	540.30	1.3107	16.60	33.96
220	9.0942	2917.98	8.5617	1.9583	545.65	1.3095	17.01	34.95
230	9.2794	2937.59	8.6011	1.9634	550.93	1.3084	17.42	35.95
240	9.4646	2957.25	8.6398	1.9687	556.15	1.3072	17.82	36.96
250	9.6496	2976.96	8.6778	1.9740	561.30	1.3060	18.23	37.99
260	9.8347	2996.73	8.7153	1.9795	566.40	1.3048	18.65	39.03
270	10.0197	3016.55	8.7521	1.9852	571.44	1.3036	19.06	40.08
280	10.2047	3036.44	8.7884	1.9909	576.43	1.3024	19.47	41.14
290	10.3897	3056.37	8.8241	1.9967	581.36	1.3012	19.88	42.22
300	10.5746	3076.37	8.8593	2.0026	586.24	1.3000	20.30	43.30
310	10.7595	3096.43	8.8940	2.0086	591.07	1.2988	20.71	44.40
320	10.9444	3116.54	8.9282	2.0146	595.86	1.2976	21.13	45.50
330	11.1293	3136.72	8.9619	2.0207	600.59	1.2964	21.54	46.62
340	11.3141	3156.96	8.9952	2.0269	605.28	1.2952	21.96	47.74

Table 4 Single-Phase Region (Regions 1 to 3 of IAPWS-IF97) – Continuation
Homogenes Zustandsgebiet (Bereiche 1 bis 3 der IAPWS-IF97) – Fortsetzung

p = 0.025 MPa								
t [°C]	v [m³ kg⁻¹]	h [kJ kg⁻¹]	s [kJ kg⁻¹ K⁻¹]	c_p [kJ kg⁻¹ K⁻¹]	w [m s⁻¹]	κ	$\eta \times 10^6$ [kg m⁻¹ s⁻¹]	$\lambda \times 10^3$ [W m⁻¹ K⁻¹]
350	11.4990	3177.26	9.0280	2.0331	609.92	1.2940	22.38	48.88
360	11.6838	3197.62	9.0605	2.0394	614.52	1.2928	22.79	50.02
370	11.8686	3218.04	9.0925	2.0457	619.08	1.2917	23.21	51.18
380	12.0534	3238.53	9.1241	2.0521	623.59	1.2905	23.62	52.34
390	12.2382	3259.09	9.1553	2.0585	628.06	1.2893	24.04	53.51
400	12.4230	3279.70	9.1862	2.0650	632.50	1.2881	24.45	54.69
410	12.6078	3300.39	9.2167	2.0715	636.89	1.2869	24.87	55.88
420	12.7926	3321.13	9.2468	2.0780	641.25	1.2857	25.28	57.08
430	12.9773	3341.95	9.2766	2.0846	645.57	1.2846	25.70	58.28
440	13.1621	3362.83	9.3061	2.0913	649.85	1.2834	26.11	59.49
450	13.3468	3383.77	9.3353	2.0979	654.10	1.2822	26.52	60.71
460	13.5316	3404.79	9.3641	2.1046	658.31	1.2811	26.93	61.94
470	13.7163	3425.87	9.3927	2.1114	662.49	1.2799	27.34	63.18
480	13.9010	3447.01	9.4210	2.1182	666.64	1.2788	27.76	64.42
490	14.0857	3468.23	9.4489	2.1250	670.75	1.2776	28.16	65.67
500	14.2704	3489.51	9.4767	2.1318	674.83	1.2765	28.57	66.93
510	14.4552	3510.86	9.5041	2.1387	678.88	1.2753	28.98	68.19
520	14.6399	3532.29	9.5313	2.1455	682.90	1.2742	29.39	69.46
530	14.8246	3553.78	9.5582	2.1524	686.89	1.2731	29.80	70.74
540	15.0093	3575.33	9.5849	2.1594	690.85	1.2720	30.20	72.02
550	15.1940	3596.96	9.6113	2.1663	694.79	1.2708	30.61	73.31
560	15.3787	3618.66	9.6375	2.1733	698.69	1.2697	31.01	74.61
570	15.5633	3640.43	9.6635	2.1803	702.57	1.2686	31.41	75.91
580	15.7480	3662.27	9.6892	2.1873	706.42	1.2675	31.81	77.22
590	15.9327	3684.18	9.7148	2.1943	710.25	1.2665	32.21	78.53
600	16.1174	3706.15	9.7401	2.2014	714.05	1.2654	32.61	79.85
610	16.3021	3728.20	9.7652	2.2084	717.82	1.2643	33.01	81.18
620	16.4867	3750.32	9.7901	2.2155	721.57	1.2632	33.41	82.51
630	16.6714	3772.51	9.8148	2.2225	725.30	1.2622	33.81	83.84
640	16.8561	3794.77	9.8393	2.2296	729.00	1.2611	34.20	85.19
650	17.0408	3817.10	9.8636	2.2366	732.68	1.2601	34.60	86.53
660	17.2254	3839.51	9.8878	2.2437	736.34	1.2590	34.99	87.88
670	17.4101	3861.98	9.9117	2.2508	739.97	1.2580	35.38	89.24
680	17.5947	3884.52	9.9355	2.2579	743.58	1.2570	35.77	90.60
690	17.7794	3907.14	9.9591	2.2649	747.17	1.2560	36.16	91.97
700	17.9641	3929.82	9.9825	2.2720	750.74	1.2550	36.55	93.34
720	18.3334	3975.40	10.0289	2.2861	757.82	1.2530	37.33	96.10
740	18.7027	4021.27	10.0746	2.3003	764.82	1.2511	38.09	98.87
760	19.0720	4067.41	10.1197	2.3144	771.75	1.2492	38.86	101.66
780	19.4413	4113.84	10.1642	2.3285	778.60	1.2473	39.62	104.47
800	19.8106	4160.55	10.2082	2.3426	785.37	1.2454	40.37	107.30

Table 4 Single-Phase Region (Regions 1 to 3 of IAPWS-IF97) – Continuation
Homogenes Zustandsgebiet (Bereiche 1 bis 3 der IAPWS-IF97) – Fortsetzung

				p = 0.05 MPa				
t [°C]	v [m^3 kg^{-1}]	h [kJ kg^{-1}]	s [kJ kg^{-1} K^{-1}]	c_p [kJ kg^{-1} K^{-1}]	w [m s^{-1}]	κ	$\eta\times 10^6$ [kg m^{-1} s^{-1}]	$\lambda\times 10^3$ [W m^{-1} K^{-1}]
0	0.001000	0.01	– 0.0002	4.2197	1402.36	39325	1791.66	561.05
5	0.001000	21.07	0.0763	4.2052	1426.11	40673	1518.21	570.55
10	0.001000	42.07	0.1511	4.1956	1447.47	41890	1305.93	580.02
15	0.001001	63.03	0.2245	4.1893	1466.51	42973	1137.59	589.35
20	0.001002	83.96	0.2965	4.1850	1483.34	43925	1001.63	598.44
25	0.001003	104.88	0.3672	4.1820	1498.09	44751	890.09	607.17
30	0.001004	125.79	0.4368	4.1802	1510.90	45456	797.35	615.48
35	0.001006	146.68	0.5052	4.1791	1521.90	46046	719.32	623.31
40	0.001008	167.58	0.5724	4.1787	1531.22	46526	652.97	630.61
45	0.001010	188.47	0.6386	4.1789	1538.96	46903	596.06	637.37
50	0.001012	209.37	0.7038	4.1797	1545.24	47183	546.84	643.58
60	0.001017	251.18	0.8312	4.1829	1553.77	47472	466.39	654.38
70	0.001023	293.03	0.9550	4.1882	1557.49	47436	403.89	663.12
80	0.001029	334.95	1.0754	4.1956	1556.97	47114	354.34	670.00
t_s = 81.32				**Saturation**				
Liquid	0.001030	340.48	1.0910	4.1968	1556.61	47053	348.59	670.78
Vapour	3.2401	2645.21	7.5930	2.0155	462.16	1.3184	11.64	23.14
90	3.3235	2662.59	7.6414	1.9896	468.28	1.3196	11.95	23.75
100	3.4188	2682.40	7.6952	1.9732	474.97	1.3197	12.31	24.48
110	3.5136	2702.07	7.7472	1.9630	481.45	1.3194	12.68	25.23
120	3.6081	2721.67	7.7977	1.9562	487.78	1.3189	13.06	26.02
130	3.7022	2741.21	7.8468	1.9519	493.98	1.3182	13.44	26.83
140	3.7962	2760.71	7.8946	1.9494	500.06	1.3174	13.82	27.67
150	3.8899	2780.20	7.9412	1.9484	506.03	1.3166	14.21	28.52
160	3.9834	2799.68	7.9867	1.9486	511.90	1.3157	14.60	29.40
170	4.0768	2819.18	8.0312	1.9498	517.67	1.3147	14.99	30.29
180	4.1700	2838.68	8.0747	1.9518	523.36	1.3137	15.39	31.21
190	4.2632	2858.21	8.1174	1.9544	528.96	1.3126	15.79	32.14
200	4.3563	2877.77	8.1591	1.9576	534.48	1.3115	16.19	33.09
210	4.4493	2897.37	8.2001	1.9612	539.93	1.3104	16.60	34.05
220	4.5422	2917.00	8.2403	1.9652	545.31	1.3093	17.00	35.03
230	4.6351	2936.67	8.2798	1.9696	550.61	1.3082	17.41	36.02
240	4.7279	2956.39	8.3186	1.9742	555.86	1.3070	17.82	37.03
250	4.8207	2976.16	8.3568	1.9790	561.03	1.3059	18.23	38.05
260	4.9134	2995.97	8.3943	1.9840	566.15	1.3047	18.64	39.09
270	5.0061	3015.84	8.4312	1.9893	571.21	1.3035	19.05	40.13
280	5.0988	3035.76	8.4676	1.9946	576.21	1.3023	19.47	41.19
290	5.1915	3055.73	8.5034	2.0001	581.16	1.3011	19.88	42.26
300	5.2841	3075.76	8.5386	2.0057	586.05	1.3000	20.30	43.34
310	5.3767	3095.85	8.5733	2.0115	590.89	1.2988	20.71	44.43
320	5.4693	3115.99	8.6076	2.0173	595.69	1.2976	21.13	45.54
330	5.5619	3136.19	8.6414	2.0232	600.43	1.2964	21.54	46.65
340	5.6544	3156.46	8.6747	2.0292	605.13	1.2952	21.96	47.77

Table 4 Single-Phase Region (Regions 1 to 3 of IAPWS-IF97) – Continuation
Homogenes Zustandsgebiet (Bereiche 1 bis 3 der IAPWS-IF97) – Fortsetzung

p = **0.05 MPa**

t [°C]	v [m³ kg⁻¹]	h [kJ kg⁻¹]	s [kJ kg⁻¹ K⁻¹]	c_p [kJ kg⁻¹ K⁻¹]	w [m s⁻¹]	κ	$\eta \times 10^6$ [kg m⁻¹ s⁻¹]	$\lambda \times 10^3$ [W m⁻¹ K⁻¹]
350	5.7470	3176.78	8.7076	2.0353	609.78	1.2940	22.37	48.91
360	5.8395	3197.16	8.7400	2.0414	614.39	1.2928	22.79	50.05
370	5.9320	3217.61	8.7721	2.0476	618.95	1.2916	23.21	51.20
380	6.0245	3238.11	8.8037	2.0539	623.47	1.2904	23.62	52.36
390	6.1170	3258.68	8.8349	2.0602	627.95	1.2893	24.04	53.53
400	6.2095	3279.32	8.8658	2.0665	632.39	1.2881	24.45	54.71
410	6.3020	3300.01	8.8964	2.0730	636.79	1.2869	24.87	55.90
420	6.3944	3320.78	8.9265	2.0794	641.15	1.2857	25.28	57.10
430	6.4869	3341.60	8.9564	2.0860	645.48	1.2846	25.70	58.30
440	6.5793	3362.49	8.9859	2.0925	649.76	1.2834	26.11	59.51
450	6.6718	3383.45	9.0150	2.0991	654.02	1.2822	26.52	60.73
460	6.7642	3404.48	9.0439	2.1058	658.23	1.2811	26.93	61.96
470	6.8566	3425.57	9.0725	2.1124	662.42	1.2799	27.34	63.20
480	6.9490	3446.73	9.1008	2.1192	666.57	1.2788	27.76	64.44
490	7.0415	3467.95	9.1288	2.1259	670.68	1.2776	28.17	65.69
500	7.1339	3489.24	9.1565	2.1327	674.77	1.2765	28.57	66.94
510	7.2263	3510.61	9.1839	2.1395	678.82	1.2753	28.98	68.21
520	7.3187	3532.04	9.2111	2.1464	682.84	1.2742	29.39	69.48
530	7.4111	3553.53	9.2381	2.1532	686.84	1.2731	29.80	70.75
540	7.5035	3575.10	9.2648	2.1601	690.80	1.2720	30.20	72.04
550	7.5959	3596.74	9.2912	2.1671	694.74	1.2708	30.61	73.33
560	7.6882	3618.44	9.3174	2.1740	698.64	1.2697	31.01	74.62
570	7.7806	3640.22	9.3434	2.1810	702.52	1.2686	31.41	75.93
580	7.8730	3662.06	9.3691	2.1879	706.38	1.2675	31.81	77.23
590	7.9654	3683.98	9.3947	2.1949	710.21	1.2665	32.22	78.55
600	8.0578	3705.96	9.4200	2.2019	714.01	1.2654	32.62	79.87
610	8.1501	3728.01	9.4451	2.2090	717.78	1.2643	33.01	81.19
620	8.2425	3750.14	9.4700	2.2160	721.54	1.2632	33.41	82.52
630	8.3349	3772.33	9.4947	2.2230	725.26	1.2622	33.81	83.86
640	8.4272	3794.60	9.5193	2.2301	728.97	1.2611	34.20	85.20
650	8.5196	3816.94	9.5436	2.2371	732.65	1.2601	34.60	86.55
660	8.6120	3839.34	9.5677	2.2442	736.31	1.2591	34.99	87.90
670	8.7043	3861.82	9.5917	2.2512	739.94	1.2580	35.38	89.25
680	8.7967	3884.37	9.6155	2.2583	743.56	1.2570	35.77	90.62
690	8.8890	3906.99	9.6391	2.2653	747.15	1.2560	36.16	91.98
700	8.9814	3929.67	9.6625	2.2724	750.72	1.2550	36.55	93.35
720	9.1661	3975.26	9.7089	2.2865	757.80	1.2530	37.33	96.11
740	9.3508	4021.13	9.7546	2.3006	764.81	1.2511	38.10	98.89
760	9.5355	4067.29	9.7997	2.3147	771.73	1.2492	38.86	101.68
780	9.7201	4113.72	9.8442	2.3288	778.58	1.2473	39.62	104.49
800	9.9048	4160.44	9.8882	2.3428	785.36	1.2454	40.37	107.31

Table 4 Single-Phase Region (Regions 1 to 3 of IAPWS-IF97) – Continuation
Homogenes Zustandsgebiet (Bereiche 1 bis 3 der IAPWS-IF97) – Fortsetzung

				p = 0.075 MPa				
t [°C]	v [m^3 kg^{-1}]	h [kJ kg^{-1}]	s [kJ kg^{-1} K^{-1}]	c_p [kJ kg^{-1} K^{-1}]	w [m s^{-1}]	κ	$\eta \times 10^6$ [kg m^{-1} s^{-1}]	$\lambda \times 10^3$ [W m^{-1} K^{-1}]
0	0.001000	0.03	−0.0001	4.2196	1402.40	26218	1791.59	561.06
5	0.001000	21.09	0.0763	4.2051	1426.15	27117	1518.17	570.56
10	0.001000	42.09	0.1511	4.1955	1447.51	27928	1305.90	580.03
15	0.001001	63.05	0.2245	4.1892	1466.55	28650	1137.57	589.37
20	0.001002	83.99	0.2965	4.1849	1483.38	29285	1001.62	598.45
25	0.001003	104.90	0.3672	4.1820	1498.13	29836	890.09	607.18
30	0.001004	125.81	0.4368	4.1801	1510.94	30306	797.35	615.49
35	0.001006	146.71	0.5051	4.1790	1521.94	30699	719.32	623.32
40	0.001008	167.60	0.5724	4.1786	1531.26	31019	652.98	630.62
45	0.001010	188.49	0.6386	4.1788	1539.00	31271	596.07	637.38
50	0.001012	209.39	0.7038	4.1796	1545.28	31457	546.85	643.60
60	0.001017	251.20	0.8312	4.1828	1553.81	31650	466.40	654.39
70	0.001023	293.05	0.9550	4.1882	1557.54	31626	403.89	663.13
80	0.001029	334.97	1.0754	4.1956	1557.02	31412	354.35	670.02
90	0.001036	376.97	1.1927	4.2051	1552.72	31030	314.41	675.26
t_s = 91.76				**Saturation**				
Liquid	0.001037	384.37	1.2130	4.2070	1551.61	30947	308.19	676.03
Vapour	2.2171	2662.39	7.4557	2.0474	467.93	1.3168	11.99	24.20
100	2.2704	2679.14	7.5011	2.0202	473.70	1.3178	12.29	24.77
110	2.3345	2699.23	7.5542	2.0001	480.38	1.3180	12.66	25.50
120	2.3982	2719.16	7.6056	1.9868	486.84	1.3177	13.04	26.25
130	2.4617	2738.98	7.6553	1.9774	493.15	1.3172	13.42	27.04
140	2.5248	2758.72	7.7037	1.9710	499.32	1.3166	13.81	27.85
150	2.5878	2778.41	7.7508	1.9668	505.37	1.3159	14.20	28.69
160	2.6506	2798.06	7.7967	1.9643	511.30	1.3151	14.59	29.55
170	2.7132	2817.70	7.8415	1.9633	517.13	1.3142	14.98	30.43
180	2.7757	2837.33	7.8853	1.9635	522.87	1.3133	15.38	31.33
190	2.8381	2856.97	7.9282	1.9646	528.52	1.3123	15.78	32.25
200	2.9004	2876.63	7.9702	1.9666	534.08	1.3113	16.18	33.19
210	2.9627	2896.31	8.0113	1.9691	539.56	1.3102	16.59	34.14
220	3.0248	2916.01	8.0517	1.9722	544.97	1.3091	17.00	35.11
230	3.0869	2935.75	8.0913	1.9758	550.30	1.3080	17.40	36.10
240	3.1490	2955.53	8.1303	1.9798	555.56	1.3069	17.81	37.10
250	3.2110	2975.35	8.1685	1.9840	560.76	1.3057	18.22	38.11
260	3.2730	2995.21	8.2061	1.9886	565.90	1.3046	18.64	39.14
270	3.3350	3015.12	8.2431	1.9934	570.98	1.3034	19.05	40.18
280	3.3969	3035.08	8.2795	1.9984	575.99	1.3022	19.46	41.24
290	3.4588	3055.09	8.3154	2.0036	580.95	1.3011	19.88	42.30
300	3.5206	3075.15	8.3507	2.0089	585.86	1.2999	20.29	43.38
310	3.5825	3095.27	8.3855	2.0144	590.72	1.2987	20.71	44.47
320	3.6443	3115.44	8.4198	2.0200	595.52	1.2975	21.12	45.57
330	3.7061	3135.67	8.4536	2.0257	600.27	1.2963	21.54	46.68
340	3.7679	3155.95	8.4870	2.0315	604.98	1.2952	21.96	47.81

Table 4 Single-Phase Region (Regions 1 to 3 of IAPWS-IF97) – Continuation
Homogenes Zustandsgebiet (Bereiche 1 bis 3 der IAPWS-IF97) – Fortsetzung

$p = 0.075$ MPa

t [°C]	v [m^3 kg^{-1}]	h [kJ kg^{-1}]	s [kJ kg^{-1} K^{-1}]	c_p [kJ kg^{-1} K^{-1}]	w [m s^{-1}]	κ	$\eta \times 10^6$ [kg m^{-1} s^{-1}]	$\lambda \times 10^3$ [W m^{-1} K^{-1}]
350	3.8296	3176.30	8.5199	2.0374	609.64	1.2940	22.37	48.94
360	3.8914	3196.70	8.5523	2.0434	614.25	1.2928	22.79	50.08
370	3.9531	3217.17	8.5844	2.0495	618.82	1.2916	23.20	51.23
380	4.0149	3237.69	8.6161	2.0556	623.35	1.2904	23.62	52.39
390	4.0766	3258.28	8.6474	2.0618	627.84	1.2892	24.04	53.56
400	4.1383	3278.93	8.6783	2.0681	632.28	1.2881	24.45	54.74
410	4.2000	3299.64	8.7088	2.0744	636.69	1.2869	24.87	55.92
420	4.2617	3320.42	8.7390	2.0808	641.06	1.2857	25.28	57.12
430	4.3234	3341.26	8.7689	2.0873	645.39	1.2846	25.69	58.32
440	4.3851	3362.16	8.7984	2.0938	649.68	1.2834	26.11	59.53
450	4.4467	3383.13	8.8276	2.1003	653.93	1.2822	26.52	60.75
460	4.5084	3404.17	8.8565	2.1069	658.16	1.2811	26.93	61.98
470	4.5701	3425.27	8.8851	2.1135	662.34	1.2799	27.34	63.21
480	4.6317	3446.44	8.9134	2.1202	666.50	1.2788	27.76	64.46
490	4.6934	3467.67	8.9414	2.1269	670.62	1.2776	28.17	65.71
500	4.7550	3488.98	8.9691	2.1336	674.70	1.2765	28.57	66.96
510	4.8167	3510.35	8.9966	2.1404	678.76	1.2753	28.98	68.22
520	4.8783	3531.78	9.0238	2.1472	682.79	1.2742	29.39	69.49
530	4.9399	3553.29	9.0507	2.1540	686.78	1.2731	29.80	70.77
540	5.0015	3574.87	9.0774	2.1609	690.75	1.2720	30.20	72.05
550	5.0632	3596.51	9.1039	2.1678	694.69	1.2709	30.61	73.34
560	5.1248	3618.22	9.1301	2.1747	698.60	1.2697	31.01	74.64
570	5.1864	3640.00	9.1561	2.1816	702.48	1.2686	31.41	75.94
580	5.2480	3661.85	9.1818	2.1886	706.34	1.2676	31.82	77.25
590	5.3096	3683.78	9.2074	2.1956	710.17	1.2665	32.22	78.56
600	5.3712	3705.77	9.2327	2.2025	713.97	1.2654	32.62	79.88
610	5.4328	3727.83	9.2578	2.2095	717.75	1.2643	33.01	81.21
620	5.4944	3749.96	9.2827	2.2165	721.50	1.2633	33.41	82.54
630	5.5560	3772.16	9.3075	2.2236	725.23	1.2622	33.81	83.87
640	5.6176	3794.43	9.3320	2.2306	728.94	1.2611	34.20	85.21
650	5.6792	3816.77	9.3563	2.2376	732.62	1.2601	34.60	86.56
660	5.7408	3839.18	9.3805	2.2446	736.28	1.2591	34.99	87.91
670	5.8024	3861.66	9.4044	2.2517	739.92	1.2580	35.38	89.27
680	5.8640	3884.21	9.4282	2.2587	743.53	1.2570	35.78	90.63
690	5.9256	3906.84	9.4518	2.2658	747.13	1.2560	36.16	92.00
700	5.9872	3929.53	9.4753	2.2728	750.70	1.2550	36.55	93.37
720	6.1103	3975.13	9.5216	2.2869	757.78	1.2530	37.33	96.12
740	6.2335	4021.00	9.5674	2.3009	764.79	1.2511	38.10	98.90
760	6.3566	4067.16	9.6125	2.3150	771.72	1.2492	38.86	101.69
780	6.4798	4113.60	9.6570	2.3290	778.57	1.2473	39.62	104.50
800	6.6029	4160.32	9.7010	2.3431	785.35	1.2455	40.38	107.32

Table 4 Single-Phase Region (Regions 1 to 3 of IAPWS-IF97) – Continuation
Homogenes Zustandsgebiet (Bereiche 1 bis 3 der IAPWS-IF97) – Fortsetzung

				p = 0.1 MPa				
t [°C]	v [m^3 kg^{-1}]	h [kJ kg^{-1}]	s [kJ kg^{-1} K^{-1}]	c_p [kJ kg^{-1} K^{-1}]	w [m s^{-1}]	κ	$\eta \times 10^6$ [kg m^{-1} s^{-1}]	$\lambda \times 10^3$ [W m^{-1} K^{-1}]
0	0.001000	0.06	– 0.0001	4.2194	1402.44	19665	1791.53	561.08
5	0.001000	21.12	0.0763	4.2050	1426.19	20339	1518.13	570.57
10	0.001000	42.12	0.1511	4.1955	1447.55	20947	1305.88	580.05
15	0.001001	63.08	0.2245	4.1891	1466.59	21489	1137.55	589.38
20	0.001002	84.01	0.2965	4.1848	1483.42	21965	1001.61	598.46
25	0.001003	104.93	0.3672	4.1819	1498.17	22378	890.08	607.19
30	0.001004	125.83	0.4368	4.1800	1510.98	22731	797.35	615.50
35	0.001006	146.73	0.5051	4.1790	1521.98	23026	719.32	623.33
40	0.001008	167.62	0.5724	4.1786	1531.30	23266	652.98	630.63
45	0.001010	188.52	0.6386	4.1788	1539.04	23454	596.07	637.40
50	0.001012	209.41	0.7038	4.1796	1545.32	23594	546.85	643.61
60	0.001017	251.22	0.8312	4.1828	1553.86	23739	466.40	654.40
70	0.001023	293.07	0.9550	4.1881	1557.58	23721	403.90	663.14
80	0.001029	334.99	1.0754	4.1955	1557.06	23560	354.35	670.03
90	0.001036	376.99	1.1926	4.2050	1552.77	23274	314.41	675.28
t_s = 99.61				**Saturation**				
Liquid	0.001043	417.44	1.3026	4.2161	1545.45	22896	282.92	678.97
Vapour	1.6940	2674.95	7.3588	2.0759	472.05	1.3154	12.26	25.05
100	1.6960	2675.77	7.3610	2.0741	472.34	1.3155	12.27	25.08
110	1.7448	2696.32	7.4154	2.0399	479.27	1.3165	12.64	25.77
120	1.7932	2716.61	7.4676	2.0187	485.89	1.3166	13.02	26.49
130	1.8413	2736.72	7.5181	2.0039	492.31	1.3163	13.41	27.25
140	1.8891	2756.70	7.5671	1.9933	498.57	1.3158	13.79	28.04
150	1.9367	2776.59	7.6147	1.9857	504.70	1.3152	14.18	28.86
160	1.9841	2796.42	7.6610	1.9805	510.70	1.3145	14.58	29.70
170	2.0314	2816.21	7.7062	1.9772	516.59	1.3137	14.97	30.56
180	2.0785	2835.97	7.7503	1.9755	522.38	1.3129	15.37	31.45
190	2.1256	2855.72	7.7934	1.9751	528.07	1.3119	15.77	32.36
200	2.1725	2875.48	7.8356	1.9757	533.67	1.3110	16.18	33.28
210	2.2194	2895.24	7.8769	1.9772	539.19	1.3099	16.58	34.23
220	2.2661	2915.02	7.9174	1.9793	544.62	1.3089	16.99	35.19
230	2.3129	2934.83	7.9572	1.9821	549.98	1.3078	17.40	36.17
240	2.3596	2954.66	7.9962	1.9854	555.27	1.3067	17.81	37.17
250	2.4062	2974.54	8.0346	1.9891	560.49	1.3056	18.22	38.17
260	2.4528	2994.45	8.0723	1.9932	565.65	1.3045	18.63	39.20
270	2.4994	3014.40	8.1094	1.9975	570.74	1.3033	19.05	40.24
280	2.5459	3034.40	8.1458	2.0022	575.77	1.3022	19.46	41.29
290	2.5924	3054.45	8.1818	2.0070	580.75	1.3010	19.87	42.35
300	2.6389	3074.54	8.2171	2.0121	585.67	1.2998	20.29	43.42
310	2.6853	3094.69	8.2520	2.0173	590.54	1.2987	20.71	44.51
320	2.7318	3114.89	8.2863	2.0227	595.35	1.2975	21.12	45.61
330	2.7782	3135.14	8.3202	2.0282	600.11	1.2963	21.54	46.72
340	2.8246	3155.45	8.3536	2.0338	604.83	1.2951	21.95	47.84

Table 4 Single-Phase Region (Regions 1 to 3 of IAPWS-IF97) – Continuation
Homogenes Zustandsgebiet (Bereiche 1 bis 3 der IAPWS-IF97) – Fortsetzung

				p = 0.1 MPa				
t [°C]	v [$m^3\,kg^{-1}$]	h [$kJ\,kg^{-1}$]	s [$kJ\,kg^{-1}\,K^{-1}$]	c_p [$kJ\,kg^{-1}\,K^{-1}$]	w [$m\,s^{-1}$]	κ	$\eta \times 10^6$ [$kg\,m^{-1}\,s^{-1}$]	$\lambda \times 10^3$ [$W\,m^{-1}\,K^{-1}$]
350	2.8710	3175.82	8.3865	2.0396	609.50	1.2939	22.37	48.97
360	2.9173	3196.24	8.4190	2.0454	614.12	1.2928	22.79	50.11
370	2.9637	3216.73	8.4511	2.0514	618.70	1.2916	23.20	51.26
380	3.0101	3237.27	8.4828	2.0574	623.23	1.2904	23.62	52.42
390	3.0564	3257.87	8.5141	2.0635	627.73	1.2892	24.04	53.58
400	3.1027	3278.54	8.5451	2.0697	632.18	1.2881	24.45	54.76
410	3.1490	3299.27	8.5756	2.0759	636.59	1.2869	24.87	55.95
420	3.1954	3320.06	8.6059	2.0822	640.96	1.2857	25.28	57.14
430	3.2417	3340.91	8.6357	2.0886	645.30	1.2845	25.69	58.34
440	3.2879	3361.83	8.6653	2.0950	649.59	1.2834	26.11	59.55
450	3.3342	3382.81	8.6945	2.1015	653.85	1.2822	26.52	60.77
460	3.3805	3403.86	8.7234	2.1080	658.08	1.2811	26.93	62.00
470	3.4268	3424.97	8.7520	2.1146	662.27	1.2799	27.34	63.23
480	3.4731	3446.15	8.7803	2.1212	666.43	1.2788	27.76	64.47
490	3.5193	3467.40	8.8083	2.1279	670.55	1.2776	28.17	65.72
500	3.5656	3488.71	8.8361	2.1345	674.64	1.2765	28.57	66.98
510	3.6118	3510.09	8.8635	2.1413	678.70	1.2753	28.98	68.24
520	3.6581	3531.53	8.8907	2.1480	682.73	1.2742	29.39	69.51
530	3.7043	3553.05	8.9177	2.1548	686.73	1.2731	29.80	70.79
540	3.7506	3574.63	8.9444	2.1617	690.70	1.2720	30.20	72.07
550	3.7968	3596.28	8.9709	2.1685	694.64	1.2709	30.61	73.36
560	3.8430	3618.00	8.9971	2.1754	698.55	1.2698	31.01	74.65
570	3.8893	3639.79	9.0231	2.1823	702.44	1.2687	31.41	75.96
580	3.9355	3661.65	9.0489	2.1892	706.29	1.2676	31.82	77.26
590	3.9817	3683.58	9.0744	2.1962	710.12	1.2665	32.22	78.58
600	4.0279	3705.57	9.0998	2.2031	713.93	1.2654	32.62	79.90
610	4.0742	3727.64	9.1249	2.2101	717.71	1.2643	33.02	81.22
620	4.1204	3749.77	9.1498	2.2171	721.47	1.2633	33.41	82.55
630	4.1666	3771.98	9.1745	2.2241	725.20	1.2622	33.81	83.89
640	4.2128	3794.26	9.1991	2.2311	728.90	1.2612	34.21	85.23
650	4.2590	3816.60	9.2234	2.2381	732.59	1.2601	34.60	86.57
660	4.3052	3839.02	9.2476	2.2451	736.25	1.2591	34.99	87.93
670	4.3514	3861.50	9.2715	2.2521	739.89	1.2581	35.39	89.28
680	4.3976	3884.06	9.2953	2.2591	743.51	1.2570	35.78	90.64
690	4.4438	3906.69	9.3189	2.2662	747.10	1.2560	36.17	92.01
700	4.4900	3929.38	9.3424	2.2732	750.68	1.2550	36.55	93.38
720	4.5824	3974.99	9.3888	2.2872	757.76	1.2531	37.33	96.14
740	4.6748	4020.87	9.4345	2.3013	764.77	1.2511	38.10	98.91
760	4.7672	4067.04	9.4796	2.3153	771.70	1.2492	38.86	101.70
780	4.8596	4113.48	9.5241	2.3293	778.55	1.2473	39.62	104.51
800	4.9520	4160.21	9.5681	2.3434	785.34	1.2455	40.38	107.33

Table 4 Single-Phase Region (Regions 1 to 3 of IAPWS-IF97) – Continuation
Homogenes Zustandsgebiet (Bereiche 1 bis 3 der IAPWS-IF97) – Fortsetzung

p = 0.125 MPa

t [°C]	v [m^3 kg^{-1}]	h [kJ kg^{-1}]	s [kJ kg^{-1} K^{-1}]	c_p [kJ kg^{-1} K^{-1}]	w [m s^{-1}]	κ	$\eta \times 10^6$ [kg m^{-1} s^{-1}]	$\lambda \times 10^3$ [W m^{-1} K^{-1}]
0	0.001000	0.09	– 0.0001	4.2193	1402.48	15733	1791.47	561.09
5	0.001000	21.14	0.0762	4.2048	1426.23	16272	1518.09	570.59
10	0.001000	42.14	0.1511	4.1954	1447.59	16759	1305.85	580.06
15	0.001001	63.10	0.2245	4.1890	1466.63	17192	1137.53	589.39
20	0.001002	84.04	0.2965	4.1847	1483.46	17573	1001.60	598.47
25	0.001003	104.95	0.3672	4.1818	1498.21	17904	890.08	607.21
30	0.001004	125.86	0.4368	4.1800	1511.02	18186	797.34	615.52
35	0.001006	146.75	0.5051	4.1789	1522.02	18422	719.32	623.34
40	0.001008	167.65	0.5724	4.1785	1531.34	18614	652.98	630.65
45	0.001010	188.54	0.6386	4.1787	1539.08	18765	596.07	637.41
50	0.001012	209.43	0.7037	4.1795	1545.36	18877	546.86	643.62
60	0.001017	251.24	0.8312	4.1827	1553.90	18992	466.41	654.41
70	0.001023	293.09	0.9549	4.1880	1557.63	18978	403.91	663.15
80	0.001029	335.01	1.0753	4.1955	1557.11	18850	354.36	670.04
90	0.001036	377.01	1.1926	4.2050	1552.82	18621	314.42	675.29
100	0.001043	419.12	1.3070	4.2166	1545.14	18304	281.75	679.11
t_s = 105.79				**Saturation**				
Liquid	0.001048	444.30	1.3739	4.2246	1539.08	18078	265.02	680.78
Vapour	1.3749	2684.89	7.2840	2.1025	475.25	1.3142	12.47	25.77
110	1.3909	2693.33	7.3061	2.0841	478.11	1.3148	12.63	26.04
120	1.4302	2714.00	7.3594	2.0525	484.92	1.3154	13.01	26.73
130	1.4690	2734.41	7.4106	2.0314	491.46	1.3153	13.39	27.46
140	1.5077	2754.65	7.4602	2.0162	497.82	1.3150	13.78	28.23
150	1.5460	2774.75	7.5083	2.0051	504.03	1.3146	14.17	29.03
160	1.5842	2794.76	7.5550	1.9970	510.10	1.3139	14.57	29.85
170	1.6223	2814.70	7.6005	1.9914	516.05	1.3132	14.96	30.70
180	1.6602	2834.59	7.6449	1.9878	521.89	1.3124	15.36	31.57
190	1.6980	2854.46	7.6883	1.9857	527.62	1.3116	15.76	32.47
200	1.7357	2874.31	7.7307	1.9850	533.26	1.3107	16.17	33.38
210	1.7734	2894.16	7.7722	1.9853	538.81	1.3097	16.58	34.32
220	1.8109	2914.02	7.8129	1.9866	544.28	1.3087	16.98	35.27
230	1.8484	2933.90	7.8528	1.9885	549.66	1.3076	17.39	36.24
240	1.8859	2953.80	7.8920	1.9911	554.98	1.3066	17.80	37.23
250	1.9233	2973.72	7.9304	1.9942	560.22	1.3055	18.22	38.24
260	1.9607	2993.68	7.9682	1.9978	565.39	1.3043	18.63	39.25
270	1.9980	3013.68	8.0054	2.0017	570.51	1.3032	19.04	40.29
280	2.0353	3033.72	8.0419	2.0060	575.55	1.3021	19.46	41.33
290	2.0726	3053.80	8.0779	2.0105	580.54	1.3009	19.87	42.39
300	2.1098	3073.93	8.1133	2.0153	585.48	1.2998	20.29	43.46
310	2.1470	3094.11	8.1482	2.0202	590.36	1.2986	20.70	44.55
320	2.1843	3114.33	8.1826	2.0254	595.18	1.2974	21.12	45.64
330	2.2214	3134.61	8.2165	2.0307	599.96	1.2963	21.54	46.75
340	2.2586	3154.95	8.2500	2.0361	604.68	1.2951	21.95	47.87

Table 4 Single-Phase Region (Regions 1 to 3 of IAPWS-IF97) – Continuation
Homogenes Zustandsgebiet (Bereiche 1 bis 3 der IAPWS-IF97) – Fortsetzung

p = 0.125 MPa

t [°C]	v [m^3 kg^{-1}]	h [kJ kg^{-1}]	s [kJ kg^{-1} K^{-1}]	c_p [kJ kg^{-1} K^{-1}]	w [m s^{-1}]	κ	$\eta \times 10^6$ [kg m^{-1} s^{-1}]	$\lambda \times 10^3$ [W m^{-1} K^{-1}]
350	2.2958	3175.34	8.2830	2.0417	609.36	1.2939	22.37	49.00
360	2.3329	3195.78	8.3155	2.0474	613.99	1.2927	22.79	50.13
370	2.3700	3216.29	8.3476	2.0532	618.57	1.2916	23.20	51.28
380	2.4072	3236.85	8.3794	2.0592	623.12	1.2904	23.62	52.44
390	2.4443	3257.47	8.4107	2.0652	627.61	1.2892	24.03	53.61
400	2.4814	3278.15	8.4417	2.0712	632.07	1.2880	24.45	54.78
410	2.5185	3298.89	8.4722	2.0774	636.49	1.2869	24.87	55.97
420	2.5555	3319.70	8.5025	2.0836	640.87	1.2857	25.28	57.16
430	2.5926	3340.57	8.5324	2.0899	645.20	1.2845	25.69	58.36
440	2.6297	3361.50	8.5619	2.0963	649.51	1.2834	26.11	59.57
450	2.6667	3382.49	8.5912	2.1027	653.77	1.2822	26.52	60.79
460	2.7038	3403.55	8.6201	2.1091	658.00	1.2811	26.93	62.02
470	2.7408	3424.67	8.6487	2.1156	662.19	1.2799	27.34	63.25
480	2.7779	3445.86	8.6770	2.1222	666.36	1.2788	27.76	64.49
490	2.8149	3467.12	8.7051	2.1288	670.48	1.2776	28.17	65.74
500	2.8519	3488.44	8.7328	2.1355	674.58	1.2765	28.57	67.00
510	2.8889	3509.83	8.7603	2.1422	678.64	1.2753	28.98	68.26
520	2.9260	3531.28	8.7875	2.1489	682.67	1.2742	29.39	69.53
530	2.9630	3552.81	8.8145	2.1556	686.67	1.2731	29.80	70.80
540	3.0000	3574.40	8.8412	2.1624	690.65	1.2720	30.20	72.09
550	3.0370	3596.06	8.8677	2.1693	694.59	1.2709	30.61	73.37
560	3.0740	3617.78	8.8939	2.1761	698.50	1.2698	31.01	74.67
570	3.1110	3639.58	8.9199	2.1830	702.39	1.2687	31.42	75.97
580	3.1480	3661.44	8.9457	2.1899	706.25	1.2676	31.82	77.28
590	3.1850	3683.37	8.9713	2.1968	710.08	1.2665	32.22	78.59
600	3.2220	3705.38	8.9966	2.2037	713.89	1.2654	32.62	79.91
610	3.2590	3727.45	9.0217	2.2107	717.67	1.2643	33.02	81.24
620	3.2960	3749.59	9.0467	2.2176	721.43	1.2633	33.41	82.57
630	3.3329	3771.80	9.0714	2.2246	725.16	1.2622	33.81	83.90
640	3.3699	3794.08	9.0959	2.2316	728.87	1.2612	34.21	85.24
650	3.4069	3816.43	9.1203	2.2386	732.56	1.2601	34.60	86.59
660	3.4439	3838.85	9.1444	2.2456	736.22	1.2591	34.99	87.94
670	3.4809	3861.34	9.1684	2.2526	739.86	1.2581	35.39	89.30
680	3.5178	3883.91	9.1922	2.2596	743.48	1.2571	35.78	90.66
690	3.5548	3906.54	9.2158	2.2666	747.08	1.2560	36.17	92.02
700	3.5918	3929.24	9.2393	2.2736	750.65	1.2550	36.56	93.40
720	3.6657	3974.85	9.2857	2.2876	757.74	1.2531	37.33	96.15
740	3.7396	4020.74	9.3314	2.3016	764.75	1.2511	38.10	98.92
760	3.8136	4066.91	9.3765	2.3156	771.68	1.2492	38.86	101.72
780	3.8875	4113.37	9.4211	2.3296	778.54	1.2473	39.62	104.52
800	3.9614	4160.10	9.4650	2.3436	785.32	1.2455	40.38	107.35

Table 4 Single-Phase Region (Regions 1 to 3 of IAPWS-IF97) – Continuation
Homogenes Zustandsgebiet (Bereiche 1 bis 3 der IAPWS-IF97) – Fortsetzung

				p = 0.15 MPa				
t [°C]	v [m^3 kg^{-1}]	h [kJ kg^{-1}]	s [kJ kg^{-1} K^{-1}]	c_p [kJ kg^{-1} K^{-1}]	w [m s^{-1}]	κ	$\eta \times 10^6$ [kg m^{-1} s^{-1}]	$\lambda \times 10^3$ [W m^{-1} K^{-1}]
0	0.001000	0.11	– 0.0001	4.2192	1402.52	13111	1791.41	561.10
5	0.001000	21.17	0.0762	4.2047	1426.27	13561	1518.04	570.60
10	0.001000	42.17	0.1511	4.1953	1447.63	13967	1305.82	580.07
15	0.001001	63.13	0.2244	4.1889	1466.67	14328	1137.51	589.40
20	0.001002	84.06	0.2965	4.1846	1483.50	14645	1001.58	598.48
25	0.001003	104.97	0.3672	4.1818	1498.25	14921	890.07	607.22
30	0.001004	125.88	0.4367	4.1799	1511.06	15156	797.34	615.53
35	0.001006	146.77	0.5051	4.1788	1522.06	15352	719.32	623.35
40	0.001008	167.67	0.5724	4.1784	1531.38	15512	652.98	630.66
45	0.001010	188.56	0.6386	4.1787	1539.12	15638	596.08	637.42
50	0.001012	209.45	0.7037	4.1794	1545.40	15731	546.86	643.63
60	0.001017	251.26	0.8311	4.1827	1553.94	15828	466.41	654.42
70	0.001023	293.12	0.9549	4.1880	1557.67	15816	403.91	663.17
80	0.001029	335.03	1.0753	4.1954	1557.16	15709	354.37	670.05
90	0.001036	377.03	1.1926	4.2049	1552.87	15518	314.43	675.30
100	0.001043	419.14	1.3070	4.2165	1545.19	15254	281.76	679.12
110	0.001052	461.37	1.4187	4.2303	1534.45	14927	254.70	681.69
t_s = 111.35				**Saturation**				
Liquid	0.001053	467.08	1.4335	4.2324	1532.78	14878	251.41	681.95
Vapour	1.1594	2693.11	7.2229	2.1278	477.84	1.3130	12.66	26.40
120	1.1880	2711.34	7.2698	2.0889	483.91	1.3141	12.99	26.98
130	1.2208	2732.07	7.3219	2.0602	490.60	1.3143	13.38	27.68
140	1.2533	2752.57	7.3721	2.0399	497.06	1.3142	13.77	28.42
150	1.2856	2772.89	7.4207	2.0250	503.35	1.3139	14.16	29.20
160	1.3176	2793.08	7.4679	2.0140	509.49	1.3134	14.55	30.00
170	1.3495	2813.18	7.5138	2.0059	515.50	1.3128	14.95	30.84
180	1.3813	2833.21	7.5584	2.0003	521.39	1.3120	15.35	31.69
190	1.4130	2853.19	7.6021	1.9966	527.17	1.3112	15.76	32.58
200	1.4445	2873.14	7.6447	1.9944	532.85	1.3104	16.16	33.48
210	1.4760	2893.08	7.6864	1.9936	538.43	1.3094	16.57	34.41
220	1.5074	2913.02	7.7272	1.9939	543.93	1.3085	16.98	35.35
230	1.5388	2932.96	7.7673	1.9950	549.35	1.3074	17.39	36.32
240	1.5701	2952.92	7.8066	1.9969	554.68	1.3064	17.80	37.30
250	1.6013	2972.90	7.8451	1.9994	559.95	1.3053	18.21	38.30
260	1.6326	2992.91	7.8830	2.0025	565.14	1.3042	18.62	39.31
270	1.6637	3012.96	7.9202	2.0060	570.27	1.3031	19.04	40.34
280	1.6949	3033.03	7.9569	2.0098	575.33	1.3020	19.45	41.38
290	1.7260	3053.15	7.9929	2.0140	580.34	1.3009	19.87	42.44
300	1.7571	3073.31	8.0284	2.0185	585.29	1.2997	20.28	43.51
310	1.7882	3093.52	8.0634	2.0232	590.18	1.2985	20.70	44.59
320	1.8192	3113.78	8.0978	2.0281	595.01	1.2974	21.12	45.68
330	1.8503	3134.09	8.1318	2.0332	599.80	1.2962	21.53	46.78
340	1.8813	3154.44	8.1652	2.0385	604.53	1.2951	21.95	47.90

Table 4 Single-Phase Region (Regions 1 to 3 of IAPWS-IF97) – Continuation
Homogenes Zustandsgebiet (Bereiche 1 bis 3 der IAPWS-IF97) – Fortsetzung

$p = 0.15$ MPa								
t [°C]	v [m³ kg⁻¹]	h [kJ kg⁻¹]	s [kJ kg⁻¹ K⁻¹]	c_p [kJ kg⁻¹ K⁻¹]	w [m s⁻¹]	κ	$\eta \times 10^6$ [kg m⁻¹ s⁻¹]	$\lambda \times 10^3$ [W m⁻¹ K⁻¹]
350	1.9123	3174.86	8.1983	2.0439	609.22	1.2939	22.37	49.03
360	1.9433	3195.32	8.2308	2.0495	613.86	1.2927	22.78	50.16
370	1.9743	3215.85	8.2630	2.0551	618.45	1.2915	23.20	51.31
380	2.0052	3236.43	8.2947	2.0609	623.00	1.2904	23.62	52.47
390	2.0362	3257.06	8.3261	2.0668	627.50	1.2892	24.03	53.63
400	2.0671	3277.76	8.3571	2.0728	631.97	1.2880	24.45	54.81
410	2.0981	3298.52	8.3877	2.0789	636.39	1.2869	24.86	55.99
420	2.1290	3319.34	8.4180	2.0850	640.77	1.2857	25.28	57.18
430	2.1599	3340.22	8.4479	2.0912	645.11	1.2845	25.69	58.38
440	2.1908	3361.16	8.4774	2.0975	649.42	1.2834	26.11	59.59
450	2.2217	3382.17	8.5067	2.1039	653.69	1.2822	26.52	60.81
460	2.2526	3403.24	8.5356	2.1103	657.92	1.2811	26.93	62.04
470	2.2835	3424.38	8.5643	2.1167	662.12	1.2799	27.34	63.27
480	2.3144	3445.58	8.5926	2.1232	666.29	1.2788	27.76	64.51
490	2.3453	3466.84	8.6206	2.1298	670.42	1.2776	28.17	65.76
500	2.3762	3488.17	8.6484	2.1364	674.51	1.2765	28.58	67.01
510	2.4070	3509.57	8.6759	2.1430	678.58	1.2754	28.98	68.28
520	2.4379	3531.03	8.7031	2.1497	682.61	1.2742	29.39	69.54
530	2.4687	3552.56	8.7301	2.1564	686.62	1.2731	29.80	70.82
540	2.4996	3574.16	8.7568	2.1632	690.59	1.2720	30.20	72.10
550	2.5305	3595.83	8.7833	2.1700	694.54	1.2709	30.61	73.39
560	2.5613	3617.56	8.8096	2.1768	698.46	1.2698	31.01	74.69
570	2.5922	3639.36	8.8356	2.1836	702.35	1.2687	31.42	75.99
580	2.6230	3661.23	8.8614	2.1905	706.21	1.2676	31.82	77.29
590	2.6538	3683.17	8.8869	2.1974	710.04	1.2665	32.22	78.61
600	2.6847	3705.18	8.9123	2.2043	713.85	1.2654	32.62	79.93
610	2.7155	3727.26	8.9374	2.2112	717.64	1.2643	33.02	81.25
620	2.7463	3749.41	8.9624	2.2182	721.40	1.2633	33.42	82.58
630	2.7772	3771.62	8.9871	2.2251	725.13	1.2622	33.81	83.92
640	2.8080	3793.91	9.0116	2.2321	728.84	1.2612	34.21	85.26
650	2.8388	3816.27	9.0360	2.2390	732.53	1.2601	34.60	86.60
660	2.8696	3838.69	9.0602	2.2460	736.19	1.2591	34.99	87.95
670	2.9005	3861.19	9.0841	2.2530	739.84	1.2581	35.39	89.31
680	2.9313	3883.75	9.1079	2.2600	743.46	1.2571	35.78	90.67
690	2.9621	3906.39	9.1316	2.2670	747.05	1.2561	36.17	92.04
700	2.9929	3929.09	9.1550	2.2740	750.63	1.2551	36.56	93.41
720	3.0546	3974.71	9.2014	2.2880	757.72	1.2531	37.33	96.16
740	3.1162	4020.61	9.2472	2.3020	764.73	1.2511	38.10	98.94
760	3.1778	4066.79	9.2923	2.3159	771.67	1.2492	38.86	101.73
780	3.2394	4113.25	9.3368	2.3299	778.53	1.2474	39.62	104.53
800	3.3010	4159.99	9.3808	2.3439	785.31	1.2455	40.38	107.36

Table 4 Single-Phase Region (Regions 1 to 3 of IAPWS-IF97) – Continuation
Homogenes Zustandsgebiet (Bereiche 1 bis 3 der IAPWS-IF97) – Fortsetzung

				p = 0.175 MPa				
t [°C]	v [m^3 kg^{-1}]	h [kJ kg^{-1}]	s [kJ kg^{-1} K^{-1}]	c_p [kJ kg^{-1} K^{-1}]	w [m s^{-1}]	κ	$\eta\times 10^6$ [kg m^{-1} s^{-1}]	$\lambda\times 10^3$ [W m^{-1} K^{-1}]
0	0.001000	0.14	– 0.0001	4.2191	1402.56	11239	1791.34	561.12
5	0.001000	21.19	0.0762	4.2046	1426.31	11624	1518.00	570.61
10	0.001000	42.19	0.1511	4.1952	1447.67	11972	1305.79	580.08
15	0.001001	63.15	0.2244	4.1888	1466.71	12282	1137.50	589.41
20	0.001002	84.08	0.2965	4.1846	1483.54	12554	1001.57	598.49
25	0.001003	105.00	0.3672	4.1817	1498.29	12790	890.06	607.23
30	0.001004	125.90	0.4367	4.1798	1511.10	12991	797.34	615.54
35	0.001006	146.80	0.5051	4.1788	1522.10	13160	719.32	623.36
40	0.001008	167.69	0.5724	4.1784	1531.42	13297	652.98	630.67
45	0.001010	188.58	0.6386	4.1786	1539.17	13405	596.08	637.43
50	0.001012	209.48	0.7037	4.1794	1545.45	13485	546.87	643.64
60	0.001017	251.28	0.8311	4.1826	1553.99	13568	466.42	654.44
70	0.001023	293.14	0.9549	4.1879	1557.72	13557	403.92	663.18
80	0.001029	335.05	1.0753	4.1954	1557.20	13466	354.37	670.07
90	0.001036	377.05	1.1926	4.2049	1552.92	13302	314.43	675.32
100	0.001043	419.15	1.3070	4.2165	1545.24	13076	281.76	679.14
110	0.001052	461.39	1.4186	4.2303	1534.50	12795	254.71	681.71
t_s = 116.04				**Saturation**				
Liquid	0.001057	486.97	1.4849	4.2397	1526.64	12602	240.56	682.72
Vapour	1.0036	2700.13	7.1714	2.1519	480.02	1.3119	12.82	26.97
120	1.0150	2708.60	7.1931	2.1295	482.86	1.3126	12.97	27.23
130	1.0435	2729.69	7.2461	2.0905	489.72	1.3133	13.36	27.90
140	1.0716	2750.46	7.2969	2.0645	496.29	1.3134	13.75	28.62
150	1.0995	2771.00	7.3461	2.0456	502.67	1.3132	14.15	29.37
160	1.1272	2791.38	7.3937	2.0313	508.88	1.3128	14.54	30.16
170	1.1547	2811.64	7.4399	2.0208	514.95	1.3123	14.94	30.97
180	1.1821	2831.81	7.4849	2.0130	520.89	1.3116	15.34	31.82
190	1.2094	2851.91	7.5288	2.0076	526.72	1.3109	15.75	32.69
200	1.2365	2871.97	7.5717	2.0041	532.44	1.3101	16.15	33.58
210	1.2636	2892.00	7.6135	2.0021	538.06	1.3092	16.56	34.50
220	1.2906	2912.01	7.6545	2.0013	543.58	1.3082	16.97	35.44
230	1.3176	2932.03	7.6947	2.0016	549.03	1.3073	17.38	36.39
240	1.3445	2952.05	7.7341	2.0028	554.39	1.3062	17.79	37.37
250	1.3714	2972.08	7.7728	2.0047	559.67	1.3052	18.21	38.36
260	1.3982	2992.14	7.8108	2.0072	564.89	1.3041	18.62	39.37
270	1.4250	3012.23	7.8481	2.0102	570.03	1.3030	19.03	40.39
280	1.4518	3032.35	7.8848	2.0137	575.11	1.3019	19.45	41.43
290	1.4785	3052.50	7.9209	2.0175	580.13	1.3008	19.86	42.48
300	1.5052	3072.70	7.9565	2.0217	585.09	1.2996	20.28	43.55
310	1.5319	3092.94	7.9915	2.0261	590.00	1.2985	20.70	44.63
320	1.5585	3113.22	8.0260	2.0308	594.84	1.2973	21.11	45.72
330	1.5852	3133.56	8.0600	2.0357	599.64	1.2962	21.53	46.82
340	1.6118	3153.94	8.0935	2.0408	604.38	1.2950	21.95	47.93

Table 4 Single-Phase Region (Regions 1 to 3 of IAPWS-IF97) – Continuation
Homogenes Zustandsgebiet (Bereiche 1 bis 3 der IAPWS-IF97) – Fortsetzung

p = 0.175 MPa

t [°C]	v [m³ kg⁻¹]	h [kJ kg⁻¹]	s [kJ kg⁻¹ K⁻¹]	c_p [kJ kg⁻¹ K⁻¹]	w [m s⁻¹]	κ	$\eta\times10^6$ [kg m⁻¹ s⁻¹]	$\lambda\times10^3$ [W m⁻¹ K⁻¹]
350	1.6384	3174.37	8.1265	2.0461	609.08	1.2939	22.37	49.06
360	1.6650	3194.86	8.1592	2.0515	613.72	1.2927	22.78	50.19
370	1.6916	3215.40	8.1913	2.0571	618.32	1.2915	23.20	51.34
380	1.7181	3236.00	8.2231	2.0627	622.88	1.2904	23.62	52.49
390	1.7447	3256.66	8.2545	2.0685	627.39	1.2892	24.03	53.66
400	1.7712	3277.37	8.2855	2.0744	631.86	1.2880	24.45	54.83
410	1.7978	3298.15	8.3161	2.0803	636.29	1.2869	24.86	56.01
420	1.8243	3318.98	8.3464	2.0864	640.67	1.2857	25.28	57.21
430	1.8508	3339.88	8.3764	2.0925	645.02	1.2845	25.69	58.41
440	1.8774	3360.83	8.4059	2.0988	649.33	1.2834	26.11	59.61
450	1.9039	3381.85	8.4352	2.1050	653.61	1.2822	26.52	60.83
460	1.9304	3402.93	8.4642	2.1114	657.85	1.2811	26.93	62.06
470	1.9569	3424.08	8.4928	2.1178	662.05	1.2799	27.34	63.29
480	1.9834	3445.29	8.5212	2.1242	666.22	1.2788	27.76	64.53
490	2.0098	3466.56	8.5492	2.1308	670.35	1.2776	28.17	65.78
500	2.0363	3487.90	8.5770	2.1373	674.45	1.2765	28.58	67.03
510	2.0628	3509.31	8.6045	2.1439	678.52	1.2754	28.98	68.29
520	2.0893	3530.78	8.6318	2.1506	682.56	1.2742	29.39	69.56
530	2.1157	3552.32	8.6587	2.1572	686.56	1.2731	29.80	70.84
540	2.1422	3573.93	8.6855	2.1640	690.54	1.2720	30.20	72.12
550	2.1686	3595.60	8.7120	2.1707	694.49	1.2709	30.61	73.41
560	2.1951	3617.34	8.7382	2.1775	698.41	1.2698	31.01	74.70
570	2.2216	3639.15	8.7642	2.1843	702.30	1.2687	31.42	76.00
580	2.2480	3661.03	8.7900	2.1912	706.17	1.2676	31.82	77.31
590	2.2744	3682.97	8.8156	2.1980	710.00	1.2665	32.22	78.62
600	2.3009	3704.99	8.8410	2.2049	713.81	1.2654	32.62	79.94
610	2.3273	3727.07	8.8661	2.2118	717.60	1.2644	33.02	81.27
620	2.3538	3749.22	8.8911	2.2187	721.36	1.2633	33.42	82.60
630	2.3802	3771.45	8.9158	2.2256	725.10	1.2622	33.81	83.93
640	2.4066	3793.74	8.9404	2.2326	728.81	1.2612	34.21	85.27
650	2.4331	3816.10	8.9647	2.2395	732.50	1.2602	34.60	86.62
660	2.4595	3838.53	8.9889	2.2465	736.17	1.2591	35.00	87.97
670	2.4859	3861.03	9.0129	2.2535	739.81	1.2581	35.39	89.32
680	2.5123	3883.60	9.0367	2.2604	743.43	1.2571	35.78	90.69
690	2.5388	3906.24	9.0603	2.2674	747.03	1.2561	36.17	92.05
700	2.5652	3928.94	9.0837	2.2744	750.61	1.2551	36.56	93.42
720	2.6180	3974.57	9.1302	2.2883	757.70	1.2531	37.33	96.18
740	2.6708	4020.48	9.1759	2.3023	764.72	1.2512	38.10	98.95
760	2.7237	4066.66	9.2211	2.3163	771.65	1.2493	38.86	101.74
780	2.7765	4113.13	9.2656	2.3302	778.51	1.2474	39.62	104.55
800	2.8293	4159.87	9.3096	2.3442	785.30	1.2455	40.38	107.37

Table 4 Single-Phase Region (Regions 1 to 3 of IAPWS-IF97) – Continuation
Homogenes Zustandsgebiet (Bereiche 1 bis 3 der IAPWS-IF97) – Fortsetzung

t [°C]	v [m³ kg⁻¹]	h [kJ kg⁻¹]	s [kJ kg⁻¹ K⁻¹]	c_p [kJ kg⁻¹ K⁻¹]	w [m s⁻¹]	κ	$\eta\times10^6$ [kg m⁻¹ s⁻¹]	$\lambda\times10^3$ [W m⁻¹ K⁻¹]
				p = 0.2 MPa				
0	0.001000	0.16	– 0.0001	4.2189	1402.59	9835	1791.28	561.13
5	0.001000	21.22	0.0762	4.2045	1426.35	10172	1517.96	570.63
10	0.001000	42.21	0.1511	4.1951	1447.71	10476	1305.77	580.09
15	0.001001	63.17	0.2244	4.1888	1466.75	10747	1137.48	589.43
20	0.001002	84.11	0.2965	4.1845	1483.57	10985	1001.56	598.51
25	0.001003	105.02	0.3672	4.1816	1498.33	11192	890.06	607.24
30	0.001004	125.92	0.4367	4.1798	1511.14	11368	797.34	615.55
35	0.001006	146.82	0.5051	4.1787	1522.14	11516	719.32	623.38
40	0.001008	167.71	0.5724	4.1783	1531.46	11636	652.99	630.68
45	0.001010	188.60	0.6385	4.1785	1539.21	11730	596.08	637.44
50	0.001012	209.50	0.7037	4.1793	1545.49	11800	546.87	643.65
60	0.001017	251.31	0.8311	4.1825	1554.03	11872	466.42	654.45
70	0.001023	293.16	0.9549	4.1879	1557.76	11864	403.93	663.19
80	0.001029	335.07	1.0753	4.1953	1557.25	11783	354.38	670.08
90	0.001036	377.07	1.1926	4.2048	1552.96	11640	314.44	675.33
100	0.001043	419.17	1.3069	4.2164	1545.29	11442	281.77	679.15
110	0.001052	461.40	1.4186	4.2302	1534.56	11197	254.71	681.72
120	0.001060	503.79	1.5278	4.2464	1521.01	10909	232.05	683.19
t_s = 120.21				Saturation				
Liquid	0.001061	504.68	1.5301	4.2467	1520.69	10902	231.61	683.21
Vapour	0.8857	2706.24	7.1269	2.1752	481.88	1.3108	12.96	27.49
130	0.9104	2727.25	7.1796	2.1232	488.81	1.3122	13.34	28.12
140	0.9353	2748.31	7.2312	2.0902	495.51	1.3126	13.74	28.81
150	0.9599	2769.09	7.2809	2.0667	501.97	1.3125	14.13	29.54
160	0.9843	2789.66	7.3290	2.0492	508.26	1.3122	14.53	30.31
170	1.0085	2810.09	7.3756	2.0359	514.39	1.3118	14.93	31.11
180	1.0326	2830.39	7.4209	2.0261	520.39	1.3112	15.33	31.94
190	1.0566	2850.62	7.4650	2.0189	526.26	1.3105	15.74	32.80
200	1.0805	2870.78	7.5081	2.0139	532.02	1.3098	16.15	33.68
210	1.1043	2890.90	7.5502	2.0106	537.68	1.3089	16.55	34.59
220	1.1281	2911.00	7.5914	2.0089	543.23	1.3080	16.96	35.52
230	1.1517	2931.08	7.6317	2.0083	548.70	1.3071	17.37	36.47
240	1.1753	2951.17	7.6712	2.0087	554.09	1.3061	17.79	37.44
250	1.1989	2971.26	7.7100	2.0100	559.40	1.3050	18.20	38.42
260	1.2224	2991.37	7.7481	2.0120	564.63	1.3040	18.61	39.42
270	1.2459	3011.50	7.7855	2.0145	569.80	1.3029	19.03	40.44
280	1.2694	3031.66	7.8223	2.0176	574.89	1.3018	19.45	41.48
290	1.2928	3051.85	7.8584	2.0211	579.93	1.3007	19.86	42.53
300	1.3162	3072.08	7.8940	2.0250	584.90	1.2996	20.28	43.59
310	1.3396	3092.35	7.9291	2.0291	589.81	1.2984	20.70	44.66
320	1.3630	3112.67	7.9636	2.0336	594.67	1.2973	21.11	45.75
330	1.3863	3133.03	7.9977	2.0383	599.48	1.2961	21.53	46.85
340	1.4097	3153.43	8.0312	2.0432	604.23	1.2950	21.95	47.96

Table 4 Single-Phase Region (Regions 1 to 3 of IAPWS-IF97) – Continuation
Homogenes Zustandsgebiet (Bereiche 1 bis 3 der IAPWS-IF97) – Fortsetzung

				p = 0.2 MPa				
t [°C]	v [m³ kg⁻¹]	h [kJ kg⁻¹]	s [kJ kg⁻¹ K⁻¹]	c_p [kJ kg⁻¹ K⁻¹]	w [m s⁻¹]	κ	$\eta \times 10^6$ [kg m⁻¹ s⁻¹]	$\lambda \times 10^3$ [W m⁻¹ K⁻¹]
350	1.4330	3173.89	8.0643	2.0483	608.93	1.2938	22.36	49.09
360	1.4563	3194.40	8.0970	2.0535	613.59	1.2927	22.78	50.22
370	1.4795	3214.96	8.1292	2.0590	618.20	1.2915	23.20	51.36
380	1.5028	3235.58	8.1610	2.0645	622.76	1.2903	23.61	52.52
390	1.5261	3256.25	8.1924	2.0702	627.28	1.2892	24.03	53.68
400	1.5493	3276.98	8.2235	2.0760	631.75	1.2880	24.45	54.85
410	1.5726	3297.77	8.2541	2.0818	636.19	1.2868	24.86	56.04
420	1.5958	3318.62	8.2844	2.0878	640.58	1.2857	25.28	57.23
430	1.6190	3339.53	8.3144	2.0939	644.93	1.2845	25.69	58.43
440	1.6423	3360.50	8.3440	2.1000	649.25	1.2834	26.11	59.63
450	1.6655	3381.53	8.3733	2.1062	653.53	1.2822	26.52	60.85
460	1.6887	3402.62	8.4022	2.1125	657.77	1.2811	26.93	62.07
470	1.7119	3423.78	8.4309	2.1189	661.97	1.2799	27.34	63.31
480	1.7351	3445.00	8.4593	2.1253	666.14	1.2788	27.76	64.55
490	1.7583	3466.29	8.4873	2.1317	670.28	1.2776	28.17	65.79
500	1.7814	3487.64	8.5151	2.1382	674.39	1.2765	28.58	67.05
510	1.8046	3509.05	8.5426	2.1448	678.46	1.2754	28.98	68.31
520	1.8278	3530.53	8.5699	2.1514	682.50	1.2742	29.39	69.58
530	1.8510	3552.08	8.5969	2.1580	686.51	1.2731	29.80	70.85
540	1.8741	3573.69	8.6236	2.1647	690.49	1.2720	30.21	72.13
550	1.8973	3595.37	8.6501	2.1715	694.44	1.2709	30.61	73.42
560	1.9204	3617.12	8.6764	2.1782	698.36	1.2698	31.01	74.72
570	1.9436	3638.94	8.7024	2.1850	702.26	1.2687	31.42	76.02
580	1.9667	3660.82	8.7282	2.1918	706.12	1.2676	31.82	77.32
590	1.9899	3682.77	8.7538	2.1986	709.96	1.2665	32.22	78.64
600	2.0130	3704.79	8.7792	2.2055	713.78	1.2654	32.62	79.96
610	2.0362	3726.88	8.8043	2.2124	717.56	1.2644	33.02	81.28
620	2.0593	3749.04	8.8293	2.2193	721.33	1.2633	33.42	82.61
630	2.0825	3771.27	8.8540	2.2262	725.06	1.2623	33.81	83.95
640	2.1056	3793.56	8.8786	2.2331	728.78	1.2612	34.21	85.29
650	2.1287	3815.93	8.9029	2.2400	732.47	1.2602	34.60	86.63
660	2.1519	3838.36	8.9271	2.2470	736.14	1.2591	35.00	87.98
670	2.1750	3860.87	8.9511	2.2539	739.78	1.2581	35.39	89.34
680	2.1981	3883.44	8.9749	2.2609	743.40	1.2571	35.78	90.70
690	2.2212	3906.09	8.9985	2.2678	747.01	1.2561	36.17	92.07
700	2.2444	3928.80	9.0220	2.2748	750.59	1.2551	36.56	93.44
720	2.2906	3974.43	9.0684	2.2887	757.68	1.2531	37.33	96.19
740	2.3368	4020.35	9.1142	2.3026	764.70	1.2512	38.10	98.96
760	2.3831	4066.54	9.1593	2.3166	771.64	1.2493	38.87	101.75
780	2.4293	4113.01	9.2039	2.3305	778.50	1.2474	39.63	104.56
800	2.4755	4159.76	9.2479	2.3445	785.29	1.2455	40.38	107.38

Table 4 **Single-Phase Region** (Regions 1 to 3 of IAPWS-IF97) – Continuation
Homogenes Zustandsgebiet (Bereiche 1 bis 3 der IAPWS-IF97) – Fortsetzung

p = 0.25 MPa								
t [°C]	v [m^3 kg^{-1}]	h [kJ kg^{-1}]	s [kJ kg^{-1} K^{-1}]	c_p [kJ kg^{-1} K^{-1}]	w [m s^{-1}]	κ	$\eta\times 10^6$ [kg m^{-1} s^{-1}]	$\lambda\times 10^3$ [W m^{-1} K^{-1}]
0	0.001000	0.21	– 0.0001	4.2187	1402.67	7869	1791.15	561.16
5	0.001000	21.27	0.0762	4.2043	1426.43	8139	1517.88	570.65
10	0.001000	42.26	0.1511	4.1949	1447.79	8382	1305.71	580.12
15	0.001001	63.22	0.2244	4.1886	1466.83	8599	1137.44	589.45
20	0.001002	84.15	0.2965	4.1843	1483.65	8789	1001.54	598.53
25	0.001003	105.07	0.3672	4.1815	1498.41	8954	890.05	607.26
30	0.001004	125.97	0.4367	4.1796	1511.22	9096	797.33	615.57
35	0.001006	146.86	0.5051	4.1786	1522.22	9214	719.32	623.40
40	0.001008	167.76	0.5723	4.1782	1531.54	9310	652.99	630.70
45	0.001010	188.65	0.6385	4.1784	1539.29	9385	596.09	637.46
50	0.001012	209.54	0.7037	4.1792	1545.57	9441	546.88	643.68
60	0.001017	251.35	0.8311	4.1824	1554.11	9499	466.44	654.47
70	0.001023	293.20	0.9549	4.1878	1557.85	9492	403.94	663.22
80	0.001029	335.11	1.0753	4.1952	1557.34	9428	354.39	670.10
90	0.001036	377.11	1.1925	4.2047	1553.06	9314	314.45	675.36
100	0.001043	419.21	1.3069	4.2163	1545.40	9155	281.79	679.18
110	0.001052	461.44	1.4186	4.2301	1534.66	8959	254.73	681.75
120	0.001060	503.82	1.5278	4.2462	1521.12	8728	232.06	683.22
t_s = 127.41				**Saturation**				
Liquid	0.001067	535.35	1.6072	4.2598	1509.38	8538	217.56	683.65
Vapour	0.7187	2716.50	7.0524	2.2197	484.95	1.3089	13.21	28.42
130	0.7240	2722.21	7.0666	2.1995	486.86	1.3095	13.31	28.58
140	0.7444	2743.92	7.1198	2.1456	493.89	1.3108	13.71	29.21
150	0.7644	2765.19	7.1707	2.1112	500.57	1.3111	14.11	29.90
160	0.7842	2786.17	7.2197	2.0862	507.01	1.3111	14.51	30.63
170	0.8039	2806.93	7.2671	2.0673	513.27	1.3108	14.91	31.39
180	0.8234	2827.53	7.3130	2.0529	519.37	1.3104	15.32	32.19
190	0.8428	2848.00	7.3577	2.0421	525.34	1.3098	15.72	33.03
200	0.8621	2868.38	7.4013	2.0340	531.18	1.3092	16.13	33.89
210	0.8813	2888.69	7.4437	2.0282	536.91	1.3084	16.54	34.77
220	0.9004	2908.95	7.4852	2.0243	542.53	1.3076	16.95	35.68
230	0.9195	2929.18	7.5259	2.0219	548.06	1.3067	17.36	36.62
240	0.9385	2949.39	7.5656	2.0208	553.49	1.3057	17.78	37.57
250	0.9574	2969.60	7.6046	2.0208	558.85	1.3048	18.19	38.55
260	0.9764	2989.81	7.6429	2.0217	564.12	1.3038	18.61	39.54
270	0.9952	3010.04	7.6805	2.0233	569.32	1.3027	19.02	40.55
280	1.0141	3030.28	7.7174	2.0255	574.45	1.3016	19.44	41.57
290	1.0329	3050.55	7.7537	2.0283	579.51	1.3005	19.85	42.62
300	1.0517	3070.85	7.7895	2.0315	584.51	1.2994	20.27	43.67
310	1.0705	3091.18	7.8246	2.0352	589.45	1.2983	20.69	44.74
320	1.0892	3111.55	7.8593	2.0391	594.33	1.2972	21.11	45.82
330	1.1079	3131.97	7.8934	2.0434	599.16	1.2961	21.53	46.92
340	1.1267	3152.42	7.9270	2.0479	603.93	1.2949	21.94	48.03

Table 4 Single-Phase Region (Regions 1 to 3 of IAPWS-IF97) – Continuation
Homogenes Zustandsgebiet (Bereiche 1 bis 3 der IAPWS-IF97) – Fortsetzung

p = 0.25 MPa								
t [°C]	v [m^3 kg^{-1}]	h [kJ kg^{-1}]	s [kJ kg^{-1} K^{-1}]	c_p [kJ kg^{-1} K^{-1}]	w [m s^{-1}]	κ	$\eta \times 10^6$ [kg m^{-1} s^{-1}]	$\lambda \times 10^3$ [W m^{-1} K^{-1}]
350	1.1454	3172.93	7.9602	2.0527	608.65	1.2938	22.36	49.15
360	1.1640	3193.48	7.9929	2.0577	613.32	1.2926	22.78	50.28
370	1.1827	3214.08	8.0252	2.0628	617.94	1.2915	23.20	51.42
380	1.2014	3234.73	8.0571	2.0681	622.52	1.2903	23.61	52.57
390	1.2200	3255.44	8.0885	2.0735	627.05	1.2891	24.03	53.73
400	1.2387	3276.20	8.1196	2.0791	631.54	1.2880	24.45	54.90
410	1.2573	3297.02	8.1503	2.0848	635.98	1.2868	24.86	56.08
420	1.2759	3317.90	8.1807	2.0906	640.39	1.2857	25.28	57.27
430	1.2945	3338.84	8.2106	2.0965	644.75	1.2845	25.69	58.47
440	1.3131	3359.83	8.2403	2.1025	649.08	1.2834	26.11	59.67
450	1.3317	3380.89	8.2696	2.1086	653.36	1.2822	26.52	60.89
460	1.3503	3402.00	8.2986	2.1148	657.61	1.2811	26.93	62.11
470	1.3689	3423.18	8.3273	2.1210	661.83	1.2799	27.34	63.34
480	1.3875	3444.42	8.3557	2.1273	666.00	1.2788	27.76	64.58
490	1.4060	3465.73	8.3838	2.1337	670.15	1.2776	28.17	65.83
500	1.4246	3487.10	8.4116	2.1401	674.26	1.2765	28.58	67.08
510	1.4432	3508.53	8.4392	2.1466	678.34	1.2754	28.99	68.34
520	1.4617	3530.03	8.4664	2.1531	682.38	1.2742	29.39	69.61
530	1.4803	3551.59	8.4935	2.1597	686.40	1.2731	29.80	70.88
540	1.4988	3573.22	8.5202	2.1663	690.39	1.2720	30.21	72.17
550	1.5174	3594.92	8.5467	2.1729	694.34	1.2709	30.61	73.45
560	1.5359	3616.68	8.5730	2.1796	698.27	1.2698	31.02	74.75
570	1.5545	3638.51	8.5991	2.1863	702.17	1.2687	31.42	76.05
580	1.5730	3660.41	8.6249	2.1931	706.04	1.2676	31.82	77.35
590	1.5915	3682.37	8.6505	2.1999	709.88	1.2665	32.22	78.67
600	1.6101	3704.41	8.6759	2.2067	713.70	1.2655	32.62	79.99
610	1.6286	3726.51	8.7010	2.2135	717.49	1.2644	33.02	81.31
620	1.6471	3748.68	8.7260	2.2203	721.26	1.2633	33.42	82.64
630	1.6656	3770.91	8.7507	2.2272	725.00	1.2623	33.82	83.97
640	1.6842	3793.22	8.7753	2.2341	728.72	1.2612	34.21	85.31
650	1.7027	3815.59	8.7997	2.2410	732.41	1.2602	34.61	86.66
660	1.7212	3838.04	8.8239	2.2479	736.08	1.2592	35.00	88.01
670	1.7397	3860.55	8.8479	2.2548	739.73	1.2581	35.39	89.37
680	1.7582	3883.13	8.8717	2.2617	743.35	1.2571	35.78	90.73
690	1.7767	3905.79	8.8953	2.2686	746.96	1.2561	36.17	92.09
700	1.7952	3928.51	8.9188	2.2756	750.54	1.2551	36.56	93.46
720	1.8322	3974.16	8.9652	2.2894	757.64	1.2532	37.33	96.22
740	1.8693	4020.08	9.0110	2.3033	764.66	1.2512	38.10	98.99
760	1.9063	4066.29	9.0562	2.3172	771.61	1.2493	38.87	101.78
780	1.9432	4112.77	9.1007	2.3311	778.47	1.2474	39.63	104.58
800	1.9802	4159.53	9.1447	2.3450	785.26	1.2456	40.38	107.41

Table 4 Single-Phase Region (Regions 1 to 3 of IAPWS-IF97) – Continuation
Homogenes Zustandsgebiet (Bereiche 1 bis 3 der IAPWS-IF97) – Fortsetzung

				p = 0.3 MPa				
t	v	h	s	c_p	w	κ	$\eta\times10^6$	$\lambda\times10^3$
[°C]	[m^3 kg^{-1}]	[kJ kg^{-1}]	[kJ kg^{-1} K^{-1}]	[kJ kg^{-1} K^{-1}]	[m s^{-1}]		[kg m^{-1} s^{-1}]	[W m^{-1} K^{-1}]
0	0.001000	0.26	– 0.0001	4.2184	1402.75	6558	1791.03	561.19
5	0.001000	21.32	0.0762	4.2041	1426.51	6783	1517.80	570.68
10	0.001000	42.31	0.1511	4.1947	1447.87	6986	1305.66	580.14
15	0.001001	63.27	0.2244	4.1884	1466.90	7166	1137.41	589.47
20	0.001002	84.20	0.2964	4.1842	1483.73	7325	1001.52	598.55
25	0.001003	105.11	0.3672	4.1813	1498.49	7463	890.04	607.29
30	0.001004	126.01	0.4367	4.1795	1511.30	7580	797.33	615.59
35	0.001006	146.91	0.5051	4.1784	1522.30	7679	719.32	623.42
40	0.001008	167.80	0.5723	4.1781	1531.62	7759	652.99	630.73
45	0.001010	188.69	0.6385	4.1783	1539.37	7822	596.10	637.49
50	0.001012	209.58	0.7037	4.1791	1545.65	7868	546.89	643.70
60	0.001017	251.39	0.8311	4.1823	1554.20	7917	466.45	654.50
70	0.001023	293.24	0.9548	4.1877	1557.94	7911	403.95	663.24
80	0.001029	335.15	1.0752	4.1951	1557.44	7858	354.41	670.13
90	0.001036	377.15	1.1925	4.2046	1553.16	7762	314.47	675.38
100	0.001043	419.25	1.3069	4.2162	1545.50	7631	281.80	679.21
110	0.001051	461.48	1.4185	4.2300	1534.77	7467	254.74	681.78
120	0.001060	503.86	1.5277	4.2461	1521.23	7275	232.08	683.25
130	0.001070	546.41	1.6346	4.2648	1505.08	7058	212.90	683.71
t_s = 133.53				**Saturation**				
Liquid	0.001073	561.46	1.6718	4.2720	1498.80	6977	206.84	683.66
Vapour	0.6058	2724.89	6.9916	2.2618	487.39	1.3071	13.42	29.24
140	0.6170	2739.36	7.0269	2.2099	492.17	1.3087	13.68	29.62
150	0.6340	2761.18	7.0791	2.1593	499.11	1.3097	14.08	30.26
160	0.6508	2782.60	7.1291	2.1254	505.73	1.3099	14.49	30.95
170	0.6674	2803.72	7.1773	2.1002	512.12	1.3098	14.89	31.68
180	0.6839	2824.62	7.2239	2.0809	518.34	1.3096	15.30	32.45
190	0.7002	2845.35	7.2692	2.0661	524.41	1.3091	15.70	33.25
200	0.7164	2865.95	7.3132	2.0548	530.33	1.3086	16.11	34.09
210	0.7326	2886.46	7.3561	2.0463	536.13	1.3079	16.53	34.96
220	0.7486	2906.89	7.3979	2.0401	541.82	1.3071	16.94	35.85
230	0.7646	2927.26	7.4388	2.0359	547.40	1.3063	17.35	36.77
240	0.7806	2947.61	7.4789	2.0332	552.89	1.3054	17.77	37.71
250	0.7965	2967.93	7.5181	2.0318	558.29	1.3045	18.18	38.67
260	0.8123	2988.25	7.5566	2.0315	563.60	1.3035	18.60	39.65
270	0.8281	3008.57	7.5943	2.0322	568.84	1.3025	19.01	40.65
280	0.8439	3028.89	7.6314	2.0336	574.00	1.3015	19.43	41.67
290	0.8596	3049.24	7.6679	2.0356	579.10	1.3004	19.85	42.71
300	0.8753	3069.61	7.7037	2.0382	584.12	1.2993	20.27	43.75
310	0.8910	3090.01	7.7390	2.0413	589.09	1.2982	20.68	44.82
320	0.9067	3110.43	7.7737	2.0447	593.99	1.2971	21.10	45.90
330	0.9224	3130.90	7.8079	2.0486	598.84	1.2960	21.52	46.99
340	0.9380	3151.41	7.8417	2.0527	603.63	1.2948	21.94	48.09

Table 4 Single-Phase Region (Regions 1 to 3 of IAPWS-IF97) – Continuation
Homogenes Zustandsgebiet (Bereiche 1 bis 3 der IAPWS-IF97) – Fortsetzung

				p = 0.3 MPa				
t [°C]	v [m³ kg⁻¹]	h [kJ kg⁻¹]	s [kJ kg⁻¹ K⁻¹]	c_p [kJ kg⁻¹ K⁻¹]	w [m s⁻¹]	κ	$\eta \times 10^6$ [kg m⁻¹ s⁻¹]	$\lambda \times 10^3$ [W m⁻¹ K⁻¹]
350	0.9536	3171.96	7.8749	2.0571	608.37	1.2937	22.36	49.21
360	0.9692	3192.55	7.9077	2.0618	613.05	1.2926	22.78	50.33
370	0.9848	3213.19	7.9400	2.0667	617.69	1.2914	23.19	51.47
380	1.0004	3233.89	7.9720	2.0717	622.28	1.2903	23.61	52.62
390	1.0160	3254.63	8.0035	2.0769	626.83	1.2891	24.03	53.78
400	1.0315	3275.42	8.0346	2.0823	631.33	1.2880	24.44	54.95
410	1.0471	3296.27	8.0654	2.0878	635.78	1.2868	24.86	56.13
420	1.0626	3317.18	8.0957	2.0934	640.20	1.2856	25.28	57.31
430	1.0782	3338.14	8.1258	2.0992	644.57	1.2845	25.69	58.51
440	1.0937	3359.17	8.1554	2.1050	648.90	1.2833	26.11	59.72
450	1.1092	3380.25	8.1848	2.1110	653.20	1.2822	26.52	60.93
460	1.1247	3401.39	8.2138	2.1170	657.46	1.2810	26.93	62.15
470	1.1402	3422.59	8.2426	2.1231	661.68	1.2799	27.34	63.38
480	1.1557	3443.85	8.2710	2.1293	665.86	1.2788	27.76	64.62
490	1.1712	3465.17	8.2991	2.1356	670.02	1.2776	28.17	65.86
500	1.1867	3486.56	8.3269	2.1419	674.13	1.2765	28.58	67.12
510	1.2022	3508.01	8.3545	2.1483	678.22	1.2754	28.99	68.38
520	1.2177	3529.53	8.3818	2.1548	682.27	1.2743	29.39	69.64
530	1.2332	3551.11	8.4089	2.1613	686.29	1.2731	29.80	70.92
540	1.2486	3572.75	8.4356	2.1678	690.28	1.2720	30.21	72.20
550	1.2641	3594.46	8.4622	2.1744	694.24	1.2709	30.61	73.49
560	1.2796	3616.24	8.4885	2.1810	698.17	1.2698	31.02	74.78
570	1.2950	3638.08	8.5145	2.1877	702.08	1.2687	31.42	76.08
580	1.3105	3659.99	8.5404	2.1944	705.95	1.2676	31.82	77.39
590	1.3260	3681.97	8.5660	2.2011	709.80	1.2666	32.22	78.70
600	1.3414	3704.02	8.5914	2.2078	713.62	1.2655	32.62	80.02
610	1.3569	3726.13	8.6165	2.2146	717.42	1.2644	33.02	81.34
620	1.3723	3748.31	8.6415	2.2214	721.19	1.2634	33.42	82.67
630	1.3877	3770.56	8.6663	2.2282	724.93	1.2623	33.82	84.00
640	1.4032	3792.87	8.6909	2.2351	728.65	1.2613	34.21	85.34
650	1.4186	3815.26	8.7152	2.2419	732.35	1.2602	34.61	86.69
660	1.4341	3837.71	8.7394	2.2488	736.02	1.2592	35.00	88.04
670	1.4495	3860.23	8.7634	2.2557	739.67	1.2582	35.39	89.40
680	1.4649	3882.83	8.7873	2.2626	743.30	1.2572	35.78	90.76
690	1.4804	3905.49	8.8109	2.2695	746.91	1.2562	36.17	92.12
700	1.4958	3928.21	8.8344	2.2764	750.49	1.2552	36.56	93.49
720	1.5267	3973.88	8.8808	2.2902	757.60	1.2532	37.34	96.24
740	1.5575	4019.82	8.9266	2.3040	764.63	1.2512	38.11	99.02
760	1.5884	4066.04	8.9718	2.3178	771.57	1.2493	38.87	101.80
780	1.6192	4112.54	9.0164	2.3317	778.44	1.2475	39.63	104.61
800	1.6500	4159.31	9.0604	2.3456	785.24	1.2456	40.38	107.43

Table 4 Single-Phase Region (Regions 1 to 3 of IAPWS-IF97) – Continuation
Homogenes Zustandsgebiet (Bereiche 1 bis 3 der IAPWS-IF97) – Fortsetzung

				p = 0.35 MPa				
t [°C]	v [m^3 kg^{-1}]	h [kJ kg^{-1}]	s [kJ kg^{-1} K^{-1}]	c_p [kJ kg^{-1} K^{-1}]	w [m s^{-1}]	κ	$\eta \times 10^6$ [kg m^{-1} s^{-1}]	$\lambda \times 10^3$ [W m^{-1} K^{-1}]
0	0.001000	0.31	− 0.0001	4.2182	1402.83	5622	1790.90	561.21
5	0.001000	21.37	0.0762	4.2039	1426.59	5815	1517.71	570.70
10	0.001000	42.36	0.1511	4.1945	1447.95	5989	1305.60	580.17
15	0.001001	63.32	0.2244	4.1882	1466.98	6143	1137.37	589.50
20	0.001002	84.25	0.2964	4.1840	1483.81	6279	1001.50	598.57
25	0.001003	105.16	0.3672	4.1812	1498.57	6398	890.02	607.31
30	0.001004	126.06	0.4367	4.1793	1511.38	6498	797.32	615.62
35	0.001006	146.95	0.5050	4.1783	1522.38	6583	719.32	623.44
40	0.001008	167.84	0.5723	4.1779	1531.70	6651	653.00	630.75
45	0.001010	188.73	0.6385	4.1782	1539.45	6705	596.10	637.51
50	0.001012	209.63	0.7036	4.1790	1545.74	6745	546.90	643.72
60	0.001017	251.43	0.8310	4.1822	1554.29	6787	466.46	654.52
70	0.001023	293.28	0.9548	4.1876	1558.03	6782	403.96	663.27
80	0.001029	335.19	1.0752	4.1950	1557.53	6736	354.42	670.16
90	0.001036	377.18	1.1925	4.2045	1553.26	6654	314.48	675.41
100	0.001043	419.29	1.3068	4.2161	1545.60	6541	281.81	679.23
110	0.001051	461.51	1.4185	4.2299	1534.88	6401	254.75	681.81
120	0.001060	503.89	1.5277	4.2460	1521.35	6237	232.09	683.28
130	0.001070	546.44	1.6346	4.2646	1505.21	6051	212.92	683.75
t_s = 138.86				**Saturation**				
Liquid	0.001079	584.31	1.7275	4.2835	1488.84	5871	198.28	683.39
Vapour	0.5242	2731.97	6.9401	2.3019	489.40	1.3055	13.61	29.98
140	0.5259	2734.58	6.9464	2.2894	490.28	1.3059	13.65	30.04
150	0.5408	2757.05	7.0002	2.2129	497.59	1.3080	14.06	30.63
160	0.5555	2778.94	7.0513	2.1671	504.42	1.3087	14.46	31.27
170	0.5699	2800.44	7.1004	2.1346	510.96	1.3089	14.87	31.97
180	0.5842	2821.66	7.1477	2.1100	517.30	1.3087	15.28	32.71
190	0.5984	2842.66	7.1935	2.0910	523.46	1.3084	15.69	33.48
200	0.6124	2863.49	7.2381	2.0762	529.47	1.3079	16.10	34.30
210	0.6263	2884.19	7.2814	2.0649	535.35	1.3074	16.51	35.14
220	0.6402	2904.80	7.3236	2.0564	541.11	1.3067	16.92	36.02
230	0.6540	2925.33	7.3648	2.0502	546.75	1.3059	17.34	36.92
240	0.6678	2945.81	7.4051	2.0459	552.29	1.3051	17.75	37.85
250	0.6815	2966.25	7.4445	2.0431	557.73	1.3042	18.17	38.80
260	0.6951	2986.67	7.4832	2.0416	563.09	1.3033	18.59	39.77
270	0.7087	3007.09	7.5211	2.0412	568.36	1.3023	19.01	40.76
280	0.7223	3027.50	7.5584	2.0417	573.56	1.3013	19.42	41.77
290	0.7358	3047.92	7.5950	2.0430	578.68	1.3002	19.84	42.80
300	0.7494	3068.36	7.6310	2.0449	583.73	1.2992	20.26	43.84
310	0.7629	3088.82	7.6663	2.0474	588.72	1.2981	20.68	44.90
320	0.7763	3109.31	7.7012	2.0504	593.65	1.2970	21.10	45.97
330	0.7898	3129.83	7.7355	2.0538	598.52	1.2959	21.52	47.05
340	0.8032	3150.39	7.7693	2.0576	603.33	1.2948	21.94	48.15

Table 4 Single-Phase Region (Regions 1 to 3 of IAPWS-IF97) – Continuation
Homogenes Zustandsgebiet (Bereiche 1 bis 3 der IAPWS-IF97) – Fortsetzung

$p = 0.35$ **MPa**

t [°C]	v [m^3 kg^{-1}]	h [kJ kg^{-1}]	s [kJ kg^{-1} K^{-1}]	c_p [kJ kg^{-1} K^{-1}]	w [m s^{-1}]	κ	$\eta \times 10^6$ [kg m^{-1} s^{-1}]	$\lambda \times 10^3$ [W m^{-1} K^{-1}]
350	0.8167	3170.99	7.8026	2.0616	608.08	1.2936	22.35	49.27
360	0.8301	3191.62	7.8355	2.0660	612.79	1.2925	22.77	50.39
370	0.8435	3212.31	7.8679	2.0705	617.44	1.2914	23.19	51.53
380	0.8569	3233.04	7.8999	2.0753	622.04	1.2902	23.61	52.67
390	0.8702	3253.81	7.9314	2.0803	626.60	1.2891	24.03	53.83
400	0.8836	3274.64	7.9626	2.0855	631.11	1.2879	24.44	54.99
410	0.8969	3295.52	7.9934	2.0908	635.58	1.2868	24.86	56.17
420	0.9103	3316.46	8.0238	2.0963	640.00	1.2856	25.28	57.36
430	0.9236	3337.45	8.0539	2.1019	644.39	1.2845	25.69	58.55
440	0.9370	3358.50	8.0836	2.1076	648.73	1.2833	26.11	59.76
450	0.9503	3379.60	8.1130	2.1134	653.03	1.2822	26.52	60.97
460	0.9636	3400.77	8.1421	2.1193	657.30	1.2810	26.93	62.19
470	0.9769	3421.99	8.1708	2.1253	661.53	1.2799	27.34	63.42
480	0.9902	3443.27	8.1993	2.1314	665.72	1.2788	27.76	64.65
490	1.0035	3464.62	8.2274	2.1376	669.88	1.2776	28.17	65.90
500	1.0168	3486.02	8.2553	2.1438	674.01	1.2765	28.58	67.15
510	1.0301	3507.49	8.2829	2.1501	678.10	1.2754	28.99	68.41
520	1.0434	3529.03	8.3102	2.1564	682.15	1.2743	29.39	69.68
530	1.0567	3550.62	8.3373	2.1629	686.18	1.2731	29.80	70.95
540	1.0699	3572.28	8.3641	2.1693	690.18	1.2720	30.21	72.23
550	1.0832	3594.01	8.3906	2.1759	694.14	1.2709	30.61	73.52
560	1.0965	3615.80	8.4169	2.1824	698.08	1.2698	31.02	74.81
570	1.1097	3637.66	8.4430	2.1890	701.99	1.2687	31.42	76.11
580	1.1230	3659.58	8.4689	2.1957	705.87	1.2677	31.82	77.42
590	1.1363	3681.57	8.4945	2.2023	709.72	1.2666	32.22	78.73
600	1.1495	3703.63	8.5199	2.2090	713.54	1.2655	32.62	80.05
610	1.1628	3725.75	8.5451	2.2158	717.34	1.2644	33.02	81.37
620	1.1760	3747.94	8.5701	2.2225	721.12	1.2634	33.42	82.70
630	1.1893	3770.20	8.5948	2.2293	724.87	1.2623	33.82	84.03
640	1.2025	3792.53	8.6194	2.2361	728.59	1.2613	34.21	85.37
650	1.2157	3814.92	8.6438	2.2429	732.29	1.2602	34.61	86.72
660	1.2290	3837.39	8.6680	2.2497	735.97	1.2592	35.00	88.07
670	1.2422	3859.92	8.6920	2.2566	739.62	1.2582	35.39	89.42
680	1.2555	3882.52	8.7159	2.2634	743.25	1.2572	35.79	90.78
690	1.2687	3905.19	8.7395	2.2703	746.86	1.2562	36.18	92.15
700	1.2819	3927.92	8.7630	2.2771	750.45	1.2552	36.56	93.52
720	1.3084	3973.60	8.8095	2.2909	757.56	1.2532	37.34	96.27
740	1.3349	4019.56	8.8553	2.3047	764.59	1.2513	38.11	99.04
760	1.3613	4065.79	8.9005	2.3185	771.54	1.2494	38.87	101.83
780	1.3878	4112.30	8.9451	2.3323	778.42	1.2475	39.63	104.63
800	1.4142	4159.08	8.9891	2.3461	785.22	1.2457	40.39	107.45

Table 4 Single-Phase Region (Regions 1 to 3 of IAPWS-IF97) – Continuation
Homogenes Zustandsgebiet (Bereiche 1 bis 3 der IAPWS-IF97) – Fortsetzung

p = 0.4 MPa

t [°C]	v [m^3 kg^{-1}]	h [kJ kg^{-1}]	s [kJ kg^{-1} K^{-1}]	c_p [kJ kg^{-1} K^{-1}]	w [m s^{-1}]	κ	$\eta \times 10^6$ [kg m^{-1} s^{-1}]	$\lambda \times 10^3$ [W m^{-1} K^{-1}]
0	0.001000	0.37	− 0.0001	4.2179	1402.91	4920	1790.78	561.24
5	0.001000	21.42	0.0762	4.2037	1426.66	5089	1517.63	570.73
10	0.001000	42.41	0.1510	4.1943	1448.03	5241	1305.55	580.19
15	0.001001	63.36	0.2244	4.1881	1467.06	5376	1137.34	589.52
20	0.001002	84.29	0.2964	4.1839	1483.89	5495	1001.48	598.60
25	0.001003	105.21	0.3672	4.1810	1498.64	5599	890.01	607.33
30	0.001004	126.11	0.4367	4.1792	1511.46	5687	797.32	615.64
35	0.001006	147.00	0.5050	4.1782	1522.46	5760	719.32	623.47
40	0.001008	167.89	0.5723	4.1778	1531.78	5821	653.00	630.77
45	0.001010	188.78	0.6385	4.1781	1539.53	5868	596.11	637.53
50	0.001012	209.67	0.7036	4.1789	1545.82	5903	546.90	643.75
60	0.001017	251.47	0.8310	4.1821	1554.37	5939	466.47	654.54
70	0.001023	293.32	0.9548	4.1874	1558.12	5935	403.97	663.29
80	0.001029	335.23	1.0752	4.1949	1557.62	5895	354.43	670.18
90	0.001036	377.22	1.1924	4.2044	1553.35	5823	314.49	675.44
100	0.001043	419.32	1.3068	4.2160	1545.70	5725	281.83	679.26
110	0.001051	461.55	1.4185	4.2297	1534.99	5602	254.77	681.84
120	0.001060	503.93	1.5276	4.2459	1521.46	5458	232.10	683.31
130	0.001070	546.48	1.6345	4.2645	1505.33	5296	212.93	683.78
140	0.001080	589.23	1.7392	4.2859	1486.73	5117	196.55	683.32
t_s = 143.61				**Saturation**				
Liquid	0.001084	604.72	1.7766	4.2944	1479.43	5049	191.22	682.94
Vapour	0.4624	2738.06	6.8954	2.3403	491.09	1.3039	13.77	30.66
150	0.4709	2752.78	6.9305	2.2749	495.98	1.3060	14.03	31.00
160	0.4839	2775.19	6.9828	2.2121	503.07	1.3074	14.44	31.60
170	0.4968	2797.09	7.0328	2.1708	509.78	1.3078	14.85	32.26
180	0.5094	2818.64	7.0809	2.1403	516.24	1.3079	15.26	32.97
190	0.5219	2839.92	7.1274	2.1167	522.51	1.3077	15.67	33.72
200	0.5343	2860.99	7.1724	2.0984	528.61	1.3073	16.08	34.51
210	0.5467	2881.90	7.2161	2.0841	534.56	1.3068	16.50	35.33
220	0.5589	2902.69	7.2587	2.0732	540.38	1.3062	16.91	36.19
230	0.5710	2923.37	7.3002	2.0649	546.08	1.3055	17.33	37.08
240	0.5831	2943.99	7.3408	2.0588	551.68	1.3048	17.74	37.99
250	0.5952	2964.56	7.3805	2.0546	557.17	1.3039	18.16	38.93
260	0.6072	2985.09	7.4194	2.0519	562.57	1.3030	18.58	39.89
270	0.6192	3005.60	7.4575	2.0504	567.88	1.3021	19.00	40.87
280	0.6311	3026.10	7.4949	2.0500	573.11	1.3011	19.42	41.87
290	0.6430	3046.60	7.5316	2.0505	578.26	1.3001	19.84	42.89
300	0.6549	3067.11	7.5677	2.0518	583.34	1.2990	20.25	43.92
310	0.6667	3087.64	7.6032	2.0537	588.36	1.2980	20.67	44.97
320	0.6786	3108.19	7.6381	2.0561	593.31	1.2969	21.09	46.04
330	0.6904	3128.76	7.6725	2.0591	598.19	1.2958	21.51	47.12
340	0.7022	3149.37	7.7064	2.0624	603.02	1.2947	21.93	48.22

Table 4 Single-Phase Region (Regions 1 to 3 of IAPWS-IF97) – Continuation
Homogenes Zustandsgebiet (Bereiche 1 bis 3 der IAPWS-IF97) – Fortsetzung

				$p = 0.4$ MPa				
t	v	h	s	c_p	w	κ	$\eta \times 10^6$	$\lambda \times 10^3$
[°C]	[m^3 kg^{-1}]	[kJ kg^{-1}]	[kJ kg^{-1} K^{-1}]	[kJ kg^{-1} K^{-1}]	[m s^{-1}]		[kg m^{-1} s^{-1}]	[W m^{-1} K^{-1}]
350	0.7139	3170.01	7.7398	2.0661	607.80	1.2936	22.35	49.33
360	0.7257	3190.69	7.7728	2.0702	612.52	1.2924	22.77	50.45
370	0.7375	3211.42	7.8052	2.0745	617.18	1.2913	23.19	51.58
380	0.7492	3232.18	7.8373	2.0790	621.80	1.2902	23.61	52.72
390	0.7609	3253.00	7.8689	2.0838	626.37	1.2890	24.02	53.88
400	0.7726	3273.86	7.9001	2.0887	630.90	1.2879	24.44	55.04
410	0.7843	3294.77	7.9310	2.0938	635.38	1.2868	24.86	56.22
420	0.7960	3315.74	7.9614	2.0991	639.81	1.2856	25.27	57.40
430	0.8077	3336.76	7.9915	2.1046	644.20	1.2845	25.69	58.59
440	0.8194	3357.83	8.0213	2.1101	648.56	1.2833	26.10	59.80
450	0.8311	3378.96	8.0507	2.1158	652.87	1.2822	26.52	61.01
460	0.8428	3400.15	8.0798	2.1216	657.14	1.2810	26.93	62.23
470	0.8544	3421.39	8.1086	2.1275	661.38	1.2799	27.34	63.45
480	0.8661	3442.69	8.1371	2.1334	665.58	1.2788	27.76	64.69
490	0.8777	3464.06	8.1652	2.1395	669.75	1.2776	28.17	65.93
500	0.8894	3485.49	8.1931	2.1456	673.88	1.2765	28.58	67.19
510	0.9010	3506.97	8.2207	2.1519	677.98	1.2754	28.99	68.44
520	0.9126	3528.52	8.2481	2.1581	682.04	1.2743	29.40	69.71
530	0.9243	3550.14	8.2752	2.1645	686.07	1.2732	29.80	70.98
540	0.9359	3571.81	8.3020	2.1709	690.07	1.2720	30.21	72.26
550	0.9475	3593.55	8.3286	2.1773	694.04	1.2709	30.61	73.55
560	0.9591	3615.36	8.3549	2.1838	697.99	1.2698	31.02	74.84
570	0.9708	3637.23	8.3810	2.1904	701.90	1.2688	31.42	76.14
580	0.9824	3659.17	8.4069	2.1969	705.78	1.2677	31.83	77.45
590	0.9940	3681.17	8.4325	2.2036	709.64	1.2666	32.23	78.76
600	1.0056	3703.24	8.4579	2.2102	713.47	1.2655	32.63	80.08
610	1.0172	3725.37	8.4831	2.2169	717.27	1.2645	33.03	81.40
620	1.0288	3747.58	8.5081	2.2236	721.05	1.2634	33.42	82.73
630	1.0404	3769.85	8.5329	2.2303	724.80	1.2623	33.82	84.06
640	1.0520	3792.18	8.5575	2.2371	728.53	1.2613	34.22	85.40
650	1.0636	3814.59	8.5819	2.2439	732.23	1.2603	34.61	86.75
660	1.0752	3837.06	8.6061	2.2506	735.91	1.2592	35.00	88.10
670	1.0868	3859.60	8.6302	2.2575	739.57	1.2582	35.40	89.45
680	1.0984	3882.21	8.6540	2.2643	743.20	1.2572	35.79	90.81
690	1.1099	3904.89	8.6777	2.2711	746.81	1.2562	36.18	92.18
700	1.1215	3927.63	8.7012	2.2779	750.40	1.2552	36.57	93.55
720	1.1447	3973.33	8.7476	2.2916	757.52	1.2532	37.34	96.30
740	1.1679	4019.30	8.7935	2.3054	764.55	1.2513	38.11	99.07
760	1.1910	4065.54	8.8387	2.3191	771.51	1.2494	38.87	101.86
780	1.2142	4112.06	8.8833	2.3329	778.39	1.2475	39.63	104.66
800	1.2373	4158.85	8.9273	2.3466	785.19	1.2457	40.39	107.48

Table 4 Single-Phase Region (Regions 1 to 3 of IAPWS-IF97) – Continuation
Homogenes Zustandsgebiet (Bereiche 1 bis 3 der IAPWS-IF97) – Fortsetzung

p = 0.45 MPa

t [°C]	v [m³ kg⁻¹]	h [kJ kg⁻¹]	s [kJ kg⁻¹ K⁻¹]	c_p [kJ kg⁻¹ K⁻¹]	w [m s⁻¹]	κ	$\eta \times 10^6$ [kg m⁻¹ s⁻¹]	$\lambda \times 10^3$ [W m⁻¹ K⁻¹]
0	0.001000	0.42	– 0.0001	4.2177	1402.99	4374	1790.65	561.27
5	0.001000	21.47	0.0762	4.2034	1426.74	4524	1517.55	570.75
10	0.001000	42.46	0.1510	4.1941	1448.10	4659	1305.49	580.22
15	0.001001	63.41	0.2244	4.1879	1467.14	4779	1137.30	589.54
20	0.001002	84.34	0.2964	4.1837	1483.97	4885	1001.46	598.62
25	0.001003	105.25	0.3671	4.1809	1498.72	4977	890.00	607.35
30	0.001004	126.15	0.4367	4.1791	1511.54	5055	797.31	615.66
35	0.001006	147.04	0.5050	4.1781	1522.54	5121	719.32	623.49
40	0.001008	167.93	0.5723	4.1777	1531.87	5174	653.01	630.79
45	0.001010	188.82	0.6384	4.1779	1539.62	5216	596.12	637.56
50	0.001012	209.71	0.7036	4.1787	1545.90	5248	546.91	643.77
60	0.001017	251.52	0.8310	4.1820	1554.46	5280	466.48	654.57
70	0.001023	293.36	0.9547	4.1873	1558.21	5276	403.99	663.31
80	0.001029	335.27	1.0751	4.1948	1557.71	5240	354.45	670.21
90	0.001036	377.26	1.1924	4.2042	1553.45	5177	314.51	675.46
100	0.001043	419.36	1.3067	4.2158	1545.81	5089	281.84	679.29
110	0.001051	461.59	1.4184	4.2296	1535.10	4980	254.78	681.87
120	0.001060	503.96	1.5276	4.2457	1521.58	4852	232.12	683.34
130	0.001070	546.51	1.6345	4.2644	1505.45	4708	212.94	683.81
140	0.001080	589.26	1.7392	4.2858	1486.86	4550	196.56	683.35
t_s = 147.91				**Saturation**				
Liquid	0.001088	623.22	1.8206	4.3049	1470.48	4415	185.24	682.37
Vapour	0.4139	2743.39	6.8560	2.3772	492.54	1.3025	13.92	31.28
150	0.4164	2748.33	6.8677	2.3496	494.21	1.3034	14.01	31.39
160	0.4282	2771.34	6.9214	2.2617	501.66	1.3059	14.42	31.94
170	0.4398	2793.68	6.9724	2.2091	508.56	1.3068	14.83	32.56
180	0.4512	2815.57	7.0213	2.1718	515.16	1.3070	15.24	33.23
190	0.4625	2837.14	7.0683	2.1434	521.54	1.3070	15.65	33.95
200	0.4736	2858.46	7.1139	2.1213	527.73	1.3067	16.07	34.72
210	0.4847	2879.58	7.1581	2.1039	533.76	1.3063	16.48	35.52
220	0.4956	2900.55	7.2010	2.0904	539.65	1.3058	16.90	36.36
230	0.5065	2921.40	7.2429	2.0799	545.42	1.3051	17.32	37.23
240	0.5173	2942.16	7.2837	2.0721	551.06	1.3044	17.73	38.13
250	0.5281	2962.85	7.3237	2.0663	556.60	1.3036	18.15	39.05
260	0.5388	2983.49	7.3627	2.0623	562.04	1.3028	18.57	40.00
270	0.5495	3004.10	7.4010	2.0598	567.39	1.3019	18.99	40.97
280	0.5602	3024.69	7.4386	2.0584	572.65	1.3009	19.41	41.97
290	0.5708	3045.27	7.4755	2.0581	577.84	1.2999	19.83	42.98
300	0.5814	3065.86	7.5117	2.0587	582.95	1.2989	20.25	44.01
310	0.5920	3086.45	7.5473	2.0600	587.99	1.2979	20.67	45.05
320	0.6025	3107.06	7.5824	2.0619	592.96	1.2968	21.09	46.11
330	0.6130	3127.69	7.6169	2.0644	597.87	1.2957	21.51	47.19
340	0.6236	3148.35	7.6508	2.0673	602.72	1.2946	21.93	48.28

Table 4 Single-Phase Region (Regions 1 to 3 of IAPWS-IF97) – Continuation
Homogenes Zustandsgebiet (Bereiche 1 bis 3 der IAPWS-IF97) – Fortsetzung

p = **0.45 MPa**

t [°C]	v [m³ kg⁻¹]	h [kJ kg⁻¹]	s [kJ kg⁻¹ K⁻¹]	c_p [kJ kg⁻¹ K⁻¹]	w [m s⁻¹]	κ	$\eta \times 10^6$ [kg m⁻¹ s⁻¹]	$\lambda \times 10^3$ [W m⁻¹ K⁻¹]
350	0.6341	3169.04	7.6843	2.0707	607.51	1.2935	22.35	49.39
360	0.6445	3189.76	7.7173	2.0744	612.25	1.2924	22.77	50.50
370	0.6550	3210.53	7.7498	2.0784	616.93	1.2913	23.19	51.63
380	0.6655	3231.33	7.7819	2.0827	621.56	1.2901	23.60	52.77
390	0.6759	3252.18	7.8136	2.0872	626.15	1.2890	24.02	53.93
400	0.6863	3273.08	7.8449	2.0919	630.68	1.2879	24.44	55.09
410	0.6968	3294.02	7.8758	2.0969	635.17	1.2867	24.86	56.26
420	0.7072	3315.01	7.9063	2.1020	639.62	1.2856	25.27	57.44
430	0.7176	3336.06	7.9364	2.1072	644.02	1.2845	25.69	58.64
440	0.7280	3357.16	7.9662	2.1127	648.38	1.2833	26.10	59.84
450	0.7384	3378.31	7.9957	2.1182	652.71	1.2822	26.52	61.05
460	0.7488	3399.52	8.0248	2.1239	656.99	1.2810	26.93	62.27
470	0.7591	3420.79	8.0536	2.1296	661.23	1.2799	27.35	63.49
480	0.7695	3442.12	8.0821	2.1355	665.44	1.2788	27.76	64.73
490	0.7799	3463.50	8.1103	2.1415	669.61	1.2776	28.17	65.97
500	0.7902	3484.95	8.1383	2.1475	673.75	1.2765	28.58	67.22
510	0.8006	3506.45	8.1659	2.1536	677.85	1.2754	28.99	68.48
520	0.8110	3528.02	8.1933	2.1598	681.92	1.2743	29.40	69.74
530	0.8213	3549.65	8.2204	2.1661	685.96	1.2732	29.80	71.02
540	0.8316	3571.34	8.2472	2.1724	689.97	1.2721	30.21	72.30
550	0.8420	3593.10	8.2738	2.1788	693.95	1.2710	30.62	73.58
560	0.8523	3614.92	8.3001	2.1852	697.89	1.2699	31.02	74.87
570	0.8627	3636.80	8.3262	2.1917	701.81	1.2688	31.42	76.17
580	0.8730	3658.75	8.3521	2.1982	705.70	1.2677	31.83	77.48
590	0.8833	3680.77	8.3778	2.2048	709.56	1.2666	32.23	78.79
600	0.8936	3702.85	8.4032	2.2114	713.39	1.2655	32.63	80.11
610	0.9040	3725.00	8.4284	2.2180	717.20	1.2645	33.03	81.43
620	0.9143	3747.21	8.4534	2.2247	720.98	1.2634	33.43	82.76
630	0.9246	3769.49	8.4783	2.2314	724.73	1.2624	33.82	84.09
640	0.9349	3791.84	8.5029	2.2381	728.46	1.2613	34.22	85.43
650	0.9452	3814.25	8.5273	2.2448	732.17	1.2603	34.61	86.78
660	0.9555	3836.73	8.5515	2.2516	735.85	1.2593	35.01	88.13
670	0.9659	3859.28	8.5755	2.2583	739.51	1.2583	35.40	89.48
680	0.9762	3881.90	8.5994	2.2651	743.15	1.2572	35.79	90.84
690	0.9865	3904.59	8.6231	2.2719	746.77	1.2562	36.18	92.20
700	0.9968	3927.34	8.6466	2.2787	750.36	1.2552	36.57	93.57
720	1.0174	3973.05	8.6931	2.2924	757.48	1.2533	37.34	96.33
740	1.0380	4019.03	8.7389	2.3060	764.52	1.2514	38.11	99.10
760	1.0586	4065.29	8.7841	2.3197	771.48	1.2495	38.88	101.88
780	1.0791	4111.82	8.8287	2.3334	778.36	1.2476	39.64	104.68
800	1.0997	4158.63	8.8728	2.3472	785.17	1.2457	40.39	107.50

Table 4 Single-Phase Region (Regions 1 to 3 of IAPWS-IF97) – Continuation
Homogenes Zustandsgebiet (Bereiche 1 bis 3 der IAPWS-IF97) – Fortsetzung

t [°C]	v [m³ kg⁻¹]	h [kJ kg⁻¹]	s [kJ kg⁻¹ K⁻¹]	c_p [kJ kg⁻¹ K⁻¹]	w [m s⁻¹]	κ	$\eta\times10^6$ [kg m⁻¹ s⁻¹]	$\lambda\times10^3$ [W m⁻¹ K⁻¹]
				p = 0.5 MPa				
0	0.001000	0.47	− 0.0001	4.2174	1403.06	3937	1790.53	561.30
5	0.001000	21.52	0.0762	4.2032	1426.82	4072	1517.46	570.78
10	0.001000	42.51	0.1510	4.1939	1448.18	4194	1305.44	580.24
15	0.001001	63.46	0.2244	4.1877	1467.22	4302	1137.27	589.57
20	0.001002	84.39	0.2964	4.1836	1484.05	4397	1001.43	598.64
25	0.001003	105.30	0.3671	4.1807	1498.80	4480	889.99	607.38
30	0.001004	126.20	0.4366	4.1789	1511.62	4550	797.31	615.68
35	0.001006	147.09	0.5050	4.1779	1522.62	4609	719.32	623.51
40	0.001008	167.98	0.5722	4.1776	1531.95	4658	653.01	630.82
45	0.001010	188.87	0.6384	4.1778	1539.70	4695	596.12	637.58
50	0.001012	209.76	0.7036	4.1786	1545.99	4723	546.92	643.79
60	0.001017	251.56	0.8310	4.1819	1554.54	4752	466.49	654.59
70	0.001023	293.40	0.9547	4.1872	1558.30	4749	404.00	663.34
80	0.001029	335.31	1.0751	4.1946	1557.81	4717	354.46	670.23
90	0.001036	377.30	1.1923	4.2041	1553.55	4660	314.52	675.49
100	0.001043	419.40	1.3067	4.2157	1545.91	4581	281.85	679.32
110	0.001051	461.62	1.4184	4.2295	1535.20	4483	254.79	681.89
120	0.001060	504.00	1.5275	4.2456	1521.69	4368	232.13	683.37
130	0.001070	546.54	1.6344	4.2642	1505.57	4238	212.95	683.84
140	0.001080	589.29	1.7391	4.2856	1486.98	4095	196.58	683.39
150	0.001090	632.27	1.8419	4.3102	1466.05	3941	182.47	682.06
t_s = 151.84				**Saturation**				
Liquid	0.001093	640.19	1.8606	4.3151	1461.95	3912	180.09	681.72
Vapour	0.3748	2748.11	6.8206	2.4127	493.80	1.3011	14.06	31.87
160	0.3837	2767.38	6.8655	2.3176	500.18	1.3042	14.39	32.29
170	0.3943	2790.19	6.9176	2.2500	507.32	1.3056	14.81	32.86
180	0.4047	2812.45	6.9672	2.2048	514.07	1.3061	15.22	33.50
190	0.4149	2834.32	7.0150	2.1711	520.56	1.3062	15.64	34.19
200	0.4250	2855.90	7.0611	2.1448	526.84	1.3061	16.05	34.93
210	0.4351	2877.24	7.1057	2.1242	532.96	1.3058	16.47	35.71
220	0.4450	2898.40	7.1491	2.1080	538.92	1.3053	16.89	36.53
230	0.4549	2919.41	7.1912	2.0954	544.74	1.3047	17.30	37.39
240	0.4647	2940.31	7.2324	2.0856	550.44	1.3041	17.72	38.27
250	0.4744	2961.13	7.2726	2.0783	556.03	1.3033	18.14	39.18
260	0.4841	2981.88	7.3119	2.0730	561.51	1.3025	18.56	40.12
270	0.4938	3002.59	7.3503	2.0693	566.90	1.3016	18.98	41.08
280	0.5034	3023.28	7.3881	2.0670	572.20	1.3007	19.40	42.06
290	0.5130	3043.94	7.4251	2.0659	577.42	1.2998	19.82	43.07
300	0.5226	3064.60	7.4614	2.0657	582.55	1.2988	20.24	44.09
310	0.5321	3085.26	7.4972	2.0664	587.62	1.2977	20.66	45.13
320	0.5417	3105.93	7.5323	2.0678	592.61	1.2967	21.08	46.19
330	0.5512	3126.61	7.5669	2.0698	597.55	1.2956	21.50	47.26
340	0.5607	3147.32	7.6010	2.0723	602.41	1.2945	21.92	48.35

Table 4 Single-Phase Region (Regions 1 to 3 of IAPWS-IF97) – Continuation
Homogenes Zustandsgebiet (Bereiche 1 bis 3 der IAPWS-IF97) – Fortsetzung

p = 0.5 MPa								
t [°C]	v [m³ kg⁻¹]	h [kJ kg⁻¹]	s [kJ kg⁻¹ K⁻¹]	c_p [kJ kg⁻¹ K⁻¹]	w [m s⁻¹]	κ	$\eta \times 10^6$ [kg m⁻¹ s⁻¹]	$\lambda \times 10^3$ [W m⁻¹ K⁻¹]
350	0.5701	3168.06	7.6345	2.0753	607.22	1.2934	22.34	49.45
360	0.5796	3188.83	7.6676	2.0786	611.98	1.2923	22.76	50.56
370	0.5890	3209.63	7.7002	2.0823	616.68	1.2912	23.18	51.69
380	0.5985	3230.48	7.7323	2.0864	621.32	1.2901	23.60	52.83
390	0.6079	3251.36	7.7641	2.0906	625.92	1.2890	24.02	53.98
400	0.6173	3272.29	7.7954	2.0952	630.47	1.2878	24.44	55.14
410	0.6267	3293.27	7.8263	2.0999	634.97	1.2867	24.86	56.31
420	0.6361	3314.29	7.8569	2.1048	639.43	1.2856	25.27	57.49
430	0.6455	3335.36	7.8871	2.1099	643.84	1.2844	25.69	58.68
440	0.6548	3356.49	7.9169	2.1152	648.21	1.2833	26.10	59.88
450	0.6642	3377.67	7.9464	2.1206	652.54	1.2822	26.52	61.09
460	0.6736	3398.90	7.9756	2.1261	656.83	1.2810	26.93	62.30
470	0.6829	3420.19	8.0044	2.1318	661.09	1.2799	27.35	63.53
480	0.6923	3441.54	8.0329	2.1376	665.30	1.2788	27.76	64.76
490	0.7016	3462.94	8.0612	2.1434	669.48	1.2776	28.17	66.01
500	0.7109	3484.41	8.0891	2.1494	673.62	1.2765	28.58	67.26
510	0.7203	3505.93	8.1168	2.1554	677.73	1.2754	28.99	68.51
520	0.7296	3527.52	8.1442	2.1615	681.81	1.2743	29.40	69.78
530	0.7389	3549.16	8.1713	2.1677	685.85	1.2732	29.81	71.05
540	0.7483	3570.87	8.1981	2.1740	689.86	1.2721	30.21	72.33
550	0.7576	3592.64	8.2247	2.1803	693.85	1.2710	30.62	73.61
560	0.7669	3614.48	8.2511	2.1866	697.80	1.2699	31.02	74.91
570	0.7762	3636.38	8.2772	2.1931	701.72	1.2688	31.43	76.20
580	0.7855	3658.34	8.3031	2.1995	705.61	1.2677	31.83	77.51
590	0.7948	3680.37	8.3288	2.2060	709.48	1.2666	32.23	78.82
600	0.8041	3702.46	8.3543	2.2126	713.31	1.2656	32.63	80.14
610	0.8134	3724.62	8.3795	2.2192	717.12	1.2645	33.03	81.46
620	0.8227	3746.84	8.4045	2.2258	720.91	1.2634	33.43	82.79
630	0.8320	3769.13	8.4293	2.2324	724.67	1.2624	33.82	84.12
640	0.8413	3791.49	8.4539	2.2391	728.40	1.2614	34.22	85.46
650	0.8506	3813.91	8.4784	2.2458	732.11	1.2603	34.61	86.80
660	0.8598	3836.41	8.5026	2.2525	735.80	1.2593	35.01	88.15
670	0.8691	3858.97	8.5266	2.2592	739.46	1.2583	35.40	89.51
680	0.8784	3881.59	8.5505	2.2660	743.10	1.2573	35.79	90.87
690	0.8877	3904.28	8.5742	2.2727	746.72	1.2563	36.18	92.23
700	0.8970	3927.05	8.5977	2.2795	750.31	1.2553	36.57	93.60
720	0.9155	3972.77	8.6442	2.2931	757.44	1.2533	37.34	96.35
740	0.9341	4018.77	8.6901	2.3067	764.48	1.2514	38.11	99.12
760	0.9526	4065.04	8.7353	2.3204	771.45	1.2495	38.88	101.91
780	0.9711	4111.58	8.7799	2.3340	778.33	1.2476	39.64	104.71
800	0.9897	4158.40	8.8240	2.3477	785.14	1.2458	40.39	107.53

Table 4 Single-Phase Region (Regions 1 to 3 of IAPWS-IF97) – Continuation
Homogenes Zustandsgebiet (Bereiche 1 bis 3 der IAPWS-IF97) – Fortsetzung

				p = 0.6 MPa				
t	v	h	s	c_p	w	κ	$\eta\times10^6$	$\lambda\times10^3$
[°C]	[m^3 kg^{-1}]	[kJ kg^{-1}]	[kJ kg^{-1} K^{-1}]	[kJ kg^{-1} K^{-1}]	[m s^{-1}]		[kg m^{-1} s^{-1}]	[W m^{-1} K^{-1}]
0	0.001000	0.57	– 0.0001	4.2169	1403.22	3282	1790.27	561.35
5	0.001000	21.62	0.0762	4.2028	1426.98	3394	1517.30	570.83
10	0.001000	42.60	0.1510	4.1936	1448.34	3495	1305.33	580.29
15	0.001001	63.56	0.2244	4.1874	1467.38	3586	1137.20	589.61
20	0.001002	84.48	0.2964	4.1832	1484.21	3665	1001.39	598.69
25	0.001003	105.39	0.3671	4.1805	1498.96	3734	889.97	607.42
30	0.001004	126.29	0.4366	4.1787	1511.78	3793	797.30	615.73
35	0.001006	147.18	0.5050	4.1777	1522.78	3842	719.32	623.56
40	0.001008	168.07	0.5722	4.1773	1532.11	3882	653.02	630.86
45	0.001010	188.95	0.6384	4.1776	1539.86	3914	596.14	637.62
50	0.001012	209.84	0.7035	4.1784	1546.15	3937	546.94	643.84
60	0.001017	251.64	0.8309	4.1817	1554.72	3961	466.51	654.64
70	0.001022	293.48	0.9547	4.1870	1558.47	3959	404.02	663.39
80	0.001029	335.39	1.0750	4.1944	1557.99	3932	354.49	670.28
90	0.001036	377.38	1.1923	4.2039	1553.74	3884	314.55	675.54
100	0.001043	419.47	1.3066	4.2155	1546.11	3819	281.88	679.37
110	0.001051	461.70	1.4183	4.2293	1535.42	3737	254.82	681.95
120	0.001060	504.07	1.5275	4.2453	1521.92	3641	232.16	683.42
130	0.001070	546.61	1.6343	4.2639	1505.81	3533	212.98	683.90
140	0.001080	589.35	1.7390	4.2853	1487.24	3414	196.60	683.45
150	0.001090	632.33	1.8418	4.3099	1466.32	3286	182.49	682.12
t_s = 158.83				**Saturation**				
Liquid	0.001101	670.50	1.9311	4.3345	1445.94	3166	171.58	680.24
Vapour	0.3156	2756.14	6.7592	2.4800	495.88	1.2987	14.30	32.95
160	0.3167	2759.02	6.7658	2.4597	496.87	1.2994	14.35	33.00
170	0.3258	2782.97	6.8205	2.3427	504.69	1.3029	14.76	33.49
180	0.3347	2806.04	6.8720	2.2759	511.81	1.3042	15.18	34.05
190	0.3435	2828.55	6.9211	2.2295	518.55	1.3047	15.60	34.68
200	0.3521	2850.66	6.9684	2.1942	525.04	1.3048	16.02	35.37
210	0.3606	2872.46	7.0139	2.1666	531.32	1.3047	16.44	36.10
220	0.3690	2894.01	7.0581	2.1446	537.43	1.3044	16.86	36.88
230	0.3774	2915.37	7.1010	2.1273	543.38	1.3039	17.28	37.70
240	0.3857	2936.57	7.1427	2.1136	549.19	1.3034	17.70	38.56
250	0.3939	2957.65	7.1834	2.1030	554.88	1.3027	18.12	39.44
260	0.4021	2978.64	7.2231	2.0949	560.45	1.3020	18.54	40.36
270	0.4102	2999.56	7.2620	2.0888	565.91	1.3012	18.97	41.30
280	0.4183	3020.42	7.3001	2.0845	571.28	1.3003	19.39	42.26
290	0.4264	3041.25	7.3374	2.0817	576.56	1.2994	19.81	43.25
300	0.4344	3062.06	7.3740	2.0800	581.76	1.2985	20.23	44.26
310	0.4424	3082.86	7.4100	2.0794	586.88	1.2975	20.65	45.29
320	0.4504	3103.65	7.4453	2.0797	591.92	1.2965	21.07	46.34
330	0.4584	3124.45	7.4801	2.0807	596.89	1.2954	21.50	47.40
340	0.4663	3145.27	7.5143	2.0823	601.80	1.2944	21.92	48.48

Table 4 Single-Phase Region (Regions 1 to 3 of IAPWS-IF97) – Continuation
Homogenes Zustandsgebiet (Bereiche 1 bis 3 der IAPWS-IF97) – Fortsetzung

				p = 0.6 MPa				
t [°C]	v [m^3 kg^{-1}]	h [kJ kg^{-1}]	s [kJ kg^{-1} K^{-1}]	c_p [kJ kg^{-1} K^{-1}]	w [m s^{-1}]	κ	$\eta \times 10^6$ [kg m^{-1} s^{-1}]	$\lambda \times 10^3$ [W m^{-1} K^{-1}]
350	0.4743	3166.10	7.5480	2.0845	606.65	1.2933	22.34	49.57
360	0.4822	3186.96	7.5812	2.0872	611.43	1.2922	22.76	50.68
370	0.4901	3207.85	7.6140	2.0903	616.16	1.2911	23.18	51.80
380	0.4980	3228.77	7.6463	2.0938	620.84	1.2900	23.60	52.93
390	0.5059	3249.72	7.6781	2.0976	625.46	1.2889	24.02	54.08
400	0.5137	3270.72	7.7095	2.1017	630.04	1.2878	24.44	55.23
410	0.5216	3291.76	7.7405	2.1060	634.56	1.2867	24.85	56.40
420	0.5294	3312.84	7.7712	2.1106	639.04	1.2855	25.27	57.58
430	0.5373	3333.97	7.8014	2.1154	643.47	1.2844	25.69	58.76
440	0.5451	3355.15	7.8314	2.1203	647.86	1.2833	26.10	59.96
450	0.5530	3376.38	7.8609	2.1255	652.21	1.2822	26.52	61.17
460	0.5608	3397.66	7.8901	2.1307	656.52	1.2810	26.93	62.38
470	0.5686	3418.99	7.9190	2.1362	660.79	1.2799	27.35	63.60
480	0.5764	3440.38	7.9476	2.1417	665.02	1.2788	27.76	64.84
490	0.5842	3461.83	7.9759	2.1473	669.21	1.2776	28.17	66.08
500	0.5920	3483.33	8.0039	2.1531	673.37	1.2765	28.58	67.32
510	0.5998	3504.89	8.0316	2.1590	677.49	1.2754	28.99	68.58
520	0.6076	3526.51	8.0591	2.1649	681.58	1.2743	29.40	69.84
530	0.6154	3548.19	8.0862	2.1709	685.63	1.2732	29.81	71.11
540	0.6232	3569.93	8.1131	2.1770	689.66	1.2721	30.21	72.39
550	0.6309	3591.73	8.1398	2.1832	693.65	1.2710	30.62	73.68
560	0.6387	3613.59	8.1662	2.1895	697.61	1.2699	31.02	74.97
570	0.6465	3635.52	8.1923	2.1958	701.54	1.2688	31.43	76.27
580	0.6542	3657.51	8.2183	2.2021	705.44	1.2677	31.83	77.57
590	0.6620	3679.56	8.2439	2.2085	709.31	1.2667	32.23	78.88
600	0.6698	3701.68	8.2694	2.2150	713.16	1.2656	32.63	80.20
610	0.6775	3723.86	8.2947	2.2214	716.98	1.2645	33.03	81.52
620	0.6853	3746.11	8.3197	2.2280	720.77	1.2635	33.43	82.85
630	0.6930	3768.42	8.3446	2.2345	724.53	1.2624	33.83	84.18
640	0.7008	3790.80	8.3692	2.2411	728.28	1.2614	34.22	85.52
650	0.7085	3813.24	8.3937	2.2477	731.99	1.2604	34.62	86.86
660	0.7163	3835.75	8.4179	2.2544	735.68	1.2593	35.01	88.21
670	0.7240	3858.33	8.4420	2.2610	739.35	1.2583	35.40	89.57
680	0.7318	3880.97	8.4659	2.2677	743.00	1.2573	35.80	90.93
690	0.7395	3903.68	8.4896	2.2744	746.62	1.2563	36.19	92.29
700	0.7473	3926.46	8.5131	2.2811	750.22	1.2553	36.57	93.66
720	0.7627	3972.22	8.5596	2.2946	757.36	1.2534	37.35	96.41
740	0.7782	4018.24	8.6055	2.3081	764.41	1.2515	38.12	99.18
760	0.7937	4064.54	8.6508	2.3216	771.38	1.2496	38.88	101.96
780	0.8091	4111.11	8.6954	2.3352	778.28	1.2477	39.64	104.76
800	0.8246	4157.95	8.7395	2.3488	785.10	1.2459	40.40	107.58

Table 4 **Single-Phase Region** (Regions 1 to 3 of IAPWS-IF97) – Continuation
Homogenes Zustandsgebiet (Bereiche 1 bis 3 der IAPWS-IF97) – Fortsetzung

p = 0.7 MPa

t [°C]	v [m³ kg⁻¹]	h [kJ kg⁻¹]	s [kJ kg⁻¹ K⁻¹]	c_p [kJ kg⁻¹ K⁻¹]	w [m s⁻¹]	κ	$\eta \times 10^6$ [kg m⁻¹ s⁻¹]	$\lambda \times 10^3$ [W m⁻¹ K⁻¹]
0	0.001000	0.67	− 0.0001	4.2164	1403.38	2813	1790.02	561.41
5	0.001000	21.72	0.0762	4.2024	1427.14	2910	1517.13	570.88
10	0.001000	42.70	0.1510	4.1932	1448.50	2997	1305.22	580.34
15	0.001001	63.65	0.2244	4.1871	1467.54	3074	1137.13	589.66
20	0.001002	84.58	0.2964	4.1829	1484.37	3142	1001.35	598.73
25	0.001003	105.48	0.3671	4.1802	1499.12	3201	889.94	607.47
30	0.001004	126.38	0.4366	4.1784	1511.94	3252	797.29	615.77
35	0.001006	147.27	0.5049	4.1774	1522.95	3294	719.32	623.60
40	0.001008	168.15	0.5722	4.1771	1532.27	3328	653.03	630.91
45	0.001010	189.04	0.6383	4.1773	1540.03	3355	596.15	637.67
50	0.001012	209.93	0.7035	4.1782	1546.32	3375	546.96	643.88
60	0.001017	251.73	0.8309	4.1814	1554.89	3396	466.54	654.69
70	0.001022	293.56	0.9546	4.1868	1558.65	3394	404.05	663.44
80	0.001029	335.47	1.0750	4.1942	1558.18	3371	354.51	670.33
90	0.001036	377.46	1.1922	4.2037	1553.94	3330	314.57	675.59
100	0.001043	419.55	1.3065	4.2153	1546.32	3274	281.91	679.43
110	0.001051	461.77	1.4182	4.2290	1535.63	3204	254.85	682.01
120	0.001060	504.14	1.5274	4.2451	1522.15	3122	232.18	683.48
130	0.001069	546.68	1.6342	4.2637	1506.05	3029	213.01	683.96
140	0.001080	589.42	1.7389	4.2850	1487.49	2928	196.63	683.52
150	0.001090	632.39	1.8417	4.3095	1466.59	2818	182.52	682.19
160	0.001102	675.62	1.9427	4.3376	1443.41	2701	170.26	680.01
t_s = 164.95				**Saturation**				
Liquid	0.001108	697.14	1.9921	4.3530	1431.09	2640	164.77	678.62
Vapour	0.2728	2762.75	6.7070	2.5433	497.53	1.2964	14.51	33.92
170	0.2769	2775.36	6.7356	2.4582	501.80	1.2993	14.72	34.13
180	0.2847	2799.38	6.7892	2.3561	509.44	1.3021	15.14	34.62
190	0.2924	2822.60	6.8399	2.2927	516.48	1.3031	15.57	35.18
200	0.3000	2845.29	6.8884	2.2468	523.19	1.3035	15.99	35.81
210	0.3074	2867.57	6.9350	2.2113	529.65	1.3036	16.41	36.50
220	0.3148	2889.54	6.9800	2.1832	535.91	1.3034	16.83	37.24
230	0.3220	2911.26	7.0236	2.1607	541.99	1.3031	17.26	38.02
240	0.3292	2932.77	7.0659	2.1428	547.92	1.3027	17.68	38.85
250	0.3364	2954.12	7.1071	2.1287	553.71	1.3021	18.10	39.71
260	0.3434	2975.35	7.1473	2.1176	559.37	1.3015	18.53	40.60
270	0.3505	2996.48	7.1866	2.1090	564.92	1.3008	18.95	41.52
280	0.3575	3017.54	7.2250	2.1026	570.36	1.3000	19.37	42.47
290	0.3645	3038.54	7.2626	2.0979	575.70	1.2991	19.80	43.44
300	0.3714	3059.50	7.2995	2.0947	580.96	1.2982	20.22	44.43
310	0.3783	3080.44	7.3357	2.0927	586.13	1.2973	20.64	45.45
320	0.3852	3101.36	7.3713	2.0918	591.22	1.2963	21.06	46.48
330	0.3921	3122.28	7.4063	2.0918	596.24	1.2953	21.49	47.54
340	0.3989	3143.20	7.4407	2.0925	601.19	1.2942	21.91	48.61

Table 4 Single-Phase Region (Regions 1 to 3 of IAPWS-IF97) – Continuation
Homogenes Zustandsgebiet (Bereiche 1 bis 3 der IAPWS-IF97) – Fortsetzung

				p = 0.7 MPa				
t	v	h	s	c_p	w	κ	$\eta \times 10^6$	$\lambda \times 10^3$
[°C]	[$m^3\,kg^{-1}$]	[$kJ\,kg^{-1}$]	[$kJ\,kg^{-1}\,K^{-1}$]	[$kJ\,kg^{-1}\,K^{-1}$]	[$m\,s^{-1}$]		[$kg\,m^{-1}\,s^{-1}$]	[$W\,m^{-1}\,K^{-1}$]
350	0.4058	3164.13	7.4745	2.0939	606.07	1.2932	22.33	49.69
360	0.4126	3185.08	7.5079	2.0959	610.89	1.2921	22.75	50.79
370	0.4194	3206.05	7.5408	2.0984	615.65	1.2910	23.17	51.91
380	0.4262	3227.05	7.5732	2.1013	620.36	1.2899	23.59	53.04
390	0.4330	3248.08	7.6051	2.1046	625.01	1.2888	24.01	54.18
400	0.4398	3269.14	7.6366	2.1083	629.60	1.2877	24.43	55.33
410	0.4465	3290.24	7.6678	2.1122	634.15	1.2866	24.85	56.49
420	0.4533	3311.39	7.6985	2.1164	638.65	1.2855	25.27	57.66
430	0.4600	3332.57	7.7288	2.1209	643.11	1.2844	25.69	58.85
440	0.4668	3353.81	7.7588	2.1255	647.52	1.2833	26.10	60.04
450	0.4735	3375.08	7.7884	2.1303	651.88	1.2821	26.52	61.25
460	0.4802	3396.41	7.8177	2.1354	656.21	1.2810	26.93	62.46
470	0.4869	3417.79	7.8467	2.1405	660.49	1.2799	27.35	63.68
480	0.4936	3439.22	7.8753	2.1458	664.74	1.2788	27.76	64.91
490	0.5003	3460.71	7.9037	2.1513	668.94	1.2777	28.17	66.15
500	0.5070	3482.25	7.9317	2.1569	673.11	1.2765	28.58	67.40
510	0.5137	3503.85	7.9595	2.1625	677.25	1.2754	28.99	68.65
520	0.5204	3525.50	7.9870	2.1683	681.35	1.2743	29.40	69.91
530	0.5271	3547.21	8.0142	2.1742	685.41	1.2732	29.81	71.18
540	0.5338	3568.99	8.0411	2.1801	689.45	1.2721	30.22	72.46
550	0.5405	3590.82	8.0678	2.1862	693.45	1.2710	30.62	73.74
560	0.5472	3612.71	8.0942	2.1923	697.42	1.2699	31.03	75.03
570	0.5538	3634.66	8.1204	2.1985	701.36	1.2689	31.43	76.33
580	0.5605	3656.68	8.1464	2.2047	705.27	1.2678	31.83	77.63
590	0.5672	3678.76	8.1721	2.2110	709.15	1.2667	32.24	78.94
600	0.5738	3700.90	8.1976	2.2173	713.00	1.2656	32.64	80.26
610	0.5805	3723.10	8.2229	2.2237	716.83	1.2646	33.04	81.58
620	0.5871	3745.37	8.2480	2.2301	720.63	1.2635	33.43	82.91
630	0.5938	3767.71	8.2728	2.2366	724.40	1.2625	33.83	84.24
640	0.6004	3790.11	8.2975	2.2431	728.15	1.2615	34.23	85.58
650	0.6071	3812.57	8.3220	2.2497	731.87	1.2604	34.62	86.92
660	0.6137	3835.10	8.3462	2.2562	735.57	1.2594	35.02	88.27
670	0.6204	3857.69	8.3703	2.2628	739.25	1.2584	35.41	89.62
680	0.6270	3880.36	8.3942	2.2694	742.90	1.2574	35.80	90.98
690	0.6337	3903.08	8.4179	2.2760	746.52	1.2564	36.19	92.35
700	0.6403	3925.88	8.4415	2.2827	750.13	1.2554	36.58	93.71
720	0.6536	3971.66	8.4881	2.2960	757.28	1.2534	37.35	96.46
740	0.6669	4017.72	8.5340	2.3094	764.34	1.2515	38.12	99.23
760	0.6801	4064.04	8.5792	2.3229	771.32	1.2496	38.89	102.01
780	0.6934	4110.63	8.6239	2.3364	778.22	1.2478	39.65	104.81
800	0.7066	4157.50	8.6680	2.3499	785.05	1.2459	40.40	107.62

Table 4 Single-Phase Region (Regions 1 to 3 of IAPWS-IF97) – Continuation
Homogenes Zustandsgebiet (Bereiche 1 bis 3 der IAPWS-IF97) – Fortsetzung

t [°C]	v [m^3 kg^{-1}]	h [kJ kg^{-1}]	s [kJ kg^{-1} K^{-1}]	c_p [kJ kg^{-1} K^{-1}]	w [m s^{-1}]	κ	$\eta \times 10^6$ [kg m^{-1} s^{-1}]	$\lambda \times 10^3$ [W m^{-1} K^{-1}]
				p = 0.8 MPa				
0	0.001000	0.77	– 0.0001	4.2160	1403.53	2462	1789.77	561.46
5	0.001000	21.81	0.0762	4.2020	1427.30	2547	1516.97	570.93
10	0.001000	42.80	0.1510	4.1928	1448.66	2623	1305.11	580.39
15	0.001001	63.75	0.2243	4.1867	1467.69	2691	1137.06	589.71
20	0.001001	84.67	0.2963	4.1826	1484.52	2750	1001.31	598.78
25	0.001003	105.58	0.3671	4.1799	1499.28	2802	889.92	607.51
30	0.001004	126.47	0.4366	4.1781	1512.10	2846	797.28	615.82
35	0.001006	147.36	0.5049	4.1772	1523.11	2883	719.32	623.65
40	0.001008	168.24	0.5721	4.1768	1532.43	2913	653.03	630.95
45	0.001010	189.13	0.6383	4.1771	1540.19	2937	596.16	637.72
50	0.001012	210.02	0.7034	4.1779	1546.48	2954	546.97	643.93
60	0.001017	251.81	0.8308	4.1812	1555.06	2972	466.56	654.73
70	0.001022	293.65	0.9545	4.1866	1558.83	2970	404.07	663.49
80	0.001029	335.55	1.0749	4.1940	1558.37	2950	354.54	670.38
90	0.001036	377.53	1.1921	4.2035	1554.13	2915	314.60	675.65
100	0.001043	419.62	1.3065	4.2150	1546.52	2866	281.93	679.48
110	0.001051	461.84	1.4181	4.2288	1535.85	2804	254.87	682.06
120	0.001060	504.21	1.5273	4.2448	1522.37	2733	232.21	683.54
130	0.001069	546.75	1.6341	4.2634	1506.29	2652	213.03	684.02
140	0.001079	589.48	1.7388	4.2847	1487.75	2563	196.65	683.58
150	0.001090	632.45	1.8416	4.3092	1466.86	2466	182.54	682.26
160	0.001102	675.68	1.9426	4.3373	1443.70	2364	170.29	680.09
170	0.001114	719.21	2.0419	4.3695	1418.29	2256	159.55	677.06
t_s = 170.41				**Saturation**				
Liquid	0.001115	721.02	2.0460	4.3709	1417.19	2252	159.14	676.92
Vapour	0.2403	2768.30	6.6615	2.6032	498.87	1.2944	14.70	34.82
180	0.2472	2792.44	6.7154	2.4499	506.90	1.2994	15.10	35.20
190	0.2541	2816.46	6.7678	2.3621	514.34	1.3013	15.53	35.70
200	0.2609	2839.77	6.8176	2.3030	521.29	1.3021	15.96	36.27
210	0.2675	2862.57	6.8653	2.2586	527.95	1.3024	16.38	36.91
220	0.2740	2884.97	6.9112	2.2236	534.36	1.3025	16.81	37.60
230	0.2805	2907.06	6.9555	2.1956	540.58	1.3023	17.23	38.35
240	0.2869	2928.90	6.9985	2.1732	546.63	1.3020	17.66	39.14
250	0.2932	2950.54	7.0403	2.1553	552.52	1.3015	18.08	39.97
260	0.2995	2972.02	7.0810	2.1411	558.28	1.3010	18.51	40.84
270	0.3057	2993.37	7.1206	2.1298	563.91	1.3003	18.93	41.74
280	0.3119	3014.63	7.1594	2.1211	569.42	1.2996	19.36	42.67
290	0.3180	3035.80	7.1974	2.1145	574.84	1.2988	19.78	43.63
300	0.3242	3056.92	7.2345	2.1097	580.15	1.2979	20.21	44.61
310	0.3302	3078.00	7.2710	2.1063	585.37	1.2970	20.63	45.61
320	0.3363	3099.05	7.3068	2.1042	590.51	1.2961	21.06	46.63
330	0.3424	3120.09	7.3420	2.1031	595.58	1.2951	21.48	47.68
340	0.3484	3141.12	7.3765	2.1029	600.57	1.2941	21.90	48.74

Table 4 Single-Phase Region (Regions 1 to 3 of IAPWS-IF97) – Continuation
Homogenes Zustandsgebiet (Bereiche 1 bis 3 der IAPWS-IF97) – Fortsetzung

				p = 0.8 MPa				
t [°C]	v [m³ kg⁻¹]	h [kJ kg⁻¹]	s [kJ kg⁻¹ K⁻¹]	c_p [kJ kg⁻¹ K⁻¹]	w [m s⁻¹]	κ	$\eta \times 10^6$ [kg m⁻¹ s⁻¹]	$\lambda \times 10^3$ [W m⁻¹ K⁻¹]
350	0.3544	3162.15	7.4106	2.1035	605.49	1.2930	22.32	49.82
360	0.3604	3183.19	7.4441	2.1048	610.34	1.2920	22.75	50.91
370	0.3664	3204.25	7.4771	2.1066	615.14	1.2909	23.17	52.02
380	0.3724	3225.33	7.5096	2.1090	619.87	1.2899	23.59	53.14
390	0.3783	3246.43	7.5416	2.1117	624.55	1.2888	24.01	54.28
400	0.3843	3267.56	7.5733	2.1149	629.17	1.2877	24.43	55.42
410	0.3902	3288.73	7.6045	2.1184	633.74	1.2866	24.85	56.58
420	0.3961	3309.93	7.6353	2.1223	638.26	1.2855	25.27	57.75
430	0.4021	3331.17	7.6657	2.1264	642.74	1.2844	25.68	58.93
440	0.4080	3352.46	7.6958	2.1307	647.17	1.2832	26.10	60.12
450	0.4139	3373.79	7.7255	2.1352	651.55	1.2821	26.52	61.33
460	0.4198	3395.16	7.7548	2.1400	655.89	1.2810	26.93	62.54
470	0.4257	3416.59	7.7839	2.1449	660.19	1.2799	27.35	63.76
480	0.4316	3438.06	7.8126	2.1500	664.45	1.2788	27.76	64.98
490	0.4374	3459.59	7.8410	2.1553	668.67	1.2777	28.17	66.22
500	0.4433	3481.17	7.8690	2.1606	672.86	1.2766	28.58	67.47
510	0.4492	3502.80	7.8969	2.1661	677.00	1.2754	28.99	68.72
520	0.4551	3524.49	7.9244	2.1717	681.12	1.2743	29.40	69.98
530	0.4609	3546.24	7.9516	2.1774	685.19	1.2732	29.81	71.25
540	0.4668	3568.04	7.9786	2.1833	689.24	1.2721	30.22	72.52
550	0.4726	3589.90	8.0053	2.1892	693.25	1.2711	30.62	73.81
560	0.4785	3611.82	8.0318	2.1951	697.23	1.2700	31.03	75.10
570	0.4843	3633.81	8.0580	2.2012	701.18	1.2689	31.43	76.39
580	0.4902	3655.85	8.0840	2.2073	705.10	1.2678	31.84	77.70
590	0.4960	3677.95	8.1098	2.2135	708.99	1.2667	32.24	79.00
600	0.5019	3700.12	8.1353	2.2197	712.85	1.2657	32.64	80.32
610	0.5077	3722.35	8.1606	2.2260	716.68	1.2646	33.04	81.64
620	0.5135	3744.64	8.1857	2.2323	720.49	1.2636	33.44	82.97
630	0.5194	3766.99	8.2106	2.2387	724.27	1.2625	33.84	84.30
640	0.5252	3789.41	8.2353	2.2451	728.02	1.2615	34.23	85.64
650	0.5310	3811.90	8.2598	2.2516	731.75	1.2605	34.63	86.98
660	0.5368	3834.44	8.2841	2.2581	735.46	1.2595	35.02	88.33
670	0.5427	3857.06	8.3082	2.2646	739.14	1.2584	35.41	89.68
680	0.5485	3879.74	8.3321	2.2711	742.80	1.2574	35.80	91.04
690	0.5543	3902.48	8.3558	2.2777	746.43	1.2564	36.19	92.40
700	0.5601	3925.29	8.3794	2.2843	750.04	1.2555	36.58	93.77
720	0.5717	3971.11	8.4260	2.2975	757.20	1.2535	37.36	96.52
740	0.5834	4017.19	8.4719	2.3108	764.27	1.2516	38.13	99.28
760	0.5950	4063.54	8.5172	2.3242	771.26	1.2497	38.89	102.06
780	0.6066	4110.16	8.5619	2.3376	778.17	1.2478	39.65	104.86
800	0.6182	4157.04	8.6060	2.3510	785.00	1.2460	40.41	107.67

Table 4 Single-Phase Region (Regions 1 to 3 of IAPWS-IF97) – Continuation
Homogenes Zustandsgebiet (Bereiche 1 bis 3 der IAPWS-IF97) – Fortsetzung

p = 0.9 MPa

t [°C]	v [m³ kg⁻¹]	h [kJ kg⁻¹]	s [kJ kg⁻¹ K⁻¹]	c_p [kJ kg⁻¹ K⁻¹]	w [m s⁻¹]	κ	$\eta \times 10^6$ [kg m⁻¹ s⁻¹]	$\lambda \times 10^3$ [W m⁻¹ K⁻¹]
0	0.001000	0.87	– 0.0001	4.2155	1403.69	2189	1789.52	561.52
5	0.001000	21.91	0.0762	4.2015	1427.45	2264	1516.80	570.98
10	0.001000	42.90	0.1510	4.1924	1448.82	2332	1305.01	580.43
15	0.001001	63.84	0.2243	4.1864	1467.85	2392	1136.99	589.75
20	0.001001	84.76	0.2963	4.1823	1484.68	2445	1001.26	598.83
25	0.001003	105.67	0.3670	4.1796	1499.44	2491	889.90	607.56
30	0.001004	126.56	0.4365	4.1779	1512.25	2530	797.27	615.86
35	0.001006	147.45	0.5049	4.1769	1523.27	2563	719.32	623.69
40	0.001007	168.33	0.5721	4.1766	1532.59	2590	653.04	631.00
45	0.001010	189.22	0.6382	4.1769	1540.35	2611	596.18	637.76
50	0.001012	210.10	0.7034	4.1777	1546.65	2627	546.99	643.98
60	0.001017	251.89	0.8307	4.1810	1555.23	2643	466.58	654.78
70	0.001022	293.73	0.9545	4.1863	1559.01	2641	404.10	663.53
80	0.001029	335.63	1.0748	4.1938	1558.55	2623	354.56	670.43
90	0.001036	377.61	1.1921	4.2032	1554.33	2592	314.63	675.70
100	0.001043	419.70	1.3064	4.2148	1546.73	2548	281.96	679.53
110	0.001051	461.91	1.4180	4.2285	1536.07	2494	254.90	682.12
120	0.001060	504.28	1.5272	4.2446	1522.60	2430	232.24	683.60
130	0.001069	546.81	1.6340	4.2631	1506.53	2358	213.06	684.09
140	0.001079	589.55	1.7387	4.2844	1488.01	2279	196.68	683.64
150	0.001090	632.51	1.8415	4.3089	1467.13	2193	182.57	682.33
160	0.001102	675.74	1.9424	4.3369	1443.98	2102	170.31	680.16
170	0.001114	719.27	2.0418	4.3691	1418.60	2006	159.58	677.13
t_s = 175.36				**Saturation**				
Liquid	0.001121	742.72	2.0944	4.3882	1404.08	1953	154.36	675.16
Vapour	0.2149	2773.04	6.6212	2.6602	499.98	1.2926	14.87	35.65
180	0.2179	2785.15	6.6481	2.5646	504.12	1.2959	15.07	35.81
190	0.2242	2810.11	6.7026	2.4396	512.08	1.2993	15.50	36.22
200	0.2304	2834.10	6.7538	2.3634	519.34	1.3007	15.93	36.73
210	0.2364	2857.45	6.8027	2.3085	526.20	1.3013	16.35	37.32
220	0.2423	2880.31	6.8495	2.2660	532.79	1.3015	16.78	37.97
230	0.2482	2902.79	6.8947	2.2321	539.15	1.3014	17.21	38.68
240	0.2539	2924.97	6.9383	2.2049	545.32	1.3012	17.64	39.44
250	0.2596	2946.91	6.9806	2.1830	551.32	1.3009	18.06	40.24
260	0.2653	2968.65	7.0218	2.1654	557.18	1.3004	18.49	41.08
270	0.2708	2990.23	7.0619	2.1514	562.89	1.2998	18.92	41.96
280	0.2764	3011.68	7.1010	2.1403	568.48	1.2992	19.34	42.87
290	0.2819	3033.04	7.1393	2.1317	573.96	1.2984	19.77	43.81
300	0.2874	3054.32	7.1768	2.1251	579.34	1.2976	20.20	44.78
310	0.2928	3075.55	7.2135	2.1202	584.61	1.2967	20.62	45.77
320	0.2983	3096.73	7.2495	2.1168	589.81	1.2958	21.05	46.78
330	0.3037	3117.89	7.2849	2.1146	594.91	1.2949	21.47	47.82
340	0.3091	3139.03	7.3196	2.1135	599.94	1.2939	21.89	48.87

Table 4 Single-Phase Region (Regions 1 to 3 of IAPWS-IF97) – Continuation
Homogenes Zustandsgebiet (Bereiche 1 bis 3 der IAPWS-IF97) – Fortsetzung

				p = 0.9 MPa				
t [°C]	v [$m^3\,kg^{-1}$]	h [$kJ\,kg^{-1}$]	s [$kJ\,kg^{-1}\,K^{-1}$]	c_p [$kJ\,kg^{-1}\,K^{-1}$]	w [$m\,s^{-1}$]	κ	$\eta \times 10^6$ [$kg\,m^{-1}\,s^{-1}$]	$\lambda \times 10^3$ [$W\,m^{-1}\,K^{-1}$]
350	0.3145	3160.16	7.3538	2.1132	604.90	1.2929	22.32	49.94
360	0.3198	3181.30	7.3875	2.1137	609.79	1.2919	22.74	51.03
370	0.3252	3202.44	7.4206	2.1149	614.62	1.2908	23.16	52.13
380	0.3305	3223.60	7.4532	2.1167	619.38	1.2898	23.59	53.25
390	0.3358	3244.77	7.4854	2.1189	624.09	1.2887	24.01	54.38
400	0.3411	3265.98	7.5172	2.1216	628.73	1.2876	24.43	55.52
410	0.3464	3287.21	7.5485	2.1247	633.33	1.2865	24.85	56.68
420	0.3517	3308.47	7.5794	2.1282	637.87	1.2854	25.27	57.84
430	0.3570	3329.77	7.6099	2.1319	642.37	1.2843	25.68	59.02
440	0.3623	3351.11	7.6400	2.1359	646.82	1.2832	26.10	60.21
450	0.3675	3372.49	7.6698	2.1402	651.22	1.2821	26.52	61.41
460	0.3728	3393.91	7.6992	2.1447	655.58	1.2810	26.93	62.61
470	0.3780	3415.38	7.7283	2.1493	659.89	1.2799	27.35	63.83
480	0.3833	3436.90	7.7571	2.1542	664.17	1.2788	27.76	65.06
490	0.3885	3458.47	7.7855	2.1592	668.40	1.2777	28.17	66.29
500	0.3938	3480.09	7.8136	2.1644	672.60	1.2766	28.58	67.54
510	0.3990	3501.76	7.8415	2.1697	676.76	1.2755	29.00	68.79
520	0.4042	3523.48	7.8691	2.1752	680.89	1.2744	29.40	70.05
530	0.4094	3545.26	7.8963	2.1807	684.97	1.2733	29.81	71.31
540	0.4147	3567.10	7.9234	2.1864	689.03	1.2722	30.22	72.59
550	0.4199	3588.99	7.9501	2.1921	693.05	1.2711	30.63	73.87
560	0.4251	3610.94	7.9766	2.1980	697.04	1.2700	31.03	75.16
570	0.4303	3632.95	8.0029	2.2039	701.00	1.2689	31.44	76.46
580	0.4355	3655.02	8.0289	2.2099	704.93	1.2678	31.84	77.76
590	0.4407	3677.15	8.0547	2.2160	708.82	1.2668	32.24	79.07
600	0.4459	3699.34	8.0803	2.2221	712.69	1.2657	32.64	80.38
610	0.4511	3721.59	8.1056	2.2283	716.54	1.2647	33.04	81.70
620	0.4563	3743.90	8.1307	2.2345	720.35	1.2636	33.44	83.03
630	0.4615	3766.28	8.1556	2.2408	724.14	1.2626	33.84	84.36
640	0.4667	3788.72	8.1803	2.2471	727.90	1.2616	34.24	85.70
650	0.4718	3811.22	8.2049	2.2535	731.63	1.2605	34.63	87.04
660	0.4770	3833.79	8.2292	2.2599	735.34	1.2595	35.02	88.39
670	0.4822	3856.42	8.2533	2.2664	739.03	1.2585	35.42	89.74
680	0.4874	3879.12	8.2772	2.2728	742.69	1.2575	35.81	91.10
690	0.4926	3901.88	8.3010	2.2793	746.33	1.2565	36.20	92.46
700	0.4977	3924.70	8.3246	2.2859	749.95	1.2555	36.59	93.83
720	0.5081	3970.55	8.3712	2.2990	757.12	1.2536	37.36	96.57
740	0.5184	4016.66	8.4172	2.3122	764.20	1.2517	38.13	99.34
760	0.5288	4063.04	8.4625	2.3254	771.20	1.2498	38.90	102.12
780	0.5391	4109.68	8.5072	2.3387	778.11	1.2479	39.66	104.91
800	0.5494	4156.59	8.5513	2.3521	784.96	1.2461	40.41	107.72

Table 4 **Single-Phase Region** (Regions 1 to 3 of IAPWS-IF97) – Continuation
Homogenes Zustandsgebiet (Bereiche 1 bis 3 der IAPWS-IF97) – Fortsetzung

$p = 1$ MPa

t [°C]	v [m^3 kg^{-1}]	h [kJ kg^{-1}]	s [kJ kg^{-1} K^{-1}]	c_p [kJ kg^{-1} K^{-1}]	w [m s^{-1}]	κ	$\eta \times 10^6$ [kg m^{-1} s^{-1}]	$\lambda \times 10^3$ [W m^{-1} K^{-1}]
0	0.001000	0.98	– 0.0001	4.2150	1403.85	1971	1789.27	561.57
5	0.001000	22.01	0.0762	4.2011	1427.61	2038	1516.64	571.03
10	0.001000	42.99	0.1510	4.1921	1448.97	2099	1304.90	580.48
15	0.001000	63.94	0.2243	4.1860	1468.01	2154	1136.92	589.80
20	0.001001	84.86	0.2963	4.1820	1484.84	2201	1001.22	598.87
25	0.001003	105.76	0.3670	4.1793	1499.60	2243	889.87	607.60
30	0.001004	126.65	0.4365	4.1776	1512.41	2278	797.26	615.91
35	0.001006	147.54	0.5048	4.1766	1523.43	2307	719.32	623.74
40	0.001007	168.42	0.5720	4.1763	1532.76	2331	653.05	631.04
45	0.001009	189.30	0.6382	4.1766	1540.52	2350	596.19	637.81
50	0.001012	210.19	0.7033	4.1775	1546.82	2364	547.01	644.02
60	0.001017	251.98	0.8307	4.1808	1555.40	2379	466.60	654.83
70	0.001022	293.81	0.9544	4.1861	1559.19	2378	404.12	663.58
80	0.001029	335.71	1.0748	4.1935	1558.74	2362	354.59	670.48
90	0.001035	377.69	1.1920	4.2030	1554.52	2333	314.65	675.75
100	0.001043	419.77	1.3063	4.2146	1546.93	2294	281.99	679.59
110	0.001051	461.99	1.4179	4.2283	1536.28	2245	254.93	682.18
120	0.001060	504.35	1.5271	4.2443	1522.83	2187	232.26	683.66
130	0.001069	546.88	1.6339	4.2629	1506.77	2123	213.08	684.15
140	0.001079	589.61	1.7386	4.2841	1488.26	2052	196.70	683.71
150	0.001090	632.57	1.8414	4.3086	1467.40	1975	182.59	682.40
160	0.001102	675.80	1.9423	4.3366	1444.27	1893	170.34	680.23
170	0.001114	719.32	2.0417	4.3687	1418.91	1807	159.60	677.21
t_s = 179.89				**Saturation**				
Liquid	0.001127	762.68	2.1384	4.4051	1391.64	1718	150.24	673.38
Vapour	0.1943	2777.12	6.5850	2.7150	500.89	1.2910	15.02	36.43
180	0.1944	2777.43	6.5857	2.7119	501.00	1.2911	15.03	36.43
190	0.2003	2803.52	6.6426	2.5285	509.69	1.2969	15.46	36.77
200	0.2060	2828.27	6.6955	2.4288	517.32	1.2991	15.89	37.21
210	0.2115	2852.20	6.7455	2.3614	524.41	1.3000	16.33	37.74
220	0.2170	2875.55	6.7934	2.3105	531.18	1.3005	16.76	38.35
230	0.2223	2898.45	6.8393	2.2702	537.69	1.3006	17.19	39.02
240	0.2276	2920.98	6.8837	2.2378	543.99	1.3005	17.62	39.74
250	0.2327	2943.22	6.9266	2.2116	550.11	1.3002	18.05	40.52
260	0.2379	2965.23	6.9683	2.1905	556.06	1.2999	18.47	41.33
270	0.2430	2987.05	7.0088	2.1735	561.86	1.2994	18.90	42.19
280	0.2480	3008.71	7.0484	2.1600	567.53	1.2988	19.33	43.08
290	0.2530	3030.25	7.0870	2.1492	573.08	1.2981	19.76	44.01
300	0.2580	3051.70	7.1247	2.1408	578.51	1.2973	20.18	44.96
310	0.2629	3073.08	7.1617	2.1344	583.85	1.2965	20.61	45.94
320	0.2678	3094.40	7.1979	2.1297	589.09	1.2956	21.04	46.94
330	0.2727	3115.68	7.2335	2.1263	594.25	1.2947	21.46	47.96
340	0.2776	3136.93	7.2685	2.1242	599.32	1.2938	21.89	49.00

Table 4 Single-Phase Region (Regions 1 to 3 of IAPWS-IF97) – Continuation
Homogenes Zustandsgebiet (Bereiche 1 bis 3 der IAPWS-IF97) – Fortsetzung

t [°C]	v [m³ kg⁻¹]	h [kJ kg⁻¹]	s [kJ kg⁻¹ K⁻¹]	c_p [kJ kg⁻¹ K⁻¹]	w [m s⁻¹]	κ	$\eta \times 10^6$ [kg m⁻¹ s⁻¹]	$\lambda \times 10^3$ [W m⁻¹ K⁻¹]
				p = 1 MPa				
350	0.2825	3158.16	7.3028	2.1231	604.32	1.2928	22.31	50.07
360	0.2873	3179.39	7.3366	2.1228	609.24	1.2918	22.74	51.15
370	0.2922	3200.62	7.3699	2.1233	614.10	1.2907	23.16	52.24
380	0.2970	3221.86	7.4026	2.1245	618.89	1.2897	23.58	53.35
390	0.3018	3243.11	7.4349	2.1262	623.62	1.2886	24.00	54.48
400	0.3066	3264.39	7.4668	2.1284	628.30	1.2876	24.42	55.62
410	0.3114	3285.68	7.4982	2.1311	632.92	1.2865	24.84	56.77
420	0.3162	3307.01	7.5292	2.1341	637.48	1.2854	25.26	57.93
430	0.3209	3328.37	7.5598	2.1375	642.00	1.2843	25.68	59.11
440	0.3257	3349.76	7.5900	2.1412	646.47	1.2832	26.10	60.29
450	0.3304	3371.19	7.6198	2.1451	650.89	1.2821	26.52	61.49
460	0.3352	3392.66	7.6493	2.1494	655.26	1.2810	26.93	62.69
470	0.3399	3414.18	7.6785	2.1538	659.59	1.2799	27.35	63.91
480	0.3447	3435.74	7.7073	2.1584	663.88	1.2788	27.76	65.13
490	0.3494	3457.35	7.7358	2.1632	668.13	1.2777	28.17	66.37
500	0.3541	3479.00	7.7640	2.1682	672.34	1.2766	28.59	67.61
510	0.3588	3500.71	7.7919	2.1733	676.52	1.2755	29.00	68.86
520	0.3635	3522.47	7.8195	2.1786	680.65	1.2744	29.41	70.12
530	0.3683	3544.28	7.8468	2.1840	684.75	1.2733	29.82	71.38
540	0.3730	3566.15	7.8739	2.1895	688.82	1.2722	30.22	72.66
550	0.3777	3588.07	7.9007	2.1951	692.85	1.2711	30.63	73.94
560	0.3824	3610.05	7.9272	2.2008	696.85	1.2700	31.04	75.22
570	0.3870	3632.09	7.9535	2.2066	700.82	1.2690	31.44	76.52
580	0.3917	3654.19	7.9795	2.2125	704.76	1.2679	31.84	77.82
590	0.3964	3676.34	8.0054	2.2185	708.66	1.2668	32.25	79.13
600	0.4011	3698.56	8.0309	2.2245	712.54	1.2658	32.65	80.44
610	0.4058	3720.83	8.0563	2.2306	716.39	1.2647	33.05	81.76
620	0.4105	3743.17	8.0815	2.2367	720.21	1.2637	33.45	83.09
630	0.4151	3765.57	8.1064	2.2429	724.00	1.2626	33.84	84.42
640	0.4198	3788.03	8.1311	2.2492	727.77	1.2616	34.24	85.76
650	0.4245	3810.55	8.1557	2.2555	731.51	1.2606	34.63	87.10
660	0.4292	3833.14	8.1800	2.2618	735.23	1.2596	35.03	88.45
670	0.4338	3855.79	8.2041	2.2682	738.92	1.2586	35.42	89.80
680	0.4385	3878.50	8.2281	2.2746	742.59	1.2576	35.81	91.15
690	0.4432	3901.28	8.2519	2.2810	746.24	1.2566	36.20	92.52
700	0.4478	3924.12	8.2755	2.2875	749.86	1.2556	36.59	93.88
720	0.4572	3970.00	8.3221	2.3005	757.04	1.2536	37.37	96.63
740	0.4665	4016.14	8.3681	2.3135	764.13	1.2517	38.14	99.39
760	0.4758	4062.54	8.4135	2.3267	771.13	1.2498	38.90	102.17
780	0.4851	4109.21	8.4582	2.3399	778.06	1.2480	39.66	104.96
800	0.4944	4156.14	8.5024	2.3532	784.91	1.2462	40.41	107.77

Table 4 Single-Phase Region (Regions 1 to 3 of IAPWS-IF97) – Continuation
Homogenes Zustandsgebiet (Bereiche 1 bis 3 der IAPWS-IF97) – Fortsetzung

				p = 1.25 MPa				
t [°C]	v [m³ kg⁻¹]	h [kJ kg⁻¹]	s [kJ kg⁻¹ K⁻¹]	c_p [kJ kg⁻¹ K⁻¹]	w [m s⁻¹]	κ	$\eta\times10^6$ [kg m⁻¹ s⁻¹]	$\lambda\times10^3$ [W m⁻¹ K⁻¹]
0	0.001000	1.23	− 0.0001	4.2137	1404.24	1578	1788.65	561.71
5	0.000999	22.26	0.0762	4.2000	1428.01	1632	1516.23	571.16
10	0.001000	43.24	0.1510	4.1911	1449.37	1680	1304.63	580.60
15	0.001000	64.18	0.2243	4.1852	1468.41	1724	1136.75	589.92
20	0.001001	85.09	0.2962	4.1812	1485.24	1762	1001.12	598.99
25	0.001002	105.99	0.3669	4.1786	1499.99	1795	889.82	607.71
30	0.001004	126.88	0.4364	4.1769	1512.81	1823	797.24	616.02
35	0.001005	147.76	0.5047	4.1760	1523.83	1847	719.32	623.85
40	0.001007	168.64	0.5719	4.1757	1533.16	1866	653.07	631.16
45	0.001009	189.52	0.6381	4.1760	1540.93	1881	596.22	637.92
50	0.001012	210.40	0.7032	4.1769	1547.23	1893	547.05	644.14
60	0.001017	252.19	0.8306	4.1802	1555.83	1904	466.66	654.95
70	0.001022	294.01	0.9543	4.1856	1559.63	1903	404.19	663.71
80	0.001028	335.91	1.0746	4.1930	1559.20	1891	354.66	670.61
90	0.001035	377.88	1.1918	4.2025	1555.01	1868	314.72	675.88
100	0.001043	419.96	1.3061	4.2140	1547.44	1836	282.05	679.73
110	0.001051	462.17	1.4177	4.2277	1536.82	1797	254.99	682.32
120	0.001060	504.52	1.5269	4.2437	1523.39	1751	232.33	683.81
130	0.001069	547.05	1.6337	4.2622	1507.37	1700	213.15	684.30
140	0.001079	589.78	1.7384	4.2834	1488.90	1643	196.77	683.87
150	0.001090	632.73	1.8411	4.3078	1468.08	1581	182.66	682.56
160	0.001102	675.94	1.9420	4.3357	1445.00	1516	170.40	680.40
170	0.001114	719.46	2.0413	4.3677	1419.68	1447	159.67	677.40
180	0.001127	763.31	2.1392	4.4044	1392.14	1375	150.20	673.52
t_s = 189.82				**Saturation**				
Liquid	0.001141	806.75	2.2340	4.4460	1362.93	1302	141.93	668.85
Vapour	0.1570	2785.17	6.5074	2.8439	502.59	1.2873	15.36	38.20
190	0.1571	2785.69	6.5085	2.8382	502.77	1.2875	15.37	38.20
200	0.1620	2812.88	6.5666	2.6258	511.85	1.2940	15.81	38.46
210	0.1667	2838.52	6.6202	2.5102	519.73	1.2966	16.25	38.84
220	0.1712	2863.21	6.6708	2.4320	527.02	1.2978	16.69	39.32
230	0.1757	2887.22	6.7190	2.3728	533.94	1.2983	17.13	39.88
240	0.1800	2910.70	6.7652	2.3258	540.58	1.2986	17.56	40.52
250	0.1843	2933.77	6.8098	2.2879	546.99	1.2986	18.00	41.22
260	0.1886	2956.49	6.8528	2.2571	553.20	1.2984	18.43	41.97
270	0.1927	2978.93	6.8945	2.2320	559.23	1.2981	18.86	42.77
280	0.1969	3001.14	6.9350	2.2116	565.10	1.2977	19.30	43.61
290	0.2010	3023.17	6.9745	2.1952	570.83	1.2972	19.73	44.49
300	0.2050	3045.05	7.0130	2.1819	576.43	1.2965	20.16	45.41
310	0.2091	3066.82	7.0506	2.1713	581.91	1.2958	20.59	46.35
320	0.2131	3088.49	7.0875	2.1630	587.28	1.2950	21.01	47.32
330	0.2170	3110.09	7.1236	2.1566	592.56	1.2942	21.44	48.32
340	0.2210	3131.63	7.1590	2.1518	597.74	1.2933	21.87	49.34

Table 4 Single-Phase Region (Regions 1 to 3 of IAPWS-IF97) – Continuation
Homogenes Zustandsgebiet (Bereiche 1 bis 3 der IAPWS-IF97) – Fortsetzung

p = 1.25 MPa								
t [°C]	v [m^3 kg^{-1}]	h [kJ kg^{-1}]	s [kJ kg^{-1} K^{-1}]	c_p [kJ kg^{-1} K^{-1}]	w [m s^{-1}]	κ	$\eta\times 10^6$ [kg m^{-1} s^{-1}]	$\lambda\times 10^3$ [W m^{-1} K^{-1}]
350	0.2249	3153.13	7.1938	2.1483	602.83	1.2924	22.30	50.38
360	0.2289	3174.60	7.2280	2.1461	607.85	1.2915	22.72	51.45
370	0.2328	3196.05	7.2616	2.1448	612.79	1.2905	23.15	52.53
380	0.2367	3217.50	7.2947	2.1443	617.66	1.2895	23.57	53.62
390	0.2406	3238.94	7.3273	2.1446	622.46	1.2884	23.99	54.74
400	0.2444	3260.39	7.3594	2.1456	627.20	1.2874	24.42	55.86
410	0.2483	3281.85	7.3910	2.1471	631.88	1.2864	24.84	57.00
420	0.2522	3303.33	7.4222	2.1492	636.50	1.2853	25.26	58.16
430	0.2560	3324.84	7.4530	2.1516	641.07	1.2842	25.68	59.32
440	0.2599	3346.37	7.4834	2.1545	645.59	1.2831	26.10	60.50
450	0.2637	3367.93	7.5135	2.1577	650.05	1.2821	26.52	61.69
460	0.2675	3389.52	7.5431	2.1612	654.47	1.2810	26.93	62.89
470	0.2713	3411.15	7.5724	2.1650	658.84	1.2799	27.35	64.10
480	0.2751	3432.82	7.6014	2.1690	663.17	1.2788	27.76	65.32
490	0.2789	3454.53	7.6300	2.1733	667.46	1.2777	28.18	66.55
500	0.2827	3476.29	7.6583	2.1778	671.70	1.2766	28.59	67.79
510	0.2865	3498.09	7.6864	2.1824	675.91	1.2755	29.00	69.03
520	0.2903	3519.94	7.7141	2.1872	680.07	1.2744	29.41	70.29
530	0.2941	3541.83	7.7415	2.1922	684.20	1.2733	29.82	71.55
540	0.2979	3563.78	7.7687	2.1973	688.29	1.2723	30.23	72.82
550	0.3017	3585.78	7.7956	2.2026	692.35	1.2712	30.64	74.10
560	0.3054	3607.83	7.8222	2.2080	696.38	1.2701	31.04	75.39
570	0.3092	3629.94	7.8486	2.2135	700.37	1.2690	31.45	76.68
580	0.3130	3652.10	7.8747	2.2190	704.33	1.2680	31.85	77.98
590	0.3168	3674.32	7.9006	2.2247	708.25	1.2669	32.25	79.29
600	0.3205	3696.60	7.9263	2.2305	712.15	1.2659	32.66	80.60
610	0.3243	3718.93	7.9517	2.2363	716.02	1.2648	33.06	81.92
620	0.3280	3741.33	7.9769	2.2422	719.86	1.2638	33.45	83.24
630	0.3318	3763.78	8.0019	2.2482	723.67	1.2628	33.85	84.57
640	0.3355	3786.29	8.0267	2.2542	727.46	1.2617	34.25	85.91
650	0.3393	3808.86	8.0513	2.2603	731.22	1.2607	34.64	87.25
660	0.3430	3831.50	8.0757	2.2665	734.95	1.2597	35.04	88.59
670	0.3468	3854.19	8.0999	2.2727	738.66	1.2587	35.43	89.94
680	0.3505	3876.95	8.1239	2.2789	742.34	1.2577	35.82	91.30
690	0.3543	3899.77	8.1477	2.2851	746.00	1.2567	36.21	92.66
700	0.3580	3922.65	8.1713	2.2914	749.63	1.2557	36.60	94.03
720	0.3655	3968.61	8.2181	2.3041	756.83	1.2538	37.38	96.77
740	0.3729	4014.82	8.2641	2.3170	763.95	1.2519	38.15	99.53
760	0.3804	4061.29	8.3095	2.3299	770.98	1.2500	38.91	102.30
780	0.3879	4108.01	8.3543	2.3429	777.92	1.2482	39.67	105.09
800	0.3953	4155.00	8.3985	2.3560	784.79	1.2464	40.43	107.90

Table 4 Single-Phase Region (Regions 1 to 3 of IAPWS-IF97) – Continuation
Homogenes Zustandsgebiet (Bereiche 1 bis 3 der IAPWS-IF97) – Fortsetzung

				p = 1.5 MPa				
t	v	h	s	c_p	w	κ	$\eta \times 10^6$	$\lambda \times 10^3$
[°C]	[$m^3\,kg^{-1}$]	[$kJ\,kg^{-1}$]	[$kJ\,kg^{-1}\,K^{-1}$]	[$kJ\,kg^{-1}\,K^{-1}$]	[$m\,s^{-1}$]		[$kg\,m^{-1}\,s^{-1}$]	[$W\,m^{-1}\,K^{-1}$]
0	0.000999	1.48	– 0.0001	4.2125	1404.63	1316	1788.03	561.85
5	0.000999	22.51	0.0762	4.1990	1428.40	1361	1515.82	571.29
10	0.001000	43.48	0.1510	4.1902	1449.76	1401	1304.36	580.72
15	0.001000	64.42	0.2242	4.1844	1468.80	1437	1136.58	590.03
20	0.001001	85.33	0.2962	4.1805	1485.63	1469	1001.01	599.10
25	0.001002	106.22	0.3669	4.1779	1500.39	1497	889.76	607.83
30	0.001004	127.11	0.4363	4.1762	1513.21	1520	797.21	616.13
35	0.001005	147.99	0.5046	4.1754	1524.23	1540	719.32	623.96
40	0.001007	168.86	0.5719	4.1751	1533.57	1556	653.09	631.27
45	0.001009	189.74	0.6380	4.1755	1541.34	1569	596.26	638.04
50	0.001011	210.62	0.7031	4.1763	1547.65	1578	547.09	644.25
60	0.001016	252.40	0.8304	4.1797	1556.26	1588	466.71	655.07
70	0.001022	294.22	0.9541	4.1850	1560.08	1587	404.25	663.83
80	0.001028	336.10	1.0744	4.1925	1559.67	1576	354.72	670.74
90	0.001035	378.07	1.1916	4.2019	1555.50	1558	314.79	676.01
100	0.001043	420.15	1.3059	4.2134	1547.95	1531	282.12	679.86
110	0.001051	462.35	1.4175	4.2271	1537.36	1499	255.06	682.46
120	0.001060	504.70	1.5266	4.2431	1523.96	1461	232.39	683.96
130	0.001069	547.22	1.6334	4.2615	1507.97	1418	213.21	684.46
140	0.001079	589.94	1.7381	4.2827	1489.53	1370	196.83	684.03
150	0.001090	632.88	1.8408	4.3070	1468.76	1319	182.72	682.73
160	0.001101	676.09	1.9417	4.3348	1445.72	1265	170.46	680.58
170	0.001114	719.59	2.0410	4.3667	1420.45	1207	159.73	677.58
180	0.001127	763.44	2.1389	4.4033	1392.96	1147	150.26	673.72
190	0.001141	807.68	2.2354	4.4455	1363.25	1085	141.85	668.97
t_s = 198.30				**Saturation**				
Liquid	0.001154	844.72	2.3147	4.4855	1336.89	1032	135.54	664.32
Vapour	0.1317	2791.01	6.4431	2.9645	503.66	1.2841	15.66	39.79
200	0.1324	2796.02	6.4537	2.9088	505.43	1.2859	15.73	39.80
210	0.1366	2823.91	6.5120	2.6938	514.60	1.2922	16.18	40.01
220	0.1406	2850.19	6.5658	2.5718	522.60	1.2947	16.63	40.35
230	0.1445	2875.46	6.6166	2.4873	530.01	1.2959	17.07	40.79
240	0.1483	2900.00	6.6649	2.4226	537.04	1.2965	17.51	41.33
250	0.1520	2923.96	6.7111	2.3710	543.77	1.2969	17.95	41.94
260	0.1556	2947.45	6.7556	2.3291	550.26	1.2970	18.39	42.63
270	0.1592	2970.57	6.7986	2.2949	556.53	1.2969	18.83	43.36
280	0.1628	2993.37	6.8402	2.2669	562.62	1.2966	19.26	44.15
290	0.1662	3015.92	6.8806	2.2441	568.53	1.2962	19.70	44.99
300	0.1697	3038.27	6.9199	2.2254	574.30	1.2957	20.13	45.86
310	0.1731	3060.44	6.9582	2.2103	579.94	1.2951	20.56	46.77
320	0.1765	3082.48	6.9957	2.1980	585.45	1.2945	20.99	47.72
330	0.1799	3104.41	7.0324	2.1883	590.84	1.2937	21.42	48.69
340	0.1832	3126.25	7.0683	2.1806	596.13	1.2929	21.85	49.69

Table 4 Single-Phase Region (Regions 1 to 3 of IAPWS-IF97) – Continuation
Homogenes Zustandsgebiet (Bereiche 1 bis 3 der IAPWS-IF97) – Fortsetzung

p = 1.5 MPa

t [°C]	v [m^3 kg^{-1}]	h [kJ kg^{-1}]	s [kJ kg^{-1} K^{-1}]	c_p [kJ kg^{-1} K^{-1}]	w [m s^{-1}]	κ	$\eta \times 10^6$ [kg m^{-1} s^{-1}]	$\lambda \times 10^3$ [W m^{-1} K^{-1}]
350	0.1866	3148.03	7.1035	2.1746	601.33	1.2920	22.28	50.71
360	0.1899	3169.75	7.1381	2.1701	606.44	1.2911	22.71	51.75
370	0.1932	3191.43	7.1721	2.1669	611.46	1.2902	23.14	52.82
380	0.1965	3213.09	7.2055	2.1648	616.41	1.2892	23.56	53.90
390	0.1998	3234.73	7.2384	2.1636	621.29	1.2883	23.99	55.00
400	0.2030	3256.37	7.2708	2.1632	626.09	1.2873	24.41	56.11
410	0.2063	3278.00	7.3027	2.1635	630.83	1.2862	24.83	57.24
420	0.2095	3299.64	7.3341	2.1645	635.51	1.2852	25.26	58.39
430	0.2127	3321.29	7.3651	2.1660	640.14	1.2841	25.68	59.55
440	0.2160	3342.96	7.3957	2.1680	644.70	1.2831	26.10	60.72
450	0.2192	3364.65	7.4259	2.1704	649.21	1.2820	26.52	61.90
460	0.2224	3386.37	7.4558	2.1732	653.68	1.2809	26.93	63.09
470	0.2256	3408.12	7.4852	2.1763	658.09	1.2799	27.35	64.30
480	0.2288	3429.90	7.5143	2.1797	662.46	1.2788	27.77	65.51
490	0.2320	3451.71	7.5431	2.1834	666.78	1.2777	28.18	66.73
500	0.2352	3473.57	7.5716	2.1874	671.06	1.2766	28.59	67.97
510	0.2383	3495.46	7.5997	2.1916	675.29	1.2755	29.01	69.21
520	0.2415	3517.40	7.6275	2.1959	679.49	1.2745	29.42	70.46
530	0.2447	3539.38	7.6551	2.2005	683.65	1.2734	29.83	71.72
540	0.2479	3561.41	7.6823	2.2052	687.77	1.2723	30.24	72.99
550	0.2510	3583.49	7.7093	2.2101	691.85	1.2713	30.64	74.27
560	0.2542	3605.61	7.7360	2.2152	695.90	1.2702	31.05	75.55
570	0.2573	3627.79	7.7625	2.2203	699.92	1.2691	31.46	76.84
580	0.2605	3650.02	7.7887	2.2256	703.90	1.2681	31.86	78.14
590	0.2636	3672.30	7.8147	2.2310	707.85	1.2670	32.26	79.44
600	0.2668	3694.64	7.8404	2.2365	711.76	1.2660	32.66	80.76
610	0.2699	3717.03	7.8659	2.2421	715.65	1.2649	33.06	82.07
620	0.2731	3739.48	7.8912	2.2477	719.51	1.2639	33.46	83.40
630	0.2762	3761.99	7.9162	2.2535	723.34	1.2629	33.86	84.72
640	0.2793	3784.55	7.9411	2.2593	727.14	1.2619	34.26	86.06
650	0.2825	3807.17	7.9657	2.2652	730.92	1.2609	34.65	87.40
660	0.2856	3829.86	7.9902	2.2711	734.67	1.2599	35.05	88.74
670	0.2887	3852.60	8.0144	2.2771	738.39	1.2589	35.44	90.09
680	0.2919	3875.40	8.0384	2.2832	742.09	1.2579	35.83	91.45
690	0.2950	3898.26	8.0623	2.2893	745.76	1.2569	36.22	92.81
700	0.2981	3921.18	8.0860	2.2954	749.41	1.2559	36.61	94.17
720	0.3044	3967.22	8.1328	2.3078	756.63	1.2540	37.39	96.91
740	0.3106	4013.50	8.1789	2.3204	763.77	1.2521	38.16	99.67
760	0.3168	4060.03	8.2244	2.3330	770.82	1.2502	38.92	102.44
780	0.3231	4106.82	8.2693	2.3458	777.79	1.2484	39.68	105.22
800	0.3293	4153.87	8.3135	2.3587	784.68	1.2466	40.44	108.02

Table 4 Single-Phase Region (Regions 1 to 3 of IAPWS-IF97) – Continuation
Homogenes Zustandsgebiet (Bereiche 1 bis 3 der IAPWS-IF97) – Fortsetzung

				p = 1.75 MPa				
t [°C]	v [$m^3\,kg^{-1}$]	h [$kJ\,kg^{-1}$]	s [$kJ\,kg^{-1}\,K^{-1}$]	c_p [$kJ\,kg^{-1}\,K^{-1}$]	w [$m\,s^{-1}$]	κ	$\eta \times 10^6$ [$kg\,m^{-1}\,s^{-1}$]	$\lambda \times 10^3$ [$W\,m^{-1}\,K^{-1}$]
0	0.000999	1.74	0.0000	4.2113	1405.02	1128	1787.41	561.99
5	0.000999	22.76	0.0762	4.1979	1428.80	1167	1515.41	571.42
10	0.001000	43.73	0.1509	4.1892	1450.16	1202	1304.09	580.85
15	0.001000	64.66	0.2242	4.1835	1469.20	1233	1136.40	590.15
20	0.001001	85.56	0.2961	4.1797	1486.03	1260	1000.91	599.22
25	0.001002	106.46	0.3668	4.1772	1500.79	1284	889.70	607.94
30	0.001004	127.34	0.4363	4.1756	1513.61	1304	797.19	616.25
35	0.001005	148.21	0.5046	4.1747	1524.63	1321	719.33	624.07
40	0.001007	169.08	0.5718	4.1745	1533.97	1335	653.11	631.38
45	0.001009	189.96	0.6379	4.1749	1541.75	1345	596.29	638.15
50	0.001011	210.83	0.7030	4.1758	1548.06	1354	547.14	644.37
60	0.001016	252.61	0.8303	4.1791	1556.69	1362	466.77	655.18
70	0.001022	294.42	0.9540	4.1845	1560.52	1361	404.31	663.95
80	0.001028	336.30	1.0743	4.1919	1560.13	1352	354.79	670.87
90	0.001035	378.27	1.1914	4.2014	1555.98	1336	314.85	676.15
100	0.001043	420.34	1.3057	4.2129	1548.46	1314	282.19	680.00
110	0.001051	462.53	1.4173	4.2265	1537.89	1286	255.13	682.60
120	0.001059	504.88	1.5264	4.2424	1524.53	1253	232.46	684.11
130	0.001069	547.39	1.6332	4.2608	1508.57	1216	213.28	684.61
140	0.001079	590.10	1.7378	4.2819	1490.17	1176	196.90	684.19
150	0.001090	633.04	1.8405	4.3062	1469.43	1132	182.78	682.90
160	0.001101	676.24	1.9414	4.3339	1446.44	1085	170.52	680.76
170	0.001114	719.73	2.0407	4.3657	1421.21	1036	159.79	677.77
180	0.001127	763.56	2.1385	4.4022	1393.78	985.19	150.32	673.91
190	0.001141	807.79	2.2350	4.4443	1364.13	931.95	141.91	669.17
200	0.001156	852.47	2.3305	4.4929	1332.22	877.081	134.37	663.50
t_s = 205.73				**Saturation**				
Liquid	0.001166	878.32	2.3848	4.5242	1312.90	845.00	130.39	659.80
Vapour	0.1134	2795.28	6.3878	3.0794	504.32	1.2813	15.91	41.23
210	0.1151	2808.11	6.4145	2.9400	508.77	1.2856	16.11	41.25
220	0.1187	2836.40	6.4724	2.7386	517.84	1.2909	16.56	41.43
230	0.1222	2863.14	6.5261	2.6167	525.87	1.2932	17.01	41.75
240	0.1256	2888.85	6.5767	2.5294	533.34	1.2944	17.46	42.18
250	0.1289	2913.79	6.6248	2.4615	540.43	1.2950	17.90	42.70
260	0.1321	2938.12	6.6709	2.4069	547.21	1.2954	18.35	43.30
270	0.1353	2961.96	6.7152	2.3624	553.75	1.2955	18.79	43.98
280	0.1384	2985.39	6.7580	2.3260	560.06	1.2955	19.23	44.71
290	0.1414	3008.50	6.7994	2.2960	566.18	1.2952	19.67	45.50
300	0.1445	3031.33	6.8396	2.2715	572.13	1.2949	20.10	46.33
310	0.1474	3053.94	6.8787	2.2513	577.92	1.2944	20.54	47.21
320	0.1504	3076.37	6.9168	2.2348	583.57	1.2938	20.97	48.12
330	0.1533	3098.65	6.9541	2.2214	589.10	1.2932	21.40	49.06
340	0.1563	3120.81	6.9905	2.2105	594.51	1.2924	21.84	50.04

Table 4 Single-Phase Region (Regions 1 to 3 of IAPWS-IF97) – Continuation
Homogenes Zustandsgebiet (Bereiche 1 bis 3 der IAPWS-IF97) – Fortsetzung

				p = 1.75 MPa				
t [°C]	v [m^3 kg^{-1}]	h [kJ kg^{-1}]	s [kJ kg^{-1} K^{-1}]	c_p [kJ kg^{-1} K^{-1}]	w [m s^{-1}]	κ	$\eta \times 10^6$ [kg m^{-1} s^{-1}]	$\lambda \times 10^3$ [W m^{-1} K^{-1}]
350	0.1592	3142.87	7.0262	2.2018	599.81	1.2917	22.27	51.04
360	0.1620	3164.85	7.0612	2.1950	605.01	1.2908	22.70	52.06
370	0.1649	3186.77	7.0955	2.1897	610.13	1.2899	23.13	53.11
380	0.1678	3208.65	7.1293	2.1858	615.15	1.2890	23.55	54.18
390	0.1706	3230.49	7.1625	2.1830	620.10	1.2881	23.98	55.26
400	0.1734	3252.31	7.1951	2.1812	624.98	1.2871	24.41	56.37
410	0.1762	3274.12	7.2273	2.1803	629.78	1.2861	24.83	57.49
420	0.1790	3295.92	7.2590	2.1801	634.52	1.2851	25.25	58.62
430	0.1818	3317.72	7.2902	2.1806	639.19	1.2841	25.68	59.77
440	0.1846	3339.53	7.3210	2.1817	643.81	1.2830	26.10	60.93
450	0.1874	3361.36	7.3514	2.1833	648.37	1.2820	26.52	62.11
460	0.1902	3383.20	7.3814	2.1853	652.88	1.2809	26.93	63.29
470	0.1929	3405.07	7.4110	2.1878	657.33	1.2799	27.35	64.49
480	0.1957	3426.96	7.4403	2.1906	661.74	1.2788	27.77	65.70
490	0.1984	3448.88	7.4692	2.1937	666.10	1.2777	28.18	66.92
500	0.2012	3470.83	7.4978	2.1971	670.41	1.2767	28.60	68.15
510	0.2039	3492.82	7.5260	2.2008	674.68	1.2756	29.01	69.39
520	0.2066	3514.85	7.5540	2.2047	678.91	1.2745	29.42	70.64
530	0.2094	3536.92	7.5816	2.2089	683.09	1.2735	29.83	71.90
540	0.2121	3559.03	7.6090	2.2132	687.24	1.2724	30.24	73.16
550	0.2148	3581.18	7.6361	2.2177	691.35	1.2713	30.65	74.44
560	0.2176	3603.38	7.6629	2.2224	695.42	1.2703	31.06	75.72
570	0.2203	3625.63	7.6894	2.2272	699.46	1.2692	31.46	77.01
580	0.2230	3647.93	7.7157	2.2322	703.47	1.2682	31.87	78.30
590	0.2257	3670.28	7.7417	2.2373	707.44	1.2671	32.27	79.61
600	0.2284	3692.68	7.7675	2.2425	711.38	1.2661	32.67	80.91
610	0.2311	3715.13	7.7931	2.2479	715.28	1.2651	33.07	82.23
620	0.2338	3737.63	7.8185	2.2533	719.16	1.2640	33.47	83.55
630	0.2365	3760.19	7.8436	2.2588	723.01	1.2630	33.87	84.88
640	0.2392	3782.81	7.8685	2.2644	726.83	1.2620	34.27	86.21
650	0.2419	3805.48	7.8932	2.2701	730.62	1.2610	34.66	87.55
660	0.2446	3828.21	7.9177	2.2758	734.38	1.2600	35.06	88.89
670	0.2473	3851.00	7.9419	2.2817	738.12	1.2590	35.45	90.24
680	0.2500	3873.85	7.9660	2.2875	741.83	1.2580	35.84	91.60
690	0.2527	3896.75	7.9899	2.2935	745.52	1.2570	36.23	92.95
700	0.2553	3919.71	8.0137	2.2994	749.18	1.2561	36.62	94.32
720	0.2607	3965.82	8.0606	2.3116	756.43	1.2542	37.40	97.05
740	0.2661	4012.18	8.1068	2.3238	763.59	1.2523	38.17	99.81
760	0.2714	4058.78	8.1523	2.3362	770.67	1.2504	38.93	102.57
780	0.2768	4105.63	8.1972	2.3488	777.65	1.2486	39.69	105.36
800	0.2821	4152.73	8.2415	2.3615	784.56	1.2468	40.45	108.15

Table 4 **Single-Phase Region** (Regions 1 to 3 of IAPWS-IF97) – Continuation
Homogenes Zustandsgebiet (Bereiche 1 bis 3 der IAPWS-IF97) – Fortsetzung

				$p = 2$ MPa				
t [°C]	v [m^3 kg^{-1}]	h [kJ kg^{-1}]	s [kJ kg^{-1} K^{-1}]	c_p [kJ kg^{-1} K^{-1}]	w [m s^{-1}]	κ	$\eta \times 10^6$ [kg m^{-1} s^{-1}]	$\lambda \times 10^3$ [W m^{-1} K^{-1}]
0	0.000999	1.99	0.0000	4.2100	1405.42	988.40	1786.79	562.12
5	0.000999	23.01	0.0762	4.1969	1429.19	1022	1515.00	571.55
10	0.000999	43.97	0.1509	4.1883	1450.56	1052	1303.82	580.97
15	0.001000	64.89	0.2242	4.1827	1469.59	1079	1136.23	590.27
20	0.001001	85.80	0.2961	4.1789	1486.42	1103	1000.80	599.33
25	0.001002	106.69	0.3667	4.1764	1501.18	1124	889.64	608.05
30	0.001004	127.56	0.4362	4.1749	1514.01	1142	797.17	616.36
35	0.001005	148.44	0.5045	4.1741	1525.03	1156	719.33	624.19
40	0.001007	169.31	0.5717	4.1739	1534.38	1168	653.13	631.50
45	0.001009	190.18	0.6378	4.1743	1542.16	1178	596.32	638.27
50	0.001011	211.05	0.7029	4.1752	1548.48	1185	547.18	644.49
60	0.001016	252.82	0.8302	4.1786	1557.12	1192	466.83	655.30
70	0.001022	294.63	0.9538	4.1840	1560.97	1192	404.37	664.07
80	0.001028	336.50	1.0741	4.1914	1560.60	1184	354.85	670.99
90	0.001035	378.46	1.1913	4.2008	1556.47	1170	314.92	676.28
100	0.001042	420.53	1.3055	4.2123	1548.97	1150	282.25	680.14
110	0.001051	462.71	1.4171	4.2259	1538.43	1126	255.19	682.75
120	0.001059	505.05	1.5262	4.2418	1525.09	1097	232.52	684.25
130	0.001069	547.56	1.6329	4.2601	1509.17	1065	213.34	684.77
140	0.001079	590.26	1.7376	4.2812	1490.80	1030	196.96	684.36
150	0.001089	633.19	1.8403	4.3053	1470.11	991.86	182.85	683.07
160	0.001101	676.38	1.9411	4.3330	1447.15	951.08	170.59	680.94
170	0.001113	719.87	2.0404	4.3647	1421.98	908.11	159.85	677.95
180	0.001127	763.69	2.1382	4.4011	1394.60	863.22	150.39	674.11
190	0.001141	807.91	2.2347	4.4430	1365.00	816.67	141.97	669.38
200	0.001156	852.57	2.3301	4.4914	1333.16	768.70	134.43	663.72
210	0.001173	897.76	2.4246	4.5476	1299.04	719.55	127.63	657.06
$t_s = 212.38$				**Saturation**				
Liquid	0.001177	908.62	2.4470	4.5623	1290.55	707.68	126.10	655.32
Vapour	0.0996	2798.38	6.3392	3.1904	504.66	1.2788	16.14	42.57
220	0.1022	2821.67	6.3868	2.9487	512.58	1.2858	16.50	42.58
230	0.1054	2850.17	6.4440	2.7665	521.47	1.2901	16.95	42.75
240	0.1085	2877.21	6.4972	2.6481	529.47	1.2920	17.41	43.06
250	0.1115	2903.23	6.5474	2.5602	536.96	1.2931	17.86	43.49
260	0.1144	2928.47	6.5952	2.4909	544.07	1.2938	18.30	44.01
270	0.1173	2953.09	6.6410	2.4349	550.89	1.2941	18.75	44.61
280	0.1200	2977.21	6.6850	2.3890	557.44	1.2943	19.19	45.29
290	0.1228	3000.90	6.7274	2.3512	563.78	1.2942	19.64	46.02
300	0.1255	3024.25	6.7685	2.3201	569.91	1.2940	20.08	46.81
310	0.1282	3047.32	6.8084	2.2944	575.87	1.2937	20.51	47.65
320	0.1308	3070.16	6.8472	2.2733	581.67	1.2932	20.95	48.53
330	0.1334	3092.80	6.8851	2.2559	587.33	1.2926	21.39	49.45
340	0.1360	3115.28	6.9221	2.2417	592.86	1.2920	21.82	50.39

Table 4 Single-Phase Region (Regions 1 to 3 of IAPWS-IF97) – Continuation
Homogenes Zustandsgebiet (Bereiche 1 bis 3 der IAPWS-IF97) – Fortsetzung

				p = 2 MPa				
t [°C]	v [$m^3\,kg^{-1}$]	h [$kJ\,kg^{-1}$]	s [$kJ\,kg^{-1}\,K^{-1}$]	c_p [$kJ\,kg^{-1}\,K^{-1}$]	w [$m\,s^{-1}$]	κ	$\eta \times 10^6$ [$kg\,m^{-1}\,s^{-1}$]	$\lambda \times 10^3$ [$W\,m^{-1}\,K^{-1}$]
350	0.1386	3137.64	6.9582	2.2301	598.27	1.2913	22.25	51.37
360	0.1411	3159.89	6.9937	2.2207	603.57	1.2905	22.69	52.38
370	0.1437	3182.06	7.0284	2.2133	608.77	1.2897	23.12	53.41
380	0.1462	3204.16	7.0625	2.2074	613.88	1.2888	23.55	54.46
390	0.1487	3226.21	7.0960	2.2030	618.91	1.2879	23.97	55.53
400	0.1512	3248.23	7.1290	2.1997	623.85	1.2869	24.40	56.62
410	0.1537	3270.21	7.1614	2.1974	628.72	1.2860	24.83	57.73
420	0.1562	3292.18	7.1933	2.1961	633.52	1.2850	25.25	58.86
430	0.1586	3314.14	7.2248	2.1955	638.25	1.2840	25.67	60.00
440	0.1611	3336.09	7.2558	2.1957	642.92	1.2830	26.10	61.15
450	0.1635	3358.05	7.2863	2.1964	647.52	1.2819	26.52	62.32
460	0.1660	3380.02	7.3165	2.1976	652.08	1.2809	26.94	63.50
470	0.1684	3402.01	7.3463	2.1994	656.57	1.2799	27.35	64.69
480	0.1708	3424.01	7.3757	2.2015	661.02	1.2788	27.77	65.90
490	0.1733	3446.04	7.4048	2.2041	665.41	1.2777	28.19	67.11
500	0.1757	3468.09	7.4335	2.2069	669.76	1.2767	28.60	68.34
510	0.1781	3490.18	7.4619	2.2101	674.06	1.2756	29.02	69.57
520	0.1805	3512.30	7.4899	2.2136	678.32	1.2746	29.43	70.82
530	0.1829	3534.45	7.5177	2.2173	682.54	1.2735	29.84	72.07
540	0.1853	3556.64	7.5451	2.2212	686.71	1.2725	30.25	73.33
550	0.1877	3578.88	7.5723	2.2254	690.85	1.2714	30.66	74.61
560	0.1901	3601.15	7.5992	2.2297	694.95	1.2704	31.07	75.89
570	0.1925	3623.47	7.6258	2.2342	699.01	1.2693	31.47	77.17
580	0.1949	3645.84	7.6522	2.2389	703.04	1.2683	31.88	78.47
590	0.1972	3668.25	7.6783	2.2437	707.03	1.2672	32.28	79.77
600	0.1996	3690.71	7.7042	2.2486	710.99	1.2662	32.68	81.08
610	0.2020	3713.22	7.7298	2.2537	714.92	1.2652	33.08	82.39
620	0.2044	3735.78	7.7552	2.2589	718.81	1.2642	33.48	83.71
630	0.2067	3758.40	7.7804	2.2642	722.68	1.2631	33.88	85.04
640	0.2091	3781.07	7.8054	2.2695	726.51	1.2621	34.28	86.37
650	0.2115	3803.79	7.8301	2.2750	730.32	1.2611	34.67	87.70
660	0.2138	3826.57	7.8547	2.2806	734.10	1.2601	35.07	89.05
670	0.2162	3849.40	7.8790	2.2862	737.85	1.2592	35.46	90.39
680	0.2185	3872.29	7.9032	2.2919	741.58	1.2582	35.85	91.75
690	0.2209	3895.24	7.9271	2.2976	745.28	1.2572	36.25	93.10
700	0.2233	3918.24	7.9509	2.3035	748.96	1.2562	36.63	94.46
720	0.2280	3964.43	7.9978	2.3153	756.23	1.2543	37.41	97.20
740	0.2327	4010.86	8.0441	2.3273	763.42	1.2525	38.18	99.95
760	0.2374	4057.52	8.0897	2.3394	770.51	1.2506	38.94	102.71
780	0.2421	4104.43	8.1347	2.3518	777.52	1.2488	39.70	105.49
800	0.2467	4151.59	8.1791	2.3642	784.44	1.2470	40.46	108.28

Table 4 Single-Phase Region (Regions 1 to 3 of IAPWS-IF97) – Continuation
Homogenes Zustandsgebiet (Bereiche 1 bis 3 der IAPWS-IF97) – Fortsetzung

				p = 2.5 MPa				
t	v	h	s	c_p	w	κ	$\eta \times 10^6$	$\lambda \times 10^3$
[°C]	[m^3 kg^{-1}]	[kJ kg^{-1}]	[kJ kg^{-1} K^{-1}]	[kJ kg^{-1} K^{-1}]	[m s^{-1}]		[kg m^{-1} s^{-1}]	[W m^{-1} K^{-1}]
0	0.000999	2.50	0.0000	4.2076	1406.21	791.80	1785.56	562.40
5	0.000999	23.50	0.0762	4.1947	1429.98	818.87	1514.19	571.80
10	0.000999	44.45	0.1509	4.1864	1451.35	843.28	1303.29	581.21
15	0.001000	65.37	0.2241	4.1810	1470.39	865.00	1135.89	590.50
20	0.001001	86.27	0.2960	4.1774	1487.22	884.11	1000.60	599.56
25	0.001002	107.15	0.3666	4.1750	1501.98	900.68	889.53	608.28
30	0.001003	128.02	0.4360	4.1736	1514.81	914.84	797.13	616.58
35	0.001005	148.89	0.5043	4.1728	1525.84	926.70	719.33	624.41
40	0.001007	169.75	0.5715	4.1727	1535.19	936.38	653.17	631.72
45	0.001009	190.61	0.6376	4.1731	1542.98	943.99	596.39	638.49
50	0.001011	211.48	0.7026	4.1740	1549.31	949.67	547.27	644.72
60	0.001016	253.24	0.8299	4.1775	1557.98	955.63	466.94	655.54
70	0.001022	295.04	0.9535	4.1829	1561.86	955.11	404.50	664.32
80	0.001028	336.90	1.0738	4.1903	1561.52	948.88	354.98	671.25
90	0.001035	378.85	1.1909	4.1997	1557.44	937.66	315.05	676.54
100	0.001042	420.90	1.3051	4.2112	1549.99	922.05	282.39	680.41
110	0.001050	463.08	1.4167	4.2247	1539.50	902.60	255.33	683.03
120	0.001059	505.40	1.5257	4.2405	1526.23	879.80	232.66	684.55
130	0.001068	547.90	1.6325	4.2588	1510.36	854.07	213.47	685.07
140	0.001078	590.59	1.7371	4.2797	1492.07	825.76	197.09	684.68
150	0.001089	633.50	1.8397	4.3037	1471.45	795.19	182.97	683.41
160	0.001101	676.67	1.9405	4.3312	1448.59	762.62	170.71	681.29
170	0.001113	720.14	2.0397	4.3627	1423.51	728.31	159.98	678.32
180	0.001126	763.94	2.1375	4.3989	1396.24	692.46	150.51	674.50
190	0.001140	808.14	2.2339	4.4405	1366.76	655.28	142.09	669.79
200	0.001156	852.77	2.3293	4.4885	1335.04	616.97	134.56	664.15
210	0.001172	897.93	2.4237	4.5442	1301.06	577.71	127.76	657.53
220	0.001190	943.69	2.5175	4.6093	1264.73	537.69	121.56	649.84
t_s = 223.96				**Saturation**				
Liquid	0.001197	961.98	2.5544	4.6380	1249.68	521.68	119.26	646.48
Vapour	0.0799	2802.04	6.2560	3.4044	504.68	1.2743	16.55	45.03
230	0.0817	2821.86	6.2956	3.1700	511.47	1.2808	16.84	44.93
240	0.0844	2852.28	6.3555	2.9367	521.08	1.2863	17.30	44.96
250	0.0870	2880.86	6.4106	2.7887	529.58	1.2888	17.76	45.16
260	0.0896	2908.19	6.4624	2.6806	537.46	1.2903	18.22	45.50
270	0.0920	2934.55	6.5114	2.5959	544.90	1.2912	18.68	45.95
280	0.0944	2960.16	6.5581	2.5276	551.99	1.2917	19.13	46.49
290	0.0967	2985.15	6.6029	2.4717	558.78	1.2920	19.58	47.12
300	0.0989	3009.63	6.6460	2.4256	565.32	1.2921	20.02	47.82
310	0.1012	3033.69	6.6876	2.3875	571.64	1.2921	20.47	48.57
320	0.1034	3057.40	6.7279	2.3559	577.76	1.2918	20.91	49.38
330	0.1055	3080.82	6.7671	2.3296	583.70	1.2915	21.35	50.24
340	0.1077	3104.01	6.8052	2.3078	589.48	1.2910	21.79	51.13

Table 4 Single-Phase Region (Regions 1 to 3 of IAPWS-IF97) – Continuation
Homogenes Zustandsgebiet (Bereiche 1 bis 3 der IAPWS-IF97) – Fortsetzung

p = 2.5 MPa

t [°C]	v [m³ kg⁻¹]	h [kJ kg⁻¹]	s [kJ kg⁻¹ K⁻¹]	c_p [kJ kg⁻¹ K⁻¹]	w [m s⁻¹]	κ	$\eta \times 10^6$ [kg m⁻¹ s⁻¹]	$\lambda \times 10^3$ [W m⁻¹ K⁻¹]
350	0.1098	3126.99	6.8424	2.2898	595.12	1.2904	22.23	52.06
360	0.1119	3149.81	6.8787	2.2749	600.63	1.2898	22.66	53.03
370	0.1140	3172.50	6.9142	2.2626	606.02	1.2891	23.10	54.02
380	0.1160	3195.07	6.9491	2.2525	611.30	1.2883	23.53	55.04
390	0.1181	3217.55	6.9832	2.2444	616.48	1.2875	23.96	56.08
400	0.1201	3239.96	7.0168	2.2379	621.57	1.2866	24.39	57.15
410	0.1221	3262.32	7.0497	2.2328	626.57	1.2857	24.82	58.24
420	0.1242	3284.63	7.0822	2.2290	631.49	1.2848	25.25	59.34
430	0.1262	3306.90	7.1141	2.2262	636.34	1.2838	25.67	60.46
440	0.1282	3329.15	7.1455	2.2243	641.11	1.2828	26.10	61.60
450	0.1301	3351.39	7.1765	2.2232	645.82	1.2819	26.52	62.75
460	0.1321	3373.62	7.2070	2.2228	650.46	1.2809	26.94	63.92
470	0.1341	3395.85	7.2371	2.2231	655.04	1.2798	27.36	65.10
480	0.1361	3418.08	7.2668	2.2239	659.56	1.2788	27.78	66.29
490	0.1380	3440.33	7.2962	2.2251	664.04	1.2778	28.20	67.50
500	0.1400	3462.59	7.3251	2.2269	668.45	1.2768	28.61	68.71
510	0.1419	3484.86	7.3538	2.2290	672.82	1.2757	29.03	69.94
520	0.1439	3507.17	7.3821	2.2315	677.14	1.2747	29.44	71.18
530	0.1458	3529.50	7.4100	2.2343	681.42	1.2736	29.85	72.43
540	0.1478	3551.85	7.4377	2.2374	685.65	1.2726	30.26	73.69
550	0.1497	3574.24	7.4651	2.2408	689.84	1.2716	30.67	74.95
560	0.1516	3596.67	7.4922	2.2444	693.99	1.2705	31.08	76.23
570	0.1536	3619.13	7.5190	2.2482	698.10	1.2695	31.49	77.51
580	0.1555	3641.64	7.5455	2.2522	702.17	1.2685	31.89	78.80
590	0.1574	3664.18	7.5718	2.2565	706.21	1.2675	32.30	80.10
600	0.1593	3686.76	7.5978	2.2608	710.21	1.2664	32.70	81.40
610	0.1612	3709.40	7.6235	2.2654	714.18	1.2654	33.10	82.71
620	0.1631	3732.07	7.6491	2.2701	718.11	1.2644	33.50	84.03
630	0.1650	3754.80	7.6744	2.2749	722.01	1.2634	33.90	85.35
640	0.1670	3777.57	7.6995	2.2798	725.89	1.2624	34.30	86.68
650	0.1689	3800.39	7.7243	2.2849	729.73	1.2614	34.70	88.02
660	0.1708	3823.27	7.7490	2.2900	733.54	1.2604	35.09	89.36
670	0.1727	3846.20	7.7734	2.2953	737.32	1.2595	35.48	90.70
680	0.1746	3869.17	7.7976	2.3006	741.08	1.2585	35.88	92.05
690	0.1765	3892.21	7.8217	2.3060	744.80	1.2575	36.27	93.40
700	0.1783	3915.30	7.8455	2.3115	748.51	1.2566	36.66	94.76
720	0.1821	3961.64	7.8927	2.3227	755.84	1.2547	37.43	97.49
740	0.1859	4008.21	7.9391	2.3342	763.07	1.2528	38.20	100.23
760	0.1897	4055.01	7.9848	2.3458	770.21	1.2510	38.97	102.99
780	0.1934	4102.04	8.0299	2.3577	777.25	1.2492	39.73	105.76
800	0.1972	4149.32	8.0744	2.3698	784.22	1.2474	40.48	108.55

Table 4 **Single-Phase Region** (Regions 1 to 3 of IAPWS-IF97) – Continuation
Homogenes Zustandsgebiet (Bereiche 1 bis 3 der IAPWS-IF97) – Fortsetzung

				p = 3 MPa				
t [°C]	v [m^3 kg^{-1}]	h [kJ kg^{-1}]	s [kJ kg^{-1} K^{-1}]	c_p [kJ kg^{-1} K^{-1}]	w [m s^{-1}]	κ	$\eta\times 10^6$ [kg m^{-1} s^{-1}]	$\lambda\times 10^3$ [W m^{-1} K^{-1}]
0	0.000999	3.01	0.0000	4.2052	1406.99	660.74	1784.34	562.68
5	0.000999	24.00	0.0762	4.1926	1430.77	683.32	1513.38	572.06
10	0.000999	44.94	0.1508	4.1846	1452.14	703.67	1302.76	581.45
15	0.001000	65.85	0.2240	4.1793	1471.18	721.78	1135.55	590.74
20	0.001000	86.74	0.2959	4.1759	1488.01	737.71	1000.39	599.79
25	0.001002	107.61	0.3665	4.1736	1502.77	751.53	889.42	608.51
30	0.001003	128.47	0.4359	4.1722	1515.61	763.34	797.08	616.81
35	0.001005	149.33	0.5041	4.1716	1526.64	773.23	719.34	624.64
40	0.001007	170.19	0.5713	4.1715	1536.00	781.31	653.22	631.95
45	0.001009	191.05	0.6373	4.1719	1543.80	787.67	596.46	638.72
50	0.001011	211.91	0.7024	4.1729	1550.14	792.41	547.36	644.95
60	0.001016	253.66	0.8296	4.1764	1558.83	797.41	467.05	655.78
70	0.001021	295.45	0.9532	4.1818	1562.75	797.01	404.62	664.56
80	0.001028	337.30	1.0734	4.1892	1562.45	791.86	355.11	671.50
90	0.001035	379.24	1.1905	4.1986	1558.41	782.54	315.19	676.80
100	0.001042	421.28	1.3048	4.2100	1551.01	769.57	282.52	680.68
110	0.001050	463.44	1.4163	4.2235	1540.57	753.40	255.46	683.31
120	0.001059	505.76	1.5253	4.2393	1527.35	734.45	232.79	684.85
130	0.001068	548.24	1.6320	4.2574	1511.56	713.05	213.60	685.38
140	0.001078	590.91	1.7365	4.2783	1493.33	689.50	197.22	685.00
150	0.001089	633.81	1.8391	4.3022	1472.80	664.07	183.10	683.75
160	0.001100	676.97	1.9399	4.3295	1450.02	636.98	170.84	681.64
170	0.001113	720.42	2.0391	4.3608	1425.03	608.43	160.10	678.69
180	0.001126	764.20	2.1368	4.3966	1397.87	578.61	150.63	674.89
190	0.001140	808.37	2.2332	4.4379	1368.50	547.68	142.22	670.20
200	0.001155	852.98	2.3285	4.4856	1336.92	515.81	134.68	664.59
210	0.001171	898.10	2.4229	4.5409	1303.07	483.14	127.88	657.99
220	0.001189	943.83	2.5165	4.6053	1266.89	449.85	121.69	650.34
230	0.001209	990.25	2.6097	4.6810	1228.28	416.05	116.02	641.53
t_s = 233.86				**Saturation**				
Liquid	0.001217	1008.37	2.6456	4.7138	1212.71	402.91	113.94	637.80
Vapour	0.0667	2803.26	6.1858	3.6123	504.07	1.2705	16.90	47.26
240	0.0682	2824.56	6.2275	3.3435	511.35	1.2775	17.20	47.08
250	0.0706	2856.55	6.2893	3.0772	521.44	1.2834	17.67	47.00
260	0.0729	2886.42	6.3458	2.9070	530.33	1.2862	18.14	47.11
270	0.0751	2914.84	6.3987	2.7826	538.52	1.2879	18.60	47.38
280	0.0772	2942.16	6.4485	2.6854	546.22	1.2890	19.06	47.78
290	0.0792	2968.61	6.4959	2.6071	553.54	1.2897	19.52	48.28
300	0.0812	2994.35	6.5412	2.5431	560.52	1.2902	19.97	48.87
310	0.0831	3019.51	6.5847	2.4902	567.23	1.2904	20.42	49.54
320	0.0850	3044.18	6.6267	2.4463	573.70	1.2904	20.87	50.27
330	0.0869	3068.46	6.6673	2.4098	579.95	1.2903	21.32	51.06
340	0.0887	3092.40	6.7066	2.3794	586.00	1.2900	21.76	51.90

Table 4 Single-Phase Region (Regions 1 to 3 of IAPWS-IF97) – Continuation
Homogenes Zustandsgebiet (Bereiche 1 bis 3 der IAPWS-IF97) – Fortsetzung

				$p = 3$ MPa				
t	v	h	s	c_p	w	κ	$\eta \times 10^6$	$\lambda \times 10^3$
[°C]	[$m^3\,kg^{-1}$]	[$kJ\,kg^{-1}$]	[$kJ\,kg^{-1}\,K^{-1}$]	[$kJ\,kg^{-1}\,K^{-1}$]	[$m\,s^{-1}$]		[$kg\,m^{-1}\,s^{-1}$]	[$W\,m^{-1}\,K^{-1}$]
350	0.0906	3116.06	6.7449	2.3539	591.89	1.2896	22.20	52.78
360	0.0924	3139.49	6.7822	2.3327	597.62	1.2891	22.64	53.70
370	0.0941	3162.73	6.8186	2.3150	603.21	1.2885	23.08	54.66
380	0.0959	3185.80	6.8542	2.3003	608.67	1.2878	23.52	55.64
390	0.0976	3208.74	6.8891	2.2881	614.01	1.2870	23.95	56.66
400	0.0994	3231.57	6.9233	2.2780	619.25	1.2863	24.38	57.69
410	0.1011	3254.31	6.9568	2.2698	624.39	1.2854	24.81	58.76
420	0.1028	3276.97	6.9897	2.2632	629.44	1.2846	25.24	59.84
430	0.1045	3299.58	7.0221	2.2579	634.40	1.2837	25.67	60.94
440	0.1062	3322.14	7.0540	2.2539	639.28	1.2827	26.10	62.06
450	0.1079	3344.66	7.0853	2.2508	644.09	1.2818	26.52	63.20
460	0.1096	3367.16	7.1162	2.2487	648.82	1.2808	26.95	64.35
470	0.1112	3389.64	7.1467	2.2474	653.49	1.2798	27.37	65.52
480	0.1129	3412.10	7.1767	2.2467	658.10	1.2788	27.79	66.70
490	0.1145	3434.57	7.2063	2.2467	662.65	1.2778	28.21	67.89
500	0.1162	3457.04	7.2356	2.2472	667.14	1.2768	28.62	69.10
510	0.1178	3479.52	7.2645	2.2483	671.58	1.2758	29.04	70.32
520	0.1195	3502.01	7.2930	2.2497	675.96	1.2748	29.45	71.55
530	0.1211	3524.51	7.3212	2.2516	680.30	1.2738	29.87	72.79
540	0.1227	3547.04	7.3491	2.2538	684.59	1.2728	30.28	74.04
550	0.1244	3569.59	7.3767	2.2564	688.83	1.2717	30.69	75.30
560	0.1260	3592.17	7.4039	2.2593	693.03	1.2707	31.10	76.57
570	0.1276	3614.78	7.4309	2.2624	697.19	1.2697	31.51	77.85
580	0.1292	3637.42	7.4576	2.2658	701.31	1.2687	31.91	79.14
590	0.1308	3660.09	7.4840	2.2694	705.39	1.2677	32.32	80.43
600	0.1324	3682.81	7.5102	2.2732	709.43	1.2667	32.72	81.73
610	0.1341	3705.56	7.5361	2.2772	713.44	1.2657	33.12	83.04
620	0.1357	3728.35	7.5617	2.2814	717.41	1.2647	33.52	84.36
630	0.1373	3751.19	7.5872	2.2857	721.35	1.2637	33.92	85.68
640	0.1389	3774.07	7.6124	2.2902	725.26	1.2627	34.32	87.00
650	0.1405	3796.99	7.6373	2.2948	729.13	1.2617	34.72	88.34
660	0.1420	3819.96	7.6621	2.2996	732.97	1.2608	35.11	89.67
670	0.1436	3842.98	7.6866	2.3044	736.79	1.2598	35.51	91.01
680	0.1452	3866.05	7.7109	2.3094	740.57	1.2588	35.90	92.36
690	0.1468	3889.17	7.7351	2.3145	744.33	1.2579	36.29	93.71
700	0.1484	3912.34	7.7590	2.3196	748.06	1.2569	36.68	95.07
720	0.1516	3958.84	7.8063	2.3302	755.44	1.2551	37.46	97.79
740	0.1547	4005.55	7.8529	2.3411	762.72	1.2532	38.23	100.53
760	0.1579	4052.49	7.8987	2.3523	769.90	1.2514	38.99	103.28
780	0.1610	4099.64	7.9440	2.3637	776.99	1.2496	39.75	106.04
800	0.1642	4147.03	7.9885	2.3753	783.99	1.2478	40.50	108.81

Table 4 Single-Phase Region (Regions 1 to 3 of IAPWS-IF97) – Continuation
Homogenes Zustandsgebiet (Bereiche 1 bis 3 der IAPWS-IF97) – Fortsetzung

p = 3.5 MPa

t [°C]	v [m³ kg⁻¹]	h [kJ kg⁻¹]	s [kJ kg⁻¹ K⁻¹]	c_p [kJ kg⁻¹ K⁻¹]	w [m s⁻¹]	κ	$\eta \times 10^6$ [kg m⁻¹ s⁻¹]	$\lambda \times 10^3$ [W m⁻¹ K⁻¹]
0	0.000998	3.51	0.0001	4.2027	1407.78	567.13	1783.12	562.95
5	0.000998	24.50	0.0762	4.1906	1431.57	586.49	1512.57	572.32
10	0.000999	45.43	0.1508	4.1828	1452.94	603.94	1302.23	581.70
15	0.000999	66.33	0.2239	4.1777	1471.97	619.48	1135.22	590.97
20	0.001000	87.21	0.2958	4.1743	1488.80	633.14	1000.19	600.02
25	0.001001	108.07	0.3663	4.1722	1503.57	645.00	889.31	608.73
30	0.001003	128.93	0.4357	4.1709	1516.40	655.13	797.04	617.04
35	0.001004	149.78	0.5040	4.1703	1527.44	663.62	719.35	624.86
40	0.001006	170.63	0.5711	4.1703	1536.81	670.55	653.26	632.18
45	0.001008	191.49	0.6371	4.1708	1544.62	676.01	596.53	638.95
50	0.001011	212.34	0.7022	4.1718	1550.97	680.09	547.45	645.18
60	0.001016	254.08	0.8294	4.1753	1559.69	684.39	467.16	656.02
70	0.001021	295.85	0.9529	4.1807	1563.64	684.08	404.75	664.81
80	0.001027	337.70	1.0731	4.1881	1563.38	679.70	355.25	671.75
90	0.001034	379.62	1.1902	4.1975	1559.38	671.74	315.32	677.06
100	0.001042	421.65	1.3044	4.2089	1552.03	660.65	282.65	680.95
110	0.001050	463.81	1.4158	4.2223	1541.64	646.83	255.59	683.60
120	0.001058	506.11	1.5248	4.2380	1528.48	630.62	232.92	685.14
130	0.001068	548.58	1.6315	4.2561	1512.75	612.31	213.73	685.69
140	0.001078	591.24	1.7360	4.2768	1494.60	592.17	197.34	685.32
150	0.001088	634.12	1.8386	4.3006	1474.14	570.41	183.22	684.08
160	0.001100	677.26	1.9393	4.3277	1451.44	547.23	170.96	682.00
170	0.001112	720.69	2.0385	4.3588	1426.55	522.81	160.23	679.06
180	0.001125	764.45	2.1361	4.3945	1399.49	497.29	150.76	675.28
190	0.001139	808.60	2.2325	4.4354	1370.24	470.82	142.34	670.61
200	0.001155	853.18	2.3277	4.4827	1338.78	443.55	134.81	665.03
210	0.001171	898.28	2.4220	4.5375	1305.07	415.59	128.01	658.46
220	0.001189	943.96	2.5156	4.6014	1269.04	387.10	121.82	650.84
230	0.001208	990.34	2.6087	4.6763	1230.60	358.17	116.15	642.07
240	0.001229	1037.54	2.7016	4.7651	1189.60	328.93	110.89	632.03
t_s = 242.56				**Saturation**				
Liquid	0.001235	1049.78	2.7254	4.7905	1178.66	321.40	109.60	629.23
Vapour	0.0571	2802.74	6.1245	3.8172	503.02	1.2670	17.22	49.36
250	0.0588	2829.73	6.1765	3.4689	512.16	1.2756	17.58	49.05
260	0.0609	2862.90	6.2393	3.1876	522.52	1.2812	18.06	48.88
270	0.0629	2893.80	6.2967	3.0028	531.68	1.2842	18.53	48.94
280	0.0648	2923.11	6.3502	2.8665	540.11	1.2860	19.00	49.16
290	0.0667	2951.22	6.4006	2.7598	548.01	1.2872	19.47	49.52
300	0.0685	2978.38	6.4484	2.6738	555.50	1.2880	19.93	49.99
310	0.0702	3004.75	6.4940	2.6034	562.64	1.2886	20.38	50.56
320	0.0719	3030.49	6.5377	2.5453	569.48	1.2889	20.84	51.21
330	0.0736	3055.69	6.5799	2.4969	576.06	1.2890	21.29	51.93
340	0.0752	3080.45	6.6206	2.4566	582.41	1.2889	21.73	52.70

Table 4 Single-Phase Region (Regions 1 to 3 of IAPWS-IF97) – Continuation
Homogenes Zustandsgebiet (Bereiche 1 bis 3 der IAPWS-IF97) – Fortsetzung

				p = 3.5 MPa				
t	v	h	s	c_p	w	κ	$\eta \times 10^6$	$\lambda \times 10^3$
[°C]	[m^3 kg^{-1}]	[kJ kg^{-1}]	[kJ kg^{-1} K^{-1}]	[kJ kg^{-1} K^{-1}]	[m s^{-1}]		[kg m^{-1} s^{-1}]	[W m^{-1} K^{-1}]
350	0.0768	3104.84	6.6601	2.4228	588.56	1.2886	22.18	53.53
360	0.0784	3128.92	6.6984	2.3945	594.52	1.2883	22.62	54.40
370	0.0800	3152.75	6.7357	2.3707	600.32	1.2878	23.06	55.32
380	0.0815	3176.35	6.7721	2.3508	605.98	1.2872	23.50	56.27
390	0.0830	3199.77	6.8077	2.3341	611.50	1.2866	23.94	57.25
400	0.0846	3223.04	6.8426	2.3201	616.89	1.2859	24.38	58.26
410	0.0861	3246.18	6.8767	2.3085	622.17	1.2851	24.81	59.29
420	0.0876	3269.22	6.9102	2.2988	627.35	1.2843	25.24	60.35
430	0.0890	3292.17	6.9430	2.2909	632.44	1.2835	25.67	61.43
440	0.0905	3315.04	6.9753	2.2845	637.43	1.2826	26.10	62.53
450	0.0920	3337.86	7.0071	2.2793	642.34	1.2817	26.53	63.65
460	0.0934	3360.63	7.0384	2.2753	647.17	1.2808	26.95	64.79
470	0.0949	3383.37	7.0692	2.2723	651.93	1.2798	27.38	65.95
480	0.0963	3406.08	7.0995	2.2701	656.63	1.2789	27.80	67.12
490	0.0978	3428.78	7.1295	2.2687	661.25	1.2779	28.22	68.30
500	0.0992	3451.46	7.1590	2.2680	665.82	1.2769	28.64	69.50
510	0.1006	3474.14	7.1882	2.2679	670.32	1.2759	29.05	70.71
520	0.1020	3496.82	7.2169	2.2683	674.77	1.2749	29.47	71.93
530	0.1035	3519.50	7.2454	2.2692	679.17	1.2739	29.88	73.16
540	0.1049	3542.20	7.2734	2.2705	683.52	1.2729	30.30	74.41
550	0.1063	3564.92	7.3012	2.2723	687.82	1.2719	30.71	75.66
560	0.1077	3587.65	7.3287	2.2743	692.07	1.2709	31.12	76.93
570	0.1091	3610.40	7.3558	2.2768	696.28	1.2699	31.53	78.20
580	0.1105	3633.19	7.3827	2.2795	700.44	1.2689	31.93	79.49
590	0.1119	3655.99	7.4092	2.2824	704.57	1.2679	32.34	80.78
600	0.1133	3678.83	7.4356	2.2857	708.65	1.2669	32.74	82.07
610	0.1146	3701.71	7.4616	2.2891	712.70	1.2659	33.15	83.38
620	0.1160	3724.62	7.4874	2.2928	716.71	1.2650	33.55	84.69
630	0.1174	3747.56	7.5129	2.2966	720.69	1.2640	33.95	86.01
640	0.1188	3770.55	7.5383	2.3006	724.63	1.2630	34.34	87.33
650	0.1202	3793.58	7.5633	2.3048	728.54	1.2620	34.74	88.66
660	0.1215	3816.65	7.5882	2.3091	732.41	1.2611	35.14	89.99
670	0.1229	3839.76	7.6128	2.3136	736.26	1.2601	35.53	91.33
680	0.1243	3862.92	7.6373	2.3182	740.07	1.2592	35.92	92.68
690	0.1256	3886.13	7.6615	2.3229	743.86	1.2582	36.31	94.03
700	0.1270	3909.38	7.6855	2.3278	747.61	1.2573	36.70	95.38
720	0.1297	3956.03	7.7329	2.3378	755.05	1.2554	37.48	98.09
740	0.1325	4002.89	7.7797	2.3481	762.37	1.2536	38.25	100.82
760	0.1352	4049.96	7.8257	2.3587	769.60	1.2518	39.02	103.57
780	0.1379	4097.24	7.8710	2.3697	776.73	1.2500	39.77	106.32
800	0.1406	4144.75	7.9157	2.3809	783.77	1.2483	40.53	109.09

Table 4 Single-Phase Region (Regions 1 to 3 of IAPWS-IF97) – Continuation
Homogenes Zustandsgebiet (Bereiche 1 bis 3 der IAPWS-IF97) – Fortsetzung

				p = 4 MPa				
t [°C]	v [m^3 kg^{-1}]	h [kJ kg^{-1}]	s [kJ kg^{-1} K^{-1}]	c_p [kJ kg^{-1} K^{-1}]	w [m s^{-1}]	κ	$\eta \times 10^6$ [kg m^{-1} s^{-1}]	$\lambda \times 10^3$ [W m^{-1} K^{-1}]
0	0.000998	4.02	0.0001	4.2003	1408.57	496.92	1781.91	563.23
5	0.000998	24.99	0.0762	4.1885	1432.36	513.88	1511.78	572.57
10	0.000998	45.91	0.1507	4.1809	1453.73	529.16	1301.71	581.94
15	0.000999	66.80	0.2239	4.1760	1472.77	542.76	1134.88	591.20
20	0.001000	87.68	0.2957	4.1728	1489.60	554.72	999.98	600.25
25	0.001001	108.53	0.3662	4.1708	1504.36	565.10	889.20	608.96
30	0.001003	129.38	0.4356	4.1696	1517.20	573.97	797.00	617.26
35	0.001004	150.23	0.5038	4.1690	1528.25	581.40	719.36	625.09
40	0.001006	171.08	0.5709	4.1691	1537.62	587.48	653.30	632.40
45	0.001008	191.92	0.6369	4.1696	1545.44	592.27	596.60	639.18
50	0.001010	212.77	0.7020	4.1706	1551.80	595.84	547.53	645.41
60	0.001015	254.50	0.8291	4.1742	1560.55	599.64	467.28	656.25
70	0.001021	296.26	0.9526	4.1796	1564.53	599.38	404.87	665.05
80	0.001027	338.09	1.0728	4.1871	1564.31	595.57	355.38	672.01
90	0.001034	380.01	1.1898	4.1964	1560.35	588.64	315.45	677.33
100	0.001041	422.03	1.3040	4.2078	1553.04	578.97	282.79	681.23
110	0.001050	464.17	1.4154	4.2212	1542.71	566.91	255.72	683.88
120	0.001058	506.46	1.5244	4.2368	1529.61	552.75	233.05	685.44
130	0.001068	548.92	1.6310	4.2548	1513.94	536.76	213.86	686.00
140	0.001077	591.56	1.7355	4.2754	1495.85	519.17	197.47	685.64
150	0.001088	634.43	1.8380	4.2990	1475.47	500.17	183.35	684.42
160	0.001100	677.56	1.9388	4.3260	1452.86	479.92	171.09	682.35
170	0.001112	720.97	2.0378	4.3569	1428.07	458.59	160.35	679.44
180	0.001125	764.71	2.1354	4.3923	1401.11	436.30	150.88	675.67
190	0.001139	808.83	2.2317	4.4330	1371.97	413.18	142.47	671.02
200	0.001154	853.39	2.3269	4.4799	1340.63	389.35	134.93	665.46
210	0.001170	898.45	2.4212	4.5342	1307.06	364.93	128.13	658.92
220	0.001188	944.10	2.5147	4.5975	1271.18	340.03	121.95	651.33
230	0.001207	990.44	2.6077	4.6717	1232.91	314.76	116.28	642.60
240	0.001228	1037.58	2.7005	4.7595	1192.10	289.22	111.03	632.61
250	0.001252	1085.69	2.7933	4.8646	1148.50	263.45	106.12	621.22
t_s = 250.36				**Saturation**				
Liquid	0.001253	1087.43	2.7967	4.8688	1146.89	262.53	105.95	620.78
Vapour	0.0498	2800.90	6.0697	4.0217	501.64	1.2639	17.51	51.34
260	0.0518	2837.19	6.1384	3.5536	513.78	1.2746	17.98	50.85
270	0.0537	2871.20	6.2016	3.2702	524.26	1.2797	18.47	50.64
280	0.0555	2902.88	6.2594	3.0774	533.58	1.2826	18.94	50.66
290	0.0572	2932.91	6.3132	2.9332	542.18	1.2844	19.41	50.85
300	0.0589	2961.65	6.3638	2.8199	550.23	1.2857	19.88	51.19
310	0.0605	2989.38	6.4118	2.7285	557.85	1.2866	20.34	51.65
320	0.0620	3016.28	6.4575	2.6536	565.10	1.2872	20.80	52.20
330	0.0635	3042.49	6.5014	2.5915	572.04	1.2876	21.26	52.84
340	0.0650	3068.14	6.5435	2.5399	578.71	1.2877	21.71	53.54

Table 4 Single-Phase Region (Regions 1 to 3 of IAPWS-IF97) – Continuation
Homogenes Zustandsgebiet (Bereiche 1 bis 3 der IAPWS-IF97) – Fortsetzung

$p = 4$ MPa								
t [°C]	v [m^3 kg^{-1}]	h [kJ kg^{-1}]	s [kJ kg^{-1} K^{-1}]	c_p [kJ kg^{-1} K^{-1}]	w [m s^{-1}]	κ	$\eta \times 10^6$ [kg m^{-1} s^{-1}]	$\lambda \times 10^3$ [W m^{-1} K^{-1}]
350	0.0665	3093.32	6.5843	2.4967	585.14	1.2877	22.16	54.31
360	0.0679	3118.10	6.6237	2.4604	591.35	1.2875	22.61	55.13
370	0.0693	3142.55	6.6620	2.4299	597.38	1.2871	23.05	56.00
380	0.0707	3166.71	6.6993	2.4042	603.23	1.2867	23.49	56.91
390	0.0721	3190.64	6.7357	2.3825	608.93	1.2862	23.93	57.86
400	0.0734	3214.37	6.7712	2.3642	614.49	1.2855	24.37	58.84
410	0.0748	3237.94	6.8059	2.3488	619.92	1.2849	24.81	59.85
420	0.0761	3261.36	6.8400	2.3359	625.24	1.2841	25.24	60.88
430	0.0774	3284.66	6.8734	2.3251	630.45	1.2833	25.67	61.94
440	0.0787	3307.87	6.9061	2.3162	635.56	1.2825	26.10	63.02
450	0.0800	3330.99	6.9383	2.3088	640.58	1.2816	26.53	64.12
460	0.0813	3354.05	6.9700	2.3027	645.51	1.2807	26.96	65.25
470	0.0826	3377.05	7.0012	2.2979	650.36	1.2798	27.38	66.39
480	0.0839	3400.01	7.0318	2.2941	655.14	1.2789	27.81	67.54
490	0.0852	3422.94	7.0621	2.2913	659.84	1.2780	28.23	68.72
500	0.0864	3445.84	7.0919	2.2892	664.48	1.2770	28.65	69.90
510	0.0877	3468.72	7.1213	2.2879	669.06	1.2760	29.07	71.10
520	0.0890	3491.60	7.1503	2.2872	673.58	1.2751	29.49	72.32
530	0.0902	3514.47	7.1790	2.2871	678.04	1.2741	29.90	73.55
540	0.0915	3537.34	7.2073	2.2875	682.44	1.2731	30.31	74.78
550	0.0927	3560.22	7.2353	2.2883	686.80	1.2721	30.73	76.03
560	0.0939	3583.11	7.2629	2.2896	691.10	1.2711	31.14	77.29
570	0.0952	3606.01	7.2902	2.2913	695.36	1.2701	31.55	78.56
580	0.0964	3628.93	7.3172	2.2933	699.58	1.2692	31.95	79.84
590	0.0976	3651.88	7.3440	2.2956	703.75	1.2682	32.36	81.13
600	0.0989	3674.85	7.3704	2.2982	707.87	1.2672	32.76	82.42
610	0.1001	3697.84	7.3966	2.3011	711.96	1.2662	33.17	83.72
620	0.1013	3720.87	7.4226	2.3042	716.01	1.2652	33.57	85.03
630	0.1025	3743.93	7.4482	2.3076	720.03	1.2643	33.97	86.35
640	0.1037	3767.02	7.4737	2.3111	724.00	1.2633	34.37	87.67
650	0.1049	3790.15	7.4989	2.3149	727.94	1.2624	34.76	88.99
660	0.1062	3813.32	7.5238	2.3188	731.85	1.2614	35.16	90.32
670	0.1074	3836.53	7.5486	2.3229	735.73	1.2605	35.55	91.66
680	0.1086	3859.78	7.5731	2.3271	739.57	1.2595	35.95	93.00
690	0.1098	3883.07	7.5974	2.3315	743.39	1.2586	36.34	94.35
700	0.1110	3906.41	7.6215	2.3360	747.17	1.2577	36.73	95.69
720	0.1134	3953.22	7.6691	2.3453	754.65	1.2558	37.51	98.40
740	0.1158	4000.23	7.7160	2.3551	762.03	1.2540	38.28	101.13
760	0.1182	4047.43	7.7621	2.3652	769.30	1.2522	39.04	103.86
780	0.1205	4094.84	7.8075	2.3757	776.47	1.2505	39.80	106.61
800	0.1229	4142.46	7.8523	2.3865	783.55	1.2487	40.55	109.37

Table 4 Single-Phase Region (Regions 1 to 3 of IAPWS-IF97) – Continuation
Homogenes Zustandsgebiet (Bereiche 1 bis 3 der IAPWS-IF97) – Fortsetzung

				p = 4.5 MPa				
t [°C]	v [m³ kg⁻¹]	h [kJ kg⁻¹]	s [kJ kg⁻¹ K⁻¹]	c_p [kJ kg⁻¹ K⁻¹]	w [m s⁻¹]	κ	$\eta \times 10^6$ [kg m⁻¹ s⁻¹]	$\lambda \times 10^3$ [W m⁻¹ K⁻¹]
0	0.000998	4.53	0.0001	4.1979	1409.37	442.32	1780.71	563.51
5	0.000998	25.49	0.0761	4.1864	1433.16	457.40	1510.98	572.83
10	0.000998	46.40	0.1507	4.1791	1454.53	470.99	1301.19	582.18
15	0.000999	67.28	0.2238	4.1744	1473.56	483.08	1134.55	591.44
20	0.001000	88.14	0.2956	4.1713	1490.39	493.72	999.78	600.48
25	0.001001	109.00	0.3661	4.1694	1505.16	502.95	889.10	609.19
30	0.001002	129.84	0.4354	4.1683	1518.00	510.84	796.96	617.49
35	0.001004	150.68	0.5036	4.1678	1529.05	517.46	719.36	625.32
40	0.001006	171.52	0.5707	4.1679	1538.43	522.87	653.35	632.63
45	0.001008	192.36	0.6367	4.1684	1546.26	527.13	596.67	639.41
50	0.001010	213.20	0.7017	4.1695	1552.63	530.32	547.62	645.64
60	0.001015	254.91	0.8288	4.1731	1561.40	533.71	467.39	656.49
70	0.001021	296.67	0.9523	4.1786	1565.42	533.51	405.00	665.30
80	0.001027	338.49	1.0725	4.1860	1565.23	530.15	355.51	672.26
90	0.001034	380.40	1.1895	4.1953	1561.32	524.01	315.59	677.59
100	0.001041	422.41	1.3036	4.2066	1554.06	515.44	282.92	681.50
110	0.001049	464.54	1.4150	4.2200	1543.78	504.74	255.86	684.17
120	0.001058	506.81	1.5239	4.2355	1530.73	492.18	233.18	685.73
130	0.001067	549.26	1.6305	4.2534	1515.12	478.00	213.99	686.31
140	0.001077	591.89	1.7350	4.2740	1497.11	462.39	197.60	685.97
150	0.001088	634.74	1.8375	4.2974	1476.80	445.53	183.48	684.76
160	0.001099	677.85	1.9382	4.3242	1454.28	427.57	171.21	682.70
170	0.001111	721.24	2.0372	4.3549	1429.58	408.64	160.47	679.80
180	0.001124	764.96	2.1348	4.3901	1402.72	388.86	151.01	676.06
190	0.001138	809.06	2.2310	4.4305	1373.69	368.34	142.59	671.44
200	0.001154	853.59	2.3261	4.4771	1342.48	347.19	135.06	665.90
210	0.001170	898.63	2.4203	4.5310	1309.04	325.52	128.26	659.38
220	0.001187	944.24	2.5138	4.5937	1273.32	303.42	122.08	651.83
230	0.001207	990.54	2.6067	4.6672	1235.21	280.00	116.41	643.14
240	0.001228	1037.63	2.6994	4.7540	1194.59	258.32	111.16	633.20
250	0.001251	1085.67	2.7921	4.8578	1151.21	235.46	106.26	621.86
t_s = 257.44				**Saturation**				
Liquid	0.001270	1122.14	2.8613	4.9492	1116.93	218.35	102.79	612.42
Vapour	0.0441	2798.00	6.0198	4.2279	500.02	1.2610	17.78	53.30
260	0.0446	2808.60	6.0397	4.0583	503.69	1.2649	17.91	53.08
270	0.0465	2846.71	6.1106	3.6066	516.08	1.2741	18.40	52.52
280	0.0482	2881.29	6.1737	3.3285	526.57	1.2787	18.89	52.29
290	0.0498	2913.55	6.2315	3.1328	535.99	1.2814	19.37	52.29
300	0.0514	2944.10	6.2852	2.9844	544.69	1.2833	19.84	52.47
310	0.0529	2973.34	6.3358	2.8671	552.84	1.2846	20.31	52.80
320	0.0543	3001.52	6.3837	2.7723	560.55	1.2855	20.77	53.25
330	0.0557	3028.84	6.4294	2.6944	567.88	1.2861	21.23	53.79
340	0.0571	3055.45	6.4732	2.6298	574.89	1.2865	21.69	54.43

Table 4 Single-Phase Region (Regions 1 to 3 of IAPWS-IF97) – Continuation
Homogenes Zustandsgebiet (Bereiche 1 bis 3 der IAPWS-IF97) – Fortsetzung

p = 4.5 MPa								
t [°C]	v [m³ kg⁻¹]	h [kJ kg⁻¹]	s [kJ kg⁻¹ K⁻¹]	c_p [kJ kg⁻¹ K⁻¹]	w [m s⁻¹]	κ	$\eta \times 10^6$ [kg m⁻¹ s⁻¹]	$\lambda \times 10^3$ [W m⁻¹ K⁻¹]
350	0.0584	3081.47	6.5153	2.5759	581.62	1.2867	22.14	55.13
360	0.0597	3107.00	6.5559	2.5308	588.10	1.2866	22.59	55.90
370	0.0610	3132.11	6.5953	2.4927	594.36	1.2864	23.04	56.72
380	0.0623	3156.88	6.6335	2.4606	600.42	1.2861	23.49	57.58
390	0.0635	3181.34	6.6706	2.4335	606.31	1.2857	23.93	58.49
400	0.0648	3205.56	6.7069	2.4105	612.04	1.2852	24.37	59.44
410	0.0660	3229.56	6.7423	2.3910	617.63	1.2846	24.81	60.42
420	0.0672	3253.39	6.7769	2.3746	623.09	1.2839	25.24	61.43
430	0.0684	3277.06	6.8108	2.3607	628.43	1.2832	25.68	62.46
440	0.0696	3300.61	6.8441	2.3490	633.66	1.2824	26.11	63.52
450	0.0708	3324.05	6.8767	2.3392	638.79	1.2816	26.54	64.61
460	0.0719	3347.40	6.9088	2.3310	643.83	1.2807	26.97	65.71
470	0.0731	3370.67	6.9403	2.3242	648.77	1.2798	27.40	66.84
480	0.0742	3393.89	6.9713	2.3187	653.64	1.2789	27.82	67.98
490	0.0754	3417.05	7.0019	2.3143	658.43	1.2780	28.24	69.14
500	0.0765	3440.18	7.0320	2.3109	663.14	1.2771	28.66	70.32
510	0.0777	3463.27	7.0617	2.3083	667.79	1.2762	29.08	71.51
520	0.0788	3486.34	7.0910	2.3064	672.38	1.2752	29.50	72.72
530	0.0799	3509.40	7.1199	2.3052	676.90	1.2742	29.92	73.93
540	0.0810	3532.45	7.1484	2.3047	681.37	1.2733	30.33	75.17
550	0.0821	3555.50	7.1765	2.3046	685.78	1.2723	30.75	76.41
560	0.0833	3578.55	7.2044	2.3051	690.14	1.2713	31.16	77.66
570	0.0844	3601.60	7.2319	2.3060	694.45	1.2704	31.57	78.93
580	0.0855	3624.67	7.2591	2.3073	698.71	1.2694	31.97	80.20
590	0.0866	3647.75	7.2860	2.3090	702.92	1.2684	32.38	81.48
600	0.0877	3670.85	7.3126	2.3110	707.09	1.2675	32.79	82.77
610	0.0888	3693.97	7.3389	2.3133	711.22	1.2665	33.19	84.07
620	0.0898	3717.11	7.3650	2.3158	715.31	1.2655	33.59	85.38
630	0.0909	3740.29	7.3908	2.3187	719.36	1.2646	33.99	86.69
640	0.0920	3763.49	7.4163	2.3217	723.38	1.2636	34.39	88.01
650	0.0931	3786.72	7.4416	2.3250	727.35	1.2627	34.79	89.33
660	0.0942	3809.99	7.4667	2.3285	731.29	1.2618	35.18	90.66
670	0.0953	3833.29	7.4915	2.3322	735.20	1.2608	35.58	91.99
680	0.0963	3856.63	7.5162	2.3361	739.07	1.2599	35.97	93.33
690	0.0974	3880.01	7.5406	2.3401	742.92	1.2590	36.36	94.67
700	0.0985	3903.44	7.5647	2.3442	746.73	1.2580	36.75	96.02
720	0.1006	3950.41	7.6125	2.3529	754.26	1.2562	37.53	98.72
740	0.1028	3997.56	7.6595	2.3621	761.69	1.2544	38.30	101.44
760	0.1049	4044.89	7.7058	2.3718	769.00	1.2527	39.07	104.17
780	0.1070	4092.43	7.7514	2.3818	776.21	1.2509	39.82	106.91
800	0.1092	4140.17	7.7963	2.3921	783.33	1.2492	40.58	109.66

Table 4 Single-Phase Region (Regions 1 to 3 of IAPWS-IF97) – Continuation
Homogenes Zustandsgebiet (Bereiche 1 bis 3 der IAPWS-IF97) – Fortsetzung

				$p = 5$ MPa				
t [°C]	v [m^3 kg^{-1}]	h [kJ kg^{-1}]	s [kJ kg^{-1} K^{-1}]	c_p [kJ kg^{-1} K^{-1}]	w [m s^{-1}]	κ	$\eta \times 10^6$ [kg m^{-1} s^{-1}]	$\lambda \times 10^3$ [W m^{-1} K^{-1}]
0	0.000998	5.03	0.0001	4.1955	1410.16	398.63	1779.51	563.78
5	0.000998	25.98	0.0761	4.1843	1433.95	412.22	1510.19	573.08
10	0.000998	46.88	0.1506	4.1773	1455.32	424.45	1300.68	582.43
15	0.000999	67.76	0.2237	4.1727	1474.36	435.34	1134.22	591.67
20	0.001000	88.61	0.2955	4.1698	1491.19	444.92	999.59	600.71
25	0.001001	109.46	0.3660	4.1680	1505.96	453.24	888.99	609.41
30	0.001002	130.29	0 .4353	4.1670	1518.80	460.35	796.92	617.71
35	0.001004	151.13	0.5034	4.1665	1529.86	466.31	719.37	625.54
40	0.001006	171.96	0.5705	4.1667	1539.24	471.18	653.39	632.86
45	0.001008	192.80	0.6365	4.1673	1547.08	475.03	596.74	639.64
50	0.001010	213.63	0.7015	4.1684	1553.46	477.90	547.71	645.87
60	0.001015	255.33	0.8286	4.1720	1562.26	480.97	467.50	656.73
70	0.001020	297.08	0.9520	4.1775	1566.31	480.81	405.13	665.54
80	0.001027	338.89	1.0721	4.1849	1566.16	477.81	355.64	672.51
90	0.001034	380.78	1.1891	4.1942	1562.28	472.30	315.72	677.85
100	0.001041	422.78	1.3032	4.2055	1555.07	464.61	283.05	681.77
110	0.001049	464.90	1.4146	4.2188	1544.84	455.01	255.99	684.45
120	0.001058	507.17	1.5235	4.2343	1531.85	443.73	233.31	686.03
130	0.001067	549.60	1.6301	4.2521	1516.30	430.99	214.12	686.62
140	0.001077	592.22	1.7345	4.2725	1498.36	416.97	197.73	686.29
150	0.001087	635.06	1.8369	4.2959	1478.13	401.82	183.60	685.09
160	0.001099	678.14	1.9376	4.3225	1455.69	385.68	171.34	683.05
170	0.001111	721.52	2.0366	4.3530	1431.08	368.67	160.60	680.17
180	0.001124	765.22	2.1341	4.3879	1404.32	350.90	151.13	676.45
190	0.001138	809.29	2.2303	4.4280	1375.41	332.47	142.71	671.84
200	0.001153	853.80	2.3254	4.4743	1344.32	313.47	135.18	666.33
210	0.001169	898.80	2.4195	4.5277	1311.02	293.99	128.39	659.85
220	0.001187	944.38	2.5129	4.5899	1275.44	274.13	122.21	652.32
230	0.001206	990.64	2.6057	4.6626	1237.50	253.98	116.54	643.67
240	0.001227	1037.68	2.6983	4.7485	1197.07	233.60	111.30	633.78
250	0.001250	1085.66	2.7909	4.8511	1153.91	213.06	106.40	622.50
260	0.001275	1134.77	2.8839	4.9755	1107.65	192.38	101.77	609.67
$t_s = 263.94$				**Saturation**				
Liquid	0.001286	1154.50	2.9207	5.0322	1088.43	184.19	100.00	604.15
Vapour	0.0394	2794.23	5.9737	4.4378	498.18	1.2584	18.03	55.20
270	0.0406	2819.84	6.0211	4.0460	506.87	1.2666	18.34	54.65
280	0.0423	2858.08	6.0909	3.6350	518.94	1.2740	18.83	54.09
290	0.0439	2893.00	6.1535	3.3662	529.38	1.2780	19.32	53.85
300	0.0453	2925.64	6.2109	3.1714	538.84	1.2806	19.80	53.85
310	0.0468	2956.58	6.2645	3.0218	547.60	1.2824	20.27	54.03
320	0.0481	2986.18	6.3148	2.9028	555.81	1.2837	20.74	54.36
330	0.0494	3014.71	6.3625	2.8063	563.57	1.2846	21.21	54.81
340	0.0507	3042.36	6.4080	2.7269	570.94	1.2852	21.67	55.36

Table 4 Single-Phase Region (Regions 1 to 3 of IAPWS-IF97) – Continuation
Homogenes Zustandsgebiet (Bereiche 1 bis 3 der IAPWS-IF97) – Fortsetzung

$p = 5$ MPa

t [°C]	v [m^3 kg^{-1}]	h [kJ kg^{-1}]	s [kJ kg^{-1} K^{-1}]	c_p [kJ kg^{-1} K^{-1}]	w [m s^{-1}]	κ	$\eta \times 10^6$ [kg m^{-1} s^{-1}]	$\lambda \times 10^3$ [W m^{-1} K^{-1}]
350	0.0520	3069.29	6.4515	2.6610	577.99	1.2856	22.13	55.99
360	0.0532	3095.62	6.4934	2.6058	584.75	1.2858	22.58	56.69
370	0.0544	3121.44	6.5339	2.5594	591.26	1.2857	23.03	57.46
380	0.0556	3146.83	6.5731	2.5203	597.55	1.2855	23.48	58.28
390	0.0567	3171.86	6.6111	2.4871	603.64	1.2852	23.93	59.15
400	0.0578	3196.59	6.6481	2.4590	609.56	1.2848	24.37	60.06
410	0.0590	3221.06	6.6842	2.4351	615.31	1.2843	24.81	61.01
420	0.0601	3245.31	6.7194	2.4148	620.92	1.2837	25.25	61.99
430	0.0612	3269.37	6.7539	2.3976	626.39	1.2830	25.68	63.00
440	0.0622	3293.27	6.7877	2.3829	631.75	1.2823	26.12	64.04
450	0.0633	3317.03	6.8208	2.3705	636.99	1.2815	26.55	65.10
460	0.0644	3340.68	6.8532	2.3601	642.13	1.2807	26.98	66.19
470	0.0654	3364.24	6.8851	2.3513	647.17	1.2799	27.41	67.30
480	0.0665	3387.71	6.9165	2.3440	652.13	1.2790	27.83	68.43
490	0.0675	3411.12	6.9474	2.3379	657.00	1.2781	28.26	69.58
500	0.0686	3434.48	6.9778	2.3330	661.80	1.2772	28.68	70.74
510	0.0696	3457.79	7.0078	2.3291	666.52	1.2763	29.10	71.92
520	0.0706	3481.06	7.0373	2.3260	671.17	1.2754	29.52	73.12
530	0.0717	3504.31	7.0664	2.3237	675.76	1.2744	29.94	74.33
540	0.0727	3527.54	7.0952	2.3221	680.29	1.2735	30.35	75.56
550	0.0737	3550.75	7.1235	2.3212	684.75	1.2725	30.77	76.79
560	0.0747	3573.96	7.1516	2.3208	689.17	1.2716	31.18	78.04
570	0.0757	3597.17	7.1793	2.3209	693.53	1.2706	31.59	79.30
580	0.0767	3620.38	7.2066	2.3215	697.84	1.2697	32.00	80.57
590	0.0777	3643.60	7.2337	2.3224	702.10	1.2687	32.40	81.85
600	0.0787	3666.83	7.2604	2.3238	706.31	1.2678	32.81	83.13
610	0.0797	3690.08	7.2869	2.3255	710.49	1.2668	33.21	84.43
620	0.0807	3713.34	7.3131	2.3275	714.61	1.2659	33.62	85.73
630	0.0817	3736.63	7.3390	2.3299	718.70	1.2649	34.02	87.04
640	0.0827	3759.94	7.3647	2.3324	722.75	1.2640	34.42	88.35
650	0.0836	3783.28	7.3901	2.3353	726.76	1.2630	34.81	89.67
660	0.0846	3806.65	7.4153	2.3383	730.74	1.2621	35.21	91.00
670	0.0856	3830.05	7.4402	2.3416	734.67	1.2612	35.60	92.33
680	0.0866	3853.48	7.4650	2.3450	738.58	1.2603	36.00	93.67
690	0.0875	3876.95	7.4894	2.3487	742.45	1.2593	36.39	95.00
700	0.0885	3900.45	7.5137	2.3525	746.29	1.2584	36.78	96.35
720	0.0905	3947.58	7.5617	2.3606	753.88	1.2566	37.56	99.04
740	0.0924	3994.88	7.6088	2.3692	761.35	1.2549	38.33	101.75
760	0.0943	4042.35	7.6552	2.3783	768.71	1.2531	39.09	104.47
780	0.0962	4090.02	7.7009	2.3879	775.96	1.2514	39.85	107.21
800	0.0982	4137.87	7.7459	2.3978	783.12	1.2496	40.60	109.95

Table 4 Single-Phase Region (Regions 1 to 3 of IAPWS-IF97) – Continuation
Homogenes Zustandsgebiet (Bereiche 1 bis 3 der IAPWS-IF97) – Fortsetzung

				p = 6 MPa				
t [°C]	v [m³ kg⁻¹]	h [kJ kg⁻¹]	s [kJ kg⁻¹ K⁻¹]	c_p [kJ kg⁻¹ K⁻¹]	w [m s⁻¹]	κ	$\eta \times 10^6$ [kg m⁻¹ s⁻¹]	$\lambda \times 10^3$ [W m⁻¹ K⁻¹]
0	0.000997	6.04	0.0002	4.1908	1411.75	333.11	1777.13	564.34
5	0.000997	26.97	0.0761	4.1803	1435.55	344.45	1508.63	573.60
10	0.000998	47.85	0.1505	4.1737	1456.92	354.65	1299.65	582.91
15	0.000998	68.71	0.2235	4.1695	1475.95	363.74	1133.57	592.14
20	0.000999	89.55	0.2952	4.1668	1492.78	371.73	999.19	601.16
25	0.001000	110.38	0.3657	4.1652	1507.55	378.67	888.78	609.87
30	0.001002	131.20	0.4350	4.1643	1520.40	384.60	796.85	618.16
35	0.001003	152.02	0.5031	4.1641	1531.47	389.58	719.39	625.99
40	0.001005	172.84	0.5701	4.1643	1540.87	393.65	653.48	633.31
45	0.001007	193.67	0.6361	4.1650	1548.72	396.87	596.88	640.10
50	0.001009	214.49	0.7010	4.1661	1555.13	399.28	547.89	646.34
60	0.001014	256.17	0.8280	4.1698	1563.97	401.87	467.73	657.21
70	0.001020	297.90	0.9514	4.1754	1568.08	401.76	405.38	666.04
80	0.001026	339.69	1.0715	4.1828	1568.01	399.29	355.90	673.02
90	0.001033	381.56	1.1884	4.1921	1564.21	394.74	315.99	678.38
100	0.001040	423.53	1.3024	4.2033	1557.09	388.37	283.32	682.32
110	0.001048	465.63	1.4138	4.2165	1546.96	380.41	256.25	685.02
120	0.001057	507.87	1.5226	4.2318	1534.09	371.05	233.57	686.62
130	0.001066	550.28	1.6291	4.2495	1518.66	360.47	214.38	687.23
140	0.001076	592.87	1.7335	4.2697	1500.85	348.83	197.98	686.93
150	0.001087	635.68	1.8358	4.2927	1480.77	336.25	183.85	685.76
160	0.001098	678.73	1.9364	4.3191	1458.50	322.85	171.58	683.76
170	0.001110	722.07	2.0353	4.3492	1434.07	308.72	160.85	680.91
180	0.001123	765.73	2.1327	4.3837	1407.51	293.96	151.37	677.22
190	0.001137	809.76	2.2289	4.4232	1378.82	278.65	142.96	672.66
200	0.001152	854.22	2.3238	4.4687	1347.98	262.87	135.43	667.19
210	0.001168	899.16	2.4178	4.5213	1314.94	246.69	128.64	660.77
220	0.001186	944.67	2.5110	4.5823	1279.66	230.19	122.46	653.31
230	0.001205	990.84	2.6037	4.6537	1242.04	213.45	116.80	644.74
240	0.001225	1037.79	2.6961	4.7378	1201.98	196.52	111.57	634.94
250	0.001248	1085.65	2.7885	4.8379	1159.25	179.46	106.68	623.77
260	0.001273	1134.61	2.8812	4.9589	1113.52	162.29	102.06	611.09
270	0.001302	1184.92	2.9747	5.1082	1064.22	145.00	97.65	596.70
t_s = 275.59				**Saturation**				
Liquid	0.001319	1213.73	3.0274	5.2080	1034.81	135.28	95.25	587.84
Vapour	0.0324	2784.56	5.8901	4.8768	494.01	1.2535	18.51	59.01
280	0.0332	2805.25	5.9276	4.5160	501.07	1.2604	18.73	58.40
290	0.0348	2847.50	6.0033	3.9812	514.56	1.2694	19.24	57.48
300	0.0362	2885.49	6.0702	3.6378	526.04	1.2743	19.73	56.98
310	0.0375	2920.58	6.1309	3.3929	536.29	1.2775	20.22	56.79
320	0.0388	2953.55	6.1870	3.2078	545.69	1.2797	20.70	56.82
330	0.0400	2984.87	6.2393	3.0626	554.43	1.2813	21.17	57.03
340	0.0411	3014.89	6.2887	2.9459	562.64	1.2825	21.64	57.38

Table 4 Single-Phase Region (Regions 1 to 3 of IAPWS-IF97) – Continuation
Homogenes Zustandsgebiet (Bereiche 1 bis 3 der IAPWS-IF97) – Fortsetzung

$p = 6$ MPa								
t [°C]	v [m^3 kg^{-1}]	h [kJ kg^{-1}]	s [kJ kg^{-1} K^{-1}]	c_p [kJ kg^{-1} K^{-1}]	w [m s^{-1}]	κ	$\eta \times 10^6$ [kg m^{-1} s^{-1}]	$\lambda \times 10^3$ [W m^{-1} K^{-1}]
350	0.0423	3043.86	6.3356	2.8504	570.41	1.2834	22.10	57.84
360	0.0433	3071.96	6.3803	2.7714	577.79	1.2839	22.56	58.41
370	0.0444	3099.33	6.4232	2.7054	584.85	1.2842	23.02	59.05
380	0.0454	3126.10	6.4645	2.6499	591.62	1.2843	23.48	59.77
390	0.0464	3152.36	6.5044	2.6030	598.14	1.2842	23.93	60.55
400	0.0474	3178.18	6.5431	2.5632	604.44	1.2840	24.37	61.38
410	0.0484	3203.64	6.5806	2.5293	610.54	1.2837	24.82	62.26
420	0.0494	3228.79	6.6171	2.5004	616.47	1.2832	25.26	63.18
430	0.0503	3253.66	6.6528	2.4757	622.23	1.2827	25.70	64.13
440	0.0512	3278.31	6.6876	2.4545	627.84	1.2821	26.14	65.12
450	0.0522	3302.76	6.7216	2.4364	633.32	1.2814	26.57	66.14
460	0.0531	3327.05	6.7550	2.4209	638.68	1.2807	27.01	67.19
470	0.0540	3351.19	6.7877	2.4077	643.93	1.2799	27.44	68.27
480	0.0549	3375.21	6.8198	2.3964	649.07	1.2791	27.87	69.36
490	0.0558	3399.12	6.8513	2.3868	654.12	1.2783	28.29	70.48
500	0.0567	3422.95	6.8824	2.3787	659.07	1.2775	28.72	71.62
510	0.0576	3446.70	6.9129	2.3719	663.95	1.2766	29.14	72.78
520	0.0584	3470.39	6.9429	2.3663	668.74	1.2757	29.56	73.96
530	0.0593	3494.03	6.9726	2.3616	673.46	1.2748	29.98	75.15
540	0.0602	3517.63	7.0018	2.3579	678.11	1.2739	30.39	76.36
550	0.0610	3541.19	7.0306	2.3550	682.70	1.2730	30.81	77.59
560	0.0619	3564.73	7.0590	2.3528	687.22	1.2721	31.22	78.82
570	0.0627	3588.25	7.0870	2.3512	691.69	1.2712	31.63	80.07
580	0.0636	3611.76	7.1148	2.3503	696.09	1.2702	32.04	81.33
590	0.0644	3635.26	7.1421	2.3498	700.45	1.2693	32.45	82.60
600	0.0653	3658.76	7.1692	2.3499	704.75	1.2684	32.86	83.88
610	0.0661	3682.26	7.1960	2.3504	709.01	1.2674	33.26	85.16
620	0.0669	3705.76	7.2224	2.3512	713.22	1.2665	33.67	86.46
630	0.0678	3729.28	7.2486	2.3525	717.38	1.2656	34.07	87.76
640	0.0686	3752.81	7.2745	2.3541	721.50	1.2647	34.47	89.07
650	0.0694	3776.36	7.3002	2.3559	725.58	1.2638	34.87	90.38
660	0.0703	3799.93	7.3256	2.3581	729.62	1.2629	35.26	91.70
670	0.0711	3823.53	7.3507	2.3605	733.63	1.2619	35.66	93.03
680	0.0719	3847.15	7.3756	2.3632	737.59	1.2610	36.05	94.35
690	0.0727	3870.79	7.4003	2.3661	741.52	1.2601	36.44	95.69
700	0.0735	3894.47	7.4248	2.3692	745.42	1.2593	36.83	97.02
720	0.0752	3941.92	7.4730	2.3760	753.11	1.2575	37.61	99.71
740	0.0768	3989.51	7.5205	2.3834	760.68	1.2557	38.38	102.40
760	0.0784	4037.26	7.5672	2.3915	768.13	1.2540	39.14	105.11
780	0.0800	4085.18	7.6131	2.4001	775.46	1.2523	39.90	107.82
800	0.0816	4133.27	7.6583	2.4092	782.69	1.2506	40.66	110.55

Table 4 Single-Phase Region (Regions 1 to 3 of IAPWS-IF97) – Continuation
Homogenes Zustandsgebiet (Bereiche 1 bis 3 der IAPWS-IF97) – Fortsetzung

				p = 7 MPa				
t [°C]	v [m³ kg⁻¹]	h [kJ kg⁻¹]	s [kJ kg⁻¹ K⁻¹]	c_p [kJ kg⁻¹ K⁻¹]	w [m s⁻¹]	κ	$\eta \times 10^6$ [kg m⁻¹ s⁻¹]	$\lambda \times 10^3$ [W m⁻¹ K⁻¹]
0	0.000997	7.05	0.0002	4.1861	1413.35	286.31	1774.79	564.89
5	0.000997	27.96	0.0761	4.1762	1437.15	296.04	1507.08	574.11
10	0.000997	48.82	0.1504	4.1701	1458.52	304.80	1298.64	583.40
15	0.000998	69.66	0.2234	4.1663	1477.55	312.59	1132.93	592.61
20	0.000999	90.48	0.2950	4.1639	1494.38	319.45	998.81	601.62
25	0.001000	111.30	0.3654	4.1625	1509.15	325.41	888.58	610.32
30	0.001001	132.11	0.4346	4.1618	1522.01	330.50	796.77	618.62
35	0.001003	152.92	0.5027	4.1616	1533.08	334.77	719.42	626.45
40	0.001005	173.73	0.5697	4.1619	1542.49	338.27	653.57	633.77
45	0.001007	194.54	0.6356	4.1627	1550.36	341.04	597.03	640.56
50	0.001009	215.36	0.7006	4.1639	1556.79	343.12	548.07	646.80
60	0.001014	257.01	0.8275	4.1677	1565.69	345.36	467.96	657.68
70	0.001020	298.71	0.9509	4.1733	1569.86	345.30	405.63	666.53
80	0.001026	340.48	1.0708	4.1807	1569.85	343.21	356.17	673.53
90	0.001033	382.33	1.1877	4.1899	1566.14	339.34	316.26	678.90
100	0.001040	424.29	1.3017	4.2011	1559.11	333.91	283.59	682.86
110	0.001048	466.36	1.4129	4.2142	1549.08	327.12	256.52	685.58
120	0.001057	508.58	1.5217	4.2294	1536.31	319.13	233.83	687.21
130	0.001066	550.96	1.6281	4.2468	1521.01	310.10	214.63	687.85
140	0.001076	593.52	1.7324	4.2668	1503.34	300.16	198.23	687.57
150	0.001086	636.30	1.8347	4.2897	1483.40	289.42	184.10	686.44
160	0.001097	679.33	1.9352	4.3157	1461.29	277.97	171.83	684.46
170	0.001109	722.63	2.0341	4.3454	1437.04	265.90	161.09	681.65
180	0.001122	766.25	2.1314	4.3794	1410.68	253.29	151.62	678.00
190	0.001136	810.23	2.2274	4.4184	1382.21	240.21	143.21	673.48
200	0.001151	854.64	2.3223	4.4632	1351.60	226.72	135.68	668.06
210	0.001167	899.52	2.4161	4.5149	1318.83	212.90	128.88	661.68
220	0.001184	944.96	2.5092	4.5749	1283.83	198.80	122.71	654.29
230	0.001203	991.05	2.6018	4.6449	1246.54	184.49	117.06	645.79
240	0.001224	1037.90	2.6939	4.7272	1206.84	170.02	111.83	636.08
250	0.001246	1085.65	2.7861	4.8250	1164.53	155.45	106.96	625.04
260	0.001271	1134.47	2.8785	4.9428	1119.31	140.78	102.36	612.49
270	0.001299	1184.60	2.9717	5.0874	1070.67	126.04	97.96	598.28
280	0.001331	1236.34	3.0661	5.2698	1017.77	111.17	93.70	582.20
t_s = 285.83				**Saturation**				
Liquid	0.001352	1267.44	3.1220	5.4004	984.51	102.43	91.25	571.88
Vapour	0.0274	2772.57	5.8146	5.3540	489.28	1.2491	18.96	62.91
290	0.0280	2793.98	5.8528	4.9360	496.60	1.2563	19.17	62.12
300	0.0295	2839.83	5.9335	4.2919	511.27	1.2661	19.68	60.82
310	0.0308	2880.57	6.0040	3.8817	523.61	1.2715	20.17	60.06
320	0.0320	2917.86	6.0674	3.5913	534.56	1.2751	20.66	59.68
330	0.0332	2952.63	6.1255	3.3740	544.52	1.2777	21.14	59.56
340	0.0342	2985.50	6.1796	3.2052	553.73	1.2796	21.62	59.66

Table 4 Single-Phase Region (Regions 1 to 3 of IAPWS-IF97) – Continuation
Homogenes Zustandsgebiet (Bereiche 1 bis 3 der IAPWS-IF97) – Fortsetzung

			p = 7 MPa					
t [°C]	v [m^3 kg^{-1}]	h [kJ kg^{-1}]	s [kJ kg^{-1} K^{-1}]	c_p [kJ kg^{-1} K^{-1}]	w [m s^{-1}]	κ	$\eta \times 10^6$ [kg m^{-1} s^{-1}]	$\lambda \times 10^3$ [W m^{-1} K^{-1}]
350	0.0353	3016.85	6.2303	3.0704	562.33	1.2810	22.09	59.92
360	0.0363	3046.99	6.2783	2.9608	570.43	1.2820	22.56	60.31
370	0.0372	3076.13	6.3240	2.8703	578.11	1.2827	23.02	60.80
380	0.0382	3104.44	6.3677	2.7949	585.42	1.2831	23.48	61.40
390	0.0391	3132.07	6.4096	2.7314	592.42	1.2833	23.94	62.07
400	0.0400	3159.10	6.4501	2.6778	599.14	1.2833	24.39	62.80
410	0.0408	3185.65	6.4892	2.6321	605.62	1.2831	24.84	63.60
420	0.0417	3211.77	6.5272	2.5932	611.88	1.2828	25.28	64.45
430	0.0425	3237.53	6.5641	2.5599	617.95	1.2824	25.72	65.34
440	0.0434	3262.98	6.6000	2.5313	623.84	1.2819	26.17	66.27
450	0.0442	3288.17	6.6351	2.5067	629.58	1.2814	26.60	67.24
460	0.0450	3313.13	6.6694	2.4855	635.17	1.2808	27.04	68.25
470	0.0458	3337.89	6.7029	2.4673	640.63	1.2801	27.47	69.28
480	0.0466	3362.48	6.7358	2.4516	645.97	1.2793	27.90	70.35
490	0.0474	3386.93	6.7681	2.4381	651.19	1.2786	28.33	71.44
500	0.0482	3411.25	6.7997	2.4265	656.32	1.2778	28.76	72.55
510	0.0489	3435.46	6.8308	2.4166	661.35	1.2770	29.18	73.68
520	0.0497	3459.59	6.8614	2.4081	666.29	1.2761	29.60	74.84
530	0.0505	3483.63	6.8916	2.4009	671.15	1.2753	30.02	76.01
540	0.0512	3507.61	6.9212	2.3949	675.93	1.2744	30.44	77.21
550	0.0520	3531.53	6.9505	2.3899	680.63	1.2735	30.86	78.41
560	0.0527	3555.41	6.9793	2.3857	685.27	1.2726	31.27	79.64
570	0.0535	3579.25	7.0078	2.3824	689.84	1.2717	31.68	80.87
580	0.0542	3603.06	7.0358	2.3798	694.35	1.2708	32.10	82.12
590	0.0549	3626.85	7.0635	2.3779	698.80	1.2699	32.50	83.38
600	0.0557	3650.62	7.0909	2.3765	703.19	1.2690	32.91	84.65
610	0.0564	3674.38	7.1180	2.3757	707.53	1.2681	33.32	85.93
620	0.0571	3698.14	7.1447	2.3754	711.83	1.2672	33.72	87.21
630	0.0578	3721.89	7.1712	2.3755	716.07	1.2663	34.12	88.51
640	0.0586	3745.65	7.1973	2.3760	720.26	1.2654	34.52	89.81
650	0.0593	3769.41	7.2232	2.3769	724.41	1.2645	34.92	91.12
660	0.0600	3793.19	7.2488	2.3782	728.52	1.2637	35.32	92.43
670	0.0607	3816.98	7.2742	2.3798	732.59	1.2628	35.71	93.75
680	0.0614	3840.78	7.2993	2.3816	736.61	1.2619	36.11	95.07
690	0.0621	3864.61	7.3242	2.3837	740.60	1.2610	36.50	96.40
700	0.0628	3888.46	7.3488	2.3861	744.55	1.2601	36.89	97.73
720	0.0643	3936.23	7.3974	2.3916	752.35	1.2584	37.67	100.40
740	0.0657	3984.13	7.4452	2.3978	760.02	1.2567	38.44	103.08
760	0.0671	4032.15	7.4921	2.4048	767.56	1.2550	39.20	105.77
780	0.0685	4080.32	7.5383	2.4124	774.98	1.2533	39.96	108.46
800	0.0698	4128.65	7.5837	2.4206	782.28	1.2516	40.71	111.17

Table 4 Single-Phase Region (Regions 1 to 3 of IAPWS-IF97) – Continuation
Homogenes Zustandsgebiet (Bereiche 1 bis 3 der IAPWS-IF97) – Fortsetzung

				p = 8 MPa				
t [°C]	v [m³ kg⁻¹]	h [kJ kg⁻¹]	s [kJ kg⁻¹ K⁻¹]	c_p [kJ kg⁻¹ K⁻¹]	w [m s⁻¹]	κ	$\eta \times 10^6$ [kg m⁻¹ s⁻¹]	$\lambda \times 10^3$ [W m⁻¹ K⁻¹]
0	0.000996	8.06	0.0003	4.1814	1414.95	251.22	1772.46	565.44
5	0.000996	28.94	0.0760	4.1722	1438.75	259.74	1505.55	574.62
10	0.000997	49.79	0.1503	4.1665	1460.12	267.41	1297.64	583.89
15	0.000997	70.61	0.2232	4.1630	1479.15	274.24	1132.29	593.08
20	0.000998	91.42	0.2948	4.1609	1495.98	280.24	998.43	602.08
25	0.000999	112.22	0.3652	4.1597	1510.75	285.46	888.38	610.77
30	0.001001	133.02	0.4343	4.1592	1523.61	289.92	796.71	619.07
35	0.001003	153.81	0.5024	4.1591	1534.69	293.67	719.45	626.90
40	0.001004	174.61	0.5693	4.1596	1544.12	296.74	653.67	634.22
45	0.001006	195.41	0.6352	4.1604	1552.00	299.17	597.17	641.02
50	0.001009	216.22	0.7001	4.1617	1558.45	301.00	548.26	647.27
60	0.001014	257.85	0.8270	4.1656	1567.40	302.99	468.19	658.16
70	0.001019	299.53	0.9503	4.1712	1571.63	302.96	405.88	667.02
80	0.001025	341.28	1.0702	4.1786	1571.70	301.15	356.43	674.03
90	0.001032	383.11	1.1870	4.1878	1568.07	297.79	316.52	679.43
100	0.001039	425.04	1.3009	4.1989	1561.13	293.07	283.86	683.41
110	0.001047	467.09	1.4121	4.2119	1551.20	287.15	256.78	686.15
120	0.001056	509.28	1.5208	4.2269	1538.54	280.19	234.10	687.80
130	0.001065	551.64	1.6272	4.2442	1523.35	272.32	214.89	688.46
140	0.001075	594.18	1.7314	4.2640	1505.81	263.65	198.49	688.22
150	0.001086	636.93	1.8337	4.2866	1486.02	254.29	184.36	687.11
160	0.001097	679.92	1.9341	4.3123	1464.06	244.30	172.08	685.16
170	0.001109	723.19	2.0328	4.3417	1439.99	233.78	161.34	682.38
180	0.001122	766.77	2.1301	4.3752	1413.83	222.78	151.87	678.77
190	0.001135	810.71	2.2260	4.4137	1385.57	211.37	143.45	674.29
200	0.001150	855.06	2.3207	4.4578	1355.20	199.61	135.92	668.92
210	0.001166	899.89	2.4145	4.5087	1322.68	187.55	129.13	662.60
220	0.001183	945.26	2.5074	4.5677	1287.97	175.25	122.97	655.27
230	0.001202	991.27	2.5998	4.6363	1251.00	162.77	117.32	646.84
240	0.001222	1038.03	2.6918	4.7169	1211.65	150.15	112.10	637.23
250	0.001245	1085.66	2.7837	4.8125	1169.76	137.43	107.23	626.29
260	0.001269	1134.34	2.8759	4.9272	1125.03	124.65	102.64	613.88
270	0.001297	1184.29	2.9687	5.0674	1077.01	111.80	98.27	599.83
280	0.001328	1235.81	3.0627	5.2431	1024.95	98.86	94.03	583.97
290	0.001364	1289.33	3.1586	5.4713	967.72	85.80	89.86	566.10
t_s = 295.01				**Saturation**				
Liquid	0.001385	1317.08	3.2077	5.6140	936.73	79.21	87.77	556.34
Vapour	0.0235	2758.61	5.7448	5.8831	484.07	1.2450	19.40	67.06
300	0.0243	2786.38	5.7935	5.2870	493.56	1.2541	19.65	65.79
310	0.0256	2835.27	5.8781	4.5559	509.11	1.2640	20.15	64.11
320	0.0268	2878.35	5.9514	4.0905	522.16	1.2697	20.65	63.10
330	0.0280	2917.53	6.0169	3.7620	533.69	1.2736	21.14	62.53
340	0.0290	2953.87	6.0766	3.5176	544.12	1.2764	21.62	62.27

Table 4 Single-Phase Region (Regions 1 to 3 of IAPWS-IF97) – Continuation
Homogenes Zustandsgebiet (Bereiche 1 bis 3 der IAPWS-IF97) – Fortsetzung

				p = 8 MPa				
t [°C]	v [m³ kg⁻¹]	h [kJ kg⁻¹]	s [kJ kg⁻¹ K⁻¹]	c_p [kJ kg⁻¹ K⁻¹]	w [m s⁻¹]	κ	$\eta \times 10^6$ [kg m⁻¹ s⁻¹]	$\lambda \times 10^3$ [W m⁻¹ K⁻¹]
350	0.0300	2988.06	6.1319	3.3288	553.71	1.2785	22.09	62.26
360	0.0309	3020.57	6.1837	3.1789	562.64	1.2800	22.57	62.43
370	0.0318	3051.73	6.2325	3.0573	571.02	1.2811	23.03	62.74
380	0.0327	3081.79	6.2789	2.9572	578.94	1.2818	23.50	63.18
390	0.0335	3110.93	6.3232	2.8738	586.46	1.2823	23.95	63.72
400	0.0343	3139.31	6.3657	2.8037	593.65	1.2825	24.41	64.35
410	0.0351	3167.04	6.4066	2.7444	600.54	1.2826	24.86	65.05
420	0.0359	3194.23	6.4461	2.6939	607.17	1.2825	25.31	65.81
430	0.0367	3220.95	6.4843	2.6507	613.56	1.2822	25.76	66.63
440	0.0375	3247.26	6.5215	2.6136	619.75	1.2819	26.20	67.50
450	0.0382	3273.23	6.5577	2.5817	625.75	1.2814	26.64	68.41
460	0.0389	3298.91	6.5929	2.5541	631.59	1.2809	27.08	69.37
470	0.0397	3324.33	6.6274	2.5303	637.27	1.2803	27.51	70.36
480	0.0404	3349.53	6.6611	2.5097	642.82	1.2796	27.94	71.38
490	0.0411	3374.53	6.6940	2.4919	648.24	1.2789	28.37	72.44
500	0.0418	3399.37	6.7264	2.4765	653.54	1.2782	28.80	73.52
510	0.0425	3424.07	6.7581	2.4631	658.73	1.2774	29.23	74.63
520	0.0431	3448.64	6.7893	2.4516	663.82	1.2766	29.65	75.76
530	0.0438	3473.11	6.8199	2.4416	668.82	1.2758	30.07	76.91
540	0.0445	3497.48	6.8501	2.4331	673.73	1.2750	30.49	78.08
550	0.0452	3521.77	6.8798	2.4258	678.56	1.2741	30.91	79.27
560	0.0458	3546.00	6.9091	2.4196	683.31	1.2733	31.33	80.48
570	0.0465	3570.17	6.9379	2.4144	687.99	1.2724	31.74	81.70
580	0.0472	3594.29	6.9663	2.4101	692.60	1.2715	32.15	82.94
590	0.0478	3618.37	6.9944	2.4066	697.15	1.2707	32.56	84.19
600	0.0485	3642.42	7.0221	2.4038	701.64	1.2698	32.97	85.45
610	0.0491	3666.45	7.0495	2.4016	706.06	1.2689	33.37	86.72
620	0.0498	3690.46	7.0765	2.4000	710.44	1.2680	33.78	88.00
630	0.0504	3714.45	7.1032	2.3990	714.76	1.2671	34.18	89.28
640	0.0510	3738.44	7.1296	2.3984	719.03	1.2663	34.58	90.58
650	0.0517	3762.42	7.1557	2.3983	723.25	1.2654	34.98	91.88
660	0.0523	3786.40	7.1816	2.3986	727.43	1.2645	35.38	93.18
670	0.0529	3810.39	7.2072	2.3992	731.56	1.2636	35.77	94.49
680	0.0536	3834.39	7.2325	2.4003	735.65	1.2628	36.16	95.81
690	0.0542	3858.40	7.2575	2.4016	739.69	1.2619	36.56	97.13
700	0.0548	3882.42	7.2823	2.4032	743.70	1.2610	36.95	98.45
720	0.0561	3930.53	7.3313	2.4073	751.61	1.2593	37.72	101.11
740	0.0573	3978.72	7.3793	2.4123	759.37	1.2576	38.49	103.77
760	0.0585	4027.03	7.4265	2.4182	767.00	1.2560	39.26	106.45
780	0.0598	4075.46	7.4729	2.4249	774.50	1.2543	40.01	109.13
800	0.0610	4124.02	7.5186	2.4322	781.88	1.2526	40.77	111.81

Table 4 Single-Phase Region (Regions 1 to 3 of IAPWS-IF97) – Continuation
Homogenes Zustandsgebiet (Bereiche 1 bis 3 der IAPWS-IF97) – Fortsetzung

p = 9 MPa								
t [°C]	v [m³ kg⁻¹]	h [kJ kg⁻¹]	s [kJ kg⁻¹ K⁻¹]	c_p [kJ kg⁻¹ K⁻¹]	w [m s⁻¹]	κ	$\eta \times 10^6$ [kg m⁻¹ s⁻¹]	$\lambda \times 10^3$ [W m⁻¹ K⁻¹]
0	0.000996	9.06	0.0003	4.1768	1416.55	223.92	1770.17	565.99
5	0.000996	29.93	0.0760	4.1682	1440.36	231.51	1504.04	575.14
10	0.000996	50.75	0.1502	4.1630	1461.72	238.33	1296.66	584.37
15	0.000997	71.56	0.2230	4.1599	1480.75	244.41	1131.67	593.55
20	0.000998	92.35	0.2946	4.1580	1497.57	249.75	998.06	602.54
25	0.000999	113.14	0.3649	4.1570	1512.35	254.39	888.19	611.23
30	0.001000	133.92	0.4340	4.1566	1525.22	258.37	796.64	619.52
35	0.001002	154.71	0.5020	4.1567	1536.31	261.70	719.48	627.35
40	0.001004	175.49	0.5689	4.1572	1545.75	264.44	653.77	634.68
45	0.001006	196.28	0.6348	4.1582	1553.65	266.61	597.32	641.48
50	0.001008	217.07	0.6996	4.1595	1560.11	268.24	548.44	647.73
60	0.001013	258.69	0.8265	4.1634	1569.11	270.03	468.42	658.63
70	0.001019	300.35	0.9497	4.1691	1573.40	270.02	406.14	667.51
80	0.001025	342.08	1.0695	4.1765	1573.54	268.44	356.69	674.54
90	0.001032	383.88	1.1863	4.1856	1569.99	265.48	316.79	679.95
100	0.001039	425.79	1.3001	4.1967	1563.14	261.30	284.12	683.95
110	0.001047	467.82	1.4113	4.2096	1553.30	256.07	257.05	686.72
120	0.001055	509.99	1.5199	4.2245	1540.75	249.91	234.36	688.39
130	0.001065	552.32	1.6262	4.2417	1525.69	242.94	215.15	689.08
140	0.001074	594.83	1.7304	4.2612	1508.27	235.26	198.74	688.86
150	0.001085	637.56	1.8326	4.2836	1488.62	226.96	184.61	687.78
160	0.001096	680.52	1.9329	4.3090	1466.82	218.11	172.33	685.86
170	0.001108	723.75	2.0316	4.3380	1442.93	208.79	161.58	683.12
180	0.001121	767.29	2.1288	4.3711	1416.95	199.05	152.11	679.54
190	0.001134	811.19	2.2246	4.4090	1388.90	188.94	143.69	675.10
200	0.001149	855.49	2.3192	4.4525	1358.76	178.51	136.17	669.77
210	0.001165	900.26	2.4128	4.5026	1326.50	167.83	129.38	663.51
220	0.001182	945.57	2.5057	4.5605	1292.07	156.93	123.22	656.24
230	0.001201	991.50	2.5979	4.6279	1255.41	145.87	117.57	647.89
240	0.001221	1038.16	2.6897	4.7069	1216.41	134.68	112.36	638.36
250	0.001243	1085.68	2.7814	4.8002	1174.92	123.41	107.50	627.53
260	0.001267	1134.23	2.8733	4.9120	1130.68	112.09	102.93	615.25
270	0.001295	1184.00	2.9658	5.0480	1083.26	100.71	98.57	601.37
280	0.001325	1235.30	3.0594	5.2176	1032.00	89.28	94.36	585.71
290	0.001361	1288.52	3.1547	5.4359	975.83	77.75	90.22	568.11
300	0.001402	1344.27	3.2529	5.7305	913.54	66.12	86.06	548.39
t_s = 303.35				**Saturation**				
Liquid	0.001418	1363.65	3.2866	5.8542	891.17	62.22	84.65	541.29
Vapour	0.0205	2742.88	5.6790	6.4762	478.44	1.2411	19.83	71.56
310	0.0214	2782.61	5.7475	5.5578	491.95	1.2537	20.16	69.39
320	0.0227	2833.89	5.8348	4.7670	508.10	1.2631	20.66	67.34
330	0.0238	2878.87	5.9100	4.2599	521.69	1.2688	21.15	66.08
340	0.0249	2919.58	5.9769	3.9023	533.66	1.2728	21.63	65.33

Table 4 Single-Phase Region (Regions 1 to 3 of IAPWS-IF97) – Continuation
Homogenes Zustandsgebiet (Bereiche 1 bis 3 der IAPWS-IF97) – Fortsetzung

p = 9 MPa

t [°C]	v [m^3 kg^{-1}]	h [kJ kg^{-1}]	s [kJ kg^{-1} K^{-1}]	c_p [kJ kg^{-1} K^{-1}]	w [m s^{-1}]	κ	$\eta\times10^6$ [kg m^{-1} s^{-1}]	$\lambda\times10^3$ [W m^{-1} K^{-1}]
350	0.0258	2957.22	6.0378	3.6370	544.46	1.2757	22.11	64.95
360	0.0267	2992.53	6.0940	3.4328	554.36	1.2779	22.59	64.83
370	0.0276	3026.01	6.1465	3.2710	563.55	1.2795	23.06	64.91
380	0.0284	3058.05	6.1959	3.1400	572.15	1.2806	23.52	65.16
390	0.0292	3088.89	6.2428	3.0323	580.26	1.2814	23.98	65.54
400	0.0300	3118.75	6.2875	2.9425	587.95	1.2819	24.44	66.03
410	0.0307	3147.79	6.3303	2.8671	595.29	1.2821	24.90	66.62
420	0.0314	3176.13	6.3715	2.8031	602.31	1.2822	25.35	67.28
430	0.0321	3203.88	6.4113	2.7486	609.06	1.2821	25.80	68.02
440	0.0328	3231.13	6.4497	2.7018	615.56	1.2819	26.24	68.81
450	0.0335	3257.94	6.4871	2.6617	621.85	1.2815	26.68	69.66
460	0.0342	3284.38	6.5234	2.6270	627.95	1.2811	27.12	70.56
470	0.0349	3310.50	6.5588	2.5970	633.87	1.2806	27.56	71.50
480	0.0355	3336.33	6.5933	2.5710	639.63	1.2800	27.99	72.48
490	0.0362	3361.93	6.6271	2.5484	645.24	1.2793	28.42	73.49
500	0.0368	3387.31	6.6601	2.5287	650.72	1.2787	28.85	74.54
510	0.0374	3412.51	6.6925	2.5116	656.08	1.2779	29.28	75.61
520	0.0380	3437.55	6.7243	2.4967	661.33	1.2772	29.71	76.72
530	0.0387	3462.45	6.7555	2.4838	666.48	1.2764	30.13	77.85
540	0.0393	3487.23	6.7861	2.4726	671.52	1.2756	30.55	79.00
550	0.0399	3511.91	6.8163	2.4629	676.48	1.2748	30.97	80.17
560	0.0405	3536.49	6.8460	2.4545	681.35	1.2740	31.38	81.36
570	0.0411	3561.00	6.8752	2.4473	686.14	1.2731	31.80	82.57
580	0.0417	3585.44	6.9040	2.4411	690.86	1.2723	32.21	83.79
590	0.0423	3609.83	6.9324	2.4359	695.50	1.2714	32.62	85.03
600	0.0429	3634.16	6.9605	2.4316	700.08	1.2706	33.03	86.28
610	0.0434	3658.46	6.9881	2.4280	704.60	1.2697	33.43	87.54
620	0.0440	3682.73	7.0155	2.4251	709.06	1.2689	33.84	88.81
630	0.0446	3706.97	7.0424	2.4228	713.46	1.2680	34.24	90.09
640	0.0452	3731.18	7.0691	2.4211	717.80	1.2671	34.64	91.37
650	0.0458	3755.39	7.0955	2.4199	722.10	1.2663	35.04	92.66
660	0.0463	3779.59	7.1216	2.4192	726.34	1.2654	35.44	93.96
670	0.0469	3803.78	7.1473	2.4190	730.54	1.2646	35.83	95.27
680	0.0475	3827.97	7.1728	2.4191	734.69	1.2637	36.23	96.57
690	0.0480	3852.16	7.1981	2.4197	738.79	1.2629	36.62	97.89
700	0.0486	3876.36	7.2231	2.4205	742.86	1.2620	37.01	99.20
720	0.0497	3924.80	7.2724	2.4232	750.87	1.2603	37.78	101.84
740	0.0508	3973.30	7.3207	2.4270	758.73	1.2586	38.55	104.49
760	0.0519	4021.88	7.3682	2.4318	766.45	1.2570	39.32	107.15
780	0.0530	4070.57	7.4149	2.4374	774.03	1.2553	40.07	109.81
800	0.0541	4119.38	7.4608	2.4438	781.49	1.2537	40.82	112.48

Table 4 Single-Phase Region (Regions 1 to 3 of IAPWS-IF97) – Continuation
Homogenes Zustandsgebiet (Bereiche 1 bis 3 der IAPWS-IF97) – Fortsetzung

				p = 10 MPa				
t [°C]	v [m^3 kg^{-1}]	h [kJ kg^{-1}]	s [kJ kg^{-1} K^{-1}]	c_p [kJ kg^{-1} K^{-1}]	w [m s^{-1}]	κ	$\eta \times 10^6$ [kg m^{-1} s^{-1}]	$\lambda \times 10^3$ [W m^{-1} K^{-1}]
0	0.000995	10.07	0.0003	4.1723	1418.16	202.09	1767.90	566.55
5	0.000995	30.91	0.0759	4.1643	1441.97	208.92	1502.54	575.65
10	0.000996	51.72	0.1501	4.1595	1463.33	215.07	1295.68	584.86
15	0.000996	72.51	0.2229	4.1567	1482.35	220.54	1131.06	594.02
20	0.000997	93.29	0.2944	4.1551	1499.18	225.36	997.70	603.00
25	0.000999	114.06	0.3646	4.1543	1513.95	229.54	888.01	611.68
30	0.001000	134.83	0.4337	4.1541	1526.82	233.12	796.58	619.97
35	0.001002	155.60	0.5017	4.1543	1537.92	236.13	719.51	627.81
40	0.001003	176.37	0.5685	4.1549	1547.37	238.60	653.87	635.13
45	0.001006	197.15	0.6344	4.1559	1555.29	240.56	597.47	641.93
50	0.001008	217.93	0.6992	4.1573	1561.78	242.04	548.63	648.19
60	0.001013	259.53	0.8259	4.1613	1570.83	243.66	468.65	659.11
70	0.001018	301.17	0.9491	4.1670	1575.18	243.67	406.39	668.00
80	0.001024	342.87	1.0689	4.1744	1575.38	242.27	356.96	675.05
90	0.001031	384.66	1.1856	4.1835	1571.91	239.62	317.06	680.48
100	0.001038	426.55	1.2994	4.1945	1565.15	235.89	284.39	684.50
110	0.001046	468.56	1.4105	4.2073	1555.41	231.20	257.31	687.28
120	0.001055	510.70	1.5190	4.2221	1542.96	225.68	234.62	688.98
130	0.001064	553.00	1.6253	4.2391	1528.01	219.43	215.41	689.69
140	0.001074	595.49	1.7294	4.2585	1510.72	212.54	198.99	689.50
150	0.001084	638.18	1.8315	4.2806	1491.21	205.10	184.86	688.45
160	0.001095	681.11	1.9318	4.3057	1469.57	197.16	172.58	686.56
170	0.001107	724.31	2.0304	4.3343	1445.84	188.80	161.83	683.85
180	0.001120	767.81	2.1274	4.3670	1420.05	180.06	152.35	680.31
190	0.001134	811.66	2.2232	4.4044	1392.21	170.99	143.94	675.91
200	0.001148	855.92	2.3177	4.4472	1362.30	161.64	136.41	670.63
210	0.001164	900.63	2.4112	4.4965	1330.29	152.05	129.63	664.41
220	0.001181	945.87	2.5039	4.5535	1296.13	142.27	123.47	657.20
230	0.001199	991.73	2.5959	4.6196	1259.78	132.34	117.83	648.93
240	0.001219	1038.30	2.6876	4.6970	1221.12	122.30	112.62	639.48
250	0.001241	1085.72	2.7791	4.7883	1180.03	112.19	107.78	628.76
260	0.001265	1134.13	2.8708	4.8972	1136.25	102.03	103.21	616.61
270	0.001292	1183.74	2.9629	5.0293	1089.42	91.84	98.87	602.89
280	0.001323	1234.82	3.0561	5.1931	1038.92	81.61	94.68	587.43
290	0.001357	1287.75	3.1510	5.4023	983.78	71.30	90.57	570.08
300	0.001398	1343.10	3.2484	5.6816	922.76	60.91	86.46	550.68
310	0.001447	1401.77	3.3498	6.0782	854.92	50.51	82.23	529.11
t_s = 311.00				**Saturation**				
Liquid	0.001453	1407.87	3.3603	6.1275	847.74	49.47	81.79	526.83
Vapour	0.0180	2725.47	5.6159	7.1472	472.44	1.2377	20.27	76.54
320	0.0193	2782.66	5.7131	5.7468	491.71	1.2546	20.70	72.87
330	0.0204	2835.67	5.8017	4.9228	508.20	1.2632	21.19	70.48
340	0.0215	2882.06	5.8780	4.3885	522.16	1.2688	21.67	69.00

Table 4 Single-Phase Region (Regions 1 to 3 of IAPWS-IF97) – Continuation
Homogenes Zustandsgebiet (Bereiche 1 bis 3 der IAPWS-IF97) – Fortsetzung

				p = 10 MPa				
t [°C]	v [m³ kg⁻¹]	h [kJ kg⁻¹]	s [kJ kg⁻¹ K⁻¹]	c_p [kJ kg⁻¹ K⁻¹]	w [m s⁻¹]	κ	$\eta\times 10^6$ [kg m⁻¹ s⁻¹]	$\lambda\times 10^3$ [W m⁻¹ K⁻¹]
350	0.0224	2923.96	5.9458	4.0118	534.45	1.2728	22.15	68.09
360	0.0233	2962.61	6.0073	3.7324	545.52	1.2757	22.63	67.58
370	0.0242	2998.82	6.0641	3.5174	555.64	1.2779	23.10	67.36
380	0.0250	3033.11	6.1170	3.3471	565.02	1.2794	23.56	67.37
390	0.0257	3065.87	6.1668	3.2092	573.79	1.2806	24.03	67.55
400	0.0264	3097.38	6.2139	3.0958	582.04	1.2813	24.49	67.88
410	0.0271	3127.85	6.2589	3.0013	589.86	1.2818	24.94	68.33
420	0.0278	3157.45	6.3019	2.9217	597.31	1.2820	25.40	68.87
430	0.0285	3186.32	6.3432	2.8542	604.44	1.2821	25.84	69.51
440	0.0291	3214.57	6.3831	2.7965	611.28	1.2820	26.29	70.21
450	0.0298	3242.28	6.4217	2.7470	617.87	1.2817	26.73	70.99
460	0.0304	3269.53	6.4591	2.7043	624.24	1.2814	27.18	71.82
470	0.0310	3296.38	6.4955	2.6674	630.41	1.2810	27.61	72.70
480	0.0316	3322.89	6.5310	2.6354	636.39	1.2804	28.05	73.63
490	0.0322	3349.11	6.5655	2.6076	642.21	1.2799	28.48	74.60
500	0.0328	3375.06	6.5993	2.5833	647.89	1.2792	28.91	75.61
510	0.0334	3400.78	6.6324	2.5622	653.42	1.2786	29.34	76.65
520	0.0340	3426.31	6.6648	2.5437	658.83	1.2779	29.76	77.72
530	0.0345	3451.67	6.6965	2.5275	664.12	1.2771	30.19	78.83
540	0.0351	3476.87	6.7277	2.5134	669.31	1.2764	30.61	79.95
550	0.0357	3501.94	6.7584	2.5011	674.39	1.2756	31.03	81.11
560	0.0362	3526.90	6.7885	2.4904	679.39	1.2748	31.44	82.28
570	0.0368	3551.75	6.8182	2.4811	684.29	1.2740	31.86	83.47
580	0.0373	3576.52	6.8474	2.4730	689.11	1.2731	32.27	84.68
590	0.0378	3601.22	6.8761	2.4660	693.86	1.2723	32.68	85.90
600	0.0384	3625.84	6.9045	2.4600	698.54	1.2715	33.09	87.14
610	0.0389	3650.42	6.9325	2.4549	703.14	1.2706	33.50	88.39
620	0.0394	3674.95	6.9601	2.4507	707.68	1.2698	33.90	89.65
630	0.0400	3699.43	6.9874	2.4471	712.16	1.2689	34.30	90.92
640	0.0405	3723.89	7.0143	2.4442	716.59	1.2681	34.70	92.19
650	0.0410	3748.32	7.0409	2.4420	720.95	1.2672	35.10	93.48
660	0.0415	3772.73	7.0672	2.4402	725.26	1.2664	35.50	94.77
670	0.0421	3797.13	7.0932	2.4390	729.53	1.2655	35.90	96.06
680	0.0426	3821.51	7.1189	2.4383	733.74	1.2647	36.29	97.37
690	0.0431	3845.89	7.1444	2.4380	737.91	1.2639	36.68	98.67
700	0.0436	3870.27	7.1696	2.4380	742.03	1.2630	37.07	99.98
720	0.0446	3919.05	7.2192	2.4393	750.15	1.2614	37.85	102.60
740	0.0456	3967.85	7.2678	2.4418	758.10	1.2597	38.62	105.24
760	0.0466	4016.72	7.3156	2.4454	765.91	1.2581	39.38	107.88
780	0.0476	4065.68	7.3625	2.4501	773.58	1.2564	40.13	110.52
800	0.0486	4114.73	7.4087	2.4555	781.12	1.2548	40.88	113.17

Table 4 **Single-Phase Region** (Regions 1 to 3 of IAPWS-IF97) – Continuation
Homogenes Zustandsgebiet (Bereiche 1 bis 3 der IAPWS-IF97) – Fortsetzung

				p = 11 MPa				
t [°C]	v [m³ kg⁻¹]	h [kJ kg⁻¹]	s [kJ kg⁻¹ K⁻¹]	c_p [kJ kg⁻¹ K⁻¹]	w [m s⁻¹]	κ	$\eta \times 10^6$ [kg m⁻¹ s⁻¹]	$\lambda \times 10^3$ [W m⁻¹ K⁻¹]
0	0.000995	11.07	0.0004	4.1678	1419.78	184.23	1765.65	567.10
5	0.000995	31.89	0.0759	4.1603	1443.59	190.45	1501.07	576.17
10	0.000995	52.68	0.1500	4.1560	1464.94	196.04	1294.72	585.35
15	0.000996	73.45	0.2227	4.1536	1483.96	201.02	1130.45	594.49
20	0.000997	94.22	0.2941	4.1523	1500.78	205.40	997.34	603.46
25	0.000998	114.98	0.3644	4.1517	1515.56	209.21	887.82	612.14
30	0.001000	135.74	0.4334	4.1516	1528.43	212.47	796.53	620.43
35	0.001001	156.49	0.5013	4.1519	1539.54	215.21	719.55	628.26
40	0.001003	177.26	0.5682	4.1526	1549.00	217.46	653.97	635.59
45	0.001005	198.02	0.6339	4.1537	1556.94	219.25	597.62	642.39
50	0.001007	218.79	0.6987	4.1551	1563.44	220.60	548.81	648.66
60	0.001012	260.36	0.8254	4.1592	1572.54	222.09	468.88	659.59
70	0.001018	301.98	0.9485	4.1649	1576.95	222.12	406.64	668.49
80	0.001024	343.67	1.0682	4.1723	1577.22	220.86	357.22	675.55
90	0.001031	385.43	1.1849	4.1814	1573.83	218.47	317.33	681.00
100	0.001038	427.30	1.2986	4.1923	1567.15	215.09	284.66	685.04
110	0.001046	469.29	1.4096	4.2050	1557.50	210.85	257.58	687.85
120	0.001054	511.41	1.5182	4.2197	1545.16	205.85	234.88	689.57
130	0.001063	553.69	1.6244	4.2366	1530.32	200.19	215.66	690.30
140	0.001073	596.15	1.7284	4.2557	1513.16	193.95	199.25	690.14
150	0.001084	638.81	1.8304	4.2776	1493.79	187.21	185.10	689.11
160	0.001095	681.71	1.9306	4.3024	1472.30	180.02	172.82	687.26
170	0.001107	724.87	2.0291	4.3307	1448.74	172.44	162.07	684.58
180	0.001119	768.34	2.1261	4.3630	1423.13	164.52	152.60	681.08
190	0.001133	812.15	2.2217	4.3998	1395.50	156.30	144.18	676.72
200	0.001147	856.35	2.3162	4.4420	1365.81	147.82	136.65	671.48
210	0.001163	901.01	2.4096	4.4906	1334.04	139.13	129.87	665.32
220	0.001180	946.19	2.5021	4.5466	1300.16	130.27	123.71	658.17
230	0.001198	991.97	2.5940	4.6115	1264.10	121.27	118.08	649.96
240	0.001218	1038.45	2.6855	4.6873	1225.79	112.17	112.88	640.60
250	0.001239	1085.76	2.7768	4.7766	1185.08	103.01	108.05	629.98
260	0.001263	1134.04	2.8682	4.8828	1141.76	93.80	103.50	617.96
270	0.001290	1183.49	2.9601	5.0112	1095.49	84.57	99.17	604.39
280	0.001320	1234.36	3.0529	5.1695	1045.71	75.31	95.00	589.13
290	0.001354	1287.02	3.1473	5.3703	991.56	66.01	90.92	572.01
300	0.001394	1341.98	3.2440	5.6357	931.80	56.63	86.85	552.91
310	0.001442	1400.08	3.3445	6.0070	865.35	47.23	82.68	531.73
t_s = 318.08				**Saturation**				
Liquid	0.001489	1450.28	3.4300	6.4427	806.15	39.69	79.13	513.02
Vapour	0.0160	2706.39	5.5545	7.9168	466.11	1.2349	20.72	82.15
320	0.0163	2721.07	5.5793	7.4062	471.35	1.2408	20.80	80.81
330	0.0176	2786.37	5.6885	5.8578	492.74	1.2565	21.27	76.19
340	0.0187	2840.45	5.7775	5.0240	509.38	1.2642	21.74	73.51

Table 4 Single-Phase Region (Regions 1 to 3 of IAPWS-IF97) – Continuation
Homogenes Zustandsgebiet (Bereiche 1 bis 3 der IAPWS-IF97) – Fortsetzung

p = 11 MPa

t [°C]	v [m^3 kg^{-1}]	h [kJ kg^{-1}]	s [kJ kg^{-1} K^{-1}]	c_p [kJ kg^{-1} K^{-1}]	w [m s^{-1}]	κ	$\eta \times 10^6$ [kg m^{-1} s^{-1}]	$\lambda \times 10^3$ [W m^{-1} K^{-1}]
350	0.0196	2887.79	5.8541	4.4777	523.55	1.2696	22.21	71.83
360	0.0205	2930.53	5.9221	4.0917	536.03	1.2735	22.68	70.78
370	0.0213	2969.95	5.9839	3.8048	547.26	1.2763	23.15	70.16
380	0.0221	3006.84	6.0408	3.5836	557.53	1.2784	23.62	69.85
390	0.0228	3041.77	6.0939	3.4080	567.03	1.2798	24.08	69.79
400	0.0235	3075.12	6.1438	3.2658	575.91	1.2809	24.54	69.92
410	0.0242	3107.17	6.1911	3.1485	584.26	1.2816	25.00	70.19
420	0.0249	3138.15	6.2361	3.0507	592.17	1.2820	25.45	70.60
430	0.0255	3168.23	6.2792	2.9682	599.70	1.2822	25.90	71.12
440	0.0261	3197.55	6.3206	2.8980	606.91	1.2822	26.35	71.72
450	0.0267	3226.23	6.3605	2.8381	613.82	1.2821	26.79	72.40
460	0.0273	3254.34	6.3991	2.7865	620.47	1.2818	27.24	73.16
470	0.0279	3281.98	6.4366	2.7419	626.90	1.2815	27.67	73.97
480	0.0284	3309.20	6.4730	2.7033	633.12	1.2810	28.11	74.85
490	0.0290	3336.06	6.5084	2.6697	639.15	1.2805	28.54	75.77
500	0.0296	3362.61	6.5430	2.6405	645.02	1.2799	28.97	76.73
510	0.0301	3388.89	6.5767	2.6149	650.74	1.2793	29.40	77.73
520	0.0306	3414.92	6.6098	2.5925	656.31	1.2786	29.83	78.77
530	0.0312	3440.74	6.6421	2.5729	661.76	1.2779	30.25	79.85
540	0.0317	3466.39	6.6738	2.5557	667.09	1.2772	30.67	80.95
550	0.0322	3491.87	6.7050	2.5406	672.31	1.2764	31.09	82.08
560	0.0327	3517.20	6.7356	2.5273	677.42	1.2756	31.51	83.23
570	0.0332	3542.42	6.7657	2.5158	682.44	1.2748	31.92	84.40
580	0.0337	3567.52	6.7953	2.5056	687.38	1.2740	32.34	85.60
590	0.0342	3592.53	6.8244	2.4968	692.22	1.2732	32.75	86.81
600	0.0347	3617.46	6.8531	2.4891	697.00	1.2724	33.16	88.03
610	0.0352	3642.32	6.8814	2.4824	701.69	1.2716	33.56	89.27
620	0.0357	3667.11	6.9094	2.4767	706.32	1.2708	33.97	90.52
630	0.0362	3691.86	6.9369	2.4718	710.88	1.2699	34.37	91.78
640	0.0367	3716.55	6.9641	2.4677	715.38	1.2691	34.77	93.04
650	0.0371	3741.21	6.9910	2.4643	719.82	1.2682	35.17	94.32
660	0.0376	3765.84	7.0175	2.4615	724.20	1.2674	35.57	95.60
670	0.0381	3790.45	7.0437	2.4593	728.53	1.2666	35.96	96.89
680	0.0386	3815.03	7.0696	2.4577	732.80	1.2657	36.36	98.18
690	0.0390	3839.60	7.0953	2.4565	737.03	1.2649	36.75	99.48
700	0.0395	3864.16	7.1207	2.4558	741.21	1.2641	37.14	100.78
720	0.0404	3913.27	7.1706	2.4555	749.43	1.2624	37.91	103.39
740	0.0414	3962.39	7.2196	2.4568	757.49	1.2608	38.68	106.00
760	0.0423	4011.55	7.2676	2.4592	765.39	1.2592	39.44	108.63
780	0.0432	4060.77	7.3148	2.4628	773.14	1.2576	40.20	111.25
800	0.0441	4110.07	7.3612	2.4674	780.75	1.2560	40.95	113.88

Table 4 Single-Phase Region (Regions 1 to 3 of IAPWS-IF97) – Continuation
Homogenes Zustandsgebiet (Bereiche 1 bis 3 der IAPWS-IF97) – Fortsetzung

p = 12 MPa

t [°C]	v [m³ kg⁻¹]	h [kJ kg⁻¹]	s [kJ kg⁻¹ K⁻¹]	c_p [kJ kg⁻¹ K⁻¹]	w [m s⁻¹]	κ	$\eta \times 10^6$ [kg m⁻¹ s⁻¹]	$\lambda \times 10^3$ [W m⁻¹ K⁻¹]
0	0.000994	12.07	0.0004	4.1633	1421.40	169.34	1763.43	567.65
5	0.000994	32.87	0.0758	4.1565	1445.21	175.05	1499.61	576.68
10	0.000995	53.64	0.1499	4.1526	1466.56	180.19	1293.78	585.84
15	0.000995	74.40	0.2225	4.1505	1485.57	184.75	1129.85	594.97
20	0.000996	95.15	0.2939	4.1494	1502.39	188.77	996.99	603.92
25	0.000998	115.90	0.3641	4.1490	1517.16	192.26	887.65	612.59
30	0.000999	136.64	0.4331	4.1491	1530.04	195.26	796.48	620.88
35	0.001001	157.39	0.5010	4.1495	1541.16	197.78	719.58	628.71
40	0.001003	178.14	0.5678	4.1503	1550.63	199.84	654.08	636.05
45	0.001005	198.89	0.6335	4.1515	1558.58	201.49	597.77	642.85
50	0.001007	219.65	0.6983	4.1530	1565.11	202.73	549.00	649.12
60	0.001012	261.20	0.8249	4.1571	1574.25	204.11	469.11	660.06
70	0.001017	302.80	0.9479	4.1629	1578.72	204.15	406.90	668.98
80	0.001023	344.46	1.0676	4.1703	1579.06	203.02	357.49	676.06
90	0.001030	386.21	1.1842	4.1793	1575.74	200.85	317.59	681.53
100	0.001038	428.06	1.2978	4.1902	1569.15	197.77	284.93	685.59
110	0.001045	470.02	1.4088	4.2028	1559.60	193.89	257.84	688.41
120	0.001054	512.12	1.5173	4.2174	1547.35	189.33	235.14	690.15
130	0.001063	554.37	1.6234	4.2340	1532.63	184.16	215.92	690.92
140	0.001073	596.81	1.7274	4.2530	1515.59	178.46	199.50	690.78
150	0.001083	639.44	1.8294	4.2746	1496.35	172.30	185.35	689.78
160	0.001094	682.31	1.9295	4.2992	1475.01	165.73	173.07	687.96
170	0.001106	725.44	2.0279	4.3271	1451.61	158.80	162.32	685.31
180	0.001118	768.86	2.1248	4.3590	1426.19	151.56	152.84	681.84
190	0.001132	812.63	2.2204	4.3953	1398.75	144.05	144.42	677.53
200	0.001146	856.79	2.3147	4.4369	1369.29	136.31	136.90	672.33
210	0.001162	901.39	2.4080	4.4847	1337.76	128.36	130.12	666.22
220	0.001179	946.51	2.5004	4.5397	1304.15	120.26	123.96	659.13
230	0.001197	992.21	2.5921	4.6035	1268.39	112.04	118.33	650.99
240	0.001216	1038.61	2.6834	4.6779	1230.40	103.72	113.14	641.71
250	0.001238	1085.81	2.7745	4.7652	1190.07	95.35	108.31	631.19
260	0.001262	1133.97	2.8657	4.8688	1147.20	86.94	103.77	619.29
270	0.001288	1183.26	2.9573	4.9936	1101.47	78.51	99.46	605.88
280	0.001317	1233.94	3.0498	5.1467	1052.38	70.07	95.31	590.80
290	0.001351	1286.33	3.1436	5.3398	999.16	61.59	91.26	573.92
300	0.001390	1340.93	3.2397	5.5924	940.65	53.06	87.23	555.11
310	0.001436	1398.49	3.3393	5.9411	875.62	44.49	83.12	534.28
320	0.001494	1460.31	3.4444	6.4621	803.26	36.00	78.79	511.38
t_s = 324.68				**Saturation**				
Liquid	0.001526	1491.33	3.4965	6.8126	765.59	32.00	76.62	499.89
Vapour	0.0143	2685.58	5.4941	8.8189	459.46	1.2329	21.18	88.57
330	0.0150	2728.14	5.5650	7.3313	474.22	1.2474	21.40	84.19
340	0.0162	2793.47	5.6725	5.8968	494.92	1.2591	21.85	79.31

Table 4 Single-Phase Region (Regions 1 to 3 of IAPWS-IF97) – Continuation
Homogenes Zustandsgebiet (Bereiche 1 bis 3 der IAPWS-IF97) – Fortsetzung

p = 12 MPa								
t [°C]	v [m³ kg⁻¹]	h [kJ kg⁻¹]	s [kJ kg⁻¹ K⁻¹]	c_p [kJ kg⁻¹ K⁻¹]	w [m s⁻¹]	κ	$\eta\times10^6$ [kg m⁻¹ s⁻¹]	$\lambda\times10^3$ [W m⁻¹ K⁻¹]
350	0.0172	2848.01	5.7607	5.0746	511.54	1.2661	22.31	76.41
360	0.0181	2895.87	5.8369	4.5309	525.77	1.2711	22.77	74.56
370	0.0189	2939.15	5.9047	4.1447	538.33	1.2748	23.23	73.39
380	0.0197	2979.09	5.9664	3.8563	549.63	1.2774	23.70	72.68
390	0.0204	3016.49	6.0232	3.6329	559.97	1.2793	24.16	72.29
400	0.0211	3051.90	6.0762	3.4551	569.54	1.2806	24.61	72.17
410	0.0218	3085.70	6.1261	3.3106	578.47	1.2815	25.07	72.24
420	0.0224	3118.19	6.1733	3.1912	586.89	1.2821	25.52	72.48
430	0.0230	3149.59	6.2182	3.0914	594.85	1.2825	25.97	72.85
440	0.0236	3180.07	6.2613	3.0071	602.43	1.2826	25.42	73.34
450	0.0242	3209.77	6.3027	2.9353	609.68	1.2826	25.86	73.92
460	0.0247	3238.81	6.3425	2.8737	616.64	1.2824	27.30	74.59
470	0.0253	3267.28	6.3811	2.8207	623.34	1.2821	27.74	75.33
480	0.0258	3295.25	6.4185	2.7748	629.81	1.2817	28.18	76.13
490	0.0263	3322.79	6.4548	2.7349	636.07	1.2813	28.61	77.00
500	0.0268	3349.97	6.4902	2.7002	642.14	1.2807	29.04	77.91
510	0.0273	3376.81	6.5247	2.6698	648.04	1.2801	29.47	78.87
520	0.0278	3403.37	6.5584	2.6432	653.79	1.2795	29.90	79.87
530	0.0283	3429.69	6.5914	2.6199	659.39	1.2788	30.32	80.91
540	0.0288	3455.78	6.6237	2.5993	664.86	1.2781	30.74	81.99
550	0.0293	3481.68	6.6553	2.5813	670.22	1.2774	31.16	83.09
560	0.0298	3507.41	6.6864	2.5654	675.46	1.2766	31.58	84.22
570	0.0303	3533.00	6.7169	2.5514	680.60	1.2758	31.99	85.37
580	0.0307	3558.45	6.7469	2.5391	685.64	1.2751	32.41	86.55
590	0.0312	3583.78	6.7764	2.5283	690.60	1.2743	32.82	87.74
600	0.0317	3609.02	6.8055	2.5188	695.46	1.2734	33.23	88.95
610	0.0321	3634.16	6.8341	2.5105	700.25	1.2726	33.63	90.18
620	0.0326	3659.23	6.8624	2.5033	704.97	1.2718	34.04	91.42
630	0.0330	3684.23	6.8902	2.4970	709.61	1.2710	34.44	92.66
640	0.0335	3709.18	6.9177	2.4917	714.18	1.2702	34.84	93.92
650	0.0339	3734.07	6.9448	2.4871	718.70	1.2693	35.24	95.19
660	0.0344	3758.92	6.9716	2.4832	723.15	1.2685	35.64	96.46
670	0.0348	3783.73	6.9980	2.4800	727.54	1.2677	36.03	97.74
680	0.0352	3808.52	7.0241	2.4773	731.88	1.2669	36.42	99.02
690	0.0357	3833.28	7.0500	2.4753	736.17	1.2660	36.82	100.31
700	0.0361	3858.03	7.0756	2.4737	740.40	1.2652	37.21	101.60
720	0.0370	3907.48	7.1259	2.4720	748.73	1.2636	37.98	104.20
740	0.0378	3956.92	7.1751	2.4718	756.89	1.2620	38.75	106.79
760	0.0387	4006.36	7.2235	2.4731	764.87	1.2603	39.51	109.40
780	0.0395	4055.85	7.2709	2.4756	772.71	1.2587	40.26	112.01
800	0.0404	4105.40	7.3175	2.4793	780.40	1.2571	41.01	114.61

Table 4 Single-Phase Region (Regions 1 to 3 of IAPWS-IF97) – Continuation
Homogenes Zustandsgebiet (Bereiche 1 bis 3 der IAPWS-IF97) – Fortsetzung

t [°C]	v [m³ kg⁻¹]	h [kJ kg⁻¹]	s [kJ kg⁻¹ K⁻¹]	c_p [kJ kg⁻¹ K⁻¹]	w [m s⁻¹]	κ	$\eta\times10^6$ [kg m⁻¹ s⁻¹]	$\lambda\times10^3$ [W m⁻¹ K⁻¹]
				p = 13 MPa				
0	0.000994	13.07	0.0004	4.1589	1423.02	156.75	1761.24	568.21
5	0.000994	33.85	0.0758	4.1526	1446.83	162.03	1498.17	577.19
10	0.000994	54.60	0.1497	4.1492	1468.18	166.77	1292.84	586.32
15	0.000995	75.35	0.2224	4.1474	1487.18	170.99	1129.27	595.44
20	0.000996	96.08	0.2937	4.1466	1504.00	174.70	996.64	604.39
25	0.000997	116.81	0.3638	4.1464	1518.77	177.93	887.48	613.05
30	0.000999	137.54	0.4328	4.1466	1531.65	180.70	796.43	621.33
35	0.001000	158.28	0.5006	4.1471	1542.78	183.02	719.63	629.17
40	0.001002	179.02	0.5674	4.1480	1552.26	184.94	654.18	636.50
45	0.001004	199.76	0.6331	4.1493	1560.23	186.46	597.93	643.31
50	0.001006	220.51	0.6978	4.1508	1566.77	187.62	549.19	649.59
60	0.001011	262.04	0.8244	4.1551	1575.96	188.90	469.35	660.54
70	0.001017	303.62	0.9473	4.1608	1580.49	188.95	407.15	669.47
80	0.001023	345.26	1.0670	4.1682	1580.90	187.92	357.75	676.57
90	0.001030	386.99	1.1835	4.1773	1577.66	185.93	317.86	682.05
100	0.001037	428.81	1.2971	4.1880	1571.14	183.10	285.19	686.13
110	0.001045	470.75	1.4080	4.2006	1561.69	179.55	258.11	688.98
120	0.001053	512.83	1.5164	4.2150	1549.54	175.35	235.40	690.74
130	0.001062	555.06	1.6225	4.2315	1534.93	170.59	216.18	691.53
140	0.001072	597.47	1.7264	4.2503	1518.01	165.35	199.75	691.41
150	0.001082	640.07	1.8283	4.2717	1498.91	159.68	185.60	690.45
160	0.001093	682.91	1.9283	4.2960	1477.71	153.64	173.31	688.65
170	0.001105	726.00	2.0267	4.3236	1454.47	147.26	162.56	686.04
180	0.001118	769.39	2.1235	4.3550	1429.23	140.60	153.08	682.61
190	0.001131	813.12	2.2190	4.3908	1401.99	133.69	144.66	678.33
200	0.001145	857.23	2.3132	4.4318	1372.74	126.56	137.14	673.18
210	0.001161	901.78	2.4064	4.4789	1341.45	119.25	130.36	667.11
220	0.001177	946.83	2.4986	4.5330	1308.10	111.79	124.21	660.08
230	0.001195	992.47	2.5903	4.5957	1272.63	104.22	118.58	652.01
240	0.001215	1038.78	2.6814	4.6686	1234.97	96.57	113.40	642.81
250	0.001236	1085.88	2.7723	4.7540	1195.01	88.86	108.58	632.39
260	0.001260	1133.91	2.8632	4.8552	1152.57	81.12	104.05	620.61
270	0.001286	1183.05	2.9546	4.9766	1107.36	73.37	99.75	607.35
280	0.001315	1233.53	3.0466	5.1248	1058.94	65.61	95.63	592.45
290	0.001348	1285.67	3.1401	5.3106	1006.61	57.83	91.60	575.79
300	0.001386	1339.92	3.2355	5.5515	949.29	50.02	87.61	557.26
310	0.001431	1396.99	3.3342	5.8798	885.71	42.17	83.55	536.78
320	0.001487	1458.02	3.4380	6.3606	815.02	34.37	79.30	514.30
330	0.001559	1525.23	3.5503	7.1610	733.04	26.51	74.63	489.65
t_s = 330.86				**Saturation**				
Liquid	0.001566	1531.40	3.5606	7.2579	724.84	25.80	74.20	487.41
Vapour	0.0128	2662.89	5.4339	9.9072	452.50	1.2320	21.68	96.02
340	0.0140	2738.92	5.5589	7.1954	478.10	1.2533	22.02	87.21

Table 4 Single-Phase Region (Regions 1 to 3 of IAPWS-IF97) – Continuation
Homogenes Zustandsgebiet (Bereiche 1 bis 3 der IAPWS-IF97) – Fortsetzung

				p = 13 MPa				
t	v	h	s	c_p	w	κ	$\eta \times 10^6$	$\lambda \times 10^3$
[°C]	[m^3 kg^{-1}]	[kJ kg^{-1}]	[kJ kg^{-1} K^{-1}]	[kJ kg^{-1} K^{-1}]	[m s^{-1}]		[kg m^{-1} s^{-1}]	[W m^{-1} K^{-1}]
350	0.0151	2803.61	5.6635	5.8727	498.14	1.2625	22.44	82.20
360	0.0161	2858.09	5.7503	5.0817	514.59	1.2688	22.89	79.15
370	0.0169	2906.10	5.8255	4.5530	528.76	1.2734	23.34	77.18
380	0.0177	2949.64	5.8927	4.1742	541.28	1.2766	23.79	75.92
390	0.0184	2989.90	5.9539	3.8893	552.57	1.2789	24.25	75.12
400	0.0190	3027.64	6.0104	3.6672	562.91	1.2806	24.70	74.68
410	0.0197	3063.39	6.0631	3.4896	572.49	1.2817	25.15	74.50
420	0.0203	3097.54	6.1127	3.3448	581.45	1.2825	25.60	74.54
430	0.0209	3130.37	6.1598	3.2248	589.88	1.2830	26.05	74.74
440	0.0214	3162.10	6.2046	3.1242	597.87	1.2832	26.50	75.08
450	0.0220	3192.90	6.2475	3.0391	605.48	1.2833	26.94	75.55
460	0.0225	3222.92	6.2887	2.9664	612.75	1.2832	27.38	76.11
470	0.0230	3252.27	6.3285	2.9040	619.74	1.2829	27.82	76.76
480	0.0235	3281.03	6.3669	2.8501	626.46	1.2826	28.25	77.49
490	0.0240	3309.29	6.4042	2.8033	632.95	1.2822	28.69	78.29
500	0.0245	3337.12	6.4404	2.7626	639.23	1.2817	29.12	79.15
510	0.0250	3364.56	6.4757	2.7270	645.33	1.2811	29.55	80.06
520	0.0255	3391.67	6.5101	2.6959	651.25	1.2805	29.97	81.02
530	0.0259	3418.49	6.5437	2.6685	657.02	1.2798	30.40	82.03
540	0.0264	3445.05	6.5765	2.6445	662.64	1.2792	30.82	83.07
550	0.0269	3471.39	6.6087	2.6233	668.13	1.2784	31.24	84.14
560	0.0273	3497.53	6.6403	2.6046	673.50	1.2777	31.65	85.25
570	0.0278	3523.49	6.6713	2.5880	678.76	1.2769	32.07	86.38
580	0.0282	3549.29	6.7017	2.5734	683.92	1.2762	32.48	87.54
590	0.0286	3574.96	6.7316	2.5606	688.98	1.2754	32.89	88.71
600	0.0291	3600.51	6.7610	2.5492	693.94	1.2746	33.30	89.91
610	0.0295	3625.95	6.7900	2.5392	698.82	1.2738	33.71	91.12
620	0.0299	3651.30	6.8185	2.5304	703.62	1.2729	34.11	92.35
630	0.0303	3676.56	6.8467	2.5227	708.35	1.2721	34.51	93.58
640	0.0308	3701.76	6.8744	2.5160	713.00	1.2713	34.91	94.83
650	0.0312	3726.89	6.9018	2.5102	717.59	1.2705	35.31	96.09
660	0.0316	3751.96	6.9288	2.5051	722.11	1.2697	35.71	97.35
670	0.0320	3776.99	6.9555	2.5009	726.57	1.2689	36.10	98.62
680	0.0324	3801.98	6.9818	2.4973	730.97	1.2680	36.50	99.89
690	0.0328	3826.94	7.0079	2.4943	735.32	1.2672	36.89	101.17
700	0.0332	3851.87	7.0336	2.4919	739.61	1.2664	37.28	102.45
720	0.0340	3901.67	7.0843	2.4886	748.05	1.2648	38.05	105.03
740	0.0348	3951.42	7.1339	2.4871	756.30	1.2632	38.82	107.61
760	0.0356	4001.16	7.1825	2.4872	764.37	1.2616	39.58	110.19
780	0.0364	4050.92	7.2302	2.4886	772.29	1.2600	40.33	112.78
800	0.0372	4100.72	7.2771	2.4913	780.05	1.2584	41.08	115.37

Table 4 Single-Phase Region (Regions 1 to 3 of IAPWS-IF97) – Continuation
Homogenes Zustandsgebiet (Bereiche 1 bis 3 der IAPWS-IF97) – Fortsetzung

p = 14 MPa								
t [°C]	v [$m^3\,kg^{-1}$]	h [$kJ\,kg^{-1}$]	s [$kJ\,kg^{-1}\,K^{-1}$]	c_p [$kJ\,kg^{-1}\,K^{-1}$]	w [$m\,s^{-1}$]	κ	$\eta \times 10^6$ [$kg\,m^{-1}\,s^{-1}$]	$\lambda \times 10^3$ [$W\,m^{-1}\,K^{-1}$]
0	0.000993	14.07	0.0004	4.1545	1424.65	145.96	1759.07	568.76
5	0.000993	34.83	0.0757	4.1488	1448.46	150.86	1496.74	577.71
10	0.000994	55.56	0.1496	4.1458	1469.80	155.27	1291.92	586.81
15	0.000995	76.29	0.2222	4.1444	1488.80	159.19	1128.69	595.91
20	0.000996	97.01	0.2935	4.1438	1505.61	162.64	996.30	604.85
25	0.000997	117.73	0.3636	4.1438	1520.38	165.64	887.31	613.51
30	0.000998	138.45	0.4325	4.1441	1533.27	168.22	796.39	621.79
35	0.001000	159.17	0.5003	4.1448	1544.40	170.38	719.67	629.62
40	0.001002	179.90	0.5670	4.1458	1553.89	172.16	654.29	636.96
45	0.001004	200.63	0.6327	4.1471	1561.87	173.58	598.09	643.77
50	0.001006	221.37	0.6973	4.1487	1568.43	174.66	549.38	650.05
60	0.001011	262.88	0.8238	4.1530	1577.67	175.87	469.58	661.02
70	0.001016	304.43	0.9468	4.1588	1582.26	175.93	407.41	669.96
80	0.001023	346.06	1.0663	4.1662	1582.73	174.98	358.02	677.07
90	0.001029	387.76	1.1828	4.1752	1579.57	173.15	318.13	682.58
100	0.001037	429.57	1.2963	4.1859	1573.14	170.54	285.46	686.67
110	0.001044	471.49	1.4072	4.1983	1563.77	167.25	258.37	689.54
120	0.001053	513.54	1.5155	4.2127	1551.73	163.37	235.66	691.33
130	0.001062	555.75	1.6215	4.2290	1537.22	158.97	216.43	692.14
140	0.001071	598.13	1.7254	4.2476	1520.42	154.11	200.00	692.05
150	0.001082	640.71	1.8272	4.2688	1501.45	148.87	185.85	691.11
160	0.001093	683.51	1.9272	4.2928	1480.39	143.27	173.56	689.35
170	0.001104	726.57	2.0255	4.3200	1457.32	137.37	162.80	686.77
180	0.001117	769.93	2.1222	4.3511	1432.25	131.20	153.32	683.37
190	0.001130	813.61	2.2176	4.3864	1405.20	124.80	144.90	679.13
200	0.001144	857.67	2.3117	4.4269	1376.16	118.20	137.38	674.02
210	0.001160	902.17	2.4048	4.4732	1345.11	111.44	130.60	668.01
220	0.001176	947.16	2.4969	4.5265	1312.02	104.53	124.45	661.03
230	0.001194	992.72	2.5884	4.5880	1276.84	97.52	118.83	653.02
240	0.001213	1038.95	2.6794	4.6595	1239.50	90.43	113.66	643.91
250	0.001235	1085.95	2.7701	4.7431	1199.90	83.30	108.84	633.58
260	0.001258	1133.86	2.8608	4.8419	1157.88	76.14	104.33	621.92
270	0.001283	1182.85	2.9518	4.9600	1113.18	68.96	100.04	608.80
280	0.001312	1233.15	3.0436	5.1037	1065.39	61.79	95.93	594.08
290	0.001345	1285.05	3.1366	5.2826	1013.90	54.61	91.93	577.63
300	0.001382	1338.97	3.2315	5.5128	957.73	47.41	87.97	559.37
310	0.001426	1395.56	3.3293	5.8226	895.59	40.18	83.97	539.22
320	0.001480	1455.87	3.4319	6.2683	826.50	32.98	79.79	517.15
330	0.001549	1521.80	3.5421	6.9830	748.38	25.83	75.26	493.08
t_s = 336.67				**Saturation**				
Liquid	0.001610	1570.88	3.6230	7.8117	682.84	20.69	71.84	475.45
Vapour	0.0115	2638.09	5.3730	11.2598	445.18	1.2322	22.21	104.81
340	0.0120	2672.38	5.4291	9.4878	457.16	1.2442	22.29	99.19

Table 4 Single-Phase Region (Regions 1 to 3 of IAPWS-IF97) – Continuation
Homogenes Zustandsgebiet (Bereiche 1 bis 3 der IAPWS-IF97) – Fortsetzung

				p = 14 MPa				
t	v	h	s	c_p	w	κ	$\eta \times 10^6$	$\lambda \times 10^3$
[°C]	[m^3 kg^{-1}]	[kJ kg^{-1}]	[kJ kg^{-1} K^{-1}]	[kJ kg^{-1} K^{-1}]	[m s^{-1}]		[kg m^{-1} s^{-1}]	[W m^{-1} K^{-1}]
350	0.0132	2752.92	5.5595	7.0059	482.90	1.2588	22.64	89.88
360	0.0142	2816.39	5.6605	5.7978	502.28	1.2665	23.05	84.84
370	0.0151	2870.38	5.7452	5.0539	518.43	1.2721	23.47	81.72
380	0.0159	2918.26	5.8190	4.5496	532.40	1.2761	23.91	79.68
390	0.0166	2961.83	5.8853	4.1840	544.80	1.2789	24.36	78.35
400	0.0172	3002.23	5.9457	3.9064	556.02	1.2808	24.80	77.50
410	0.0179	3040.16	6.0017	3.6884	566.31	1.2822	25.25	77.01
420	0.0185	3076.14	6.0539	3.5132	575.85	1.2831	25.70	76.79
430	0.0190	3110.53	6.1032	3.3697	584.79	1.2837	26.14	76.79
440	0.0196	3143.61	6.1499	3.2504	593.21	1.2840	26.59	76.97
450	0.0201	3175.60	6.1945	3.1500	601.20	1.2841	27.03	77.29
460	0.0206	3206.66	6.2371	3.0648	608.81	1.2841	27.47	77.74
470	0.0211	3236.94	6.2782	2.9920	616.09	1.2839	27.90	78.29
480	0.0216	3266.54	6.3177	2.9293	623.08	1.2836	28.34	78.93
490	0.0221	3295.55	6.3560	2.8750	629.81	1.2832	28.77	79.66
500	0.0225	3324.06	6.3931	2.8278	636.31	1.2828	29.20	80.46
510	0.0230	3352.13	6.4292	2.7867	642.60	1.2822	29.63	81.31
520	0.0235	3379.81	6.4643	2.7506	648.71	1.2816	30.05	82.23
530	0.0239	3407.15	6.4986	2.7190	654.64	1.2810	30.48	83.19
540	0.0243	3434.20	6.5320	2.6911	660.42	1.2803	30.90	84.19
550	0.0248	3460.99	6.5648	2.6666	666.05	1.2796	31.32	85.24
560	0.0252	3487.54	6.5968	2.6449	671.55	1.2789	31.73	86.32
570	0.0256	3513.89	6.6283	2.6256	676.93	1.2781	32.15	87.43
580	0.0260	3540.06	6.6591	2.6086	682.20	1.2774	32.56	88.56
590	0.0264	3566.07	6.6894	2.5936	687.37	1.2766	32.97	89.72
600	0.0268	3591.94	6.7192	2.5803	692.43	1.2758	33.38	90.90
610	0.0272	3617.68	6.7486	2.5685	697.40	1.2750	33.78	92.09
620	0.0276	3643.31	6.7774	2.5580	702.29	1.2742	34.19	93.30
630	0.0280	3668.85	6.8058	2.5488	707.10	1.2734	34.59	94.53
640	0.0284	3694.29	6.8339	2.5407	711.83	1.2725	34.99	95.76
650	0.0288	3719.67	6.8615	2.5336	716.49	1.2717	35.39	97.01
660	0.0292	3744.97	6.8888	2.5274	721.08	1.2709	35.78	98.26
670	0.0296	3770.22	6.9157	2.5221	725.61	1.2701	36.18	99.52
680	0.0300	3795.41	6.9422	2.5175	730.07	1.2693	36.57	100.79
690	0.0304	3820.57	6.9685	2.5135	734.48	1.2685	36.96	102.06
700	0.0308	3845.69	6.9944	2.5103	738.83	1.2676	37.35	103.33
720	0.0315	3895.84	7.0455	2.5054	747.37	1.2660	38.12	105.88
740	0.0323	3945.91	7.0954	2.5025	755.72	1.2644	38.89	108.45
760	0.0330	3995.95	7.1443	2.5013	763.89	1.2628	39.65	111.01
780	0.0337	4045.98	7.1922	2.5017	771.88	1.2612	40.40	113.58
800	0.0345	4096.02	7.2393	2.5033	779.72	1.2596	41.14	116.14

Table 4 **Single-Phase Region** (Regions 1 to 3 of IAPWS-IF97) – Continuation
Homogenes Zustandsgebiet (Bereiche 1 bis 3 der IAPWS-IF97) – Fortsetzung

				p = 15 MPa				
t [°C]	v [m^3 kg^{-1}]	h [kJ kg^{-1}]	s [kJ kg^{-1} K^{-1}]	c_p [kJ kg^{-1} K^{-1}]	w [m s^{-1}]	κ	$\eta \times 10^6$ [kg m^{-1} s^{-1}]	$\lambda \times 10^3$ [W m^{-1} K^{-1}]
0	0.000993	15.07	0.0004	4.1501	1426.28	136.61	1756.92	569.31
5	0.000993	35.81	0.0757	4.1451	1450.09	141.19	1495.34	578.22
10	0.000993	56.52	0.1495	4.1425	1471.42	145.31	1291.01	587.30
15	0.000994	77.23	0.2220	4.1414	1490.41	148.97	1128.12	596.38
20	0.000995	97.94	0.2932	4.1410	1507.22	152.19	995.97	605.31
25	0.000996	118.64	0.3633	4.1412	1521.99	155.00	887.15	613.96
30	0.000998	139.35	0.4322	4.1417	1534.88	157.40	796.35	622.24
35	0.000999	160.06	0.4999	4.1424	1546.02	159.43	719.72	630.08
40	0.001001	180.78	0.5666	4.1435	1555.53	161.09	654.41	637.42
45	0.001003	201.50	0.6322	4.1449	1563.52	162.42	598.24	644.23
50	0.001006	222.23	0.6969	4.1466	1570.10	163.43	549.58	650.52
60	0.001011	263.71	0.8233	4.1509	1579.38	164.57	469.82	661.49
70	0.001016	305.25	0.9462	4.1568	1584.02	164.64	407.67	670.45
80	0.001022	346.85	1.0657	4.1642	1584.57	163.77	358.29	677.58
90	0.001029	388.54	1.1821	4.1731	1581.47	162.07	318.40	683.10
100	0.001036	430.32	1.2956	4.1838	1575.13	159.65	285.73	687.22
110	0.001044	472.22	1.4064	4.1961	1565.85	156.59	258.63	690.11
120	0.001052	514.25	1.5147	4.2103	1553.90	152.98	235.92	691.92
130	0.001061	556.43	1.6206	4.2266	1539.50	148.89	216.69	692.75
140	0.001071	598.79	1.7244	4.2450	1522.82	144.37	200.26	692.69
150	0.001081	641.34	1.8262	4.2659	1503.98	139.49	186.10	691.78
160	0.001092	684.12	1.9261	4.2896	1483.06	134.28	173.80	690.05
170	0.001104	727.14	2.0243	4.3166	1460.14	128.79	163.04	687.50
180	0.001116	770.46	2.1209	4.3472	1435.25	123.05	153.56	684.13
190	0.001129	814.10	2.2162	4.3821	1408.39	117.10	145.14	679.93
200	0.001143	858.12	2.3102	4.4219	1379.56	110.96	137.62	674.87
210	0.001159	902.56	2.4032	4.4675	1348.74	104.66	130.84	668.90
220	0.001175	947.49	2.4952	4.5200	1315.90	98.23	124.70	661.98
230	0.001193	992.99	2.5865	4.5804	1281.00	91.71	119.08	654.03
240	0.001212	1039.13	2.6774	4.6506	1243.97	85.11	113.91	645.00
250	0.001233	1086.04	2.7679	4.7325	1204.73	78.47	109.10	634.76
260	0.001256	1133.83	2.8584	4.8289	1163.13	71.81	104.60	623.22
270	0.001281	1182.68	2.9491	4.9440	1118.92	65.14	100.33	610.23
280	0.001310	1232.79	3.0406	5.0833	1071.74	58.47	96.24	595.68
290	0.001342	1284.45	3.1331	5.2558	1021.04	51.81	92.26	579.45
300	0.001378	1338.06	3.2275	5.4760	965.96	45.13	88.33	561.44
310	0.001421	1394.21	3.3246	5.7692	905.25	38.44	84.37	541.60
320	0.001473	1453.85	3.4260	6.1839	837.75	31.76	80.27	519.92
330	0.001539	1518.64	3.5343	6.8289	762.35	25.17	75.86	496.36
340	0.001631	1592.27	3.6553	8.0647	666.58	18.16	70.77	470.19
t_s = 342.16				**Saturation**				
Liquid	0.001657	1610.15	3.6844	8.5252	639.77	16.47	69.50	463.92
Vapour	0.0103	2610.86	5.3108	12.9821	437.44	1.2337	22.79	115.34

Table 4 Single-Phase Region (Regions 1 to 3 of IAPWS-IF97) – Continuation
Homogenes Zustandsgebiet (Bereiche 1 bis 3 der IAPWS-IF97) – Fortsetzung

p = 15 MPa

t [°C]	v [m^3 kg^{-1}]	h [kJ kg^{-1}]	s [kJ kg^{-1} K^{-1}]	c_p [kJ kg^{-1} K^{-1}]	w [m s^{-1}]	κ	$\eta\times 10^6$ [kg m^{-1} s^{-1}]	$\lambda\times 10^3$ [W m^{-1} K^{-1}]
350	0.0115	2693.00	5.4435	8.7885	464.90	1.2551	22.94	100.79
360	0.0126	2769.56	5.5654	6.7740	488.52	1.2645	23.26	92.19
370	0.0135	2831.40	5.6624	5.6868	507.20	1.2710	23.65	87.24
380	0.0143	2884.61	5.7445	5.0003	522.93	1.2758	24.07	84.12
390	0.0150	2932.11	5.8166	4.5264	536.62	1.2791	24.49	82.05
400	0.0157	2975.55	5.8817	4.1778	548.83	1.2814	24.93	80.68
410	0.0163	3015.93	5.9412	3.9102	559.91	1.2829	25.37	79.80
420	0.0169	3053.94	5.9965	3.6985	570.10	1.2840	25.81	79.27
430	0.0174	3090.04	6.0482	3.5273	579.58	1.2847	26.25	79.03
440	0.0180	3124.58	6.0970	3.3864	588.46	1.2851	26.69	79.01
450	0.0185	3157.84	6.1433	3.2687	596.85	1.2852	27.13	79.17
460	0.0190	3190.02	6.1875	3.1695	604.81	1.2852	27.56	79.48
470	0.0195	3221.28	6.2298	3.0851	612.40	1.2851	28.00	79.91
480	0.0199	3251.76	6.2706	3.0126	619.67	1.2848	28.43	80.46
490	0.0204	3281.57	6.3099	2.9501	626.65	1.2845	28.86	81.10
500	0.0208	3310.79	6.3479	2.8960	633.37	1.2840	29.29	81.83
510	0.0213	3339.51	6.3848	2.8487	639.87	1.2835	29.72	82.62
520	0.0217	3367.79	6.4207	2.8074	646.16	1.2829	30.14	83.48
530	0.0221	3395.68	6.4556	2.7712	652.26	1.2823	30.56	84.40
540	0.0225	3423.22	6.4897	2.7393	658.20	1.2816	30.98	85.37
550	0.0229	3450.47	6.5230	2.7112	663.97	1.2809	31.40	86.38
560	0.0233	3477.46	6.5556	2.6863	669.61	1.2802	31.82	87.43
570	0.0237	3504.21	6.5875	2.6643	675.11	1.2794	32.23	88.51
580	0.0241	3530.75	6.6188	2.6447	680.50	1.2787	32.64	89.62
590	0.0245	3557.11	6.6496	2.6274	685.77	1.2779	33.05	90.76
600	0.0249	3583.31	6.6797	2.6120	690.93	1.2771	33.46	91.92
610	0.0253	3609.36	6.7094	2.5983	696.00	1.2763	33.86	93.10
620	0.0257	3635.28	6.7386	2.5862	700.98	1.2755	34.27	94.30
630	0.0261	3661.09	6.7673	2.5754	705.87	1.2747	34.67	95.51
640	0.0264	3686.79	6.7956	2.5659	710.68	1.2738	35.07	96.73
650	0.0268	3712.41	6.8235	2.5575	715.41	1.2730	35.46	97.96
660	0.0272	3737.95	6.8510	2.5501	720.07	1.2722	35.86	99.20
670	0.0275	3763.41	6.8782	2.5436	724.66	1.2714	36.25	100.45
680	0.0279	3788.82	6.9050	2.5379	729.19	1.2706	36.65	101.71
690	0.0283	3814.17	6.9314	2.5330	733.66	1.2698	37.04	102.97
700	0.0286	3839.48	6.9576	2.5288	738.07	1.2689	37.42	104.23
720	0.0293	3889.99	7.0090	2.5223	746.72	1.2673	38.20	106.76
740	0.0300	3940.39	7.0592	2.5180	755.16	1.2657	38.96	109.31
760	0.0307	3990.72	7.1084	2.5156	763.41	1.2641	39.72	111.85
780	0.0314	4041.03	7.1566	2.5148	771.49	1.2625	40.47	114.40
800	0.0321	4091.33	7.2039	2.5155	779.40	1.2609	41.21	116.94

Table 4 Single-Phase Region (Regions 1 to 3 of IAPWS-IF97) – Continuation
Homogenes Zustandsgebiet (Bereiche 1 bis 3 der IAPWS-IF97) – Fortsetzung

p = 16 MPa

t [°C]	v [m^3 kg^{-1}]	h [kJ kg^{-1}]	s [kJ kg^{-1} K^{-1}]	c_p [kJ kg^{-1} K^{-1}]	w [m s^{-1}]	κ	$\eta \times 10^6$ [kg m^{-1} s^{-1}]	$\lambda \times 10^3$ [W m^{-1} K^{-1}]
0	0.000992	16.07	0.0005	4.1458	1427.92	128.43	1754.80	569.86
5	0.000992	36.78	0.0756	4.1413	1451.73	132.73	1493.95	578.73
10	0.000993	57.48	0.1494	4.1392	1473.05	136.59	1290.11	587.79
15	0.000994	78.18	0.2218	4.1384	1492.04	140.03	1127.56	596.85
20	0.000995	98.87	0.2930	4.1382	1508.84	143.05	995.65	605.77
25	0.000996	119.56	0.3630	4.1386	1523.61	145.68	886.99	614.42
30	0.000997	140.25	0.4318	4.1392	1536.50	147.94	796.31	622.69
35	0.000999	160.95	0.4996	4.1401	1547.64	149.84	719.77	630.53
40	0.001001	181.66	0.5662	4.1413	1557.16	151.41	654.52	637.87
45	0.001003	202.37	0.6318	4.1427	1565.17	152.66	598.40	644.69
50	0.001005	223.08	0.6964	4.1445	1571.77	153.61	549.77	650.98
60	0.001010	264.55	0.8228	4.1489	1581.09	154.68	470.05	661.97
70	0.001016	306.07	0.9456	4.1548	1585.79	154.76	407.92	670.94
80	0.001022	347.65	1.0650	4.1621	1586.40	153.95	358.55	678.09
90	0.001028	389.31	1.1814	4.1711	1583.38	152.37	318.67	683.62
100	0.001036	431.08	1.2948	4.1817	1577.11	150.11	286.00	687.76
110	0.001043	472.95	1.4056	4.1939	1567.92	147.26	258.90	690.67
120	0.001052	514.96	1.5138	4.2080	1556.07	143.89	236.18	692.50
130	0.001061	557.12	1.6197	4.2241	1541.78	140.07	216.94	693.36
140	0.001070	599.45	1.7234	4.2423	1525.21	135.85	200.51	693.33
150	0.001080	641.97	1.8251	4.2630	1506.49	131.29	186.35	692.44
160	0.001091	684.72	1.9250	4.2865	1485.72	126.42	174.05	690.74
170	0.001103	727.72	2.0231	4.3131	1462.95	121.29	163.29	688.22
180	0.001115	770.99	2.1197	4.3434	1438.23	115.92	153.80	684.89
190	0.001128	814.60	2.2148	4.3778	1411.56	110.36	145.38	680.73
200	0.001143	858.57	2.3088	4.4171	1382.93	104.62	137.86	675.71
210	0.001158	902.96	2.4016	4.4620	1352.34	98.73	131.08	669.78
220	0.001174	947.83	2.4935	4.5136	1319.75	92.72	124.94	662.92
230	0.001192	993.25	2.5847	4.5730	1285.12	86.62	119.33	655.04
240	0.001211	1039.32	2.6754	4.6418	1248.41	80.46	114.16	646.08
250	0.001231	1086.13	2.7657	4.7220	1209.51	74.25	109.36	635.93
260	0.001254	1133.81	2.8560	4.8163	1168.31	68.02	104.87	624.50
270	0.001279	1182.51	2.9465	4.9284	1124.58	61.79	100.61	611.65
280	0.001307	1232.45	3.0376	5.0636	1077.99	55.56	96.54	597.27
290	0.001339	1283.89	3.1297	5.2301	1028.05	49.34	92.58	581.23
300	0.001375	1337.20	3.2236	5.4411	973.99	43.13	88.69	563.47
310	0.001417	1392.93	3.3200	5.7190	914.67	36.91	84.77	543.93
320	0.001467	1451.94	3.4203	6.1063	848.79	30.69	80.74	522.63
330	0.001530	1515.71	3.5269	6.6932	775.52	24.56	76.43	499.53
340	0.001616	1587.27	3.6445	7.7439	687.82	18.29	71.55	474.23
t_s = 347.36				**Saturation**				
Liquid	0.001710	1649.67	3.7457	9.4729	597.67	13.06	67.14	452.83
Vapour	0.0093	2580.80	5.2463	15.2072	429.20	1.2369	23.44	128.06

Table 4 Single-Phase Region (Regions 1 to 3 of IAPWS-IF97) – Continuation
Homogenes Zustandsgebiet (Bereiche 1 bis 3 der IAPWS-IF97) – Fortsetzung

p = 16 MPa

t [°C]	v [m^3 kg^{-1}]	h [kJ kg^{-1}]	s [kJ kg^{-1} K^{-1}]	c_p [kJ kg^{-1} K^{-1}]	w [m s^{-1}]	κ	$\eta \times 10^6$ [kg m^{-1} s^{-1}]	$\lambda \times 10^3$ [W m^{-1} K^{-1}]
350	0.0098	2616.99	5.3045	12.4134	441.49	1.2475	23.41	118.93
360	0.0111	2715.63	5.4616	8.1928	472.82	1.2634	23.57	102.12
370	0.0120	2788.30	5.5755	6.5172	494.85	1.2705	23.89	94.18
380	0.0129	2848.27	5.6680	5.5538	512.76	1.2760	24.26	89.43
390	0.0136	2900.49	5.7474	4.9289	527.98	1.2798	24.66	86.36
400	0.0143	2947.46	5.8177	4.4882	541.32	1.2824	25.08	84.30
410	0.0149	2990.62	5.8814	4.1588	553.28	1.2841	25.51	82.92
420	0.0155	3030.88	5.9399	3.9032	564.18	1.2853	25.94	82.02
430	0.0160	3068.85	5.9943	3.6993	574.24	1.2860	26.37	81.48
440	0.0165	3104.99	6.0453	3.5333	583.62	1.2864	26.80	81.22
450	0.0170	3139.61	6.0935	3.3960	592.43	1.2866	27.24	81.19
460	0.0175	3172.98	6.1393	3.2809	600.76	1.2866	27.67	81.34
470	0.0180	3205.29	6.1831	3.1835	608.68	1.2865	28.10	81.64
480	0.0185	3236.70	6.2251	3.1004	616.23	1.2862	28.53	82.08
490	0.0189	3267.34	6.2655	3.0289	623.47	1.2859	28.96	82.63
500	0.0193	3297.31	6.3045	2.9671	630.42	1.2854	29.39	83.27
510	0.0197	3326.71	6.3423	2.9134	637.13	1.2849	29.81	84.00
520	0.0202	3355.60	6.3790	2.8664	643.61	1.2844	30.23	84.80
530	0.0206	3384.05	6.4146	2.8253	649.89	1.2837	30.65	85.67
540	0.0210	3412.12	6.4494	2.7891	655.98	1.2831	31.07	86.59
550	0.0214	3439.85	6.4832	2.7572	661.91	1.2824	31.49	87.56
560	0.0217	3467.28	6.5164	2.7290	667.68	1.2816	31.90	88.58
570	0.0221	3494.44	6.5488	2.7039	673.31	1.2809	32.32	89.63
580	0.0225	3521.37	6.5805	2.6817	678.81	1.2801	32.73	90.72
590	0.0229	3548.08	6.6117	2.6620	684.18	1.2793	33.14	91.83
600	0.0232	3574.61	6.6422	2.6444	689.45	1.2785	33.54	92.98
610	0.0236	3600.98	6.6722	2.6288	694.61	1.2777	33.95	94.14
620	0.0240	3627.20	6.7018	2.6149	699.68	1.2769	34.35	95.32
630	0.0243	3653.28	6.7308	2.6025	704.65	1.2761	34.75	96.51
640	0.0247	3679.25	6.7594	2.5915	709.54	1.2752	35.15	97.72
650	0.0250	3705.11	6.7876	2.5817	714.35	1.2744	35.55	98.94
660	0.0254	3730.89	6.8153	2.5731	719.08	1.2736	35.94	100.17
670	0.0257	3756.58	6.8427	2.5654	723.74	1.2728	36.33	101.41
680	0.0261	3782.20	6.8697	2.5587	728.33	1.2719	36.73	102.65
690	0.0264	3807.76	6.8964	2.5528	732.85	1.2711	37.12	103.90
700	0.0267	3833.26	6.9228	2.5477	737.32	1.2703	37.50	105.15
720	0.0274	3884.12	6.9745	2.5395	746.07	1.2687	38.27	107.67
740	0.0281	3934.85	7.0251	2.5337	754.61	1.2670	39.04	110.19
760	0.0287	3985.49	7.0746	2.5300	762.95	1.2654	39.79	112.71
780	0.0294	4036.06	7.1231	2.5281	771.11	1.2638	40.54	115.24
800	0.0301	4086.62	7.1706	2.5277	779.09	1.2622	41.28	117.76

Table 4 **Single-Phase Region** (Regions 1 to 3 of IAPWS-IF97) – Continuation
Homogenes Zustandsgebiet (Bereiche 1 bis 3 der IAPWS-IF97) – Fortsetzung

				p = 17 MPa				
t	v	h	s	c_p	w	κ	$\eta \times 10^6$	$\lambda \times 10^3$
[°C]	[$m^3\ kg^{-1}$]	[$kJ\ kg^{-1}$]	[$kJ\ kg^{-1}\ K^{-1}$]	[$kJ\ kg^{-1}\ K^{-1}$]	[$m\ s^{-1}$]		[$kg\ m^{-1}\ s^{-1}$]	[$W\ m^{-1}\ K^{-1}$]
0	0.000992	17.06	0.0005	4.1416	1429.56	121.21	1752.71	570.41
5	0.000992	37.76	0.0756	4.1376	1453.37	125.26	1492.57	579.24
10	0.000992	58.44	0.1493	4.1359	1474.68	128.90	1289.22	588.27
15	0.000993	79.12	0.2216	4.1354	1493.66	132.13	1127.01	597.32
20	0.000994	99.79	0.2928	4.1355	1510.45	134.98	995.33	606.23
25	0.000995	120.47	0.3627	4.1360	1525.22	137.46	886.84	614.87
30	0.000997	141.15	0.4315	4.1368	1538.11	139.59	796.28	623.15
35	0.000999	161.84	0.4992	4.1378	1549.26	141.38	719.82	630.98
40	0.001001	182.53	0.5658	4.1391	1558.79	142.86	654.64	638.33
45	0.001003	203.23	0.6314	4.1406	1566.82	144.04	598.57	645.16
50	0.001005	223.94	0.6960	4.1424	1573.43	144.94	549.97	651.45
60	0.001010	265.38	0.8223	4.1469	1582.81	145.96	470.29	662.45
70	0.001015	306.88	0.9450	4.1528	1587.56	146.04	408.18	671.43
80	0.001021	348.45	1.0644	4.1601	1588.23	145.30	358.82	678.59
90	0.001028	390.09	1.1807	4.1691	1585.28	143.82	318.94	684.15
100	0.001035	431.83	1.2941	4.1796	1579.09	141.71	286.26	688.30
110	0.001043	473.69	1.4048	4.1918	1569.99	139.03	259.16	691.23
120	0.001051	515.67	1.5129	4.2057	1558.24	135.87	236.44	693.09
130	0.001060	557.81	1.6188	4.2217	1544.05	132.29	217.20	693.97
140	0.001070	600.11	1.7224	4.2397	1527.59	128.33	200.76	693.96
150	0.001080	642.61	1.8241	4.2602	1509.00	124.05	186.59	693.11
160	0.001091	685.33	1.9238	4.2834	1488.37	119.48	174.29	691.43
170	0.001102	728.29	2.0219	4.3097	1465.75	114.66	163.53	688.95
180	0.001114	771.53	2.1184	4.3396	1441.19	109.63	154.04	685.65
190	0.001128	815.09	2.2135	4.3735	1414.70	104.40	145.62	681.52
200	0.001142	859.02	2.3073	4.4123	1386.28	99.02	138.10	676.54
210	0.001157	903.36	2.4000	4.4565	1355.91	93.49	131.32	670.67
220	0.001173	948.17	2.4918	4.5073	1323.57	87.86	125.18	663.86
230	0.001190	993.53	2.5829	4.5657	1289.21	82.13	119.57	656.04
240	0.001209	1039.51	2.6734	4.6332	1252.80	76.34	114.41	647.15
250	0.001230	1086.23	2.7635	4.7118	1214.24	70.52	109.62	637.10
260	0.001252	1133.80	2.8536	4.8040	1173.43	64.67	105.14	625.77
270	0.001277	1182.36	2.9439	4.9132	1130.17	58.83	100.89	613.06
280	0.001305	1232.13	3.0346	5.0445	1084.14	52.99	96.83	598.84
290	0.001336	1283.35	3.1264	5.2054	1034.92	47.17	92.90	583.00
300	0.001371	1336.38	3.2197	5.4078	981.82	41.36	89.04	565.46
310	0.001412	1391.71	3.3154	5.6719	923.83	35.55	85.16	546.22
320	0.001461	1450.14	3.4148	6.0346	859.60	29.75	81.19	525.26
330	0.001522	1512.98	3.5198	6.5721	788.20	24.01	76.98	502.60
340	0.001603	1582.79	3.6346	7.4849	706.04	18.29	72.28	478.01
350	0.001727	1666.59	3.7701	9.6869	591.09	11.90	66.44	449.63

Table 4 Single-Phase Region (Regions 1 to 3 of IAPWS-IF97) – Continuation
Homogenes Zustandsgebiet (Bereiche 1 bis 3 der IAPWS-IF97) – Fortsetzung

p = 17 MPa									
t [°C]	v [$m^3\,kg^{-1}$]	h [$kJ\,kg^{-1}$]	s [$kJ\,kg^{-1}\,K^{-1}$]	c_p [$kJ\,kg^{-1}\,K^{-1}$]	w [$m\,s^{-1}$]	κ	$\eta \times 10^6$ [$kg\,m^{-1}\,s^{-1}$]	$\lambda \times 10^3$ [$W\,m^{-1}\,K^{-1}$]
t_s = 352.29				Saturation				
Liquid	0.001769	1690.04	3.8077	10.8176	556.15	10.2832	64.72	442.29
Vapour	0.0084	2547.44	5.1785	18.3060	420.27	1.2414	24.18	143.92
360	0.0096	2650.94	5.3431	10.5468	454.04	1.2629	24.02	116.71
370	0.0107	2739.76	5.4823	7.6571	481.09	1.2710	24.20	103.18
380	0.0116	2808.65	5.5887	6.2541	501.74	1.2768	24.50	95.92
390	0.0124	2866.66	5.6768	5.4104	518.80	1.2811	24.87	91.42
400	0.0130	2917.78	5.7533	4.8463	533.46	1.2840	25.26	88.45
410	0.0137	2964.10	5.8217	4.4391	546.41	1.2858	25.67	86.43
420	0.0142	3006.88	5.8838	4.1301	558.09	1.2870	26.08	85.06
430	0.0148	3046.93	5.9412	3.8874	568.78	1.2877	26.51	84.16
440	0.0153	3084.79	5.9947	3.6923	578.69	1.2881	26.93	83.62
450	0.0158	3120.89	6.0449	3.5325	587.95	1.2883	27.36	83.36
460	0.0163	3155.53	6.0925	3.3996	596.66	1.2883	27.79	83.33
470	0.0167	3188.95	6.1378	3.2878	604.92	1.2882	28.21	83.49
480	0.0172	3221.34	6.1811	3.1928	612.77	1.2879	28.64	83.79
490	0.0176	3252.85	6.2227	3.1115	620.27	1.2875	29.06	84.24
500	0.0180	3283.61	6.2627	3.0414	627.47	1.2871	29.49	84.79
510	0.0184	3313.71	6.3014	2.9806	634.39	1.2866	29.91	85.44
520	0.0188	3343.25	6.3389	2.9276	641.06	1.2860	30.33	86.18
530	0.0192	3372.29	6.3752	2.8812	647.52	1.2853	30.75	86.99
540	0.0196	3400.89	6.4106	2.8405	653.77	1.2847	31.17	87.86
550	0.0199	3429.11	6.4451	2.8046	659.85	1.2839	31.58	88.79
560	0.0203	3457.00	6.4788	2.7728	665.76	1.2832	32.00	89.77
570	0.0207	3484.58	6.5117	2.7446	671.51	1.2824	32.41	90.80
580	0.0210	3511.90	6.5439	2.7196	677.13	1.2816	32.82	91.86
590	0.0214	3538.98	6.5755	2.6974	682.61	1.2808	33.23	92.95
600	0.0218	3565.86	6.6064	2.6776	687.98	1.2800	33.63	94.07
610	0.0221	3592.54	6.6368	2.6599	693.24	1.2792	34.04	95.21
620	0.0224	3619.06	6.6667	2.6441	698.39	1.2784	34.44	96.37
630	0.0228	3645.43	6.6961	2.6301	703.45	1.2775	34.84	97.55
640	0.0231	3671.67	6.7249	2.6175	708.42	1.2767	35.23	98.75
650	0.0235	3697.79	6.7534	2.6063	713.30	1.2759	35.63	99.95
660	0.0238	3723.80	6.7814	2.5964	718.10	1.2750	36.03	101.17
670	0.0241	3749.72	6.8090	2.5875	722.82	1.2742	36.42	102.40
680	0.0244	3775.55	6.8363	2.5797	727.48	1.2734	36.81	103.63
690	0.0248	3801.31	6.8632	2.5728	732.06	1.2725	37.20	104.86
700	0.0251	3827.01	6.8897	2.5667	736.58	1.2717	37.58	106.11
720	0.0257	3878.24	6.9418	2.5568	745.44	1.2701	38.35	108.60
740	0.0264	3929.30	6.9927	2.5495	754.08	1.2684	39.11	111.10
760	0.0270	3980.24	7.0425	2.5445	762.50	1.2668	39.87	113.60
780	0.0276	4031.09	7.0913	2.5414	770.74	1.2652	40.62	116.10
800	0.0282	4081.90	7.1391	2.5401	778.80	1.2636	41.36	118.60

Table 4 Single-Phase Region (Regions 1 to 3 of IAPWS-IF97) – Continuation
Homogenes Zustandsgebiet (Bereiche 1 bis 3 der IAPWS-IF97) – Fortsetzung

p = 18 MPa								
t [°C]	v [$m^3 kg^{-1}$]	h [$kJ kg^{-1}$]	s [$kJ kg^{-1} K^{-1}$]	c_p [$kJ kg^{-1} K^{-1}$]	w [$m s^{-1}$]	κ	$\eta \times 10^6$ [$kg m^{-1} s^{-1}$]	$\lambda \times 10^3$ [$W m^{-1} K^{-1}$]
0	0.000991	18.05	0.0005	4.1373	1431.21	114.80	1750.64	570.96
5	0.000991	38.73	0.0755	4.1340	1455.01	118.62	1491.22	579.76
10	0.000992	59.40	0.1491	4.1327	1476.32	122.06	1288.35	588.76
15	0.000993	80.06	0.2215	4.1324	1495.28	125.12	1126.46	597.79
20	0.000994	100.72	0.2926	4.1328	1512.07	127.81	995.02	606.69
25	0.000995	121.39	0.3625	4.1335	1526.84	130.16	886.69	615.33
30	0.000997	142.06	0.4312	4.1344	1539.73	132.17	796.25	623.60
35	0.000998	162.73	0.4989	4.1355	1550.89	133.86	719.88	631.44
40	0.001000	183.41	0.5654	4.1369	1560.43	135.26	654.76	638.79
45	0.001002	204.10	0.6310	4.1384	1568.47	136.38	598.73	645.62
50	0.001004	224.80	0.6955	4.1403	1575.10	137.24	550.16	651.91
60	0.001009	266.22	0.8218	4.1449	1584.52	138.21	470.53	662.92
70	0.001015	307.70	0.9444	4.1508	1589.32	138.30	408.44	671.92
80	0.001021	349.24	1.0638	4.1582	1590.06	137.60	359.09	679.10
90	0.001027	390.87	1.1800	4.1670	1587.18	136.22	319.21	684.67
100	0.001035	432.59	1.2933	4.1775	1581.07	134.23	286.53	688.84
110	0.001042	474.42	1.4040	4.1896	1572.06	131.72	259.43	691.80
120	0.001051	516.39	1.5121	4.2035	1560.40	128.74	236.70	693.67
130	0.001060	558.50	1.6178	4.2192	1546.31	125.37	217.45	694.58
140	0.001069	600.78	1.7214	4.2371	1529.97	121.64	201.01	694.60
150	0.001079	643.25	1.8230	4.2574	1511.50	117.61	186.84	693.77
160	0.001090	685.93	1.9227	4.2803	1491.00	113.31	174.53	692.13
170	0.001101	728.86	2.0207	4.3063	1468.53	108.77	163.77	689.67
180	0.001114	772.07	2.1171	4.3358	1444.13	104.03	154.28	686.41
190	0.001127	815.59	2.2121	4.3693	1417.82	99.11	145.86	682.32
200	0.001141	859.47	2.3058	4.4075	1389.60	94.04	138.33	677.38
210	0.001156	903.76	2.3985	4.4511	1359.45	88.84	131.56	671.55
220	0.001172	948.52	2.4902	4.5011	1327.35	83.53	125.42	664.79
230	0.001189	993.81	2.5811	4.5585	1293.26	78.14	119.82	657.04
240	0.001208	1039.72	2.6714	4.6248	1257.14	72.68	114.66	648.22
250	0.001228	1086.34	2.7614	4.7018	1218.92	67.20	109.88	638.26
260	0.001251	1133.80	2.8513	4.7919	1178.50	61.69	105.40	627.04
270	0.001275	1182.23	2.9413	4.8984	1135.68	56.19	101.17	614.45
280	0.001302	1231.83	3.0317	5.0260	1090.21	50.70	97.13	600.38
290	0.001333	1282.84	3.1231	5.1816	1041.67	45.22	93.22	584.73
300	0.001368	1335.60	3.2160	5.3761	989.47	39.77	89.38	567.43
310	0.001408	1390.56	3.3110	5.6275	932.73	34.33	85.54	548.45
320	0.001456	1448.44	3.4095	5.9683	870.13	28.90	81.63	527.83
330	0.001514	1510.43	3.5131	6.4631	800.55	23.51	77.50	505.59
340	0.001591	1578.71	3.6253	7.2691	722.19	18.21	72.96	481.58
350	0.001703	1658.65	3.7546	8.9989	619.19	12.51	67.51	454.52

Table 4 Single-Phase Region (Regions 1 to 3 of IAPWS-IF97) – Continuation
Homogenes Zustandsgebiet (Bereiche 1 bis 3 der IAPWS-IF97) – Fortsetzung

				p = 18 MPa				
t [°C]	v [m^3 kg^{-1}]	h [kJ kg^{-1}]	s [kJ kg^{-1} K^{-1}]	c_p [kJ kg^{-1} K^{-1}]	w [m s^{-1}]	κ	$\eta \times 10^6$ [kg m^{-1} s^{-1}]	$\lambda \times 10^3$ [W m^{-1} K^{-1}]
t_s = 356.99				**Saturation**				
Liquid	0.001839	1732.04	3.8717	12.8388	513.14	7.9524	62.17	432.20
Vapour	0.0075	2509.59	5.1056	22.9543	410.35	1.2474	25.05	164.48
360	0.0081	2566.03	5.1950	15.8202	428.85	1.2599	24.77	142.51
370	0.0095	2683.67	5.3795	9.3270	465.42	1.2733	24.63	115.41
380	0.0104	2764.89	5.5048	7.1712	489.69	1.2786	24.82	104.04
390	0.0112	2830.24	5.6041	5.9991	509.01	1.2832	25.12	97.46
400	0.0119	2886.31	5.6881	5.2645	525.21	1.2862	25.47	93.25
410	0.0125	2936.27	5.7618	4.7572	539.29	1.2881	25.85	90.41
420	0.0131	2981.89	5.8281	4.3826	551.82	1.2892	26.25	88.46
430	0.0137	3024.21	5.8887	4.0939	563.20	1.2899	26.66	87.12
440	0.0142	3063.96	5.9448	3.8649	573.67	1.2902	27.08	86.24
450	0.0147	3101.65	5.9973	3.6792	583.40	1.2903	27.49	85.72
460	0.0151	3137.66	6.0468	3.5262	592.52	1.2903	27.91	85.47
470	0.0156	3172.26	6.0936	3.3983	601.13	1.2901	28.34	85.45
480	0.0160	3205.69	6.1383	3.2902	609.29	1.2898	28.76	85.62
490	0.0164	3238.12	6.1811	3.1981	617.07	1.2894	29.18	85.94
500	0.0168	3269.69	6.2222	3.1190	624.51	1.2889	29.60	86.39
510	0.0172	3300.53	6.2618	3.0506	631.65	1.2884	30.02	86.96
520	0.0176	3330.73	6.3002	2.9911	638.52	1.2878	30.44	87.62
530	0.0180	3360.38	6.3373	2.9391	645.16	1.2871	30.86	88.37
540	0.0183	3389.54	6.3734	2.8935	651.58	1.2864	31.27	89.19
550	0.0187	3418.27	6.4085	2.8533	657.80	1.2857	31.68	90.07
560	0.0191	3446.62	6.4427	2.8179	663.85	1.2849	32.10	91.01
570	0.0194	3474.64	6.4762	2.7864	669.73	1.2841	32.51	92.00
580	0.0198	3502.36	6.5089	2.7585	675.47	1.2833	32.91	93.03
590	0.0201	3529.82	6.5408	2.7336	681.06	1.2825	33.32	94.10
600	0.0204	3557.04	6.5722	2.7114	686.53	1.2816	33.72	95.19
610	0.0208	3584.05	6.6030	2.6916	691.89	1.2808	34.13	96.31
620	0.0211	3610.88	6.6332	2.6739	697.13	1.2800	34.53	97.46
630	0.0214	3637.54	6.6629	2.6581	702.27	1.2791	34.93	98.62
640	0.0217	3664.05	6.6920	2.6440	707.31	1.2783	35.32	99.80
650	0.0221	3690.42	6.7208	2.6313	712.27	1.2774	35.72	100.99
660	0.0224	3716.68	6.7491	2.6200	717.14	1.2766	36.11	102.20
670	0.0227	3742.83	6.7769	2.6099	721.93	1.2757	36.50	103.41
680	0.0230	3768.88	6.8044	2.6010	726.64	1.2749	36.89	104.63
690	0.0233	3794.85	6.8315	2.5930	731.29	1.2740	37.28	105.85
700	0.0236	3820.74	6.8583	2.5859	735.86	1.2732	37.67	107.08
720	0.0242	3872.34	6.9107	2.5743	744.83	1.2715	38.43	109.55
740	0.0248	3923.73	6.9620	2.5655	753.56	1.2699	39.19	112.03
760	0.0254	3974.98	7.0121	2.5591	762.07	1.2683	39.95	114.50
780	0.0260	4026.11	7.0611	2.5549	770.39	1.2666	40.69	116.98
800	0.0266	4077.18	7.1091	2.5525	778.51	1.2650	41.43	119.46

Table 4 Single-Phase Region (Regions 1 to 3 of IAPWS-IF97) – Continuation
Homogenes Zustandsgebiet (Bereiche 1 bis 3 der IAPWS-IF97) – Fortsetzung

				p = 19 MPa				
t	v	h	s	c_p	w	κ	$\eta \times 10^6$	$\lambda \times 10^3$
[°C]	[$m^3\ kg^{-1}$]	[$kJ\ kg^{-1}$]	[$kJ\ kg^{-1}\ K^{-1}$]	[$kJ\ kg^{-1}\ K^{-1}$]	[$m\ s^{-1}$]		[$kg\ m^{-1}\ s^{-1}$]	[$W\ m^{-1}\ K^{-1}$]
0	0.000991	19.04	0.0005	4.1332	1432.86	109.06	1748.59	571.51
5	0.000991	39.70	0.0754	4.1304	1456.65	112.69	1489.88	580.27
10	0.000992	60.35	0.1490	4.1294	1477.95	115.95	1287.49	589.25
15	0.000992	81.00	0.2213	4.1295	1496.91	118.85	1125.93	598.26
20	0.000993	101.65	0.2923	4.1301	1513.69	121.40	994.72	607.15
25	0.000995	122.30	0.3622	4.1310	1528.46	123.62	886.55	615.78
30	0.000996	142.96	0.4309	4.1320	1541.35	125.53	796.22	624.06
35	0.000998	163.62	0.4985	4.1333	1552.52	127.14	719.94	631.89
40	0.001000	184.29	0.5650	4.1347	1562.07	128.47	654.88	639.24
45	0.001002	204.97	0.6305	4.1363	1570.12	129.53	598.90	646.08
50	0.001004	225.65	0.6951	4.1382	1576.77	130.34	550.36	652.38
60	0.001009	267.06	0.8212	4.1429	1586.23	131.27	470.77	663.40
70	0.001014	308.51	0.9439	4.1488	1591.09	131.36	408.70	672.41
80	0.001020	350.04	1.0631	4.1562	1591.89	130.72	359.35	679.60
90	0.001027	391.64	1.1793	4.1650	1589.08	129.41	319.47	685.19
100	0.001034	433.34	1.2926	4.1754	1583.05	127.54	286.80	689.39
110	0.001042	475.16	1.4032	4.1874	1574.12	125.17	259.69	692.36
120	0.001050	517.10	1.5112	4.2012	1562.55	122.37	236.96	694.26
130	0.001059	559.19	1.6169	4.2168	1548.57	119.18	217.71	695.19
140	0.001068	601.44	1.7205	4.2345	1532.34	115.66	201.26	695.23
150	0.001079	643.89	1.8220	4.2546	1513.99	111.85	187.09	694.43
160	0.001089	686.54	1.9216	4.2773	1493.62	107.79	174.78	692.82
170	0.001101	729.44	2.0195	4.3029	1471.29	103.50	164.01	690.40
180	0.001113	772.61	2.1159	4.3321	1447.06	99.02	154.52	687.17
190	0.001126	816.09	2.2108	4.3652	1420.92	94.38	146.10	683.11
200	0.001140	859.93	2.3044	4.4028	1392.90	89.58	138.57	678.21
210	0.001155	904.17	2.3969	4.4458	1362.96	84.67	131.80	672.43
220	0.001171	948.87	2.4885	4.4950	1331.10	79.65	125.66	665.72
230	0.001188	994.09	2.5793	4.5514	1297.27	74.56	120.06	658.03
240	0.001207	1039.92	2.6695	4.6165	1261.44	69.41	114.91	649.29
250	0.001227	1086.46	2.7593	4.6920	1223.55	64.22	110.14	639.41
260	0.001249	1133.81	2.8489	4.7802	1183.50	59.03	105.67	628.29
270	0.001273	1182.11	2.9387	4.8841	1141.13	53.83	101.45	615.83
280	0.001300	1231.55	3.0289	5.0081	1096.18	48.64	97.42	601.91
290	0.001330	1282.36	3.1199	5.1587	1048.30	43.48	93.53	586.44
300	0.001364	1334.85	3.2123	5.3458	996.96	38.34	89.72	569.36
310	0.001404	1389.45	3.3067	5.5856	941.38	33.23	85.92	550.64
320	0.001450	1446.83	3.4043	5.9066	880.37	28.13	82.05	530.34
330	0.001507	1508.03	3.5066	6.3642	812.62	23.07	78.01	508.49
340	0.001580	1574.97	3.6167	7.0846	737.04	18.10	73.60	485.00
350	0.001683	1651.88	3.7410	8.4936	643.78	12.96	68.46	459.03
360	0.001873	1755.11	3.9053	13.6384	501.01	7.05	61.08	427.75

Table 4 Single-Phase Region (Regions 1 to 3 of IAPWS-IF97) – Continuation
Homogenes Zustandsgebiet (Bereiche 1 bis 3 der IAPWS-IF97) – Fortsetzung

$p = 19$ MPa								
t [°C]	v [$\mathrm{m^3\,kg^{-1}}$]	h [$\mathrm{kJ\,kg^{-1}}$]	s [$\mathrm{kJ\,kg^{-1}\,K^{-1}}$]	c_p [$\mathrm{kJ\,kg^{-1}\,K^{-1}}$]	w [$\mathrm{m\,s^{-1}}$]	κ	$\eta \times 10^6$ [$\mathrm{kg\,m^{-1}\,s^{-1}}$]	$\lambda \times 10^3$ [$\mathrm{W\,m^{-1}\,K^{-1}}$]
$t_s = 361.47$				**Saturation**				
Liquid	0.001925	1776.94	3.9397	16.2346	468.76	6.0065	59.40	423.20
Vapour	0.0067	2465.52	5.0247	30.5865	398.68	1.2536	25.11	192.39
370	0.0082	2616.04	5.2606	12.1689	446.59	1.2773	25.26	133.52
380	0.0093	2715.73	5.4145	8.4188	476.36	1.2822	25.23	114.51
390	0.0102	2790.66	5.5283	6.7384	498.48	1.2862	25.43	104.78
400	0.0109	2852.77	5.6213	5.7605	516.53	1.2893	25.73	98.85
410	0.0115	2906.98	5.7013	5.1209	531.88	1.2911	26.07	94.94
420	0.0121	2955.79	5.7722	4.6649	545.37	1.2921	26.45	92.25
430	0.0127	3000.65	5.8365	4.3211	557.49	1.2926	26.84	90.39
440	0.0132	3042.47	5.8955	4.0525	568.56	1.2928	27.24	89.11
450	0.0136	3081.88	5.9504	3.8372	578.79	1.2928	27.65	88.27
460	0.0141	3119.34	6.0019	3.6613	588.34	1.2926	28.06	87.77
470	0.0145	3155.20	6.0504	3.5153	597.31	1.2924	28.47	87.55
480	0.0149	3189.72	6.0966	3.3928	605.79	1.2920	28.89	87.55
490	0.0154	3223.12	6.1406	3.2889	613.85	1.2915	29.30	87.73
500	0.0157	3255.55	6.1829	3.2000	621.54	1.2910	29.72	88.07
510	0.0161	3287.16	6.2235	3.1233	628.91	1.2904	30.14	88.54
520	0.0165	3318.05	6.2627	3.0569	635.99	1.2898	30.55	89.13
530	0.0169	3348.32	6.3006	2.9989	642.81	1.2891	30.97	89.81
540	0.0172	3378.06	6.3374	2.9482	649.39	1.2883	31.38	90.57
550	0.0176	3407.31	6.3732	2.9035	655.77	1.2876	31.79	91.40
560	0.0179	3436.14	6.4080	2.8641	661.96	1.2868	32.20	92.30
570	0.0183	3464.61	6.4419	2.8292	667.97	1.2859	32.61	93.25
580	0.0186	3492.74	6.4751	2.7982	673.82	1.2851	33.01	94.25
590	0.0189	3520.58	6.5076	2.7706	679.53	1.2843	33.42	95.28
600	0.0192	3548.16	6.5393	2.7460	685.10	1.2834	33.82	96.35
610	0.0196	3575.51	6.5705	2.7240	690.55	1.2825	34.22	97.45
620	0.0199	3602.65	6.6010	2.7043	695.88	1.2817	34.62	98.58
630	0.0202	3629.60	6.6310	2.6866	701.10	1.2808	35.02	99.72
640	0.0205	3656.39	6.6605	2.6708	706.23	1.2799	35.42	100.88
650	0.0208	3683.03	6.6895	2.6567	711.26	1.2790	35.81	102.06
660	0.0211	3709.53	6.7181	2.6440	716.20	1.2782	36.20	103.25
670	0.0214	3735.91	6.7462	2.6327	721.05	1.2773	36.59	104.45
680	0.0217	3762.18	6.7739	2.6225	725.83	1.2765	36.98	105.65
690	0.0220	3788.36	6.8013	2.6135	730.53	1.2756	37.37	106.87
700	0.0223	3814.46	6.8282	2.6054	735.16	1.2748	37.75	108.08
720	0.0229	3866.43	6.8811	2.5920	744.23	1.2731	38.52	110.53
740	0.0235	3918.16	6.9326	2.5816	753.05	1.2714	39.28	112.98
760	0.0240	3969.71	6.9830	2.5739	761.65	1.2697	40.03	115.43
780	0.0246	4021.13	7.0323	2.5684	770.04	1.2681	40.77	117.89
800	0.0252	4072.46	7.0806	2.5650	778.24	1.2664	41.51	120.33

Table 4 Single-Phase Region (Regions 1 to 3 of IAPWS-IF97) – Continuation
Homogenes Zustandsgebiet (Bereiche 1 bis 3 der IAPWS-IF97) – Fortsetzung

				p = 20 MPa				
t [°C]	v [$m^3\ kg^{-1}$]	h [$kJ\ kg^{-1}$]	s [$kJ\ kg^{-1}\ K^{-1}$]	c_p [$kJ\ kg^{-1}\ K^{-1}$]	w [$m\ s^{-1}$]	κ	$\eta \times 10^6$ [$kg\ m^{-1}\ s^{-1}$]	$\lambda \times 10^3$ [$W\ m^{-1}\ K^{-1}$]
0	0.000990	20.03	0.0005	4.1290	1434.52	103.89	1746.57	572.06
5	0.000991	40.67	0.0753	4.1268	1458.30	107.35	1488.56	580.78
10	0.000991	61.30	0.1489	4.1262	1479.59	110.45	1286.64	589.74
15	0.000992	81.94	0.2211	4.1266	1498.54	113.20	1125.40	598.73
20	0.000993	102.57	0.2921	4.1274	1515.32	115.63	994.42	607.61
25	0.000994	123.21	0.3619	4.1285	1530.08	117.74	886.41	616.24
30	0.000996	143.86	0.4306	4.1297	1542.98	119.55	796.20	624.51
35	0.000997	164.51	0.4981	4.1310	1554.14	121.09	720.00	632.35
40	0.000999	185.17	0.5646	4.1325	1563.70	122.35	655.00	639.70
45	0.001001	205.83	0.6301	4.1342	1571.77	123.36	599.06	646.54
50	0.001003	226.51	0.6946	4.1361	1578.44	124.14	550.56	652.84
60	0.001008	267.89	0.8207	4.1409	1587.94	125.03	471.01	663.87
70	0.001014	309.33	0.9433	4.1468	1592.85	125.13	408.96	672.90
80	0.001020	350.83	1.0625	4.1542	1593.72	124.52	359.62	680.11
90	0.001027	392.42	1.1786	4.1630	1590.98	123.29	319.74	685.72
100	0.001034	434.10	1.2918	4.1734	1585.03	121.52	287.06	689.93
110	0.001041	475.89	1.4024	4.1853	1576.18	119.28	259.95	692.92
120	0.001050	517.81	1.5104	4.1989	1564.70	116.63	237.22	694.84
130	0.001058	559.88	1.6160	4.2144	1550.82	113.61	217.96	695.80
140	0.001068	602.11	1.7195	4.2320	1534.69	110.28	201.51	695.87
150	0.001078	644.52	1.8209	4.2518	1516.46	106.67	187.33	695.09
160	0.001089	687.15	1.9205	4.2742	1496.22	102.82	175.02	693.51
170	0.001100	730.02	2.0183	4.2996	1474.04	98.76	164.25	691.12
180	0.001112	773.16	2.1146	4.3284	1449.96	94.52	154.76	687.92
190	0.001125	816.60	2.2094	4.3610	1424.00	90.11	146.33	683.90
200	0.001139	860.39	2.3030	4.3982	1396.17	85.57	138.81	679.04
210	0.001154	904.58	2.3954	4.4405	1366.45	80.91	132.04	673.31
220	0.001170	949.22	2.4868	4.4889	1334.81	76.16	125.90	666.65
230	0.001187	994.38	2.5775	4.5444	1301.25	71.34	120.31	659.02
240	0.001205	1040.14	2.6675	4.6084	1265.71	66.46	115.16	650.34
250	0.001225	1086.58	2.7572	4.6824	1228.14	61.55	110.39	640.55
260	0.001247	1133.83	2.8466	4.7687	1188.45	56.62	105.93	629.53
270	0.001271	1182.01	2.9362	4.8701	1146.51	51.70	101.72	617.19
280	0.001298	1231.29	3.0261	4.9908	1102.08	46.79	97.71	603.42
290	0.001328	1281.91	3.1167	5.1366	1054.82	41.91	93.84	588.13
300	0.001361	1334.14	3.2087	5.3168	1004.28	37.05	90.05	571.26
310	0.001400	1388.40	3.3025	5.5459	949.79	32.23	86.29	552.80
320	0.001445	1445.30	3.3993	5.8491	890.30	27.43	82.47	532.79
330	0.001500	1505.79	3.5004	6.2740	824.41	22.66	78.50	511.31
340	0.001569	1571.52	3.6085	6.9239	751.10	17.97	74.21	488.30
350	0.001665	1645.95	3.7288	8.1062	664.96	13.28	69.31	463.20
360	0.001825	1740.13	3.8787	11.4601	542.74	8.07	62.81	433.43

Table 4 Single-Phase Region (Regions 1 to 3 of IAPWS-IF97) – Continuation
Homogenes Zustandsgebiet (Bereiche 1 bis 3 der IAPWS-IF97) – Fortsetzung

p = 20 MPa

t [°C]	v [m³ kg⁻¹]	h [kJ kg⁻¹]	s [kJ kg⁻¹ K⁻¹]	c_p [kJ kg⁻¹ K⁻¹]	w [m s⁻¹]	κ	$\eta \times 10^6$ [kg m⁻¹ s⁻¹]	$\lambda \times 10^3$ [W m⁻¹ K⁻¹]
t_s = 365.75				**Saturation**				
Liquid	0.002039	1827.17	4.0155	23.1784	422.31	4.3740	56.22	417.64
Vapour	0.0059	2411.53	4.9301	45.5883	384.56	1.2620	27.50	234.27
370	0.0069	2526.48	5.1095	18.6597	421.11	1.2806	26.32	165.66
380	0.0083	2659.19	5.3144	10.2212	461.33	1.2886	25.78	128.52
390	0.0092	2747.17	5.4482	7.6925	487.08	1.2908	25.82	113.84
400	0.0099	2816.84	5.5525	6.3601	507.34	1.2935	26.03	105.46
410	0.0106	2876.05	5.6398	5.5410	524.18	1.2951	26.33	100.13
420	0.0112	2928.51	5.7160	4.9820	538.73	1.2957	26.67	96.52
430	0.0117	2976.18	5.7843	4.5718	551.66	1.2960	27.04	94.00
440	0.0122	3020.26	5.8466	4.2568	563.37	1.2959	27.42	92.24
450	0.0127	3061.53	5.9041	4.0074	574.13	1.2957	27.81	91.03
460	0.0132	3100.57	5.9577	3.8056	584.12	1.2954	28.21	90.24
470	0.0136	3137.77	6.0081	3.6395	593.47	1.2950	28.62	89.79
480	0.0140	3173.45	6.0558	3.5010	602.28	1.2945	29.03	89.60
490	0.0144	3207.86	6.1012	3.3841	610.63	1.2939	29.44	89.63
500	0.0148	3241.19	6.1445	3.2845	618.58	1.2933	29.85	89.85
510	0.0152	3273.59	6.1862	3.1990	626.18	1.2927	30.26	90.21
520	0.0155	3305.21	6.2263	3.1251	633.46	1.2920	30.67	90.70
530	0.0159	3336.13	6.2650	3.0608	640.47	1.2912	31.08	91.31
540	0.0162	3366.45	6.3026	3.0046	647.23	1.2904	31.49	92.01
550	0.0166	3396.24	6.3390	2.9552	653.76	1.2896	31.90	92.78
560	0.0169	3425.57	6.3744	2.9116	660.09	1.2888	32.31	93.63
570	0.0172	3454.49	6.4089	2.8731	666.23	1.2879	32.71	94.54
580	0.0176	3483.05	6.4426	2.8389	672.20	1.2870	33.12	95.50
590	0.0179	3511.28	6.4755	2.8084	678.02	1.2862	33.52	96.51
600	0.0182	3539.23	6.5077	2.7812	683.69	1.2853	33.92	97.55
610	0.0185	3566.91	6.5392	2.7569	689.24	1.2844	34.32	98.63
620	0.0188	3594.37	6.5701	2.7352	694.66	1.2835	34.72	99.73
630	0.0191	3621.63	6.6005	2.7157	699.96	1.2826	35.12	100.85
640	0.0194	3648.69	6.6303	2.6982	705.16	1.2817	35.51	102.00
650	0.0197	3675.59	6.6596	2.6824	710.26	1.2808	35.90	103.16
660	0.0200	3702.35	6.6884	2.6683	715.27	1.2799	36.29	104.33
670	0.0203	3728.97	6.7168	2.6557	720.19	1.2790	36.68	105.51
680	0.0206	3755.46	6.7447	2.6443	725.03	1.2781	37.07	106.71
690	0.0208	3781.86	6.7723	2.6342	729.79	1.2772	37.46	107.90
700	0.0211	3808.15	6.7994	2.6251	734.48	1.2764	37.84	109.11
720	0.0217	3860.50	6.8527	2.6098	743.65	1.2746	38.60	111.53
740	0.0222	3912.57	6.9046	2.5979	752.56	1.2729	39.36	113.95
760	0.0228	3964.43	6.9553	2.5887	761.25	1.2713	40.11	116.38
780	0.0233	4016.13	7.0048	2.5821	769.72	1.2696	40.85	118.81
800	0.0239	4067.73	7.0534	2.5775	777.99	1.2679	41.59	121.24

Table 4 Single-Phase Region (Regions 1 to 3 of IAPWS-IF97) – Continuation
Homogenes Zustandsgebiet (Bereiche 1 bis 3 der IAPWS-IF97) – Fortsetzung

				p = 21 MPa				
t [°C]	v [m^3 kg^{-1}]	h [kJ kg^{-1}]	s [kJ kg^{-1} K^{-1}]	c_p [kJ kg^{-1} K^{-1}]	w [m s^{-1}]	κ	$\eta\times10^6$ [kg m^{-1} s^{-1}]	$\lambda\times10^3$ [W m^{-1} K^{-1}]
0	0.000990	21.02	0.0005	4.1249	1436.18	99.22	1744.57	572.61
5	0.000990	41.64	0.0753	4.1232	1459.96	102.51	1487.26	581.29
10	0.000991	62.26	0.1487	4.1231	1481.23	105.47	1285.80	590.22
15	0.000991	82.87	0.2209	4.1237	1500.17	108.09	1124.89	599.20
20	0.000992	103.50	0.2919	4.1248	1516.94	110.41	994.13	608.07
25	0.000994	124.12	0.3616	4.1260	1531.70	112.42	886.28	616.70
30	0.000995	144.76	0.4303	4.1273	1544.60	114.15	796.19	624.96
35	0.000997	165.40	0.4978	4.1288	1555.77	115.61	720.07	632.80
40	0.000999	186.04	0.5643	4.1303	1565.34	116.82	655.13	640.16
45	0.001001	206.70	0.6297	4.1321	1573.42	117.79	599.23	647.00
50	0.001003	227.36	0.6941	4.1341	1580.10	118.53	550.76	653.31
60	0.001008	268.73	0.8202	4.1389	1589.65	119.38	471.25	664.35
70	0.001013	310.15	0.9427	4.1449	1594.62	119.48	409.22	673.39
80	0.001019	351.63	1.0619	4.1522	1595.54	118.91	359.89	680.61
90	0.001026	393.20	1.1779	4.1610	1592.87	117.75	320.01	686.24
100	0.001033	434.86	1.2911	4.1713	1587.00	116.08	287.33	690.47
110	0.001041	476.63	1.4016	4.1832	1578.23	113.95	260.22	693.48
120	0.001049	518.53	1.5095	4.1967	1566.85	111.43	237.48	695.43
130	0.001058	560.57	1.6151	4.2121	1553.06	108.57	218.22	696.41
140	0.001067	602.77	1.7185	4.2294	1537.04	105.40	201.76	696.50
150	0.001077	645.16	1.8199	4.2491	1518.93	101.98	187.58	695.75
160	0.001088	687.76	1.9194	4.2712	1498.82	98.32	175.26	694.20
170	0.001099	730.60	2.0172	4.2963	1476.78	94.47	164.49	691.84
180	0.001111	773.70	2.1133	4.3247	1452.85	90.44	154.99	688.67
190	0.001124	817.11	2.2081	4.3570	1427.07	86.25	146.57	684.69
200	0.001138	860.85	2.3015	4.3936	1399.42	81.94	139.04	679.87
210	0.001153	904.99	2.3939	4.4353	1369.90	77.52	132.27	674.18
220	0.001169	949.58	2.4852	4.4830	1338.50	73.00	126.14	667.57
230	0.001186	994.68	2.5757	4.5376	1305.19	68.42	120.55	660.00
240	0.001204	1040.36	2.6656	4.6004	1269.93	63.78	115.41	651.39
250	0.001224	1086.72	2.7551	4.6730	1232.67	59.12	110.64	641.68
260	0.001246	1133.86	2.8444	4.7574	1193.35	54.44	106.19	630.76
270	0.001269	1181.92	2.9337	4.8565	1151.82	49.77	101.99	618.54
280	0.001296	1231.05	3.0233	4.9739	1107.88	45.11	97.99	604.92
290	0.001325	1281.47	3.1136	5.1153	1061.23	40.48	94.14	589.80
300	0.001358	1333.46	3.2051	5.2890	1011.45	35.88	90.38	573.13
310	0.001396	1387.40	3.2984	5.5082	957.98	31.31	86.65	554.91
320	0.001440	1443.85	3.3944	5.7953	899.91	26.78	82.88	535.19
330	0.001493	1503.67	3.4944	6.1914	835.88	22.28	78.97	514.06
340	0.001560	1568.32	3.6007	6.7821	764.68	17.85	74.80	491.50
350	0.001649	1640.67	3.7177	7.7974	683.51	13.49	70.09	467.10
360	0.001789	1728.76	3.8579	10.2538	575.48	8.82	64.17	438.88

Table 4 Single-Phase Region (Regions 1 to 3 of IAPWS-IF97) – Continuation
Homogenes Zustandsgebiet (Bereiche 1 bis 3 der IAPWS-IF97) – Fortsetzung

				p = 21 MPa				
t [°C]	v [m³ kg⁻¹]	h [kJ kg⁻¹]	s [kJ kg⁻¹ K⁻¹]	c_p [kJ kg⁻¹ K⁻¹]	w [m s⁻¹]	κ	$\eta \times 10^6$ [kg m⁻¹ s⁻¹]	$\lambda \times 10^3$ [W m⁻¹ K⁻¹]
t_s = 369.83				**Saturation**				
Liquid	0.002212	1889.43	4.1093	45.0287	372.33	2.9846	52.14	427.21
Vapour	0.0050	2337.72	4.8065	89.2672	365.83	1.2774	29.58	313.56
370	0.0051	2351.56	4.8280	74.1535	370.03	1.2787	29.30	293.27
380	0.0072	2591.72	5.1993	13.1781	443.49	1.2980	26.58	148.65
390	0.0083	2698.69	5.3619	8.9597	474.64	1.2979	26.32	125.28
400	0.0091	2778.04	5.4807	7.1006	497.56	1.2990	26.41	113.37
410	0.0098	2843.27	5.5769	6.0327	516.15	1.3001	26.63	106.11
420	0.0104	2899.92	5.6593	5.3404	531.89	1.3003	26.93	101.32
430	0.0109	2950.74	5.7321	4.8492	545.70	1.3001	27.26	98.01
440	0.0114	2997.30	5.7978	4.4796	558.10	1.2996	27.62	95.67
450	0.0119	3040.60	5.8581	4.1910	569.42	1.2991	28.00	94.03
460	0.0123	3081.31	5.9140	3.9600	579.87	1.2985	28.38	92.91
470	0.0128	3119.94	5.9664	3.7714	589.61	1.2979	28.78	92.19
480	0.0132	3156.85	6.0157	3.6151	598.77	1.2973	29.18	91.79
490	0.0135	3192.33	6.0625	3.4839	607.42	1.2966	29.58	91.64
500	0.0139	3226.59	6.1071	3.3728	615.63	1.2959	29.99	91.71
510	0.0143	3259.83	6.1498	3.2777	623.46	1.2952	30.39	91.96
520	0.0146	3292.19	6.1909	3.1957	630.95	1.2944	30.80	92.35
530	0.0150	3323.79	6.2305	3.1247	638.15	1.2936	31.21	92.87
540	0.0153	3354.71	6.2687	3.0627	645.08	1.2927	31.61	93.50
550	0.0157	3385.06	6.3058	3.0083	651.76	1.2919	32.02	94.22
560	0.0160	3414.90	6.3419	2.9604	658.23	1.2910	32.42	95.02
570	0.0163	3444.29	6.3769	2.9180	664.51	1.2901	32.83	95.88
580	0.0166	3473.28	6.4111	2.8805	670.60	1.2891	33.23	96.80
590	0.0169	3501.91	6.4445	2.8470	676.53	1.2882	33.63	97.78
600	0.0172	3530.23	6.4771	2.8172	682.30	1.2873	34.03	98.79
610	0.0175	3558.27	6.5090	2.7905	687.94	1.2863	34.43	99.84
620	0.0178	3586.05	6.5403	2.7666	693.45	1.2854	34.82	100.92
630	0.0181	3613.61	6.5710	2.7451	698.84	1.2844	35.22	102.02
640	0.0184	3640.96	6.6011	2.7259	704.12	1.2835	35.61	103.14
650	0.0187	3668.13	6.6307	2.7086	709.29	1.2826	36.00	104.29
660	0.0190	3695.14	6.6598	2.6930	714.37	1.2817	36.39	105.44
670	0.0192	3722.00	6.6884	2.6790	719.36	1.2807	36.78	106.61
680	0.0195	3748.72	6.7166	2.6664	724.25	1.2798	37.16	107.79
690	0.0198	3775.33	6.7444	2.6551	729.07	1.2789	37.55	108.97
700	0.0201	3801.83	6.7718	2.6450	733.82	1.2781	37.93	110.16
720	0.0206	3854.55	6.8254	2.6278	743.08	1.2763	38.69	112.55
740	0.0211	3906.97	6.8776	2.6143	752.09	1.2745	39.45	114.95
760	0.0217	3959.14	6.9286	2.6037	760.86	1.2728	40.19	117.36
780	0.0222	4011.13	6.9785	2.5958	769.40	1.2711	40.93	119.76
800	0.0227	4062.99	7.0273	2.5902	777.74	1.2694	41.67	122.16

Table 4 Single-Phase Region (Regions 1 to 3 of IAPWS-IF97) – Continuation
Homogenes Zustandsgebiet (Bereiche 1 bis 3 der IAPWS-IF97) – Fortsetzung

p = 22 MPa

t [°C]	v [m³ kg⁻¹]	h [kJ kg⁻¹]	s [kJ kg⁻¹ K⁻¹]	c_p [kJ kg⁻¹ K⁻¹]	w [m s⁻¹]	κ	$\eta \times 10^6$ [kg m⁻¹ s⁻¹]	$\lambda \times 10^3$ [W m⁻¹ K⁻¹]
0	0.000989	22.01	0.0005	4.1209	1437.84	94.98	1742.60	573.16
5	0.000990	42.61	0.0752	4.1197	1461.61	98.12	1485.97	581.81
10	0.000990	63.21	0.1486	4.1199	1482.88	100.94	1284.97	590.71
15	0.000991	83.81	0.2207	4.1209	1501.81	103.45	1124.38	599.68
20	0.000992	104.42	0.2916	4.1221	1518.57	105.66	993.84	608.54
25	0.000993	125.03	0.3613	4.1235	1533.33	107.58	886.15	617.15
30	0.000995	145.65	0.4299	4.1250	1546.23	109.24	796.17	625.42
35	0.000997	166.28	0.4974	4.1265	1557.40	110.63	720.14	633.26
40	0.000998	186.92	0.5639	4.1282	1566.98	111.79	655.25	640.61
45	0.001000	207.56	0.6293	4.1300	1575.07	112.72	599.41	647.46
50	0.001003	228.22	0.6937	4.1321	1581.77	113.43	550.96	653.77
60	0.001008	269.56	0.8197	4.1369	1591.36	114.25	471.49	664.83
70	0.001013	310.96	0.9421	4.1429	1596.38	114.35	409.48	673.88
80	0.001019	352.43	1.0612	4.1503	1597.37	113.82	360.15	681.12
90	0.001026	393.97	1.1772	4.1591	1594.77	112.72	320.28	686.76
100	0.001033	435.61	1.2904	4.1693	1588.97	111.13	287.60	691.01
110	0.001040	477.36	1.4008	4.1811	1580.28	109.11	260.48	694.05
120	0.001049	519.24	1.5087	4.1945	1568.99	106.71	237.74	696.01
130	0.001057	561.26	1.6142	4.2097	1555.30	103.99	218.47	697.02
140	0.001067	603.44	1.7175	4.2269	1539.39	100.97	202.01	697.13
150	0.001077	645.80	1.8189	4.2463	1521.39	97.71	187.82	696.41
160	0.001087	688.37	1.9183	4.2683	1501.40	94.23	175.50	694.89
170	0.001099	731.18	2.0160	4.2931	1479.50	90.56	164.73	692.56
180	0.001111	774.25	2.1121	4.3211	1455.73	86.72	155.23	689.43
190	0.001124	817.61	2.2068	4.3529	1430.11	82.74	146.80	685.48
200	0.001137	861.32	2.3001	4.3891	1402.64	78.64	139.28	680.70
210	0.001152	905.41	2.3923	4.4302	1373.33	74.43	132.51	675.05
220	0.001168	949.94	2.4836	4.4771	1342.16	70.13	126.38	668.49
230	0.001184	994.98	2.5740	4.5309	1309.09	65.76	120.79	660.98
240	0.001203	1040.59	2.6637	4.5926	1274.11	61.35	115.65	652.44
250	0.001222	1086.86	2.7530	4.6638	1237.16	56.91	110.89	642.81
260	0.001244	1133.90	2.8421	4.7464	1198.19	52.46	106.45	631.99
270	0.001267	1181.84	2.9312	4.8432	1157.07	48.01	102.26	619.88
280	0.001293	1230.82	3.0205	4.9576	1113.61	43.58	98.28	606.40
290	0.001322	1281.06	3.1105	5.0947	1067.55	39.18	94.44	591.44
300	0.001355	1332.82	3.2016	5.2623	1018.49	34.80	90.70	574.97
310	0.001392	1386.45	3.2944	5.4724	965.96	30.47	87.00	556.98
320	0.001435	1442.47	3.3896	5.7448	909.21	26.18	83.27	537.54
330	0.001487	1501.67	3.4886	6.1153	846.98	21.93	79.43	516.74
340	0.001551	1565.33	3.5933	6.6555	777.89	17.74	75.35	494.59
350	0.001635	1635.90	3.7074	7.5429	700.27	13.63	70.82	470.81
360	0.001760	1719.47	3.8404	9.4598	602.89	9.39	65.32	443.93
370	0.002029	1842.65	4.0333	18.3525	452.83	4.59	56.64	411.16

Table 4 Single-Phase Region (Regions 1 to 3 of IAPWS-IF97) – Continuation
Homogenes Zustandsgebiet (Bereiche 1 bis 3 der IAPWS-IF97) – Fortsetzung

p = 22 MPa								
t [°C]	v [m^3 kg^{-1}]	h [kJ kg^{-1}]	s [kJ kg^{-1} K^{-1}]	c_p [kJ kg^{-1} K^{-1}]	w [m s^{-1}]	κ	$\eta \times 10^6$ [kg m^{-1} s^{-1}]	$\lambda \times 10^3$ [W m^{-1} K^{-1}]
t_s = 373.71				**Saturation**				
Liquid	0.002750	2021.91	4.3109	1164.94	315.24	1.6423	43.29	876.77
Vapour	0.0036	2169.25	4.5386	1399.93	327.19	1.3484	35.53	900.15
380	0.0061	2504.56	5.0556	19.4188	420.44	1.3118	27.88	182.13
390	0.0074	2643.66	5.2671	10.7225	460.90	1.3088	26.99	140.13
400	0.0083	2735.76	5.4050	8.0332	487.13	1.3066	26.87	122.95
410	0.0090	2808.37	5.5121	6.6169	507.74	1.3064	26.99	113.08
420	0.0096	2869.89	5.6015	5.7489	524.85	1.3059	27.22	106.77
430	0.0101	2924.25	5.6794	5.1572	539.62	1.3050	27.52	102.47
440	0.0106	2973.55	5.7491	4.7230	552.76	1.3041	27.85	99.44
450	0.0111	3019.05	5.8124	4.3894	564.66	1.3031	28.20	97.28
460	0.0116	3061.57	5.8708	4.1251	575.59	1.3022	28.57	95.77
470	0.0120	3101.72	5.9252	3.9113	585.74	1.3014	28.95	94.75
480	0.0124	3139.92	5.9763	3.7354	595.25	1.3005	29.34	94.11
490	0.0128	3176.52	6.0246	3.5887	604.20	1.2997	29.74	93.77
500	0.0131	3211.77	6.0704	3.4649	612.68	1.2988	30.13	93.68
510	0.0135	3245.88	6.1143	3.3595	620.75	1.2980	30.53	93.79
520	0.0138	3279.01	6.1563	3.2689	628.46	1.2971	30.94	94.08
530	0.0142	3311.30	6.1968	3.1906	635.85	1.2962	31.34	94.51
540	0.0145	3342.86	6.2358	3.1225	642.95	1.2952	31.74	95.06
550	0.0148	3373.78	6.2736	3.0628	649.79	1.2943	32.14	95.71
560	0.0151	3404.14	6.3103	3.0104	656.41	1.2933	32.54	96.45
570	0.0155	3434.01	6.3459	2.9640	662.81	1.2924	32.94	97.27
580	0.0158	3463.44	6.3806	2.9230	669.02	1.2914	33.34	98.15
590	0.0161	3492.48	6.4145	2.8864	675.06	1.2904	33.74	99.08
600	0.0163	3521.18	6.4475	2.8538	680.94	1.2894	34.14	100.06
610	0.0166	3549.57	6.4798	2.8247	686.67	1.2884	34.53	101.08
620	0.0169	3577.68	6.5115	2.7986	692.27	1.2874	34.93	102.14
630	0.0172	3605.55	6.5425	2.7751	697.74	1.2864	35.32	103.22
640	0.0175	3633.19	6.5730	2.7540	703.10	1.2855	35.71	104.32
650	0.0178	3660.64	6.6029	2.7351	708.34	1.2845	36.10	105.44
660	0.0180	3687.90	6.6322	2.7180	713.49	1.2835	36.49	106.58
670	0.0183	3715.00	6.6611	2.7026	718.54	1.2826	36.88	107.73
680	0.0186	3741.96	6.6896	2.6888	723.50	1.2816	37.26	108.89
690	0.0188	3768.78	6.7176	2.6763	728.37	1.2807	37.64	110.06
700	0.0191	3795.49	6.7451	2.6651	733.17	1.2798	38.02	111.24
720	0.0196	3848.59	6.7992	2.6460	742.54	1.2780	38.78	113.60
740	0.0201	3901.35	6.8518	2.6308	751.63	1.2762	39.53	115.97
760	0.0206	3953.85	6.9031	2.6188	760.48	1.2744	40.28	118.35
780	0.0211	4006.13	6.9532	2.6096	769.10	1.2727	41.02	120.73
800	0.0216	4058.25	7.0022	2.6029	777.51	1.2710	41.75	123.10

Table 4 Single-Phase Region (Regions 1 to 3 of IAPWS-IF97) – Continuation
Homogenes Zustandsgebiet (Bereiche 1 bis 3 der IAPWS-IF97) – Fortsetzung

$p = 23$ MPa

t [°C]	v [m^3 kg^{-1}]	h [kJ kg^{-1}]	s [kJ kg^{-1} K^{-1}]	c_p [kJ kg^{-1} K^{-1}]	w [m s^{-1}]	κ	$\eta\times 10^6$ [kg m^{-1} s^{-1}]	$\lambda\times 10^3$ [W m^{-1} K^{-1}]
0	0.000989	23.00	0.0004	4.1169	1439.51	91.10	1740.65	573.71
5	0.000989	43.58	0.0751	4.1162	1463.27	94.11	1484.70	582.32
10	0.000990	64.16	0.1485	4.1168	1484.53	96.81	1284.16	591.20
15	0.000991	84.75	0.2205	4.1181	1503.45	99.21	1123.88	600.15
20	0.000992	105.34	0.2914	4.1195	1520.20	101.33	993.56	609.00
25	0.000993	125.94	0.3611	4.1211	1534.95	103.17	886.03	617.61
30	0.000994	146.55	0.4296	4.1227	1547.85	104.75	796.16	625.87
35	0.000996	167.17	0.4971	4.1243	1559.03	106.09	720.21	633.71
40	0.000998	187.79	0.5635	4.1261	1568.62	107.20	655.38	641.07
45	0.001000	208.43	0.6288	4.1279	1576.73	108.09	599.58	647.92
50	0.001002	229.07	0.6932	4.1300	1583.44	108.77	551.17	654.24
60	0.001007	270.40	0.8192	4.1349	1593.07	109.57	471.73	665.30
70	0.001013	311.78	0.9415	4.1410	1598.15	109.67	409.74	674.37
80	0.001019	353.22	1.0606	4.1484	1599.19	109.16	360.42	681.62
90	0.001025	394.75	1.1766	4.1571	1596.66	108.12	320.55	687.28
100	0.001032	436.37	1.2896	4.1673	1590.93	106.61	287.87	691.55
110	0.001040	478.10	1.4000	4.1790	1582.33	104.68	260.74	694.61
120	0.001048	519.95	1.5078	4.1923	1571.12	102.40	237.99	696.60
130	0.001057	561.95	1.6133	4.2074	1557.53	99.80	218.72	697.62
140	0.001066	604.11	1.7166	4.2244	1541.72	96.93	202.26	697.77
150	0.001076	646.45	1.8178	4.2436	1523.84	93.82	188.07	697.07
160	0.001087	688.99	1.9172	4.2653	1503.97	90.50	175.74	695.57
170	0.001098	731.76	2.0148	4.2898	1482.21	87.00	164.96	693.27
180	0.001110	774.80	2.1109	4.3175	1458.59	83.34	155.47	690.18
190	0.001123	818.12	2.2054	4.3490	1433.13	79.54	147.04	686.27
200	0.001136	861.79	2.2987	4.3846	1405.85	75.62	139.51	681.52
210	0.001151	905.83	2.3908	4.4252	1376.74	71.60	132.74	675.92
220	0.001167	950.31	2.4819	4.4714	1345.78	67.50	126.62	669.41
230	0.001183	995.28	2.5722	4.5242	1312.96	63.34	121.03	661.95
240	0.001201	1040.82	2.6618	4.5849	1278.25	59.13	115.89	653.48
250	0.001221	1087.01	2.7510	4.6547	1241.61	54.89	111.14	643.92
260	0.001242	1133.95	2.8399	4.7357	1202.97	50.65	106.71	633.20
270	0.001266	1181.77	2.9287	4.8302	1162.25	46.41	102.53	621.21
280	0.001291	1230.61	3.0178	4.9417	1119.26	42.18	98.56	607.86
290	0.001320	1280.68	3.1075	5.0748	1073.76	37.98	94.74	593.07
300	0.001352	1332.20	3.1982	5.2367	1025.40	33.82	91.02	576.79
310	0.001388	1385.54	3.2905	5.4382	973.75	29.70	87.35	559.02
320	0.001430	1441.16	3.3850	5.6972	918.21	25.63	83.66	539.83
330	0.001480	1499.78	3.4830	6.0449	857.71	21.60	79.88	519.35
340	0.001542	1562.55	3.5862	6.5417	790.73	17.63	75.89	497.60

Table 4 Single-Phase Region (Regions 1 to 3 of IAPWS-IF97) – Continuation
Homogenes Zustandsgebiet (Bereiche 1 bis 3 der IAPWS-IF97) – Fortsetzung

				p = 23 MPa				
t [°C]	v [m^3 kg^{-1}]	h [kJ kg^{-1}]	s [kJ kg^{-1} K^{-1}]	c_p [kJ kg^{-1} K^{-1}]	w [m s^{-1}]	κ	$\eta \times 10^6$ [kg m^{-1} s^{-1}]	$\lambda \times 10^3$ [W m^{-1} K^{-1}]
350	0.001622	1631.55	3.6978	7.3278	715.91	13.7397	71.51	474.35
360	0.001736	1711.59	3.8252	8.8860	626.70	9.8359	65.34	448.59
370	0.001945	1818.84	3.9932	13.8083	501.72	5.6260	59.06	416.56
380	0.004782	2364.90	4.8334	43.2888	384.28	1.3425	30.73	262.21
390	0.006504	2579.46	5.1599	13.4179	445.15	1.3246	27.91	160.35
400	0.007477	2689.22	5.3242	9.2285	475.96	1.3172	27.45	134.72
410	0.008233	2771.01	5.4449	7.3217	498.93	1.3146	27.42	121.26
420	0.008872	2838.27	5.5426	6.2188	517.58	1.3129	27.57	112.97
430	0.009436	2896.64	5.6262	5.5007	533.42	1.3111	27.81	107.44
440	0.009948	2948.96	5.7001	4.9894	547.34	1.3094	28.10	103.58
450	0.010420	2996.84	5.7668	4.6037	559.86	1.3078	28.43	100.83
460	0.010862	3041.31	5.8279	4.3020	571.30	1.3065	28.78	98.86
470	0.011278	3083.08	5.8845	4.0600	581.87	1.3052	29.14	97.49
480	0.011674	3122.66	5.9374	3.8624	591.73	1.3041	29.52	96.58
490	0.012052	3160.44	5.9872	3.6985	601.00	1.3031	29.90	96.02
500	0.012415	3196.72	6.0345	3.5610	609.75	1.3020	30.29	95.75
510	0.012766	3231.73	6.0795	3.4445	618.07	1.3010	30.68	95.72
520	0.013106	3265.67	6.1225	3.3447	625.99	1.3000	31.08	95.88
530	0.013435	3298.67	6.1639	3.2587	633.57	1.2990	31.48	96.21
540	0.013757	3330.88	6.2037	3.1841	640.85	1.2980	31.87	96.67
550	0.014070	3362.39	6.2422	3.1189	647.85	1.2969	32.27	97.25
560	0.014377	3393.28	6.2795	3.0616	654.60	1.2959	32.67	97.93
570	0.014677	3423.64	6.3158	3.0111	661.14	1.2948	33.06	98.70
580	0.014972	3453.52	6.3510	2.9664	667.47	1.2938	33.46	99.53
590	0.015262	3482.98	6.3853	2.9266	673.62	1.2927	33.86	100.43
600	0.015547	3512.07	6.4188	2.8912	679.60	1.2916	34.25	101.37
610	0.015827	3540.82	6.4516	2.8595	685.43	1.2906	34.64	102.36
620	0.016104	3569.27	6.4836	2.8311	691.11	1.2896	35.04	103.39
630	0.016377	3597.45	6.5150	2.8055	696.67	1.2885	35.43	104.45
640	0.016646	3625.39	6.5457	2.7826	702.10	1.2875	35.82	105.53
650	0.016913	3653.11	6.5759	2.7619	707.42	1.2865	36.20	106.63
660	0.017176	3680.64	6.6056	2.7433	712.63	1.2855	36.59	107.75
670	0.017437	3707.98	6.6347	2.7265	717.74	1.2845	36.98	108.88
680	0.017695	3735.17	6.6634	2.7114	722.76	1.2835	37.36	110.03
690	0.017951	3762.22	6.6916	2.6977	727.69	1.2826	37.74	111.18
700	0.018204	3789.13	6.7195	2.6854	732.54	1.2816	38.12	112.34
720	0.018705	3842.62	6.7739	2.6644	742.01	1.2798	38.88	114.67
740	0.019199	3895.73	6.8268	2.6475	751.19	1.2779	39.63	117.02
760	0.019686	3948.54	6.8784	2.6340	760.12	1.2761	40.37	119.37
780	0.020167	4001.11	6.9288	2.6236	768.82	1.2743	41.11	121.71
800	0.020642	4053.50	6.9781	2.6157	777.29	1.2726	41.84	124.06

Table 4 Single-Phase Region (Regions 1 to 3 of IAPWS-IF97) – Continuation
Homogenes Zustandsgebiet (Bereiche 1 bis 3 der IAPWS-IF97) – Fortsetzung

				p = 24 MPa				
t	v	h	s	c_p	w	κ	$\eta \times 10^6$	$\lambda \times 10^3$
[°C]	[m^3 kg^{-1}]	[kJ kg^{-1}]	[kJ kg^{-1} K^{-1}]	[kJ kg^{-1} K^{-1}]	[m s^{-1}]		[kg m^{-1} s^{-1}]	[W m^{-1} K^{-1}]
0	0.000988	23.98	0.0004	4.1129	1441.18	87.55	1738.72	574.26
5	0.000989	44.54	0.0750	4.1127	1464.93	90.44	1483.44	582.83
10	0.000989	65.11	0.1483	4.1137	1486.18	93.03	1283.36	591.68
15	0.000990	85.68	0.2203	4.1152	1505.09	95.33	1123.39	600.62
20	0.000991	106.26	0.2911	4.1169	1521.83	97.36	993.29	609.46
25	0.000993	126.85	0.3608	4.1186	1536.58	99.12	885.91	618.06
30	0.000994	147.45	0.4293	4.1204	1549.48	100.64	796.16	626.33
35	0.000996	168.05	0.4967	4.1221	1560.67	101.93	720.28	634.17
40	0.000998	188.67	0.5631	4.1239	1570.26	102.99	655.52	641.53
45	0.001000	209.29	0.6284	4.1259	1578.38	103.84	599.75	648.38
50	0.001002	229.93	0.6928	4.1280	1585.11	104.50	551.37	654.70
60	0.001007	271.23	0.8187	4.1330	1594.78	105.27	471.97	665.78
70	0.001012	312.59	0.9410	4.1391	1599.91	105.38	410.00	674.86
80	0.001018	354.02	1.0600	4.1464	1601.01	104.90	360.69	682.13
90	0.001025	395.53	1.1759	4.1551	1598.55	103.91	320.82	687.80
100	0.001032	437.13	1.2889	4.1653	1592.90	102.47	288.13	692.09
110	0.001039	478.84	1.3992	4.1769	1584.38	100.63	261.01	695.17
120	0.001048	520.67	1.5070	4.1901	1573.26	98.45	238.25	697.18
130	0.001056	562.64	1.6124	4.2050	1559.76	95.96	218.98	698.23
140	0.001066	604.78	1.7156	4.2219	1544.05	93.22	202.51	698.40
150	0.001076	647.09	1.8168	4.2409	1526.28	90.25	188.31	697.73
160	0.001086	689.60	1.9161	4.2624	1506.54	87.07	175.99	696.26
170	0.001097	732.35	2.0137	4.2866	1484.90	83.73	165.20	693.99
180	0.001109	775.35	2.1096	4.3140	1461.43	80.23	155.70	690.93
190	0.001122	818.64	2.2041	4.3450	1436.13	76.60	147.27	687.05
200	0.001135	862.26	2.2973	4.3802	1409.03	72.85	139.75	682.35
210	0.001150	906.26	2.3893	4.4202	1380.12	69.01	132.98	676.78
220	0.001166	950.68	2.4803	4.4657	1349.38	65.09	126.85	670.32
230	0.001182	995.59	2.5705	4.5177	1316.80	61.11	121.27	662.92
240	0.001200	1041.06	2.6600	4.5773	1282.35	57.09	116.14	654.51
250	0.001220	1087.17	2.7490	4.6458	1246.00	53.04	111.39	645.04
260	0.001241	1134.01	2.8376	4.7252	1207.71	48.98	106.96	634.40
270	0.001264	1181.72	2.9263	4.8176	1167.37	44.93	102.79	622.52
280	0.001289	1230.42	3.0151	4.9262	1124.84	40.90	98.84	609.30
290	0.001317	1280.31	3.1045	5.0555	1079.88	36.89	95.03	594.67
300	0.001349	1331.62	3.1948	5.2120	1032.19	32.91	91.34	578.58
310	0.001385	1384.67	3.2866	5.4056	981.37	28.98	87.69	561.02
320	0.001426	1439.91	3.3805	5.6524	926.94	25.11	84.04	542.09
330	0.001475	1497.99	3.4776	5.9796	868.05	21.29	80.31	521.89
340	0.001534	1559.93	3.5794	6.4387	803.19	17.52	76.40	500.52

Table 4 Single-Phase Region (Regions 1 to 3 of IAPWS-IF97) – Continuation
Homogenes Zustandsgebiet (Bereiche 1 bis 3 der IAPWS-IF97) – Fortsetzung

				p = 24 MPa				
t	v	h	s	c_p	w	κ	$\eta \times 10^6$	$\lambda \times 10^3$
[°C]	[m^3 kg^{-1}]	[kJ kg^{-1}]	[kJ kg^{-1} K^{-1}]	[kJ kg^{-1} K^{-1}]	[m s^{-1}]		[kg m^{-1} s^{-1}]	[W m^{-1} K^{-1}]
350	0.001610	1627.56	3.6888	7.1423	730.87	13.8251	72.15	477.77
360	0.001715	1704.72	3.8116	8.4459	647.89	10.1966	67.25	452.93
370	0.001891	1802.54	3.9649	11.7728	538.24	6.3833	60.80	422.66
380	0.002612	2025.16	4.3076	67.8457	355.44	2.0153	45.44	411.72
390	0.005613	2500.75	5.0320	18.2003	426.49	1.3501	29.30	190.25
400	0.006731	2637.37	5.2366	10.8035	463.97	1.3325	28.19	149.39
410	0.007540	2730.75	5.3744	8.1820	489.68	1.3250	27.93	130.93
420	0.008205	2804.86	5.4821	6.7651	510.09	1.3214	27.97	120.06
430	0.008781	2867.81	5.5723	5.8859	527.11	1.3183	28.14	113.00
440	0.009299	2923.48	5.6509	5.2815	541.87	1.3156	28.38	108.14
450	0.009774	2973.96	5.7212	4.8354	555.03	1.3133	28.68	104.68
460	0.010215	3020.53	5.7852	4.4912	566.99	1.3113	29.00	102.20
470	0.010629	3064.02	5.8441	4.2179	577.99	1.3097	29.35	100.43
480	0.011021	3105.06	5.8990	3.9963	588.23	1.3082	29.71	99.20
490	0.011394	3144.08	5.9504	3.8138	597.81	1.3068	30.08	98.40
500	0.011753	3181.43	5.9991	3.6614	606.84	1.3056	30.46	97.93
510	0.012098	3217.39	6.0453	3.5328	615.40	1.3044	30.84	97.74
520	0.012431	3252.15	6.0894	3.4232	623.54	1.3032	31.23	97.77
530	0.012754	3285.90	6.1317	3.3290	631.32	1.3021	31.62	97.98
540	0.013069	3318.78	6.1723	3.2474	638.77	1.3009	32.01	98.35
550	0.013375	3350.89	6.2116	3.1764	645.93	1.2998	32.41	98.86
560	0.013674	3382.33	6.2496	3.1141	652.83	1.2986	32.80	99.47
570	0.013967	3413.19	6.2864	3.0593	659.50	1.2975	33.19	100.18
580	0.014255	3443.54	6.3222	3.0107	665.95	1.2963	33.58	100.96
590	0.014537	3473.43	6.3570	2.9676	672.21	1.2952	33.98	101.82
600	0.014814	3502.91	6.3910	2.9292	678.29	1.2940	34.37	102.72
610	0.015087	3532.02	6.4241	2.8949	684.21	1.2929	34.76	103.68
620	0.015355	3560.82	6.4565	2.8641	689.98	1.2918	35.15	104.68
630	0.015620	3589.32	6.4883	2.8364	695.62	1.2907	35.54	105.71
640	0.015882	3617.55	6.5194	2.8116	701.12	1.2897	35.92	106.77
650	0.016140	3645.56	6.5499	2.7892	706.51	1.2886	36.31	107.85
660	0.016396	3673.34	6.5798	2.7690	711.79	1.2875	36.70	108.95
670	0.016648	3700.94	6.6092	2.7507	716.96	1.2865	37.08	110.06
680	0.016898	3728.37	6.6381	2.7343	722.04	1.2855	37.46	111.19
690	0.017146	3755.63	6.6666	2.7194	727.03	1.2845	37.84	112.32
700	0.017391	3782.76	6.6946	2.7060	731.93	1.2835	38.22	113.47
720	0.017876	3836.64	6.7494	2.6829	741.50	1.2816	38.97	115.77
740	0.018353	3890.10	6.8027	2.6643	750.77	1.2797	39.72	118.08
760	0.018823	3943.23	6.8547	2.6493	759.78	1.2778	40.46	120.40
780	0.019287	3996.10	6.9053	2.6376	768.54	1.2760	41.19	122.72
800	0.019746	4048.76	6.9549	2.6286	777.08	1.2742	41.92	125.04

Table 4 **Single-Phase Region** (Regions 1 to 3 of IAPWS-IF97) – Continuation
Homogenes Zustandsgebiet (Bereiche 1 bis 3 der IAPWS-IF97) – Fortsetzung

				p = 25 MPa				
t	v	h	s	c_p	w	κ	$\eta \times 10^6$	$\lambda \times 10^3$
[°C]	[$m^3\ kg^{-1}$]	[$kJ\ kg^{-1}$]	[$kJ\ kg^{-1}\ K^{-1}$]	[$kJ\ kg^{-1}\ K^{-1}$]	[$m\ s^{-1}$]		[$kg\ m^{-1}\ s^{-1}$]	[$W\ m^{-1}\ K^{-1}$]
0	0.000988	24.96	0.0004	4.1090	1442.86	84.29	1736.82	574.80
5	0.000988	45.51	0.0749	4.1093	1466.60	87.06	1482.20	583.34
10	0.000989	66.06	0.1482	4.1107	1487.83	89.55	1282.57	592.17
15	0.000990	86.62	0.2201	4.1125	1506.73	91.76	1122.91	601.09
20	0.000991	107.18	0.2909	4.1144	1523.47	93.70	993.02	609.92
25	0.000992	127.76	0.3605	4.1162	1538.21	95.40	885.80	618.52
30	0.000994	148.35	0.4290	4.1181	1551.11	96.86	796.15	626.78
35	0.000995	168.94	0.4964	4.1199	1562.30	98.09	720.36	634.62
40	0.000997	189.54	0.5627	4.1218	1571.91	99.12	655.65	641.98
45	0.000999	210.16	0.6280	4.1238	1580.03	99.94	599.93	648.84
50	0.001001	230.78	0.6923	4.1260	1586.78	100.57	551.58	655.17
60	0.001006	272.07	0.8181	4.1311	1596.50	101.32	472.22	666.26
70	0.001012	313.41	0.9404	4.1372	1601.67	101.43	410.26	675.35
80	0.001018	354.82	1.0593	4.1445	1602.84	100.98	360.96	682.63
90	0.001024	396.30	1.1752	4.1532	1600.44	100.03	321.09	688.33
100	0.001031	437.88	1.2881	4.1633	1594.86	98.65	288.40	692.63
110	0.001039	479.57	1.3984	4.1748	1586.42	96.90	261.27	695.73
120	0.001047	521.38	1.5061	4.1879	1575.38	94.81	238.51	697.76
130	0.001056	563.34	1.6115	4.2027	1561.98	92.44	219.23	698.84
140	0.001065	605.45	1.7147	4.2194	1546.37	89.81	202.75	699.03
150	0.001075	647.73	1.8158	4.2383	1528.71	86.96	188.56	698.39
160	0.001085	690.22	1.9150	4.2595	1509.09	83.92	176.23	696.95
170	0.001097	732.93	2.0125	4.2834	1487.59	80.72	165.44	694.71
180	0.001108	775.90	2.1084	4.3105	1464.25	77.37	155.94	691.67
190	0.001121	819.15	2.2028	4.3411	1439.12	73.89	147.51	687.83
200	0.001135	862.73	2.2959	4.3758	1412.19	70.31	139.98	683.17
210	0.001149	906.69	2.3878	4.4152	1383.47	66.63	133.21	677.65
220	0.001164	951.06	2.4787	4.4601	1352.95	62.88	127.09	671.23
230	0.001181	995.91	2.5688	4.5113	1320.60	59.07	121.50	663.89
240	0.001199	1041.31	2.6581	4.5698	1286.42	55.21	116.38	655.54
250	0.001218	1087.33	2.7469	4.6371	1250.36	51.34	111.64	646.14
260	0.001239	1134.08	2.8355	4.7149	1212.39	47.45	107.22	635.60
270	0.001262	1181.67	2.9239	4.8052	1172.43	43.57	103.06	623.82
280	0.001287	1230.24	3.0125	4.9112	1130.34	39.71	99.11	610.74
290	0.001315	1279.96	3.1016	5.0369	1085.91	35.87	95.32	596.26
300	0.001346	1331.06	3.1915	5.1883	1038.86	32.08	91.65	580.34
310	0.001381	1383.84	3.2828	5.3745	988.82	28.32	88.03	563.00
320	0.001421	1438.72	3.3761	5.6099	935.42	24.62	84.42	544.30
330	0.001469	1496.28	3.4723	5.9188	878.00	20.99	80.74	524.38
340	0.001526	1557.48	3.5729	6.3448	815.21	17.42	76.90	503.35

Table 4 Single-Phase Region (Regions 1 to 3 of IAPWS-IF97) – Continuation
Homogenes Zustandsgebiet (Bereiche 1 bis 3 der IAPWS-IF97) – Fortsetzung

p = 25 MPa

t [°C]	v [m^3 kg^{-1}]	h [kJ kg^{-1}]	s [kJ kg^{-1} K^{-1}]	c_p [kJ kg^{-1} K^{-1}]	w [m s^{-1}]	κ	$\eta \times 10^6$ [kg m^{-1} s^{-1}]	$\lambda \times 10^3$ [W m^{-1} K^{-1}]
350	0.001599	1623.86	3.6803	6.9800	745.33	13.8983	72.76	481.07
360	0.001697	1698.64	3.7993	8.0944	667.09	10.4897	68.08	457.01
370	0.001850	1789.93	3.9424	10.5651	568.12	6.9776	62.21	428.37
380	0.002218	1935.67	4.1670	23.1840	426.79	3.2844	52.39	392.66
390	0.004647	2395.53	4.8656	28.4614	403.34	1.4003	31.70	239.85
400	0.006005	2578.59	5.1399	13.0025	450.94	1.3546	29.17	168.15
410	0.006884	2687.10	5.2999	9.2401	480.01	1.3389	28.56	142.43
420	0.007579	2769.45	5.4196	7.4062	502.36	1.3319	28.43	128.20
430	0.008172	2837.67	5.5174	6.3206	520.68	1.3270	28.51	119.23
440	0.008697	2897.06	5.6013	5.6026	536.34	1.3230	28.70	113.16
450	0.009175	2950.38	5.6755	5.0860	550.18	1.3196	28.95	108.88
460	0.009617	2999.20	5.7426	4.6938	562.68	1.3169	29.25	105.79
470	0.010029	3044.53	5.8040	4.3856	574.13	1.3147	29.57	103.57
480	0.010418	3087.11	5.8609	4.1377	584.74	1.3127	29.92	101.99
490	0.010788	3127.44	5.9141	3.9347	594.64	1.3110	30.27	100.91
500	0.011142	3165.92	5.9642	3.7661	603.96	1.3095	30.64	100.23
510	0.011482	3202.85	6.0117	3.6245	612.76	1.3081	31.01	99.85
520	0.011810	3238.48	6.0569	3.5044	621.13	1.3067	31.39	99.74
530	0.012127	3272.99	6.1001	3.4014	629.10	1.3054	31.77	99.83
540	0.012435	3306.55	6.1416	3.3126	636.72	1.3041	32.16	100.10
550	0.012735	3339.28	6.1816	3.2354	644.04	1.3028	32.55	100.52
560	0.013028	3371.29	6.2203	3.1679	651.09	1.3015	32.94	101.06
570	0.013314	3402.67	6.2577	3.1085	657.88	1.3003	33.32	101.70
580	0.013595	3433.49	6.2941	3.0560	664.45	1.2990	33.71	102.44
590	0.013870	3463.81	6.3294	3.0094	670.82	1.2978	34.10	103.24
600	0.014140	3493.69	6.3638	2.9679	677.01	1.2966	34.49	104.11
610	0.014405	3523.18	6.3974	2.9308	683.02	1.2954	34.88	105.04
620	0.014667	3552.32	6.4302	2.8976	688.88	1.2942	35.27	106.00
630	0.014925	3581.14	6.4623	2.8678	694.59	1.2931	35.65	107.00
640	0.015179	3609.69	6.4937	2.8410	700.17	1.2919	36.04	108.04
650	0.015430	3637.97	6.5246	2.8168	705.63	1.2908	36.42	109.09
660	0.015678	3666.03	6.5548	2.7949	710.98	1.2897	36.80	110.17
670	0.015923	3693.88	6.5845	2.7752	716.21	1.2886	37.18	111.27
680	0.016165	3721.54	6.6136	2.7574	721.35	1.2876	37.56	112.37
690	0.016406	3749.03	6.6423	2.7413	726.39	1.2865	37.94	113.49
700	0.016643	3776.37	6.6706	2.7267	731.35	1.2855	38.32	114.62
720	0.017113	3830.64	6.7258	2.7016	741.01	1.2835	39.07	116.89
740	0.017575	3884.47	6.7794	2.6812	750.37	1.2815	39.81	119.17
760	0.018030	3937.92	6.8317	2.6647	759.45	1.2796	40.55	121.46
780	0.018479	3991.08	6.8826	2.6517	768.29	1.2777	41.28	123.75
800	0.018922	4044.00	6.9324	2.6415	776.89	1.2759	42.01	126.04

Table 4 Single-Phase Region (Regions 1 to 3 of IAPWS-IF97) – Continuation
Homogenes Zustandsgebiet (Bereiche 1 bis 3 der IAPWS-IF97) – Fortsetzung

				p = 26 MPa				
t [°C]	v [$m^3 kg^{-1}$]	h [$kJ kg^{-1}$]	s [$kJ kg^{-1} K^{-1}$]	c_p [$kJ kg^{-1} K^{-1}$]	w [$m s^{-1}$]	κ	$\eta \times 10^6$ [$kg m^{-1} s^{-1}$]	$\lambda \times 10^3$ [$W m^{-1} K^{-1}$]
0	0.000988	25.95	0.0004	4.1051	1444.54	81.27	1734.94	575.35
5	0.000988	46.47	0.0749	4.1059	1468.27	83.94	1480.98	583.85
10	0.000988	67.01	0.1480	4.1076	1489.49	86.33	1281.79	592.65
15	0.000989	87.55	0.2199	4.1097	1508.38	88.46	1122.43	601.56
20	0.000990	108.10	0.2907	4.1118	1525.10	90.33	992.76	610.38
25	0.000992	128.67	0.3602	4.1138	1539.84	91.96	885.69	618.98
30	0.000993	149.24	0.4287	4.1158	1552.74	93.37	796.15	627.24
35	0.000995	169.83	0.4960	4.1178	1563.94	94.56	720.44	635.08
40	0.000997	190.42	0.5623	4.1197	1573.55	95.54	655.79	642.44
45	0.000999	211.02	0.6276	4.1218	1581.69	96.34	600.11	649.30
50	0.001001	231.64	0.6919	4.1240	1588.45	96.95	551.79	655.63
60	0.001006	272.90	0.8176	4.1291	1598.21	97.67	472.46	666.73
70	0.001011	314.22	0.9398	4.1353	1603.43	97.78	410.52	675.84
80	0.001017	355.61	1.0587	4.1426	1604.66	97.36	361.23	683.14
90	0.001024	397.08	1.1745	4.1513	1602.32	96.45	321.36	688.85
100	0.001031	438.64	1.2874	4.1613	1596.82	95.14	288.67	693.17
110	0.001038	480.31	1.3976	4.1727	1588.45	93.45	261.53	696.29
120	0.001047	522.10	1.5053	4.1857	1577.50	91.45	238.77	698.34
130	0.001055	564.03	1.6106	4.2004	1564.19	89.18	219.48	699.44
140	0.001064	606.12	1.7137	4.2170	1548.68	86.66	203.00	699.66
150	0.001074	648.38	1.8148	4.2356	1531.13	83.93	188.80	699.05
160	0.001085	690.84	1.9140	4.2566	1511.63	81.02	176.47	697.63
170	0.001096	733.52	2.0114	4.2803	1490.26	77.94	165.68	695.42
180	0.001108	776.45	2.1072	4.3070	1467.07	74.73	156.17	692.42
190	0.001120	819.67	2.2015	4.3373	1442.09	71.39	147.74	688.61
200	0.001134	863.21	2.2945	4.3715	1415.33	67.95	140.21	683.98
210	0.001148	907.12	2.3863	4.4104	1386.80	64.43	133.44	678.51
220	0.001163	951.44	2.4771	4.4546	1356.49	60.83	127.32	672.14
230	0.001180	996.23	2.5671	4.5049	1324.37	57.17	121.74	664.85
240	0.001198	1041.56	2.6563	4.5625	1290.44	53.48	116.62	656.57
250	0.001217	1087.51	2.7449	4.6286	1254.67	49.76	111.88	647.24
260	0.001237	1134.16	2.8333	4.7048	1217.02	46.03	107.47	636.78
270	0.001260	1181.64	2.9215	4.7932	1177.43	42.31	103.32	625.11
280	0.001285	1230.08	3.0099	4.8966	1135.77	38.61	99.38	612.15
290	0.001312	1279.64	3.0987	5.0188	1091.86	34.94	95.61	597.82
300	0.001343	1330.53	3.1882	5.1654	1045.42	31.30	91.95	582.08
310	0.001378	1383.05	3.2791	5.3447	996.13	27.70	88.36	564.94
320	0.001417	1437.58	3.3718	5.5697	943.68	24.17	84.78	546.47
330	0.001464	1494.67	3.4672	5.8618	887.58	20.70	81.15	526.82
340	0.001519	1555.16	3.5667	6.2589	826.76	17.31	77.38	506.11

Table 4 Single-Phase Region (Regions 1 to 3 of IAPWS-IF97) – Continuation
Homogenes Zustandsgebiet (Bereiche 1 bis 3 der IAPWS-IF97) – Fortsetzung

p = 26 MPa

t [°C]	v [m^3 kg^{-1}]	h [kJ kg^{-1}]	s [kJ kg^{-1} K^{-1}]	c_p [kJ kg^{-1} K^{-1}]	w [m s^{-1}]	κ	$\eta \times 10^6$ [kg m^{-1} s^{-1}]	$\lambda \times 10^3$ [W m^{-1} K^{-1}]
350	0.001588	1620.44	3.6723	6.8365	759.34	13.9613	73.35	484.26
360	0.001681	1693.16	3.7880	7.8050	684.72	10.7302	68.85	460.85
370	0.001818	1779.58	3.9234	9.7460	593.74	7.4598	63.40	433.63
380	0.002087	1901.05	4.1107	16.2115	474.65	4.1520	55.47	398.47
390	0.003552	2242.70	4.6290	47.3156	379.01	1.5556	36.76	318.55
400	0.005287	2510.55	5.0304	16.2374	436.89	1.3886	30.53	192.82
410	0.006256	2639.46	5.2206	10.5538	469.97	1.3578	29.34	156.17
420	0.006990	2731.76	5.3548	8.1627	494.42	1.3450	28.98	137.58
430	0.007602	2806.08	5.4612	6.8141	514.15	1.3374	28.94	126.23
440	0.008137	2869.65	5.5510	5.9567	530.78	1.3316	29.06	118.71
450	0.008619	2926.06	5.6296	5.3574	545.33	1.3270	29.26	113.45
460	0.009062	2977.30	5.7000	4.9105	558.38	1.3233	29.52	109.67
470	0.009474	3024.60	5.7640	4.5635	570.28	1.3203	29.82	106.93
480	0.009861	3068.80	5.8231	4.2866	581.27	1.3178	30.14	104.96
490	0.010228	3110.51	5.8781	4.0614	591.50	1.3157	30.48	103.57
500	0.010578	3150.16	5.9298	3.8754	601.10	1.3138	30.83	102.64
510	0.010913	3188.11	5.9785	3.7198	610.16	1.3121	31.19	102.07
520	0.011236	3224.64	6.0249	3.5883	618.74	1.3105	31.56	101.79
530	0.011548	3259.95	6.0691	3.4761	626.91	1.3090	31.94	101.76
540	0.011850	3294.21	6.1115	3.3796	634.71	1.3075	32.32	101.92
550	0.012144	3327.58	6.1523	3.2959	642.19	1.3061	32.70	102.24
560	0.012431	3360.17	6.1917	3.2229	649.38	1.3047	33.08	102.70
570	0.012711	3392.07	6.2297	3.1587	656.30	1.3033	33.46	103.28
580	0.012985	3423.37	6.2666	3.1021	662.99	1.3019	33.85	103.96
590	0.013254	3454.13	6.3025	3.0520	669.47	1.3006	34.23	104.72
600	0.013517	3484.42	6.3374	3.0073	675.75	1.2993	34.62	105.54
610	0.013776	3514.29	6.3714	2.9674	681.86	1.2980	35.00	106.43
620	0.014031	3543.78	6.4046	2.9317	687.80	1.2967	35.39	107.36
630	0.014282	3572.94	6.4371	2.8996	693.59	1.2955	35.77	108.33
640	0.014530	3601.79	6.4688	2.8707	699.25	1.2943	36.15	109.34
650	0.014774	3630.36	6.5000	2.8447	704.78	1.2931	36.53	110.37
660	0.015015	3658.69	6.5305	2.8212	710.19	1.2919	36.91	111.43
670	0.015253	3686.79	6.5604	2.8000	715.48	1.2908	37.29	112.50
680	0.015489	3714.70	6.5899	2.7807	720.68	1.2897	37.67	113.59
690	0.015722	3742.42	6.6188	2.7633	725.77	1.2886	38.05	114.69
700	0.015953	3769.97	6.6473	2.7476	730.78	1.2875	38.42	115.80
720	0.016409	3824.64	6.7029	2.7204	740.54	1.2854	39.17	118.03
740	0.016856	3878.82	6.7569	2.6982	749.98	1.2834	39.91	120.28
760	0.017297	3932.60	6.8094	2.6803	759.14	1.2814	40.65	122.54
780	0.017732	3986.05	6.8607	2.6659	768.05	1.2795	41.38	124.80
800	0.018162	4039.25	6.9107	2.6545	776.71	1.2776	42.10	127.05

Table 4 Single-Phase Region (Regions 1 to 3 of IAPWS-IF97) – Continuation
Homogenes Zustandsgebiet (Bereiche 1 bis 3 der IAPWS-IF97) – Fortsetzung

				p = 27 MPa				
t	v	h	s	c_p	w	κ	$\eta\times 10^6$	$\lambda\times 10^3$
[°C]	[m^3 kg^{-1}]	[kJ kg^{-1}]	[kJ kg^{-1} K^{-1}]	[kJ kg^{-1} K^{-1}]	[m s^{-1}]		[kg m^{-1} s^{-1}]	[W m^{-1} K^{-1}]
0	0.000987	26.93	0.0004	4.1012	1446.22	78.48	1733.09	575.89
5	0.000987	47.43	0.0748	4.1025	1469.94	81.05	1479.77	584.36
10	0.000988	67.95	0.1479	4.1046	1491.15	83.36	1281.02	593.14
15	0.000989	88.48	0.2198	4.1070	1510.02	85.41	1121.97	602.03
20	0.000990	109.02	0.2904	4.1093	1526.74	87.21	992.51	610.84
25	0.000991	129.57	0.3599	4.1115	1541.48	88.78	885.59	619.43
30	0.000993	150.14	0.4283	4.1136	1554.37	90.14	796.16	627.69
35	0.000994	170.71	0.4956	4.1156	1565.57	91.28	720.52	635.53
40	0.000996	191.29	0.5619	4.1177	1575.19	92.24	655.93	642.90
45	0.000998	211.89	0.6271	4.1198	1583.35	93.00	600.29	649.76
50	0.001001	232.49	0.6914	4.1221	1590.13	93.59	552.00	656.10
60	0.001005	273.74	0.8171	4.1272	1599.92	94.29	472.71	667.21
70	0.001011	315.04	0.9393	4.1334	1605.19	94.41	410.79	676.33
80	0.001017	356.41	1.0581	4.1407	1606.48	94.00	361.50	683.64
90	0.001023	397.86	1.1738	4.1493	1604.21	93.14	321.63	689.37
100	0.001030	439.40	1.2867	4.1593	1598.77	91.88	288.93	693.71
110	0.001038	481.05	1.3968	4.1707	1590.49	90.27	261.80	696.85
120	0.001046	522.82	1.5044	4.1836	1579.62	88.35	239.03	698.92
130	0.001055	564.72	1.6097	4.1981	1566.40	86.16	219.74	700.05
140	0.001064	606.79	1.7128	4.2145	1550.99	83.74	203.25	700.29
150	0.001074	649.02	1.8138	4.2330	1533.54	81.12	189.04	699.70
160	0.001084	691.45	1.9129	4.2538	1514.16	78.32	176.71	698.32
170	0.001095	734.11	2.0102	4.2772	1492.92	75.37	165.91	696.14
180	0.001107	777.01	2.1060	4.3036	1469.87	72.28	156.40	693.17
190	0.001120	820.19	2.2002	4.3334	1445.04	69.08	147.97	689.39
200	0.001133	863.69	2.2931	4.3672	1418.45	65.78	140.44	684.80
210	0.001147	907.55	2.3849	4.4056	1390.11	62.39	133.68	679.36
220	0.001162	951.82	2.4756	4.4491	1360.00	58.93	127.55	673.05
230	0.001179	996.55	2.5654	4.4987	1328.12	55.42	121.97	665.80
240	0.001196	1041.81	2.6544	4.5553	1294.44	51.87	116.86	657.59
250	0.001215	1087.68	2.7430	4.6202	1258.94	48.30	112.13	648.33
260	0.001236	1134.25	2.8311	4.6949	1221.60	44.72	107.72	637.96
270	0.001258	1181.62	2.9192	4.7814	1182.37	41.15	103.58	626.40
280	0.001283	1229.93	3.0073	4.8823	1141.13	37.59	99.65	613.55
290	0.001310	1279.33	3.0958	5.0012	1097.71	34.07	95.90	599.37
300	0.001340	1330.03	3.1850	5.1433	1051.87	30.58	92.26	583.80
310	0.001374	1382.30	3.2754	5.3161	1003.30	27.13	88.69	566.85
320	0.001413	1436.49	3.3676	5.5315	951.73	23.74	85.14	548.60
330	0.001458	1493.13	3.4622	5.8084	896.83	20.43	81.55	529.20
340	0.001512	1552.97	3.5606	6.1798	837.81	17.19	77.84	508.79

Table 4 Single-Phase Region (Regions 1 to 3 of IAPWS-IF97) – Continuation
Homogenes Zustandsgebiet (Bereiche 1 bis 3 der IAPWS-IF97) – Fortsetzung

				p = 27 MPa				
t [°C]	v [m^3 kg^{-1}]	h [kJ kg^{-1}]	s [kJ kg^{-1} K^{-1}]	c_p [kJ kg^{-1} K^{-1}]	w [m s^{-1}]	κ	$\eta \times 10^6$ [kg m^{-1} s^{-1}]	$\lambda \times 10^3$ [W m^{-1} K^{-1}]
350	0.001579	1617.24	3.6646	6.7084	772.86	14.0128	73.91	487.35
360	0.001666	1688.19	3.7775	7.5614	701.06	10.9287	59.56	464.51
370	0.001790	1770.78	3.9069	9.1448	616.33	7.8594	54.45	438.48
380	0.002008	1878.90	4.0737	13.3100	511.31	4.8216	57.57	405.45
390	0.002734	2096.01	4.4031	37.5530	389.20	2.0517	44.25	359.13
400	0.004566	2429.83	4.9032	21.1421	422.18	1.4456	32.52	225.94
410	0.005653	2586.98	5.1351	12.2342	459.66	1.3844	30.33	172.78
420	0.006433	2691.52	5.2870	9.0550	486.30	1.3615	29.64	148.40
430	0.007067	2772.93	5.4037	7.3768	507.54	1.3500	29.44	134.10
440	0.007614	2841.18	5.5001	6.3483	525.20	1.3418	29.46	124.82
450	0.008101	2900.98	5.5833	5.6513	540.48	1.3355	29.60	118.43
460	0.008546	2954.83	5.6573	5.1423	554.10	1.3306	29.81	113.85
470	0.008958	3004.22	5.7242	4.7521	566.46	1.3267	30.08	110.53
480	0.009344	3050.14	5.7856	4.4435	577.84	1.3235	30.38	108.12
490	0.009708	3093.29	5.8425	4.1941	588.40	1.3209	30.70	106.39
500	0.010054	3134.17	5.8958	3.9893	598.28	1.3186	31.03	105.18
510	0.010385	3173.18	5.9459	3.8187	607.59	1.3165	31.38	104.40
520	0.010704	3210.63	5.9934	3.6752	616.39	1.3147	31.74	103.94
530	0.011011	3246.76	6.0387	3.5531	624.76	1.3129	32.11	103.76
540	0.011308	3281.75	6.0820	3.4484	632.74	1.3112	32.48	103.80
550	0.011597	3315.78	6.1236	3.3579	640.37	1.3096	32.85	104.03
560	0.011878	3348.95	6.1636	3.2791	647.70	1.3081	33.23	104.41
570	0.012153	3381.39	6.2023	3.2100	654.76	1.3065	33.61	104.91
580	0.012421	3413.18	6.2398	3.1492	661.57	1.3050	33.99	105.53
590	0.012684	3444.40	6.2762	3.0953	668.15	1.3036	34.37	106.23
600	0.012941	3475.11	6.3116	3.0473	674.53	1.3021	34.75	107.01
610	0.013194	3505.36	6.3460	3.0046	680.73	1.3008	35.13	107.86
620	0.013443	3535.21	6.3796	2.9663	686.75	1.2994	35.51	108.75
630	0.013688	3564.70	6.4125	2.9319	692.62	1.2981	35.89	109.70
640	0.013929	3593.86	6.4446	2.9009	698.35	1.2968	36.27	110.67
650	0.014167	3622.73	6.4760	2.8730	703.95	1.2955	36.65	111.68
660	0.014402	3651.33	6.5068	2.8478	709.42	1.2943	37.03	112.71
670	0.014634	3679.69	6.5371	2.8250	714.78	1.2931	37.41	113.76
680	0.014863	3707.84	6.5667	2.8043	720.03	1.2919	37.78	114.83
690	0.015090	3735.78	6.5959	2.7856	725.18	1.2908	38.16	115.91
700	0.015314	3763.55	6.6246	2.7687	730.23	1.2896	38.53	117.00
720	0.015757	3818.63	6.6806	2.7394	740.08	1.2874	39.27	119.20
740	0.016192	3873.17	6.7350	2.7154	749.61	1.2853	40.01	121.42
760	0.016620	3927.27	6.7879	2.6959	758.85	1.2833	40.74	123.64
780	0.017042	3981.03	6.8394	2.6801	767.82	1.2813	41.47	125.87
800	0.017458	4034.50	6.8897	2.6676	776.55	1.2793	42.19	128.09

Table 4 Single-Phase Region (Regions 1 to 3 of IAPWS-IF97) – Continuation
Homogenes Zustandsgebiet (Bereiche 1 bis 3 der IAPWS-IF97) – Fortsetzung

p = 28 MPa

t [°C]	v [m³ kg⁻¹]	h [kJ kg⁻¹]	s [kJ kg⁻¹ K⁻¹]	c_p [kJ kg⁻¹ K⁻¹]	w [m s⁻¹]	κ	$\eta\times 10^6$ [kg m⁻¹ s⁻¹]	$\lambda\times 10^3$ [W m⁻¹ K⁻¹]
0	0.000987	27.91	0.0003	4.0974	1447.91	75.89	1731.25	576.44
5	0.000987	48.40	0.0747	4.0992	1471.62	78.37	1478.58	584.86
10	0.000988	68.90	0.1477	4.1017	1492.81	80.60	1280.26	593.62
15	0.000988	89.41	0.2196	4.1043	1511.67	82.57	1121.51	602.50
20	0.000989	109.94	0.2902	4.1068	1528.38	84.31	992.26	611.30
25	0.000991	130.48	0.3597	4.1091	1543.11	85.83	885.49	619.89
30	0.000992	151.03	0.4280	4.1113	1556.01	87.14	796.17	628.14
35	0.000994	171.59	0.4953	4.1135	1567.21	88.25	720.61	635.98
40	0.000996	192.17	0.5615	4.1156	1576.84	89.16	656.07	643.35
45	0.000998	212.75	0.6267	4.1178	1585.00	89.91	600.47	650.22
50	0.001000	233.34	0.6910	4.1201	1591.80	90.48	552.21	656.56
60	0.001005	274.57	0.8166	4.1253	1601.63	91.16	472.95	667.68
70	0.001010	315.85	0.9387	4.1315	1606.96	91.27	411.05	676.81
80	0.001016	357.20	1.0575	4.1388	1608.29	90.89	361.77	684.14
90	0.001023	398.63	1.1732	4.1474	1606.09	90.06	321.90	689.89
100	0.001030	440.16	1.2859	4.1573	1600.73	88.85	289.20	694.25
110	0.001037	481.79	1.3960	4.1686	1592.52	87.30	262.06	697.41
120	0.001046	523.53	1.5036	4.1814	1581.74	85.46	239.28	699.50
130	0.001054	565.42	1.6088	4.1959	1568.60	83.36	219.99	700.65
140	0.001063	607.46	1.7118	4.2121	1553.29	81.03	203.50	700.92
150	0.001073	649.67	1.8128	4.2304	1535.95	78.51	189.29	700.36
160	0.001084	692.07	1.9118	4.2509	1516.68	75.82	176.94	699.00
170	0.001095	734.70	2.0091	4.2741	1495.56	72.98	166.15	696.85
180	0.001106	777.56	2.1047	4.3002	1472.65	70.01	156.64	693.91
190	0.001119	820.71	2.1989	4.3297	1447.97	66.93	148.20	690.17
200	0.001132	864.17	2.2918	4.3630	1421.55	63.75	140.67	685.62
210	0.001146	907.99	2.3834	4.4008	1393.39	60.49	133.91	680.22
220	0.001161	952.20	2.4740	4.4437	1363.49	57.17	127.78	673.95
230	0.001178	996.88	2.5637	4.4925	1331.82	53.79	122.21	666.76
240	0.001195	1042.08	2.6526	4.5482	1298.39	50.38	117.09	658.60
250	0.001214	1087.87	2.7410	4.6120	1263.17	46.94	112.37	649.42
260	0.001234	1134.35	2.8290	4.6852	1226.14	43.50	107.97	639.13
270	0.001257	1181.61	2.9168	4.7699	1187.25	40.06	103.84	627.66
280	0.001281	1229.79	3.0047	4.8684	1146.42	36.64	99.92	614.94
290	0.001308	1279.04	3.0929	4.9842	1103.49	33.26	96.18	600.90
300	0.001338	1329.55	3.1818	5.1219	1058.22	29.90	92.56	585.49
310	0.001371	1381.57	3.2718	5.2887	1010.34	26.59	89.01	568.73
320	0.001409	1435.45	3.3634	5.4951	959.59	23.34	85.50	550.70
330	0.001453	1491.66	3.4574	5.7582	905.78	20.16	81.95	531.53
340	0.001506	1550.90	3.5548	6.1067	848.38	17.07	78.30	511.41

Table 4 Single-Phase Region (Regions 1 to 3 of IAPWS-IF97) – Continuation
Homogenes Zustandsgebiet (Bereiche 1 bis 3 der IAPWS-IF97) – Fortsetzung

p = 28 MPa

t [°C]	v [$m^3\,kg^{-1}$]	h [$kJ\,kg^{-1}$]	s [$kJ\,kg^{-1}\,K^{-1}$]	c_p [$kJ\,kg^{-1}\,K^{-1}$]	w [$m\,s^{-1}$]	κ	$\eta \times 10^6$ [$kg\,m^{-1}\,s^{-1}$]	$\lambda \times 10^3$ [$W\,m^{-1}\,K^{-1}$]
350	0.001570	1614.25	3.6573	6.5932	785.84	14.0510	74.44	490.34
360	0.001652	1683.64	3.7677	7.3526	716.33	11.0935	70.24	468.00
370	0.001766	1763.10	3.8922	8.6798	636.68	8.1956	65.39	442.98
380	0.001952	1862.42	4.0454	11.6724	541.56	5.3658	59.21	412.01
390	0.002398	2022.34	4.2882	23.4816	428.02	2.7287	49.29	369.69
400	0.003855	2334.42	4.7552	27.3105	409.44	1.5533	35.51	267.40
410	0.005071	2529.06	5.0424	14.3338	449.43	1.4227	31.60	192.68
420	0.005904	2648.46	5.2160	10.1031	478.15	1.3830	30.43	160.85
430	0.006563	2738.07	5.3444	8.0188	500.89	1.3652	30.01	142.94
440	0.007123	2811.60	5.4483	6.7824	519.63	1.3538	29.91	131.57
450	0.007617	2875.11	5.5367	5.9700	535.67	1.3454	29.97	123.84
460	0.008065	2931.77	5.6145	5.3900	549.86	1.3389	30.14	118.36
470	0.008477	2983.38	5.6845	4.9518	562.68	1.3339	30.36	114.38
480	0.008862	3031.12	5.7483	4.6086	574.44	1.3298	30.63	111.47
490	0.009224	3075.78	5.8072	4.3330	585.34	1.3266	30.93	109.36
500	0.009568	3117.94	5.8621	4.1079	595.51	1.3238	31.25	107.86
510	0.009895	3158.06	5.9136	3.9213	605.06	1.3213	31.58	106.83
520	0.010210	3196.47	5.9624	3.7649	614.09	1.3191	31.93	106.19
530	0.010512	3233.44	6.0087	3.6324	622.65	1.3171	32.29	105.85
540	0.010805	3269.18	6.0529	3.5191	630.80	1.3152	32.65	105.76
550	0.011089	3303.87	6.0953	3.4214	638.60	1.3134	33.01	105.88
560	0.011365	3337.65	6.1361	3.3366	646.07	1.3116	33.38	106.16
570	0.011634	3370.64	6.1755	3.2624	653.25	1.3100	33.76	106.59
580	0.011897	3402.93	6.2136	3.1971	660.17	1.3083	34.13	107.14
590	0.012154	3434.61	6.2505	3.1393	666.86	1.3067	34.51	107.79
600	0.012407	3465.74	6.2863	3.0880	673.34	1.3052	34.89	108.52
610	0.012654	3496.39	6.3212	3.0423	679.63	1.3036	35.26	109.33
620	0.012897	3526.60	6.3552	3.0013	685.74	1.3022	35.64	110.18
630	0.013136	3556.43	6.3885	2.9645	691.68	1.3008	36.02	111.09
640	0.013371	3585.90	6.4209	2.9315	697.48	1.2994	36.39	112.04
650	0.013603	3615.07	6.4527	2.9016	703.15	1.2980	36.77	113.02
660	0.013832	3643.95	6.4838	2.8747	708.68	1.2967	37.15	114.02
670	0.014058	3672.57	6.5143	2.8503	714.10	1.2954	37.52	115.05
680	0.014282	3700.96	6.5442	2.8282	719.40	1.2942	37.89	116.10
690	0.014503	3729.14	6.5737	2.8081	724.61	1.2930	38.27	117.15
700	0.014721	3757.13	6.6026	2.7900	729.71	1.2918	38.64	118.23
720	0.015152	3812.60	6.6590	2.7586	739.65	1.2895	39.38	120.39
740	0.015574	3867.51	6.7137	2.7327	749.26	1.2873	40.11	122.57
760	0.015990	3921.94	6.7669	2.7116	758.57	1.2852	40.84	124.76
780	0.016400	3976.00	6.8188	2.6945	767.61	1.2831	41.56	126.96
800	0.016805	4029.74	6.8693	2.6808	776.40	1.2811	42.28	129.15

Table 4 Single-Phase Region (Regions 1 to 3 of IAPWS-IF97) – Continuation
Homogenes Zustandsgebiet (Bereiche 1 bis 3 der IAPWS-IF97) – Fortsetzung

p = 29 MPa

t [°C]	v [m^3 kg^{-1}]	h [kJ kg^{-1}]	s [kJ kg^{-1} K^{-1}]	c_p [kJ kg^{-1} K^{-1}]	w [m s^{-1}]	κ	$\eta \times 10^6$ [kg m^{-1} s^{-1}]	$\lambda \times 10^3$ [W m^{-1} K^{-1}]
0	0.000986	28.88	0.0003	4.0936	1449.60	73.48	1729.45	576.98
5	0.000986	49.36	0.0746	4.0959	1473.30	75.88	1477.41	585.37
10	0.000987	69.84	0.1476	4.0987	1494.48	78.02	1279.52	594.11
15	0.000988	90.34	0.2194	4.1016	1513.33	79.93	1121.06	602.96
20	0.000989	110.86	0.2899	4.1043	1530.03	81.62	992.02	611.76
25	0.000990	131.39	0.3594	4.1068	1544.75	83.08	885.40	620.34
30	0.000992	151.93	0.4277	4.1091	1557.64	84.35	796.18	628.60
35	0.000994	172.48	0.4949	4.1113	1568.85	85.42	720.69	636.44
40	0.000996	193.04	0.5611	4.1135	1578.49	86.31	656.21	643.81
45	0.000998	213.61	0.6263	4.1158	1586.66	87.02	600.65	650.68
50	0.001000	234.20	0.6905	4.1181	1593.47	87.58	552.42	657.02
60	0.001005	275.40	0.8161	4.1234	1603.34	88.24	473.20	668.16
70	0.001010	316.67	0.9381	4.1296	1608.72	88.36	411.31	677.30
80	0.001016	358.00	1.0568	4.1370	1610.11	87.99	362.04	684.65
90	0.001022	399.41	1.1725	4.1455	1607.97	87.20	322.17	690.41
100	0.001029	440.91	1.2852	4.1554	1602.68	86.04	289.47	694.79
110	0.001037	482.52	1.3953	4.1666	1594.55	84.55	262.32	697.97
120	0.001045	524.25	1.5028	4.1793	1583.85	82.77	239.54	700.09
130	0.001054	566.12	1.6079	4.1936	1570.80	80.75	220.24	701.25
140	0.001063	608.13	1.7109	4.2097	1555.59	78.51	203.75	701.55
150	0.001073	650.32	1.8118	4.2278	1538.35	76.08	189.53	701.01
160	0.001083	692.69	1.9107	4.2481	1519.19	73.49	177.18	699.68
170	0.001094	735.29	2.0079	4.2710	1498.20	70.76	166.38	697.56
180	0.001106	778.12	2.1035	4.2968	1475.42	67.90	156.87	694.65
190	0.001118	821.23	2.1976	4.3259	1450.89	64.93	148.44	690.95
200	0.001131	864.65	2.2904	4.3588	1424.64	61.87	140.90	686.43
210	0.001145	908.43	2.3819	4.3961	1396.66	58.73	134.14	681.07
220	0.001160	952.59	2.4724	4.4384	1366.95	55.52	128.02	674.84
230	0.001177	997.21	2.5620	4.4865	1335.50	52.27	122.44	667.71
240	0.001194	1042.35	2.6508	4.5413	1302.31	48.98	117.33	659.61
250	0.001213	1088.06	2.7390	4.6039	1267.36	45.67	112.61	650.49
260	0.001233	1134.45	2.8269	4.6757	1230.63	42.36	108.22	640.29
270	0.001255	1181.62	2.9145	4.7586	1192.08	39.05	104.09	628.92
280	0.001279	1229.67	3.0022	4.8549	1151.64	35.76	100.19	616.32
290	0.001305	1278.77	3.0901	4.9677	1109.18	32.50	96.46	602.41
300	0.001335	1329.09	3.1787	5.1013	1064.47	29.27	92.86	587.16
310	0.001368	1380.89	3.2683	5.2623	1017.25	26.09	89.33	570.59
320	0.001405	1434.46	3.3594	5.4604	967.29	22.96	85.84	552.76
330	0.001448	1490.27	3.4527	5.7108	914.46	19.91	82.33	533.82
340	0.001499	1548.93	3.5491	6.0387	858.50	16.95	78.73	513.96

Table 4 Single-Phase Region (Regions 1 to 3 of IAPWS-IF97) – Continuation
Homogenes Zustandsgebiet (Bereiche 1 bis 3 der IAPWS-IF97) – Fortsetzung

p = 29 MPa								
t [°C]	v [m^3 kg^{-1}]	h [kJ kg^{-1}]	s [kJ kg^{-1} K^{-1}]	c_p [kJ kg^{-1} K^{-1}]	w [m s^{-1}]	κ	$\eta \times 10^6$ [kg m^{-1} s^{-1}]	$\lambda \times 10^3$ [W m^{-1} K^{-1}]
350	0.001561	1611.44	3.6503	6.4887	798.21	14.0742	74.96	493.24
360	0.001639	1679.45	3.7585	7.1709	730.70	11.2307	70.88	471.35
370	0.001746	1756.30	3.8789	8.3066	655.26	8.4815	66.25	447.20
380	0.001909	1849.25	4.0223	10.6004	567.61	5.8210	60.58	418.00
390	0.002234	1982.05	4.2239	17.4122	465.23	3.3403	52.45	378.82
400	0.003231	2234.27	4.6012	29.8466	406.40	1.7630	39.64	306.40
410	0.004511	2465.14	4.9418	16.9324	439.90	1.4792	33.25	216.45
420	0.005399	2602.20	5.1411	11.3603	470.18	1.4118	31.38	175.27
430	0.006088	2701.40	5.2833	8.7469	494.26	1.3837	30.67	152.87
440	0.006662	2780.85	5.3955	7.2640	514.09	1.3680	30.42	139.01
450	0.007164	2848.43	5.4896	6.3158	530.90	1.3567	30.39	129.73
460	0.007615	2908.10	5.5716	5.6546	545.67	1.3483	30.49	123.21
470	0.008029	2962.08	5.6447	5.1630	558.96	1.3419	30.67	118.49
480	0.008413	3011.73	5.7111	4.7820	571.10	1.3369	30.91	115.03
490	0.008773	3057.98	5.7721	4.4782	582.32	1.3328	31.18	112.50
500	0.009114	3101.48	5.8287	4.2313	592.78	1.3295	31.48	110.67
510	0.009439	3142.75	5.8817	4.0277	602.58	1.3266	31.80	109.38
520	0.009749	3182.15	5.9317	3.8576	611.83	1.3240	32.13	108.53
530	0.010048	3219.99	5.9791	3.7140	620.58	1.3217	32.47	108.02
540	0.010337	3256.50	6.0243	3.5916	628.91	1.3195	32.83	107.79
550	0.010616	3291.88	6.0676	3.4864	636.86	1.3174	33.18	107.79
560	0.010887	3326.28	6.1091	3.3953	644.47	1.3155	33.55	107.98
570	0.011152	3359.82	6.1491	3.3157	651.78	1.3136	33.91	108.33
580	0.011410	3392.62	6.1878	3.2458	658.82	1.3118	34.28	108.81
590	0.011662	3424.77	6.2253	3.1841	665.61	1.3100	34.65	109.40
600	0.011909	3456.33	6.2616	3.1293	672.19	1.3083	35.03	110.08
610	0.012151	3487.37	6.2970	3.0805	678.56	1.3067	35.40	110.83
620	0.012389	3517.96	6.3314	3.0368	684.75	1.3051	35.77	111.65
630	0.012622	3548.12	6.3650	2.9976	690.78	1.3036	36.15	112.52
640	0.012852	3577.92	6.3978	2.9624	696.64	1.3021	36.52	113.44
650	0.013079	3607.38	6.4299	2.9305	702.37	1.3007	36.89	114.39
660	0.013303	3636.54	6.4613	2.9018	707.97	1.2993	37.27	115.36
670	0.013523	3665.43	6.4921	2.8758	713.44	1.2979	37.64	116.37
680	0.013741	3694.07	6.5223	2.8522	718.80	1.2966	38.01	117.39
690	0.013956	3722.48	6.5520	2.8308	724.06	1.2953	38.38	118.43
700	0.014169	3750.69	6.5811	2.8114	729.21	1.2941	38.75	119.48
720	0.014589	3806.57	6.6380	2.7778	739.24	1.2917	39.48	121.60
740	0.015000	3861.84	6.6931	2.7501	748.93	1.2894	40.22	123.75
760	0.015405	3916.61	6.7466	2.7274	758.31	1.2872	40.94	125.91
780	0.015803	3970.97	6.7987	2.7089	767.42	1.2850	41.66	128.07
800	0.016196	4024.99	6.8495	2.6940	776.26	1.2829	42.38	130.22

Table 4 Single-Phase Region (Regions 1 to 3 of IAPWS-IF97) – Continuation
Homogenes Zustandsgebiet (Bereiche 1 bis 3 der IAPWS-IF97) – Fortsetzung

p = 30 MPa

t [°C]	v [m^3 kg^{-1}]	h [kJ kg^{-1}]	s [kJ kg^{-1} K^{-1}]	c_p [kJ kg^{-1} K^{-1}]	w [m s^{-1}]	κ	$\eta \times 10^6$ [kg m^{-1} s^{-1}]	$\lambda \times 10^3$ [W m^{-1} K^{-1}]
0	0.000986	29.86	0.0003	4.0899	1451.30	71.23	1727.66	577.53
5	0.000986	50.32	0.0745	4.0926	1474.98	73.55	1476.25	585.88
10	0.000987	70.79	0.1474	4.0958	1496.15	75.63	1278.79	594.59
15	0.000988	91.27	0.2192	4.0989	1514.98	77.47	1120.62	603.43
20	0.000989	111.78	0.2897	4.1018	1531.67	79.10	991.79	612.22
25	0.000990	132.29	0.3591	4.1044	1546.39	80.52	885.31	620.80
30	0.000992	152.82	0.4274	4.1069	1559.28	81.74	796.19	629.05
35	0.000993	173.36	0.4946	4.1092	1570.49	82.78	720.79	636.89
40	0.000995	193.91	0.5607	4.1115	1580.13	83.64	656.36	644.27
45	0.000997	214.47	0.6259	4.1138	1588.32	84.33	600.84	651.14
50	0.000999	235.05	0.6900	4.1162	1595.15	84.87	552.63	657.49
60	0.001004	276.24	0.8156	4.1215	1605.06	85.52	473.45	668.63
70	0.001010	317.48	0.9376	4.1278	1610.48	85.63	411.58	677.79
80	0.001016	358.80	1.0562	4.1351	1611.93	85.29	362.31	685.15
90	0.001022	400.19	1.1718	4.1436	1609.85	84.53	322.44	690.93
100	0.001029	441.67	1.2845	4.1534	1604.63	83.41	289.74	695.33
110	0.001037	483.26	1.3945	4.1646	1596.57	81.97	262.58	698.52
120	0.001045	524.97	1.5019	4.1772	1585.95	80.27	239.80	700.67
130	0.001053	566.81	1.6070	4.1914	1573.00	78.32	220.49	701.86
140	0.001062	608.80	1.7099	4.2073	1557.88	76.16	203.99	702.18
150	0.001072	650.96	1.8107	4.2252	1540.74	73.82	189.77	701.67
160	0.001082	693.31	1.9097	4.2453	1521.69	71.32	177.42	700.36
170	0.001093	735.88	2.0068	4.2680	1500.82	68.68	166.62	698.27
180	0.001105	778.68	2.1023	4.2935	1478.18	65.92	157.10	695.40
190	0.001117	821.76	2.1964	4.3222	1453.80	63.06	148.67	691.72
200	0.001130	865.14	2.2890	4.3547	1427.70	60.10	141.13	687.24
210	0.001144	908.87	2.3805	4.3915	1399.90	57.08	134.37	681.92
220	0.001159	952.99	2.4709	4.4332	1370.38	53.99	128.25	675.74
230	0.001176	997.55	2.5603	4.4805	1339.15	50.85	122.67	668.65
240	0.001193	1042.62	2.6490	4.5344	1306.20	47.68	117.57	660.61
250	0.001211	1088.26	2.7371	4.5959	1271.51	44.49	112.85	651.57
260	0.001231	1134.57	2.8248	4.6664	1235.08	41.29	108.46	641.44
270	0.001253	1181.63	2.9122	4.7476	1196.86	38.10	104.35	630.17
280	0.001277	1229.56	2.9997	4.8417	1156.80	34.93	100.45	617.68
290	0.001303	1278.51	3.0874	4.9516	1114.79	31.79	96.74	603.91
300	0.001332	1328.66	3.1756	5.0814	1070.62	28.68	93.15	588.81
310	0.001365	1380.23	3.2648	5.2370	1024.05	25.62	89.65	572.42
320	0.001401	1433.51	3.3554	5.4272	974.84	22.60	86.19	554.79
330	0.001444	1488.93	3.4481	5.6659	922.91	19.67	82.71	536.07
340	0.001493	1547.07	3.5437	5.9753	868.21	16.83	79.16	516.45

Table 4 Single-Phase Region (Regions 1 to 3 of IAPWS-IF97) – Continuation
Homogenes Zustandsgebiet (Bereiche 1 bis 3 der IAPWS-IF97) – Fortsetzung

p = 30 MPa								
t [°C]	v [m^3 kg^{-1}]	h [kJ kg^{-1}]	s [kJ kg^{-1} K^{-1}]	c_p [kJ kg^{-1} K^{-1}]	w [m s^{-1}]	κ	$\eta \times 10^6$ [kg m^{-1} s^{-1}]	$\lambda \times 10^3$ [W m^{-1} K^{-1}]
350	0.001553	1608.80	3.6435	6.3935	809.96	14.0820	75.46	496.05
360	0.001628	1675.57	3.7498	7.0110	744.31	11.3452	71.49	474.57
370	0.001727	1750.19	3.8667	7.9986	672.42	8.7269	67.05	451.17
380	0.001873	1838.26	4.0026	9.8342	590.65	6.2087	61.77	423.48
390	0.002133	1955.23	4.1802	14.3444	497.30	3.8646	54.72	387.25
400	0.002796	2152.37	4.4750	25.7975	419.63	2.0990	43.95	331.51
410	0.003984	2395.84	4.8342	19.6849	432.54	1.5654	35.40	243.37
420	0.004921	2552.87	5.0625	12.7301	462.80	1.4509	32.53	191.41
430	0.005638	2662.82	5.2201	9.5627	487.80	1.4067	31.44	163.94
440	0.006228	2748.86	5.3416	7.7964	508.64	1.3847	30.99	147.20
450	0.006738	2820.91	5.4419	6.6908	526.23	1.3699	30.85	136.13
460	0.007193	2883.84	5.5284	5.9370	541.56	1.3591	30.88	128.43
470	0.007608	2940.32	5.6049	5.3861	555.31	1.3510	31.01	122.88
480	0.007992	2991.99	5.6740	4.9638	567.83	1.3447	31.20	118.81
490	0.008352	3039.89	5.7372	4.6297	579.37	1.3398	31.45	115.81
500	0.008690	3084.79	5.7956	4.3597	590.11	1.3357	31.72	113.62
510	0.009012	3127.24	5.8502	4.1378	600.16	1.3322	32.02	112.05
520	0.009320	3167.67	5.9015	3.9532	609.62	1.3292	32.34	110.97
530	0.009615	3206.41	5.9500	3.7979	618.57	1.3265	32.67	110.27
540	0.009899	3243.71	5.9962	3.6660	627.07	1.3241	33.01	109.89
550	0.010175	3279.79	6.0403	3.5529	635.17	1.3217	33.36	109.77
560	0.010442	3314.82	6.0826	3.4552	642.92	1.3196	33.72	109.86
570	0.010701	3348.94	6.1233	3.3701	650.36	1.3175	34.08	110.12
580	0.010955	3382.25	6.1626	3.2955	657.51	1.3155	34.44	110.52
590	0.011202	3414.87	6.2006	3.2296	664.40	1.3135	34.80	111.04
600	0.011444	3446.87	6.2374	3.1713	671.07	1.3117	35.17	111.67
610	0.011681	3478.32	6.2732	3.1193	677.53	1.3099	35.54	112.38
620	0.011914	3509.28	6.3081	3.0728	683.80	1.3082	35.91	113.15
630	0.012143	3539.79	6.3421	3.0311	689.90	1.3065	36.28	113.99
640	0.012368	3569.91	6.3752	2.9936	695.84	1.3049	36.65	114.87
650	0.012590	3599.68	6.4077	2.9598	701.63	1.3034	37.02	115.78
660	0.012808	3629.12	6.4394	2.9292	707.28	1.3019	37.39	116.74
670	0.013024	3658.27	6.4705	2.9016	712.81	1.3005	37.76	117.71
680	0.013236	3687.16	6.5009	2.8765	718.23	1.2991	38.13	118.71
690	0.013446	3715.81	6.5308	2.8537	723.53	1.2977	38.50	119.73
700	0.013654	3744.24	6.5602	2.8331	728.73	1.2964	38.86	120.75
720	0.014063	3800.53	6.6175	2.7972	738.85	1.2939	39.59	122.84
740	0.014464	3856.17	6.6729	2.7676	748.62	1.2915	40.32	124.95
760	0.014858	3911.27	6.7268	2.7433	758.07	1.2892	41.04	127.07
780	0.015246	3965.93	6.7792	2.7234	767.24	1.2870	41.76	129.19
800	0.015629	4020.23	6.8303	2.7072	776.14	1.2848	42.47	131.31

Table 4 Single-Phase Region (Regions 1 to 3 of IAPWS-IF97) – Continuation
Homogenes Zustandsgebiet (Bereiche 1 bis 3 der IAPWS-IF97) – Fortsetzung

p = 31 MPa

t [°C]	v [m^3 kg^{-1}]	h [kJ kg^{-1}]	s [kJ kg^{-1} K^{-1}]	c_p [kJ kg^{-1} K^{-1}]	w [m s^{-1}]	κ	$\eta \times 10^6$ [kg m^{-1} s^{-1}]	$\lambda \times 10^3$ [W m^{-1} K^{-1}]
0	0.000985	30.84	0.0002	4.0862	1453.00	69.13	1725.90	578.07
5	0.000986	51.27	0.0744	4.0894	1476.67	71.37	1475.11	586.39
10	0.000986	71.73	0.1473	4.0929	1497.82	73.38	1278.07	595.07
15	0.000987	92.20	0.2189	4.0963	1516.64	75.17	1120.19	603.90
20	0.000988	112.69	0.2894	4.0993	1533.32	76.74	991.56	612.68
25	0.000990	133.20	0.3588	4.1021	1548.03	78.12	885.22	621.25
30	0.000991	153.71	0.4270	4.1047	1560.92	79.30	796.21	629.50
35	0.000993	174.24	0.4942	4.1071	1572.13	80.31	720.88	637.35
40	0.000995	194.78	0.5603	4.1095	1581.78	81.14	656.50	644.72
45	0.000997	215.34	0.6255	4.1118	1589.98	81.82	601.02	651.60
50	0.000999	235.90	0.6896	4.1143	1596.82	82.34	552.85	657.95
60	0.001004	277.07	0.8151	4.1197	1606.77	82.97	473.70	669.11
70	0.001009	318.30	0.9370	4.1259	1612.24	83.09	411.84	678.28
80	0.001015	359.59	1.0556	4.1332	1613.75	82.76	362.58	685.65
90	0.001022	400.97	1.1711	4.1417	1611.73	82.03	322.71	691.45
100	0.001029	442.43	1.2838	4.1515	1606.58	80.95	290.00	695.86
110	0.001036	484.00	1.3937	4.1626	1598.59	79.57	262.85	699.08
120	0.001044	525.69	1.5011	4.1751	1588.06	77.92	240.06	701.24
130	0.001053	567.51	1.6061	4.1891	1575.18	76.04	220.75	702.46
140	0.001062	609.48	1.7090	4.2049	1560.16	73.96	204.24	702.80
150	0.001071	651.61	1.8098	4.2227	1543.12	71.70	190.01	702.32
160	0.001082	693.94	1.9086	4.2426	1524.19	69.28	177.66	701.04
170	0.001093	736.47	2.0057	4.2649	1503.43	66.74	166.85	698.98
180	0.001104	779.25	2.1011	4.2901	1480.92	64.07	157.34	696.14
190	0.001116	822.29	2.1951	4.3186	1456.68	61.31	148.90	692.50
200	0.001130	865.63	2.2877	4.3506	1430.74	58.46	141.36	688.05
210	0.001144	909.31	2.3790	4.3869	1403.11	55.53	134.60	682.77
220	0.001158	953.38	2.4693	4.4280	1373.79	52.55	128.48	676.63
230	0.001174	997.89	2.5587	4.4746	1342.78	49.52	122.91	669.60
240	0.001192	1042.90	2.6472	4.5276	1310.06	46.46	117.80	661.61
250	0.001210	1088.47	2.7352	4.5881	1275.63	43.38	113.09	652.63
260	0.001230	1134.69	2.8227	4.6573	1239.48	40.29	108.71	642.59
270	0.001252	1181.65	2.9100	4.7369	1201.58	37.21	104.60	631.41
280	0.001275	1229.47	2.9972	4.8288	1161.90	34.15	100.72	619.03
290	0.001301	1278.28	3.0846	4.9360	1120.32	31.12	97.01	605.39
300	0.001330	1328.25	3.1726	5.0621	1076.68	28.12	93.44	590.45
310	0.001362	1379.60	3.2614	5.2126	1030.74	25.17	89.96	574.22
320	0.001398	1432.61	3.3515	5.3955	982.25	22.27	86.52	556.78
330	0.001439	1487.66	3.4436	5.6234	931.16	19.44	83.08	538.28
340	0.001487	1545.29	3.5383	5.9159	877.55	16.70	79.58	518.89

Table 4 Single-Phase Region (Regions 1 to 3 of IAPWS-IF97) – Continuation
Homogenes Zustandsgebiet (Bereiche 1 bis 3 der IAPWS-IF97) – Fortsetzung

p = 31 MPa								
t [°C]	v [$m^3\ kg^{-1}$]	h [$kJ\ kg^{-1}$]	s [$kJ\ kg^{-1}\ K^{-1}$]	c_p [$kJ\ kg^{-1}\ K^{-1}$]	w [$m\ s^{-1}$]	κ	$\eta \times 10^6$ [$kg\ m^{-1}\ s^{-1}$]	$\lambda \times 10^3$ [$W\ m^{-1}\ K^{-1}$]
350	0.001545	1606.30	3.6370	6.3060	821.09	14.0751	75.94	498.78
360	0.001617	1671.95	3.7415	6.8687	757.24	11.4408	72.07	477.67
370	0.001710	1744.65	3.8554	7.7389	688.42	8.9390	67.79	454.94
380	0.001843	1828.81	3.9853	9.2539	611.42	6.5435	62.82	428.52
390	0.002061	1935.27	4.1469	12.5118	525.18	4.3163	56.51	394.83
400	0.002533	2095.80	4.3871	20.6241	443.35	2.5037	47.51	347.16
410	0.003513	2324.74	4.7246	21.5620	429.82	1.6965	38.06	270.65
420	0.004470	2500.72	4.9805	14.2355	456.48	1.5036	33.93	209.47
430	0.005213	2622.34	5.1548	10.4639	481.69	1.4357	32.33	176.22
440	0.005819	2715.63	5.2865	8.3796	503.36	1.4045	31.64	156.18
450	0.006338	2792.54	5.3937	7.0966	521.67	1.3851	31.36	143.08
460	0.006798	2858.96	5.4849	6.2381	537.56	1.3713	31.30	134.03
470	0.007215	2918.10	5.5650	5.6213	551.74	1.3611	31.37	127.55
480	0.007599	2971.88	5.6369	5.1542	564.63	1.3534	31.52	122.81
490	0.007957	3021.52	5.7024	4.7876	576.49	1.3474	31.73	119.30
500	0.008294	3067.87	5.7627	4.4928	587.50	1.3425	31.98	116.72
510	0.008613	3111.55	5.8189	4.2517	597.79	1.3384	32.26	114.84
520	0.008917	3153.04	5.8715	4.0518	607.47	1.3349	32.56	113.51
530	0.009209	3192.70	5.9212	3.8842	616.61	1.3318	32.88	112.62
540	0.009490	3230.81	5.9684	3.7423	625.28	1.3290	33.21	112.08
550	0.009762	3267.61	6.0134	3.6209	633.53	1.3263	33.55	111.82
560	0.010025	3303.29	6.0564	3.5163	641.42	1.3239	33.89	111.79
570	0.010280	3337.98	6.0978	3.4254	648.97	1.3216	34.24	111.96
580	0.010529	3371.83	6.1377	3.3459	656.24	1.3194	34.60	112.28
590	0.010772	3404.93	6.1763	3.2758	663.23	1.3172	34.96	112.74
600	0.011010	3437.38	6.2137	3.2138	669.99	1.3152	35.32	113.30
610	0.011243	3469.23	6.2500	3.1586	676.54	1.3133	35.68	113.96
620	0.011471	3500.57	6.2853	3.1092	682.88	1.3114	36.05	114.69
630	0.011695	3531.43	6.3196	3.0650	689.05	1.3096	36.42	115.48
640	0.011915	3561.88	6.3532	3.0252	695.06	1.3079	36.78	116.33
650	0.012132	3591.95	6.3859	2.9894	700.91	1.3063	37.15	117.21
660	0.012346	3621.68	6.4179	2.9569	706.63	1.3047	37.52	118.14
670	0.012557	3651.10	6.4493	2.9276	712.21	1.3031	37.88	119.08
680	0.012764	3680.24	6.4800	2.9010	717.68	1.3017	38.25	120.06
690	0.012970	3709.13	6.5102	2.8768	723.03	1.3002	38.61	121.05
700	0.013173	3737.78	6.5398	2.8549	728.28	1.2989	38.98	122.05
720	0.013572	3794.49	6.5975	2.8168	738.48	1.2962	39.71	124.10
740	0.013963	3850.50	6.6533	2.7853	748.33	1.2937	40.43	126.17
760	0.014347	3905.94	6.7075	2.7593	757.85	1.2913	41.15	128.25
780	0.014726	3960.90	6.7602	2.7379	767.08	1.2890	41.86	130.34
800	0.015098	4015.48	6.8115	2.7206	776.04	1.2867	42.57	132.42

Table 4 Single-Phase Region (Regions 1 to 3 of IAPWS-IF97) – Continuation
Homogenes Zustandsgebiet (Bereiche 1 bis 3 der IAPWS-IF97) – Fortsetzung

p = 32 MPa

t [°C]	v [m^3 kg^{-1}]	h [kJ kg^{-1}]	s [kJ kg^{-1} K^{-1}]	c_p [kJ kg^{-1} K^{-1}]	w [m s^{-1}]	κ	$\eta \times 10^6$ [kg m^{-1} s^{-1}]	$\lambda \times 10^3$ [W m^{-1} K^{-1}]
0	0.000985	31.81	0.0002	4.0825	1454.70	67.15	1724.15	578.61
5	0.000985	52.23	0.0743	4.0862	1478.36	69.33	1473.99	586.89
10	0.000986	72.67	0.1471	4.0900	1499.49	71.28	1277.35	595.56
15	0.000987	93.13	0.2187	4.0936	1518.30	73.01	1119.77	604.37
20	0.000988	113.61	0.2892	4.0969	1534.96	74.54	991.33	613.14
25	0.000989	134.10	0.3585	4.0999	1549.67	75.87	885.15	621.71
30	0.000991	154.60	0.4267	4.1025	1562.56	77.02	796.24	629.96
35	0.000992	175.12	0.4939	4.1050	1573.78	77.99	720.98	637.80
40	0.000994	195.65	0.5599	4.1074	1583.43	78.80	656.65	645.18
45	0.000996	216.20	0.6250	4.1099	1591.64	79.46	601.21	652.06
50	0.000999	236.75	0.6891	4.1124	1598.50	79.97	553.07	658.42
60	0.001003	277.90	0.8145	4.1178	1608.48	80.58	473.95	669.58
70	0.001009	319.11	0.9364	4.1241	1614.00	80.70	412.11	678.77
80	0.001015	360.39	1.0550	4.1314	1615.56	80.38	362.85	686.16
90	0.001021	401.74	1.1705	4.1399	1613.61	79.68	322.98	691.97
100	0.001028	443.19	1.2830	4.1496	1608.52	78.65	290.27	696.40
110	0.001036	484.74	1.3929	4.1606	1600.61	77.31	263.11	699.64
120	0.001044	526.41	1.5003	4.1730	1590.15	75.72	240.31	701.82
130	0.001052	568.20	1.6053	4.1869	1577.37	73.90	221.00	703.06
140	0.001061	610.15	1.7080	4.2026	1562.44	71.89	204.49	703.43
150	0.001071	652.26	1.8088	4.2201	1545.50	69.71	190.26	702.97
160	0.001081	694.56	1.9076	4.2398	1526.67	67.37	177.90	701.72
170	0.001092	737.07	2.0046	4.2620	1506.04	64.91	167.09	699.69
180	0.001103	779.81	2.0999	4.2869	1483.65	62.34	157.57	696.88
190	0.001116	822.81	2.1938	4.3149	1459.56	59.67	149.13	693.27
200	0.001129	866.12	2.2863	4.3466	1433.77	56.91	141.59	688.86
210	0.001143	909.76	2.3776	4.3824	1406.31	54.09	134.82	683.62
220	0.001158	953.78	2.4678	4.4229	1377.18	51.20	128.70	677.52
230	0.001173	998.24	2.5570	4.4688	1346.37	48.28	123.14	670.53
240	0.001190	1043.18	2.6455	4.5210	1313.88	45.32	118.03	662.61
250	0.001209	1088.68	2.7333	4.5804	1279.70	42.34	113.33	653.69
260	0.001228	1134.82	2.8206	4.6484	1243.83	39.36	108.95	643.73
270	0.001250	1181.68	2.9077	4.7263	1206.26	36.38	104.85	632.64
280	0.001273	1229.38	2.9947	4.8163	1166.93	33.42	100.98	620.37
290	0.001299	1278.06	3.0819	4.9208	1125.78	30.49	97.28	606.85
300	0.001327	1327.86	3.1696	5.0434	1082.65	27.60	93.73	592.06
310	0.001359	1379.00	3.2581	5.1891	1037.31	24.75	90.27	576.00
320	0.001394	1431.74	3.3477	5.3652	989.53	21.95	86.85	558.75
330	0.001435	1486.44	3.4392	5.5830	939.22	19.22	83.45	540.45
340	0.001482	1543.60	3.5332	5.8602	886.59	16.58	79.99	521.29

Table 4 Single-Phase Region (Regions 1 to 3 of IAPWS-IF97) – Continuation
Homogenes Zustandsgebiet (Bereiche 1 bis 3 der IAPWS-IF97) – Fortsetzung

				p = 32 MPa				
t [°C]	v [m^3 kg^{-1}]	h [kJ kg^{-1}]	s [kJ kg^{-1} K^{-1}]	c_p [kJ kg^{-1} K^{-1}]	w [m s^{-1}]	κ	$\eta \times 10^6$ [kg m^{-1} s^{-1}]	$\lambda \times 10^3$ [W m^{-1} K^{-1}]
350	0.001538	1603.94	3.6308	6.2253	831.65	14.0554	76.41	501.43
360	0.001607	1668.57	3.7336	6.7412	769.58	11.5206	72.62	480.68
370	0.001695	1739.58	3.8449	7.5161	703.44	9.1234	68.49	458.52
380	0.001817	1820.54	3.9698	8.7959	630.41	6.8356	63.77	433.20
390	0.002006	1919.40	4.1199	11.2892	549.87	4.7096	58.01	401.66
400	0.002367	2056.81	4.3255	16.9320	469.62	2.9119	50.27	358.93
410	0.003127	2258.47	4.6228	21.5042	434.13	1.8835	41.03	294.57
420	0.004053	2446.50	4.8961	15.7049	452.00	1.5752	35.61	228.97
430	0.004812	2579.90	5.0873	11.4309	476.37	1.4737	33.38	189.72
440	0.005434	2681.17	5.2303	9.0081	498.35	1.4283	32.38	165.96
450	0.005961	2763.32	5.3448	7.5331	517.30	1.4028	31.93	150.58
460	0.006426	2833.48	5.4411	6.5584	533.70	1.3852	31.76	140.04
470	0.006845	2895.43	5.5251	5.8688	548.29	1.3725	31.76	132.53
480	0.007229	2951.43	5.5999	5.3530	561.53	1.3631	31.86	127.04
490	0.007586	3002.87	5.6678	4.9517	573.69	1.3558	32.04	122.98
500	0.007922	3050.73	5.7301	4.6307	584.97	1.3499	32.26	119.97
510	0.008239	3095.68	5.7879	4.3693	595.50	1.3451	32.51	117.75
520	0.008540	3138.26	5.8419	4.1533	605.38	1.3410	32.79	116.16
530	0.008829	3178.87	5.8928	3.9728	614.71	1.3374	33.10	115.05
540	0.009107	3217.81	5.9410	3.8203	623.54	1.3342	33.41	114.33
550	0.009375	3255.35	5.9868	3.6904	631.94	1.3312	33.74	113.93
560	0.009634	3291.68	6.0307	3.5786	639.96	1.3285	34.08	113.79
570	0.009886	3326.97	6.0728	3.4818	647.64	1.3259	34.42	113.85
580	0.010131	3361.36	6.1134	3.3971	655.01	1.3235	34.77	114.09
590	0.010370	3394.95	6.1525	3.3227	662.11	1.3211	35.12	114.48
600	0.010603	3427.84	6.1904	3.2569	668.95	1.3189	35.48	114.98
610	0.010831	3460.11	6.2271	3.1984	675.58	1.3168	35.83	115.58
620	0.011055	3491.83	6.2629	3.1461	682.00	1.3148	36.19	116.26
630	0.011275	3523.05	6.2976	3.0993	688.24	1.3128	36.56	117.01
640	0.011491	3553.83	6.3315	3.0572	694.31	1.3110	36.92	117.82
650	0.011704	3584.21	6.3646	3.0192	700.23	1.3092	37.28	118.67
660	0.011913	3614.22	6.3969	2.9849	706.00	1.3075	37.65	119.56
670	0.012119	3643.91	6.4286	2.9538	711.64	1.3059	38.01	120.48
680	0.012322	3673.31	6.4596	2.9256	717.16	1.3043	38.37	121.43
690	0.012523	3702.44	6.4900	2.9001	722.56	1.3028	38.74	122.40
700	0.012721	3731.32	6.5198	2.8768	727.85	1.3014	39.10	123.38
720	0.013111	3788.44	6.5779	2.8364	738.14	1.2986	39.82	125.38
740	0.013494	3844.82	6.6341	2.8030	748.06	1.2960	40.54	127.41
760	0.013869	3900.60	6.6887	2.7753	757.65	1.2935	41.25	129.45
780	0.014237	3955.87	6.7416	2.7525	766.94	1.2910	41.96	131.50
800	0.014601	4010.73	6.7932	2.7340	775.95	1.2887	42.67	133.55

Table 4 Single-Phase Region (Regions 1 to 3 of IAPWS-IF97) – Continuation
Homogenes Zustandsgebiet (Bereiche 1 bis 3 der IAPWS-IF97) – Fortsetzung

				p = 33 MPa				
t [°C]	v [m³ kg⁻¹]	h [kJ kg⁻¹]	s [kJ kg⁻¹ K⁻¹]	c_p [kJ kg⁻¹ K⁻¹]	w [m s⁻¹]	κ	$\eta \times 10^6$ [kg m⁻¹ s⁻¹]	$\lambda \times 10^3$ [W m⁻¹ K⁻¹]
0	0.000984	32.78	0.0002	4.0788	1456.41	65.30	1722.44	579.15
5	0.000985	53.19	0.0742	4.0830	1480.05	67.41	1472.88	587.40
10	0.000985	73.61	0.1470	4.0872	1501.17	69.30	1276.66	596.04
15	0.000986	94.06	0.2185	4.0910	1519.96	70.98	1119.35	604.84
20	0.000987	114.52	0.2889	4.0945	1536.61	72.46	991.12	613.60
25	0.000989	135.00	0.3582	4.0976	1551.31	73.76	885.07	622.16
30	0.000990	155.50	0.4264	4.1004	1564.20	74.87	796.26	630.41
35	0.000992	176.01	0.4935	4.1030	1575.42	75.82	721.08	638.25
40	0.000994	196.53	0.5596	4.1054	1585.08	76.60	656.81	645.64
45	0.000996	217.06	0.6246	4.1079	1593.30	77.24	601.40	652.52
50	0.000998	237.61	0.6887	4.1104	1600.17	77.74	553.28	658.88
60	0.001003	278.74	0.8140	4.1159	1610.19	78.34	474.20	670.06
70	0.001008	319.93	0.9359	4.1222	1615.76	78.46	412.37	679.25
80	0.001014	361.18	1.0544	4.1296	1617.38	78.16	363.12	686.66
90	0.001021	402.52	1.1698	4.1380	1615.49	77.48	323.25	692.48
100	0.001028	443.95	1.2823	4.1477	1610.46	76.48	290.54	696.94
110	0.001035	485.48	1.3922	4.1586	1602.63	75.19	263.37	700.20
120	0.001043	527.13	1.4995	4.1709	1592.25	73.65	240.57	702.40
130	0.001052	568.90	1.6044	4.1847	1579.55	71.90	221.25	703.66
140	0.001061	610.83	1.7071	4.2002	1564.71	69.95	204.73	704.06
150	0.001070	652.91	1.8078	4.2176	1547.87	67.84	190.50	703.63
160	0.001080	695.19	1.9065	4.2371	1529.15	65.58	178.14	702.40
170	0.001091	737.66	2.0035	4.2590	1508.63	63.20	167.32	700.40
180	0.001103	780.37	2.0988	4.2836	1486.37	60.71	157.80	697.61
190	0.001115	823.35	2.1926	4.3113	1462.41	58.12	149.35	694.04
200	0.001128	866.61	2.2850	4.3426	1436.78	55.46	141.82	689.66
210	0.001142	910.21	2.3762	4.3779	1409.49	52.73	135.05	684.46
220	0.001157	954.19	2.4662	4.4178	1380.54	49.94	128.93	678.41
230	0.001172	998.59	2.5554	4.4631	1349.94	47.11	123.37	671.47
240	0.001189	1043.47	2.6437	4.5145	1317.67	44.24	118.27	663.60
250	0.001207	1088.90	2.7314	4.5729	1283.74	41.36	113.56	654.75
260	0.001227	1134.95	2.8186	4.6396	1248.15	38.48	109.19	644.86
270	0.001248	1181.72	2.9055	4.7160	1210.88	35.60	105.10	633.86
280	0.001271	1229.31	2.9923	4.8040	1171.91	32.74	101.23	621.70
290	0.001297	1277.85	3.0793	4.9060	1131.17	29.90	97.55	608.30
300	0.001325	1327.49	3.1666	5.0253	1088.53	27.11	94.01	593.65
310	0.001356	1378.43	3.2547	5.1664	1043.77	24.35	90.57	577.75
320	0.001391	1430.91	3.3440	5.3361	996.67	21.65	87.18	560.68
330	0.001430	1485.28	3.4349	5.5446	947.13	19.01	83.81	542.58
340	0.001476	1541.98	3.5281	5.8076	895.36	16.46	80.39	523.63

Table 4 Single-Phase Region (Regions 1 to 3 of IAPWS-IF97) – Continuation
Homogenes Zustandsgebiet (Bereiche 1 bis 3 der IAPWS-IF97) – Fortsetzung

				p = 33 MPa				
t	v	h	s	c_p	w	κ	$\eta \times 10^6$	$\lambda \times 10^3$
[°C]	[m^3 kg^{-1}]	[kJ kg^{-1}]	[kJ kg^{-1} K^{-1}]	[kJ kg^{-1} K^{-1}]	[m s^{-1}]		[kg m^{-1} s^{-1}]	[W m^{-1} K^{-1}]
350	0.001531	1601.69	3.6247	6.1505	841.70	14.0253	76.86	504.02
360	0.001597	1665.40	3.7261	6.6259	781.41	11.5870	73.16	483.59
370	0.001681	1734.91	3.8350	7.3223	717.62	9.2845	69.15	461.95
380	0.001794	1813.17	3.9557	8.4232	647.97	7.0923	64.65	437.57
390	0.001962	1906.25	4.0971	10.4108	572.10	5.0555	59.29	407.84
400	0.002254	2028.31	4.2797	14.4965	495.04	3.2952	52.46	368.76
410	0.002837	2202.63	4.5367	19.7924	445.66	2.1218	43.96	313.57
420	0.003677	2391.90	4.8118	16.8338	450.32	1.6711	37.55	248.92
430	0.004439	2536.26	5.0187	12.3705	472.11	1.5216	34.58	204.06
440	0.005071	2645.56	5.1730	9.6702	493.77	1.4570	33.21	176.52
450	0.005607	2733.27	5.2952	7.9974	513.16	1.4233	32.56	158.66
460	0.006076	2807.39	5.3970	6.8977	530.01	1.4011	32.27	146.47
470	0.006497	2872.32	5.4850	6.1286	544.98	1.3853	32.19	137.82
480	0.006882	2930.63	5.5629	5.5602	558.54	1.3738	32.23	131.51
490	0.007238	2983.95	5.6333	5.1218	570.99	1.3650	32.36	126.85
500	0.007572	3033.36	5.6976	4.7733	582.52	1.3580	32.55	123.37
510	0.007887	3079.63	5.7571	4.4906	593.28	1.3523	32.78	120.79
520	0.008186	3123.34	5.8125	4.2577	603.36	1.3476	33.04	118.91
530	0.008472	3164.92	5.8646	4.0636	612.87	1.3434	33.32	117.58
540	0.008747	3204.71	5.9139	3.9002	621.87	1.3398	33.62	116.67
550	0.009011	3243.00	5.9607	3.7613	630.41	1.3364	33.94	116.12
560	0.009267	3280.01	6.0054	3.6421	638.56	1.3334	34.27	115.85
570	0.009515	3315.90	6.0482	3.5390	646.36	1.3305	34.60	115.80
580	0.009756	3350.83	6.0894	3.4491	653.83	1.3278	34.94	115.95
590	0.009991	3384.92	6.1291	3.3702	661.02	1.3252	35.29	116.26
600	0.010221	3418.26	6.1675	3.3005	667.96	1.3228	35.64	116.70
610	0.010445	3450.95	6.2047	3.2386	674.66	1.3205	35.99	117.24
620	0.010665	3483.06	6.2409	3.1834	681.16	1.3183	36.34	117.87
630	0.010881	3514.64	6.2761	3.1339	687.47	1.3162	36.70	118.58
640	0.011093	3545.75	6.3103	3.0894	693.60	1.3142	37.06	119.35
650	0.011301	3576.44	6.3437	3.0493	699.58	1.3123	37.42	120.16
660	0.011506	3606.75	6.3764	3.0131	705.40	1.3105	37.78	121.02
670	0.011708	3636.72	6.4083	2.9803	711.10	1.3088	38.14	121.91
680	0.011907	3666.37	6.4396	2.9505	716.66	1.3071	38.50	122.83
690	0.012104	3695.74	6.4703	2.9235	722.11	1.3055	38.86	123.77
700	0.012298	3724.85	6.5003	2.8989	727.45	1.3040	39.22	124.73
720	0.012679	3782.38	6.5589	2.8562	737.81	1.3010	39.94	126.68
740	0.013053	3839.14	6.6154	2.8208	747.81	1.2983	40.65	128.67
760	0.013419	3895.26	6.6703	2.7915	757.46	1.2957	41.36	130.67
780	0.013779	3950.84	6.7236	2.7672	766.81	1.2931	42.07	132.68
800	0.014134	4005.98	6.7754	2.7474	775.88	1.2907	42.77	134.70

Table 4 Single-Phase Region (Regions 1 to 3 of IAPWS-IF97) – Continuation
Homogenes Zustandsgebiet (Bereiche 1 bis 3 der IAPWS-IF97) – Fortsetzung

p = 34 MPa

t [°C]	v [m³ kg⁻¹]	h [kJ kg⁻¹]	s [kJ kg⁻¹ K⁻¹]	c_p [kJ kg⁻¹ K⁻¹]	w [m s⁻¹]	κ	$\eta\times10^6$ [kg m⁻¹ s⁻¹]	$\lambda\times10^3$ [W m⁻¹ K⁻¹]
0	0.000984	33.75	0.0001	4.0753	1458.12	63.56	1720.74	579.69
5	0.000984	54.14	0.0741	4.0798	1481.75	65.61	1471.78	587.90
10	0.000985	74.55	0.1468	4.0843	1502.84	67.45	1275.97	596.52
15	0.000986	94.98	0.2183	4.0884	1521.62	69.08	1118.94	605.31
20	0.000987	115.44	0.2887	4.0921	1538.27	70.51	990.91	614.06
25	0.000988	135.90	0.3579	4.0953	1552.96	71.77	885.00	622.62
30	0.000990	156.39	0.4261	4.0982	1565.84	72.85	796.29	630.86
35	0.000992	176.89	0.4931	4.1009	1577.06	73.77	721.18	638.71
40	0.000993	197.40	0.5592	4.1035	1586.74	74.54	656.96	646.09
45	0.000996	217.92	0.6242	4.1060	1594.96	75.16	601.59	652.98
50	0.000998	238.46	0.6882	4.1086	1601.85	75.64	553.50	659.34
60	0.001003	279.57	0.8135	4.1141	1611.91	76.23	474.45	670.53
70	0.001008	320.74	0.9353	4.1204	1617.52	76.35	412.64	679.74
80	0.001014	361.98	1.0537	4.1277	1619.19	76.06	363.39	687.16
90	0.001020	403.30	1.1691	4.1362	1617.36	75.41	323.52	693.00
100	0.001027	444.71	1.2816	4.1458	1612.41	74.44	290.80	697.47
110	0.001035	486.22	1.3914	4.1566	1604.64	73.20	263.63	700.75
120	0.001043	527.84	1.4986	4.1688	1594.34	71.71	240.83	702.98
130	0.001051	569.60	1.6035	4.1826	1581.72	70.01	221.50	704.26
140	0.001060	611.50	1.7062	4.1979	1566.97	68.13	204.98	704.68
150	0.001070	653.57	1.8068	4.2151	1550.23	66.08	190.74	704.28
160	0.001080	695.81	1.9054	4.2344	1531.62	63.90	178.37	703.08
170	0.001091	738.26	2.0023	4.2560	1511.21	61.59	167.56	701.11
180	0.001102	780.94	2.0976	4.2804	1489.08	59.18	158.03	698.35
190	0.001114	823.88	2.1913	4.3078	1465.26	56.67	149.58	694.81
200	0.001127	867.11	2.2836	4.3386	1439.78	54.09	142.05	690.47
210	0.001141	910.66	2.3747	4.3735	1412.65	51.44	135.28	685.31
220	0.001156	954.59	2.4647	4.4129	1383.88	48.74	129.16	679.29
230	0.001171	998.94	2.5538	4.4575	1353.48	46.00	123.60	672.40
240	0.001188	1043.76	2.6420	4.5080	1321.43	43.23	118.50	664.59
250	0.001206	1089.12	2.7295	4.5655	1287.75	40.44	113.80	655.80
260	0.001226	1135.10	2.8166	4.6310	1252.42	37.64	109.43	645.98
270	0.001247	1181.77	2.9033	4.7059	1215.46	34.86	105.35	635.07
280	0.001270	1229.25	2.9899	4.7920	1176.83	32.09	101.49	623.01
290	0.001295	1277.66	3.0766	4.8916	1136.49	29.34	97.82	609.74
300	0.001322	1327.14	3.1637	5.0077	1094.32	26.64	94.29	595.22
310	0.001353	1377.88	3.2515	5.1445	1050.13	23.98	90.87	579.49
320	0.001387	1430.12	3.3403	5.3082	1003.70	21.36	87.50	562.59
330	0.001426	1484.16	3.4307	5.5080	954.89	18.81	84.16	544.68
340	0.001471	1540.44	3.5232	5.7580	903.92	16.34	80.78	525.94

Table 4 Single-Phase Region (Regions 1 to 3 of IAPWS-IF97) – Continuation
Homogenes Zustandsgebiet (Bereiche 1 bis 3 der IAPWS-IF97) – Fortsetzung

				p = 34 MPa				
t	v	h	s	c_p	w	κ	$\eta \times 10^6$	$\lambda \times 10^3$
[°C]	[m^3 kg^{-1}]	[kJ kg^{-1}]	[kJ kg^{-1} K^{-1}]	[kJ kg^{-1} K^{-1}]	[m s^{-1}]		[kg m^{-1} s^{-1}]	[W m^{-1} K^{-1}]
350	0.001524	1599.57	3.6188	6.0807	851.32	13.9874	77.31	506.55
360	0.001588	1662.42	3.7189	6.5210	792.78	11.6422	73.68	486.43
370	0.001668	1730.59	3.8257	7.1517	731.08	9.4259	69.78	465.24
380	0.001773	1806.54	3.9429	8.1126	664.34	7.3196	65.46	441.68
390	0.001925	1895.03	4.0773	9.7456	592.38	5.3622	60.44	413.49
400	0.002170	2006.25	4.2437	12.8373	518.75	3.6467	54.26	377.32
410	0.002627	2158.31	4.4678	17.5600	462.20	2.3920	46.60	328.53
420	0.003352	2339.36	4.7309	17.3079	452.36	1.7956	39.70	268.05
430	0.004094	2491.92	4.9495	13.2510	469.31	1.5822	35.95	219.16
440	0.004731	2608.99	5.1149	10.3545	489.86	1.4919	34.15	187.84
450	0.005273	2702.45	5.2451	8.4825	509.36	1.4472	33.26	167.30
460	0.005746	2780.74	5.3526	7.2545	526.55	1.4192	32.82	153.31
470	0.006169	2848.77	5.4448	6.4004	541.84	1.3997	32.65	143.42
480	0.006555	2909.50	5.5260	5.7755	555.69	1.3856	32.62	136.23
490	0.006911	2964.77	5.5989	5.2977	568.40	1.3750	32.71	130.90
500	0.007243	3015.79	5.6653	4.9203	580.17	1.3668	32.86	126.93
510	0.007556	3063.42	5.7265	4.6153	591.14	1.3602	33.06	123.97
520	0.007853	3108.28	5.7834	4.3649	601.42	1.3546	33.30	121.78
530	0.008137	3150.85	5.8368	4.1567	611.10	1.3499	33.56	120.19
540	0.008408	3191.52	5.8871	3.9819	620.26	1.3457	33.85	119.09
550	0.008670	3230.58	5.9348	3.8336	628.94	1.3420	34.15	118.37
560	0.008922	3268.26	5.9804	3.7067	637.22	1.3385	34.46	117.96
570	0.009167	3304.77	6.0239	3.5972	645.12	1.3353	34.79	117.81
580	0.009404	3340.26	6.0658	3.5019	652.70	1.3324	35.12	117.86
590	0.009636	3374.85	6.1061	3.4184	659.99	1.3295	35.46	118.09
600	0.009862	3408.66	6.1450	3.3447	667.01	1.3269	35.80	118.46
610	0.010082	3441.77	6.1827	3.2793	673.79	1.3244	36.15	118.94
620	0.010298	3474.27	6.2193	3.2210	680.36	1.3220	36.50	119.52
630	0.010510	3506.21	6.2549	3.1688	686.73	1.3197	36.85	120.18
640	0.010718	3537.66	6.2895	3.1219	692.92	1.3176	37.20	120.90
650	0.010922	3568.67	6.3233	3.0797	698.96	1.3155	37.56	121.68
660	0.011123	3599.27	6.3563	3.0415	704.84	1.3136	37.91	122.51
670	0.011321	3629.51	6.3885	3.0069	710.58	1.3117	38.27	123.37
680	0.011517	3659.42	6.4200	2.9756	716.19	1.3100	38.63	124.25
690	0.011709	3689.03	6.4509	2.9471	721.69	1.3083	38.99	125.17
700	0.011899	3718.37	6.4812	2.9212	727.07	1.3066	39.34	126.10
720	0.012273	3776.33	6.5402	2.8761	737.51	1.3035	40.05	128.00
740	0.012638	3833.46	6.5972	2.8387	747.58	1.3007	40.76	129.95
760	0.012996	3889.92	6.6523	2.8077	757.30	1.2979	41.47	131.91
780	0.013348	3945.81	6.7059	2.7820	766.71	1.2953	42.18	133.88
800	0.013694	4001.23	6.7580	2.7609	775.82	1.2927	42.87	135.86

Table 4 Single-Phase Region (Regions 1 to 3 of IAPWS-IF97) – Continuation
Homogenes Zustandsgebiet (Bereiche 1 bis 3 der IAPWS-IF97) – Fortsetzung

p = 35 MPa								
t [°C]	v [m^3 kg^{-1}]	h [kJ kg^{-1}]	s [kJ kg^{-1} K^{-1}]	c_p [kJ kg^{-1} K^{-1}]	w [m s^{-1}]	κ	$\eta \times 10^6$ [kg m^{-1} s^{-1}]	$\lambda \times 10^3$ [W m^{-1} K^{-1}]
0	0.000983	34.72	0.0001	4.0717	1459.84	61.92	1719.06	580.23
5	0.000984	55.09	0.0740	4.0767	1483.45	63.91	1470.70	588.41
10	0.000984	75.49	0.1466	4.0815	1504.53	65.69	1275.29	597.00
15	0.000985	95.91	0.2181	4.0859	1523.29	67.28	1118.54	605.77
20	0.000987	116.35	0.2884	4.0897	1539.92	68.68	990.70	614.52
25	0.000988	136.81	0.3576	4.0931	1554.60	69.90	884.93	623.07
30	0.000989	157.28	0.4257	4.0961	1567.49	70.95	796.33	631.32
35	0.000991	177.77	0.4928	4.0989	1578.71	71.84	721.28	639.16
40	0.000993	198.27	0.5588	4.1015	1588.39	72.59	657.12	646.55
45	0.000995	218.78	0.6238	4.1041	1596.63	73.19	601.79	653.43
50	0.000997	239.31	0.6878	4.1067	1603.52	73.66	553.72	659.81
60	0.001002	280.40	0.8130	4.1123	1613.62	74.24	474.70	671.00
70	0.001008	321.56	0.9347	4.1186	1619.28	74.36	412.91	680.23
80	0.001013	362.78	1.0531	4.1259	1621.00	74.08	363.66	687.66
90	0.001020	404.08	1.1685	4.1343	1619.23	73.46	323.79	693.52
100	0.001027	445.47	1.2809	4.1439	1614.35	72.52	291.07	698.01
110	0.001034	486.96	1.3906	4.1547	1606.65	71.32	263.90	701.31
120	0.001042	528.56	1.4978	4.1668	1596.43	69.88	241.08	703.56
130	0.001051	570.30	1.6026	4.1804	1583.89	68.23	221.75	704.87
140	0.001060	612.18	1.7052	4.1956	1569.23	66.40	205.22	705.31
150	0.001069	654.22	1.8058	4.2126	1552.59	64.42	190.98	704.93
160	0.001079	696.44	1.9044	4.2317	1534.08	62.30	178.61	703.76
170	0.001090	738.86	2.0012	4.2531	1513.79	60.07	167.79	701.81
180	0.001101	781.51	2.0964	4.2772	1491.78	57.73	158.26	699.09
190	0.001113	824.41	2.1900	4.3042	1468.09	55.30	149.81	695.58
200	0.001126	867.61	2.2823	4.3347	1442.75	52.80	142.27	691.27
210	0.001140	911.12	2.3733	4.3691	1415.79	50.24	135.50	686.15
220	0.001155	955.00	2.4632	4.4080	1387.20	47.62	129.39	680.18
230	0.001170	999.30	2.5521	4.4519	1356.99	44.96	123.82	673.33
240	0.001187	1044.06	2.6402	4.5017	1325.16	42.27	118.73	665.57
250	0.001205	1089.35	2.7276	4.5582	1291.72	39.57	114.03	656.84
260	0.001224	1135.25	2.8145	4.6225	1256.66	36.86	109.67	647.10
270	0.001245	1181.83	2.9011	4.6960	1219.99	34.16	105.59	636.28
280	0.001268	1229.20	2.9875	4.7803	1181.69	31.47	101.75	624.31
290	0.001292	1277.48	3.0740	4.8775	1141.74	28.82	98.09	611.16
300	0.001320	1326.81	3.1608	4.9906	1100.03	26.20	94.57	596.78
310	0.001350	1377.36	3.2483	5.1233	1056.39	23.62	91.16	581.20
320	0.001384	1429.36	3.3367	5.2814	1010.60	21.09	87.82	564.47
330	0.001422	1483.10	3.4265	5.4731	962.51	18.61	84.50	546.75
340	0.001466	1538.97	3.5184	5.7111	912.29	16.22	81.16	528.20

Table 4 Single-Phase Region (Regions 1 to 3 of IAPWS-IF97) – Continuation
Homogenes Zustandsgebiet (Bereiche 1 bis 3 der IAPWS-IF97) – Fortsetzung

p = 35 MPa

t [°C]	v [m^3 kg^{-1}]	h [kJ kg^{-1}]	s [kJ kg^{-1} K^{-1}]	c_p [kJ kg^{-1} K^{-1}]	w [m s^{-1}]	κ	$\eta \times 10^6$ [kg m^{-1} s^{-1}]	$\lambda \times 10^3$ [W m^{-1} K^{-1}]
350	0.001517	1597.54	3.6131	6.0154	860.58	13.9444	77.74	509.02
360	0.001579	1659.61	3.7119	6.4251	803.72	11.6877	74.18	489.19
370	0.001656	1726.57	3.8169	7.0001	743.90	9.5504	70.38	468.41
380	0.001755	1800.51	3.9309	7.8489	679.71	7.5220	66.21	445.56
390	0.001893	1885.27	4.0597	9.2219	611.08	5.6361	61.46	418.69
400	0.002106	1988.43	4.2140	11.6505	540.68	3.9668	55.79	384.93
410	0.002474	2123.59	4.4133	15.5328	481.15	2.6732	48.87	340.78
420	0.003082	2291.32	4.6570	17.0301	458.49	1.9489	41.93	285.39
430	0.003782	2447.74	4.8811	13.9618	468.42	1.6576	37.48	234.58
440	0.004413	2571.64	5.0561	11.0053	486.97	1.5353	35.20	199.73
450	0.004959	2670.97	5.1945	8.9762	506.01	1.4752	34.03	176.45
460	0.005436	2753.55	5.3079	7.6253	523.37	1.4398	33.43	160.58
470	0.005861	2824.83	5.4045	6.6833	538.90	1.4158	33.14	149.34
480	0.006246	2888.06	5.4890	5.9984	552.99	1.3987	33.05	141.19
490	0.006602	2945.34	5.5646	5.4791	565.94	1.3861	33.07	135.16
500	0.006933	2998.02	5.6331	5.0715	577.93	1.3764	33.18	130.65
510	0.007245	3047.03	5.6961	4.7434	589.10	1.3686	33.35	127.27
520	0.007540	3093.08	5.7546	4.4747	599.56	1.3622	33.56	124.75
530	0.007821	3136.68	5.8092	4.2519	609.41	1.3568	33.81	122.90
540	0.008089	3178.24	5.8606	4.0652	618.71	1.3521	34.08	121.59
550	0.008348	3218.08	5.9093	3.9073	627.54	1.3478	34.36	120.70
560	0.008597	3256.46	5.9557	3.7724	635.93	1.3440	34.67	120.15
570	0.008838	3293.59	6.0000	3.6563	643.95	1.3405	34.98	119.87
580	0.009073	3329.64	6.0425	3.5554	651.62	1.3372	35.30	119.83
590	0.009301	3364.74	6.0834	3.4671	659.00	1.3341	35.63	119.96
600	0.009523	3399.02	6.1229	3.3894	666.10	1.3312	35.97	120.26
610	0.009740	3432.56	6.1611	3.3205	672.96	1.3284	36.31	120.68
620	0.009953	3465.45	6.1981	3.2591	679.59	1.3258	36.65	121.20
630	0.010161	3497.76	6.2341	3.2041	686.03	1.3234	37.00	121.81
640	0.010365	3529.55	6.2691	3.1548	692.28	1.3211	37.35	122.49
650	0.010566	3560.87	6.3032	3.1103	698.37	1.3189	37.70	123.23
660	0.010763	3591.77	6.3365	3.0701	704.30	1.3168	38.05	124.02
670	0.010957	3622.29	6.3690	3.0338	710.10	1.3148	38.41	124.85
680	0.011149	3652.46	6.4008	3.0008	715.76	1.3129	38.76	125.70
690	0.011337	3682.31	6.4320	2.9708	721.29	1.3111	39.11	126.59
700	0.011524	3711.88	6.4625	2.9435	726.71	1.3094	39.47	127.49
720	0.011889	3770.27	6.5219	2.8961	737.24	1.3061	40.18	129.35
740	0.012247	3827.78	6.5793	2.8567	747.37	1.3031	40.88	131.25
760	0.012598	3884.58	6.6348	2.8239	757.16	1.3002	41.58	133.17
780	0.012942	3940.78	6.6887	2.7968	766.62	1.2975	42.28	135.10
800	0.013280	3996.48	6.7411	2.7744	775.78	1.2948	42.98	137.04

Table 4 Single-Phase Region (Regions 1 to 3 of IAPWS-IF97) – Continuation
Homogenes Zustandsgebiet (Bereiche 1 bis 3 der IAPWS-IF97) – Fortsetzung

				p = 36 MPa				
t [°C]	v [m³ kg⁻¹]	h [kJ kg⁻¹]	s [kJ kg⁻¹ K⁻¹]	c_p [kJ kg⁻¹ K⁻¹]	w [m s⁻¹]	κ	$\eta \times 10^6$ [kg m⁻¹ s⁻¹]	$\lambda \times 10^3$ [W m⁻¹ K⁻¹]
0	0.000983	35.69	0.0000	4.0682	1461.55	60.37	1717.41	580.77
5	0.000983	56.05	0.0739	4.0736	1485.15	62.31	1469.64	588.91
10	0.000984	76.43	0.1465	4.0788	1506.21	64.04	1274.63	597.48
15	0.000985	96.83	0.2179	4.0833	1524.95	65.58	1118.15	606.24
20	0.000986	117.26	0.2882	4.0874	1541.58	66.94	990.51	614.97
25	0.000988	137.71	0.3574	4.0909	1556.25	68.13	884.87	623.52
30	0.000989	158.17	0.4254	4.0940	1569.13	69.15	796.36	631.77
35	0.000991	178.65	0.4924	4.0968	1580.36	70.02	721.39	639.61
40	0.000993	199.14	0.5584	4.0995	1590.04	70.75	657.27	647.00
45	0.000995	219.64	0.6233	4.1021	1598.29	71.34	601.98	653.89
50	0.000997	240.16	0.6873	4.1048	1605.20	71.80	553.94	660.27
60	0.001002	281.23	0.8125	4.1104	1615.33	72.36	474.95	671.48
70	0.001007	322.37	0.9342	4.1168	1621.03	72.48	413.17	680.71
80	0.001013	363.57	1.0525	4.1241	1622.81	72.22	363.93	688.16
90	0.001019	404.86	1.1678	4.1325	1621.11	71.61	324.06	694.04
100	0.001026	446.23	1.2802	4.1420	1616.28	70.71	291.34	698.54
110	0.001034	487.70	1.3898	4.1527	1608.66	69.54	264.16	701.86
120	0.001042	529.28	1.4970	4.1647	1598.51	68.14	241.34	704.13
130	0.001050	571.00	1.6018	4.1782	1586.06	66.55	222.00	705.46
140	0.001059	612.85	1.7043	4.1933	1571.49	64.78	205.47	705.93
150	0.001069	654.87	1.8048	4.2102	1554.94	62.86	191.22	705.58
160	0.001079	697.06	1.9034	4.2291	1536.53	60.80	178.85	704.44
170	0.001089	739.46	2.0001	4.2502	1516.35	58.63	168.02	702.52
180	0.001101	782.08	2.0952	4.2740	1494.46	56.36	158.49	699.82
190	0.001113	824.95	2.1888	4.3008	1470.91	54.01	150.04	696.35
200	0.001126	868.10	2.2810	4.3309	1445.72	51.58	142.50	692.07
210	0.001139	911.58	2.3719	4.3648	1418.91	49.09	135.73	686.98
220	0.001154	955.42	2.4617	4.4031	1390.50	46.55	129.61	681.05
230	0.001169	999.66	2.5505	4.4464	1360.48	43.97	124.05	674.25
240	0.001186	1044.36	2.6385	4.4954	1328.86	41.37	118.96	666.55
250	0.001204	1089.59	2.7258	4.5510	1295.65	38.74	114.27	657.88
260	0.001223	1135.41	2.8125	4.6142	1260.85	36.12	109.91	648.21
270	0.001243	1181.90	2.8989	4.6863	1224.47	33.49	105.84	637.47
280	0.001266	1229.17	2.9852	4.7688	1186.50	30.89	102.00	625.61
290	0.001290	1277.32	3.0714	4.8638	1146.92	28.32	98.35	612.57
300	0.001317	1326.49	3.1580	4.9740	1105.65	25.78	94.85	598.32
310	0.001347	1376.86	3.2451	5.1029	1062.55	23.28	91.46	582.88
320	0.001380	1428.63	3.3331	5.2556	1017.39	20.83	88.13	566.33
330	0.001418	1482.08	3.4225	5.4397	970.00	18.43	84.84	548.78
340	0.001461	1537.57	3.5137	5.6667	920.50	16.11	81.53	530.42

Table 4 Single-Phase Region (Regions 1 to 3 of IAPWS-IF97) – Continuation
Homogenes Zustandsgebiet (Bereiche 1 bis 3 der IAPWS-IF97) – Fortsetzung

p = 36 MPa								
t	v	h	s	c_p	w	κ	$\eta \times 10^6$	$\lambda \times 10^3$
[°C]	[m³ kg⁻¹]	[kJ kg⁻¹]	[kJ kg⁻¹ K⁻¹]	[kJ kg⁻¹ K⁻¹]	[m s⁻¹]		[kg m⁻¹ s⁻¹]	[W m⁻¹ K⁻¹]
350	0.001511	1595.61	3.6076	5.9542	869.55	13.8984	78.16	511.45
360	0.001571	1656.95	3.7053	6.3369	814.30	11.7251	74.66	491.87
370	0.001644	1722.82	3.8085	6.8642	756.17	9.6603	70.95	471.46
380	0.001738	1795.00	3.9198	7.6215	694.23	7.7030	66.93	449.25
390	0.001865	1876.62	4.0438	8.7973	628.47	5.8821	62.40	423.52
400	0.002053	1973.55	4.1888	10.7634	561.02	4.2583	57.12	391.78
410	0.002360	2096.00	4.3694	13.8993	500.75	2.9511	50.82	351.20
420	0.002865	2249.28	4.5921	16.1898	468.35	2.1268	44.12	300.57
430	0.003505	2404.82	4.8149	14.3950	469.83	1.7495	39.14	249.81
440	0.004121	2534.26	4.9977	11.5780	485.24	1.5873	36.36	211.94
450	0.004665	2639.00	5.1436	9.4642	503.25	1.5080	34.88	186.07
460	0.005143	2725.89	5.2630	8.0038	520.54	1.4634	34.09	168.24
470	0.005570	2800.50	5.3641	6.9757	536.20	1.4339	33.68	155.58
480	0.005956	2866.34	5.4521	6.2284	550.48	1.4133	33.50	146.40
490	0.006311	2925.68	5.5303	5.6655	563.63	1.3982	33.46	139.61
500	0.006641	2980.06	5.6011	5.2264	575.81	1.3868	33.53	134.52
510	0.006951	3030.50	5.6660	4.8745	587.16	1.3777	33.66	130.70
520	0.007244	3077.76	5.7259	4.5871	597.80	1.3704	33.85	127.83
530	0.007522	3122.41	5.7819	4.3492	607.80	1.3642	34.07	125.71
540	0.007788	3164.87	5.8344	4.1502	617.24	1.3588	34.32	124.17
550	0.008044	3205.51	5.8841	3.9822	626.20	1.3541	34.59	123.09
560	0.008290	3244.60	5.9313	3.8391	634.71	1.3498	34.88	122.39
570	0.008529	3282.36	5.9764	3.7162	642.83	1.3459	35.18	121.99
580	0.008760	3318.98	6.0195	3.6096	650.60	1.3422	35.49	121.84
590	0.008985	3354.60	6.0610	3.5165	658.06	1.3388	35.82	121.89
600	0.009204	3389.35	6.1011	3.4345	665.24	1.3356	36.14	122.10
610	0.009417	3423.32	6.1398	3.3620	672.17	1.3327	36.48	122.45
620	0.009627	3456.61	6.1772	3.2975	678.87	1.3298	36.82	122.92
630	0.009831	3489.29	6.2136	3.2397	685.37	1.3272	37.16	123.48
640	0.010032	3521.43	6.2490	3.1879	691.68	1.3247	37.50	124.11
650	0.010229	3553.07	6.2835	3.1412	697.82	1.3223	37.85	124.81
660	0.010423	3584.27	6.3171	3.0990	703.80	1.3201	38.20	125.56
670	0.010614	3615.06	6.3499	3.0608	709.64	1.3180	38.55	126.35
680	0.010801	3645.49	6.3820	3.0261	715.35	1.3160	38.90	127.18
690	0.010987	3675.59	6.4134	2.9947	720.93	1.3141	39.25	128.03
700	0.011169	3705.40	6.4442	2.9660	726.39	1.3122	39.60	128.91
720	0.011528	3764.20	6.5040	2.9162	736.99	1.3088	40.30	130.72
740	0.011878	3822.10	6.5618	2.8747	747.19	1.3056	41.00	132.57
760	0.012221	3879.24	6.6176	2.8403	757.03	1.3026	41.70	134.45
780	0.012558	3935.75	6.6718	2.8116	766.55	1.2997	42.39	136.34
800	0.012889	3991.74	6.7244	2.7880	775.75	1.2969	43.09	138.23

Table 4 Single-Phase Region (Regions 1 to 3 of IAPWS-IF97) – Continuation
Homogenes Zustandsgebiet (Bereiche 1 bis 3 der IAPWS-IF97) – Fortsetzung

p = 37 MPa

t [°C]	v [m^3 kg^{-1}]	h [kJ kg^{-1}]	s [kJ kg^{-1} K^{-1}]	c_p [kJ kg^{-1} K^{-1}]	w [m s^{-1}]	κ	$\eta \times 10^6$ [kg m^{-1} s^{-1}]	$\lambda \times 10^3$ [W m^{-1} K^{-1}]
0	0.000982	36.66	0.0000	4.0647	1463.27	58.90	1715.78	581.30
5	0.000983	57.00	0.0737	4.0706	1486.85	60.79	1468.59	589.41
10	0.000984	77.36	0.1463	4.0760	1507.89	62.48	1273.97	597.96
15	0.000985	97.76	0.2177	4.0808	1526.62	63.98	1117.77	606.71
20	0.000986	118.17	0.2879	4.0850	1543.23	65.30	990.31	615.43
25	0.000987	138.61	0.3571	4.0887	1557.90	66.45	884.82	623.98
30	0.000989	159.06	0.4251	4.0919	1570.78	67.45	796.40	632.22
35	0.000990	179.52	0.4920	4.0948	1582.01	68.30	721.50	640.07
40	0.000992	200.01	0.5580	4.0976	1591.70	69.01	657.44	647.46
45	0.000994	220.50	0.6229	4.1002	1599.96	69.58	602.18	654.35
50	0.000997	241.01	0.6869	4.1029	1606.88	70.03	554.17	660.73
60	0.001001	282.07	0.8120	4.1086	1617.05	70.58	475.21	671.95
70	0.001007	323.18	0.9336	4.1150	1622.79	70.70	413.44	681.20
80	0.001013	364.37	1.0519	4.1223	1624.63	70.45	364.20	688.66
90	0.001019	405.63	1.1671	4.1307	1622.98	69.87	324.33	694.55
100	0.001026	446.99	1.2795	4.1401	1618.22	68.99	291.61	699.08
110	0.001033	488.44	1.3891	4.1508	1610.67	67.86	264.42	702.42
120	0.001041	530.01	1.4962	4.1627	1600.60	66.51	241.60	704.71
130	0.001050	571.70	1.6009	4.1761	1588.22	64.96	222.25	706.06
140	0.001058	613.53	1.7034	4.1910	1573.74	63.24	205.71	706.55
150	0.001068	655.52	1.8038	4.2077	1557.28	61.37	191.46	706.23
160	0.001078	697.69	1.9023	4.2264	1538.98	59.38	179.08	705.11
170	0.001089	740.06	1.9990	4.2474	1518.91	57.27	168.25	703.22
180	0.001100	782.65	2.0941	4.2709	1497.14	55.07	158.72	700.56
190	0.001112	825.49	2.1876	4.2973	1473.71	52.79	150.27	697.11
200	0.001125	868.61	2.2797	4.3270	1448.66	50.43	142.72	692.87
210	0.001138	912.04	2.3705	4.3605	1422.01	48.01	135.95	687.82
220	0.001153	955.83	2.4602	4.3983	1393.77	45.55	129.84	681.93
230	0.001168	1000.02	2.5489	4.4410	1363.94	43.04	124.28	675.18
240	0.001185	1044.67	2.6368	4.4892	1332.54	40.51	119.19	667.52
250	0.001202	1089.83	2.7240	4.5439	1299.56	37.96	114.50	658.92
260	0.001221	1135.57	2.8106	4.6060	1265.01	35.41	110.15	649.31
270	0.001242	1181.98	2.8968	4.6768	1228.91	32.87	106.08	638.66
280	0.001264	1229.14	2.9828	4.7576	1191.26	30.34	102.25	626.89
290	0.001288	1277.17	3.0689	4.8505	1152.04	27.84	98.61	613.96
300	0.001315	1326.20	3.1552	4.9579	1111.20	25.38	95.12	599.84
310	0.001344	1376.39	3.2420	5.0831	1068.60	22.96	91.75	584.55
320	0.001377	1427.94	3.3296	5.2308	1024.06	20.58	88.44	568.16
330	0.001414	1481.10	3.4185	5.4078	977.36	18.26	85.18	550.78
340	0.001456	1536.22	3.5091	5.6244	928.56	16.00	81.90	532.61

Table 4 Single-Phase Region (Regions 1 to 3 of IAPWS-IF97) – Continuation
Homogenes Zustandsgebiet (Bereiche 1 bis 3 der IAPWS-IF97) – Fortsetzung

$p = 37$ MPa

t [°C]	v [m^3 kg^{-1}]	h [kJ kg^{-1}]	s [kJ kg^{-1} K^{-1}]	c_p [kJ kg^{-1} K^{-1}]	w [m s^{-1}]	κ	$\eta\times10^6$ [kg m^{-1} s^{-1}]	$\lambda\times10^3$ [W m^{-1} K^{-1}]
350	0.001505	1593.77	3.6022	5.8966	878.31	13.8515	78.57	513.83
360	0.001563	1654.44	3.6988	6.2554	824.53	11.7553	75.13	494.50
370	0.001633	1719.30	3.8004	6.7414	767.95	9.7577	71.51	474.43
380	0.001722	1789.91	3.9094	7.4229	708.01	7.8656	67.60	452.78
390	0.001841	1868.88	4.0293	8.4448	644.77	6.1042	63.27	428.04
400	0.002009	1960.82	4.1669	10.0756	579.96	4.5241	58.31	398.02
410	0.002272	2073.55	4.3331	12.6238	520.10	3.2180	52.51	360.32
420	0.002694	2213.50	4.5364	15.0901	481.03	2.3218	46.17	313.76
430	0.003264	2364.30	4.7524	14.4914	473.74	1.8585	40.88	264.39
440	0.003854	2497.25	4.9402	12.0538	484.90	1.6490	37.62	224.37
450	0.004391	2606.78	5.0928	9.9393	501.28	1.5465	35.81	196.11
460	0.004869	2697.88	5.2179	8.3801	518.16	1.4903	34.80	176.26
470	0.005296	2775.85	5.3236	7.2743	533.80	1.4542	34.25	162.13
480	0.005682	2844.35	5.4151	6.4645	548.18	1.4294	33.98	151.85
490	0.006036	2905.81	5.4962	5.8564	561.48	1.4115	33.88	144.26
500	0.006365	2961.93	5.5693	5.3848	573.82	1.3981	33.89	138.56
510	0.006673	3013.83	5.6360	5.0083	585.34	1.3876	33.99	134.27
520	0.006964	3062.32	5.6975	4.7017	596.13	1.3791	34.14	131.03
530	0.007241	3108.04	5.7548	4.4483	606.28	1.3720	34.34	128.61
540	0.007504	3151.43	5.8085	4.2368	615.85	1.3660	34.57	126.83
550	0.007757	3192.88	5.8592	4.0585	624.93	1.3607	34.82	125.56
560	0.008001	3232.69	5.9072	3.9069	633.55	1.3559	35.10	124.69
570	0.008236	3271.09	5.9531	3.7769	641.77	1.3515	35.39	124.16
580	0.008464	3308.28	5.9969	3.6644	649.63	1.3475	35.69	123.90
590	0.008686	3344.43	6.0390	3.5663	657.17	1.3438	35.00	123.85
600	0.008902	3379.65	6.0796	3.4802	664.43	1.3403	35.32	123.98
610	0.009112	3414.06	6.1188	3.4039	671.42	1.3371	35.65	124.27
620	0.009318	3447.76	6.1567	3.3362	678.19	1.3340	36.98	124.67
630	0.009520	3480.81	6.1935	3.2756	684.74	1.3312	37.32	125.18
640	0.009717	3513.29	6.2293	3.2212	691.11	1.3285	37.66	125.76
650	0.009911	3545.25	6.2641	3.1722	697.30	1.3259	38.00	126.42
660	0.010101	3576.75	6.2980	3.1280	703.33	1.3235	38.34	127.13
670	0.010289	3607.83	6.3312	3.0880	709.22	1.3213	38.69	127.89
680	0.010473	3638.52	6.3635	3.0516	714.97	1.3191	39.03	128.68
690	0.010655	3668.87	6.3952	3.0187	720.59	1.3171	39.38	129.50
700	0.010834	3698.90	6.4262	2.9886	726.09	1.3152	39.73	130.35
720	0.011186	3758.14	6.4865	2.9364	736.76	1.3115	40.42	132.11
740	0.011529	3816.42	6.5446	2.8929	747.03	1.3082	41.12	133.91
760	0.011866	3873.90	6.6008	2.8567	756.93	1.3050	41.81	135.74
780	0.012196	3930.72	6.6553	2.8266	766.49	1.3020	42.50	137.59
800	0.012520	3987.00	6.7082	2.8016	775.74	1.2991	43.19	139.44

Table 4 Single-Phase Region (Regions 1 to 3 of IAPWS-IF97) – Continuation
Homogenes Zustandsgebiet (Bereiche 1 bis 3 der IAPWS-IF97) – Fortsetzung

				p = 38 MPa				
t [°C]	v [$m^3\,kg^{-1}$]	h [$kJ\,kg^{-1}$]	s [$kJ\,kg^{-1}\,K^{-1}$]	c_p [$kJ\,kg^{-1}\,K^{-1}$]	w [$m\,s^{-1}$]	κ	$\eta \times 10^6$ [$kg\,m^{-1}\,s^{-1}$]	$\lambda \times 10^3$ [$W\,m^{-1}\,K^{-1}$]
0	0.000982	37.63	– 0.0001	4.0612	1465.00	57.51	1714.17	581.84
5	0.000982	57.95	0.0736	4.0675	1488.56	59.35	1467.56	589.92
10	0.000983	78.30	0.1461	4.0733	1509.58	60.99	1273.33	598.44
15	0.000984	98.68	0.2175	4.0783	1528.29	62.46	1117.40	607.17
20	0.000985	119.08	0.2877	4.0827	1544.89	63.74	990.13	615.89
25	0.000987	139.51	0.3568	4.0865	1559.55	64.87	884.76	624.43
30	0.000988	159.95	0.4248	4.0898	1572.43	65.84	796.45	632.67
35	0.000990	180.40	0.4917	4.0928	1583.66	66.67	721.61	640.52
40	0.000992	200.87	0.5576	4.0956	1593.35	67.36	657.60	647.91
45	0.000994	221.36	0.6225	4.0983	1601.62	67.92	602.38	654.81
50	0.000996	241.86	0.6864	4.1011	1608.56	68.36	554.39	661.20
60	0.001001	282.90	0.8115	4.1068	1618.76	68.89	475.46	672.42
70	0.001006	324.00	0.9330	4.1132	1624.55	69.02	413.71	681.69
80	0.001012	365.16	1.0513	4.1205	1626.44	68.78	364.48	689.16
90	0.001019	406.41	1.1665	4.1288	1624.85	68.21	324.60	695.07
100	0.001025	447.75	1.2787	4.1383	1620.15	67.37	291.87	699.61
110	0.001033	489.18	1.3883	4.1488	1612.67	66.27	264.68	702.97
120	0.001041	530.73	1.4954	4.1607	1602.67	64.96	241.85	705.29
130	0.001049	572.40	1.6000	4.1740	1590.38	63.45	222.50	706.66
140	0.001058	614.21	1.7025	4.1888	1575.98	61.78	205.96	707.18
150	0.001067	656.18	1.8029	4.2053	1559.62	59.97	191.70	706.88
160	0.001077	698.32	1.9013	4.2238	1541.41	58.03	179.32	705.79
170	0.001088	740.66	1.9979	4.2445	1521.45	55.99	168.48	703.93
180	0.001099	783.22	2.0929	4.2678	1499.80	53.85	158.95	701.29
190	0.001111	826.03	2.1863	4.2939	1476.50	51.63	150.49	697.88
200	0.001124	869.11	2.2784	4.3232	1451.60	49.33	142.95	693.67
210	0.001137	912.50	2.3691	4.3563	1425.10	46.99	136.18	688.65
220	0.001152	956.25	2.4587	4.3936	1397.03	44.59	130.06	682.80
230	0.001167	1000.39	2.5473	4.4356	1367.38	42.16	124.50	676.10
240	0.001184	1044.98	2.6351	4.4832	1336.18	39.70	119.41	668.49
250	0.001201	1090.08	2.7221	4.5370	1303.43	37.22	114.73	659.94
260	0.001220	1135.74	2.8086	4.5980	1269.13	34.74	110.38	650.41
270	0.001240	1182.06	2.8947	4.6674	1233.31	32.27	106.32	639.83
280	0.001262	1229.13	2.9805	4.7466	1195.96	29.82	102.50	628.16
290	0.001286	1277.04	3.0664	4.8375	1157.10	27.39	98.87	615.34
300	0.001313	1325.92	3.1524	4.9422	1116.67	25.00	95.39	601.35
310	0.001342	1375.94	3.2389	5.0639	1074.57	22.65	92.03	586.20
320	0.001374	1427.27	3.3262	5.2068	1030.61	20.34	88.75	569.96
330	0.001410	1480.16	3.4146	5.3772	984.59	18.09	85.51	552.76
340	0.001452	1534.94	3.5047	5.5843	936.50	15.90	82.26	534.76

Table 4 Single-Phase Region (Regions 1 to 3 of IAPWS-IF97) – Continuation
Homogenes Zustandsgebiet (Bereiche 1 bis 3 der IAPWS-IF97) – Fortsetzung

p = 38 MPa								
t [°C]	v [m^3 kg^{-1}]	h [kJ kg^{-1}]	s [kJ kg^{-1} K^{-1}]	c_p [kJ kg^{-1} K^{-1}]	w [m s^{-1}]	κ	$\eta \times 10^6$ [kg m^{-1} s^{-1}]	$\lambda \times 10^3$ [W m^{-1} K^{-1}]
350	0.001499	1592.02	3.5970	5.8422	886.88	13.8047	78.97	516.17
360	0.001556	1652.05	3.6926	6.1799	834.45	11.7795	75.58	497.07
370	0.001623	1715.99	3.7928	6.6298	779.28	9.8442	72.04	477.30
380	0.001708	1785.20	3.8995	7.2475	721.14	8.0123	68.24	456.15
390	0.001819	1861.87	4.0160	8.1467	660.13	6.3057	64.08	432.28
400	0.001972	1949.73	4.1475	9.5262	597.69	4.7672	59.38	403.75
410	0.002201	2054.84	4.3025	11.6255	538.79	3.4709	53.99	368.45
420	0.002558	2183.36	4.4892	13.9663	495.49	2.5258	48.04	325.29
430	0.003058	2327.04	4.6950	14.2685	480.14	1.9835	42.64	278.01
440	0.003613	2461.17	4.8844	12.3896	486.14	1.7214	38.97	236.79
450	0.004138	2574.56	5.0423	10.3579	500.23	1.5914	36.81	206.39
460	0.004612	2669.65	5.1730	8.7418	516.31	1.5211	35.57	184.57
470	0.005038	2750.94	5.2831	7.5743	531.75	1.4771	34.87	168.97
480	0.005423	2822.13	5.3783	6.7051	546.13	1.4472	34.49	157.55
490	0.005777	2885.76	5.4622	6.0511	559.51	1.4260	34.32	149.10
500	0.006105	2943.64	5.5376	5.5460	571.99	1.4104	34.28	142.76
510	0.006411	2997.02	5.6062	5.1445	583.64	1.3982	34.33	137.96
520	0.006700	3046.78	5.6693	4.8183	594.57	1.3885	34.45	134.33
530	0.006974	3093.58	5.7279	4.5492	604.85	1.3805	34.62	131.60
540	0.007235	3137.91	5.7828	4.3247	614.55	1.3736	34.83	129.57
550	0.007486	3180.19	5.8345	4.1359	623.73	1.3677	35.07	128.09
560	0.007727	3220.73	5.8834	3.9756	632.46	1.3623	35.33	127.06
570	0.007959	3259.78	5.9300	3.8384	640.78	1.3575	35.60	126.39
580	0.008185	3297.56	5.9746	3.7199	648.72	1.3531	35.89	126.01
590	0.008403	3334.23	6.0173	3.6167	656.34	1.3490	36.20	125.86
600	0.008616	3369.93	6.0584	3.5262	663.67	1.3452	36.51	125.91
610	0.008824	3404.78	6.0981	3.4462	670.73	1.3417	36.83	126.12
620	0.009027	3438.89	6.1365	3.3752	677.55	1.3384	37.15	126.46
630	0.009225	3472.31	6.1737	3.3117	684.16	1.3353	37.48	126.91
640	0.009419	3505.14	6.2099	3.2548	690.58	1.3324	37.81	127.44
650	0.009610	3537.43	6.2451	3.2035	696.82	1.3297	38.15	128.05
660	0.009797	3569.23	6.2793	3.1572	702.90	1.3271	38.49	128.72
670	0.009981	3600.59	6.3128	3.1153	708.83	1.3247	38.83	129.44
680	0.010163	3631.55	6.3454	3.0773	714.62	1.3224	39.17	130.20
690	0.010341	3662.14	6.3773	3.0428	720.28	1.3202	39.52	131.00
700	0.010517	3692.41	6.4086	3.0113	725.82	1.3182	39.86	131.82
720	0.010862	3752.07	6.4693	2.9566	736.56	1.3144	40.55	133.51
740	0.011199	3810.74	6.5278	2.9111	746.89	1.3108	41.24	135.27
760	0.011529	3868.57	6.5843	2.8731	756.85	1.3075	41.93	137.05
780	0.011852	3925.70	6.6391	2.8415	766.46	1.3044	42.62	138.86
800	0.012170	3982.26	6.6923	2.8153	775.75	1.3013	43.30	140.67

Table 4 Single-Phase Region (Regions 1 to 3 of IAPWS-IF97) – Continuation
Homogenes Zustandsgebiet (Bereiche 1 bis 3 der IAPWS-IF97) – Fortsetzung

				p = 39 MPa				
t [°C]	v [m³ kg⁻¹]	h [kJ kg⁻¹]	s [kJ kg⁻¹ K⁻¹]	c_p [kJ kg⁻¹ K⁻¹]	w [m s⁻¹]	κ	$\eta \times 10^6$ [kg m⁻¹ s⁻¹]	$\lambda \times 10^3$ [W m⁻¹ K⁻¹]
0	0.000982	38.59	– 0.0002	4.0578	1466.73	56.20	1712.59	582.37
5	0.000982	58.90	0.0735	4.0645	1490.27	57.99	1466.54	590.42
10	0.000983	79.24	0.1460	4.0706	1511.27	59.59	1272.70	598.92
15	0.000984	99.60	0.2173	4.0759	1529.97	61.01	1117.03	607.64
20	0.000985	119.99	0.2874	4.0804	1546.55	62.27	989.95	616.35
25	0.000986	140.41	0.3565	4.0843	1561.20	63.37	884.72	624.88
30	0.000988	160.84	0.4244	4.0877	1574.08	64.31	796.49	633.12
35	0.000990	181.28	0.4913	4.0908	1585.31	65.12	721.73	640.97
40	0.000991	201.74	0.5572	4.0937	1595.01	65.79	657.76	648.37
45	0.000994	222.22	0.6221	4.0965	1603.29	66.34	602.58	655.27
50	0.000996	242.71	0.6860	4.0992	1610.23	66.77	554.61	661.66
60	0.001001	283.73	0.8110	4.1050	1620.47	67.30	475.72	672.90
70	0.001006	324.81	0.9325	4.1115	1626.31	67.42	413.98	682.17
80	0.001012	365.96	1.0507	4.1188	1628.25	67.19	364.75	689.66
90	0.001018	407.19	1.1658	4.1270	1626.71	66.65	324.88	695.59
100	0.001025	448.51	1.2780	4.1364	1622.08	65.82	292.14	700.15
110	0.001032	489.92	1.3876	4.1469	1614.67	64.76	264.94	703.53
120	0.001040	531.45	1.4946	4.1587	1604.75	63.48	242.11	705.86
130	0.001049	573.10	1.5992	4.1718	1592.54	62.02	222.75	707.26
140	0.001057	614.89	1.7016	4.1865	1578.22	60.40	206.20	707.80
150	0.001067	656.84	1.8019	4.2029	1561.95	58.64	191.94	707.52
160	0.001077	698.95	1.9003	4.2212	1543.84	56.75	179.55	706.46
170	0.001087	741.27	1.9968	4.2417	1523.99	54.77	168.72	704.63
180	0.001099	783.80	2.0917	4.2647	1502.45	52.68	159.17	702.02
190	0.001111	826.57	2.1851	4.2905	1479.28	50.52	150.72	698.64
200	0.001123	869.62	2.2771	4.3195	1454.51	48.30	143.17	694.47
210	0.001137	912.97	2.3677	4.3521	1428.17	46.01	136.40	689.49
220	0.001151	956.67	2.4573	4.3889	1400.26	43.68	130.29	683.68
230	0.001166	1000.76	2.5458	4.4304	1370.80	41.32	124.73	677.01
240	0.001182	1045.30	2.6334	4.4772	1339.80	38.93	119.64	669.46
250	0.001200	1090.33	2.7203	4.5301	1307.27	36.52	114.96	660.97
260	0.001219	1135.92	2.8066	4.5901	1273.22	34.11	110.62	651.50
270	0.001239	1182.16	2.8926	4.6583	1237.66	31.71	106.56	641.00
280	0.001261	1229.12	2.9782	4.7359	1200.62	29.32	102.75	629.42
290	0.001284	1276.91	3.0639	4.8247	1162.10	26.96	99.13	616.71
300	0.001311	1325.66	3.1497	4.9269	1122.06	24.63	95.66	602.84
310	0.001339	1375.51	3.2359	5.0453	1080.44	22.35	92.32	587.83
320	0.001371	1426.63	3.3228	5.1837	1037.06	20.11	89.05	571.74
330	0.001407	1479.27	3.4108	5.3479	991.70	17.93	85.83	554.70
340	0.001447	1533.70	3.5003	5.5461	944.31	15.80	82.62	536.88

Table 4 Single-Phase Region (Regions 1 to 3 of IAPWS-IF97) – Continuation
Homogenes Zustandsgebiet (Bereiche 1 bis 3 der IAPWS-IF97) – Fortsetzung

p = 39 MPa

t [°C]	v [m^3 kg^{-1}]	h [kJ kg^{-1}]	s [kJ kg^{-1} K^{-1}]	c_p [kJ kg^{-1} K^{-1}]	w [m s^{-1}]	κ	$\eta\times 10^6$ [kg m^{-1} s^{-1}]	$\lambda\times 10^3$ [W m^{-1} K^{-1}]
350	0.001494	1590.34	3.5919	5.7909	895.31	13.7591	79.36	518.47
360	0.001548	1649.78	3.6865	6.1096	844.08	11.7984	76.03	499.58
370	0.001614	1712.87	3.7854	6.5279	790.20	9.9211	72.55	480.09
380	0.001695	1780.82	3.8902	7.0913	733.70	8.1449	68.85	459.39
390	0.001799	1855.48	4.0036	7.8907	674.68	6.4891	64.83	436.29
400	0.001939	1939.91	4.1300	9.0766	614.37	4.9902	60.37	409.05
410	0.002143	2038.94	4.2760	10.8328	556.68	3.7086	55.31	375.80
420	0.002449	2157.93	4.4489	12.9385	510.88	2.7322	49.72	335.49
430	0.002886	2293.54	4.6431	13.8060	488.73	2.1220	44.37	290.54
440	0.003398	2426.60	4.8310	12.5578	489.05	1.8045	40.38	248.97
450	0.003905	2542.79	4.9928	10.6936	500.29	1.6434	37.88	216.73
460	0.004372	2641.40	5.1283	9.0783	515.10	1.5561	36.40	193.13
470	0.004795	2725.85	5.2427	7.8684	530.11	1.5027	35.53	176.06
480	0.005180	2799.72	5.3415	6.9478	544.36	1.4669	35.04	163.48
490	0.005532	2865.54	5.4283	6.2485	557.76	1.4420	34.78	154.13
500	0.005858	2925.22	5.5060	5.7095	570.32	1.4237	34.68	147.11
510	0.006163	2980.10	5.5765	5.2826	582.08	1.4097	34.69	141.79
520	0.006450	3031.14	5.6413	4.9366	593.13	1.3985	34.77	137.75
530	0.006722	3079.04	5.7013	4.6515	603.52	1.3894	34.92	134.69
540	0.006981	3124.33	5.7574	4.4140	613.33	1.3817	35.10	132.39
550	0.007229	3167.44	5.8101	4.2144	622.62	1.3750	35.32	130.70
560	0.007467	3208.72	5.8599	4.0452	631.45	1.3691	35.56	129.49
570	0.007697	3248.43	5.9073	3.9006	639.85	1.3638	35.82	128.67
580	0.007920	3286.80	5.9525	3.7759	647.88	1.3590	35.10	128.17
590	0.008136	3324.00	5.9959	3.6675	655.57	1.3545	35.40	127.92
600	0.008346	3360.19	6.0376	3.5726	662.96	1.3503	35.70	127.88
610	0.008550	3395.49	6.0778	3.4888	670.08	1.3465	37.01	128.01
620	0.008750	3430.00	6.1166	3.4144	676.96	1.3429	37.33	128.28
630	0.008946	3463.80	6.1543	3.3481	683.63	1.3396	37.65	128.67
640	0.009137	3496.98	6.1908	3.2885	690.09	1.3364	37.98	129.16
650	0.009325	3529.60	6.2263	3.2349	696.38	1.3335	38.31	129.72
660	0.009509	3561.70	6.2609	3.1866	702.50	1.3308	38.64	130.35
670	0.009690	3593.34	6.2946	3.1428	708.47	1.3282	38.98	131.03
680	0.009868	3624.57	6.3276	3.1031	714.30	1.3258	39.32	131.75
690	0.010043	3655.42	6.3598	3.0670	720.00	1.3235	39.66	132.51
700	0.010216	3685.92	6.3913	3.0341	725.57	1.3213	40.00	133.30
720	0.010555	3746.01	6.4524	2.9769	736.38	1.3173	40.68	134.94
740	0.010886	3805.06	6.5113	2.9293	746.77	1.3135	41.37	136.64
760	0.011209	3863.24	6.5681	2.8896	756.79	1.3101	42.05	138.38
780	0.011527	3920.69	6.6232	2.8565	766.45	1.3068	42.73	140.14
800	0.011838	3977.53	6.6767	2.8290	775.78	1.3036	43.41	141.91

Table 4 Single-Phase Region (Regions 1 to 3 of IAPWS-IF97) – Continuation
Homogenes Zustandsgebiet (Bereiche 1 bis 3 der IAPWS-IF97) – Fortsetzung

p = 40 MPa

t [°C]	v [m³ kg⁻¹]	h [kJ kg⁻¹]	s [kJ kg⁻¹ K⁻¹]	c_p [kJ kg⁻¹ K⁻¹]	w [m s⁻¹]	κ	$\eta \times 10^6$ [kg m⁻¹ s⁻¹]	$\lambda \times 10^3$ [W m⁻¹ K⁻¹]
0	0.000981	39.56	– 0.0002	4.0544	1468.46	54.95	1711.02	582.91
5	0.000982	59.85	0.0734	4.0615	1491.98	56.69	1465.54	590.92
10	0.000982	80.17	0.1458	4.0679	1512.97	58.26	1272.07	599.40
15	0.000983	100.52	0.2171	4.0734	1531.64	59.64	1116.67	608.10
20	0.000985	120.90	0.2872	4.0781	1548.22	60.87	989.77	616.80
25	0.000986	141.30	0.3562	4.0821	1562.86	61.94	884.67	625.34
30	0.000987	161.72	0.4241	4.0857	1575.73	62.86	796.54	633.57
35	0.000989	182.16	0.4910	4.0888	1586.96	63.65	721.85	641.42
40	0.000991	202.61	0.5568	4.0918	1596.66	64.31	657.93	648.82
45	0.000993	223.08	0.6216	4.0946	1604.95	64.84	602.78	655.73
50	0.000995	243.56	0.6855	4.0974	1611.91	65.26	554.84	662.12
60	0.001000	284.56	0.8105	4.1033	1622.19	65.78	475.97	673.37
70	0.001005	325.62	0.9319	4.1097	1628.07	65.91	414.24	682.66
80	0.001011	366.76	1.0501	4.1170	1630.06	65.68	365.02	690.16
90	0.001018	407.97	1.1651	4.1252	1628.58	65.16	325.15	696.10
100	0.001024	449.27	1.2773	4.1346	1624.01	64.36	292.41	700.68
110	0.001032	490.66	1.3868	4.1450	1616.67	63.33	265.20	704.08
120	0.001040	532.17	1.4937	4.1567	1606.82	62.08	242.36	706.44
130	0.001048	573.80	1.5983	4.1697	1594.69	60.66	223.00	707.86
140	0.001057	615.57	1.7006	4.1843	1580.46	59.09	206.45	708.42
150	0.001066	657.49	1.8009	4.2005	1564.28	57.37	192.18	708.17
160	0.001076	699.59	1.8992	4.2186	1546.27	55.54	179.79	707.14
170	0.001087	741.87	1.9957	4.2389	1526.52	53.60	168.95	705.33
180	0.001098	784.37	2.0906	4.2616	1505.10	51.58	159.40	702.75
190	0.001110	827.11	2.1839	4.2871	1482.05	49.48	150.94	699.40
200	0.001122	870.12	2.2758	4.3158	1457.42	47.31	143.40	695.26
210	0.001136	913.44	2.3663	4.3480	1431.22	45.09	136.63	690.32
220	0.001150	957.10	2.4558	4.3843	1403.48	42.82	130.51	684.55
230	0.001165	1001.14	2.5442	4.4252	1374.20	40.52	124.95	677.92
240	0.001181	1045.62	2.6317	4.4713	1343.39	38.19	119.87	670.42
250	0.001199	1090.59	2.7185	4.5234	1311.08	35.85	115.19	661.99
260	0.001217	1136.11	2.8047	4.5824	1277.27	33.51	110.85	652.59
270	0.001237	1182.26	2.8905	4.6493	1241.98	31.17	106.80	642.17
280	0.001259	1229.13	2.9760	4.7254	1205.23	28.84	102.99	630.67
290	0.001283	1276.80	3.0614	4.8123	1167.04	26.55	99.38	618.06
300	0.001308	1325.41	3.1469	4.9120	1127.39	24.29	95.93	604.32
310	0.001337	1375.10	3.2329	5.0272	1086.22	22.07	92.60	589.44
320	0.001368	1426.02	3.3195	5.1615	1043.39	19.89	89.35	573.50
330	0.001403	1478.41	3.4070	5.3198	998.69	17.77	86.15	556.62
340	0.001443	1532.52	3.4960	5.5097	951.99	15.70	82.97	538.96

Table 4 Single-Phase Region (Regions 1 to 3 of IAPWS-IF97) – Continuation
Homogenes Zustandsgebiet (Bereiche 1 bis 3 der IAPWS-IF97) – Fortsetzung

				p = 40 MPa				
t [°C]	v [m^3 kg^{-1}]	h [kJ kg^{-1}]	s [kJ kg^{-1} K^{-1}]	c_p [kJ kg^{-1} K^{-1}]	w [m s^{-1}]	κ	$\eta \times 10^6$ [kg m^{-1} s^{-1}]	$\lambda \times 10^3$ [W m^{-1} K^{-1}]
350	0.001488	1588.74	3.5870	5.7424	903.62	13.7149	79.75	520.73
360	0.001542	1647.62	3.6807	6.0440	853.45	11.8128	76.46	502.04
370	0.001605	1709.93	3.7783	6.4342	800.77	9.9896	73.04	482.81
380	0.001682	1776.72	3.8814	6.9509	745.75	8.2653	69.44	462.52
390	0.001780	1849.61	3.9921	7.6679	688.52	6.6568	65.55	440.09
400	0.001911	1931.13	4.1141	8.7012	630.14	5.1954	61.27	413.99
410	0.002093	2025.18	4.2527	10.1922	573.76	3.9314	56.50	382.50
420	0.002361	2136.30	4.4142	12.0453	526.61	2.9362	51.25	344.63
430	0.002742	2263.84	4.5969	13.2052	499.08	2.2707	46.02	301.97
440	0.003210	2394.03	4.7807	12.5536	493.66	1.8982	41.82	260.72
450	0.003693	2511.77	4.9447	10.9499	501.50	1.7027	39.01	227.11
460	0.004149	2613.32	5.0842	9.3909	514.66	1.5960	37.28	201.90
470	0.004567	2700.69	5.2026	8.1475	528.95	1.5315	36.24	183.36
480	0.004950	2777.18	5.3048	7.1892	542.92	1.4888	35.62	169.62
490	0.005300	2845.20	5.3946	6.4475	556.25	1.4594	35.27	159.34
500	0.005625	2906.69	5.4746	5.8745	568.83	1.4381	35.11	151.61
510	0.005928	2963.09	5.5471	5.4221	580.67	1.4220	35.06	145.75
520	0.006213	3015.42	5.6135	5.0563	591.81	1.4093	35.11	141.28
530	0.006483	3064.43	5.6749	4.7551	602.30	1.3990	35.22	137.87
540	0.006740	3110.69	5.7322	4.5044	612.21	1.3903	35.38	135.30
550	0.006985	3154.65	5.7859	4.2938	621.60	1.3828	35.58	133.38
560	0.007221	3196.67	5.8366	4.1156	630.51	1.3763	35.80	131.98
570	0.007449	3237.05	5.8848	3.9634	638.99	1.3704	36.05	131.01
580	0.007669	3276.01	5.9308	3.8324	647.09	1.3651	36.32	130.38
590	0.007882	3313.75	5.9747	3.7188	654.85	1.3602	36.60	130.02
600	0.008089	3350.43	6.0170	3.6194	662.31	1.3557	36.89	129.88
610	0.008291	3386.18	6.0577	3.5317	669.49	1.3515	37.19	129.93
620	0.008488	3421.10	6.0970	3.4540	676.42	1.3476	37.50	130.14
630	0.008681	3455.29	6.1351	3.3846	683.13	1.3440	37.82	130.47
640	0.008869	3488.82	6.1720	3.3225	689.65	1.3407	38.14	130.90
650	0.009054	3521.76	6.2079	3.2666	695.98	1.3375	38.47	131.41
660	0.009235	3554.17	6.2428	3.2161	702.14	1.3346	38.80	132.00
670	0.009413	3586.09	6.2768	3.1704	708.15	1.3318	39.13	132.64
680	0.009589	3617.59	6.3100	3.1289	714.02	1.3292	39.46	133.32
690	0.009761	3648.69	6.3425	3.0913	719.75	1.3268	39.80	134.05
700	0.009931	3679.42	6.3743	3.0570	725.36	1.3245	40.14	134.81
720	0.010264	3739.95	6.4358	2.9973	736.23	1.3202	40.81	136.39
740	0.010589	3799.38	6.4951	2.9476	746.68	1.3163	41.49	138.04
760	0.010906	3857.91	6.5523	2.9061	756.75	1.3127	42.17	139.73
780	0.011217	3915.68	6.6077	2.8715	766.46	1.3092	42.85	141.44
800	0.011523	3972.81	6.6614	2.8428	775.83	1.3059	43.52	143.17

Table 4 Single-Phase Region (Regions 1 to 3 of IAPWS-IF97) – Continuation
Homogenes Zustandsgebiet (Bereiche 1 bis 3 der IAPWS-IF97) – Fortsetzung

p = 42.5 MPa

t [°C]	v [m³ kg⁻¹]	h [kJ kg⁻¹]	s [kJ kg⁻¹ K⁻¹]	c_p [kJ kg⁻¹ K⁻¹]	w [m s⁻¹]	κ	$\eta \times 10^6$ [kg m⁻¹ s⁻¹]	$\lambda \times 10^3$ [W m⁻¹ K⁻¹]
0	0.000980	41.96	– 0.0004	4.0461	1472.80	52.08	1707.20	584.24
5	0.000981	62.21	0.0731	4.0542	1496.27	53.72	1463.09	592.16
10	0.000981	82.50	0.1454	4.0613	1517.21	55.20	1270.57	600.60
15	0.000982	102.82	0.2165	4.0674	1535.84	56.50	1115.81	609.26
20	0.000983	123.17	0.2865	4.0725	1552.38	57.66	989.37	617.94
25	0.000985	143.55	0.3554	4.0768	1567.00	58.66	884.59	626.47
30	0.000986	163.94	0.4233	4.0806	1579.86	59.54	796.69	634.70
35	0.000988	184.35	0.4901	4.0839	1591.09	60.28	722.15	642.55
40	0.000990	204.78	0.5558	4.0870	1600.81	60.90	658.36	649.96
45	0.000992	225.22	0.6206	4.0900	1609.12	61.41	603.29	656.87
50	0.000994	245.68	0.6844	4.0929	1616.11	61.81	555.41	663.27
60	0.000999	286.64	0.8092	4.0988	1626.47	62.30	476.62	674.55
70	0.001004	327.66	0.9305	4.1053	1632.46	62.43	414.92	683.87
80	0.001010	368.75	1.0485	4.1126	1634.57	62.23	365.70	691.41
90	0.001017	409.91	1.1635	4.1208	1633.24	61.74	325.82	697.39
100	0.001023	451.17	1.2756	4.1300	1628.83	61.00	293.07	702.01
110	0.001031	492.52	1.3849	4.1403	1621.65	60.03	265.86	705.46
120	0.001038	533.98	1.4917	4.1518	1611.98	58.88	243.00	707.87
130	0.001047	575.56	1.5962	4.1645	1600.04	57.55	223.63	709.35
140	0.001056	617.27	1.6984	4.1787	1586.02	56.07	207.06	709.97
150	0.001065	659.14	1.7985	4.1946	1570.06	54.47	192.78	709.78
160	0.001075	701.17	1.8967	4.2122	1552.30	52.75	180.37	708.82
170	0.001085	743.39	1.9930	4.2320	1532.80	50.94	169.52	707.08
180	0.001096	785.82	2.0877	4.2541	1511.66	49.05	159.97	704.58
190	0.001108	828.48	2.1808	4.2789	1488.92	47.08	151.50	701.30
200	0.001120	871.40	2.2725	4.3066	1464.62	45.05	143.95	697.24
210	0.001134	914.62	2.3629	4.3379	1438.78	42.96	137.18	692.38
220	0.001148	958.17	2.4521	4.3730	1411.44	40.84	131.06	686.71
230	0.001163	1002.10	2.5403	4.4124	1382.59	38.69	125.51	680.19
240	0.001179	1046.44	2.6276	4.4569	1352.27	36.51	120.43	672.81
250	0.001196	1091.25	2.7141	4.5070	1320.48	34.31	115.76	664.51
260	0.001214	1136.60	2.7999	4.5636	1287.25	32.12	111.43	655.27
270	0.001234	1182.55	2.8853	4.6275	1252.60	29.93	107.39	645.04
280	0.001255	1229.18	2.9704	4.7000	1216.56	27.75	103.60	633.76
290	0.001278	1276.58	3.0553	4.7824	1179.16	25.60	100.01	621.40
300	0.001303	1324.87	3.1403	4.8765	1140.42	23.49	96.58	607.94
310	0.001330	1374.16	3.2256	4.9844	1100.32	21.41	93.29	593.39
320	0.001361	1424.61	3.3113	5.1089	1058.77	19.38	90.08	577.80
330	0.001395	1476.41	3.3979	5.2541	1015.61	17.40	86.93	561.30
340	0.001432	1529.78	3.4857	5.4256	970.65	15.48	83.81	544.04

Table 4 Single-Phase Region (Regions 1 to 3 of IAPWS-IF97) – Continuation
Homogenes Zustandsgebiet (Bereiche 1 bis 3 der IAPWS-IF97) – Fortsetzung

p = 42.5 MPa

t [°C]	v [m³ kg⁻¹]	h [kJ kg⁻¹]	s [kJ kg⁻¹ K⁻¹]	c_p [kJ kg⁻¹ K⁻¹]	w [m s⁻¹]	κ	$\eta \times 10^6$ [kg m⁻¹ s⁻¹]	$\lambda \times 10^3$ [W m⁻¹ K⁻¹]
350	0.001476	1585.03	3.5751	5.6317	923.91	13.6113	80.68	526.22
360	0.001525	1642.65	3.6668	5.8972	875.85	11.8323	77.49	507.98
370	0.001584	1703.22	3.7617	6.2298	825.79	10.1303	74.22	489.33
380	0.001654	1767.55	3.8609	6.6547	773.98	8.5215	70.80	469.89
390	0.001741	1836.78	3.9661	7.2178	720.53	7.0180	67.18	448.87
400	0.001851	1912.59	4.0796	7.9848	666.24	5.6423	63.29	425.06
410	0.001998	1997.41	4.2046	9.0312	613.10	4.4278	59.05	397.08
420	0.002199	2094.18	4.3452	10.3579	565.32	3.4194	54.46	364.01
430	0.002478	2204.37	4.5030	11.6024	529.68	2.6641	49.72	326.42
440	0.002839	2323.16	4.6708	11.9488	511.80	2.1713	45.37	287.54
450	0.003248	2439.49	4.8328	11.1760	509.81	1.8825	42.00	252.44
460	0.003664	2545.20	4.9780	9.9483	517.07	1.7167	39.68	224.11
470	0.004062	2638.48	5.1044	8.7260	528.58	1.6186	38.17	202.19
480	0.004432	2720.71	5.2143	7.7518	541.07	1.5542	37.21	185.73
490	0.004776	2794.03	5.3110	6.9393	553.71	1.5104	36.61	173.11
500	0.005095	2860.04	5.3970	6.2882	566.06	1.4797	36.26	163.50
510	0.005393	2920.25	5.4744	5.7731	577.90	1.4571	36.08	156.17
520	0.005673	2975.84	5.5449	5.3583	589.14	1.4396	36.02	150.55
530	0.005938	3027.66	5.6098	5.0176	599.79	1.4256	36.04	146.22
540	0.006189	3076.38	5.6701	4.7340	609.87	1.4141	36.13	142.90
550	0.006429	3122.49	5.7265	4.4958	619.44	1.4043	36.27	140.37
560	0.006659	3166.42	5.7795	4.2944	628.52	1.3959	36.45	138.48
570	0.006880	3208.48	5.8297	4.1229	637.17	1.3884	36.65	137.09
580	0.007094	3248.96	5.8774	3.9758	645.43	1.3817	36.89	136.11
590	0.007301	3288.06	5.9230	3.8485	653.34	1.3756	37.14	135.45
600	0.007502	3325.98	5.9667	3.7376	660.93	1.3701	37.40	135.07
610	0.007697	3362.86	6.0087	3.6400	668.23	1.3649	37.68	134.91
620	0.007888	3398.82	6.0492	3.5537	675.28	1.3602	37.97	134.92
630	0.008074	3433.97	6.0883	3.4769	682.10	1.3559	38.27	135.09
640	0.008256	3468.38	6.1262	3.4080	688.71	1.3519	38.57	135.37
650	0.008434	3502.15	6.1630	3.3462	695.14	1.3481	38.88	135.76
660	0.008609	3535.33	6.1987	3.2903	701.40	1.3447	39.20	136.23
670	0.008780	3567.97	6.2335	3.2398	707.50	1.3414	39.52	136.76
680	0.008948	3600.14	6.2675	3.1940	713.45	1.3384	39.84	137.35
690	0.009114	3631.87	6.3006	3.1524	719.27	1.3356	40.16	137.99
700	0.009277	3663.20	6.3329	3.1145	724.96	1.3329	40.49	138.66
720	0.009597	3724.81	6.3956	3.0485	735.98	1.3281	41.15	140.09
740	0.009908	3785.22	6.4558	2.9935	746.57	1.3236	41.81	141.60
760	0.010212	3844.61	6.5139	2.9476	756.75	1.3195	42.48	143.16
780	0.010509	3903.17	6.5700	2.9092	766.57	1.3156	43.15	144.76
800	0.010801	3961.02	6.6244	2.8772	776.02	1.3119	43.81	146.37

Table 4 Single-Phase Region (Regions 1 to 3 of IAPWS-IF97) – Continuation
Homogenes Zustandsgebiet (Bereiche 1 bis 3 der IAPWS-IF97) – Fortsetzung

				p = 45 MPa				
t [°C]	v [m³ kg⁻¹]	h [kJ kg⁻¹]	s [kJ kg⁻¹ K⁻¹]	c_p [kJ kg⁻¹ K⁻¹]	w [m s⁻¹]	κ	$\eta \times 10^6$ [kg m⁻¹ s⁻¹]	$\lambda \times 10^3$ [W m⁻¹ K⁻¹]
0	0.000979	44.36	– 0.0006	4.0380	1477.16	49.53	1703.51	585.56
5	0.000979	64.57	0.0727	4.0471	1500.58	51.09	1460.75	593.40
10	0.000980	84.83	0.1449	4.0549	1521.46	52.48	1269.13	601.78
15	0.000981	105.12	0.2160	4.0614	1540.05	53.71	1115.00	610.42
20	0.000982	125.44	0.2859	4.0669	1556.55	54.80	988.99	619.08
25	0.000984	145.78	0.3547	4.0716	1571.15	55.75	884.53	627.59
30	0.000985	166.15	0.4224	4.0756	1584.00	56.58	796.85	635.83
35	0.000987	186.54	0.4891	4.0791	1595.23	57.28	722.48	643.68
40	0.000989	206.94	0.5548	4.0824	1604.96	57.87	658.79	651.09
45	0.000991	227.36	0.6195	4.0854	1613.29	58.35	603.81	658.01
50	0.000993	247.80	0.6832	4.0884	1620.32	58.73	555.99	664.43
60	0.000998	288.71	0.8079	4.0945	1630.75	59.21	477.26	675.73
70	0.001003	329.69	0.9291	4.1010	1636.85	59.34	415.60	685.08
80	0.001009	370.74	1.0470	4.1083	1639.09	59.16	366.39	692.65
90	0.001016	411.86	1.1619	4.1164	1637.89	58.70	326.50	698.68
100	0.001022	453.07	1.2738	4.1255	1633.63	58.01	293.74	703.34
110	0.001030	494.37	1.3830	4.1356	1626.62	57.11	266.51	706.84
120	0.001037	535.78	1.4897	4.1469	1617.13	56.03	243.64	709.30
130	0.001046	577.31	1.5940	4.1594	1605.38	54.78	224.25	710.84
140	0.001054	618.98	1.6961	4.1733	1591.56	53.39	207.67	711.52
150	0.001063	660.78	1.7961	4.1888	1575.82	51.89	193.37	711.39
160	0.001073	702.76	1.8941	4.2060	1558.28	50.28	180.95	710.49
170	0.001084	744.91	1.9904	4.2252	1539.04	48.57	170.10	708.82
180	0.001095	787.27	2.0849	4.2467	1518.17	46.79	160.53	706.39
190	0.001106	829.85	2.1778	4.2708	1495.72	44.94	152.06	703.19
200	0.001119	872.69	2.2693	4.2977	1471.74	43.03	144.51	699.21
210	0.001132	915.82	2.3595	4.3280	1446.25	41.07	137.73	694.44
220	0.001145	959.27	2.4485	4.3620	1419.29	39.08	131.61	688.86
230	0.001160	1003.07	2.5365	4.4001	1390.86	37.05	126.06	682.44
240	0.001176	1047.28	2.6235	4.4430	1360.99	35.00	120.98	675.17
250	0.001193	1091.95	2.7097	4.4912	1329.70	32.94	116.32	667.01
260	0.001211	1137.13	2.7952	4.5455	1297.02	30.88	112.00	657.92
270	0.001230	1182.88	2.8803	4.6068	1262.98	28.82	107.98	647.86
280	0.001251	1229.29	2.9649	4.6759	1227.62	26.78	104.20	636.79
290	0.001273	1276.43	3.0494	4.7542	1190.96	24.75	100.63	624.67
300	0.001298	1324.41	3.1338	4.8431	1153.06	22.77	97.23	611.48
310	0.001324	1373.34	3.2185	4.9444	1113.93	20.82	93.96	597.23
320	0.001354	1423.35	3.3035	5.0604	1073.54	18.92	90.79	581.97
330	0.001386	1474.60	3.3892	5.1943	1031.78	17.06	87.69	565.82
340	0.001423	1527.31	3.4758	5.3502	988.47	15.26	84.63	548.94

Table 4 Single-Phase Region (Regions 1 to 3 of IAPWS-IF97) – Continuation
Homogenes Zustandsgebiet (Bereiche 1 bis 3 der IAPWS-IF97) – Fortsetzung

p = 45 MPa								
t [°C]	v [m³ kg⁻¹]	h [kJ kg⁻¹]	s [kJ kg⁻¹ K⁻¹]	c_p [kJ kg⁻¹ K⁻¹]	w [m s⁻¹]	κ	$\eta \times 10^6$ [kg m⁻¹ s⁻¹]	$\lambda \times 10^3$ [W m⁻¹ K⁻¹]
350	0.001464	1581.70	3.5638	5.5342	943.48	13.5139	81.57	531.50
360	0.001511	1638.21	3.6538	5.7706	896.96	11.8340	78.47	513.66
370	0.001565	1697.31	3.7464	6.0587	849.12	10.2363	75.32	495.49
380	0.001630	1759.62	3.8425	6.4163	799.96	8.7265	72.05	476.74
390	0.001707	1825.97	3.9433	6.8740	749.57	7.3132	68.64	456.80
400	0.001804	1897.57	4.0505	7.4719	698.58	6.0130	65.03	434.72
410	0.001926	1976.02	4.1661	8.2532	648.37	4.8494	61.18	409.36
420	0.002088	2063.30	4.2930	9.2300	601.62	3.8529	57.06	379.84
430	0.002302	2160.87	4.4327	10.2690	562.72	3.0565	52.78	346.27
440	0.002581	2267.51	4.5833	10.9544	536.65	2.4796	48.62	310.45
450	0.002915	2377.29	4.7362	10.8646	525.21	2.1027	45.03	275.87
460	0.003278	2482.58	4.8808	10.1200	525.29	1.8707	42.27	245.84
470	0.003642	2578.97	5.0114	9.1498	532.14	1.7277	40.32	221.57
480	0.003992	2665.37	5.1269	8.1730	542.26	1.6368	39.00	202.41
490	0.004324	2743.09	5.2294	7.3838	553.37	1.5739	38.12	187.70
500	0.004634	2813.35	5.3209	6.6881	564.91	1.5303	37.55	176.19
510	0.004925	2877.28	5.4030	6.1183	576.39	1.4989	37.21	167.29
520	0.005199	2936.08	5.4776	5.6582	587.51	1.4753	37.02	160.42
530	0.005458	2990.72	5.5461	5.2804	598.14	1.4566	36.94	155.09
540	0.005704	3041.90	5.6094	4.9655	608.27	1.4415	36.95	150.96
550	0.005938	3090.19	5.6685	4.7004	617.92	1.4288	37.02	147.77
560	0.006163	3136.04	5.7238	4.4759	627.11	1.4180	37.14	145.33
570	0.006379	3179.82	5.7761	4.2848	635.87	1.4086	37.30	143.48
580	0.006587	3221.83	5.8256	4.1212	644.23	1.4003	37.49	142.12
590	0.006788	3262.32	5.8728	3.9799	652.24	1.3928	37.71	141.15
600	0.006983	3301.49	5.9179	3.8571	659.93	1.3860	37.94	140.49
610	0.007172	3339.51	5.9612	3.7495	667.32	1.3798	38.20	140.10
620	0.007357	3376.52	6.0029	3.6544	674.46	1.3741	38.46	139.91
630	0.007537	3412.64	6.0431	3.5699	681.36	1.3689	38.74	139.90
640	0.007713	3447.95	6.0820	3.4943	688.06	1.3641	39.02	140.02
650	0.007885	3482.55	6.1197	3.4264	694.56	1.3596	39.32	140.27
660	0.008053	3516.50	6.1562	3.3651	700.90	1.3556	39.62	140.61
670	0.008219	3549.87	6.1918	3.3097	707.07	1.3518	39.92	141.03
680	0.008381	3582.71	6.2265	3.2595	713.10	1.3483	40.23	141.52
690	0.008541	3615.08	6.2602	3.2139	718.99	1.3450	40.55	142.06
700	0.008698	3647.00	6.2932	3.1723	724.75	1.3420	40.86	142.64
720	0.009005	3709.70	6.3570	3.1000	735.91	1.3364	41.50	143.90
740	0.009304	3771.08	6.4182	3.0396	746.62	1.3314	42.15	145.27
760	0.009596	3831.35	6.4771	2.9892	756.91	1.3268	42.80	146.69
780	0.009881	3890.70	6.5340	2.9470	766.81	1.3224	43.45	148.16
800	0.010160	3949.28	6.5891	2.9118	776.34	1.3182	44.11	149.65

Table 4 Single-Phase Region (Regions 1 to 3 of IAPWS-IF97) – Continuation
Homogenes Zustandsgebiet (Bereiche 1 bis 3 der IAPWS-IF97) – Fortsetzung

p = 47.5 MPa

t [°C]	v [m³ kg⁻¹]	h [kJ kg⁻¹]	s [kJ kg⁻¹ K⁻¹]	c_p [kJ kg⁻¹ K⁻¹]	w [m s⁻¹]	κ	$\eta\times10^6$ [kg m⁻¹ s⁻¹]	$\lambda\times10^3$ [W m⁻¹ K⁻¹]
0	0.000978	46.75	– 0.0008	4.0301	1481.54	47.26	1699.96	586.87
5	0.000978	66.92	0.0724	4.0401	1504.90	48.73	1458.49	594.64
10	0.000979	87.15	0.1445	4.0486	1525.73	50.05	1267.75	602.97
15	0.000980	107.41	0.2154	4.0556	1544.27	51.22	1114.23	611.57
20	0.000981	127.70	0.2852	4.0615	1560.74	52.25	988.66	620.21
25	0.000983	148.02	0.3539	4.0665	1575.31	53.15	884.49	628.72
30	0.000984	168.36	0.4216	4.0707	1588.15	53.94	797.03	636.95
35	0.000986	188.73	0.4882	4.0744	1599.38	54.61	722.82	644.80
40	0.000988	209.11	0.5538	4.0778	1609.12	55.17	659.25	652.22
45	0.000990	229.50	0.6185	4.0809	1617.47	55.62	604.34	659.15
50	0.000992	249.92	0.6821	4.0840	1624.52	55.99	556.58	665.58
60	0.000997	290.79	0.8067	4.0902	1635.04	56.44	477.92	676.90
70	0.001002	331.72	0.9277	4.0968	1641.23	56.57	416.28	686.28
80	0.001008	372.72	1.0455	4.1040	1643.59	56.41	367.07	693.90
90	0.001014	413.80	1.1602	4.1121	1642.54	55.99	327.18	699.96
100	0.001021	454.97	1.2721	4.1211	1638.42	55.34	294.41	704.67
110	0.001028	496.23	1.3812	4.1310	1631.57	54.49	267.16	708.21
120	0.001036	537.59	1.4877	4.1421	1622.25	53.47	244.28	710.73
130	0.001044	579.07	1.5919	4.1544	1610.69	52.30	224.87	712.32
140	0.001053	620.68	1.6939	4.1680	1597.06	51.00	208.27	713.06
150	0.001062	662.44	1.7937	4.1831	1581.54	49.58	193.96	713.00
160	0.001072	704.35	1.8916	4.1999	1564.23	48.06	181.54	712.16
170	0.001082	746.44	1.9877	4.2186	1545.23	46.45	170.67	710.56
180	0.001093	788.73	2.0821	4.2395	1524.63	44.77	161.10	708.20
190	0.001105	831.24	2.1749	4.2629	1502.46	43.03	152.62	705.07
200	0.001117	874.00	2.2662	4.2891	1478.79	41.23	145.06	701.17
210	0.001130	917.03	2.3562	4.3184	1453.64	39.38	138.28	696.48
220	0.001143	960.38	2.4450	4.3513	1427.04	37.50	132.16	690.99
230	0.001158	1004.07	2.5327	4.3882	1399.01	35.59	126.61	684.68
240	0.001173	1048.15	2.6195	4.4295	1369.58	33.66	121.53	677.52
250	0.001190	1092.68	2.7054	4.4760	1338.77	31.71	116.87	669.48
260	0.001208	1137.69	2.7906	4.5282	1306.62	29.76	112.56	660.54
270	0.001226	1183.26	2.8753	4.5869	1273.16	27.82	108.55	650.65
280	0.001247	1229.46	2.9596	4.6529	1238.43	25.90	104.79	639.77
290	0.001269	1276.35	3.0436	4.7274	1202.48	23.99	101.24	627.87
300	0.001293	1324.04	3.1275	4.8116	1165.36	22.12	97.86	614.94
310	0.001319	1372.62	3.2116	4.9070	1127.12	20.28	94.62	600.97
320	0.001347	1422.22	3.2959	5.0155	1087.76	18.49	91.48	586.03
330	0.001379	1472.98	3.3808	5.1394	1047.25	16.75	88.42	570.21
340	0.001414	1525.07	3.4664	5.2820	1005.46	15.06	85.41	553.67

Table 4 Single-Phase Region (Regions 1 to 3 of IAPWS-IF97) – Continuation
Homogenes Zustandsgebiet (Bereiche 1 bis 3 der IAPWS-IF97) – Fortsetzung

				p = 47.5 MPa				
t [°C]	v [m^3 kg^{-1}]	h [kJ kg^{-1}]	s [kJ kg^{-1} K^{-1}]	c_p [kJ kg^{-1} K^{-1}]	w [m s^{-1}]	κ	$\eta \times 10^6$ [kg m^{-1} s^{-1}]	$\lambda \times 10^3$ [W m^{-1} K^{-1}]
350	0.001453	1578.70	3.5532	5.4476	962.22	13.4172	82.42	536.58
360	0.001497	1634.23	3.6415	5.6599	916.97	11.8228	79.41	519.11
370	0.001548	1692.05	3.7322	5.9129	871.05	10.3162	76.35	501.36
380	0.001608	1752.66	3.8257	6.2193	824.12	8.8924	73.22	483.17
390	0.001679	1816.69	3.9229	6.6009	776.28	7.5577	69.97	464.08
400	0.001764	1885.02	4.0252	7.0838	728.02	6.3244	66.58	443.34
410	0.001870	1958.79	4.1340	7.6942	680.36	5.2100	63.01	419.99
420	0.002005	2039.37	4.2511	8.4418	635.16	4.2360	59.25	393.21
430	0.002178	2127.92	4.3779	9.2706	595.35	3.4260	55.35	362.84
440	0.002399	2224.39	4.5141	9.9785	564.69	2.7982	51.47	329.93
450	0.002670	2326.02	4.6556	10.2520	545.98	2.3508	47.88	296.83
460	0.002977	2427.50	4.7950	9.9580	538.86	2.0535	44.89	266.41
470	0.003301	2523.91	4.9256	9.2853	540.25	1.8616	42.59	240.52
480	0.003624	2612.84	5.0445	8.4981	546.76	1.7366	40.93	219.48
490	0.003936	2693.49	5.1509	7.7093	555.70	1.6517	39.77	202.66
500	0.004234	2767.24	5.2469	7.0439	565.76	1.5914	38.97	189.45
510	0.004517	2834.60	5.3335	6.4441	576.41	1.5487	38.44	178.99
520	0.004783	2896.47	5.4120	5.9466	587.09	1.5171	38.11	170.82
530	0.005035	2953.82	5.4838	5.5367	597.50	1.4926	37.92	164.44
540	0.005275	3007.43	5.5502	5.1940	607.52	1.4729	37.83	159.45
550	0.005504	3057.88	5.6118	4.9042	617.13	1.4567	37.83	155.54
560	0.005723	3105.65	5.6695	4.6578	626.32	1.4430	37.88	152.51
570	0.005933	3151.15	5.7238	4.4476	635.12	1.4313	37.99	150.18
580	0.006136	3194.71	5.7752	4.2674	643.53	1.4209	38.14	148.41
590	0.006332	3236.59	5.8240	4.1121	651.60	1.4117	38.31	147.09
600	0.006521	3277.02	5.8706	3.9773	659.34	1.4035	38.52	146.14
610	0.006705	3316.19	5.9152	3.8594	666.80	1.3960	38.74	145.49
620	0.006884	3354.25	5.9580	3.7555	673.99	1.3892	38.98	145.08
630	0.007059	3391.34	5.9993	3.6632	680.95	1.3830	39.23	144.88
640	0.007229	3427.55	6.0392	3.5808	687.70	1.3773	39.50	144.83
650	0.007396	3462.98	6.0778	3.5069	694.27	1.3721	39.77	144.93
660	0.007559	3497.71	6.1152	3.4402	700.66	1.3673	40.06	145.13
670	0.007719	3531.81	6.1516	3.3798	706.89	1.3629	40.35	145.43
680	0.007875	3565.33	6.1869	3.3252	712.98	1.3589	40.64	145.81
690	0.008030	3598.33	6.2213	3.2755	718.93	1.3551	40.94	146.24
700	0.008181	3630.85	6.2549	3.2302	724.75	1.3517	41.25	146.73
720	0.008477	3694.64	6.3198	3.1514	736.02	1.3453	41.87	147.82
740	0.008765	3756.99	6.3820	3.0858	746.83	1.3397	42.50	149.03
760	0.009046	3818.14	6.4418	3.0309	757.22	1.3345	43.13	150.31
780	0.009320	3878.29	6.4994	2.9849	767.20	1.3296	43.77	151.65
800	0.009588	3937.59	6.5552	2.9465	776.79	1.3249	44.41	153.01

Table 4 **Single-Phase Region** (Regions 1 to 3 of IAPWS-IF97) – Continuation
Homogenes Zustandsgebiet (Bereiche 1 bis 3 der IAPWS-IF97) – Fortsetzung

p = 50 MPa

t [°C]	v [m³ kg⁻¹]	h [kJ kg⁻¹]	s [kJ kg⁻¹ K⁻¹]	c_p [kJ kg⁻¹ K⁻¹]	w [m s⁻¹]	κ	$\eta \times 10^6$ [kg m⁻¹ s⁻¹]	$\lambda \times 10^3$ [W m⁻¹ K⁻¹]
0	0.000977	49.13	−0.0010	4.0225	1485.93	45.21	1696.52	588.17
5	0.000977	69.27	0.0720	4.0333	1509.24	46.61	1456.32	595.87
10	0.000978	89.46	0.1440	4.0424	1530.01	47.86	1266.44	604.15
15	0.000979	109.69	0.2148	4.0499	1548.50	48.97	1113.52	612.72
20	0.000980	129.96	0.2845	4.0562	1564.93	49.96	988.36	621.34
25	0.000982	150.25	0.3532	4.0614	1579.48	50.81	884.49	629.84
30	0.000983	170.57	0.4208	4.0659	1592.30	51.56	797.24	638.07
35	0.000985	190.91	0.4873	4.0698	1603.53	52.20	723.17	645.93
40	0.000987	211.27	0.5528	4.0733	1613.28	52.73	659.71	653.35
45	0.000989	231.64	0.6174	4.0765	1621.65	53.17	604.88	660.29
50	0.000991	252.03	0.6810	4.0797	1628.73	53.52	557.17	666.72
60	0.000996	292.86	0.8054	4.0860	1639.32	53.95	478.57	678.08
70	0.001001	333.75	0.9264	4.0926	1645.62	54.08	416.96	687.49
80	0.001007	374.71	1.0440	4.0998	1648.09	53.94	367.76	695.14
90	0.001013	415.75	1.1586	4.1078	1647.17	53.54	327.86	701.24
100	0.001020	456.87	1.2703	4.1167	1643.20	52.94	295.07	705.99
110	0.001027	498.09	1.3793	4.1265	1636.51	52.14	267.81	709.58
120	0.001035	539.41	1.4858	4.1374	1627.36	51.18	244.91	712.15
130	0.001043	580.84	1.5898	4.1494	1615.97	50.07	225.49	713.80
140	0.001052	622.40	1.6917	4.1627	1602.54	48.84	208.88	714.60
150	0.001061	664.10	1.7914	4.1774	1587.22	47.50	194.55	714.60
160	0.001070	705.95	1.8891	4.1938	1570.14	46.06	182.11	713.83
170	0.001081	747.98	1.9851	4.2121	1551.38	44.54	171.23	712.30
180	0.001091	790.20	2.0793	4.2324	1531.03	42.95	161.66	710.00
190	0.001103	832.64	2.1719	4.2551	1509.14	41.30	153.17	706.95
200	0.001115	875.31	2.2631	4.2806	1485.77	39.60	145.61	703.13
210	0.001128	918.26	2.3529	4.3090	1460.94	37.86	138.82	698.52
220	0.001141	961.50	2.4415	4.3409	1434.69	36.07	132.70	693.12
230	0.001156	1005.09	2.5290	4.3766	1407.05	34.27	127.15	686.90
240	0.001171	1049.05	2.6155	4.4165	1378.04	32.44	122.08	679.84
250	0.001187	1093.43	2.7012	4.4613	1347.69	30.60	117.42	671.93
260	0.001204	1138.29	2.7861	4.5115	1316.04	28.76	113.12	663.13
270	0.001223	1183.68	2.8704	4.5678	1283.13	26.92	109.12	653.39
280	0.001243	1229.67	2.9543	4.6310	1249.01	25.10	105.37	642.70
290	0.001265	1276.33	3.0379	4.7020	1213.73	23.30	101.84	631.02
300	0.001288	1323.74	3.1214	4.7819	1177.35	21.53	98.48	618.32
310	0.001313	1372.00	3.2049	4.8719	1139.92	19.79	95.26	604.63
320	0.001341	1421.22	3.2885	4.9736	1101.51	18.10	92.16	589.98
330	0.001371	1471.52	3.3726	5.0889	1062.11	16.45	89.14	574.47
340	0.001405	1523.05	3.4574	5.2200	1021.67	14.86	86.17	558.24

Table 4 Single-Phase Region (Regions 1 to 3 of IAPWS-IF97) – Continuation
Homogenes Zustandsgebiet (Bereiche 1 bis 3 der IAPWS-IF97) – Fortsetzung

				p = 50 MPa				
t [°C]	v [m³ kg⁻¹]	h [kJ kg⁻¹]	s [kJ kg⁻¹ K⁻¹]	c_p [kJ kg⁻¹ K⁻¹]	w [m s⁻¹]	κ	$\eta \times 10^6$ [kg m⁻¹ s⁻¹]	$\lambda \times 10^3$ [W m⁻¹ K⁻¹]
350	0.001442	1575.98	3.5430	5.3700	980.05	13.3178	83.24	541.49
360	0.001485	1630.63	3.6300	5.5620	936.04	11.8022	80.29	524.36
370	0.001533	1687.35	3.7189	5.7866	891.79	10.3760	77.33	506.97
380	0.001588	1746.51	3.8101	6.0531	846.77	9.0278	74.30	489.25
390	0.001654	1808.60	3.9045	6.3776	801.09	7.7623	71.20	470.86
400	0.001731	1874.31	4.0028	6.7781	755.14	6.5889	67.98	451.17
410	0.001825	1944.47	4.1063	7.2713	709.70	5.5207	64.64	429.40
420	0.001940	2020.07	4.2161	7.8636	666.13	4.5737	61.15	404.77
430	0.002085	2101.99	4.3334	8.5275	626.51	3.7651	57.55	377.00
440	0.002266	2190.53	4.4585	9.1599	593.57	3.1096	53.94	346.66
450	0.002487	2284.44	4.5892	9.5672	569.96	2.6120	50.48	315.36
460	0.002745	2380.52	4.7212	9.5776	556.72	2.2580	47.41	285.39
470	0.003027	2474.69	4.8488	9.2032	552.52	2.0170	44.89	258.71
480	0.003319	2563.86	4.9680	8.6087	554.76	1.8547	42.95	236.21
490	0.003609	2646.56	5.0770	7.8969	560.89	1.7436	41.52	217.60
500	0.003889	2722.52	5.1759	7.3090	568.92	1.6644	40.50	202.98
510	0.004160	2792.70	5.2661	6.7313	578.20	1.6073	39.78	191.11
520	0.004417	2857.36	5.3482	6.2131	588.07	1.5657	39.28	181.64
530	0.004662	2917.25	5.4232	5.7787	598.00	1.5340	38.97	174.18
540	0.004896	2973.16	5.4924	5.4136	607.73	1.5088	38.78	168.30
550	0.005118	3025.70	5.5566	5.1031	617.16	1.4883	38.69	163.65
560	0.005332	3075.37	5.6166	4.8372	626.25	1.4711	38.68	160.00
570	0.005537	3122.57	5.6729	4.6091	634.99	1.4565	38.73	157.14
580	0.005734	3167.66	5.7261	4.4131	643.38	1.4438	38.82	154.94
590	0.005924	3210.92	5.7765	4.2440	651.44	1.4327	38.96	153.25
600	0.006109	3252.61	5.8245	4.0973	659.20	1.4227	39.12	151.98
610	0.006288	3292.93	5.8705	3.9692	666.67	1.4138	39.31	151.06
620	0.006461	3332.05	5.9145	3.8564	673.89	1.4057	39.52	150.42
630	0.006631	3370.10	5.9569	3.7565	680.88	1.3983	39.75	150.01
640	0.006796	3407.21	5.9977	3.6673	687.66	1.3916	39.99	149.79
650	0.006957	3443.48	6.0372	3.5873	694.26	1.3856	40.25	149.72
660	0.007115	3478.99	6.0755	3.5151	700.69	1.3800	40.51	149.79
670	0.007270	3513.81	6.1126	3.4499	706.97	1.3749	40.79	149.95
680	0.007422	3548.00	6.1487	3.3907	713.10	1.3703	41.07	150.21
690	0.007571	3581.64	6.1838	3.3370	719.09	1.3660	41.36	150.53
700	0.007718	3614.76	6.2180	3.2881	724.96	1.3620	41.65	150.92
720	0.008004	3679.64	6.2840	3.2028	736.32	1.3548	42.24	151.83
740	0.008281	3742.97	6.3471	3.1318	747.23	1.3484	42.85	152.87
760	0.008552	3804.99	6.4078	3.0725	757.69	1.3426	43.47	154.01
780	0.008816	3865.93	6.4662	3.0228	767.74	1.3372	44.10	155.20
800	0.009074	3925.96	6.5226	2.9813	777.37	1.3319	44.72	156.44

Table 4 Single-Phase Region (Regions 1 to 3 of IAPWS-IF97) – Continuation
Homogenes Zustandsgebiet (Bereiche 1 bis 3 der IAPWS-IF97) – Fortsetzung

p = 52.5 MPa

t [°C]	v [m^3 kg^{-1}]	h [kJ kg^{-1}]	s [kJ kg^{-1} K^{-1}]	c_p [kJ kg^{-1} K^{-1}]	w [m s^{-1}]	κ	$\eta \times 10^6$ [kg m^{-1} s^{-1}]	$\lambda \times 10^3$ [W m^{-1} K^{-1}]
0	0.000976	51.50	– 0.0013	4.0150	1490.35	43.36	1693.22	589.47
5	0.000976	71.61	0.0717	4.0267	1513.59	44.70	1454.24	597.09
10	0.000977	91.77	0.1435	4.0364	1534.30	45.89	1265.20	605.32
15	0.000978	111.97	0.2142	4.0444	1552.74	46.95	1112.85	613.86
20	0.000979	132.21	0.2839	4.0510	1569.13	47.88	988.09	622.47
25	0.000981	152.48	0.3524	4.0565	1583.65	48.70	884.51	630.96
30	0.000983	172.77	0.4199	4.0611	1596.46	49.41	797.47	639.19
35	0.000984	193.09	0.4864	4.0652	1607.69	50.02	723.55	647.05
40	0.000986	213.42	0.5519	4.0688	1617.44	50.53	660.19	654.47
45	0.000988	233.78	0.6163	4.0722	1625.83	50.95	605.43	661.42
50	0.000990	254.14	0.6799	4.0754	1632.94	51.28	557.78	667.87
60	0.000995	294.93	0.8042	4.0818	1643.60	51.70	479.24	679.25
70	0.001000	335.78	0.9250	4.0884	1649.99	51.83	417.64	688.69
80	0.001006	376.70	1.0425	4.0957	1652.59	51.70	368.45	696.37
90	0.001012	417.70	1.1570	4.1036	1651.79	51.33	328.54	702.51
100	0.001019	458.78	1.2686	4.1124	1647.97	50.76	295.74	707.31
110	0.001026	499.95	1.3775	4.1220	1641.43	50.01	268.46	710.95
120	0.001034	541.22	1.4838	4.1327	1632.44	49.10	245.55	713.57
130	0.001042	582.60	1.5877	4.1445	1621.24	48.05	226.11	715.27
140	0.001050	624.11	1.6895	4.1575	1607.99	46.89	209.48	716.14
150	0.001059	665.76	1.7891	4.1719	1592.88	45.62	195.14	716.20
160	0.001069	707.56	1.8867	4.1879	1576.01	44.26	182.69	715.49
170	0.001079	749.52	1.9825	4.2057	1557.48	42.82	171.80	714.02
180	0.001090	791.68	2.0765	4.2255	1537.38	41.31	162.21	711.80
190	0.001101	834.04	2.1690	4.2476	1515.77	39.74	153.72	708.82
200	0.001113	876.64	2.2600	4.2723	1492.68	38.13	146.15	705.07
210	0.001126	919.49	2.3496	4.2999	1468.17	36.47	139.36	700.54
220	0.001139	962.65	2.4380	4.3308	1442.26	34.78	133.24	695.23
230	0.001153	1006.12	2.5253	4.3653	1414.98	33.07	127.69	689.10
240	0.001168	1049.96	2.6116	4.4039	1386.37	31.34	122.62	682.15
250	0.001184	1094.22	2.6970	4.4471	1356.47	29.59	117.97	674.35
260	0.001201	1138.92	2.7816	4.4955	1325.30	27.85	113.67	665.68
270	0.001220	1184.14	2.8657	4.5495	1292.92	26.10	109.68	656.10
280	0.001239	1229.94	2.9492	4.6101	1259.38	24.38	105.94	645.58
290	0.001260	1276.37	3.0324	4.6778	1224.73	22.67	102.42	634.10
300	0.001283	1323.52	3.1154	4.7537	1189.04	20.99	99.09	621.64
310	0.001308	1371.48	3.1983	4.8389	1152.39	19.34	95.89	608.19
320	0.001335	1420.33	3.2814	4.9345	1114.83	17.74	92.82	593.82
330	0.001364	1470.21	3.3648	5.0421	1076.42	16.18	89.83	578.60
340	0.001397	1521.22	3.4487	5.1632	1037.17	14.67	86.91	562.68

Table 4 Single-Phase Region (Regions 1 to 3 of IAPWS-IF97) – Continuation
Homogenes Zustandsgebiet (Bereiche 1 bis 3 der IAPWS-IF97) – Fortsetzung

p = 52.5 MPa								
t	v	h	s	c_p	w	κ	$\eta \times 10^6$	$\lambda \times 10^3$
[°C]	[m^3 kg^{-1}]	[kJ kg^{-1}]	[kJ kg^{-1} K^{-1}]	[kJ kg^{-1} K^{-1}]	[m s^{-1}]		[kg m^{-1} s^{-1}]	[W m^{-1} K^{-1}]
350	0.001433	1573.52	3.5333	5.3000	997.00	13.2149	84.02	546.24
360	0.001473	1627.38	3.6190	5.4745	954.27	11.7746	81.15	529.42
370	0.001519	1683.11	3.7063	5.6759	911.50	10.4200	78.26	512.36
380	0.001571	1741.01	3.7957	5.9105	868.16	9.1390	75.33	495.04
390	0.001631	1801.47	3.8875	6.1907	824.34	7.9350	72.34	477.21
400	0.001702	1865.02	3.9826	6.5298	780.36	6.8154	69.27	458.38
410	0.001786	1932.30	4.0818	6.9388	736.86	5.7903	66.10	437.86
420	0.001888	2004.04	4.1861	7.4216	694.88	4.8718	62.83	414.96
430	0.002012	2080.93	4.2962	7.9625	655.92	4.0728	59.48	389.32
440	0.002165	2163.30	4.4125	8.5043	622.03	3.4049	56.09	361.20
450	0.002349	2250.64	4.5341	8.9315	595.40	2.8742	52.80	331.73
460	0.002566	2341.08	4.6583	9.1056	577.50	2.4753	49.76	302.67
470	0.002809	2431.68	4.7811	8.9644	568.25	2.1893	47.13	275.83
480	0.003068	2519.52	4.8985	8.5713	566.07	1.9892	45.00	252.41
490	0.003333	2602.68	5.0082	8.0488	568.86	1.8494	43.34	232.73
500	0.003595	2680.14	5.1090	7.4459	574.51	1.7489	42.10	216.40
510	0.003850	2752.17	5.2016	6.9528	581.99	1.6757	41.19	203.41
520	0.004097	2819.13	5.2866	6.4462	590.64	1.6219	40.54	192.75
530	0.004333	2881.30	5.3645	5.9985	599.80	1.5814	40.09	184.22
540	0.004559	2939.33	5.4363	5.6181	609.02	1.5496	39.79	177.44
550	0.004775	2993.84	5.5029	5.2922	618.10	1.5239	39.61	172.03
560	0.004983	3045.32	5.5651	5.0107	626.95	1.5026	39.52	167.74
570	0.005182	3094.18	5.6234	4.7672	635.53	1.4846	39.50	164.34
580	0.005374	3140.78	5.6783	4.5566	643.82	1.4691	39.54	161.68
590	0.005559	3185.41	5.7303	4.3744	651.82	1.4557	39.63	159.60
600	0.005739	3228.34	5.7798	4.2162	659.53	1.4438	39.76	158.00
610	0.005913	3269.80	5.8270	4.0781	666.98	1.4331	39.91	156.80
620	0.006082	3309.96	5.8722	3.9567	674.18	1.4236	40.09	155.91
630	0.006246	3348.98	5.9157	3.8491	681.17	1.4149	40.29	155.29
640	0.006407	3386.98	5.9575	3.7532	687.96	1.4071	40.51	154.88
650	0.006563	3424.08	5.9979	3.6672	694.57	1.4001	40.74	154.64
660	0.006717	3460.36	6.0370	3.5897	701.01	1.3936	40.99	154.55
670	0.006867	3495.90	6.0749	3.5196	707.31	1.3878	41.25	154.58
680	0.007014	3530.77	6.1117	3.4560	713.47	1.3824	41.51	154.71
690	0.007158	3565.04	6.1474	3.3982	719.49	1.3775	41.78	154.92
700	0.007300	3598.75	6.1823	3.3456	725.39	1.3730	42.06	155.20
720	0.007577	3664.72	6.2494	3.2540	736.83	1.3649	42.63	155.92
740	0.007845	3729.01	6.3135	3.1777	747.81	1.3578	43.22	156.79
760	0.008106	3791.91	6.3749	3.1139	758.34	1.3513	43.82	157.77
780	0.008361	3853.64	6.4341	3.0606	768.43	1.3452	44.43	158.82
800	0.008610	3914.39	6.4913	3.0160	778.09	1.3393	45.04	159.92

Table 4 Single-Phase Region (Regions 1 to 3 of IAPWS-IF97) – Continuation
Homogenes Zustandsgebiet (Bereiche 1 bis 3 der IAPWS-IF97) – Fortsetzung

				p = 55 MPa				
t	v	h	s	c_p	w	κ	$\eta \times 10^6$	$\lambda \times 10^3$
[°C]	[m^3 kg^{-1}]	[kJ kg^{-1}]	[kJ kg^{-1} K^{-1}]	[kJ kg^{-1} K^{-1}]	[m s^{-1}]		[kg m^{-1} s^{-1}]	[W m^{-1} K^{-1}]
0	0.000975	53.87	−0.0015	4.0077	1494.77	41.68	1690.04	590.76
5	0.000975	73.94	0.0713	4.0202	1517.95	42.96	1452.25	598.31
10	0.000976	94.07	0.1430	4.0305	1538.60	44.09	1264.01	606.49
15	0.000977	114.24	0.2136	4.0389	1556.99	45.10	1112.22	615.00
20	0.000978	134.46	0.2832	4.0459	1573.34	46.00	987.86	623.59
25	0.000980	154.70	0.3517	4.0516	1587.83	46.78	884.56	632.07
30	0.000982	174.97	0.4191	4.0565	1600.62	47.46	797.71	640.30
35	0.000983	195.26	0.4855	4.0607	1611.85	48.04	723.93	648.16
40	0.000985	215.58	0.5509	4.0644	1621.61	48.53	660.68	655.60
45	0.000987	235.91	0.6153	4.0679	1630.02	48.93	606.00	662.55
50	0.000989	256.26	0.6787	4.0712	1637.15	49.25	558.39	669.01
60	0.000994	297.00	0.8029	4.0777	1647.88	49.66	479.90	680.42
70	0.000999	337.81	0.9236	4.0843	1654.37	49.79	418.33	689.89
80	0.001005	378.69	1.0410	4.0915	1657.07	49.67	369.14	697.60
90	0.001011	419.64	1.1554	4.0994	1656.41	49.32	329.22	703.78
100	0.001018	460.68	1.2669	4.1081	1652.72	48.79	296.41	708.63
110	0.001025	501.81	1.3756	4.1176	1646.33	48.07	269.11	712.32
120	0.001033	543.04	1.4818	4.1281	1637.51	47.21	246.18	714.99
130	0.001041	584.37	1.5857	4.1396	1626.47	46.22	226.72	716.75
140	0.001049	625.83	1.6873	4.1524	1613.42	45.11	210.08	717.67
150	0.001058	667.42	1.7867	4.1665	1598.50	43.91	195.73	717.79
160	0.001068	709.17	1.8842	4.1821	1581.84	42.61	183.27	717.15
170	0.001078	751.07	1.9799	4.1994	1563.54	41.25	172.36	715.75
180	0.001088	793.16	2.0738	4.2187	1543.69	39.81	162.77	713.59
190	0.001099	835.45	2.1661	4.2402	1522.33	38.33	154.27	710.68
200	0.001111	877.97	2.2569	4.2642	1499.53	36.79	146.69	707.01
210	0.001124	920.75	2.3464	4.2910	1475.32	35.22	139.90	702.56
220	0.001137	963.80	2.4346	4.3209	1449.74	33.61	133.77	697.32
230	0.001151	1007.18	2.5217	4.3544	1422.82	31.98	128.22	691.29
240	0.001166	1050.90	2.6077	4.3917	1394.60	30.33	123.16	684.44
250	0.001182	1095.02	2.6929	4.4334	1365.12	28.67	118.51	676.75
260	0.001198	1139.59	2.7773	4.4800	1334.41	27.01	114.22	668.21
270	0.001216	1184.64	2.8610	4.5320	1302.54	25.36	110.23	658.77
280	0.001236	1230.25	2.9442	4.5900	1269.54	23.71	106.51	648.42
290	0.001256	1276.47	3.0270	4.6548	1235.50	22.09	103.00	637.13
300	0.001279	1323.37	3.1095	4.7270	1200.48	20.49	99.68	624.88
310	0.001303	1371.03	3.1920	4.8077	1164.54	18.93	96.51	611.68
320	0.001329	1419.55	3.2745	4.8979	1127.79	17.40	93.46	597.57
330	0.001358	1469.03	3.3572	4.9985	1090.27	15.92	90.51	582.63
340	0.001389	1519.56	3.4403	5.1109	1052.05	14.49	87.63	567.00

Table 4 Single-Phase Region (Regions 1 to 3 of IAPWS-IF97) – Continuation
Homogenes Zustandsgebiet (Bereiche 1 bis 3 der IAPWS-IF97) – Fortsetzung

				p = 55 MPa				
t [°C]	v [m^3 kg^{-1}]	h [kJ kg^{-1}]	s [kJ kg^{-1} K^{-1}]	c_p [kJ kg^{-1} K^{-1}]	w [m s^{-1}]	κ	$\eta \times 10^6$ [kg m^{-1} s^{-1}]	$\lambda \times 10^3$ [W m^{-1} K^{-1}]
350	0.001424	1571.29	3.5240	5.2363	1013.14	13.1096	84.79	550.83
360	0.001462	1624.43	3.6086	5.3958	971.75	11.7420	81.97	534.31
370	0.001506	1679.28	3.6945	5.5777	930.31	10.4516	79.14	517.54
380	0.001555	1736.07	3.7821	5.7864	888.44	9.2307	76.30	500.57
390	0.001611	1795.13	3.8718	6.0315	846.25	8.0819	73.41	483.21
400	0.001676	1856.86	3.9642	6.3232	803.99	7.0109	70.46	465.07
410	0.001753	1921.78	4.0600	6.6694	762.20	6.0257	67.45	445.57
420	0.001844	1990.44	4.1597	7.0721	721.69	5.1358	64.35	424.09
430	0.001953	2063.37	4.2642	7.5211	683.62	4.3508	61.19	400.21
440	0.002084	2140.90	4.3737	7.9819	649.51	3.6800	58.00	374.01
450	0.002242	2222.82	4.4877	8.3837	621.13	3.1292	54.87	346.25
460	0.002426	2308.05	4.6048	8.6286	599.96	2.6972	51.91	318.31
470	0.002636	2394.60	4.7220	8.6407	586.55	2.3729	49.26	291.76
480	0.002864	2480.08	4.8363	8.4203	580.22	2.1370	47.00	267.88
490	0.003103	2562.47	4.9450	8.0371	579.44	1.9675	45.18	247.27
500	0.003344	2640.57	5.0467	7.5775	582.58	1.8451	43.75	229.98
510	0.003583	2713.63	5.1405	7.0789	587.86	1.7536	42.67	215.64
520	0.003817	2782.22	5.2276	6.6315	594.92	1.6860	41.86	204.00
530	0.004043	2846.28	5.3078	6.1887	602.99	1.6352	41.27	194.46
540	0.004261	2906.18	5.3820	5.8014	611.48	1.5957	40.86	186.78
550	0.004470	2962.48	5.4508	5.4666	620.03	1.5638	40.58	180.62
560	0.004671	3015.66	5.5150	5.1745	628.50	1.5376	40.41	175.68
570	0.004865	3066.10	5.5752	4.9189	636.80	1.5157	40.32	171.73
580	0.005051	3114.14	5.6318	4.6960	644.89	1.4970	40.30	168.59
590	0.005231	3160.11	5.6854	4.5020	652.75	1.4808	40.34	166.12
600	0.005406	3204.27	5.7363	4.3330	660.36	1.4667	40.42	164.18
610	0.005575	3246.84	5.7848	4.1854	667.74	1.4542	40.54	162.67
620	0.005739	3288.03	5.8311	4.0556	674.89	1.4430	40.69	161.53
630	0.005899	3328.00	5.8756	3.9406	681.84	1.4329	40.86	160.68
640	0.006055	3366.89	5.9185	3.8382	688.60	1.4238	41.05	160.07
650	0.006207	3404.80	5.9598	3.7464	695.19	1.4156	41.26	159.66
660	0.006356	3441.85	5.9997	3.6636	701.64	1.4082	41.48	159.41
670	0.006502	3478.10	6.0383	3.5887	707.94	1.4015	41.72	159.30
680	0.006644	3513.64	6.0758	3.5208	714.10	1.3954	41.97	159.30
690	0.006784	3548.54	6.1122	3.4590	720.15	1.3899	42.22	159.39
700	0.006922	3582.84	6.1477	3.4028	726.07	1.3848	42.49	159.56
720	0.007190	3649.89	6.2159	3.3048	737.56	1.3756	43.03	160.07
740	0.007450	3715.15	6.2809	3.2233	748.58	1.3676	43.60	160.77
760	0.007703	3778.91	6.3432	3.1552	759.16	1.3604	44.18	161.59
780	0.007949	3841.43	6.4032	3.0982	769.28	1.3536	44.77	162.49
800	0.008190	3902.90	6.4610	3.0506	778.96	1.3471	45.37	163.46

Table 4 Single-Phase Region (Regions 1 to 3 of IAPWS-IF97) – Continuation
Homogenes Zustandsgebiet (Bereiche 1 bis 3 der IAPWS-IF97) – Fortsetzung

p = 57.5 MPa

t [°C]	v [m³ kg⁻¹]	h [kJ kg⁻¹]	s [kJ kg⁻¹ K⁻¹]	c_p [kJ kg⁻¹ K⁻¹]	w [m s⁻¹]	κ	$\eta\times10^6$ [kg m⁻¹ s⁻¹]	$\lambda\times10^3$ [W m⁻¹ K⁻¹]
0	0.000974	56.23	− 0.0018	4.0006	1499.21	40.15	1686.98	592.04
5	0.000974	76.27	0.0709	4.0139	1522.33	41.37	1450.35	599.51
10	0.000975	96.37	0.1425	4.0247	1542.91	42.46	1262.90	607.65
15	0.000976	116.51	0.2130	4.0336	1561.25	43.42	1111.65	616.14
20	0.000978	136.70	0.2825	4.0408	1577.55	44.28	987.67	624.71
25	0.000979	156.92	0.3509	4.0468	1592.02	45.02	884.63	633.19
30	0.000981	177.17	0.4182	4.0519	1604.79	45.67	797.98	641.41
35	0.000982	197.44	0.4846	4.0562	1616.01	46.23	724.34	649.28
40	0.000984	217.73	0.5499	4.0601	1625.78	46.70	661.18	656.72
45	0.000986	238.04	0.6142	4.0637	1634.20	47.09	606.56	663.68
50	0.000989	258.36	0.6776	4.0670	1641.35	47.40	559.01	670.15
60	0.000993	299.07	0.8017	4.0736	1652.16	47.79	480.57	681.58
70	0.000998	339.84	0.9222	4.0803	1658.74	47.92	419.02	691.08
80	0.001004	380.68	1.0395	4.0875	1661.55	47.81	369.83	698.83
90	0.001010	421.59	1.1538	4.0953	1661.01	47.49	329.90	705.05
100	0.001017	462.58	1.2652	4.1039	1657.46	46.98	297.07	709.94
110	0.001024	503.67	1.3738	4.1133	1651.22	46.31	269.76	713.68
120	0.001032	544.85	1.4799	4.1236	1642.55	45.49	246.81	716.40
130	0.001039	586.14	1.5836	4.1349	1631.68	44.54	227.34	718.21
140	0.001048	627.55	1.6851	4.1473	1618.81	43.49	210.68	719.19
150	0.001057	669.10	1.7844	4.1611	1604.08	42.34	196.31	719.38
160	0.001066	710.78	1.8818	4.1763	1587.63	41.11	183.84	718.80
170	0.001076	752.63	1.9773	4.1932	1569.56	39.81	172.92	717.47
180	0.001087	794.65	2.0711	4.2120	1549.94	38.45	163.32	715.38
190	0.001098	836.87	2.1632	4.2329	1528.85	37.03	154.81	712.54
200	0.001109	879.32	2.2539	4.2563	1506.32	35.57	147.23	708.94
210	0.001122	922.01	2.3432	4.2823	1482.41	34.07	140.43	704.56
220	0.001135	964.97	2.4312	4.3113	1457.15	32.54	134.30	699.41
230	0.001149	1008.25	2.5181	4.3437	1430.58	30.98	128.75	693.47
240	0.001163	1051.86	2.6039	4.3799	1402.73	29.41	123.69	686.71
250	0.001179	1095.86	2.6888	4.4202	1373.65	27.83	119.04	679.13
260	0.001196	1140.28	2.7729	4.4651	1343.39	26.25	114.76	670.71
270	0.001213	1185.18	2.8564	4.5151	1311.99	24.67	110.78	661.41
280	0.001232	1230.60	2.9392	4.5708	1279.53	23.11	107.07	651.22
290	0.001252	1276.61	3.0217	4.6327	1246.06	21.56	103.58	640.11
300	0.001274	1323.28	3.1038	4.7017	1211.66	20.04	100.27	628.07
310	0.001298	1370.67	3.1858	4.7783	1176.42	18.55	97.12	615.09
320	0.001323	1418.87	3.2677	4.8634	1140.41	17.09	94.10	601.23
330	0.001351	1467.97	3.3498	4.9579	1103.72	15.68	91.17	586.56
340	0.001381	1518.06	3.4322	5.0625	1066.43	14.32	88.32	571.19

Table 4 Single-Phase Region (Regions 1 to 3 of IAPWS-IF97) – Continuation
Homogenes Zustandsgebiet (Bereiche 1 bis 3 der IAPWS-IF97) – Fortsetzung

p = 57.5 MPa								
t [°C]	v [m^3 kg^{-1}]	h [kJ kg^{-1}]	s [kJ kg^{-1} K^{-1}]	c_p [kJ kg^{-1} K^{-1}]	w [m s^{-1}]	κ	$\eta \times 10^6$ [kg m^{-1} s^{-1}]	$\lambda \times 10^3$ [W m^{-1} K^{-1}]
350	0.001415	1569.26	3.5150	5.1780	1028.59	13.0042	85.53	555.30
360	0.001452	1621.75	3.5986	5.3244	988.57	11.7055	82.76	539.04
370	0.001493	1675.80	3.6833	5.4900	948.32	10.4732	79.99	522.55
380	0.001540	1731.62	3.7694	5.6771	907.77	9.3064	77.22	505.88
390	0.001593	1789.44	3.8572	5.8938	867.01	8.2075	74.42	488.92
400	0.001654	1849.62	3.9473	6.1482	826.28	7.1807	71.58	471.36
410	0.001724	1912.55	4.0401	6.4460	786.00	6.2325	68.69	452.69
420	0.001806	1978.69	4.1362	6.7882	746.84	5.3703	65.73	432.37
430	0.001904	2048.44	4.2361	7.1674	709.71	4.6018	62.74	409.96
440	0.002019	2122.09	4.3401	7.5611	675.81	3.9342	59.72	385.41
450	0.002155	2199.56	4.4480	7.9239	646.51	3.3724	56.74	359.23
460	0.002315	2280.24	4.5588	8.1887	623.14	2.9171	53.88	332.47
470	0.002497	2362.79	4.6706	8.2893	606.52	2.5623	51.25	306.47
480	0.002698	2445.38	4.7810	8.1980	596.56	2.2944	48.94	282.48
490	0.002911	2526.19	4.8876	7.9419	592.36	2.0962	46.99	261.27
500	0.003132	2603.87	4.9887	7.5811	592.58	1.9501	45.41	243.10
510	0.003354	2677.63	5.0835	7.1204	595.60	1.8396	44.17	227.64
520	0.003573	2747.01	5.1716	6.7526	600.97	1.7582	43.22	215.20
530	0.003788	2812.50	5.2536	6.3426	607.67	1.6956	42.50	204.77
540	0.003996	2873.96	5.3297	5.9585	615.18	1.6471	41.97	196.25
550	0.004197	2931.83	5.4004	5.6216	623.03	1.6083	41.59	189.35
560	0.004392	2986.53	5.4665	5.3246	630.96	1.5765	41.34	183.77
570	0.004580	3038.43	5.5284	5.0615	638.85	1.5499	41.18	179.27
580	0.004761	3087.86	5.5867	4.8292	646.64	1.5275	41.10	175.65
590	0.004936	3135.11	5.6417	4.6252	654.28	1.5083	41.08	172.76
600	0.005105	3180.45	5.6940	4.4466	661.72	1.4916	41.11	170.47
610	0.005270	3224.11	5.7437	4.2901	668.97	1.4769	41.19	168.66
620	0.005430	3266.31	5.7912	4.1524	676.03	1.4639	41.30	167.25
630	0.005585	3307.22	5.8367	4.0305	682.90	1.4522	41.44	166.17
640	0.005736	3346.97	5.8805	3.9218	689.61	1.4418	41.61	165.36
650	0.005884	3385.69	5.9227	3.8244	696.16	1.4324	41.79	164.76
660	0.006029	3423.49	5.9634	3.7365	702.57	1.4239	41.99	164.35
670	0.006170	3460.45	6.0028	3.6570	708.85	1.4162	42.21	164.09
680	0.006309	3496.65	6.0410	3.5848	715.01	1.4093	42.44	163.96
690	0.006445	3532.16	6.0781	3.5191	721.06	1.4030	42.68	163.92
700	0.006578	3567.05	6.1141	3.4593	726.98	1.3972	42.93	163.98
720	0.006838	3635.16	6.1834	3.3551	738.50	1.3870	43.45	164.29
740	0.007090	3701.37	6.2494	3.2684	749.56	1.3781	43.99	164.80
760	0.007335	3766.00	6.3126	3.1960	760.17	1.3700	44.55	165.46
780	0.007574	3829.29	6.3732	3.1356	770.31	1.3625	45.12	166.21
800	0.007807	3891.48	6.4317	3.0850	779.98	1.3553	45.71	167.03

Table 4 Single-Phase Region (Regions 1 to 3 of IAPWS-IF97) – Continuation
Homogenes Zustandsgebiet (Bereiche 1 bis 3 der IAPWS-IF97) – Fortsetzung

				p = 60 MPa				
t [°C]	v [m^3 kg^{-1}]	h [kJ kg^{-1}]	s [kJ kg^{-1} K^{-1}]	c_p [kJ kg^{-1} K^{-1}]	w [m s^{-1}]	κ	$\eta \times 10^6$ [kg m^{-1} s^{-1}]	$\lambda \times 10^3$ [W m^{-1} K^{-1}]
0	0.000972	58.59	− 0.0021	3.9937	1503.66	38.75	1684.04	593.31
5	0.000973	78.59	0.0705	4.0077	1526.71	39.92	1448.53	600.72
10	0.000974	98.66	0.1420	4.0191	1547.23	40.96	1261.84	608.81
15	0.000975	118.78	0.2124	4.0284	1565.51	41.88	1111.12	617.27
20	0.000977	138.94	0.2818	4.0359	1581.77	42.70	987.51	625.83
25	0.000978	159.14	0.3501	4.0422	1596.21	43.42	884.73	634.30
30	0.000980	179.36	0.4174	4.0474	1608.97	44.04	798.27	642.52
35	0.000981	199.61	0.4836	4.0519	1620.18	44.58	724.76	650.39
40	0.000983	219.88	0.5489	4.0559	1629.95	45.03	661.69	657.84
45	0.000985	240.17	0.6132	4.0595	1638.39	45.40	607.14	664.81
50	0.000988	260.47	0.6765	4.0630	1645.56	45.70	559.63	671.29
60	0.000992	301.13	0.8004	4.0696	1656.44	46.08	481.25	682.74
70	0.000998	341.86	0.9209	4.0763	1663.10	46.21	419.71	692.27
80	0.001003	382.66	1.0381	4.0835	1666.02	46.11	370.52	700.06
90	0.001009	423.54	1.1522	4.0913	1665.60	45.81	330.58	706.32
100	0.001016	464.49	1.2634	4.0997	1662.18	45.33	297.74	711.25
110	0.001023	505.53	1.3720	4.1090	1656.08	44.69	270.41	715.03
120	0.001030	546.67	1.4780	4.1191	1647.57	43.91	247.44	717.81
130	0.001038	587.92	1.5816	4.1302	1636.87	43.01	227.95	719.68
140	0.001047	629.28	1.6829	4.1424	1624.17	42.00	211.28	720.72
150	0.001056	670.77	1.7822	4.1558	1609.64	40.91	196.90	720.96
160	0.001065	712.40	1.8794	4.1707	1593.39	39.74	184.41	720.45
170	0.001075	754.19	1.9748	4.1872	1575.54	38.49	173.48	719.18
180	0.001085	796.15	2.0684	4.2055	1556.15	37.19	163.87	717.16
190	0.001096	838.31	2.1604	4.2258	1535.31	35.84	155.35	714.39
200	0.001108	880.67	2.2509	4.2485	1513.05	34.44	147.76	710.86
210	0.001120	923.28	2.3400	4.2738	1489.43	33.01	140.96	706.56
220	0.001133	966.16	2.4279	4.3020	1464.48	31.55	134.83	701.49
230	0.001147	1009.33	2.5145	4.3334	1438.24	30.07	129.28	695.63
240	0.001161	1052.84	2.6002	4.3684	1410.76	28.57	124.21	688.97
250	0.001176	1096.72	2.6848	4.4073	1382.07	27.06	119.57	681.49
260	0.001193	1141.00	2.7687	4.4507	1352.23	25.55	115.29	673.18
270	0.001210	1185.75	2.8518	4.4988	1321.30	24.04	111.32	664.02
280	0.001229	1231.00	2.9344	4.5523	1289.34	22.55	107.62	653.98
290	0.001249	1276.81	3.0165	4.6117	1256.42	21.07	104.14	643.04
300	0.001270	1323.25	3.0982	4.6775	1222.62	19.62	100.85	631.19
310	0.001293	1370.38	3.1797	4.7504	1188.03	18.19	97.72	618.44
320	0.001318	1418.28	3.2612	4.8309	1152.72	16.80	94.72	604.81
330	0.001345	1467.03	3.3427	4.9198	1116.81	15.46	91.82	590.39
340	0.001374	1516.71	3.4244	5.0175	1080.36	14.15	89.00	575.28

Table 4 Single-Phase Region (Regions 1 to 3 of IAPWS-IF97) – Continuation
Homogenes Zustandsgebiet (Bereiche 1 bis 3 der IAPWS-IF97) – Fortsetzung

p = 60 MPa								
t [°C]	v [m^3 kg^{-1}]	h [kJ kg^{-1}]	s [kJ kg^{-1} K^{-1}]	c_p [kJ kg^{-1} K^{-1}]	w [m s^{-1}]	κ	$\eta \times 10^6$ [kg m^{-1} s^{-1}]	$\lambda \times 10^3$ [W m^{-1} K^{-1}]
350	0.001407	1567.41	3.5064	5.1244	1043.47	12.9007	86.24	559.63
360	0.001442	1619.30	3.5890	5.2592	1004.77	11.6662	83.52	543.64
370	0.001482	1672.64	3.6726	5.4109	965.62	10.4867	80.81	527.40
380	0.001526	1727.58	3.7573	5.5800	926.26	9.3691	78.11	511.00
390	0.001576	1784.32	3.8435	5.7733	886.78	8.3153	75.38	494.38
400	0.001633	1843.15	3.9316	5.9975	847.38	7.3289	72.63	477.29
410	0.001698	1904.39	4.0219	6.2571	808.45	6.4149	69.84	459.32
420	0.001774	1968.41	4.1149	6.5524	770.53	5.5794	67.01	439.96
430	0.001861	2035.54	4.2111	6.8777	734.35	4.8286	64.15	418.78
440	0.001964	2106.01	4.3106	7.2170	700.87	4.1678	61.27	395.66
450	0.002085	2179.82	4.4134	7.5398	671.18	3.6015	58.43	370.90
460	0.002224	2256.60	4.5188	7.7999	646.41	3.1309	55.68	345.32
470	0.002384	2335.44	4.6256	7.9454	627.42	2.7526	53.10	320.03
480	0.002561	2415.00	4.7319	7.9414	614.48	2.4577	50.77	296.18
490	0.002752	2493.77	4.8358	7.7888	607.17	2.2330	48.75	274.64
500	0.002952	2570.40	4.9356	7.5220	604.53	2.0636	47.06	255.82
510	0.003156	2643.99	5.0302	7.1881	605.44	1.9357	45.69	239.77
520	0.003361	2713.90	5.1189	6.7994	608.68	1.8374	44.60	226.20
530	0.003563	2780.22	5.2020	6.4533	613.85	1.7625	43.76	215.05
540	0.003762	2842.90	5.2796	6.0859	620.17	1.7040	43.12	205.75
550	0.003955	2902.06	5.3519	5.7534	627.13	1.6574	42.64	198.15
560	0.004142	2958.09	5.4195	5.4575	634.37	1.6192	42.30	191.94
570	0.004323	3011.32	5.4830	5.1919	641.73	1.5875	42.07	186.90
580	0.004499	3062.02	5.5428	4.9539	649.10	1.5609	41.92	182.80
590	0.004669	3110.48	5.5993	4.7424	656.43	1.5382	41.85	179.50
600	0.004834	3156.95	5.6528	4.5557	663.64	1.5186	41.83	176.85
610	0.004993	3201.67	5.7038	4.3914	670.71	1.5015	41.87	174.74
620	0.005149	3244.85	5.7524	4.2465	677.62	1.4864	41.94	173.06
630	0.005300	3286.66	5.7989	4.1180	684.38	1.4730	42.05	171.74
640	0.005447	3327.25	5.8436	4.0035	691.00	1.4610	42.18	170.72
650	0.005591	3366.76	5.8867	3.9007	697.48	1.4502	42.34	169.93
660	0.005731	3405.30	5.9282	3.8080	703.84	1.4406	42.52	169.36
670	0.005869	3442.95	5.9683	3.7241	710.08	1.4319	42.71	168.94
680	0.006003	3479.81	6.0072	3.6478	716.21	1.4241	42.92	168.67
690	0.006136	3515.93	6.0449	3.5783	722.24	1.4170	43.14	168.51
700	0.006265	3551.39	6.0815	3.5151	728.16	1.4105	43.38	168.45
720	0.006518	3620.56	6.1519	3.4048	739.67	1.3991	43.87	168.55
740	0.006762	3687.71	6.2188	3.3130	750.75	1.3891	44.39	168.88
760	0.007000	3753.18	6.2828	3.2365	761.36	1.3802	44.93	169.37
780	0.007231	3817.25	6.3442	3.1726	771.50	1.3719	45.48	169.96
800	0.007457	3880.15	6.4034	3.1193	781.16	1.3639	46.05	170.64

Table 4 Single-Phase Region (Regions 1 to 3 of IAPWS-IF97) – Continuation
Homogenes Zustandsgebiet (Bereiche 1 bis 3 der IAPWS-IF97) – Fortsetzung

p = 62.5 MPa

t [°C]	v [m^3 kg^{-1}]	h [kJ kg^{-1}]	s [kJ kg^{-1} K^{-1}]	c_p [kJ kg^{-1} K^{-1}]	w [m s^{-1}]	κ	$\eta \times 10^6$ [kg m^{-1} s^{-1}]	$\lambda \times 10^3$ [W m^{-1} K^{-1}]
0	0.000971	60.93	– 0.0024	3.9869	1508.13	37.46	1681.22	594.57
5	0.000972	80.91	0.0701	4.0017	1531.10	38.58	1446.79	601.91
10	0.000973	100.95	0.1415	4.0136	1551.56	39.58	1260.84	609.96
15	0.000974	121.04	0.2118	4.0232	1569.78	40.47	1110.63	618.39
20	0.000976	141.17	0.2811	4.0311	1586.00	41.25	987.38	626.94
25	0.000977	161.35	0.3493	4.0376	1600.40	41.94	884.86	635.40
30	0.000979	181.55	0.4165	4.0430	1613.14	42.54	798.57	643.62
35	0.000980	201.77	0.4827	4.0476	1624.35	43.06	725.19	651.50
40	0.000982	222.02	0.5479	4.0517	1634.13	43.49	662.21	658.95
45	0.000984	242.29	0.6121	4.0554	1642.58	43.85	607.73	665.93
50	0.000987	262.58	0.6754	4.0589	1649.77	44.14	560.26	672.42
60	0.000991	303.20	0.7992	4.0656	1660.71	44.51	481.93	683.90
70	0.000997	343.89	0.9195	4.0724	1667.46	44.64	420.41	693.46
80	0.001002	384.65	1.0366	4.0795	1670.48	44.55	371.21	701.29
90	0.001008	425.48	1.1506	4.0872	1670.17	44.26	331.27	707.58
100	0.001015	466.40	1.2618	4.0956	1666.89	43.80	298.41	712.56
110	0.001022	507.40	1.3702	4.1047	1660.93	43.20	271.06	716.39
120	0.001029	548.49	1.4761	4.1147	1652.57	42.45	248.07	719.21
130	0.001037	589.69	1.5796	4.1256	1642.03	41.60	228.56	721.14
140	0.001045	631.01	1.6808	4.1375	1629.51	40.64	211.87	722.23
150	0.001054	672.45	1.7799	4.1506	1615.16	39.59	197.48	722.54
160	0.001064	714.03	1.8770	4.1652	1599.11	38.47	184.98	722.09
170	0.001073	755.76	1.9722	4.1812	1581.47	37.28	174.04	720.89
180	0.001084	797.66	2.0657	4.1990	1562.31	36.04	164.41	718.94
190	0.001095	839.74	2.1576	4.2189	1541.71	34.75	155.89	716.23
200	0.001106	882.04	2.2480	4.2409	1519.72	33.41	148.29	712.77
210	0.001118	924.57	2.3369	4.2655	1496.38	32.04	141.49	708.55
220	0.001131	967.36	2.4246	4.2929	1471.74	30.64	135.36	703.55
230	0.001144	1010.44	2.5110	4.3233	1445.82	29.23	129.80	697.78
240	0.001159	1053.84	2.5965	4.3572	1418.69	27.79	124.74	691.21
250	0.001174	1097.60	2.6809	4.3949	1390.38	26.35	120.10	683.83
260	0.001190	1141.75	2.7645	4.4367	1360.95	24.90	115.82	675.63
270	0.001207	1186.35	2.8474	4.4832	1330.46	23.46	111.86	666.60
280	0.001225	1231.43	2.9296	4.5346	1298.98	22.03	108.16	656.70
290	0.001245	1277.06	3.0114	4.5915	1266.59	20.62	104.69	645.92
300	0.001266	1323.28	3.0927	4.6544	1233.36	19.23	101.42	634.26
310	0.001288	1370.17	3.1738	4.7238	1199.39	17.86	98.30	621.71
320	0.001313	1417.78	3.2548	4.8002	1164.76	16.54	95.32	608.31
330	0.001339	1466.20	3.3357	4.8841	1129.57	15.25	92.45	594.12
340	0.001368	1515.49	3.4168	4.9756	1093.91	14.00	89.66	579.26

Table 4 Single-Phase Region (Regions 1 to 3 of IAPWS-IF97) – Continuation
Homogenes Zustandsgebiet (Bereiche 1 bis 3 der IAPWS-IF97) – Fortsetzung

$p = 62.5$ **MPa**

t [°C]	v [m^3 kg^{-1}]	h [kJ kg^{-1}]	s [kJ kg^{-1} K^{-1}]	c_p [kJ kg^{-1} K^{-1}]	w [m s^{-1}]	κ	$\eta \times 10^6$ [kg m^{-1} s^{-1}]	$\lambda \times 10^3$ [W m^{-1} K^{-1}]
350	0.001399	1565.73	3.4981	5.0748	1057.87	12.8003	86.94	563.85
360	0.001433	1617.07	3.5798	5.1994	1020.42	11.6247	84.26	548.10
370	0.001471	1669.76	3.6623	5.3392	982.27	10.4936	81.60	532.09
380	0.001513	1723.90	3.7459	5.4930	943.98	9.4208	78.95	515.94
390	0.001561	1779.68	3.8306	5.6666	905.65	8.4083	76.30	499.62
400	0.001614	1837.32	3.9169	5.8661	867.46	7.4588	73.63	482.93
410	0.001675	1897.10	4.0050	6.0948	829.74	6.5766	70.93	465.53
420	0.001745	1959.32	4.0955	6.3529	792.94	5.7665	68.20	446.98
430	0.001825	2024.24	4.1884	6.6356	757.69	5.0336	65.46	426.85
440	0.001918	2092.07	4.2842	6.9310	724.74	4.3822	62.70	404.96
450	0.002025	2162.83	4.3828	7.2175	694.98	3.8157	59.97	381.48
460	0.002149	2236.29	4.4836	7.4624	669.36	3.3356	57.32	357.02
470	0.002290	2311.81	4.5859	7.6266	648.69	2.9401	54.80	332.51
480	0.002447	2388.44	4.6884	7.6770	633.40	2.6232	52.49	308.99
490	0.002618	2464.93	4.7893	7.6021	623.40	2.3752	50.44	287.33
500	0.002799	2540.11	4.8871	7.4166	618.09	2.1838	48.67	268.06
510	0.002987	2613.00	4.9808	7.1530	616.60	2.0369	47.20	251.36
520	0.003177	2683.09	5.0697	6.8082	617.78	1.9221	46.00	237.01
530	0.003367	2749.69	5.1532	6.5149	621.45	1.8354	45.04	225.20
540	0.003555	2813.19	5.2318	6.1819	626.43	1.7663	44.30	215.21
550	0.003739	2873.38	5.3053	5.8600	632.34	1.7112	43.72	206.94
560	0.003918	2930.50	5.3743	5.5704	638.76	1.6660	43.30	200.15
570	0.004093	2984.87	5.4392	5.3074	645.45	1.6285	42.99	194.57
580	0.004263	3036.73	5.5003	5.0679	652.31	1.5971	42.78	190.02
590	0.004427	3086.31	5.5581	4.8519	659.22	1.5705	42.64	186.30
600	0.004587	3133.85	5.6129	4.6592	666.12	1.5477	42.58	183.30
610	0.004742	3179.57	5.6649	4.4884	672.95	1.5279	42.57	180.87
620	0.004893	3223.68	5.7146	4.3371	679.68	1.5105	42.60	178.92
630	0.005040	3266.36	5.7621	4.2028	686.29	1.4952	42.67	177.36
640	0.005183	3307.78	5.8078	4.0828	692.78	1.4815	42.77	176.12
650	0.005323	3348.06	5.8516	3.9751	699.16	1.4693	42.91	175.15
660	0.005460	3387.32	5.8939	3.8779	705.44	1.4584	43.06	174.40
670	0.005593	3425.65	5.9348	3.7897	711.62	1.4486	43.23	173.84
680	0.005724	3463.14	5.9743	3.7095	717.71	1.4397	43.42	173.42
690	0.005853	3499.86	6.0127	3.6365	723.70	1.4318	43.62	173.14
700	0.005979	3535.89	6.0499	3.5698	729.59	1.4245	43.84	172.96
720	0.006224	3606.09	6.1213	3.4536	741.08	1.4118	44.30	172.84
740	0.006462	3674.16	6.1891	3.3569	752.15	1.4008	44.79	172.98
760	0.006692	3740.47	6.2540	3.2764	762.76	1.3910	45.31	173.30
780	0.006917	3805.31	6.3161	3.2092	772.88	1.3818	45.85	173.74
800	0.007136	3868.92	6.3759	3.1532	782.51	1.3729	46.40	174.27

Table 4 **Single-Phase Region** (Regions 1 to 3 of IAPWS-IF97) – Continuation
Homogenes Zustandsgebiet (Bereiche 1 bis 3 der IAPWS-IF97) – Fortsetzung

$p = 65$ MPa

t [°C]	v [m³ kg⁻¹]	h [kJ kg⁻¹]	s [kJ kg⁻¹ K⁻¹]	c_p [kJ kg⁻¹ K⁻¹]	w [m s⁻¹]	κ	$\eta \times 10^6$ [kg m⁻¹ s⁻¹]	$\lambda \times 10^3$ [W m⁻¹ K⁻¹]
0	0.000970	63.27	– 0.0027	3.9804	1512.60	36.27	1678.52	595.82
5	0.000971	83.22	0.0696	3.9958	1535.50	37.35	1445.14	603.09
10	0.000972	103.23	0.1410	4.0082	1555.89	38.31	1259.91	611.10
15	0.000973	123.29	0.2112	4.0182	1574.06	39.16	1110.19	619.51
20	0.000975	143.41	0.2804	4.0264	1590.23	39.92	987.29	628.04
25	0.000976	163.55	0.3486	4.0330	1604.60	40.58	885.01	636.50
30	0.000978	183.73	0.4157	4.0386	1617.33	41.16	798.90	644.72
35	0.000980	203.94	0.4818	4.0434	1628.52	41.65	725.64	652.60
40	0.000981	224.17	0.5469	4.0476	1638.30	42.07	662.75	660.06
45	0.000984	244.41	0.6110	4.0514	1646.76	42.42	608.33	667.05
50	0.000986	264.68	0.6742	4.0549	1653.98	42.70	560.90	673.55
60	0.000990	305.26	0.7979	4.0617	1664.97	43.06	482.61	685.06
70	0.000996	345.91	0.9182	4.0685	1671.81	43.19	421.11	694.65
80	0.001001	386.64	1.0351	4.0756	1674.93	43.11	371.91	702.51
90	0.001007	427.43	1.1490	4.0833	1674.74	42.84	331.95	708.84
100	0.001014	468.30	1.2601	4.0916	1671.58	42.40	299.07	713.86
110	0.001021	509.26	1.3684	4.1005	1665.75	41.82	271.71	717.74
120	0.001028	550.32	1.4742	4.1103	1657.54	41.11	248.70	720.61
130	0.001036	591.47	1.5775	4.1210	1647.16	40.29	229.18	722.60
140	0.001044	632.74	1.6786	4.1327	1634.81	39.37	212.47	723.75
150	0.001053	674.13	1.7776	4.1455	1620.64	38.37	198.06	724.12
160	0.001062	715.65	1.8746	4.1597	1604.79	37.30	185.54	723.73
170	0.001072	757.33	1.9697	4.1754	1587.36	36.16	174.59	722.59
180	0.001082	799.17	2.0631	4.1928	1568.43	34.97	164.96	720.71
190	0.001093	841.19	2.1548	4.2121	1548.07	33.73	156.42	718.07
200	0.001104	883.42	2.2450	4.2335	1526.33	32.46	148.82	714.68
210	0.001116	925.87	2.3338	4.2574	1503.27	31.14	142.01	710.53
220	0.001129	968.57	2.4213	4.2840	1478.92	29.81	135.87	705.61
230	0.001142	1011.56	2.5076	4.3135	1453.33	28.45	130.32	699.91
240	0.001156	1054.85	2.5928	4.3464	1426.54	27.07	125.25	693.43
250	0.001171	1098.50	2.6770	4.3828	1398.59	25.69	120.62	686.15
260	0.001187	1142.53	2.7604	4.4233	1369.56	24.30	116.34	678.06
270	0.001204	1186.98	2.8430	4.4680	1339.50	22.92	112.39	669.14
280	0.001222	1231.90	2.9250	4.5175	1308.48	21.55	108.70	659.38
290	0.001241	1277.35	3.0064	4.5722	1276.59	20.20	105.24	648.76
300	0.001262	1323.36	3.0874	4.6324	1243.91	18.86	101.98	637.28
310	0.001284	1370.01	3.1681	4.6986	1210.53	17.56	98.88	624.92
320	0.001308	1417.36	3.2486	4.7712	1176.54	16.29	95.92	611.74
330	0.001333	1465.46	3.3290	4.8504	1142.04	15.05	93.07	597.77
340	0.001361	1514.39	3.4094	4.9363	1107.12	13.85	90.31	583.14

Table 4 Single-Phase Region (Regions 1 to 3 of IAPWS-IF97) – Continuation
Homogenes Zustandsgebiet (Bereiche 1 bis 3 der IAPWS-IF97) – Fortsetzung

$p = 65$ MPa

t [°C]	v [m^3 kg^{-1}]	h [kJ kg^{-1}]	s [kJ kg^{-1} K^{-1}]	c_p [kJ kg^{-1} K^{-1}]	w [m s^{-1}]	κ	$\eta \times 10^6$ [kg m^{-1} s^{-1}]	$\lambda \times 10^3$ [W m^{-1} K^{-1}]
350	0.001391	1564.21	3.4900	5.0287	1071.87	12.7038	87.62	567.97
360	0.001424	1615.04	3.5709	5.1443	1035.55	11.5816	84.98	552.44
370	0.001461	1667.12	3.6525	5.2737	998.33	10.4950	82.36	536.65
380	0.001501	1720.55	3.7350	5.4144	961.03	9.4633	79.77	520.72
390	0.001547	1775.46	3.8184	5.5714	923.74	8.4885	77.17	504.66
400	0.001597	1832.05	3.9031	5.7502	886.63	7.5730	74.57	488.32
410	0.001654	1890.55	3.9894	5.9537	849.99	6.7203	71.96	471.41
420	0.001719	1951.21	4.0775	6.1817	814.23	5.9344	69.32	453.53
430	0.001793	2014.25	4.1678	6.4301	779.85	5.2192	66.67	434.29
440	0.001877	2079.85	4.2604	6.6898	747.48	4.5785	64.02	413.47
450	0.001975	2148.03	4.3554	6.9446	717.86	4.0150	61.40	391.13
460	0.002086	2218.65	4.4524	7.1708	691.77	3.5298	58.84	367.72
470	0.002212	2291.26	4.5507	7.3389	669.93	3.1222	56.39	344.02
480	0.002352	2365.14	4.6495	7.4209	652.84	2.7880	54.11	320.94
490	0.002505	2439.33	4.7473	7.4007	640.61	2.5201	52.04	299.33
500	0.002669	2512.82	4.8430	7.2809	632.92	2.3088	50.23	279.78
510	0.002841	2584.68	4.9354	7.0807	629.15	2.1436	48.68	262.59
520	0.003017	2654.25	5.0236	6.8274	628.51	2.0144	47.39	247.77
530	0.003194	2721.06	5.1073	6.5275	630.33	1.9135	46.33	235.13
540	0.003372	2784.99	5.1864	6.2461	633.90	1.8335	45.49	224.57
550	0.003547	2845.91	5.2609	5.9414	638.65	1.7693	44.83	215.68
560	0.003718	2903.90	5.3309	5.6615	644.13	1.7167	44.31	208.33
570	0.003886	2959.22	5.3970	5.4057	650.05	1.6729	43.93	202.25
580	0.004050	3012.08	5.4593	5.1692	656.27	1.6362	43.65	197.24
590	0.004209	3062.67	5.5182	4.9521	662.68	1.6053	43.46	193.13
600	0.004363	3111.19	5.5741	4.7557	669.19	1.5789	43.34	189.77
610	0.004514	3157.86	5.6273	4.5800	675.72	1.5562	43.28	187.03
620	0.004660	3202.86	5.6779	4.4235	682.21	1.5364	43.28	184.80
630	0.004803	3246.38	5.7264	4.2840	688.63	1.5189	43.31	183.00
640	0.004942	3288.59	5.7729	4.1593	694.97	1.5034	43.38	181.55
650	0.005078	3329.61	5.8176	4.0471	701.23	1.4896	43.49	180.39
660	0.005211	3369.57	5.8606	3.9457	707.40	1.4773	43.61	179.48
670	0.005341	3408.56	5.9022	3.8536	713.49	1.4663	43.76	178.75
680	0.005469	3446.67	5.9424	3.7698	719.51	1.4564	43.93	178.20
690	0.005594	3483.98	5.9813	3.6933	725.45	1.4475	44.11	177.78
700	0.005716	3520.55	6.0191	3.6235	731.30	1.4394	44.31	177.48
720	0.005955	3591.77	6.0915	3.5016	742.74	1.4253	44.74	177.16
740	0.006186	3660.75	6.1603	3.4001	753.77	1.4131	45.21	177.10
760	0.006410	3727.88	6.2259	3.3156	764.35	1.4022	45.70	177.25
780	0.006628	3793.47	6.2888	3.2453	774.44	1.3921	46.22	177.53
800	0.006841	3857.78	6.3493	3.1868	784.02	1.3824	46.75	177.92

Table 4 Single-Phase Region (Regions 1 to 3 of IAPWS-IF97) – Continuation
Homogenes Zustandsgebiet (Bereiche 1 bis 3 der IAPWS-IF97) – Fortsetzung

p = 67.5 MPa

t [°C]	v [m³ kg⁻¹]	h [kJ kg⁻¹]	s [kJ kg⁻¹ K⁻¹]	c_p [kJ kg⁻¹ K⁻¹]	w [m s⁻¹]	κ	$\eta \times 10^6$ [kg m⁻¹ s⁻¹]	$\lambda \times 10^3$ [W m⁻¹ K⁻¹]
0	0.000969	65.61	– 0.0030	3.9740	1517.08	35.17	1675.93	597.06
5	0.000970	85.52	0.0692	3.9901	1539.91	36.21	1443.57	604.27
10	0.000971	105.50	0.1404	4.0030	1560.24	37.14	1259.03	612.24
15	0.000972	125.54	0.2106	4.0133	1578.34	37.96	1109.80	620.62
20	0.000974	145.63	0.2797	4.0217	1594.47	38.68	987.23	629.15
25	0.000975	165.76	0.3478	4.0286	1608.81	39.32	885.19	637.60
30	0.000977	185.92	0.4148	4.0343	1621.51	39.88	799.24	645.82
35	0.000979	206.10	0.4809	4.0392	1632.70	40.35	726.10	653.70
40	0.000981	226.31	0.5459	4.0435	1642.48	40.76	663.30	661.17
45	0.000983	246.54	0.6100	4.0474	1650.95	41.09	608.94	668.17
50	0.000985	266.78	0.6731	4.0510	1658.18	41.36	561.55	674.68
60	0.000989	307.33	0.7967	4.0579	1669.24	41.72	483.30	686.21
70	0.000995	347.94	0.9168	4.0647	1676.15	41.84	421.81	695.83
80	0.001000	388.62	1.0337	4.0718	1679.37	41.77	372.60	703.72
90	0.001006	429.38	1.1475	4.0794	1679.28	41.51	332.63	710.09
100	0.001013	470.21	1.2584	4.0875	1676.24	41.10	299.74	715.16
110	0.001020	511.13	1.3666	4.0964	1670.56	40.54	272.35	719.08
120	0.001027	552.14	1.4723	4.1060	1662.48	39.87	249.33	722.01
130	0.001035	593.25	1.5755	4.1165	1652.26	39.08	229.79	724.05
140	0.001043	634.47	1.6765	4.1279	1640.07	38.20	213.06	725.26
150	0.001052	675.82	1.7754	4.1405	1626.08	37.25	198.64	725.69
160	0.001061	717.29	1.8723	4.1543	1610.42	36.22	186.11	725.36
170	0.001071	758.91	1.9673	4.1696	1593.20	35.13	175.15	724.29
180	0.001081	800.69	2.0605	4.1866	1574.49	33.98	165.50	722.47
190	0.001091	842.64	2.1521	4.2054	1554.37	32.80	156.96	719.90
200	0.001103	884.80	2.2421	4.2263	1532.89	31.57	149.35	716.58
210	0.001114	927.18	2.3308	4.2495	1510.10	30.31	142.53	712.50
220	0.001127	969.80	2.4181	4.2753	1486.04	29.03	136.39	707.66
230	0.001140	1012.69	2.5042	4.3040	1460.75	27.72	130.83	702.04
240	0.001154	1055.89	2.5892	4.3358	1434.29	26.41	125.77	695.64
250	0.001169	1099.42	2.6732	4.3711	1406.71	25.08	121.13	688.45
260	0.001185	1143.32	2.7563	4.4102	1378.06	23.75	116.86	680.47
270	0.001201	1187.64	2.8387	4.4534	1348.40	22.42	112.91	671.67
280	0.001219	1232.41	2.9204	4.5011	1317.83	21.11	109.23	662.04
290	0.001238	1277.68	3.0015	4.5536	1286.42	19.81	105.78	651.56
300	0.001258	1323.50	3.0821	4.6113	1254.26	18.53	102.53	640.24
310	0.001280	1369.92	3.1624	4.6746	1221.45	17.27	99.45	628.08
320	0.001303	1417.01	3.2425	4.7436	1188.07	16.05	96.51	615.09
330	0.001328	1464.82	3.3224	4.8186	1154.23	14.87	93.68	601.34
340	0.001355	1513.40	3.4023	4.8995	1120.02	13.72	90.94	586.93

Table 4 Single-Phase Region (Regions 1 to 3 of IAPWS-IF97) – Continuation
Homogenes Zustandsgebiet (Bereiche 1 bis 3 der IAPWS-IF97) – Fortsetzung

p = 67.5 MPa								
t [°C]	v [m^3 kg^{-1}]	h [kJ kg^{-1}]	s [kJ kg^{-1} K^{-1}]	c_p [kJ kg^{-1} K^{-1}]	w [m s^{-1}]	κ	$\eta\times 10^6$ [kg m^{-1} s^{-1}]	$\lambda\times 10^3$ [W m^{-1} K^{-1}]
350	0.001384	1562.82	3.4822	4.9858	1085.51	12.6115	88.28	571.98
360	0.001416	1613.18	3.5624	5.0932	1050.20	11.5373	85.67	556.66
370	0.001451	1664.71	3.6431	5.2135	1013.86	10.4918	83.10	541.08
380	0.001490	1717.48	3.7246	5.3429	977.45	9.4980	80.55	525.36
390	0.001533	1771.61	3.8068	5.4858	941.11	8.5579	78.01	509.53
400	0.001581	1827.26	3.8901	5.6471	904.98	7.6738	75.48	493.48
410	0.001635	1884.63	3.9747	5.8295	869.33	6.8485	72.93	476.98
420	0.001695	1943.92	4.0608	6.0327	834.50	6.0853	70.38	459.68
430	0.001764	2005.34	4.1488	6.2532	800.94	5.3877	67.81	441.20
440	0.001842	2069.02	4.2387	6.4835	769.18	4.7584	65.26	421.31
450	0.001931	2135.00	4.3306	6.7112	739.82	4.1998	62.73	399.99
460	0.002031	2203.18	4.4242	6.9185	713.51	3.7129	60.25	377.56
470	0.002145	2273.23	4.5191	7.0828	690.89	3.2971	57.86	354.65
480	0.002271	2344.61	4.6145	7.1815	672.45	2.9498	55.62	332.08
490	0.002409	2416.58	4.7095	7.1980	658.42	2.6654	53.56	310.64
500	0.002558	2488.28	4.8028	7.1284	648.67	2.4366	51.72	290.96
510	0.002715	2558.89	4.8936	6.9822	642.81	2.2544	50.12	273.40
520	0.002878	2627.74	4.9809	6.7786	640.20	2.1099	48.75	258.08
530	0.003043	2694.38	5.0644	6.5213	640.33	1.9959	47.62	244.88
540	0.003210	2758.37	5.1436	6.2784	642.46	1.9050	46.69	233.76
550	0.003376	2819.77	5.2187	5.9994	645.96	1.8314	45.94	224.32
560	0.003539	2878.41	5.2895	5.7312	650.43	1.7710	45.35	216.43
570	0.003700	2934.47	5.3564	5.4857	655.50	1.7206	44.89	209.88
580	0.003857	2988.17	5.4197	5.2561	660.99	1.6783	44.55	204.44
590	0.004010	3039.65	5.4797	5.0416	666.82	1.6426	44.30	199.94
600	0.004160	3089.06	5.5366	4.8441	672.86	1.6123	44.13	196.23
610	0.004306	3136.59	5.5907	4.6653	679.03	1.5864	44.02	193.19
620	0.004448	3182.43	5.6423	4.5049	685.23	1.5639	43.97	190.70
630	0.004587	3226.74	5.6917	4.3613	691.42	1.5441	43.97	188.66
640	0.004722	3269.70	5.7390	4.2325	697.57	1.5267	44.01	186.99
650	0.004854	3311.44	5.7844	4.1164	703.67	1.5112	44.08	185.64
660	0.004983	3352.07	5.8282	4.0113	709.72	1.4974	44.18	184.55
670	0.005110	3391.69	5.8704	3.9156	715.70	1.4851	44.30	183.68
680	0.005234	3430.41	5.9113	3.8284	721.63	1.4740	44.45	182.98
690	0.005355	3468.29	5.9508	3.7487	727.50	1.4641	44.61	182.44
700	0.005475	3505.40	5.9892	3.6758	733.29	1.4551	44.79	182.01
720	0.005707	3577.61	6.0626	3.5484	744.65	1.4395	45.19	181.47
740	0.005931	3647.48	6.1323	3.4424	755.62	1.4261	45.63	181.23
760	0.006150	3715.42	6.1987	3.3542	766.15	1.4141	46.10	181.20
780	0.006362	3781.75	6.2623	3.2809	776.19	1.4030	46.60	181.32
800	0.006568	3846.74	6.3234	3.2199	785.71	1.3924	47.11	181.56

Table 4 Single-Phase Region (Regions 1 to 3 of IAPWS-IF97) – Continuation
Homogenes Zustandsgebiet (Bereiche 1 bis 3 der IAPWS-IF97) – Fortsetzung

				p = 70 MPa				
t	v	h	s	c_p	w	κ	$\eta \times 10^6$	$\lambda \times 10^3$
[°C]	[m^3 kg^{-1}]	[kJ kg^{-1}]	[kJ kg^{-1} K^{-1}]	[kJ kg^{-1} K^{-1}]	[m s^{-1}]		[kg m^{-1} s^{-1}]	[W m^{-1} K^{-1}]
0	0.000968	67.93	– 0.0034	3.9678	1521.57	34.16	1673.45	598.29
5	0.000969	87.82	0.0688	3.9845	1544.33	35.15	1442.08	605.44
10	0.000970	107.77	0.1399	3.9978	1564.58	36.05	1258.21	613.37
15	0.000971	127.79	0.2099	4.0085	1582.63	36.84	1109.44	621.73
20	0.000973	147.86	0.2790	4.0172	1598.71	37.54	987.20	630.24
25	0.000974	167.96	0.3470	4.0242	1613.02	38.15	885.39	638.69
30	0.000976	188.10	0.4140	4.0301	1625.69	38.69	799.61	646.91
35	0.000978	208.26	0.4799	4.0351	1636.87	39.15	726.58	654.80
40	0.000980	228.45	0.5449	4.0395	1646.66	39.54	663.86	662.28
45	0.000982	248.65	0.6089	4.0435	1655.13	39.87	609.55	669.29
50	0.000984	268.88	0.6720	4.0472	1662.39	40.13	562.20	675.81
60	0.000989	309.39	0.7955	4.0541	1673.49	40.47	484.00	687.36
70	0.000994	349.96	0.9155	4.0609	1680.48	40.60	422.51	697.01
80	0.000999	390.61	1.0322	4.0680	1683.79	40.53	373.30	704.93
90	0.001005	431.32	1.1459	4.0755	1683.82	40.29	333.31	711.35
100	0.001012	472.12	1.2567	4.0836	1680.89	39.89	300.40	716.45
110	0.001019	513.00	1.3648	4.0923	1675.33	39.36	273.00	720.42
120	0.001026	553.97	1.4704	4.1018	1667.40	38.71	249.95	723.40
130	0.001034	595.04	1.5735	4.1120	1657.33	37.96	230.39	725.50
140	0.001042	636.21	1.6744	4.1233	1645.30	37.12	213.65	726.77
150	0.001050	677.50	1.7732	4.1355	1631.49	36.20	199.21	727.26
160	0.001060	718.93	1.8699	4.1491	1616.02	35.21	186.67	726.99
170	0.001069	760.49	1.9648	4.1640	1598.99	34.16	175.70	725.98
180	0.001079	802.21	2.0579	4.1805	1580.50	33.07	166.04	724.23
190	0.001090	844.11	2.1493	4.1988	1560.61	31.93	157.49	721.73
200	0.001101	886.19	2.2392	4.2191	1539.38	30.75	149.87	718.48
210	0.001113	928.50	2.3277	4.2417	1516.86	29.54	143.05	714.47
220	0.001125	971.04	2.4149	4.2668	1493.08	28.31	136.90	709.69
230	0.001138	1013.84	2.5008	4.2947	1468.10	27.05	131.34	704.15
240	0.001152	1056.94	2.5856	4.3256	1441.97	25.78	126.28	697.84
250	0.001167	1100.36	2.6694	4.3598	1414.73	24.51	121.64	690.74
260	0.001182	1144.15	2.7523	4.3976	1386.45	23.23	117.38	682.85
270	0.001198	1188.33	2.8344	4.4393	1357.19	21.96	113.43	674.16
280	0.001216	1232.95	2.9158	4.4853	1327.05	20.69	109.76	664.66
290	0.001234	1278.05	2.9966	4.5358	1296.10	19.44	106.32	654.33
300	0.001254	1323.68	3.0769	4.5912	1264.44	18.21	103.08	643.16
310	0.001275	1369.89	3.1569	4.6516	1232.17	17.01	100.01	631.17
320	0.001298	1416.73	3.2365	4.7174	1199.38	15.83	97.08	618.38
330	0.001322	1464.26	3.3160	4.7885	1166.17	14.69	94.27	604.84
340	0.001349	1512.52	3.3953	4.8648	1132.62	13.59	91.56	590.64

Table 4 Single-Phase Region (Regions 1 to 3 of IAPWS-IF97) – Continuation
Homogenes Zustandsgebiet (Bereiche 1 bis 3 der IAPWS-IF97) – Fortsetzung

				p = 70 MPa				
t	v	h	s	c_p	w	κ	$\eta \times 10^6$	$\lambda \times 10^3$
[°C]	[$m^3 kg^{-1}$]	[$kJ kg^{-1}$]	[$kJ kg^{-1} K^{-1}$]	[$kJ kg^{-1} K^{-1}$]	[$m s^{-1}$]		[$kg m^{-1} s^{-1}$]	[$W m^{-1} K^{-1}$]
350	0.001377	1561.57	3.4747	4.9456	1098.82	12.5232	88.93	575.89
360	0.001408	1611.49	3.5541	5.0456	1064.41	11.4920	86.35	560.78
370	0.001442	1662.50	3.6341	5.1581	1028.88	10.4847	83.82	545.39
380	0.001480	1714.68	3.7146	5.2776	993.30	9.5259	81.31	529.86
390	0.001521	1768.09	3.7957	5.4083	957.83	8.6180	78.82	514.24
400	0.001566	1822.89	3.8778	5.5547	922.59	7.7629	76.34	498.45
410	0.001617	1879.25	3.9609	5.7192	887.84	6.9631	73.86	482.30
420	0.001674	1937.34	4.0453	5.9017	853.87	6.2214	71.38	465.49
430	0.001738	1997.33	4.1312	6.0990	821.07	5.5407	68.89	447.67
440	0.001810	2059.35	4.2188	6.3049	789.90	4.9233	66.41	428.58
450	0.001892	2123.43	4.3080	6.5096	760.87	4.3710	63.96	408.16
460	0.001984	2189.49	4.3987	6.6990	734.52	3.8846	61.56	386.63
470	0.002087	2257.30	4.4906	6.8558	711.40	3.4639	59.24	364.50
480	0.002202	2326.43	4.5830	6.9619	691.97	3.1068	57.04	342.48
490	0.002327	2396.31	4.6752	7.0021	676.51	2.8092	55.00	321.30
500	0.002463	2466.23	4.7662	6.9693	665.04	2.5653	53.15	301.59
510	0.002607	2535.47	4.8552	6.8670	657.31	2.3678	51.51	283.79
520	0.002757	2603.38	4.9413	6.7073	652.87	2.2088	50.09	268.07
530	0.002911	2669.48	5.0242	6.5070	651.17	2.0811	48.88	254.42
540	0.003067	2733.39	5.1032	6.2797	652.00	1.9803	47.88	242.75
550	0.003223	2795.01	5.1786	6.0369	654.19	1.8968	47.05	232.81
560	0.003378	2854.09	5.2499	5.7814	657.60	1.8286	46.39	224.43
570	0.003532	2910.72	5.3175	5.5472	661.76	1.7714	45.87	217.43
580	0.003683	2965.08	5.3816	5.3277	666.47	1.7231	45.46	211.58
590	0.003830	3017.31	5.4424	5.1190	671.62	1.6823	45.15	206.71
600	0.003975	3067.51	5.5003	4.9233	677.13	1.6478	44.92	202.67
610	0.004116	3115.83	5.5553	4.7435	682.87	1.6184	44.77	199.33
620	0.004254	3162.43	5.6078	4.5806	688.74	1.5930	44.68	196.57
630	0.004388	3207.49	5.6579	4.4339	694.66	1.5709	44.64	194.29
640	0.004520	3251.16	5.7060	4.3018	700.60	1.5514	44.64	192.42
650	0.004648	3293.57	5.7522	4.1825	706.51	1.5341	44.69	190.88
660	0.004774	3334.84	5.7967	4.0742	712.40	1.5187	44.76	189.62
670	0.004897	3375.08	5.8396	3.9754	718.26	1.5050	44.86	188.59
680	0.005018	3414.38	5.8810	3.8851	724.08	1.4927	44.98	187.75
690	0.005136	3452.81	5.9211	3.8024	729.85	1.4817	45.12	187.08
700	0.005252	3490.45	5.9600	3.7267	735.57	1.4717	45.28	186.54
720	0.005478	3563.62	6.0344	3.5942	746.81	1.4545	45.65	185.78
740	0.005697	3634.37	6.1050	3.4837	757.70	1.4397	46.06	185.34
760	0.005909	3703.10	6.1722	3.3920	768.16	1.4266	46.51	185.14
780	0.006115	3770.15	6.2364	3.3158	778.13	1.4144	46.98	185.11
800	0.006317	3835.81	6.2982	3.2526	787.58	1.4028	47.48	185.20

Table 4 Single-Phase Region (Regions 1 to 3 of IAPWS-IF97) – Continuation
Homogenes Zustandsgebiet (Bereiche 1 bis 3 der IAPWS-IF97) – Fortsetzung

p = 72.5 MPa

t [°C]	v [m^3 kg^{-1}]	h [kJ kg^{-1}]	s [kJ kg^{-1} K^{-1}]	c_p [kJ kg^{-1} K^{-1}]	w [m s^{-1}]	κ	$\eta \times 10^6$ [kg m^{-1} s^{-1}]	$\lambda \times 10^3$ [W m^{-1} K^{-1}]
0	0.000967	70.26	– 0.0037	3.9618	1526.07	33.21	1671.09	599.51
5	0.000968	90.11	0.0683	3.9790	1548.75	34.17	1440.67	606.60
10	0.000969	110.04	0.1393	3.9928	1568.93	35.03	1257.45	614.50
15	0.000970	130.03	0.2093	4.0038	1586.92	35.79	1109.13	622.83
20	0.000972	150.07	0.2783	4.0127	1602.96	36.47	987.20	631.33
25	0.000973	170.16	0.3462	4.0200	1617.23	37.06	885.62	639.78
30	0.000975	190.27	0.4131	4.0260	1629.88	37.58	799.99	648.00
35	0.000977	210.42	0.4790	4.0311	1641.05	38.03	727.08	655.90
40	0.000979	230.58	0.5439	4.0356	1650.83	38.41	664.43	663.38
45	0.000981	250.77	0.6079	4.0396	1659.32	38.72	610.18	670.40
50	0.000983	270.98	0.6709	4.0433	1666.58	38.97	562.86	676.93
60	0.000988	311.45	0.7942	4.0503	1677.74	39.31	484.69	688.51
70	0.000993	351.98	0.9141	4.0572	1684.80	39.44	423.21	698.19
80	0.000998	392.59	1.0308	4.0642	1688.20	39.37	374.00	706.14
90	0.001004	433.27	1.1443	4.0717	1688.33	39.14	334.00	712.59
100	0.001011	474.03	1.2551	4.0797	1685.52	38.77	301.07	717.74
110	0.001018	514.87	1.3631	4.0883	1680.08	38.26	273.64	721.76
120	0.001025	555.79	1.4685	4.0976	1672.29	37.64	250.58	724.79
130	0.001033	596.82	1.5716	4.1077	1662.36	36.91	231.00	726.94
140	0.001041	637.95	1.6723	4.1186	1650.49	36.10	214.24	728.27
150	0.001049	679.20	1.7710	4.1307	1636.85	35.22	199.79	728.82
160	0.001058	720.57	1.8676	4.1439	1621.56	34.27	187.23	728.62
170	0.001068	762.08	1.9624	4.1584	1604.74	33.27	176.24	727.67
180	0.001078	803.74	2.0553	4.1745	1586.46	32.21	166.58	725.98
190	0.001088	845.57	2.1466	4.1924	1566.80	31.11	158.02	723.55
200	0.001099	887.60	2.2364	4.2122	1545.81	29.98	150.39	720.36
210	0.001111	929.83	2.3247	4.2342	1523.55	28.82	143.56	716.42
220	0.001123	972.29	2.4117	4.2586	1500.05	27.63	137.41	711.72
230	0.001136	1015.01	2.4975	4.2856	1475.37	26.43	131.85	706.26
240	0.001150	1058.01	2.5821	4.3156	1449.55	25.21	126.78	700.02
250	0.001164	1101.33	2.6657	4.3488	1422.65	23.98	122.15	693.01
260	0.001179	1145.00	2.7484	4.3854	1394.73	22.75	117.88	685.22
270	0.001196	1189.05	2.8302	4.4257	1365.86	21.52	113.94	676.63
280	0.001213	1233.52	2.9114	4.4700	1336.13	20.30	110.27	667.25
290	0.001231	1278.46	2.9919	4.5187	1305.63	19.10	106.84	657.05
300	0.001250	1323.91	3.0719	4.5718	1274.45	17.92	103.62	646.04
310	0.001271	1369.91	3.1515	4.6297	1242.69	16.76	100.56	634.22
320	0.001293	1416.52	3.2307	4.6924	1210.46	15.63	97.65	621.61
330	0.001317	1463.78	3.3097	4.7600	1177.85	14.53	94.86	608.26
340	0.001343	1511.74	3.3886	4.8321	1144.94	13.46	92.17	594.26

Table 4 Single-Phase Region (Regions 1 to 3 of IAPWS-IF97) – Continuation
Homogenes Zustandsgebiet (Bereiche 1 bis 3 der IAPWS-IF97) – Fortsetzung

p = 72.5 MPa								
t	v	h	s	c_p	w	κ	$\eta \times 10^6$	$\lambda \times 10^3$
[°C]	[m^3 kg^{-1}]	[kJ kg^{-1}]	[kJ kg^{-1} K^{-1}]	[kJ kg^{-1} K^{-1}]	[m s^{-1}]		[kg m^{-1} s^{-1}]	[W m^{-1} K^{-1}]
350	0.001371	1560.43	3.4673	4.9079	1111.82	12.4385	89.56	579.72
360	0.001401	1609.95	3.5462	5.0013	1078.19	11.4460	87.01	564.79
370	0.001434	1660.48	3.6254	5.1067	1043.43	10.4743	84.51	549.59
380	0.001470	1712.10	3.7050	5.2177	1008.63	9.5480	82.05	534.24
390	0.001509	1764.87	3.7852	5.3377	973.95	8.6699	79.60	518.81
400	0.001553	1818.90	3.8660	5.4711	939.53	7.8417	77.17	503.25
410	0.001601	1874.34	3.9478	5.6204	905.59	7.0657	74.75	487.40
420	0.001655	1931.36	4.0306	5.7853	872.41	6.3443	72.33	471.00
430	0.001715	1990.09	4.1148	5.9632	840.32	5.6798	69.91	453.74
440	0.001782	2050.65	4.2003	6.1487	809.72	5.0743	67.51	435.36
450	0.001858	2113.06	4.2872	6.3336	781.05	4.5293	65.13	415.75
460	0.001942	2177.28	4.3754	6.5068	754.78	4.0453	62.79	395.04
470	0.002037	2243.12	4.4646	6.6547	731.37	3.6220	60.53	373.66
480	0.002142	2310.24	4.5543	6.7625	711.23	3.2579	58.38	352.19
490	0.002256	2378.19	4.6439	6.8173	694.66	2.9500	56.36	331.33
500	0.002380	2446.38	4.7327	6.8109	681.76	2.6934	54.51	311.70
510	0.002512	2514.20	4.8198	6.7430	672.41	2.4824	52.85	293.74
520	0.002651	2581.06	4.9047	6.6206	666.30	2.3101	51.39	277.70
530	0.002794	2646.47	4.9866	6.4558	662.98	2.1700	50.13	263.65
540	0.002940	2710.02	5.0653	6.2595	662.50	2.0593	49.06	251.54
550	0.003087	2771.64	5.1406	6.0555	663.23	1.9652	48.17	241.15
560	0.003235	2831.00	5.2123	5.8155	665.53	1.8888	47.44	232.29
570	0.003381	2888.01	5.2803	5.5917	668.78	1.8248	46.85	224.87
580	0.003525	2942.88	5.3450	5.3838	672.65	1.7705	46.38	218.64
590	0.003667	2995.71	5.4066	5.1838	677.07	1.7244	46.01	213.41
600	0.003806	3046.59	5.4652	4.9925	681.98	1.6854	45.74	209.05
610	0.003943	3095.61	5.5210	4.8138	687.24	1.6523	45.54	205.42
620	0.004076	3142.91	5.5743	4.6499	692.73	1.6239	45.40	202.40
630	0.004207	3188.66	5.6252	4.5013	698.35	1.5991	45.32	199.89
640	0.004334	3232.99	5.6740	4.3669	704.04	1.5774	45.29	197.81
650	0.004459	3276.04	5.7209	4.2450	709.75	1.5582	45.30	196.09
660	0.004581	3317.92	5.7660	4.1341	715.46	1.5412	45.35	194.66
670	0.004701	3358.75	5.8095	4.0327	721.17	1.5260	45.42	193.47
680	0.004818	3398.60	5.8516	3.9397	726.85	1.5124	45.52	192.50
690	0.004933	3437.57	5.8922	3.8543	732.52	1.5002	45.64	191.69
700	0.005046	3475.72	5.9316	3.7761	738.14	1.4892	45.78	191.03
720	0.005266	3549.82	6.0070	3.6387	749.23	1.4702	46.11	190.06
740	0.005480	3621.41	6.0784	3.5241	760.01	1.4540	46.50	189.44
760	0.005686	3690.91	6.1463	3.4289	770.37	1.4396	46.92	189.07
780	0.005887	3758.68	6.2113	3.3500	780.26	1.4263	47.37	188.88
800	0.006083	3825.00	6.2737	3.2846	789.62	1.4137	47.85	188.82

Table 4 Single-Phase Region (Regions 1 to 3 of IAPWS-IF97) – Continuation
Homogenes Zustandsgebiet (Bereiche 1 bis 3 der IAPWS-IF97) – Fortsetzung

				p = 75 MPa				
t [°C]	v [m^3 kg^{-1}]	h [kJ kg^{-1}]	s [kJ kg^{-1} K^{-1}]	c_p [kJ kg^{-1} K^{-1}]	w [m s^{-1}]	κ	$\eta \times 10^6$ [kg m^{-1} s^{-1}]	$\lambda \times 10^3$ [W m^{-1} K^{-1}]
0	0.000966	72.57	– 0.0041	3.9559	1530.57	32.32	1668.83	600.72
5	0.000967	92.40	0.0678	3.9737	1553.17	33.26	1439.34	607.75
10	0.000968	112.30	0.1387	3.9879	1573.29	34.09	1256.75	615.61
15	0.000969	132.27	0.2086	3.9992	1591.22	34.82	1108.87	623.93
20	0.000971	152.29	0.2775	4.0083	1607.20	35.47	987.24	632.42
25	0.000972	172.35	0.3454	4.0158	1621.44	36.05	885.87	640.86
30	0.000974	192.44	0.4122	4.0219	1634.07	36.55	800.39	649.09
35	0.000976	212.57	0.4781	4.0272	1645.22	36.98	727.59	656.99
40	0.000978	232.72	0.5429	4.0317	1655.00	37.35	665.01	664.47
45	0.000980	252.88	0.6068	4.0358	1663.49	37.65	610.81	671.50
50	0.000982	273.07	0.6698	4.0396	1670.78	37.90	563.53	678.05
60	0.000987	313.50	0.7930	4.0466	1681.98	38.23	485.39	689.65
70	0.000992	354.00	0.9128	4.0535	1689.12	38.35	423.92	699.36
80	0.000997	394.57	1.0293	4.0605	1692.60	38.30	374.70	707.35
90	0.001003	435.22	1.1428	4.0679	1692.82	38.08	334.68	713.84
100	0.001010	475.94	1.2534	4.0758	1690.12	37.72	301.73	719.03
110	0.001017	516.74	1.3613	4.0843	1684.81	37.23	274.29	723.09
120	0.001024	557.62	1.4667	4.0934	1677.14	36.63	251.20	726.18
130	0.001031	598.61	1.5696	4.1033	1667.36	35.94	231.61	728.38
140	0.001040	639.69	1.6703	4.1141	1655.64	35.16	214.83	729.77
150	0.001048	680.89	1.7688	4.1258	1642.17	34.31	200.36	730.38
160	0.001057	722.21	1.8653	4.1387	1627.06	33.39	187.79	730.24
170	0.001066	763.67	1.9599	4.1530	1610.43	32.43	176.79	729.36
180	0.001076	805.28	2.0528	4.1687	1592.36	31.41	167.11	727.73
190	0.001087	847.05	2.1440	4.1861	1572.92	30.35	158.54	725.37
200	0.001098	889.01	2.2336	4.2054	1552.17	29.26	150.91	722.25
210	0.001109	931.16	2.3217	4.2267	1530.17	28.14	144.07	718.37
220	0.001121	973.55	2.4086	4.2505	1506.95	27.00	137.92	713.74
230	0.001134	1016.18	2.4942	4.2768	1482.56	25.84	132.35	708.35
240	0.001148	1059.09	2.5786	4.3059	1457.05	24.66	127.28	702.19
250	0.001162	1102.31	2.6620	4.3380	1430.48	23.48	122.65	695.26
260	0.001177	1145.86	2.7445	4.3735	1402.91	22.30	118.39	687.56
270	0.001193	1189.79	2.8261	4.4125	1374.42	21.11	114.45	679.08
280	0.001210	1234.13	2.9070	4.4553	1345.09	19.94	110.79	669.81
290	0.001228	1278.91	2.9872	4.5021	1315.01	18.78	107.36	659.74
300	0.001247	1324.18	3.0669	4.5532	1284.29	17.64	104.15	648.87
310	0.001267	1369.99	3.1461	4.6087	1253.03	16.52	101.10	637.21
320	0.001289	1416.37	3.2250	4.6686	1221.33	15.43	98.20	624.77
330	0.001312	1463.38	3.3036	4.7329	1189.29	14.37	95.43	611.61
340	0.001337	1511.04	3.3820	4.8011	1157.00	13.35	92.76	597.80

Table 4 Single-Phase Region (Regions 1 to 3 of IAPWS-IF97) – Continuation
Homogenes Zustandsgebiet (Bereiche 1 bis 3 der IAPWS-IF97) – Fortsetzung

p = 75 MPa								
t [°C]	v [m³ kg⁻¹]	h [kJ kg⁻¹]	s [kJ kg⁻¹ K⁻¹]	c_p [kJ kg⁻¹ K⁻¹]	w [m s⁻¹]	κ	$\eta \times 10^6$ [kg m⁻¹ s⁻¹]	$\lambda \times 10^3$ [W m⁻¹ K⁻¹]
350	0.001364	1559.41	3.4602	4.8725	1124.51	12.3572	90.18	583.45
360	0.001394	1608.54	3.5384	4.9598	1091.58	11.3995	87.66	568.71
370	0.001425	1658.63	3.6169	5.0590	1057.55	10.4611	85.19	553.69
380	0.001460	1709.73	3.6958	5.1623	1023.47	9.5651	82.76	538.51
390	0.001498	1761.90	3.7750	5.2730	989.53	8.7147	80.36	523.25
400	0.001540	1815.23	3.8548	5.3953	955.86	7.9116	77.97	507.89
410	0.001586	1869.85	3.9354	5.5314	922.66	7.1578	75.60	492.31
420	0.001637	1925.91	4.0168	5.6813	890.20	6.4554	73.23	476.27
430	0.001693	1983.52	4.0994	5.8426	858.76	5.8065	70.88	459.49
440	0.001757	2042.78	4.1830	6.0107	828.71	5.2129	68.54	441.72
450	0.001827	2103.73	4.2679	6.1787	800.43	4.6756	66.22	422.82
460	0.001906	2166.32	4.3539	6.3374	774.30	4.1951	63.95	402.87
470	0.001993	2230.41	4.4407	6.4760	750.73	3.7711	61.75	382.18
480	0.002089	2295.74	4.5280	6.5824	730.10	3.4024	59.64	361.28
490	0.002194	2361.91	4.6153	6.6455	712.68	3.0865	57.65	340.79
500	0.002308	2428.47	4.7019	6.6576	698.62	2.8196	55.81	321.29
510	0.002430	2494.89	4.7873	6.6161	687.89	2.5968	54.14	303.25
520	0.002558	2560.63	4.8707	6.5243	680.29	2.4126	52.65	286.99
530	0.002691	2625.23	4.9517	6.3908	675.47	2.2610	51.34	272.59
540	0.002827	2688.34	5.0298	6.2265	673.05	2.1363	50.22	260.06
550	0.002966	2749.64	5.1047	6.0535	673.00	2.0362	49.27	249.30
560	0.003105	2809.13	5.1765	5.8369	674.11	1.9511	48.48	240.01
570	0.003244	2866.40	5.2448	5.6217	676.44	1.8804	47.83	232.17
580	0.003382	2921.62	5.3100	5.4251	679.49	1.8201	47.30	225.58
590	0.003518	2974.92	5.3721	5.2356	683.15	1.7686	46.88	220.02
600	0.003652	3026.35	5.4313	5.0512	687.39	1.7249	46.56	215.35
610	0.003784	3075.98	5.4878	4.8756	692.12	1.6879	46.31	211.44
620	0.003913	3123.91	5.5418	4.7124	697.19	1.6563	46.13	208.17
630	0.004039	3170.27	5.5934	4.5631	702.49	1.6289	46.02	205.44
640	0.004163	3215.21	5.6429	4.4273	707.90	1.6049	45.95	203.15
650	0.004285	3258.86	5.6904	4.3037	713.38	1.5837	45.93	201.24
660	0.004403	3301.32	5.7362	4.1908	718.90	1.5649	45.95	199.65
670	0.004520	3342.71	5.7803	4.0872	724.43	1.5481	45.99	198.32
680	0.004634	3383.10	5.8229	3.9920	729.97	1.5331	46.07	197.20
690	0.004746	3422.57	5.8641	3.9043	735.50	1.5197	46.17	196.27
700	0.004856	3461.21	5.9040	3.8237	741.01	1.5076	46.29	195.50
720	0.005071	3536.22	5.9803	3.6818	751.91	1.4867	46.59	194.31
740	0.005278	3608.63	6.0525	3.5633	762.55	1.4689	46.94	193.50
760	0.005480	3678.89	6.1212	3.4649	772.79	1.4532	47.33	192.96
780	0.005676	3747.34	6.1868	3.3834	782.57	1.4387	47.76	192.62
800	0.005866	3814.32	6.2498	3.3160	791.83	1.4250	48.22	192.42

Table 4 Single-Phase Region (Regions 1 to 3 of IAPWS-IF97) – Continuation
Homogenes Zustandsgebiet (Bereiche 1 bis 3 der IAPWS-IF97) – Fortsetzung

p = 77.5 MPa

t [°C]	v [m^3 kg^{-1}]	h [kJ kg^{-1}]	s [kJ kg^{-1} K^{-1}]	c_p [kJ kg^{-1} K^{-1}]	w [m s^{-1}]	κ	$\eta \times 10^6$ [kg m^{-1} s^{-1}]	$\lambda \times 10^3$ [W m^{-1} K^{-1}]
0	0.000965	74.88	– 0.0045	3.9502	1535.07	31.50	1666.68	601.92
5	0.000966	94.68	0.0673	3.9685	1557.60	32.40	1438.09	608.90
10	0.000967	114.56	0.1382	3.9831	1577.64	33.20	1256.10	616.72
15	0.000969	134.50	0.2080	3.9947	1595.51	33.91	1108.64	625.02
20	0.000970	154.50	0.2768	4.0040	1611.45	34.55	987.31	633.50
25	0.000971	174.54	0.3446	4.0116	1625.65	35.10	886.14	641.94
30	0.000973	194.61	0.4113	4.0179	1638.26	35.59	800.81	650.17
35	0.000975	214.72	0.4771	4.0233	1649.40	36.00	728.11	658.07
40	0.000977	234.85	0.5419	4.0279	1659.17	36.36	665.61	665.57
45	0.000979	255.00	0.6058	4.0321	1667.67	36.66	611.45	672.61
50	0.000981	275.17	0.6687	4.0359	1674.96	36.90	564.20	679.16
60	0.000986	315.56	0.7918	4.0430	1686.22	37.22	486.10	690.79
70	0.000991	356.03	0.9114	4.0499	1693.41	37.34	424.63	700.53
80	0.000997	396.56	1.0279	4.0569	1696.98	37.29	375.40	708.55
90	0.001002	437.16	1.1413	4.0642	1697.29	37.08	335.36	715.08
100	0.001009	477.84	1.2518	4.0720	1694.70	36.73	302.39	720.31
110	0.001016	518.61	1.3596	4.0804	1689.50	36.27	274.93	724.42
120	0.001023	559.45	1.4648	4.0894	1681.96	35.69	251.83	727.56
130	0.001030	600.40	1.5676	4.0991	1672.32	35.02	232.21	729.82
140	0.001038	641.44	1.6682	4.1096	1660.75	34.27	215.42	731.26
150	0.001047	682.59	1.7666	4.1211	1647.44	33.45	200.93	731.93
160	0.001056	723.86	1.8630	4.1337	1632.50	32.57	188.34	731.86
170	0.001065	765.27	1.9575	4.1476	1616.06	31.64	177.33	731.04
180	0.001075	806.82	2.0502	4.1629	1598.20	30.66	167.64	729.48
190	0.001085	848.53	2.1413	4.1799	1578.98	29.64	159.06	727.18
200	0.001096	890.42	2.2308	4.1987	1558.47	28.59	151.42	724.12
210	0.001108	932.51	2.3188	4.2195	1536.72	27.51	144.58	720.32
220	0.001120	974.82	2.4055	4.2426	1513.77	26.41	138.42	715.75
230	0.001132	1017.37	2.4909	4.2681	1489.66	25.29	132.85	710.43
240	0.001146	1060.19	2.5752	4.2964	1464.46	24.16	127.78	704.35
250	0.001160	1103.31	2.6584	4.3276	1438.22	23.02	123.15	697.50
260	0.001174	1146.75	2.7406	4.3620	1410.99	21.87	118.89	689.89
270	0.001190	1190.56	2.8220	4.3997	1382.86	20.73	114.95	681.51
280	0.001207	1234.76	2.9027	4.4411	1353.92	19.60	111.30	672.34
290	0.001224	1279.39	2.9826	4.4862	1324.25	18.48	107.88	662.40
300	0.001243	1324.50	3.0620	4.5353	1293.97	17.38	104.67	651.66
310	0.001263	1370.11	3.1409	4.5886	1263.18	16.30	101.64	640.15
320	0.001285	1416.28	3.2194	4.6459	1231.99	15.25	98.75	627.89
330	0.001307	1463.05	3.2976	4.7071	1200.50	14.22	96.00	614.90
340	0.001332	1510.44	3.3755	4.7718	1168.79	13.23	93.34	601.27

Table 4 Single-Phase Region (Regions 1 to 3 of IAPWS-IF97) – Continuation
Homogenes Zustandsgebiet (Bereiche 1 bis 3 der IAPWS-IF97) – Fortsetzung

p = 77.5 MPa

t [°C]	v [m³ kg⁻¹]	h [kJ kg⁻¹]	s [kJ kg⁻¹ K⁻¹]	c_p [kJ kg⁻¹ K⁻¹]	w [m s⁻¹]	κ	$\eta \times 10^6$ [kg m⁻¹ s⁻¹]	$\lambda \times 10^3$ [W m⁻¹ K⁻¹]
350	0.001358	1558.49	3.4533	4.8391	1136.92	12.2787	90.78	587.11
360	0.001387	1607.26	3.5309	4.9208	1104.60	11.3526	88.29	572.54
370	0.001418	1656.94	3.6088	5.0144	1071.26	10.4454	85.85	557.69
380	0.001451	1707.56	3.6869	5.1110	1037.86	9.5777	83.45	542.67
390	0.001488	1759.18	3.7653	5.2135	1004.59	8.7531	81.09	527.57
400	0.001528	1811.86	3.8441	5.3259	971.62	7.9734	78.74	512.39
410	0.001572	1865.74	3.9236	5.4505	939.10	7.2404	76.42	497.04
420	0.001620	1920.92	4.0038	5.5874	907.29	6.5560	74.11	481.31
430	0.001674	1977.52	4.0848	5.7345	876.47	5.9220	71.81	464.95
440	0.001733	2035.63	4.1669	5.8877	846.94	5.3400	69.52	447.72
450	0.001799	2095.27	4.2499	6.0411	819.03	4.8108	67.27	429.45
460	0.001872	2156.42	4.3339	6.1871	793.09	4.3347	65.05	410.19
470	0.001953	2218.96	4.4186	6.3166	769.48	3.9114	62.90	390.16
480	0.002042	2282.67	4.5038	6.4197	748.50	3.5399	60.83	369.81
490	0.002139	2347.24	4.5890	6.4872	730.44	3.2180	58.87	349.70
500	0.002244	2412.27	4.6736	6.5119	715.45	2.9428	57.04	330.39
510	0.002357	2477.32	4.7572	6.4901	703.56	2.7101	55.37	312.35
520	0.002475	2541.92	4.8392	6.4231	694.65	2.5153	53.86	295.91
530	0.002599	2605.65	4.9190	6.3164	688.48	2.3531	52.53	281.24
540	0.002727	2668.14	4.9964	6.1780	684.72	2.2184	51.36	268.36
550	0.002857	2729.03	5.0708	6.0241	683.53	2.1099	50.36	257.23
560	0.002989	2788.45	5.1426	5.8470	683.24	2.0150	49.51	247.58
570	0.003121	2845.88	5.2111	5.6405	684.64	1.9377	48.81	239.34
580	0.003253	2901.33	5.2765	5.4533	686.90	1.8716	48.23	232.39
590	0.003383	2954.97	5.3390	5.2752	689.80	1.8147	47.76	226.52
600	0.003512	3006.84	5.3987	5.0993	693.35	1.7662	47.38	221.55
610	0.003639	3056.98	5.4558	4.9286	697.50	1.7252	47.09	217.37
620	0.003764	3105.45	5.5104	4.7676	702.12	1.6901	46.87	213.86
630	0.003886	3152.37	5.5626	4.6188	707.05	1.6600	46.72	210.91
640	0.004006	3197.87	5.6127	4.4827	712.17	1.6337	46.62	208.43
650	0.004124	3242.06	5.6609	4.3582	717.40	1.6104	46.57	206.34
660	0.004239	3285.07	5.7072	4.2440	722.70	1.5898	46.55	204.58
670	0.004352	3326.97	5.7519	4.1389	728.05	1.5714	46.57	203.10
680	0.004464	3367.87	5.7950	4.0419	733.41	1.5549	46.62	201.85
690	0.004573	3407.83	5.8367	3.9522	738.79	1.5402	46.70	200.80
700	0.004680	3446.94	5.8771	3.8696	744.17	1.5269	46.80	199.91
720	0.004889	3522.83	5.9543	3.7236	754.86	1.5039	47.06	198.52
740	0.005091	3596.04	6.0273	3.6013	765.31	1.4845	47.38	197.52
760	0.005287	3667.02	6.0967	3.4998	775.41	1.4673	47.75	196.82
780	0.005478	3736.15	6.1630	3.4159	785.06	1.4516	48.16	196.32
800	0.005665	3803.75	6.2265	3.3467	794.22	1.4368	48.60	195.98

Table 4 Single-Phase Region (Regions 1 to 3 of IAPWS-IF97) – Continuation
Homogenes Zustandsgebiet (Bereiche 1 bis 3 der IAPWS-IF97) – Fortsetzung

p = 80 MPa

t [°C]	v [$m^3\,kg^{-1}$]	h [$kJ\,kg^{-1}$]	s [$kJ\,kg^{-1}\,K^{-1}$]	c_p [$kJ\,kg^{-1}\,K^{-1}$]	w [$m\,s^{-1}$]	κ	$\eta \times 10^6$ [$kg\,m^{-1}\,s^{-1}$]	$\lambda \times 10^3$ [$W\,m^{-1}\,K^{-1}$]
0	0.000964	77.18	– 0.0049	3.9446	1539.57	30.72	1664.64	603.10
5	0.000965	96.95	0.0668	3.9635	1562.03	31.60	1436.91	610.03
10	0.000966	116.81	0.1376	3.9784	1582.00	32.37	1255.51	617.83
15	0.000968	136.73	0.2073	3.9903	1599.81	33.06	1108.46	626.11
20	0.000969	156.71	0.2761	3.9998	1615.70	33.67	987.41	634.58
25	0.000971	176.73	0.3438	4.0076	1629.86	34.21	886.44	643.01
30	0.000972	196.78	0.4105	4.0140	1642.44	34.68	801.25	651.24
35	0.000974	216.86	0.4762	4.0194	1653.57	35.09	728.65	659.15
40	0.000976	236.97	0.5409	4.0242	1663.34	35.43	666.21	666.66
45	0.000978	257.11	0.6047	4.0284	1671.84	35.72	612.10	673.71
50	0.000980	277.26	0.6675	4.0322	1679.15	35.95	564.89	680.27
60	0.000985	317.62	0.7905	4.0394	1690.44	36.27	486.81	691.92
70	0.000990	358.04	0.9101	4.0463	1697.70	36.39	425.34	701.69
80	0.000996	398.54	1.0264	4.0533	1701.34	36.34	376.10	709.75
90	0.001002	439.11	1.1397	4.0605	1701.75	36.14	336.05	716.31
100	0.001008	479.75	1.2501	4.0683	1699.25	35.81	303.06	721.59
110	0.001015	520.48	1.3578	4.0765	1694.17	35.36	275.57	725.75
120	0.001022	561.29	1.4630	4.0853	1686.75	34.81	252.45	728.94
130	0.001029	602.19	1.5657	4.0948	1677.23	34.16	232.81	731.25
140	0.001037	643.19	1.6661	4.1052	1665.81	33.44	216.00	732.75
150	0.001046	684.29	1.7645	4.1164	1652.66	32.65	201.50	733.49
160	0.001054	725.52	1.8607	4.1287	1637.89	31.80	188.90	733.47
170	0.001064	766.87	1.9551	4.1423	1621.64	30.90	177.87	732.72
180	0.001074	808.37	2.0477	4.1572	1603.97	29.96	168.17	731.22
190	0.001084	850.02	2.1387	4.1738	1584.97	28.97	159.58	728.98
200	0.001095	891.85	2.2280	4.1921	1564.69	27.96	151.93	725.99
210	0.001106	933.87	2.3159	4.2124	1543.19	26.92	145.09	722.25
220	0.001118	976.10	2.4024	4.2348	1520.50	25.85	138.92	717.76
230	0.001130	1018.57	2.4877	4.2597	1496.68	24.77	133.35	712.51
240	0.001143	1061.31	2.5718	4.2872	1471.78	23.68	128.28	706.50
250	0.001157	1104.33	2.6548	4.3174	1445.85	22.58	123.64	699.73
260	0.001172	1147.67	2.7368	4.3508	1418.96	21.47	119.38	692.20
270	0.001188	1191.35	2.8180	4.3873	1391.19	20.37	115.45	683.91
280	0.001204	1235.42	2.8984	4.4273	1362.62	19.28	111.80	674.85
290	0.001221	1279.91	2.9781	4.4708	1333.35	18.20	108.39	665.02
300	0.001240	1324.85	3.0572	4.5181	1303.50	17.13	105.19	654.42
310	0.001259	1370.29	3.1358	4.5692	1273.16	16.09	102.16	643.05
320	0.001280	1416.25	3.2140	4.6241	1242.46	15.07	99.29	630.94
330	0.001303	1462.78	3.2918	4.6825	1211.48	14.08	96.55	618.13
340	0.001327	1509.91	3.3693	4.7440	1180.33	13.13	93.92	604.68

Table 4 Single-Phase Region (Regions 1 to 3 of IAPWS-IF97) – Continuation
Homogenes Zustandsgebiet (Bereiche 1 bis 3 der IAPWS-IF97) – Fortsetzung

p = 80 MPa

t [°C]	v [m^3 kg^{-1}]	h [kJ kg^{-1}]	s [kJ kg^{-1} K^{-1}]	c_p [kJ kg^{-1} K^{-1}]	w [m s^{-1}]	κ	$\eta \times 10^6$ [kg m^{-1} s^{-1}]	$\lambda \times 10^3$ [W m^{-1} K^{-1}]
350	0.001352	1557.67	3.4465	4.8076	1149.04	12.2027	91.37	590.69
360	0.001380	1606.10	3.5236	4.8842	1117.26	11.3054	88.90	576.29
370	0.001410	1655.38	3.6008	4.9727	1084.58	10.4276	86.49	561.60
380	0.001443	1705.56	3.6783	5.0634	1051.82	9.5864	84.12	546.73
390	0.001478	1756.67	3.7559	5.1586	1019.19	8.7860	81.79	531.78
400	0.001516	1808.76	3.8339	5.2623	986.85	8.0282	79.49	516.77
410	0.001558	1861.95	3.9123	5.3768	954.96	7.3147	77.21	501.61
420	0.001605	1916.33	3.9913	5.5024	923.76	6.6472	74.94	486.15
430	0.001656	1972.03	4.0711	5.6371	893.49	6.0274	72.69	470.16
440	0.001712	2029.10	4.1517	5.7774	864.44	5.4566	70.46	453.39
450	0.001774	2087.58	4.2331	5.9181	836.91	4.9356	68.26	435.70
460	0.001842	2147.44	4.3153	6.0528	811.19	4.4645	66.10	417.05
470	0.001918	2208.59	4.3982	6.1737	787.59	4.0430	63.99	397.64
480	0.002001	2270.84	4.4814	6.2727	766.40	3.6701	61.96	377.82
490	0.002091	2333.94	4.5646	6.3420	747.86	3.3440	60.03	358.11
500	0.002188	2397.56	4.6474	6.3749	732.12	3.0622	58.22	339.02
510	0.002292	2461.31	4.7294	6.3679	719.27	2.8212	56.55	321.04
520	0.002403	2524.78	4.8099	6.3205	709.24	2.6172	55.03	304.49
530	0.002518	2587.59	4.8886	6.2361	701.86	2.4454	53.67	289.60
540	0.002638	2649.40	4.9651	6.1206	696.86	2.3015	52.47	276.41
550	0.002760	2709.93	5.0391	5.9817	693.95	2.1810	51.43	264.90
560	0.002885	2768.96	5.1103	5.8423	692.89	2.0804	50.54	254.97
570	0.003010	2826.43	5.1789	5.6505	693.29	1.9960	49.78	246.36
580	0.003135	2882.02	5.2445	5.4708	694.79	1.9245	49.15	239.07
590	0.003260	2935.89	5.3072	5.3034	696.96	1.8625	48.63	232.89
600	0.003384	2988.09	5.3674	5.1372	699.81	1.8092	48.21	227.65
610	0.003506	3038.64	5.4249	4.9728	703.35	1.7639	47.88	223.21
620	0.003626	3087.57	5.4800	4.8153	707.48	1.7254	47.62	219.46
630	0.003744	3134.98	5.5328	4.6682	712.02	1.6924	47.43	216.30
640	0.003861	3180.97	5.5835	4.5326	716.83	1.6637	47.29	213.62
650	0.003975	3225.67	5.6321	4.4081	721.80	1.6384	47.21	211.36
660	0.004087	3269.17	5.6790	4.2934	726.88	1.6159	47.16	209.45
670	0.004197	3311.56	5.7242	4.1874	732.01	1.5958	47.16	207.82
680	0.004306	3352.94	5.7678	4.0891	737.19	1.5778	47.19	206.44
690	0.004412	3393.37	5.8100	3.9979	742.41	1.5616	47.24	205.26
700	0.004516	3432.92	5.8509	3.9135	747.63	1.5471	47.32	204.26
720	0.004720	3509.65	5.9290	3.7638	758.06	1.5219	47.55	202.67
740	0.004917	3583.63	6.0027	3.6381	768.30	1.5007	47.84	201.49
760	0.005108	3655.32	6.0728	3.5337	778.23	1.4820	48.18	200.63
780	0.005295	3725.10	6.1397	3.4475	787.73	1.4650	48.56	199.98
800	0.005476	3793.32	6.2039	3.3765	796.77	1.4491	48.98	199.50

Table 4 Single-Phase Region (Regions 1 to 3 of IAPWS-IF97) – Continuation
Homogenes Zustandsgebiet (Bereiche 1 bis 3 der IAPWS-IF97) – Fortsetzung

p = 82.5 MPa

t [°C]	v [m³ kg⁻¹]	h [kJ kg⁻¹]	s [kJ kg⁻¹ K⁻¹]	c_p [kJ kg⁻¹ K⁻¹]	w [m s⁻¹]	κ	$\eta \times 10^6$ [kg m⁻¹ s⁻¹]	$\lambda \times 10^3$ [W m⁻¹ K⁻¹]
0	0.000963	79.48	− 0.0053	3.9392	1544.08	30.00	1662.71	604.28
5	0.000964	99.22	0.0663	3.9585	1566.46	30.84	1435.81	611.16
10	0.000965	119.06	0.1370	3.9738	1586.36	31.60	1254.98	618.92
15	0.000967	138.96	0.2067	3.9859	1604.11	32.26	1108.31	627.18
20	0.000968	158.91	0.2753	3.9957	1619.95	32.86	987.54	635.65
25	0.000970	178.91	0.3430	4.0036	1634.07	33.38	886.77	644.08
30	0.000971	198.94	0.4096	4.0101	1646.62	33.83	801.71	652.32
35	0.000973	219.01	0.4752	4.0157	1657.74	34.23	729.20	660.23
40	0.000975	239.10	0.5399	4.0205	1667.50	34.56	666.83	667.74
45	0.000977	259.21	0.6036	4.0247	1676.00	34.84	612.76	674.80
50	0.000979	279.35	0.6664	4.0286	1683.32	35.07	565.57	681.38
60	0.000984	319.67	0.7893	4.0358	1694.65	35.37	487.52	693.06
70	0.000989	360.06	0.9088	4.0427	1701.97	35.50	426.06	702.85
80	0.000995	400.52	1.0250	4.0497	1705.68	35.45	376.80	710.94
90	0.001001	441.06	1.1382	4.0569	1706.17	35.27	336.73	717.55
100	0.001007	481.66	1.2485	4.0645	1703.77	34.95	303.72	722.86
110	0.001014	522.35	1.3561	4.0727	1698.79	34.51	276.22	727.07
120	0.001021	563.12	1.4611	4.0813	1691.49	33.98	253.07	730.31
130	0.001028	603.98	1.5638	4.0907	1682.11	33.36	233.42	732.68
140	0.001036	644.93	1.6641	4.1008	1670.83	32.66	216.59	734.24
150	0.001044	686.00	1.7623	4.1118	1657.82	31.90	202.07	735.03
160	0.001053	727.17	1.8585	4.1239	1643.23	31.08	189.45	735.08
170	0.001062	768.48	1.9528	4.1371	1627.15	30.21	178.41	734.39
180	0.001072	809.92	2.0452	4.1517	1609.68	29.29	168.70	732.96
190	0.001082	851.52	2.1360	4.1678	1590.89	28.35	160.10	730.78
200	0.001093	893.28	2.2253	4.1857	1570.84	27.36	152.44	727.86
210	0.001104	935.24	2.3130	4.2054	1549.57	26.36	145.59	724.19
220	0.001116	977.40	2.3994	4.2273	1527.15	25.33	139.42	719.76
230	0.001128	1019.79	2.4845	4.2515	1503.60	24.29	133.84	714.57
240	0.001141	1062.44	2.5684	4.2782	1478.99	23.23	128.77	708.64
250	0.001155	1105.36	2.6512	4.3076	1453.38	22.16	124.13	701.94
260	0.001170	1148.60	2.7331	4.3399	1426.82	21.10	119.87	694.50
270	0.001185	1192.17	2.8141	4.3753	1399.39	20.03	115.94	686.29
280	0.001201	1236.11	2.8942	4.4139	1371.19	18.97	112.30	677.34
290	0.001218	1280.46	2.9737	4.4560	1342.31	17.93	108.89	667.62
300	0.001236	1325.24	3.0525	4.5016	1312.86	16.90	105.70	657.14
310	0.001256	1370.50	3.1308	4.5507	1282.96	15.89	102.68	645.91
320	0.001276	1416.27	3.2086	4.6033	1252.72	14.90	99.82	633.95
330	0.001298	1462.58	3.2860	4.6590	1222.24	13.95	97.10	621.30
340	0.001322	1509.46	3.3631	4.7175	1191.61	13.02	94.48	608.01

Table 4 Single-Phase Region (Regions 1 to 3 of IAPWS-IF97) – Continuation
Homogenes Zustandsgebiet (Bereiche 1 bis 3 der IAPWS-IF97) – Fortsetzung

$p = 82.5$ **MPa**

t [°C]	v [m^3 kg^{-1}]	h [kJ kg^{-1}]	s [kJ kg^{-1} K^{-1}]	c_p [kJ kg^{-1} K^{-1}]	w [m s^{-1}]	κ	$\eta \times 10^6$ [kg m^{-1} s^{-1}]	$\lambda \times 10^3$ [W m^{-1} K^{-1}]
350	0.001347	1556.93	3.4399	4.7777	1160.87	12.1287	91.95	594.19
360	0.001374	1605.05	3.5165	4.8496	1129.58	11.2579	89.51	579.95
370	0.001403	1653.96	3.5932	4.9336	1097.54	10.4079	87.12	565.42
380	0.001434	1703.73	3.6699	5.0189	1065.38	9.5917	84.78	550.70
390	0.001468	1754.35	3.7469	5.1076	1033.34	8.8139	82.48	535.89
400	0.001506	1805.90	3.8240	5.2036	1001.59	8.0766	80.21	521.02
410	0.001546	1858.46	3.9015	5.3092	970.28	7.3814	77.97	506.05
420	0.001590	1912.12	3.9795	5.4248	939.63	6.7299	75.75	490.82
430	0.001639	1966.98	4.0581	5.5487	909.88	6.1236	73.54	475.15
440	0.001692	2023.11	4.1373	5.6778	881.28	5.5637	71.36	458.79
450	0.001751	2080.54	4.2173	5.8074	854.10	5.0508	69.21	441.60
460	0.001815	2139.24	4.2979	5.9320	828.62	4.5851	67.09	423.51
470	0.001886	2199.14	4.3791	6.0451	805.09	4.1661	65.03	404.67
480	0.001963	2260.08	4.4605	6.1396	783.76	3.7930	63.04	385.37
490	0.002047	2321.85	4.5420	6.2089	764.86	3.4641	61.14	366.06
500	0.002138	2384.16	4.6231	6.2472	748.55	3.1772	59.35	347.23
510	0.002235	2446.68	4.7035	6.2509	734.90	2.9295	57.68	329.33
520	0.002337	2509.05	4.7826	6.2189	723.91	2.7175	55.16	312.73
530	0.002445	2570.94	4.8601	6.1529	715.45	2.5373	54.78	297.66
540	0.002557	2632.01	4.9357	6.0572	709.35	2.3849	53.56	284.22
550	0.002673	2692.01	5.0090	5.9378	705.32	2.2563	52.48	272.40
560	0.002790	2750.64	5.0798	5.8120	703.06	2.1474	51.54	262.15
570	0.002909	2808.02	5.1483	5.6526	702.32	2.0551	50.74	253.24
580	0.003029	2863.66	5.2139	5.4797	703.08	1.9783	50.07	245.61
590	0.003148	2917.67	5.2768	5.3219	704.56	1.9115	49.50	239.13
600	0.003266	2970.11	5.3373	5.1656	706.72	1.8534	49.04	233.62
610	0.003384	3020.98	5.3952	5.0085	709.63	1.8039	48.67	228.93
620	0.003500	3070.29	5.4507	4.8555	713.24	1.7619	48.37	224.96
630	0.003614	3118.11	5.5040	4.7110	717.38	1.7260	48.14	221.59
640	0.003727	3164.54	5.5551	4.5769	721.86	1.6949	47.97	218.73
650	0.003837	3209.69	5.6043	4.4532	726.56	1.6674	47.85	216.30
660	0.003946	3253.64	5.6516	4.3388	731.40	1.6431	47.78	214.23
670	0.004054	3296.49	5.6973	4.2326	736.32	1.6212	47.75	212.46
680	0.004159	3338.32	5.7414	4.1336	741.30	1.6017	47.75	210.95
690	0.004262	3379.19	5.7841	4.0413	746.32	1.5840	47.79	209.65
700	0.004364	3419.16	5.8254	3.9555	751.38	1.5682	47.85	208.54
720	0.004562	3496.70	5.9042	3.8026	761.50	1.5407	48.03	206.75
740	0.004755	3571.43	5.9787	3.6737	771.50	1.5174	48.29	205.40
760	0.004941	3643.80	6.0495	3.5665	781.22	1.4971	48.61	204.38
780	0.005123	3714.22	6.1170	3.4782	790.56	1.4787	48.97	203.59
800	0.005300	3783.03	6.1817	3.4055	799.47	1.4617	49.36	202.98

Table 4 Single-Phase Region (Regions 1 to 3 of IAPWS-IF97) – Continuation
Homogenes Zustandsgebiet (Bereiche 1 bis 3 der IAPWS-IF97) – Fortsetzung

p = 85 MPa

t [°C]	v [m^3 kg^{-1}]	h [kJ kg^{-1}]	s [kJ kg^{-1} K^{-1}]	c_p [kJ kg^{-1} K^{-1}]	w [m s^{-1}]	κ	$\eta \times 10^6$ [kg m^{-1} s^{-1}]	$\lambda \times 10^3$ [W m^{-1} K^{-1}]
0	0.000962	81.77	– 0.0058	3.9340	1548.59	29.32	1660.87	605.44
5	0.000963	101.49	0.0658	3.9537	1570.89	30.14	1434.78	612.27
10	0.000964	121.30	0.1364	3.9693	1590.72	30.87	1254.50	620.01
15	0.000966	141.18	0.2060	3.9817	1608.40	31.51	1108.21	628.25
20	0.000967	161.11	0.2746	3.9916	1624.19	32.09	987.70	636.71
25	0.000969	181.09	0.3421	3.9997	1638.28	32.59	887.11	645.14
30	0.000970	201.10	0.4087	4.0063	1650.81	33.04	802.19	653.38
35	0.000972	221.15	0.4743	4.0120	1661.90	33.42	729.76	661.30
40	0.000974	241.22	0.5389	4.0168	1671.66	33.74	667.45	668.82
45	0.000976	261.32	0.6026	4.0211	1680.16	34.02	613.43	675.89
50	0.000978	281.43	0.6653	4.0251	1687.49	34.24	566.27	682.48
60	0.000983	321.72	0.7881	4.0323	1698.85	34.54	488.24	694.18
70	0.000988	362.08	0.9075	4.0392	1706.22	34.66	426.77	704.01
80	0.000994	402.51	1.0236	4.0462	1710.00	34.62	377.50	712.13
90	0.001000	443.00	1.1367	4.0533	1710.57	34.44	337.41	718.77
100	0.001006	483.57	1.2469	4.0609	1708.27	34.13	304.38	724.13
110	0.001013	524.22	1.3544	4.0689	1703.39	33.71	276.86	728.39
120	0.001020	564.95	1.4593	4.0774	1696.20	33.20	253.69	731.68
130	0.001027	605.77	1.5618	4.0866	1686.93	32.60	234.02	734.10
140	0.001035	646.69	1.6621	4.0965	1675.79	31.92	217.17	735.72
150	0.001043	687.71	1.7602	4.1073	1662.93	31.18	202.63	736.58
160	0.001052	728.84	1.8562	4.1190	1648.49	30.39	190.00	736.69
170	0.001061	770.09	1.9504	4.1320	1632.59	29.55	178.95	736.06
180	0.001071	811.48	2.0428	4.1462	1615.31	28.67	169.23	734.69
190	0.001081	853.02	2.1334	4.1619	1596.73	27.75	160.62	732.58
200	0.001091	894.72	2.2225	4.1793	1576.90	26.80	152.95	729.72
210	0.001103	936.61	2.3101	4.1986	1555.87	25.83	146.09	726.11
220	0.001114	978.70	2.3964	4.2199	1533.70	24.84	139.92	721.75
230	0.001127	1021.02	2.4813	4.2434	1510.43	23.83	134.33	716.63
240	0.001139	1063.58	2.5651	4.2694	1486.11	22.80	129.26	710.76
250	0.001153	1106.41	2.6477	4.2979	1460.79	21.77	124.62	704.14
260	0.001167	1149.55	2.7294	4.3293	1434.55	20.74	120.36	696.78
270	0.001182	1193.01	2.8102	4.3636	1407.47	19.71	116.43	688.66
280	0.001198	1236.83	2.8901	4.4010	1379.62	18.69	112.79	679.80
290	0.001215	1281.04	2.9693	4.4417	1351.12	17.67	109.39	670.19
300	0.001233	1325.67	3.0479	4.4856	1322.07	16.68	106.20	659.83
310	0.001252	1370.76	3.1259	4.5328	1292.59	15.70	103.20	648.73
320	0.001272	1416.34	3.2034	4.5832	1262.79	14.75	100.35	636.91
330	0.001294	1462.44	3.2804	4.6366	1232.77	13.82	97.63	624.41
340	0.001317	1509.08	3.3571	4.6923	1202.63	12.92	95.03	611.29

Table 4 Single-Phase Region (Regions 1 to 3 of IAPWS-IF97) – Continuation
Homogenes Zustandsgebiet (Bereiche 1 bis 3 der IAPWS-IF97) – Fortsetzung

p = 85 MPa

t [°C]	v [m³ kg⁻¹]	h [kJ kg⁻¹]	s [kJ kg⁻¹ K⁻¹]	c_p [kJ kg⁻¹ K⁻¹]	w [m s⁻¹]	κ	$\eta \times 10^6$ [kg m⁻¹ s⁻¹]	$\lambda \times 10^3$ [W m⁻¹ K⁻¹]
350	0.001341	1556.29	3.4335	4.7493	1172.42	12.0564	92.52	597.62
360	0.001368	1604.10	3.5096	4.8170	1141.57	11.2103	90.09	583.54
370	0.001396	1652.67	3.5857	4.8968	1110.15	10.3865	87.73	569.15
380	0.001427	1702.04	3.6619	4.9773	1078.56	9.5938	85.42	554.57
390	0.001460	1752.22	3.7381	5.0602	1047.08	8.8374	83.15	539.90
400	0.001495	1803.26	3.8145	5.1492	1015.87	8.1194	80.92	525.17
410	0.001534	1855.24	3.8912	5.2469	985.10	7.4413	78.71	510.36
420	0.001577	1908.23	3.9682	5.3537	954.96	6.8048	76.53	495.34
430	0.001623	1962.33	4.0457	5.4681	925.68	6.2114	74.36	479.94
440	0.001674	2017.61	4.1237	5.5872	897.49	5.6620	72.22	463.95
450	0.001729	2074.08	4.2024	5.7072	870.66	5.1572	70.11	447.20
460	0.001790	2131.74	4.2815	5.8228	845.41	4.6971	68.04	429.62
470	0.001857	2190.51	4.3612	5.9286	821.97	4.2812	66.02	411.30
480	0.001929	2250.26	4.4410	6.0185	800.58	3.9087	64.07	392.50
490	0.002008	2310.80	4.5209	6.0868	781.42	3.5781	62.20	373.59
500	0.002092	2371.90	4.6004	6.1285	764.65	3.2874	60.43	355.02
510	0.002183	2433.27	4.6793	6.1401	750.35	3.0343	58.77	337.25
520	0.002279	2494.60	4.7571	6.1200	738.55	2.8156	57.24	320.64
530	0.002380	2555.57	4.8335	6.0690	729.16	2.6281	55.85	305.44
540	0.002485	2615.89	4.9081	5.9902	722.05	2.4680	54.61	291.79
550	0.002594	2675.30	4.9807	5.8879	717.00	2.3319	53.50	279.70
560	0.002705	2733.59	5.0511	5.7676	713.79	2.2161	52.53	269.12
570	0.002818	2790.62	5.1192	5.6436	711.71	2.1149	51.69	259.95
580	0.002931	2846.24	5.1848	5.4814	711.69	2.0328	50.98	252.01
590	0.003045	2900.30	5.2478	5.3323	712.54	1.9614	50.37	245.25
600	0.003159	2952.90	5.3084	5.1858	714.03	1.8989	49.87	239.47
610	0.003272	3004.01	5.3666	5.0364	716.31	1.8451	49.45	234.54
620	0.003383	3053.63	5.4224	4.8884	719.38	1.7995	49.12	230.35
630	0.003494	3101.80	5.4761	4.7472	723.09	1.7606	48.86	226.78
640	0.003603	3148.61	5.5276	4.6155	727.24	1.7271	48.65	223.74
650	0.003710	3194.14	5.5772	4.4934	731.67	1.6976	48.51	221.14
660	0.003816	3238.50	5.6250	4.3801	736.26	1.6713	48.41	218.92
670	0.003920	3281.77	5.6711	4.2743	740.95	1.6477	48.35	217.01
680	0.004022	3324.01	5.7157	4.1752	745.72	1.6265	48.32	215.38
690	0.004123	3365.29	5.7588	4.0823	750.54	1.6074	48.34	213.96
700	0.004222	3405.68	5.8005	3.9955	755.41	1.5901	48.37	212.74
720	0.004415	3483.99	5.8801	3.8399	765.19	1.5601	48.52	210.76
740	0.004603	3559.43	5.9553	3.7080	774.90	1.5347	48.75	209.24
760	0.004785	3632.46	6.0267	3.5982	784.38	1.5126	49.04	208.06
780	0.004963	3703.49	6.0948	3.5078	793.53	1.4928	49.37	207.13
800	0.005135	3772.88	6.1601	3.4334	802.32	1.4747	49.75	206.39

Table 4 Single-Phase Region (Regions 1 to 3 of IAPWS-IF97) – Continuation
Homogenes Zustandsgebiet (Bereiche 1 bis 3 der IAPWS-IF97) – Fortsetzung

p = 87.5 MPa

t [°C]	v [m³ kg⁻¹]	h [kJ kg⁻¹]	s [kJ kg⁻¹ K⁻¹]	c_p [kJ kg⁻¹ K⁻¹]	w [m s⁻¹]	κ	$\eta \times 10^6$ [kg m⁻¹ s⁻¹]	$\lambda \times 10^3$ [W m⁻¹ K⁻¹]
0	0.000961	84.05	− 0.0062	3.9289	1553.09	28.67	1659.14	606.60
5	0.000962	103.75	0.0653	3.9491	1575.32	29.47	1433.82	613.38
10	0.000964	123.53	0.1358	3.9649	1595.08	30.18	1254.07	621.09
15	0.000965	143.39	0.2053	3.9776	1612.70	30.81	1108.15	629.32
20	0.000966	163.31	0.2738	3.9877	1628.44	31.36	987.89	637.77
25	0.000968	183.27	0.3413	3.9958	1642.48	31.85	887.48	646.20
30	0.000970	203.26	0.4078	4.0026	1654.98	32.28	802.68	654.45
35	0.000971	223.29	0.4734	4.0083	1666.06	32.66	730.34	662.37
40	0.000973	243.34	0.5379	4.0132	1675.81	32.97	668.09	669.90
45	0.000975	263.42	0.6015	4.0176	1684.31	33.24	614.11	676.98
50	0.000978	283.52	0.6642	4.0216	1691.64	33.45	566.97	683.58
60	0.000982	323.77	0.7869	4.0289	1703.04	33.75	488.96	695.31
70	0.000987	364.10	0.9061	4.0358	1710.45	33.86	427.49	705.16
80	0.000993	404.49	1.0222	4.0427	1714.30	33.83	378.21	713.32
90	0.000999	444.95	1.1351	4.0498	1714.94	33.66	338.10	720.00
100	0.001005	485.48	1.2452	4.0572	1712.72	33.36	305.05	725.40
110	0.001012	526.10	1.3526	4.0651	1707.94	32.96	277.50	729.71
120	0.001019	566.79	1.4575	4.0735	1700.86	32.46	254.31	733.05
130	0.001026	607.57	1.5599	4.0825	1691.71	31.88	234.62	735.52
140	0.001034	648.44	1.6601	4.0922	1680.69	31.22	217.75	737.20
150	0.001042	689.42	1.7581	4.1028	1667.98	30.51	203.20	738.12
160	0.001051	730.50	1.8540	4.1143	1653.69	29.74	190.55	738.29
170	0.001060	771.71	1.9481	4.1269	1637.96	28.93	179.49	737.72
180	0.001069	813.04	2.0403	4.1408	1620.87	28.08	169.75	736.42
190	0.001079	854.53	2.1309	4.1562	1602.48	27.19	161.13	734.37
200	0.001090	896.17	2.2198	4.1731	1582.87	26.27	153.46	731.58
210	0.001101	938.00	2.3073	4.1919	1562.08	25.33	146.59	728.03
220	0.001113	980.02	2.3934	4.2126	1540.16	24.37	140.41	723.73
230	0.001125	1022.26	2.4782	4.2355	1517.15	23.39	134.82	718.68
240	0.001137	1064.73	2.5618	4.2608	1493.11	22.40	129.74	712.88
250	0.001151	1107.48	2.6443	4.2885	1468.09	21.40	125.10	706.33
260	0.001165	1150.51	2.7258	4.3190	1442.17	20.40	120.84	699.04
270	0.001180	1193.87	2.8063	4.3523	1415.41	19.40	116.92	691.01
280	0.001196	1237.57	2.8860	4.3885	1387.91	18.41	113.27	682.24
290	0.001212	1281.65	2.9650	4.4278	1359.77	17.43	109.88	672.73
300	0.001230	1326.14	3.0433	4.4702	1331.10	16.46	106.70	662.48
310	0.001249	1371.06	3.1210	4.5156	1302.02	15.52	103.70	651.51
320	0.001268	1416.46	3.1982	4.5640	1272.64	14.59	100.87	639.83
330	0.001289	1462.35	3.2749	4.6151	1243.08	13.70	98.16	627.47
340	0.001312	1508.77	3.3513	4.6682	1213.40	12.83	95.57	614.50

Table 4 Single-Phase Region (Regions 1 to 3 of IAPWS-IF97) – Continuation
Homogenes Zustandsgebiet (Bereiche 1 bis 3 der IAPWS-IF97) – Fortsetzung

				p = 87.5 MPa				
t [°C]	v [m^3 kg^{-1}]	h [kJ kg^{-1}]	s [kJ kg^{-1} K^{-1}]	c_p [kJ kg^{-1} K^{-1}]	w [m s^{-1}]	κ	$\eta \times 10^6$ [kg m^{-1} s^{-1}]	$\lambda \times 10^3$ [W m^{-1} K^{-1}]
350	0.001336	1555.72	3.4272	4.7223	1183.67	11.9853	93.08	600.99
360	0.001362	1603.25	3.5029	4.7861	1153.26	11.1625	90.67	587.06
370	0.001389	1651.49	3.5785	4.8621	1122.43	10.3636	88.33	572.81
380	0.001419	1700.49	3.6541	4.9382	1091.38	9.5932	86.04	558.37
390	0.001451	1750.26	3.7297	5.0159	1060.42	8.8569	83.80	543.82
400	0.001486	1800.82	3.8054	5.0988	1029.72	8.1569	81.60	529.22
410	0.001523	1852.26	3.8812	5.1894	999.44	7.4951	79.43	514.55
420	0.001564	1904.64	3.9573	5.2882	969.77	6.8728	77.28	499.71
430	0.001608	1958.05	4.0338	5.3942	940.92	6.2915	75.15	484.55
440	0.001657	2012.54	4.1108	5.5046	913.13	5.7522	73.06	468.88
450	0.001709	2068.14	4.1882	5.6159	886.61	5.2553	70.99	452.53
460	0.001767	2124.84	4.2661	5.7236	861.59	4.8009	68.96	435.41
470	0.001830	2182.58	4.3443	5.8227	838.27	4.3886	66.97	417.58
480	0.001898	2241.25	4.4227	5.9082	816.85	4.0174	65.05	399.24
490	0.001972	2300.68	4.5011	5.9749	797.51	3.6860	63.21	380.72
500	0.002051	2360.67	4.5792	6.0184	780.38	3.3926	61.46	362.44
510	0.002136	2420.96	4.6567	6.0356	765.56	3.1352	59.81	344.82
520	0.002226	2481.29	4.7332	6.0247	753.07	2.9110	58.29	328.24
530	0.002321	2541.37	4.8085	5.9860	742.87	2.7170	56.89	312.95
540	0.002420	2600.92	4.8822	5.9213	734.87	2.5502	55.63	299.11
550	0.002522	2659.72	4.9541	5.8340	728.89	2.4072	54.50	286.79
560	0.002627	2717.54	5.0239	5.7286	724.75	2.2848	53.50	275.94
570	0.002734	2774.24	5.0915	5.6170	721.46	2.1755	52.63	266.48
580	0.002843	2829.73	5.1570	5.4759	720.62	2.0878	51.88	258.28
590	0.002951	2883.78	5.2199	5.3357	720.85	2.0121	51.23	251.23
600	0.003060	2936.45	5.2806	5.1990	721.71	1.9452	50.69	245.20
610	0.003169	2987.74	5.3390	5.0574	723.35	1.8872	50.24	240.03
620	0.003276	3037.60	5.3952	4.9147	725.86	1.8380	49.87	235.63
630	0.003383	3086.05	5.4491	4.7772	729.12	1.7961	49.57	231.86
640	0.003488	3133.17	5.5010	4.6482	732.93	1.7601	49.34	228.65
650	0.003592	3179.05	5.5510	4.5285	737.08	1.7286	49.16	225.89
660	0.003695	3223.77	5.5991	4.4169	741.42	1.7004	49.03	223.52
670	0.003796	3267.41	5.6457	4.3123	745.89	1.6751	48.95	221.48
680	0.003895	3310.03	5.6906	4.2138	750.43	1.6523	48.90	219.71
690	0.003993	3351.70	5.7341	4.1208	755.04	1.6315	48.89	218.18
700	0.004090	3392.47	5.7762	4.0334	759.70	1.6128	48.91	216.86
720	0.004278	3471.52	5.8566	3.8756	769.10	1.5801	49.02	214.68
740	0.004461	3547.65	5.9325	3.7410	778.48	1.5525	49.21	213.00
760	0.004639	3621.31	6.0045	3.6288	787.68	1.5285	49.47	211.68
780	0.004812	3692.93	6.0732	3.5363	796.62	1.5072	49.78	210.61
800	0.004981	3762.87	6.1390	3.4603	805.30	1.4879	50.14	209.74

Table 4 Single-Phase Region (Regions 1 to 3 of IAPWS-IF97) – Continuation
Homogenes Zustandsgebiet (Bereiche 1 bis 3 der IAPWS-IF97) – Fortsetzung

p = 90 MPa

t [°C]	v [m³ kg⁻¹]	h [kJ kg⁻¹]	s [kJ kg⁻¹ K⁻¹]	c_p [kJ kg⁻¹ K⁻¹]	w [m s⁻¹]	κ	$\eta\times10^6$ [kg m⁻¹ s⁻¹]	$\lambda\times10^3$ [W m⁻¹ K⁻¹]
0	0.000960	86.33	− 0.0066	3.9240	1557.59	28.07	1657.51	607.74
5	0.000961	106.00	0.0647	3.9445	1579.74	28.84	1432.94	614.48
10	0.000963	125.77	0.1352	3.9607	1599.43	29.53	1253.69	622.16
15	0.000964	145.60	0.2046	3.9735	1616.99	30.14	1108.12	630.38
20	0.000965	165.50	0.2730	3.9838	1632.68	30.68	988.11	638.82
25	0.000967	185.44	0.3405	3.9921	1646.68	31.16	887.87	647.26
30	0.000969	205.42	0.4069	3.9989	1659.16	31.57	803.19	655.50
35	0.000971	225.43	0.4724	4.0047	1670.21	31.94	730.93	663.44
40	0.000972	245.46	0.5369	4.0097	1679.95	32.25	668.74	670.98
45	0.000975	265.52	0.6005	4.0141	1688.45	32.50	614.79	678.06
50	0.000977	285.60	0.6631	4.0181	1695.79	32.71	567.67	684.67
60	0.000981	325.82	0.7857	4.0255	1707.21	33.00	489.68	696.43
70	0.000986	366.11	0.9048	4.0324	1714.67	33.12	428.21	706.31
80	0.000992	406.47	1.0207	4.0392	1718.57	33.08	378.91	714.50
90	0.000998	446.90	1.1336	4.0463	1719.29	32.92	338.78	721.22
100	0.001004	487.39	1.2436	4.0537	1717.15	32.63	305.71	726.66
110	0.001011	527.97	1.3509	4.0614	1712.45	32.24	278.14	731.02
120	0.001018	568.62	1.4557	4.0697	1705.47	31.76	254.93	734.41
130	0.001025	609.37	1.5580	4.0785	1696.43	31.20	235.21	736.94
140	0.001033	650.20	1.6581	4.0881	1685.54	30.57	218.33	738.68
150	0.001041	691.13	1.7560	4.0984	1672.96	29.87	203.76	739.65
160	0.001050	732.17	1.8518	4.1096	1658.82	29.13	191.10	739.89
170	0.001059	773.33	1.9458	4.1220	1643.25	28.34	180.02	739.39
180	0.001068	814.61	2.0379	4.1355	1626.34	27.52	170.27	738.15
190	0.001078	856.04	2.1283	4.1505	1608.15	26.66	161.64	736.16
200	0.001088	897.63	2.2171	4.1670	1588.75	25.77	153.96	733.43
210	0.001099	939.39	2.3045	4.1853	1568.18	24.86	147.08	729.95
220	0.001111	981.34	2.3904	4.2055	1546.50	23.92	140.90	725.71
230	0.001123	1023.51	2.4751	4.2278	1523.76	22.98	135.30	720.72
240	0.001136	1065.90	2.5585	4.2524	1499.99	22.02	130.22	714.99
250	0.001149	1108.56	2.6408	4.2794	1475.27	21.05	125.58	708.51
260	0.001163	1151.50	2.7221	4.3090	1449.65	20.08	121.32	701.29
270	0.001178	1194.75	2.8025	4.3412	1423.21	19.11	117.40	693.34
280	0.001193	1238.34	2.8820	4.3763	1396.05	18.15	113.76	684.66
290	0.001209	1282.29	2.9608	4.4143	1368.26	17.20	110.37	675.24
300	0.001227	1326.63	3.0388	4.4553	1339.96	16.26	107.19	665.10
310	0.001245	1371.40	3.1163	4.4990	1311.26	15.34	104.20	654.25
320	0.001265	1416.62	3.1931	4.5456	1282.29	14.45	101.37	642.70
330	0.001285	1462.32	3.2695	4.5944	1253.14	13.58	98.68	630.48
340	0.001307	1508.52	3.3455	4.6451	1223.91	12.73	96.11	617.66

Table 4 Single-Phase Region (Regions 1 to 3 of IAPWS-IF97) – Continuation
Homogenes Zustandsgebiet (Bereiche 1 bis 3 der IAPWS-IF97) – Fortsetzung

				p = 90 MPa				
t [°C]	v [$m^3\,kg^{-1}$]	h [$kJ\,kg^{-1}$]	s [$kJ\,kg^{-1}\,K^{-1}$]	c_p [$kJ\,kg^{-1}\,K^{-1}$]	w [$m\,s^{-1}$]	κ	$\eta\times10^6$ [$kg\,m^{-1}\,s^{-1}$]	$\lambda\times10^3$ [$W\,m^{-1}\,K^{-1}$]
350	0.001331	1555.23	3.4211	4.6966	1194.63	11.9149	93.63	604.30
360	0.001356	1602.48	3.4963	4.7569	1164.65	11.1146	91.24	590.51
370	0.001383	1650.41	3.5714	4.8293	1134.39	10.3394	88.92	576.40
380	0.001412	1699.07	3.6465	4.9015	1103.86	9.5900	85.65	562.08
390	0.001443	1748.44	3.7215	4.9744	1073.39	8.8727	84.44	547.65
400	0.001476	1798.57	3.7965	5.0518	1043.17	8.1899	82.26	533.17
410	0.001513	1849.50	3.8716	5.1360	1013.34	7.5433	80.12	518.63
420	0.001552	1901.32	3.9469	5.2278	984.10	6.9344	78.01	503.96
430	0.001594	1954.08	4.0225	5.3261	955.66	6.3647	75.92	489.01
440	0.001641	2007.85	4.0984	5.4287	928.21	5.8350	73.86	473.62
450	0.001691	2062.66	4.1747	5.5323	901.99	5.3458	71.83	457.62
460	0.001746	2118.49	4.2514	5.6329	877.19	4.8972	69.83	440.92
470	0.001805	2175.29	4.3284	5.7260	854.00	4.4886	67.88	423.53
480	0.001870	2232.97	4.4055	5.8072	832.59	4.1192	66.00	405.63
490	0.001939	2291.38	4.4825	5.8719	813.12	3.7878	64.18	387.49
500	0.002014	2350.34	4.5593	5.9164	795.71	3.4927	62.45	369.50
510	0.002094	2409.63	4.6355	5.9376	780.46	3.2319	60.82	352.06
520	0.002179	2469.01	4.7108	5.9338	767.39	3.0031	59.30	335.53
530	0.002268	2528.22	4.7850	5.9049	756.51	2.8038	57.90	320.18
540	0.002361	2587.02	4.8578	5.8520	747.72	2.6310	56.62	306.20
550	0.002458	2645.19	4.9288	5.7777	740.90	2.4818	55.47	293.67
560	0.002557	2702.52	4.9981	5.6856	735.90	2.3533	54.45	282.58
570	0.002658	2758.85	5.0653	5.5795	732.51	2.2427	53.55	272.86
580	0.002761	2814.10	5.1304	5.4615	729.90	2.1437	52.76	264.39
590	0.002865	2868.06	5.1933	5.3323	729.49	2.0636	52.09	257.09
600	0.002970	2920.76	5.2540	5.2063	729.74	1.9925	51.51	250.80
610	0.003074	2972.16	5.3125	5.0727	730.72	1.9303	51.02	245.41
620	0.003177	3022.20	5.3689	4.9351	732.64	1.8772	50.62	240.79
630	0.003280	3070.88	5.4231	4.8011	735.44	1.8323	50.29	236.83
640	0.003382	3118.25	5.4753	4.6753	738.89	1.7939	50.02	233.44
650	0.003482	3164.41	5.5255	4.5583	742.76	1.7603	49.82	230.53
660	0.003582	3209.44	5.5741	4.4492	746.86	1.7304	49.66	228.01
670	0.003680	3253.42	5.6209	4.3466	751.10	1.7034	49.55	225.84
680	0.003777	3296.39	5.6663	4.2493	755.42	1.6788	49.48	223.95
690	0.003872	3338.42	5.7101	4.1568	759.80	1.6565	49.44	222.31
700	0.003966	3379.54	5.7526	4.0692	764.23	1.6361	49.44	220.88
720	0.004150	3459.30	5.8337	3.9097	773.21	1.6006	49.51	218.52
740	0.004329	3536.08	5.9103	3.7727	782.21	1.5706	49.68	216.68
760	0.004502	3610.36	5.9829	3.6581	791.10	1.5446	49.91	215.21
780	0.004671	3682.54	6.0521	3.5636	799.81	1.5217	50.20	214.02
800	0.004836	3753.02	6.1184	3.4861	808.37	1.5014	50.53	213.02

Table 4 Single-Phase Region (Regions 1 to 3 of IAPWS-IF97) – Continuation
Homogenes Zustandsgebiet (Bereiche 1 bis 3 der IAPWS-IF97) – Fortsetzung

p = 92.5 MPa

t [°C]	v [m^3 kg^{-1}]	h [kJ kg^{-1}]	s [kJ kg^{-1} K^{-1}]	c_p [kJ kg^{-1} K^{-1}]	w [m s^{-1}]	κ	$\eta \times 10^6$ [kg m^{-1} s^{-1}]	$\lambda \times 10^3$ [W m^{-1} K^{-1}]
0	0.000960	88.60	– 0.0071	3.9192	1562.09	27.49	1655.98	608.87
5	0.000961	108.25	0.0642	3.9401	1584.16	28.25	1432.13	615.56
10	0.000962	128.00	0.1345	3.9565	1603.78	28.91	1253.37	623.22
15	0.000963	147.81	0.2039	3.9695	1621.28	29.51	1108.14	631.43
20	0.000965	167.69	0.2723	3.9799	1636.91	30.03	988.36	639.87
25	0.000966	187.61	0.3397	3.9884	1650.88	30.50	888.29	648.30
30	0.000968	207.57	0.4061	3.9953	1663.32	30.90	803.72	656.56
35	0.000970	227.56	0.4715	4.0012	1674.36	31.26	731.54	664.50
40	0.000972	247.58	0.5359	4.0062	1684.09	31.56	669.40	672.04
45	0.000974	267.62	0.5994	4.0107	1692.59	31.81	615.48	679.14
50	0.000976	287.68	0.6620	4.0147	1699.92	32.01	568.39	685.76
60	0.000981	327.87	0.7844	4.0221	1711.36	32.29	490.41	697.54
70	0.000986	368.12	0.9035	4.0290	1718.86	32.41	428.93	707.46
80	0.000991	408.45	1.0193	4.0358	1722.82	32.38	379.62	715.68
90	0.000997	448.84	1.1321	4.0428	1723.60	32.22	339.47	722.43
100	0.001003	489.31	1.2420	4.0501	1721.53	31.94	306.37	727.92
110	0.001010	529.84	1.3492	4.0578	1716.92	31.56	278.78	732.32
120	0.001017	570.46	1.4539	4.0659	1710.03	31.10	255.55	735.77
130	0.001024	611.16	1.5561	4.0746	1701.10	30.55	235.81	738.36
140	0.001032	651.96	1.6561	4.0839	1690.32	29.94	218.91	740.15
150	0.001040	692.85	1.7539	4.0940	1677.87	29.27	204.32	741.19
160	0.001048	733.84	1.8496	4.1050	1663.88	28.55	191.64	741.48
170	0.001057	774.95	1.9434	4.1171	1648.47	27.79	180.55	741.04
180	0.001067	816.19	2.0355	4.1303	1631.72	26.98	170.79	739.87
190	0.001077	857.56	2.1258	4.1449	1613.72	26.15	162.15	737.95
200	0.001087	899.09	2.2145	4.1611	1594.52	25.29	154.46	735.28
210	0.001098	940.79	2.3017	4.1789	1574.18	24.40	147.57	731.86
220	0.001109	982.67	2.3875	4.1986	1552.74	23.50	141.38	727.68
230	0.001121	1024.77	2.4720	4.2203	1530.25	22.58	135.78	722.76
240	0.001134	1067.09	2.5553	4.2442	1506.75	21.65	130.70	717.09
250	0.001147	1109.66	2.6375	4.2705	1482.31	20.71	126.06	710.68
260	0.001161	1152.50	2.7186	4.2992	1456.99	19.77	121.80	703.53
270	0.001175	1195.65	2.7988	4.3305	1430.86	18.84	117.87	695.66
280	0.001190	1239.12	2.8781	4.3645	1404.03	17.90	114.24	687.06
290	0.001207	1282.95	2.9566	4.4013	1376.58	16.98	110.85	677.74
300	0.001224	1327.16	3.0344	4.4408	1348.63	16.07	107.68	667.70
310	0.001242	1371.78	3.1116	4.4831	1320.30	15.18	104.70	656.96
320	0.001261	1416.83	3.1882	4.5278	1291.71	14.31	101.88	645.53
330	0.001281	1462.34	3.2643	4.5746	1262.96	13.46	99.20	633.45
340	0.001303	1508.33	3.3399	4.6231	1234.14	12.64	96.63	620.76

Table 4 Single-Phase Region (Regions 1 to 3 of IAPWS-IF97) – Continuation
Homogenes Zustandsgebiet (Bereiche 1 bis 3 der IAPWS-IF97) – Fortsetzung

p = 92.5 MPa

t [°C]	v [m³ kg⁻¹]	h [kJ kg⁻¹]	s [kJ kg⁻¹ K⁻¹]	c_p [kJ kg⁻¹ K⁻¹]	w [m s⁻¹]	κ	$\eta \times 10^6$ [kg m⁻¹ s⁻¹]	$\lambda \times 10^3$ [W m⁻¹ K⁻¹]
350	0.001326	1554.80	3.4151	4.6721	1205.28	11.8449	94.17	607.54
360	0.001350	1601.80	3.4899	4.7292	1175.75	11.0666	91.80	593.89
370	0.001377	1649.44	3.5645	4.7983	1146.05	10.3140	89.49	579.92
380	0.001405	1697.76	3.6391	4.8669	1116.02	9.5846	87.25	565.72
390	0.001435	1746.78	3.7136	4.9356	1086.02	8.8852	85.06	551.41
400	0.001468	1796.49	3.7880	5.0078	1056.23	8.2185	82.91	537.04
410	0.001503	1846.95	3.8624	5.0863	1026.83	7.5864	80.80	522.62
420	0.001540	1898.24	3.9369	5.1717	997.98	6.9902	78.71	508.09
430	0.001581	1950.41	4.0116	5.2632	969.90	6.4314	76.66	493.33
440	0.001626	2003.52	4.0866	5.3588	942.79	5.9110	74.63	478.19
450	0.001674	2057.59	4.1619	5.4555	916.84	5.4293	72.64	462.50
460	0.001726	2112.62	4.2375	5.5497	892.24	4.9863	70.68	446.17
470	0.001783	2168.56	4.3133	5.6373	869.18	4.5817	68.76	429.20
480	0.001844	2225.33	4.3892	5.7144	847.80	4.2145	66.90	411.70
490	0.001910	2282.80	4.4650	5.7770	828.24	3.8836	65.11	393.94
500	0.001980	2340.81	4.5405	5.8217	810.62	3.5874	63.40	376.24
510	0.002056	2399.16	4.6155	5.8457	795.02	3.3242	61.78	358.99
520	0.002135	2457.65	4.6897	5.8474	781.48	3.0918	60.27	342.53
530	0.002219	2516.04	4.7628	5.8264	769.99	2.8879	58.87	327.16
540	0.002307	2574.10	4.8347	5.7834	760.52	2.7099	57.59	313.06
550	0.002399	2631.64	4.9050	5.7202	752.95	2.5552	56.42	300.35
560	0.002493	2688.45	4.9736	5.6398	747.16	2.4211	55.38	289.03
570	0.002589	2744.39	5.0404	5.5452	742.98	2.3050	54.45	279.07
580	0.002687	2799.35	5.1052	5.4380	739.62	2.2009	53.64	270.36
590	0.002786	2853.14	5.1678	5.3215	738.52	2.1162	52.93	262.81
600	0.002886	2905.79	5.2285	5.2078	738.11	2.0408	52.32	256.29
610	0.002986	2957.26	5.2871	5.0832	738.43	1.9742	51.80	250.66
620	0.003086	3007.43	5.3436	4.9506	739.71	1.9171	51.36	245.83
630	0.003185	3056.28	5.3980	4.8198	742.00	1.8690	51.00	241.69
640	0.003283	3103.85	5.4504	4.6969	745.08	1.8282	50.71	238.13
650	0.003380	3150.24	5.5009	4.5830	748.67	1.7926	50.47	235.06
660	0.003477	3195.54	5.5497	4.4770	752.54	1.7609	50.29	232.40
670	0.003572	3239.80	5.5969	4.3769	756.55	1.7322	50.15	230.10
680	0.003666	3283.09	5.6425	4.2815	760.64	1.7060	50.06	228.09
690	0.003759	3325.45	5.6867	4.1902	764.79	1.6820	50.00	226.34
700	0.003851	3366.91	5.7296	4.1029	768.98	1.6601	49.97	224.81
720	0.004030	3447.33	5.8114	3.9424	777.48	1.6216	50.01	222.27
740	0.004204	3524.75	5.8886	3.8033	786.06	1.5889	50.14	220.27
760	0.004373	3599.61	5.9617	3.6864	794.60	1.5608	50.35	218.66
780	0.004538	3672.34	6.0315	3.5899	803.07	1.5362	50.61	217.34
800	0.004700	3743.32	6.0982	3.5107	811.52	1.5149	50.92	216.23

Table 4 Single-Phase Region (Regions 1 to 3 of IAPWS-IF97) – Continuation
Homogenes Zustandsgebiet (Bereiche 1 bis 3 der IAPWS-IF97) – Fortsetzung

				p = 95 MPa				
t	v	h	s	c_p	w	κ	$\eta \times 10^6$	$\lambda \times 10^3$
[°C]	[m^3 kg^{-1}]	[kJ kg^{-1}]	[kJ kg^{-1} K^{-1}]	[kJ kg^{-1} K^{-1}]	[m s^{-1}]		[kg m^{-1} s^{-1}]	[W m^{-1} K^{-1}]
0	0.000959	90.87	– 0.0076	3.9146	1566.57	26.95	1654.54	609.99
5	0.000960	110.50	0.0636	3.9357	1588.58	27.68	1431.38	616.64
10	0.000961	130.22	0.1339	3.9524	1608.12	28.33	1253.10	624.28
15	0.000962	150.02	0.2032	3.9656	1625.56	28.91	1108.19	632.47
20	0.000964	169.87	0.2715	3.9762	1641.15	29.42	988.64	640.91
25	0.000965	189.77	0.3388	3.9847	1655.07	29.87	888.72	649.35
30	0.000967	209.72	0.4052	3.9918	1667.49	30.27	804.26	657.60
35	0.000969	229.69	0.4705	3.9977	1678.50	30.61	732.16	665.55
40	0.000971	249.69	0.5349	4.0028	1688.22	30.90	670.07	673.11
45	0.000973	269.72	0.5983	4.0073	1696.71	31.15	616.18	680.22
50	0.000975	289.76	0.6609	4.0114	1704.04	31.35	569.11	686.85
60	0.000980	329.91	0.7832	4.0188	1715.50	31.62	491.14	698.65
70	0.000985	370.14	0.9022	4.0257	1723.03	31.74	429.66	708.60
80	0.000990	410.43	1.0179	4.0325	1727.04	31.71	380.32	716.86
90	0.000996	450.79	1.1306	4.0394	1727.87	31.56	340.15	723.65
100	0.001002	491.22	1.2404	4.0466	1725.88	31.29	307.03	729.18
110	0.001009	531.72	1.3476	4.0542	1721.34	30.92	279.42	733.62
120	0.001016	572.30	1.4521	4.0622	1714.54	30.47	256.16	737.12
130	0.001023	612.97	1.5542	4.0707	1705.71	29.94	236.41	739.77
140	0.001031	653.72	1.6541	4.0798	1695.04	29.35	219.48	741.62
150	0.001039	694.56	1.7518	4.0897	1682.71	28.70	204.88	742.72
160	0.001047	735.51	1.8474	4.1005	1668.85	28.00	192.18	743.08
170	0.001056	776.58	1.9412	4.1122	1653.59	27.25	181.08	742.70
180	0.001065	817.76	2.0331	4.1252	1637.01	26.48	171.31	741.59
190	0.001075	859.09	2.1233	4.1394	1619.19	25.67	162.66	739.73
200	0.001085	900.56	2.2119	4.1552	1600.19	24.83	154.95	737.12
210	0.001096	942.19	2.2989	4.1726	1580.06	23.97	148.06	733.76
220	0.001107	984.01	2.3846	4.1918	1558.85	23.10	141.86	729.65
230	0.001119	1026.04	2.4690	4.2129	1536.60	22.21	136.26	724.79
240	0.001132	1068.28	2.5521	4.2362	1513.37	21.30	131.17	719.18
250	0.001145	1110.77	2.6341	4.2618	1489.21	20.39	126.53	712.84
260	0.001158	1153.52	2.7151	4.2897	1464.18	19.48	122.27	705.76
270	0.001173	1196.57	2.7950	4.3201	1438.36	18.57	118.34	697.96
280	0.001188	1239.93	2.8742	4.3531	1411.83	17.66	114.71	689.44
290	0.001204	1283.64	2.9525	4.3887	1384.71	16.77	111.33	680.21
300	0.001221	1327.72	3.0300	4.4269	1357.11	15.88	108.16	670.27
310	0.001238	1372.19	3.1070	4.4676	1329.13	15.02	105.19	659.64
320	0.001257	1417.08	3.1833	4.5106	1300.90	14.17	102.37	648.33
330	0.001277	1462.41	3.2591	4.5556	1272.52	13.35	99.70	636.37
340	0.001298	1508.19	3.3344	4.6020	1244.08	12.55	97.15	623.82

Table 4 Single-Phase Region (Regions 1 to 3 of IAPWS-IF97) – Continuation
Homogenes Zustandsgebiet (Bereiche 1 bis 3 der IAPWS-IF97) – Fortsetzung

p = 95 MPa								
t [°C]	v [$m^3\,kg^{-1}$]	h [$kJ\,kg^{-1}$]	s [$kJ\,kg^{-1}\,K^{-1}$]	c_p [$kJ\,kg^{-1}\,K^{-1}$]	w [$m\,s^{-1}$]	κ	$\eta \times 10^6$ [$kg\,m^{-1}\,s^{-1}$]	$\lambda \times 10^3$ [$W\,m^{-1}\,K^{-1}$]
350	0.001321	1554.45	3.4092	4.6487	1215.60	11.7746	94.70	610.73
360	0.001345	1601.20	3.4836	4.7028	1186.57	11.0185	92.34	597.21
370	0.001371	1648.56	3.5578	4.7689	1157.42	10.2875	90.06	583.37
380	0.001398	1696.57	3.6319	4.8342	1127.86	9.5771	87.83	569.29
390	0.001428	1745.24	3.7058	4.8990	1098.30	8.8947	85.66	555.08
400	0.001459	1794.56	3.7797	4.9667	1068.93	8.2434	83.54	540.82
410	0.001493	1844.59	3.8534	5.0399	1039.92	7.6248	81.45	526.51
420	0.001529	1895.38	3.9272	5.1195	1011.44	7.0407	79.40	512.11
430	0.001569	1947.00	4.0012	5.2049	983.69	6.4923	77.38	497.52
440	0.001612	1999.49	4.0753	5.2942	956.87	5.9806	75.38	482.60
450	0.001658	2052.89	4.1497	5.3847	931.17	5.5061	73.42	467.20
460	0.001708	2107.18	4.2242	5.4730	906.78	5.0688	71.49	451.20
470	0.001761	2162.33	4.2989	5.5557	883.84	4.6682	69.61	434.60
480	0.001820	2218.26	4.3737	5.6289	862.50	4.3036	67.78	417.49
490	0.001882	2274.86	4.4484	5.6893	842.88	3.9736	66.01	400.08
500	0.001949	2331.99	4.5227	5.7337	825.10	3.6770	64.32	382.67
510	0.002020	2389.48	4.5966	5.7597	809.21	3.4120	62.72	365.62
520	0.002096	2447.12	4.6697	5.7657	795.27	3.1767	61.21	349.26
530	0.002175	2504.72	4.7419	5.7510	783.28	2.9690	59.81	333.88
540	0.002258	2562.07	4.8129	5.7161	773.20	2.7866	58.52	319.70
550	0.002345	2618.98	4.8824	5.6625	764.97	2.6270	57.34	306.84
560	0.002434	2675.27	4.9504	5.5923	758.47	2.4879	56.28	295.32
570	0.002526	2730.78	5.0166	5.5081	753.56	2.3668	55.34	285.13
580	0.002619	2785.40	5.0810	5.4131	750.10	2.2615	54.50	276.19
590	0.002713	2838.99	5.1435	5.3028	748.07	2.1709	53.76	268.41
600	0.002809	2891.53	5.2040	5.2031	746.92	2.0905	53.12	261.65
610	0.002905	2943.01	5.2626	5.0896	746.49	2.0191	52.57	255.80
620	0.003001	2993.27	5.3192	4.9622	747.07	1.9577	52.10	250.76
630	0.003096	3042.25	5.3738	4.8340	748.80	1.9062	51.71	246.44
640	0.003191	3089.98	5.4263	4.7134	751.47	1.8627	51.39	242.71
650	0.003285	3136.55	5.4770	4.6026	754.77	1.8252	51.13	239.49
660	0.003379	3182.06	5.5261	4.5000	758.41	1.7919	50.92	236.69
670	0.003471	3226.57	5.5735	4.4032	762.20	1.7616	50.75	234.26
680	0.003563	3270.14	5.6195	4.3104	766.06	1.7337	50.64	232.13
690	0.003653	3312.79	5.6640	4.2208	769.96	1.7081	50.56	230.27
700	0.003743	3354.56	5.7071	4.1345	773.90	1.6845	50.51	228.64
720	0.003917	3435.61	5.7896	3.9736	781.90	1.6428	50.51	225.92
740	0.004087	3513.64	5.8674	3.8326	790.00	1.6073	50.61	223.77
760	0.004253	3589.06	5.9411	3.7134	798.14	1.5769	50.79	222.03
780	0.004414	3662.31	6.0113	3.6149	806.34	1.5507	51.02	220.58
800	0.004571	3733.78	6.0785	3.5339	814.70	1.5285	51.31	219.36

Table 4 Single-Phase Region (Regions 1 to 3 of IAPWS-IF97) – Continuation
Homogenes Zustandsgebiet (Bereiche 1 bis 3 der IAPWS-IF97) – Fortsetzung

p = 97.5 MPa

t [°C]	v [m³ kg⁻¹]	h [kJ kg⁻¹]	s [kJ kg⁻¹ K⁻¹]	c_p [kJ kg⁻¹ K⁻¹]	w [m s⁻¹]	κ	$\eta \times 10^6$ [kg m⁻¹ s⁻¹]	$\lambda \times 10^3$ [W m⁻¹ K⁻¹]
0	0.000958	93.13	− 0.0081	3.9101	1571.05	26.44	1653.20	611.09
5	0.000959	112.74	0.0630	3.9315	1592.98	27.15	1430.71	617.71
10	0.000960	132.44	0.1332	3.9484	1612.46	27.78	1252.88	625.32
15	0.000961	152.21	0.2025	3.9618	1629.84	28.34	1108.29	633.51
20	0.000963	172.05	0.2707	3.9725	1645.37	28.84	988.95	641.94
25	0.000964	191.94	0.3380	3.9812	1659.26	29.28	889.18	650.38
30	0.000966	211.86	0.4043	3.9883	1671.64	29.67	804.82	658.65
35	0.000968	231.82	0.4696	3.9943	1682.63	30.00	732.79	666.60
40	0.000970	251.80	0.5339	3.9994	1692.33	30.29	670.74	674.17
45	0.000972	271.81	0.5973	4.0040	1700.82	30.52	616.89	681.29
50	0.000974	291.84	0.6598	4.0081	1708.15	30.72	569.83	687.93
60	0.000979	331.96	0.7820	4.0155	1719.62	30.99	491.88	699.76
70	0.000984	372.15	0.9009	4.0224	1727.18	31.10	430.38	709.73
80	0.000989	412.41	1.0165	4.0292	1731.23	31.07	381.03	718.03
90	0.000995	452.73	1.1291	4.0360	1732.12	30.93	340.84	724.86
100	0.001001	493.13	1.2389	4.0432	1730.18	30.67	307.69	730.43
110	0.001008	533.60	1.3459	4.0506	1725.72	30.31	280.05	734.92
120	0.001015	574.14	1.4503	4.0585	1719.00	29.87	256.78	738.47
130	0.001022	614.77	1.5524	4.0668	1710.25	29.36	237.00	741.17
140	0.001030	655.48	1.6521	4.0758	1699.69	28.78	220.06	743.08
150	0.001038	696.29	1.7497	4.0855	1687.47	28.15	205.44	744.24
160	0.001046	737.19	1.8453	4.0960	1673.74	27.47	192.73	744.67
170	0.001055	778.21	1.9389	4.1075	1658.62	26.75	181.61	744.35
180	0.001064	819.35	2.0307	4.1201	1642.20	25.99	171.82	743.30
190	0.001074	860.62	2.1208	4.1341	1624.55	25.21	163.16	741.51
200	0.001084	902.03	2.2092	4.1494	1605.74	24.40	155.45	738.96
210	0.001095	943.61	2.2962	4.1664	1585.82	23.56	148.55	735.67
220	0.001106	985.37	2.3817	4.1851	1564.83	22.71	142.35	731.61
230	0.001118	1027.32	2.4660	4.2057	1542.83	21.85	136.74	726.81
240	0.001130	1069.49	2.5489	4.2284	1519.85	20.97	131.64	721.27
250	0.001143	1111.89	2.6308	4.2533	1495.95	20.09	127.00	714.99
260	0.001156	1154.56	2.7116	4.2804	1471.20	19.20	122.74	707.98
270	0.001170	1197.51	2.7914	4.3100	1445.67	18.31	118.81	700.25
280	0.001185	1240.77	2.8703	4.3420	1419.46	17.43	115.18	691.80
290	0.001201	1284.36	2.9484	4.3765	1392.65	16.56	111.80	682.66
300	0.001218	1328.30	3.0258	4.4134	1365.37	15.70	108.64	672.81
310	0.001235	1372.63	3.1024	4.4527	1337.73	14.86	105.67	662.28
320	0.001254	1417.36	3.1785	4.4941	1309.84	14.04	102.86	651.08
330	0.001273	1462.52	3.2540	4.5373	1281.81	13.23	100.20	639.25
340	0.001294	1508.11	3.3289	4.5817	1253.72	12.46	97.66	626.82

Table 4 Single-Phase Region (Regions 1 to 3 of IAPWS-IF97) – Continuation
Homogenes Zustandsgebiet (Bereiche 1 bis 3 der IAPWS-IF97) – Fortsetzung

p = 97.5 MPa								
t [°C]	v [m^3 kg^{-1}]	h [kJ kg^{-1}]	s [kJ kg^{-1} K^{-1}]	c_p [kJ kg^{-1} K^{-1}]	w [m s^{-1}]	κ	$\eta \times 10^6$ [kg m^{-1} s^{-1}]	$\lambda \times 10^3$ [W m^{-1} K^{-1}]
350	0.001316	1554.15	3.4034	4.6263	1225.59	11.7036	95.23	613.87
360	0.001340	1600.67	3.4775	4.6779	1197.13	10.9703	92.88	600.48
370	0.001365	1647.76	3.5513	4.7410	1168.51	10.2600	90.61	586.75
380	0.001392	1695.49	3.6249	4.8033	1139.41	9.5677	88.41	572.79
390	0.001420	1743.82	3.6983	4.8645	1110.27	8.9015	86.26	558.69
400	0.001451	1792.78	3.7716	4.9280	1081.29	8.2646	84.16	544.52
410	0.001484	1842.40	3.8448	4.9965	1052.64	7.6590	82.10	530.32
420	0.001519	1892.73	3.9179	5.0709	1024.49	7.0862	80.07	516.04
430	0.001557	1943.84	3.9911	5.1507	997.05	6.5477	78.07	501.60
440	0.001598	1995.76	4.0644	5.2342	970.50	6.0444	76.11	486.87
450	0.001642	2048.53	4.1379	5.3190	945.03	5.5769	74.18	471.72
460	0.001690	2102.13	4.2115	5.4021	920.81	5.1450	72.28	456.03
470	0.001742	2156.55	4.2853	5.4802	897.99	4.7486	70.42	439.77
480	0.001797	2211.71	4.3590	5.5499	876.70	4.3866	68.62	423.01
490	0.001857	2267.51	4.4326	5.6081	857.05	4.0580	66.88	405.94
500	0.001920	2323.82	4.5059	5.6519	839.13	3.7614	65.21	388.82
510	0.001988	2380.49	4.5787	5.6792	823.01	3.4953	63.62	371.97
520	0.002059	2437.35	4.6509	5.6884	808.74	3.2578	62.12	355.73
530	0.002135	2494.20	4.7221	5.6789	796.31	3.0470	60.72	340.37
540	0.002213	2550.86	4.7922	5.6508	785.72	2.8608	59.43	326.13
550	0.002295	2607.16	4.8610	5.6053	776.90	2.6970	58.24	313.13
560	0.002380	2662.91	4.9284	5.5439	769.76	2.5533	57.17	301.44
570	0.002467	2717.99	4.9941	5.4691	764.19	2.4277	56.20	291.04
580	0.002556	2772.26	5.0581	5.3832	760.06	2.3179	55.34	281.88
590	0.002646	2825.61	5.1202	5.2795	758.38	2.2290	54.58	273.87
600	0.002738	2877.95	5.1805	5.1909	756.30	2.1427	53.91	266.90
610	0.002830	2929.39	5.2391	5.0918	754.98	2.0656	53.33	260.84
620	0.002922	2979.72	5.2958	4.9709	754.74	1.9992	52.84	255.59
630	0.003014	3028.79	5.3504	4.8445	755.81	1.9437	52.42	251.07
640	0.003106	3076.63	5.4031	4.7255	758.02	1.8974	52.07	247.18
650	0.003197	3123.34	5.4539	4.6174	761.00	1.8579	51.78	243.81
660	0.003288	3169.01	5.5032	4.5184	764.40	1.8229	51.54	240.88
670	0.003377	3213.72	5.5508	4.4254	767.98	1.7911	51.36	238.31
680	0.003466	3257.53	5.5970	4.3360	771.62	1.7617	51.22	236.07
690	0.003554	3300.45	5.6418	4.2488	775.28	1.7344	51.11	234.10
700	0.003641	3342.51	5.6853	4.1638	778.95	1.7091	51.05	232.37
720	0.003812	3424.16	5.7683	4.0033	786.40	1.6640	51.01	229.48
740	0.003977	3502.77	5.8467	3.8607	793.97	1.6256	51.08	227.18
760	0.004139	3578.73	5.9209	3.7394	801.68	1.5927	51.23	225.30
780	0.004296	3652.48	5.9916	3.6387	809.59	1.5648	51.44	223.74
800	0.004450	3724.40	6.0593	3.5558	817.88	1.5418	51.71	222.40

Table 4 Single-Phase Region (Regions 1 to 3 of IAPWS-IF97) – Continuation
Homogenes Zustandsgebiet (Bereiche 1 bis 3 der IAPWS-IF97) – Fortsetzung

p = **100 MPa**

t [°C]	v [m³ kg⁻¹]	h [kJ kg⁻¹]	s [kJ kg⁻¹ K⁻¹]	c_p [kJ kg⁻¹ K⁻¹]	w [m s⁻¹]	κ	$\eta \times 10^6$ [kg m⁻¹ s⁻¹]	$\lambda \times 10^3$ [W m⁻¹ K⁻¹]
0	0.000957	95.39	− 0.0086	3.9057	1575.53	25.95	1651.96	612.19
5	0.000958	114.97	0.0625	3.9274	1597.38	26.64	1430.11	618.77
10	0.000959	134.65	0.1326	3.9446	1616.79	27.26	1252.71	626.36
15	0.000960	154.41	0.2018	3.9581	1634.11	27.80	1108.42	634.54
20	0.000962	174.23	0.2700	3.9689	1649.59	28.29	989.29	642.97
25	0.000964	194.10	0.3372	3.9777	1663.44	28.72	889.66	651.41
30	0.000965	214.00	0.4034	3.9849	1675.79	29.09	805.40	659.68
35	0.000967	233.94	0.4686	3.9909	1686.76	29.42	733.43	667.65
40	0.000969	253.91	0.5329	3.9961	1696.44	29.70	671.43	675.22
45	0.000971	273.90	0.5962	4.0007	1704.91	29.93	617.61	682.35
50	0.000973	293.92	0.6586	4.0048	1712.25	30.12	570.56	689.01
60	0.000978	334.00	0.7808	4.0122	1723.72	30.38	492.62	700.87
70	0.000983	374.16	0.8996	4.0191	1731.31	30.49	431.11	710.87
80	0.000988	414.38	1.0151	4.0259	1735.39	30.47	381.74	719.19
90	0.000994	454.68	1.1276	4.0327	1736.32	30.33	341.52	726.06
100	0.001000	495.04	1.2373	4.0397	1734.44	30.08	308.35	731.68
110	0.001007	535.47	1.3442	4.0471	1730.04	29.73	280.69	736.22
120	0.001014	575.98	1.4486	4.0548	1723.39	29.30	257.39	739.82
130	0.001021	616.57	1.5505	4.0630	1714.73	28.80	237.59	742.57
140	0.001028	657.24	1.6502	4.0718	1704.26	28.24	220.63	744.55
150	0.001036	698.01	1.7477	4.0813	1692.15	27.63	205.99	745.77
160	0.001045	738.87	1.8431	4.0916	1678.54	26.97	193.27	746.25
170	0.001054	779.84	1.9366	4.1028	1663.55	26.27	182.13	746.00
180	0.001063	820.93	2.0283	4.1152	1647.28	25.53	172.34	745.01
190	0.001072	862.15	2.1183	4.1288	1629.80	24.77	163.66	743.28
200	0.001083	903.51	2.2066	4.1437	1611.17	23.98	155.94	740.80
210	0.001093	945.03	2.2935	4.1603	1591.45	23.17	149.03	737.56
220	0.001104	986.72	2.3789	4.1785	1570.68	22.34	142.82	733.57
230	0.001116	1028.61	2.4630	4.1987	1548.90	21.50	137.21	728.83
240	0.001128	1070.70	2.5458	4.2207	1526.17	20.65	132.11	723.35
250	0.001141	1113.03	2.6275	4.2449	1502.53	19.79	127.46	717.13
260	0.001154	1155.61	2.7081	4.2714	1478.05	18.93	123.20	710.18
270	0.001168	1198.47	2.7878	4.3001	1452.81	18.07	119.28	702.52
280	0.001183	1241.62	2.8665	4.3311	1426.89	17.21	115.64	694.15
290	0.001198	1285.10	2.9444	4.3646	1400.39	16.36	112.27	685.09
300	0.001215	1328.92	3.0215	4.4003	1373.41	15.53	109.11	675.33
310	0.001232	1373.11	3.0980	4.4382	1346.08	14.71	106.15	664.90
320	0.001250	1417.69	3.1738	4.4781	1318.52	13.90	103.35	653.81
330	0.001270	1462.68	3.2490	4.5197	1290.81	13.12	100.69	642.09
340	0.001290	1508.09	3.3236	4.5622	1263.04	12.37	98.16	629.78

Table 4 Single-Phase Region (Regions 1 to 3 of IAPWS-IF97) – Continuation
Homogenes Zustandsgebiet (Bereiche 1 bis 3 der IAPWS-IF97) – Fortsetzung

				p = 100 MPa				
t	v	h	s	c_p	w	κ	$\eta \times 10^6$	$\lambda \times 10^3$
[°C]	[$m^3\,kg^{-1}$]	[$kJ\,kg^{-1}$]	[$kJ\,kg^{-1}\,K^{-1}$]	[$kJ\,kg^{-1}\,K^{-1}$]	[$m\,s^{-1}$]		[$kg\,m^{-1}\,s^{-1}$]	[$W\,m^{-1}\,K^{-1}$]
350	0.001312	1553.92	3.3978	4.6048	1235.23	11.6316	95.74	616.96
360	0.001335	1600.21	3.4715	4.6541	1207.43	10.9221	93.41	603.69
370	0.001359	1647.05	3.5449	4.7144	1179.33	10.2315	91.15	590.08
380	0.001385	1694.50	3.6181	4.7739	1150.67	9.5566	88.97	576.22
390	0.001413	1742.53	3.6910	4.8320	1121.93	8.9058	86.84	562.23
400	0.001443	1791.14	3.7638	4.8917	1093.31	8.2826	84.76	548.16
410	0.001475	1840.37	3.8364	4.9557	1065.00	7.6892	82.72	534.05
420	0.001509	1890.27	3.9089	5.0253	1037.16	7.1273	80.72	519.88
430	0.001546	1940.90	3.9814	5.1001	1010.00	6.5981	78.75	505.57
440	0.001586	1992.29	4.0540	5.1784	983.70	6.1028	76.82	491.02
450	0.001628	2044.47	4.1267	5.2581	958.43	5.6419	74.91	476.09
460	0.001674	2097.44	4.1994	5.3363	934.37	5.2154	73.04	460.67
470	0.001723	2151.18	4.2722	5.4102	911.66	4.8230	71.21	444.73
480	0.001776	2205.62	4.3450	5.4766	890.42	4.4639	69.44	428.30
490	0.001833	2260.68	4.4176	5.5326	870.75	4.1369	67.71	411.55
500	0.001893	2316.23	4.4899	5.5757	852.73	3.8407	66.06	394.70
510	0.001957	2372.14	4.5618	5.6038	836.42	3.5740	64.49	378.06
520	0.002025	2428.25	4.6330	5.6155	821.87	3.3348	63.00	361.95
530	0.002097	2484.39	4.7033	5.6101	809.07	3.1215	61.60	346.64
540	0.002172	2540.40	4.7726	5.5877	798.03	2.9322	60.31	332.35
550	0.002250	2596.09	4.8407	5.5490	788.69	2.7647	59.12	319.25
560	0.002330	2651.33	4.9074	5.4954	780.97	2.6172	58.03	307.39
570	0.002413	2705.96	4.9726	5.4288	774.80	2.4874	57.05	296.80
580	0.002498	2759.87	5.0361	5.3512	770.05	2.3735	56.16	287.43
590	0.002585	2812.94	5.0980	5.2687	769.69	2.2920	55.38	279.25
600	0.002672	2865.07	5.1580	5.1706	766.53	2.1988	54.69	272.03
610	0.002761	2916.39	5.2165	5.0887	764.05	2.1146	54.09	265.76
620	0.002850	2966.74	5.2731	4.9770	762.80	2.0420	53.56	260.31
630	0.002938	3015.89	5.3279	4.8523	763.05	1.9817	53.12	255.60
640	0.003026	3063.81	5.3806	4.7337	764.69	1.9321	52.74	251.54
650	0.003114	3110.60	5.4316	4.6275	767.31	1.8904	52.43	248.03
660	0.003202	3156.39	5.4809	4.5322	770.47	1.8538	52.17	244.96
670	0.003289	3201.27	5.5288	4.4435	773.84	1.8205	51.96	242.27
680	0.003376	3245.27	5.5752	4.3580	777.26	1.7896	51.80	239.91
690	0.003461	3288.43	5.6202	4.2739	780.67	1.7607	51.67	237.83
700	0.003546	3330.76	5.6640	4.1910	784.08	1.7336	51.59	236.00
720	0.003712	3412.96	5.7476	4.0317	790.93	1.6851	51.51	232.93
740	0.003874	3492.13	5.8265	3.8879	797.93	1.6435	51.55	230.49
760	0.004031	3568.61	5.9013	3.7644	805.15	1.6080	51.67	228.49
780	0.004185	3642.84	5.9724	3.6614	812.75	1.5784	51.86	226.81
800	0.004336	3715.19	6.0405	3.5762	821.00	1.5547	52.10	225.37

5 High-Temperature Region (Region 5 of IAPWS-IF97) Hochtemperaturgebiet (Bereich 5 der IAPWS-IF97)

This table contains values of the following thermodynamic properties in the so-called high-temperature region (region 5 of IAPWS-IF97) for temperatures from 800 °C to 2000 °C and for pressures from the triple-point pressure p_t = 611.657 Pa to 10 MPa:

- Specific volume v
- Specific enthalpy h
- Specific entropy s
- Specific isobaric heat capacity c_p
- Speed of sound w.

These thermodynamic properties were calculated from Eq. (3.25) of the industrial standard IAPWS-IF97. At 800 °C the determined values are slightly different to those given in Table 4 calculated from Eq. (3.8). However, these very small inconsistencies are irrelevant for industrial applications.

Diese Tafel enthält Werte der folgenden thermodynamischen Eigenschaften im sogenannten Hochtemperaturgebiet (Bereich 5 der IAPWS-IF97) für Temperaturen von 800 °C bis 2000 °C und für Drücke vom Tripelpunktsdruck p_t = 611,657 Pa bis 10 MPa:

- spezifisches Volumen v
- spezifische Enthalpie h
- spezifische Entropie s
- spezifische isobare Wärmekapazität c_p
- Schallgeschwindigkeit w.

Diese Zustandsgrößen wurden mit Gl. (3.25) des Industrie-Standards IAPWS-IF97 berechnet. Die für 800 °C ermittelten Werte unterscheiden sich geringfügig von denen in Tafel 4, die mit Gl. (3.8) berechnet wurden. Diese sehr geringen Inkonsistenzen sind ohne Bedeutung für industrielle Anwendungen.

Table 5 **High-Temperature Region** (Region 5 of IAPWS-IF97) **Hochtemperaturgebiet** (Bereich 5 der IAPWS-IF97)

	$p = p_t = 0.000611657$ MPa					$p = 0.001$ MPa				
t	v	h	s	c_p	w	v	h	s	c_p	w
[°C]	[m^3 kg^{-1}]	[kJ kg^{-1}]	[kJ kg^{-1} K^{-1}]	[kJ kg^{-1} K^{-1}]	[m s^{-1}]	[m^3 kg^{-1}]	[kJ kg^{-1}]	[kJ kg^{-1} K^{-1}]	[kJ kg^{-1} K^{-1}]	[m s^{-1}]
800	809.7448	4160.68	11.9207	2.3439	785.32	495.2857	4160.68	11.6939	2.3439	785.32
825	828.6086	4219.48	11.9749	2.3606	793.72	506.8240	4219.48	11.7480	2.3606	793.72
850	847.4725	4278.71	12.0282	2.3774	802.02	518.3622	4278.71	11.8013	2.3774	802.02
875	866.3363	4338.35	12.0807	2.3942	810.21	529.9004	4338.35	11.8539	2.3942	810.21
900	885.2001	4398.42	12.1325	2.4109	818.31	541.4386	4398.42	11.9056	2.4109	818.31
925	904.0640	4458.90	12.1835	2.4276	826.31	552.9768	4458.90	11.9566	2.4276	826.31
950	922.9278	4519.80	12.2338	2.4442	834.22	564.5150	4519.79	12.0069	2.4442	834.22
975	941.7916	4581.11	12.2834	2.4606	842.05	576.0532	4581.10	12.0566	2.4606	842.05
1000	960.6554	4642.82	12.3324	2.4769	849.80	587.5914	4642.82	12.1055	2.4769	849.80
1025	979.5192	4704.95	12.3807	2.4930	857.47	599.1296	4704.95	12.1538	2.4930	857.47
1050	998.3830	4767.47	12.4284	2.5089	865.06	610.6678	4767.47	12.2015	2.5089	865.06
1075	1017.2468	4830.39	12.4755	2.5245	872.58	622.2060	4830.39	12.2486	2.5245	872.58
1100	1036.1106	4893.70	12.5221	2.5400	880.04	633.7442	4893.70	12.2952	2.5400	880.04
1125	1054.9744	4957.39	12.5680	2.5551	887.43	645.2824	4957.39	12.3411	2.5551	887.43
1150	1073.8382	5021.45	12.6134	2.5701	894.76	656.8206	5021.45	12.3866	2.5701	894.76
1175	1092.7020	5085.89	12.6583	2.5847	902.02	668.3588	5085.89	12.4314	2.5847	902.02
1200	1111.5658	5150.68	12.7027	2.5991	909.23	679.8970	5150.68	12.4758	2.5991	909.23
1225	1130.4296	5215.84	12.7465	2.6132	916.38	691.4352	5215.84	12.5197	2.6132	916.38
1250	1149.2934	5281.34	12.7899	2.6270	923.47	702.9733	5281.34	12.5630	2.6270	923.47
1275	1168.1571	5347.19	12.8328	2.6406	930.51	714.5115	5347.19	12.6059	2.6406	930.51
1300	1187.0209	5413.37	12.8752	2.6538	937.50	726.0497	5413.37	12.6483	2.6538	937.50
1325	1205.8847	5479.87	12.9171	2.6668	944.43	737.5879	5479.87	12.6903	2.6668	944.43
1350	1224.7485	5546.70	12.9586	2.6795	951.32	749.1260	5546.70	12.7317	2.6795	951.32
1375	1243.6123	5613.85	12.9997	2.6919	958.16	760.6642	5613.85	12.7728	2.6919	958.16
1400	1262.4761	5681.30	13.0403	2.7041	964.95	772.2024	5681.30	12.8134	2.7041	964.95
1425	1281.3398	5749.05	13.0805	2.7160	971.69	783.7405	5749.05	12.8536	2.7160	971.69
1450	1300.2036	5817.10	13.1203	2.7276	978.39	795.2787	5817.09	12.8934	2.7276	978.39
1475	1319.0674	5885.43	13.1596	2.7389	985.05	806.8169	5885.43	12.9328	2.7389	985.05
1500	1337.9312	5954.04	13.1986	2.7501	991.66	818.3551	5954.04	12.9717	2.7501	991.66
1525	1356.7949	6022.93	13.2372	2.7609	998.23	829.8932	6022.93	13.0103	2.7609	998.23
1550	1375.6587	6092.08	13.2754	2.7715	1004.76	841.4314	6092.08	13.0485	2.7715	1004.76
1575	1394.5225	6161.50	13.3132	2.7819	1011.25	852.9696	6161.50	13.0863	2.7819	1011.25
1600	1413.3863	6231.18	13.3506	2.7921	1017.70	864.5077	6231.18	13.1238	2.7921	1017.70
1625	1432.2500	6301.10	13.3877	2.8020	1024.11	876.0459	6301.10	13.1608	2.8020	1024.11
1650	1451.1138	6371.27	13.4245	2.8117	1030.48	887.5840	6371.27	13.1976	2.8117	1030.48
1675	1469.9776	6441.69	13.4608	2.8212	1036.81	899.1222	6441.69	13.2339	2.8212	1036.81
1700	1488.8414	6512.33	13.4969	2.8305	1043.11	910.6604	6512.33	13.2700	2.8305	1043.11
1725	1507.7051	6583.21	13.5326	2.8397	1049.37	922.1985	6583.21	13.3057	2.8397	1049.37
1750	1526.5689	6654.32	13.5679	2.8486	1055.59	933.7367	6654.32	13.3410	2.8486	1055.59
1775	1545.4327	6725.64	13.6030	2.8574	1061.77	945.2749	6725.64	13.3761	2.8574	1061.77
1800	1564.2964	6797.19	13.6377	2.8661	1067.92	956.8130	6797.19	13.4108	2.8661	1067.92
1850	1602.0240	6940.91	13.7062	2.8829	1080.12	979.8893	6940.91	13.4793	2.8829	1080.12
1900	1639.7515	7085.47	13.7735	2.8992	1092.18	1002.9657	7085.47	13.5466	2.8992	1092.18
1950	1677.4790	7230.83	13.8396	2.9152	1104.10	1026.0420	7230.83	13.6127	2.9152	1104.10
2000	1715.2066	7376.98	13.9046	2.9307	1115.89	1049.1183	7376.98	13.6777	2.9307	1115.89

Table 5 High-Temperature Region (Region 5 of IAPWS-IF97) – Continuation
Hochtemperaturgebiet (Bereich 5 der IAPWS-IF97) – Fortsetzung

t	p = 0.01 MPa					p = 0.1 MPa				
	v	h	s	c_p	w	v	h	s	c_p	w
[°C]	[m^3 kg^{-1}]	[kJ kg^{-1}]	[kJ kg^{-1} K^{-1}]	[kJ kg^{-1} K^{-1}]	[m s^{-1}]	[m^3 kg^{-1}]	[kJ kg^{-1}]	[kJ kg^{-1} K^{-1}]	[kJ kg^{-1} K^{-1}]	[m s^{-1}]
800	49.5278	4160.64	10.6311	2.3440	785.31	4.9520	4160.23	9.5681	2.3449	785.27
825	50.6817	4219.44	10.6853	2.3607	793.72	5.0674	4219.06	9.6223	2.3616	793.69
850	51.8356	4278.67	10.7386	2.3775	802.02	5.1829	4278.31	9.6757	2.3783	801.99
875	52.9894	4338.32	10.7911	2.3942	810.21	5.2983	4337.97	9.7282	2.3950	810.19
900	54.1433	4398.38	10.8429	2.4110	818.31	5.4138	4398.06	9.7800	2.4117	818.29
925	55.2972	4458.87	10.8939	2.4277	826.31	5.5292	4458.56	9.8310	2.4283	826.30
950	56.4511	4519.76	10.9442	2.4442	834.22	5.6447	4519.47	9.8813	2.4449	834.22
975	57.6049	4581.08	10.9938	2.4607	842.05	5.7601	4580.80	9.9309	2.4613	842.05
1000	58.7588	4642.80	11.0428	2.4770	849.80	5.8755	4642.53	9.9799	2.4775	849.80
1025	59.9127	4704.92	11.0911	2.4930	857.47	5.9910	4704.67	10.0282	2.4936	857.47
1050	61.0665	4767.45	11.1388	2.5089	865.06	6.1064	4767.21	10.0760	2.5094	865.07
1075	62.2204	4830.37	11.1859	2.5246	872.59	6.2218	4830.14	10.1231	2.5250	872.60
1100	63.3742	4893.67	11.2325	2.5400	880.04	6.3372	4893.46	10.1696	2.5404	880.06
1125	64.5281	4957.36	11.2784	2.5552	887.43	6.4527	4957.16	10.2156	2.5556	887.45
1150	65.6819	5021.43	11.3238	2.5701	894.76	6.5681	5021.24	10.2610	2.5705	894.78
1175	66.8358	5085.87	11.3687	2.5847	902.02	6.6835	5085.68	10.3059	2.5851	902.05
1200	67.9896	5150.67	11.4131	2.5991	909.23	6.7989	5150.49	10.3503	2.5995	909.26
1225	69.1435	5215.82	11.4569	2.6132	916.38	6.9143	5215.66	10.3941	2.6136	916.41
1250	70.2973	5281.32	11.5003	2.6270	923.47	7.0297	5281.17	10.4375	2.6274	923.51
1275	71.4512	5347.17	11.5432	2.6406	930.51	7.1451	5347.02	10.4804	2.6409	930.55
1300	72.6050	5413.35	11.5856	2.6539	937.50	7.2606	5413.21	10.5228	2.6541	937.54
1325	73.7589	5479.86	11.6275	2.6668	944.44	7.3760	5479.73	10.5647	2.6671	944.47
1350	74.9127	5546.69	11.6690	2.6795	951.32	7.4914	5546.56	10.6062	2.6798	951.36
1375	76.0665	5613.84	11.7101	2.6920	958.16	7.6068	5613.72	10.6473	2.6922	958.20
1400	77.2204	5681.29	11.7507	2.7041	964.95	7.7222	5681.17	10.6879	2.7044	964.99
1425	78.3742	5749.04	11.7909	2.7160	971.70	7.8376	5748.93	10.7281	2.7162	971.74
1450	79.5280	5817.08	11.8307	2.7276	978.40	7.9530	5816.98	10.7679	2.7278	978.44
1475	80.6819	5885.42	11.8700	2.7390	985.05	8.0684	5885.32	10.8073	2.7392	985.10
1500	81.8357	5954.03	11.9090	2.7501	991.67	8.1838	5953.94	10.8462	2.7503	991.71
1525	82.9895	6022.92	11.9476	2.7609	998.24	8.2992	6022.83	10.8848	2.7611	998.28
1550	84.1434	6092.08	11.9858	2.7715	1004.77	8.4146	6091.99	10.9230	2.7717	1004.81
1575	85.2972	6161.49	12.0236	2.7819	1011.26	8.5300	6161.42	10.9608	2.7821	1011.30
1600	86.4510	6231.17	12.0611	2.7921	1017.70	8.6454	6231.10	10.9983	2.7922	1017.75
1625	87.6049	6301.10	12.0981	2.8020	1024.11	8.7608	6301.03	11.0354	2.8022	1024.16
1650	88.7587	6371.27	12.1349	2.8117	1030.48	8.8762	6371.20	11.0721	2.8119	1030.53
1675	89.9125	6441.68	12.1712	2.8212	1036.82	8.9916	6441.62	11.1085	2.8214	1036.87
1700	91.0663	6512.33	12.2073	2.8306	1043.11	9.1069	6512.27	11.1445	2.8307	1043.16
1725	92.2202	6583.21	12.2430	2.8397	1049.37	9.2223	6583.15	11.1802	2.8398	1049.42
1750	93.3740	6654.31	12.2783	2.8487	1055.59	9.3377	6654.26	11.2156	2.8488	1055.64
1775	94.5278	6725.64	12.3134	2.8575	1061.78	9.4531	6725.59	11.2506	2.8576	1061.83
1800	95.6817	6797.18	12.3481	2.8661	1067.93	9.5685	6797.14	11.2854	2.8662	1067.98
1850	97.9893	6940.91	12.4166	2.8829	1080.13	9.7993	6940.87	11.3539	2.8830	1080.18
1900	100.2970	7085.47	12.4839	2.8993	1092.18	10.0301	7085.44	11.4212	2.8994	1092.24
1950	102.6046	7230.83	12.5500	2.9152	1104.11	10.2609	7230.80	11.4873	2.9153	1104.16
2000	104.9122	7376.98	12.6150	2.9307	1115.90	10.4916	7376.96	11.5523	2.9308	1115.95

Table 5 High-Temperature Region (Region 5 of IAPWS-IF97) – Continuation
Hochtemperaturgebiet (Bereich 5 der IAPWS-IF97) – Fortsetzung

	$p = 0.2$ MPa					$p = 0.5$ MPa				
t [°C]	v [m^3 kg^{-1}]	h [kJ kg^{-1}]	s [kJ kg^{-1} K^{-1}]	c_p [kJ kg^{-1} K^{-1}]	w [m s^{-1}]	v [m^3 kg^{-1}]	h [kJ kg^{-1}]	s [kJ kg^{-1} K^{-1}]	c_p [kJ kg^{-1} K^{-1}]	w [m s^{-1}]
800	2.4755	4159.78	9.2479	2.3460	785.23	0.9897	4158.42	8.8240	2.3491	785.10
825	2.5333	4218.63	9.3021	2.3625	793.65	1.0128	4217.35	8.8782	2.3654	793.54
850	2.5911	4277.90	9.3555	2.3792	801.96	1.0360	4276.69	8.9317	2.3819	801.88
875	2.6488	4337.59	9.4080	2.3958	810.17	1.0591	4336.44	8.9843	2.3984	810.11
900	2.7066	4397.70	9.4598	2.4125	818.28	1.0823	4396.61	9.0361	2.4149	818.23
925	2.7643	4458.21	9.5108	2.4291	826.29	1.1054	4457.19	9.0872	2.4313	826.26
950	2.8221	4519.15	9.5612	2.4456	834.21	1.1285	4518.17	9.1376	2.4476	834.20
975	2.8798	4580.49	9.6108	2.4619	842.05	1.1517	4579.57	9.1873	2.4639	842.05
1000	2.9376	4642.24	9.6598	2.4781	849.81	1.1748	4641.37	9.2363	2.4800	849.82
1025	2.9953	4704.40	9.7082	2.4941	857.48	1.1979	4703.56	9.2847	2.4959	857.51
1050	3.0531	4766.95	9.7559	2.5100	865.09	1.2211	4766.16	9.3325	2.5116	865.13
1075	3.1108	4829.89	9.8030	2.5256	872.62	1.2442	4829.14	9.3796	2.5271	872.67
1100	3.1685	4893.22	9.8496	2.5409	880.08	1.2673	4892.51	9.4262	2.5424	880.14
1125	3.2263	4956.94	9.8955	2.5561	887.48	1.2904	4956.26	9.4722	2.5575	887.55
1150	3.2840	5021.03	9.9410	2.5709	894.81	1.3135	5020.38	9.5176	2.5723	894.89
1175	3.3417	5085.48	9.9859	2.5855	902.08	1.3366	5084.87	9.5626	2.5868	902.17
1200	3.3994	5150.30	10.0302	2.5999	909.29	1.3597	5149.72	9.6070	2.6011	909.38
1225	3.4571	5215.47	10.0741	2.6139	916.44	1.3828	5214.92	9.6509	2.6151	916.54
1250	3.5149	5280.99	10.1175	2.6277	923.54	1.4059	5280.47	9.6942	2.6288	923.65
1275	3.5726	5346.86	10.1604	2.6412	930.58	1.4291	5346.36	9.7372	2.6423	930.70
1300	3.6303	5413.05	10.2028	2.6545	937.57	1.4522	5412.58	9.7796	2.6555	937.69
1325	3.6880	5479.58	10.2447	2.6674	944.51	1.4753	5479.13	9.8216	2.6684	944.64
1350	3.7457	5546.42	10.2862	2.6801	951.40	1.4984	5546.00	9.8631	2.6810	951.53
1375	3.8034	5613.58	10.3273	2.6925	958.24	1.5215	5613.18	9.9041	2.6934	958.38
1400	3.8612	5681.05	10.3679	2.7046	965.04	1.5446	5680.67	9.9448	2.7054	965.17
1425	3.9189	5748.81	10.4081	2.7165	971.79	1.5677	5748.45	9.9850	2.7173	971.93
1450	3.9766	5816.87	10.4479	2.7281	978.49	1.5907	5816.53	10.0248	2.7288	978.63
1475	4.0343	5885.21	10.4873	2.7394	985.15	1.6138	5884.89	10.0642	2.7401	985.29
1500	4.0920	5953.84	10.5263	2.7505	991.76	1.6369	5953.53	10.1032	2.7512	991.91
1525	4.1497	6022.74	10.5649	2.7613	998.33	1.6600	6022.45	10.1418	2.7620	998.49
1550	4.2074	6091.90	10.6031	2.7719	1004.86	1.6831	6091.63	10.1800	2.7726	1005.02
1575	4.2651	6161.33	10.6409	2.7823	1011.35	1.7062	6161.07	10.2178	2.7829	1011.51
1600	4.3228	6231.02	10.6783	2.7924	1017.80	1.7293	6230.77	10.2553	2.7930	1017.96
1625	4.3805	6300.95	10.7154	2.8024	1024.22	1.7524	6300.72	10.2924	2.8029	1024.38
1650	4.4382	6371.13	10.7522	2.8121	1030.59	1.7755	6370.92	10.3291	2.8126	1030.75
1675	4.4959	6441.55	10.7885	2.8216	1036.92	1.7986	6441.35	10.3655	2.8221	1037.09
1700	4.5536	6512.21	10.8246	2.8309	1043.22	1.8217	6512.02	10.4015	2.8314	1043.38
1725	4.6113	6583.10	10.8603	2.8400	1049.48	1.8448	6582.92	10.4372	2.8405	1049.65
1750	4.6691	6654.21	10.8956	2.8490	1055.70	1.8678	6654.04	10.4726	2.8494	1055.87
1775	4.7268	6725.54	10.9307	2.8577	1061.89	1.8909	6725.39	10.5077	2.8582	1062.06
1800	4.7845	6797.09	10.9654	2.8664	1068.04	1.9140	6796.95	10.5424	2.8668	1068.21
1850	4.8999	6940.83	11.0339	2.8832	1080.24	1.9602	6940.71	10.6109	2.8836	1080.41
1900	5.0153	7085.40	11.1012	2.8995	1092.29	2.0064	7085.30	10.6782	2.8999	1092.47
1950	5.1307	7230.78	11.1674	2.9154	1104.22	2.0525	7230.69	10.7444	2.9157	1104.39
2000	5.2461	7376.94	11.2324	2.9309	1116.01	2.0987	7376.87	10.8094	2.9313	1116.19

Table 5 High-Temperature Region (Region 5 of IAPWS-IF97) – Continuation
Hochtemperaturgebiet (Bereich 5 der IAPWS-IF97) – Fortsetzung

t	p = 1 MPa					p = 2 MPa				
	v	h	s	c_p	w	v	h	s	c_p	w
[°C]	[m^3 kg^{-1}]	[kJ kg^{-1}]	[kJ kg^{-1} K^{-1}]	[kJ kg^{-1} K^{-1}]	[m s^{-1}]	[m^3 kg^{-1}]	[kJ kg^{-1}]	[kJ kg^{-1} K^{-1}]	[kJ kg^{-1} K^{-1}]	[m s^{-1}]
800	0.4944	4156.15	8.5024	2.3543	784.88	0.2467	4151.61	8.1791	2.3648	784.44
825	0.5060	4215.21	8.5568	2.3703	793.37	0.2526	4210.92	8.2337	2.3800	793.01
850	0.5176	4274.67	8.6103	2.3864	801.74	0.2584	4270.61	8.2874	2.3955	801.46
875	0.5292	4334.53	8.6630	2.4026	810.00	0.2643	4330.69	8.3404	2.4111	809.79
900	0.5408	4394.80	8.7149	2.4188	818.16	0.2701	4391.17	8.3925	2.4268	818.01
925	0.5524	4455.47	8.7661	2.4350	826.22	0.2759	4452.03	8.4438	2.4425	826.13
950	0.5640	4516.55	8.8166	2.4511	834.18	0.2818	4513.29	8.4944	2.4582	834.14
975	0.5756	4578.03	8.8663	2.4672	842.06	0.2876	4574.94	8.5443	2.4738	842.07
1000	0.5872	4639.90	8.9154	2.4831	849.85	0.2934	4636.98	8.5935	2.4893	849.91
1025	0.5988	4702.18	8.9638	2.4988	857.56	0.2992	4699.40	8.6421	2.5047	857.66
1050	0.6104	4764.84	9.0117	2.5144	865.20	0.3051	4762.21	8.6900	2.5199	865.33
1075	0.6220	4827.90	9.0589	2.5297	872.76	0.3109	4825.40	8.7373	2.5349	872.93
1100	0.6335	4891.33	9.1055	2.5449	880.24	0.3167	4888.96	8.7840	2.5498	880.45
1125	0.6451	4955.14	9.1515	2.5598	887.66	0.3225	4952.89	8.8301	2.5645	887.90
1150	0.6567	5019.32	9.1970	2.5745	895.02	0.3283	5017.18	8.8757	2.5789	895.28
1175	0.6683	5083.86	9.2420	2.5889	902.31	0.3341	5081.83	8.9207	2.5931	902.60
1200	0.6798	5148.76	9.2864	2.6031	909.54	0.3399	5146.83	8.9653	2.6071	909.86
1225	0.6914	5214.01	9.3303	2.6170	916.71	0.3457	5212.18	9.0092	2.6208	917.05
1250	0.7030	5279.60	9.3738	2.6306	923.83	0.3515	5277.87	9.0527	2.6342	924.18
1275	0.7145	5345.54	9.4167	2.6440	930.88	0.3573	5343.89	9.0957	2.6474	931.26
1300	0.7261	5411.80	9.4592	2.6571	937.89	0.3631	5410.24	9.1382	2.6604	938.28
1325	0.7377	5478.39	9.5012	2.6699	944.84	0.3689	5476.90	9.1803	2.6730	945.25
1350	0.7492	5545.30	9.5427	2.6825	951.74	0.3747	5543.89	9.2219	2.6855	952.17
1375	0.7608	5612.51	9.5838	2.6948	958.60	0.3805	5611.17	9.2630	2.6976	959.04
1400	0.7724	5680.03	9.6245	2.7068	965.40	0.3863	5678.76	9.3037	2.7095	965.86
1425	0.7839	5747.85	9.6647	2.7186	972.16	0.3920	5746.65	9.3440	2.7212	972.63
1450	0.7955	5815.96	9.7045	2.7301	978.87	0.3978	5814.82	9.3838	2.7326	979.35
1475	0.8070	5884.35	9.7439	2.7413	985.54	0.4036	5883.27	9.4233	2.7437	986.03
1500	0.8186	5953.02	9.7829	2.7523	992.16	0.4094	5952.00	9.4623	2.7546	992.66
1525	0.8301	6021.97	9.8215	2.7631	998.74	0.4152	6021.00	9.5009	2.7653	999.25
1550	0.8417	6091.18	9.8597	2.7736	1005.28	0.4210	6090.27	9.5392	2.7757	1005.79
1575	0.8532	6160.65	9.8976	2.7839	1011.77	0.4268	6159.79	9.5771	2.7859	1012.30
1600	0.8648	6230.37	9.9351	2.7940	1018.23	0.4325	6229.56	9.6146	2.7959	1018.76
1625	0.8764	6300.34	9.9722	2.8038	1024.65	0.4383	6299.58	9.6517	2.8057	1025.18
1650	0.8879	6370.56	10.0089	2.8135	1031.02	0.4441	6369.85	9.6885	2.8153	1031.57
1675	0.8995	6441.02	10.0453	2.8229	1037.36	0.4499	6440.34	9.7249	2.8247	1037.91
1700	0.9110	6511.71	10.0814	2.8322	1043.66	0.4557	6511.08	9.7610	2.8339	1044.22
1725	0.9226	6582.62	10.1171	2.8413	1049.92	0.4615	6582.04	9.7967	2.8429	1050.49
1750	0.9341	6653.77	10.1525	2.8502	1056.15	0.4672	6653.22	9.8321	2.8517	1056.72
1775	0.9457	6725.13	10.1875	2.8589	1062.34	0.4730	6724.62	9.8672	2.8604	1062.91
1800	0.9572	6796.71	10.2223	2.8675	1068.50	0.4788	6796.24	9.9020	2.8689	1069.07
1850	0.9803	6940.51	10.2908	2.8842	1080.70	0.4904	6940.10	9.9705	2.8856	1081.28
1900	1.0034	7085.13	10.3581	2.9005	1092.76	0.5019	7084.79	10.0379	2.9017	1093.35
1950	1.0265	7230.55	10.4243	2.9163	1104.69	0.5135	7230.27	10.1041	2.9175	1105.28
2000	1.0496	7376.75	10.4893	2.9318	1116.48	0.5250	7376.53	10.1691	2.9329	1117.07

Table 5 High-Temperature Region (Region 5 of IAPWS-IF97) – Continuation
Hochtemperaturgebiet (Bereich 5 der IAPWS-IF97) – Fortsetzung

t	p = 5 MPa					p = 10 MPa				
	v	h	s	c_p	w	v	h	s	c_p	w
[°C]	[m^3 kg^{-1}]	[kJ kg^{-1}]	[kJ kg^{-1} K^{-1}]	[kJ kg^{-1} K^{-1}]	[m s^{-1}]	[m^3 kg^{-1}]	[kJ kg^{-1}]	[kJ kg^{-1} K^{-1}]	[kJ kg^{-1} K^{-1}]	[m s^{-1}]
800	0.0982	4137.87	7.7459	2.3971	783.15	0.0486	4114.71	7.4086	2.4535	781.28
825	0.1005	4197.96	7.8013	2.4100	791.97	0.0499	4176.16	7.4652	2.4621	790.51
850	0.1029	4258.38	7.8557	2.4234	800.65	0.0511	4237.82	7.5208	2.4715	799.55
875	0.1053	4319.13	7.9092	2.4371	809.19	0.0523	4299.74	7.5753	2.4818	808.43
900	0.1077	4380.23	7.9618	2.4511	817.60	0.0535	4361.92	7.6289	2.4926	817.15
925	0.1101	4441.69	8.0136	2.4652	825.89	0.0548	4424.37	7.6815	2.5039	825.72
950	0.1124	4503.50	8.0647	2.4794	834.07	0.0560	4487.12	7.7334	2.5156	834.15
975	0.1148	4565.66	8.1150	2.4938	842.14	0.0572	4550.16	7.7844	2.5277	842.46
1000	0.1171	4628.18	8.1646	2.5081	850.11	0.0584	4613.50	7.8346	2.5399	850.65
1025	0.1195	4691.06	8.2135	2.5224	857.99	0.0596	4677.15	7.8841	2.5522	858.72
1050	0.1219	4754.30	8.2618	2.5366	865.77	0.0608	4741.12	7.9329	2.5647	866.69
1075	0.1242	4817.89	8.3094	2.5507	873.47	0.0620	4805.39	7.9811	2.5772	874.56
1100	0.1266	4881.83	8.3564	2.5647	881.09	0.0632	4869.98	8.0285	2.5897	882.34
1125	0.1289	4946.12	8.4028	2.5785	888.63	0.0644	4934.87	8.0754	2.6022	890.02
1150	0.1312	5010.76	8.4486	2.5922	896.10	0.0656	5000.08	8.1216	2.6146	897.62
1175	0.1336	5075.74	8.4938	2.6057	903.49	0.0668	5065.60	8.1672	2.6269	905.14
1200	0.1359	5141.05	8.5385	2.6191	910.82	0.0679	5131.43	8.2123	2.6392	912.58
1225	0.1383	5206.69	8.5827	2.6322	918.08	0.0691	5197.56	8.2568	2.6513	919.94
1250	0.1406	5272.65	8.6264	2.6451	925.28	0.0703	5263.99	8.3008	2.6632	927.24
1275	0.1429	5338.94	8.6696	2.6577	932.41	0.0715	5330.72	8.3442	2.6750	934.46
1300	0.1453	5405.54	8.7122	2.6702	939.49	0.0727	5397.74	8.3872	2.6866	941.62
1325	0.1476	5472.45	8.7544	2.6824	946.51	0.0738	5465.05	8.4296	2.6981	948.72
1350	0.1499	5539.66	8.7962	2.6944	953.47	0.0750	5532.64	8.4716	2.7093	955.75
1375	0.1523	5607.17	8.8374	2.7062	960.38	0.0762	5600.52	8.5131	2.7204	962.72
1400	0.1546	5674.96	8.8783	2.7177	967.24	0.0774	5668.66	8.5541	2.7313	969.64
1425	0.1569	5743.05	8.9187	2.7290	974.04	0.0786	5737.08	8.5947	2.7420	976.51
1450	0.1592	5811.41	8.9586	2.7400	980.80	0.0797	5805.76	8.6349	2.7525	983.31
1475	0.1616	5880.05	8.9982	2.7509	987.51	0.0809	5874.70	8.6746	2.7628	990.07
1500	0.1639	5948.95	9.0373	2.7615	994.17	0.0821	5943.90	8.7139	2.7729	996.78
1525	0.1662	6018.12	9.0760	2.7719	1000.78	0.0832	6013.34	8.7528	2.7828	1003.44
1550	0.1685	6087.54	9.1144	2.7820	1007.36	0.0844	6083.04	8.7913	2.7925	1010.05
1575	0.1709	6157.22	9.1523	2.7920	1013.88	0.0856	6152.97	8.8294	2.8020	1016.61
1600	0.1732	6227.14	9.1899	2.8017	1020.37	0.0867	6223.14	8.8671	2.8114	1023.13
1625	0.1755	6297.30	9.2271	2.8113	1026.81	0.0879	6293.54	8.9044	2.8206	1029.60
1650	0.1778	6367.70	9.2640	2.8207	1033.21	0.0891	6364.16	8.9414	2.8296	1036.03
1675	0.1802	6438.34	9.3005	2.8298	1039.57	0.0903	6435.02	8.9780	2.8384	1042.41
1700	0.1825	6509.19	9.3366	2.8388	1045.89	0.0914	6506.08	9.0142	2.8471	1048.76
1725	0.1848	6580.28	9.3724	2.8477	1052.18	0.0926	6577.37	9.0501	2.8556	1055.06
1750	0.1871	6651.58	9.4079	2.8563	1058.42	0.0937	6648.86	9.0857	2.8640	1061.32
1775	0.1894	6723.09	9.4430	2.8648	1064.63	0.0949	6720.57	9.1209	2.8722	1067.55
1800	0.1918	6794.82	9.4778	2.8732	1070.79	0.0961	6792.47	9.1558	2.8803	1073.73
1850	0.1964	6938.89	9.5465	2.8896	1083.02	0.0984	6936.89	9.2247	2.8962	1085.99
1900	0.2010	7083.76	9.6139	2.9054	1095.11	0.1007	7082.09	9.2922	2.9116	1098.09
1950	0.2057	7229.43	9.6802	2.9209	1107.05	0.1031	7228.04	9.3587	2.9267	1110.05
2000	0.2103	7375.85	9.7453	2.9361	1118.85	0.1054	7374.75	9.4239	2.9415	1121.87

6 Saturation State: Kinematic Viscosity, Prandtl Number, Static Dielectric Constant, and Surface Tension

Sättigungszustand: Kinematische Viskosität, Prandtl-Zahl, statische Dielektrizitätskonstante und Oberflächenspannung

This table contains values for the saturated liquid ($'$) and saturated vapour ($''$) of the following thermophysical properties for temperatures from 0 °C to the critical temperature $t_c = 373.946$ °C:

- Kinematic viscosity $\nu = \eta \rho^{-1}$
- Prandtl Number $Pr = \eta c_p \lambda^{-1}$
- Static dielectric constant ε .

Furthermore, values of the

- Surface tension σ

are listed (except for $t = 0$ °C which is outside the range of validity of Eq. (4.8)). For given temperatures, the necessary thermodynamic properties of the saturated liquid and saturated vapour were calculated as described in the introduction of Table 1, p. 103. All properties were exclusively determined from the corresponding equations in Chaps. 3 and 4. The *thermodynamic* properties were calculated from the industrial standard IAPWS-IF97

Diese Tafel enthält für die siedende Flüssigkeit ($'$) und den gesättigten Dampf ($''$) Werte der folgenden thermophysikalischen Eigenschaften für Temperaturen von 0 °C bis zur kritischen Temperatur $t_c = 373{,}946$ °C:

- kinematische Viskosität $\nu = \eta \rho^{-1}$
- Prandtl-Zahl $Pr = \eta c_p \lambda^{-1}$
- statische Dielektrizitätskonstante ε .

Darüber hinaus werden Werte für die

- Oberflächenspannung σ

aufgelistet (mit Ausnahme für die Temperatur $t = 0$ °C, die außerhalb des Gültigkeitsbereiches von Gl. (4.8) liegt). Die benötigten thermodynamischen Zustandsgrößen der siedenden Flüssigkeit und des gesättigten Dampfes wurden, wie in der Einführung zu Tafel 1, S. 103, beschrieben, ermittelt. Alle Zustandsgrößen wurden ausschließlich mit den Gleichungen in Kap. 3 und 4 berechnet. Die *thermodynamischen* Zustandsgrößen wurden mit dem Industrie-Standard IAPWS-IF97 bestimmt.

Table 6 Saturation State: Kinematic Viscosity, Prandtl Number, Static Dielectric Constant, and Surface Tension Sättigungszustand: Kinematische Viskosität, Prandtl-Zahl, statische Dielektrizitätskonstante und Oberflächenspannung

t	ν'	ν''	Pr'	Pr''	ε'	ε''	σ
[°C]	[10^{-6} m^2 s^{-1}]						[10^{-3} N m^{-1}]
0	1.7922	1899.7895	13.4776	1.0194	87.8993	1.0001	
0.01	1.7915	1898.5265	13.4723	1.0194	87.8953	1.0001	75.6463
5	1.5184	1372.5088	11.1915	1.0185	85.9115	1.0001	74.9417
10	1.3064	1005.8068	9.4478	1.0179	83.9705	1.0001	74.2210
15	1.1387	747.0211	8.0871	1.0174	82.0736	1.0002	73.4858
20	1.0035	561.8594	7.0052	1.0170	80.2188	1.0002	72.7361
25	0.8928	427.6438	6.1312	1.0168	78.4049	1.0003	71.9722
30	0.8009	329.1567	5.4157	1.0166	76.6307	1.0004	71.1942
35	0.7237	256.0451	4.8231	1.0164	74.8953	1.0005	70.4021
40	0.6581	201.1730	4.3271	1.0162	73.1979	1.0006	69.5963
45	0.6020	159.5605	3.9082	1.0160	71.5377	1.0007	68.7768
50	0.5535	127.6915	3.5515	1.0156	69.9139	1.0009	67.9439
55	0.5113	103.0559	3.2456	1.0153	68.3259	1.0012	67.0977
60	0.4744	83.8422	2.9813	1.0148	66.7728	1.0014	66.2383
65	0.4419	68.7305	2.7517	1.0144	65.2541	1.0017	65.3659
70	0.4131	56.7494	2.5510	1.0140	63.7690	1.0021	64.4808
75	0.3875	47.1777	2.3746	1.0137	62.3169	1.0025	63.5830
80	0.3646	39.4751	2.2189	1.0136	60.8971	1.0030	62.6729
85	0.3441	33.2335	2.0809	1.0136	59.5088	1.0036	61.7505
90	0.3257	28.1424	1.9579	1.0139	58.1515	1.0043	60.8160
95	0.3091	23.9634	1.8480	1.0146	56.8245	1.0050	59.8698
100	0.2940	20.5124	1.7494	1.0157	55.5271	1.0059	58.9119
105	0.2803	17.6463	1.6607	1.0172	54.2585	1.0069	57.9425
110	0.2678	15.2527	1.5806	1.0193	53.0183	1.0079	56.9620
115	0.2565	13.2433	1.5081	1.0221	51.8055	1.0092	55.9705
120	0.2461	11.5479	1.4423	1.0255	50.6197	1.0105	54.9682
125	0.2365	10.1105	1.3825	1.0296	49.4600	1.0121	53.9553
130	0.2277	8.8864	1.3280	1.0344	48.3259	1.0138	52.9321
135	0.2197	7.8392	1.2783	1.0400	47.2165	1.0156	51.8989
140	0.2122	6.9396	1.2328	1.0463	46.1313	1.0177	50.8559
145	0.2053	6.1637	1.1912	1.0533	45.0695	1.0200	49.8033
150	0.1990	5.4919	1.1531	1.0610	44.0304	1.0225	48.7413
155	0.1931	4.9080	1.1181	1.0694	43.0134	1.0252	47.6704
160	0.1876	4.3988	1.0861	1.0783	42.0178	1.0282	46.5907
165	0.1825	3.9532	1.0567	1.0878	41.0430	1.0315	45.5025
170	0.1778	3.5619	1.0297	1.0979	40.0881	1.0350	44.4062
175	0.1734	3.2173	1.0050	1.1086	39.1526	1.0389	43.3020
180	0.1693	2.9129	0.9823	1.1198	38.2358	1.0431	42.1902
185	0.1654	2.6431	0.9616	1.1317	37.3370	1.0476	41.0713
190	0.1618	2.4035	0.9428	1.1442	36.4556	1.0525	39.9454
195	0.1585	2.1900	0.9256	1.1574	35.5909	1.0578	38.8130

Table 6 Saturation State: Kinematic Viscosity, Prandtl Number, Static Dielectric Constant, and Surface Tension – Continuation
Sättigungszustand: Kinematische Viskosität, Prandtl-Zahl, statische Dielektrizitätskonstante und Oberflächenspannung – Fortsetzung

t	ν'	ν''	Pr'	Pr''	ε'	ε''	σ
[°C]	[10^{-6} m^2 s^{-1}]						[10^{-3} N m^{-1}]
200	0.1553	1.9993	0.9100	1.1714	34.7423	1.0636	37.6745
205	0.1524	1.8285	0.8960	1.1862	33.9092	1.0698	36.5302
210	0.1496	1.6752	0.8834	1.2020	33.0907	1.0765	35.3805
215	0.1471	1.5373	0.8722	1.2187	32.2865	1.0837	34.2258
220	0.1446	1.4130	0.8624	1.2365	31.4957	1.0915	33.0666
225	0.1423	1.3006	0.8539	1.2554	30.7179	1.0999	31.9033
230	0.1402	1.1988	0.8468	1.2755	29.9522	1.1089	30.7364
235	0.1382	1.1065	0.8409	1.2969	29.1982	1.1187	29.5664
240	0.1363	1.0225	0.8363	1.3195	28.4551	1.1292	28.3938
245	0.1345	0.9460	0.8330	1.3436	27.7223	1.1405	27.2191
250	0.1328	0.8762	0.8310	1.3692	26.9992	1.1527	26.0430
255	0.1312	0.8124	0.8304	1.3964	26.2851	1.1659	24.8660
260	0.1298	0.7540	0.8313	1.4253	25.5793	1.1802	23.6889
265	0.1284	0.7003	0.8336	1.4563	24.8810	1.1957	22.5122
270	0.1270	0.6511	0.8375	1.4896	24.1896	1.2124	21.3368
275	0.1258	0.6057	0.8431	1.5254	23.5042	1.2306	20.1634
280	0.1246	0.5639	0.8505	1.5642	22.8240	1.2505	18.9929
285	0.1235	0.5252	0.8599	1.6062	22.1480	1.2721	17.8262
290	0.1225	0.4895	0.8715	1.6519	21.4753	1.2958	16.6644
295	0.1215	0.4564	0.8856	1.7017	20.8048	1.3217	15.5084
300	0.1206	0.4257	0.9025	1.7558	20.1353	1.3504	14.3596
305	0.1198	0.3972	0.9225	1.8148	19.4656	1.3820	13.2192
310	0.1190	0.3706	0.9461	1.8790	18.7941	1.4172	12.0888
315	0.1182	0.3458	0.9740	1.9493	18.1192	1.4566	10.9698
320	0.1175	0.3226	1.0072	2.0272	17.4391	1.5010	9.8644
325	0.1169	0.3009	1.0469	2.1153	16.7514	1.5514	8.7744
330	0.1163	0.2805	1.0954	2.2174	16.0531	1.6091	7.7026
335	0.1158	0.2613	1.1563	2.3395	15.3401	1.6762	6.6518
340	0.1153	0.2432	1.2353	2.4896	14.6065	1.7552	5.6255
345	0.1149	0.2260	1.3415	2.6781	13.8443	1.8499	4.6282
350	0.1146	0.2096	1.4880	2.9181	13.0430	1.9663	3.6654
355	0.1144	0.1940	1.7195	3.2957	12.1825	2.1138	2.7448
360	0.1144	0.1787	2.1076	3.8956	11.2332	2.3106	1.8772
365	0.1145	0.1634	2.9195	5.0703	10.1290	2.5970	1.0801
370	0.1154	0.1469	5.7064	8.6789	8.6717	3.1071	0.3882
371	0.1157	0.1431	7.3374	10.6292	8.2784	3.2790	0.2692
372	0.1163	0.1388	10.5108	14.2652	7.8028	3.5125	0.1601
373	0.1173	0.1336	19.4579	23.9697	7.1525	3.8855	0.0648
373.946					5.3606	5.3606	0.0000

7 Kinematic Viscosity
Kinematische Viskosität

For the single-phase region, this table contains values of the

- Kinematic viscosity $\nu = \eta \rho^{-1}$

for temperatures from 0 °C to 800 °C and for pressures from the triple-point pressure $p_t = 611.657$ Pa to 100 MPa (regions 1 to 3 of IAPWS-IF97). The dynamic viscosity η was calculated from Eq. (4.1). For given pressure and temperature, the necessary density ρ was determined from the industrial standard IAPWS-IF97. The line in the respective columns for an isobar indicates the transition from liquid to vapour state.

Diese Tafel enthält für das homogene Zustandsgebiet Werte für die

- kinematische Viskosität $\nu = \eta \rho^{-1}$

für Temperaturen von 0 °C bis 800 °C und für Drücke vom Tripelpunktsdruck p_t = 611,657 Pa bis 100 MPa (Bereiche 1 bis 3 der IAPWS-IF97). Die dynamische Viskosität η wurde aus Gl. (4.1) berechnet. Für vorgegebene Werte von Druck und Temperatur wurde die benötigte Dichte ρ mit dem Industrie-Standard IAPWS-IF97 ermittelt. Die Linie in den jeweiligen Spalten für eine Isobare kennzeichnet den Übergang vom Flüssigkeits- zum Gaszustand.

Table 7 Kinematic Viscosity $\nu\,[10^{-6}\ \mathrm{m^2\ s^{-1}}]$ Kinematische Viskosität $\nu\,[10^{-6}\ \mathrm{m^2\ s^{-1}}]$

t [°C]	p_t	p [MPa] 0.001	0.01	0.05	0.1	0.2	0.3	0.4	0.5	0.6	0.7
0	1.7922	1.7922	1.7921	1.7920	1.7918	1.7915	1.7911	1.7908	1.7904	1.7901	1.7898
5	1958.57	1.5184	1.5184	1.5183	1.5182	1.5179	1.5177	1.5175	1.5172	1.5170	1.5167
10	2020.86	1235.60	1.3064	1.3064	1.3063	1.3061	1.3059	1.3058	1.3056	1.3054	1.3052
15	2085.25	1275.03	1.1387	1.1386	1.1386	1.1385	1.1383	1.1382	1.1381	1.1380	1.1378
20	2151.70	1315.71	1.0035	1.0035	1.0034	1.0033	1.0032	1.0031	1.0031	1.0030	1.0029
25	2220.19	1357.63	0.8928	0.8928	0.8927	0.8927	0.8926	0.8925	0.8925	0.8924	0.8923
30	2290.68	1400.77	0.8009	0.8009	0.8008	0.8008	0.8007	0.8007	0.8006	0.8006	0.8006
35	2363.15	1445.11	0.7237	0.7236	0.7236	0.7236	0.7236	0.7235	0.7235	0.7235	0.7234
40	2437.57	1490.65	0.6581	0.6581	0.6581	0.6581	0.6581	0.6580	0.6580	0.6580	0.6580
45	2513.92	1537.37	0.6020	0.6020	0.6020	0.6019	0.6019	0 .6019	0.6019	0.6019	0.6019
50	2592.19	1585.26	157.89	0.5535	0.5535	0.5535	0.5535	0.5534	0.5534	0.5534	0.5534
60	2754.42	1684.51	167.88	0.4744	0.4744	0.4744	0.4744	0.4744	0.4744	0.4744	0.4744
70	2924.15	1788.35	178.31	0.4131	0.4131	0.4131	0.4131	0.4131	0.4131	0.4131	0.4131
80	3101.29	1896.72	189.20	0.3646	0.3646	0.3646	0.3647	0.3647	0.3647	0.3647	0.3647
90	3285.78	2009.58	200.52	39.7071	0.3257	0.3257	0.3257	0.3257	0.3258	0.3258	0.3258
100	3477.55	2126.89	212.28	42.0909	20.8101	0.2940	0.2940	0.2940	0.2940	0.2941	0.2941
125	3988.53	2439.47	243.61	48.4163	24.0139	11.8064	0.2365	0.2365	0.2365	0.2366	0.2366
150	4544.05	2779.27	277.64	55.2667	27.4685	13.5660	8.9286	6.6071	0.1990	0.1990	0.1990
175	5143.49	3145.94	314.34	62.6419	31.1783	15.4444	10.1980	7.5733	5.9973	4.9456	4.1934
200	5786.31	3539.14	353.69	70.5389	35.1442	17.4457	11.5450	8.5939	6.8225	5.6410	4.7965
225	6471.94	3958.52	395.65	78.9536	39.3657	19.5710	12.9721	9.6722	7.6918	6.3711	5.4275
250	7199.81	4403.73	440.20	87.8806	43.8410	21.8207	14.4802	10.8097	8.6071	7.1385	6.0893
275	7969.33	4874.43	487.28	97.3139	48.5677	24.1943	16.0696	12.0070	9.5693	7.9441	6.7831
300	8779.92	5370.24	536.88	107.2470	53.5429	26.6907	17.7398	13.2643	10.5789	8.7885	7.5096
325	9630.96	5890.79	588.95	117.6730	58.7635	29.3087	19.4904	14.5812	11.6356	9.6718	8.2691
350	10521.85	6435.71	643.45	128.5847	64.2262	32.0469	21.3205	15.9572	12.7392	10.5939	9.0614
375	11451.96	7004.62	700.35	139.9749	69.9274	34.9037	23.2291	17.3918	13.8894	11.5544	9.8866
400	12420.66	7597.14	759.62	151.8361	75.8636	37.8773	25.2152	18.8842	15.0855	12.5531	10.7442
425	13427.35	8212.90	821.20	164.1610	82.0311	40.9661	27.2778	20.4336	16.3271	13.5895	11.6340
450	14471.41	8851.51	885.07	176.9422	88.4263	44.1683	29.4157	22.0393	17.6136	14.6631	12.5556
475	15552.23	9512.60	951.19	190.1723	95.0455	47.4822	31.6277	23.7005	18.9442	15.7733	13.5084
500	16669.22	10195.81	1019.52	203.8442	101.8853	50.9059	33.9128	25.4163	20.3183	16.9197	14.4922
525	17821.77	10900.78	1090.02	217.9506	108.9421	54.4379	36.2698	27.1858	21.7354	18.1018	15.5064
550	19009.32	11627.16	1162.66	232.4845	116.2123	58.0763	38.6976	29.0083	23.1947	19.3190	16.5507
575	20231.29	12374.59	1237.41	247.4391	123.6927	61.8195	41.1951	30.8829	24.6956	20.5708	17.6245
600	21487.14	13142.74	1314.23	262.8076	131.3797	65.6658	43.7612	32.8089	26.2375	21.8566	18.7274
625	22776.32	13931.28	1393.09	278.5834	139.2702	69.6136	46.3948	34.7854	27.8198	23.1760	19.8591
650	24098.31	14739.88	1473.95	294.7602	147.3610	73.6614	49.0949	36.8117	29.4418	24.5285	21.0190
675	25452.58	15568.24	1556.79	311.3316	155.6489	77.8076	51.8605	38.8870	31.1029	25.9135	22.2068
700	26838.65	16416.03	1641.58	328.2915	164.1309	82.0507	54.6906	41.0106	32.8026	27.3306	23.4221
725	28256.01	17282.98	1728.27	345.6341	172.8041	86.3891	57.5842	43.1817	34.5403	28.7793	24.6643
750	29704.21	18168.78	1816.86	363.3536	181.6656	90.8216	60.5404	45.3997	36.3154	30.2591	25.9333
775	31182.78	19073.16	1907.30	381.4443	190.7126	95.3468	63.5582	47.6639	38.1274	31.7697	27.2285
800	32691.27	19995.84	1999.57	399.9009	199.9424	99.9632	66.6368	49.9736	39.9757	33.3105	28.5496

Table 7 Kinematic Viscosity $\nu\,[10^{-6}\ m^2\ s^{-1}]$ – Continuation
Kinematische Viskosität $\nu\,[10^{-6}\ m^2\ s^{-1}]$ – Fortsetzung

t [°C]	p [MPa] 0.8	0.9	1	1.5	2	2.5	5	7.5	10	12.5	15
0	1.7894	1.7891	1.7887	1.7870	1.7854	1.7837	1.7754	1.7673	1.7594	1.7517	1.7442
5	1.5165	1.5163	1.5160	1.5148	1.5136	1.5125	1.5066	1.5009	1.4954	1.4900	1.4847
10	1.3051	1.3049	1.3047	1.3039	1.3030	1.3022	1.2980	1.2940	1.2900	1.2862	1.2824
15	1.1377	1.1376	1.1375	1.1369	1.1362	1.1356	1.1327	1.1298	1.1269	1.1241	1.1215
20	1.0028	1.0027	1.0026	1.0022	1.0017	1.0013	0.9992	0.9971	0.9950	0.9930	0.9911
25	0.8923	0.8922	0.8921	0.8918	0.8915	0.8912	0.8897	0.8882	0.8867	0.8853	0.8839
30	0.8005	0.8005	0.8004	0.8002	0.8000	0.7998	0.7987	0.7976	0.7966	0.7956	0.7946
35	0.7234	0.7234	0.7233	0.7232	0.7230	0.7229	0.7221	0.7214	0.7207	0.7200	0.7194
40	0.6579	0.6579	0.6579	0.6578	0.6577	0.6576	0.6571	0.6566	0.6562	0.6557	0.6553
45	0.6019	0.6019	0.6018	0.6018	0.6017	0.6016	0.6013	0.6011	0.6008	0.6005	0.6003
50	0.5534	0.5534	0.5534	0.5534	0.5533	0.5533	0.5532	0.5530	0.5529	0.5528	0.5527
60	0.4744	0.4744	0.4744	0.4744	0.4744	0.4744	0.4745	0.4745	0.4746	0.4747	0.4748
70	0.4131	0.4131	0.4131	0.4132	0.4132	0.4132	0.4134	0.4136	0.4138	0.4140	0.4142
80	0.3647	0.3647	0.3647	0.3648	0.3648	0.3649	0.3651	0.3654	0.3657	0.3659	0.3662
90	0.3258	0.3258	0.3258	0.3259	0.3259	0.3260	0.3263	0.3266	0.3269	0.3273	0.3276
100	0.2941	0.2941	0.2941	0.2942	0.2942	0.2943	0.2947	0.2950	0.2953	0.2957	0.2960
125	0.2366	0.2366	0.2366	0.2367	0.2368	0.2368	0.2372	0.2376	0.2380	0.2383	0.2387
150	0.1990	0.1990	0.1991	0.1991	0.1992	0.1993	0.1997	0.2000	0.2004	0.2008	0.2012
175	3.6282	0.1734	0.1734	0.1735	0.1735	0.1736	0.1740	0.1744	0.1748	0.1751	0.1755
200	4.1627	3.6693	3.2741	2.0837	0.1554	0.1555	0.1559	0.1562	0.1566	0.1570	0.1574
225	4.7195	4.1686	3.7277	2.4023	1.7360	1.3322	0.1427	0.1431	0.1435	0.1439	0.1442
250	5.3022	4.6899	4.1999	2.7284	1.9907	1.5462	0.1330	0.1334	0.1338	0.1341	0.1345
275	5.9122	5.2348	4.6928	3.0658	2.2511	1.7612	0.7702	0.1261	0.1265	0.1269	0.1273
300	6.5504	5.8042	5.2073	3.4159	2.5194	1.9809	0.8978	0.5258	0.1209	0.1213	0.1217
325	7.2170	6.3986	5.7440	3.7796	2.7969	2.2070	1.0235	0.6236	0.4164	0.1170	0.1175
350	7.9121	7.0182	6.3030	4.1573	3.0842	2.4401	1.1500	0.7169	0.4971	0.3610	0.2633
375	8.6357	7.6628	6.8845	4.5493	3.3817	2.6809	1.2784	0.8092	0.5730	0.4295	0.3317
400	9.3876	8.3324	7.4882	4.9558	3.6895	2.9297	1.4095	0.9020	0.6474	0.4939	0.3907
425	10.1674	9.0267	8.1142	5.3765	4.0077	3.1864	1.5437	0.9958	0.7216	0.5567	0.4466
450	10.9749	9.7456	8.7621	5.8116	4.3364	3.4513	1.6813	1.0913	0.7963	0.6193	0.5012
475	11.8097	10.4886	9.4316	6.2609	4.6755	3.7244	1.8223	1.1886	0.8719	0.6820	0.5556
500	12.6715	11.2555	10.1226	6.7241	5.0250	4.0055	1.9670	1.2879	0.9487	0.7454	0.6100
525	13.5599	12.0459	10.8347	7.2013	5.3846	4.2947	2.1153	1.3893	1.0267	0.8095	0.6650
550	14.4744	12.8596	11.5677	7.6921	5.7544	4.5919	2.2673	1.4930	1.1063	0.8746	0.7205
575	15.4148	13.6961	12.3212	8.1965	6.1342	4.8969	2.4229	1.5989	1.1873	0.9408	0.7767
600	16.3806	14.5552	13.0949	8.7142	6.5239	5.2098	2.5822	1.7070	1.2699	1.0080	0.8338
625	17.3714	15.4365	13.8887	9.2451	6.9234	5.5304	2.7452	1.8174	1.3540	1.0764	0.8917
650	18.3869	16.3398	14.7020	9.7889	7.3325	5.8587	2.9117	1.9301	1.4398	1.1460	0.9506
675	19.4268	17.2646	15.5348	10.3456	7.7511	6.1945	3.0819	2.0450	1.5271	1.2168	1.0103
700	20.4907	18.2107	16.3867	10.9149	8.1791	6.5377	3.2556	2.1622	1.6161	1.2889	1.0711
725	21.5781	19.1778	17.2575	11.4967	8.6164	6.8883	3.4328	2.2817	1.7067	1.3621	1.1328
750	22.6889	20.1655	18.1468	12.0907	9.0628	7.2462	3.6135	2.4034	1.7988	1.4366	1.1954
775	23.8226	21.1736	19.0544	12.6969	9.5183	7.6112	3.7977	2.5273	1.8926	1.5123	1.2591
800	24.9790	22.2018	19.9801	13.3151	9.9826	7.9833	3.9853	2.6534	1.9880	1.5892	1.3237

Table 7 Kinematic Viscosity $\nu\,[10^{-6}\,m^2\,s^{-1}]$ – Continuation
Kinematische Viskosität $\nu\,[10^{-6}\,m^2\,s^{-1}]$ – Fortsetzung

t [°C]	p [MPa] 17.5	20	22.5	25	27.5	30	32.5	35	37.5	40	45
0	1.7369	1.7297	1.7228	1.7160	1.7093	1.7029	1.6966	1.6905	1.6845	1.6787	1.6676
5	1.4795	1.4745	1.4696	1.4648	1.4601	1.4556	1.4512	1.4469	1.4427	1.4386	1.4307
10	1.2787	1.2751	1.2717	1.2682	1.2649	1.2617	1.2586	1.2555	1.2525	1.2496	1.2440
15	1.1188	1.1163	1.1138	1.1113	1.1090	1.1066	1.1044	1.1022	1.1001	1.0980	1.0941
20	0.9892	0.9874	0.9856	0.9839	0.9822	0.9805	0.9789	0.9774	0.9759	0.9744	0.9717
25	0.8826	0.8813	0.8800	0.8788	0.8776	0.8764	0.8753	0.8742	0.8732	0.8722	0.8703
30	0.7937	0.7928	0.7919	0.7910	0.7902	0.7894	0.7887	0.7879	0.7872	0.7865	0.7853
35	0.7187	0.7181	0.7175	0.7170	0.7164	0.7159	0.7154	0.7149	0.7145	0.7140	0.7132
40	0.6549	0.6545	0.6541	0.6538	0.6535	0.6531	0.6528	0.6526	0.6523	0.6521	0.6516
45	0.6000	0.5998	0.5996	0.5994	0.5993	0.5991	0.5990	0.5988	0.5987	0.5986	0.5985
50	0.5526	0.5525	0.5524	0.5524	0.5523	0.5523	0.5522	0.5522	0.5522	0.5522	0.5523
60	0.4748	0.4749	0.4751	0.4752	0.4753	0.4754	0.4756	0.4757	0.4759	0.4760	0.4764
70	0.4144	0.4146	0.4148	0.4151	0.4153	0.4155	0.4158	0.4160	0.4162	0.4165	0.4170
80	0.3665	0.3668	0.3671	0.3673	0.3676	0.3679	0.3682	0.3685	0.3688	0.3691	0.3698
90	0.3279	0.3282	0.3285	0.3289	0.3292	0.3295	0.3299	0.3302	0.3305	0.3309	0.3316
100	0.2964	0.2967	0.2971	0.2974	0.2978	0.2981	0.2985	0.2989	0.2992	0.2996	0.3003
125	0.2391	0.2395	0.2398	0.2402	0.2406	0.2410	0.2413	0.2417	0.2421	0.2425	0.2432
150	0.2016	0.2019	0.2023	0.2027	0.2031	0.2034	0.2038	0.2042	0.2045	0.2049	0.2056
175	0.1759	0.1763	0.1766	0.1770	0.1774	0.1777	0.1781	0.1784	0.1788	0.1792	0.1799
200	0.1577	0.1581	0.1585	0.1588	0.1592	0.1595	0.1599	0.1602	0.1606	0.1610	0.1616
225	0.1446	0.1450	0.1453	0.1457	0.1460	0.1464	0.1467	0.1471	0.1474	0.1478	0.1484
250	0.1349	0.1353	0.1356	0.1360	0.1363	0.1367	0.1370	0.1374	0.1377	0.1381	0.1387
275	0.1276	0.1280	0.1284	0.1288	0.1291	0.1295	0.1298	0.1302	0.1305	0.1309	0.1315
300	0.1222	0.1226	0.1230	0.1233	0.1237	0.1241	0.1245	0.1248	0.1252	0.1255	0.1262
325	0.1180	0.1184	0.1189	0.1193	0.1197	0.1201	0.1205	0.1208	0.1212	0.1215	0.1222
350	0.1149	0.1154	0.1159	0.1163	0.1168	0.1172	0.1176	0.1180	0.1183	0.1187	0.1194
375	0.2588	0.1989	0.1171	0.1152	0.1154	0.1157	0.1160	0.1163	0.1166	0.1170	0.1176
400	0.3161	0.2590	0.2134	0.1752	0.1425	0.1229	0.1186	0.1175	0.1171	0.1171	0.1173
425	0.3676	0.3081	0.2617	0.2244	0.1939	0.1689	0.1491	0.1352	0.1272	0.1232	0.1202
450	0.4169	0.3538	0.3047	0.2657	0.2339	0.2079	0.1864	0.1688	0.1547	0.1441	0.1313
475	0.4654	0.3980	0.3457	0.3042	0.2704	0.2427	0.2196	0.2004	0.1843	0.1709	0.1512
500	0.5136	0.4416	0.3858	0.3414	0.3054	0.2757	0.2509	0.2301	0.2124	0.1975	0.1740
525	0.5620	0.4850	0.4254	0.3780	0.3395	0.3078	0.2812	0.2587	0.2396	0.2232	0.1970
550	0.6107	0.5286	0.4651	0.4145	0.3734	0.3394	0.3110	0.2869	0.2663	0.2485	0.2198
575	0.6599	0.5725	0.5049	0.4510	0.4072	0.3710	0.3405	0.3147	0.2926	0.2735	0.2424
600	0.7097	0.6169	0.5450	0.4877	0.4411	0.4025	0.3701	0.3425	0.3189	0.2984	0.2649
625	0.7601	0.6617	0.5854	0.5247	0.4752	0.4342	0.3997	0.3704	0.3452	0.3233	0.2875
650	0.8113	0.7071	0.6263	0.5620	0.5095	0.4661	0.4295	0.3983	0.3715	0.3483	0.3100
675	0.8632	0.7531	0.6677	0.5997	0.5442	0.4982	0.4595	0.4265	0.3981	0.3734	0.3327
700	0.9158	0.7997	0.7096	0.6378	0.5792	0.5306	0.4897	0.4548	0.4247	0.3986	0.3554
725	0.9693	0.8469	0.7520	0.6763	0.6146	0.5634	0.5202	0.4834	0.4516	0.4240	0.3783
750	1.0235	0.8948	0.7950	0.7154	0.6504	0.5965	0.5510	0.5122	0.4788	0.4496	0.4014
775	1.0785	0.9434	0.8386	0.7549	0.6867	0.6300	0.5822	0.5414	0.5061	0.4755	0.4247
800	1.1344	0.9927	0.8827	0.7949	0.7233	0.6638	0.6136	0.5708	0.5338	0.5015	0.4481

Table 7 Kinematic Viscosity $\nu\,[10^{-6}\ \mathrm{m^2\ s^{-1}}]$ – Continuation
Kinematische Viskosität $\nu\,[10^{-6}\ \mathrm{m^2\ s^{-1}}]$ – Fortsetzung

t	p [MPa]										
[°C]	50	55	60	65	70	75	80	85	90	95	100
0	1.6570	1.6471	1.6377	1.6288	1.6205	1.6126	1.6053	1.5984	1.5920	1.5860	1.5804
5	1.4233	1.4163	1.4097	1.4035	1.3976	1.3921	1.3870	1.3822	1.3777	1.3736	1.3697
10	1.2388	1.2338	1.2292	1.2248	1.2207	1.2168	1.2133	1.2099	1.2068	1.2040	1.2014
15	1.0904	1.0869	1.0836	1.0805	1.0777	1.0750	1.0726	1.0703	1.0682	1.0663	1.0645
20	0.9691	0.9666	0.9643	0.9622	0.9603	0.9585	0.9568	0.9553	0.9539	0.9527	0.9516
25	0.8685	0.8668	0.8653	0.8639	0.8626	0.8614	0.8604	0.8594	0.8586	0.8578	0.8572
30	0.7841	0.7830	0.7820	0.7811	0.7803	0.7796	0.7790	0.7785	0.7781	0.7777	0.7774
35	0.7125	0.7119	0.7113	0.7108	0.7104	0.7100	0.7098	0.7095	0.7094	0.7093	0.7093
40	0.6512	0.6509	0.6507	0.6505	0.6503	0.6503	0.6502	0.6503	0.6503	0.6505	0.6507
45	0.5984	0.5983	0.5983	0.5983	0.5984	0.5985	0.5987	0.5989	0.5991	0.5994	0.5998
50	0.5524	0.5525	0.5527	0.5529	0.5531	0.5534	0.5537	0.5541	0.5545	0.5549	0.5553
60	0.4767	0.4771	0.4776	0.4780	0.4785	0.4790	0.4795	0.4800	0.4806	0.4811	0.4817
70	0.4176	0.4181	0.4187	0.4193	0.4199	0.4205	0.4211	0.4218	0.4224	0.4231	0.4238
80	0.3704	0.3711	0.3717	0.3724	0.3730	0.3737	0.3744	0.3751	0.3758	0.3766	0.3773
90	0.3323	0.3330	0.3337	0.3344	0.3351	0.3358	0.3366	0.3373	0.3380	0.3388	0.3395
100	0.3010	0.3017	0.3025	0.3032	0.3039	0.3047	0.3054	0.3062	0.3069	0.3077	0.3084
125	0.2440	0.2447	0.2454	0.2462	0.2469	0.2477	0.2484	0.2491	0.2499	0.2506	0.2513
150	0.2064	0.2071	0.2078	0.2086	0.2093	0.2100	0.2107	0.2114	0.2121	0.2128	0.2135
175	0.1806	0.1813	0.1820	0.1827	0.1834	0.1841	0.1847	0.1854	0.1861	0.1867	0.1874
200	0.1623	0.1630	0.1637	0.1643	0.1650	0.1657	0.1663	0.1669	0.1676	0.1682	0.1688
225	0.1491	0.1498	0.1504	0.1511	0.1517	0.1523	0.1529	0.1535	0.1541	0.1547	0.1553
250	0.1394	0.1400	0.1407	0.1413	0.1419	0.1425	0.1431	0.1437	0.1443	0.1448	0.1454
275	0.1322	0.1328	0.1334	0.1341	0.1347	0.1352	0.1358	0.1364	0.1369	0.1375	0.1380
300	0.1268	0.1275	0.1281	0.1287	0.1293	0.1298	0.1304	0.1310	0.1315	0.1320	0.1325
325	0.1229	0.1235	0.1241	0.1247	0.1253	0.1259	0.1264	0.1270	0.1275	0.1280	0.1285
350	0.1201	0.1207	0.1213	0.1219	0.1225	0.1230	0.1236	0.1241	0.1246	0.1251	0.1256
375	0.1183	0.1189	0.1195	0.1200	0.1206	0.1211	0.1216	0.1221	0.1226	0.1231	0.1236
400	0.1177	0.1181	0.1186	0.1191	0.1196	0.1201	0.1205	0.1210	0.1214	0.1219	0.1223
425	0.1192	0.1190	0.1191	0.1193	0.1196	0.1199	0.1203	0.1207	0.1210	0.1214	0.1218
450	0.1256	0.1230	0.1218	0.1212	0.1210	0.1210	0.1211	0.1212	0.1215	0.1217	0.1220
475	0.1391	0.1321	0.1282	0.1259	0.1246	0.1238	0.1233	0.1231	0.1229	0.1229	0.1230
500	0.1575	0.1463	0.1389	0.1341	0.1309	0.1288	0.1274	0.1264	0.1258	0.1254	0.1251
525	0.1776	0.1633	0.1529	0.1454	0.1401	0.1364	0.1336	0.1317	0.1302	0.1292	0.1284
550	0.1980	0.1814	0.1686	0.1590	0.1517	0.1461	0.1419	0.1388	0.1363	0.1345	0.1330
575	0.2185	0.1999	0.1852	0.1737	0.1647	0.1576	0.1519	0.1475	0.1440	0.1412	0.1390
600	0.2390	0.2185	0.2022	0.1891	0.1786	0.1700	0.1631	0.1575	0.1530	0.1492	0.1461
625	0.2595	0.2373	0.2194	0.2049	0.1930	0.1832	0.1751	0.1684	0.1629	0.1582	0.1543
650	0.2800	0.2561	0.2367	0.2208	0.2077	0.1968	0.1876	0.1800	0.1735	0.1680	0.1633
675	0.3007	0.2750	0.2542	0.2370	0.2227	0.2107	0.2005	0.1919	0.1846	0.1783	0.1729
700	0.3214	0.2941	0.2718	0.2533	0.2378	0.2248	0.2137	0.2042	0.1961	0.1890	0.1829
725	0.3423	0.3133	0.2895	0.2697	0.2531	0.2391	0.2271	0.2168	0.2079	0.2001	0.1933
750	0.3633	0.3326	0.3073	0.2863	0.2686	0.2535	0.2406	0.2295	0.2198	0.2114	0.2040
775	0.3845	0.3520	0.3253	0.3030	0.2842	0.2681	0.2544	0.2424	0.2320	0.2229	0.2148
800	0.4058	0.3716	0.3434	0.3198	0.2999	0.2829	0.2682	0.2555	0.2443	0.2346	0.2259

8 Prandtl Number

Prandtl-Zahl

For the single-phase region, this table contains values of the

- Prandtl Number $Pr = \eta\, c_p\, \lambda^{-1}$

for temperatures from 0 °C to 800 °C and for pressures from the triple-point pressure p_t = 611.657 Pa to 100 MPa (regions 1 to 3 of IAPWS-IF97). All properties were exclusively determined from the corresponding equations in Chaps. 3 and 4. For given pressure and temperature, the necessary *thermodynamic* properties were calculated from the industrial standard IAPWS-IF97. The line in the respective columns for an isobar indicates the transition from liquid to vapour state.

Für das homogene Zustandsgebiet enthält diese Tafel Werte für die

- Prandtl-Zahl $Pr = \eta\, c_p\, \lambda^{-1}$

für Temperaturen von 0 °C bis 800 °C und für Drücke vom Tripelpunktsdruck p_t = 611,657 Pa bis 100 MPa (Bereiche 1 bis 3 der IAPWS-IF97). Alle Zustandsgrößen wurden ausschließlich mit den Gleichungen in Kap. 3 und 4 ermittelt. Für vorgegebene Werte von Druck und Temperatur wurden die benötigten *thermodynamischen* Zustandsgrößen mit dem Industrie-Standard IAPWS-IF97 berechnet. Die Linie in den jeweiligen Spalten für eine Isobare kennzeichnet den Übergang vom Flüssigkeits- zum Gaszustand.

Table 8 Prandtl Number *Pr*
Prandtl-Zahl *Pr*

t [°C]	p_t	p [MPa] 0.001	0.01	0.05	0.1	0.2	0.3	0.4	0.5	0.6	0.7
0	13.4776	13.4775	13.4771	13.4752	13.4728	13.4680	13.4632	13.4584	13.4536	13.4488	13.4440
5	1.0110	11.1915	11.1912	11.1898	11.1881	11.1848	11.1814	11.1780	11.1747	11.1713	11.1679
10	1.0063	1.0124	9.4476	9.4466	9.4454	9.4429	9.4405	9.4381	9.4356	9.4332	9.4308
15	1.0033	1.0065	8.0869	8.0862	8.0853	8.0835	8.0817	8.0799	8.0781	8.0763	8.0746
20	1.0010	1.0029	7.0051	7.0045	7.0038	7.0025	7.0011	6.9998	6.9984	6.9971	6.9957
25	0.9990	1.0002	6.1312	6.1307	6.1302	6.1292	6.1281	6.1271	6.1261	6.1250	6.1240
30	0.9971	0.9981	5.4157	5.4154	5.4149	5.4141	5.4133	5.4125	5.4117	5.4109	5.4101
35	0.9954	0.9962	4.8230	4.8228	4.8225	4.8218	4.8212	4.8205	4.8199	4.8192	4.8186
40	0.9938	0.9944	4.3271	4.3269	4.3266	4.3261	4.3256	4.3251	4.3246	4.3240	4.3235
45	0.9922	0.9927	3.9082	3.9080	3.9078	3.9074	3.9070	3.9066	3.9062	3.9058	3.9053
50	0.9906	0.9911	1.0075	3.5514	3.5512	3.5509	3.5506	3.5502	3.5499	3.5495	3.5492
60	0.9877	0.9880	0.9970	2.9812	2.9811	2.9809	2.9807	2.9804	2.9802	2.9800	2.9797
70	0.9848	0.9851	0.9912	2.5509	2.5508	2.5507	2.5505	2.5504	2.5502	2.5500	2.5499
80	0.9821	0.9823	0.9868	2.2189	2.2189	2.2187	2.2186	2.2185	2.2184	2.2183	2.2181
90	0.9796	0.9797	0.9831	1.0011	1.9579	1.9578	1.9577	1.9576	1.9575	1.9574	1.9573
100	0.9771	0.9772	0.9798	0.9926	1.0148	1.7493	1.7493	1.7492	1.7491	1.7491	1.7490
125	0.9713	0.9714	0.9728	0.9795	0.9889	1.0153	1.3825	1.3824	1.3824	1.3823	1.3823
150	0.9659	0.9660	0.9668	0.9706	0.9760	0.9887	1.0049	1.0295	1.1531	1.1530	1.1530
175	0.9609	0.9609	0.9614	0.9638	0.9671	0.9748	0.9839	0.9946	1.0072	1.0226	1.0428
200	0.9560	0.9561	0.9564	0.9580	0.9602	0.9653	0.9713	0.9780	0.9856	0.9939	1.0031
225	0.9514	0.9514	0.9517	0.9529	0.9545	0.9581	0.9622	0.9669	0.9720	0.9776	0.9837
250	0.9469	0.9469	0.9471	0.9481	0.9494	0.9521	0.9552	0.9586	0.9623	0.9663	0.9705
275	0.9426	0.9426	0.9428	0.9436	0.9446	0.9469	0.9493	0.9519	0.9547	0.9577	0.9609
300	0.9384	0.9384	0.9386	0.9393	0.9402	0.9421	0.9441	0.9462	0.9484	0.9507	0.9532
325	0.9343	0.9343	0.9345	0.9351	0.9359	0.9375	0.9392	0.9410	0.9428	0.9447	0.9467
350	0.9304	0.9304	0.9306	0.9311	0.9318	0.9332	0.9347	0.9362	0.9378	0.9393	0.9410
375	0.9266	0.9267	0.9268	0.9273	0.9279	0.9291	0.9304	0.9317	0.9331	0.9344	0.9358
400	0.9230	0.9230	0.9231	0.9236	0.9241	0.9252	0.9263	0.9275	0.9286	0.9298	0.9310
425	0.9195	0.9195	0.9196	0.9200	0.9205	0.9215	0.9225	0.9235	0.9245	0.9255	0.9266
450	0.9162	0.9162	0.9163	0.9166	0.9171	0.9179	0.9188	0.9197	0.9206	0.9215	0.9224
475	0.9130	0.9130	0.9131	0.9134	0.9138	0.9146	0.9153	0.9161	0.9169	0.9177	0.9184
500	0.9100	0.9100	0.9100	0.9103	0.9106	0.9113	0.9120	0.9127	0.9134	0.9140	0.9147
525	0.9071	0.9071	0.9071	0.9074	0.9076	0.9082	0.9088	0.9094	0.9100	0.9106	0.9112
550	0.9043	0.9043	0.9043	0.9045	0.9048	0.9053	0.9058	0.9063	0.9068	0.9073	0.9078
575	0.9016	0.9016	0.9016	0.9018	0.9020	0.9025	0.9029	0.9034	0.9038	0.9042	0.9047
600	0.8990	0.8990	0.8991	0.8992	0.8994	0.8998	0.9002	0.9005	0.9009	0.9013	0.9017
625	0.8966	0.8966	0.8966	0.8967	0.8969	0.8972	0.8975	0.8978	0.8982	0.8985	0.8988
650	0.8942	0.8942	0.8942	0.8943	0.8945	0.8947	0.8950	0.8953	0.8955	0.8958	0.8961
675	0.8919	0.8919	0.8919	0.8920	0.8921	0.8923	0.8926	0.8928	0.8930	0.8933	0.8935
700	0.8897	0.8897	0.8897	0.8898	0.8898	0.8900	0.8902	0.8904	0.8906	0.8908	0.8910
725	0.8875	0.8875	0.8875	0.8876	0.8877	0.8878	0.8880	0.8881	0.8883	0.8885	0.8886
750	0.8854	0.8854	0.8854	0.8855	0.8855	0.8857	0.8858	0.8860	0.8861	0.8862	0.8864
775	0.8834	0.8834	0.8834	0.8834	0.8835	0.8836	0.8837	0.8839	0.8840	0.8841	0.8842
800	0.8814	0.8814	0.8814	0.8815	0.8815	0.8816	0.8817	0.8818	0.8819	0.8820	0.8822

Table 8 Prandtl Number *Pr* – Continuation
Prandtl-Zahl *Pr* – Fortsetzung

t [°C]	*p* [MPa] 0.8	0.9	1	1.5	2	2.5	5	7.5	10	12.5	15
0	13.4392	13.4344	13.4296	13.4059	13.3822	13.3586	13.2427	13.1296	13.0195	12.9121	12.8075
5	11.1646	11.1612	11.1579	11.1412	11.1246	11.1081	11.0266	10.9470	10.8694	10.7936	10.7196
10	9.4284	9.4260	9.4235	9.4115	9.3995	9.3876	9.3287	9.2711	9.2149	9.1599	9.1061
15	8.0728	8.0710	8.0692	8.0603	8.0514	8.0426	7.9990	7.9564	7.9146	7.8738	7.8338
20	6.9944	6.9930	6.9917	6.9850	6.9783	6.9716	6.9387	6.9064	6.8748	6.8439	6.8136
25	6.1229	6.1219	6.1209	6.1157	6.1106	6.1054	6.0801	6.0553	6.0310	6.0072	5.9838
30	5.4093	5.4085	5.4077	5.4036	5.3996	5.3956	5.3759	5.3565	5.3375	5.3188	5.3005
35	4.8180	4.8173	4.8167	4.8135	4.8103	4.8072	4.7915	4.7762	4.7611	4.7463	4.7318
40	4.3230	4.3225	4.3220	4.3194	4.3169	4.3144	4.3018	4.2895	4.2775	4.2656	4.2540
45	3.9049	3.9045	3.9041	3.9020	3.9000	3.8979	3.8878	3.8778	3.8680	3.8584	3.8490
50	3.5489	3.5485	3.5482	3.5465	3.5448	3.5431	3.5349	3.5267	3.5187	3.5108	3.5031
60	2.9795	2.9793	2.9790	2.9779	2.9767	2.9756	2.9699	2.9643	2.9588	2.9534	2.9481
70	2.5497	2.5495	2.5494	2.5485	2.5477	2.5469	2.5429	2.5390	2.5351	2.5313	2.5275
80	2.2180	2.2179	2.2178	2.2172	2.2166	2.2160	2.2131	2.2102	2.2074	2.2046	2.2019
90	1.9572	1.9572	1.9571	1.9566	1.9562	1.9557	1.9535	1.9514	1.9493	1.9472	1.9451
100	1.7489	1.7488	1.7488	1.7484	1.7481	1.7477	1.7460	1.7443	1.7427	1.7411	1.7395
125	1.3823	1.3822	1.3822	1.3819	1.3817	1.3815	1.3804	1.3793	1.3782	1.3772	1.3762
150	1.1530	1.1529	1.1529	1.1527	1.1525	1.1523	1.1513	1.1503	1.1494	1.1485	1.1476
175	1.0709	1.0050	1.0049	1.0047	1.0044	1.0042	1.0030	1.0019	1.0008	0.9997	0.9987
200	1.0133	1.0246	1.0374	1.1497	0.9097	0.9094	0.9077	0.9061	0.9046	0.9031	0.9017
225	0.9902	0.9971	1.0046	1.0495	1.1161	1.2378	0.8514	0.8489	0.8465	0.8442	0.8421
250	0.9751	0.9799	0.9850	1.0147	1.0513	1.0969	0.8292	0.8248	0.8208	0.8170	0.8134
275	0.9642	0.9677	0.9714	0.9924	1.0177	1.0472	1.3054	0.8378	0.8299	0.8228	0.8162
300	0.9557	0.9584	0.9612	0.9767	0.9949	1.0158	1.1661	1.4732	0.8921	0.8757	0.8616
325	0.9487	0.9508	0.9530	0.9649	0.9785	0.9937	1.0964	1.2585	1.5473	1.0375	0.9936
350	0.9427	0.9444	0.9461	0.9555	0.9660	0.9776	1.0516	1.1559	1.3051	1.5386	1.9998
375	0.9372	0.9386	0.9401	0.9478	0.9561	0.9651	1.0204	1.0936	1.1875	1.3106	1.4810
400	0.9322	0.9334	0.9347	0.9411	0.9479	0.9551	0.9977	1.0514	1.1167	1.1953	1.2909
425	0.9276	0.9286	0.9297	0.9351	0.9408	0.9467	0.9804	1.0212	1.0690	1.1241	1.1871
450	0.9233	0.9242	0.9251	0.9297	0.9345	0.9395	0.9667	0.9984	1.0346	1.0751	1.1200
475	0.9192	0.9200	0.9208	0.9248	0.9289	0.9331	0.9554	0.9805	1.0084	1.0391	1.0724
500	0.9154	0.9161	0.9168	0.9202	0.9237	0.9273	0.9458	0.9660	0.9878	1.0114	1.0366
525	0.9118	0.9124	0.9130	0.9159	0.9189	0.9220	0.9374	0.9537	0.9710	0.9892	1.0084
550	0.9084	0.9089	0.9094	0.9119	0.9145	0.9170	0.9299	0.9432	0.9568	0.9709	0.9856
575	0.9051	0.9056	0.9060	0.9082	0.9103	0.9125	0.9232	0.9339	0.9447	0.9556	0.9666
600	0.9020	0.9024	0.9028	0.9046	0.9064	0.9083	0.9171	0.9257	0.9341	0.9425	0.9507
625	0.8991	0.8994	0.8997	0.9013	0.9028	0.9043	0.9116	0.9184	0.9250	0.9312	0.9373
650	0.8963	0.8966	0.8969	0.8982	0.8994	0.9007	0.9066	0.9120	0.9170	0.9216	0.9259
675	0.8937	0.8939	0.8941	0.8952	0.8963	0.8973	0.9021	0.9063	0.9100	0.9133	0.9162
700	0.8912	0.8914	0.8916	0.8925	0.8933	0.8942	0.8981	0.9013	0.9040	0.9062	0.9080
725	0.8888	0.8889	0.8891	0.8899	0.8906	0.8913	0.8944	0.8969	0.8988	0.9002	0.9011
750	0.8865	0.8866	0.8868	0.8874	0.8880	0.8886	0.8911	0.8930	0.8943	0.8950	0.8953
775	0.8843	0.8844	0.8845	0.8851	0.8856	0.8861	0.8882	0.8896	0.8904	0.8907	0.8905
800	0.8823	0.8824	0.8825	0.8829	0.8834	0.8838	0.8855	0.8866	0.8871	0.8871	0.8865

Table 8 Prandtl Number *Pr* – Continuation
Prandtl-Zahl *Pr* – Fortsetzung

t [°C]	*p* [MPa] 17.5	20	22.5	25	27.5	30	32.5	35	37.5	40	45
0	12.7056	12.6064	12.5097	12.4156	12.3239	12.2347	12.1479	12.0633	11.9811	11.9011	11.7475
5	10.6474	10.5770	10.5083	10.4413	10.3759	10.3122	10.2501	10.1896	10.1306	10.0730	9.9624
10	9.0536	9.0023	8.9522	8.9032	8.8555	8.8088	8.7632	8.7188	8.6754	8.6331	8.5515
15	7.7948	7.7565	7.7192	7.6826	7.6469	7.6120	7.5778	7.5445	7.5119	7.4801	7.4187
20	6.7840	6.7549	6.7265	6.6987	6.6715	6.6448	6.6188	6.5933	6.5684	6.5441	6.4970
25	5.9609	5.9385	5.9165	5.8950	5.8739	5.8533	5.8331	5.8133	5.7940	5.7751	5.7385
30	5.2826	5.2650	5.2478	5.2309	5.2143	5.1981	5.1823	5.1667	5.1515	5.1366	5.1077
35	4.7176	4.7036	4.6899	4.6765	4.6634	4.6505	4.6379	4.6255	4.6134	4.6015	4.5785
40	4.2425	4.2313	4.2203	4.2096	4.1990	4.1886	4.1785	4.1685	4.1588	4.1492	4.1307
45	3.8397	3.8307	3.8217	3.8130	3.8044	3.7960	3.7878	3.7797	3.7717	3.7640	3.7489
50	3.4956	3.4881	3.4808	3.4737	3.4666	3.4598	3.4530	3.4464	3.4399	3.4335	3.4212
60	2.9430	2.9379	2.9328	2.9279	2.9231	2.9184	2.9138	2.9092	2.9048	2.9004	2.8919
70	2.5239	2.5203	2.5167	2.5133	2.5099	2.5065	2.5032	2.5000	2.4969	2.4938	2.4879
80	2.1992	2.1966	2.1940	2.1915	2.1891	2.1866	2.1843	2.1819	2.1797	2.1774	2.1731
90	1.9431	1.9412	1.9393	1.9374	1.9355	1.9337	1.9320	1.9302	1.9285	1.9269	1.9237
100	1.7380	1.7364	1.7350	1.7335	1.7321	1.7307	1.7293	1.7280	1.7267	1.7254	1.7230
125	1.3752	1.3743	1.3733	1.3724	1.3715	1.3706	1.3698	1.3689	1.3681	1.3673	1.3658
150	1.1467	1.1459	1.1451	1.1443	1.1435	1.1427	1.1420	1.1413	1.1406	1.1399	1.1386
175	0.9977	0.9968	0.9958	0.9949	0.9941	0.9932	0.9924	0.9916	0.9908	0.9900	0.9885
200	0.9004	0.8991	0.8978	0.8966	0.8954	0.8943	0.8932	0.8921	0.8911	0.8901	0.8882
225	0.8401	0.8381	0.8362	0.8345	0.8328	0.8311	0.8296	0.8281	0.8266	0.8252	0.8226
250	0.8101	0.8069	0.8040	0.8012	0.7985	0.7960	0.7936	0.7913	0.7892	0.7871	0.7832
275	0.8102	0.8047	0.7996	0.7948	0.7904	0.7862	0.7823	0.7787	0.7753	0.7720	0.7660
300	0.8491	0.8381	0.8282	0.8193	0.8112	0.8039	0.7971	0.7909	0.7851	0.7797	0.7701
325	0.9592	0.9315	0.9085	0.8890	0.8722	0.8575	0.8445	0.8330	0.8225	0.8131	0.7967
350	1.3802	1.2129	1.1191	1.0558	1.0090	0.9726	0.9432	0.9187	0.8977	0.8795	0.8493
375	1.7519	2.3382	9.3822	1.9305	1.4815	1.2897	1.1780	1.1029	1.0480	1.0058	0.9439
400	1.4102	1.5701	1.8144	2.2559	3.3367	3.4204	2.2023	1.6885	1.4367	1.2878	1.1177
425	1.2597	1.3440	1.4440	1.5702	1.7413	1.9743	2.2458	2.3954	2.2304	1.9238	1.4751
450	1.1696	1.2244	1.2851	1.3526	1.4285	1.5163	1.6191	1.7311	1.8334	1.8809	1.7733
475	1.1082	1.1467	1.1879	1.2322	1.2796	1.3302	1.3839	1.4411	1.5007	1.5575	1.6171
500	1.0632	1.0912	1.1206	1.1513	1.1836	1.2172	1.2522	1.2881	1.3243	1.3602	1.4255
525	1.0284	1.0492	1.0706	1.0926	1.1151	1.1383	1.1620	1.1862	1.2107	1.2350	1.2809
550	1.0006	1.0160	1.0317	1.0476	1.0637	1.0798	1.0961	1.1125	1.1289	1.1454	1.1775
575	0.9779	0.9892	1.0006	1.0120	1.0234	1.0347	1.0459	1.0570	1.0680	1.0788	1.1001
600	0.9590	0.9671	0.9752	0.9832	0.9911	0.9988	1.0064	1.0138	1.0210	1.0281	1.0417
625	0.9431	0.9488	0.9543	0.9597	0.9648	0.9698	0.9747	0.9793	0.9838	0.9881	0.9964
650	0.9299	0.9336	0.9371	0.9404	0.9434	0.9463	0.9490	0.9516	0.9540	0.9562	0.9604
675	0.9187	0.9209	0.9229	0.9245	0.9260	0.9272	0.9283	0.9292	0.9300	0.9306	0.9317
700	0.9094	0.9104	0.9112	0.9116	0.9118	0.9118	0.9116	0.9112	0.9107	0.9102	0.9088
725	0.9016	0.9017	0.9016	0.9011	0.9003	0.8994	0.8982	0.8969	0.8955	0.8940	0.8908
750	0.8952	0.8946	0.8937	0.8926	0.8911	0.8894	0.8876	0.8856	0.8835	0.8813	0.8768
775	0.8899	0.8888	0.8875	0.8858	0.8838	0.8817	0.8793	0.8768	0.8742	0.8715	0.8661
800	0.8856	0.8842	0.8825	0.8805	0.8782	0.8757	0.8730	0.8701	0.8672	0.8642	0.8582

Table 8 Prandtl Number *Pr* – Continuation
Prandtl-Zahl *Pr* – Fortsetzung

t	*p* [MPa]										
[°C]	50	55	60	65	70	75	80	85	90	95	100
0	11.6023	11.4651	11.3355	11.2133	11.0981	10.9897	10.8877	10.7919	10.7021	10.6179	10.5393
5	9.8575	9.7581	9.6639	9.5748	9.4905	9.4109	9.3358	9.2650	9.1985	9.1359	9.0772
10	8.4739	8.4001	8.3301	8.2637	8.2007	8.1411	8.0847	8.0314	7.9811	7.9336	7.8890
15	7.3601	7.3044	7.2513	7.2009	7.1530	7.1075	7.0644	7.0235	6.9849	6.9484	6.9140
20	6.4521	6.4092	6.3684	6.3294	6.2924	6.2572	6.2238	6.1920	6.1620	6.1335	6.1066
25	5.7035	5.6700	5.6381	5.6077	5.5787	5.5510	5.5248	5.4998	5.4761	5.4537	5.4324
30	5.0801	5.0537	5.0285	5.0044	4.9814	4.9595	4.9386	4.9188	4.8999	4.8820	4.8651
35	4.5564	4.5354	4.5152	4.4959	4.4775	4.4599	4.4432	4.4273	4.4121	4.3978	4.3841
40	4.1129	4.0959	4.0796	4.0641	4.0492	4.0350	4.0215	4.0086	3.9963	3.9847	3.9737
45	3.7345	3.7206	3.7074	3.6947	3.6826	3.6710	3.6600	3.6495	3.6395	3.6301	3.6211
50	3.4094	3.3980	3.3871	3.3768	3.3668	3.3573	3.3483	3.3397	3.3315	3.3237	3.3163
60	2.8838	2.8760	2.8686	2.8614	2.8546	2.8481	2.8419	2.8360	2.8304	2.8251	2.8201
70	2.4821	2.4767	2.4714	2.4664	2.4616	2.4571	2.4527	2.4486	2.4447	2.4410	2.4374
80	2.1690	2.1650	2.1613	2.1576	2.1542	2.1509	2.1478	2.1449	2.1421	2.1394	2.1369
90	1.9206	1.9177	1.9149	1.9122	1.9097	1.9072	1.9049	1.9028	1.9007	1.8987	1.8969
100	1.7206	1.7183	1.7162	1.7142	1.7122	1.7104	1.7086	1.7069	1.7054	1.7039	1.7025
125	1.3643	1.3629	1.3616	1.3603	1.3591	1.3579	1.3568	1.3558	1.3543	1.3539	1.3530
150	1.1373	1.1361	1.1350	1.1339	1.1328	1.1318	1.1308	1.1299	1.1290	1.1281	1.1273
175	0.9871	0.9858	0.9845	0.9833	0.9821	0.9810	0.9799	0.9788	0.9773	0.9768	0.9759
200	0.8864	0.8847	0.8831	0.8816	0.8801	0.8787	0.8773	0.8760	0.8747	0.8735	0.8723
225	0.8201	0.8178	0.8156	0.8135	0.8116	0.8097	0.8080	0.8063	0.8045	0.8030	0.8015
250	0.7796	0.7763	0.7733	0.7704	0.7678	0.7653	0.7629	0.7606	0.7585	0.7565	0.7545
275	0.7607	0.7558	0.7513	0.7472	0.7435	0.7400	0.7367	0.7337	0.7303	0.7281	0.7255
300	0.7616	0.7541	0.7473	0.7413	0.7358	0.7308	0.7262	0.7220	0.7180	0.7144	0.7109
325	0.7828	0.7709	0.7605	0.7514	0.7432	0.7359	0.7293	0.7233	0.7173	0.7128	0.7081
350	0.8255	0.8060	0.7897	0.7758	0.7637	0.7531	0.7437	0.7353	0.7277	0.7209	0.7146
375	0.9001	0.8669	0.8406	0.8191	0.8011	0.7857	0.7724	0.7606	0.7503	0.7410	0.7326
400	1.0213	0.9581	0.9127	0.8781	0.8507	0.8283	0.8095	0.7934	0.7794	0.7672	0.7564
425	1.2427	1.1099	1.0246	0.9652	0.9212	0.8870	0.8595	0.8369	0.8178	0.8014	0.7872
450	1.5313	1.3286	1.1877	1.0902	1.0201	0.9677	0.9272	0.8948	0.8683	0.8462	0.8273
475	1.5845	1.4722	1.3423	1.2283	1.1374	1.0666	1.0111	0.9669	0.9312	0.9017	0.8769
500	1.4583	1.4414	1.3837	1.3071	1.2282	1.1564	1.0948	1.0431	0.9999	0.9637	0.9332
525	1.3174	1.3370	1.3293	1.2939	1.2529	1.2014	1.1499	1.1019	1.0588	1.0207	0.9875
550	1.2065	1.2281	1.2382	1.2348	1.2201	1.1964	1.1614	1.1263	1.0914	1.0583	1.0275
575	1.1204	1.1384	1.1521	1.1591	1.1575	1.1479	1.1331	1.1157	1.0937	1.0690	1.0450
600	1.0547	1.0668	1.0776	1.0862	1.0913	1.0920	1.0880	1.0799	1.0692	1.0563	1.0395
625	1.0041	1.0115	1.0184	1.0247	1.0297	1.0331	1.0341	1.0325	1.0285	1.0227	1.0165
650	0.9643	0.9681	0.9719	0.9756	0.9791	0.9822	0.9845	0.9856	0.9850	0.9826	0.9782
675	0.9326	0.9336	0.9347	0.9362	0.9380	0.9401	0.9423	0.9444	0.9461	0.9471	0.9470
700	0.9074	0.9061	0.9051	0.9046	0.9047	0.9054	0.9067	0.9085	0.9108	0.9134	0.9161
725	0.8876	0.8846	0.8820	0.8799	0.8784	0.8777	0.8778	0.8786	0.8802	0.8826	0.8858
750	0.8723	0.8681	0.8643	0.8610	0.8585	0.8567	0.8557	0.8555	0.8561	0.8576	0.8598
775	0.8608	0.8557	0.8511	0.8472	0.8439	0.8414	0.8397	0.8388	0.8386	0.8391	0.8403
800	0.8523	0.8468	0.8418	0.8374	0.8338	0.8310	0.8289	0.8276	0.8269	0.8267	0.8268

9 Static Dielectric Constant
Statische Dielektrizitätskonstante

For the single-phase region, this table contains values of the

- Static dielectric constant ε

for temperatures from 0 °C to 800 °C and for pressures from the triple-point pressure p_t = 611.657 Pa to 100 MPa (regions 1 to 3 of IAPWS-IF97). The static dielectric constant was calculated from Eq. (4.9). For given pressure and temperature, the necessary density was determined from the industrial standard IAPWS-IF97. The line in the respective columns for an isobar indicates the transition from liquid to vapour state.

Diese Tafel enthält für das homogene Zustandsgebiet Werte für die

- statische Dielektrizitätskonstante ε

für Temperaturen von 0 °C bis 800 °C und für Drücke vom Tripelpunktsdruck p_t = 611,657 Pa bis 100 MPa (Bereiche 1 bis 3 der IAPWS-IF97). Die statische Dielektrizitätskonstante wurde aus Gl. (4.9) berechnet. Für vorgegebene Werte von Druck und Temperatur wurde die benötigte Dichte mit dem Industrie-Standard IAPWS-IF97 ermittelt. Die Linie in den jeweiligen Spalten für eine Isobare kennzeichnet den Übergang vom Flüssigkeits- zum Gaszustand.

Table 9 Static Dielectric Constant ε
Statische Dielektrizitätskonstante ε

t						p [MPa]					
[°C]	p_t	0.001	0.01	0.05	0.1	0.2	0.3	0.4	0.5	0.6	0.7
0	87.8993	87.8994	87.8997	87.9014	87.9035	87.9077	87.9119	87.9160	87.9202	87.9244	87.9286
5	1.0001	85.9115	85.9118	85.9135	85.9155	85.9195	85.9236	85.9277	85.9317	85.9358	85.9399
10	1.0001	1.0001	83.9708	83.9724	83.9744	83.9784	83.9823	83.9863	83.9902	83.9942	83.9982
15	1.0001	1.0001	82.0739	82.0754	82.0774	82.0812	82.0851	82.0890	82.0929	82.0967	82.1006
20	1.0001	1.0001	80.2191	80.2206	80.2225	80.2263	80.2301	80.2339	80.2377	80.2415	80.2453
25	1.0001	1.0001	78.4051	78.4066	78.4085	78.4122	78.4159	78.4197	78.4234	78.4271	78.4308
30	1.0001	1.0001	76.6309	76.6323	76.6342	76.6379	76.6415	76.6452	76.6489	76.6526	76.6562
35	1.0001	1.0001	74.8954	74.8969	74.8987	74.9023	74.9060	74.9096	74.9132	74.9169	74.9205
40	1.0000	1.0001	73.1980	73.1994	73.2012	73.2048	73.2084	73.2120	73.2156	73.2192	73.2227
45	1.0000	1.0001	71.5377	71.5391	71.5409	71.5445	71.5480	71.5516	71.5551	71.5587	71.5622
50	1.0000	1.0001	1.0008	69.9153	69.9170	69.9205	69.9241	69.9276	69.9311	69.9347	69.9382
60	1.0000	1.0001	1.0007	66.7739	66.7756	66.7791	66.7826	66.7861	66.7896	66.7931	66.7966
70	1.0000	1.0001	1.0007	63.7697	63.7714	63.7749	63.7784	63.7819	63.7853	63.7888	63.7923
80	1.0000	1.0001	1.0006	60.8972	60.8989	60.9024	60.9058	60.9093	60.9128	60.9162	60.9197
90	1.0000	1.0001	1.0006	1.0030	58.1526	58.1561	58.1595	58.1630	58.1665	58.1700	58.1734
100	1.0000	1.0001	1.0006	1.0029	1.0058	55.5305	55.5340	55.5375	55.5410	55.5445	55.5480
125	1.0000	1.0001	1.0005	1.0025	1.0051	1.0103	49.4625	49.4661	49.4697	49.4733	49.4769
150	1.0000	1.0000	1.0004	1.0022	1.0045	1.0091	1.0139	1.0187	44.0313	44.0351	44.0389
175	1.0000	1.0000	1.0004	1.0020	1.0040	1.0081	1.0123	1.0166	1.0209	1.0253	1.0299
200	1.0000	1.0000	1.0004	1.0018	1.0036	1.0073	1.0110	1.0148	1.0187	1.0226	1.0266
225	1.0000	1.0000	1.0003	1.0016	1.0033	1.0066	1.0100	1.0134	1.0168	1.0203	1.0238
250	1.0000	1.0000	1.0003	1.0015	1.0030	1.0060	1.0091	1.0121	1.0152	1.0184	1.0216
275	1.0000	1.0000	1.0003	1.0014	1.0027	1.0055	1.0083	1.0111	1.0139	1.0167	1.0196
300	1.0000	1.0000	1.0002	1.0013	1.0025	1.0050	1.0076	1.0101	1.0127	1.0153	1.0179
325	1.0000	1.0000	1.0002	1.0012	1.0023	1.0046	1.0070	1.0093	1.0117	1.0141	1.0165
350	1.0000	1.0000	1.0002	1.0011	1.0021	1.0043	1.0064	1.0086	1.0108	1.0130	1.0152
375	1.0000	1.0000	1.0002	1.0010	1.0020	1.0040	1.0060	1.0080	1.0100	1.0120	1.0141
400	1.0000	1.0000	1.0002	1.0009	1.0018	1.0037	1.0056	1.0074	1.0093	1.0112	1.0131
425	1.0000	1.0000	1.0002	1.0009	1.0017	1.0034	1.0052	1.0069	1.0087	1.0104	1.0122
450	1.0000	1.0000	1.0002	1.0008	1.0016	1.0032	1.0048	1.0065	1.0081	1.0097	1.0114
475	1.0000	1.0000	1.0002	1.0008	1.0015	1.0030	1.0045	1.0061	1.0076	1.0091	1.0107
500	1.0000	1.0000	1.0001	1.0007	1.0014	1.0028	1.0043	1.0057	1.0071	1.0086	1.0100
525	1.0000	1.0000	1.0001	1.0007	1.0013	1.0027	1.0040	1.0054	1.0067	1.0081	1.0094
550	1.0000	1.0000	1.0001	1.0006	1.0013	1.0025	1.0038	1.0051	1.0063	1.0076	1.0089
575	1.0000	1.0000	1.0001	1.0006	1.0012	1.0024	1.0036	1.0048	1.0060	1.0072	1.0084
600	1.0000	1.0000	1.0001	1.0006	1.0011	1.0023	1.0034	1.0045	1.0057	1.0068	1.0080
625	1.0000	1.0000	1.0001	1.0005	1.0011	1.0021	1.0032	1.0043	1.0054	1.0065	1.0076
650	1.0000	1.0000	1.0001	1.0005	1.0010	1.0020	1.0031	1.0041	1.0051	1.0061	1.0072
675	1.0000	1.0000	1.0001	1.0005	1.0010	1.0019	1.0029	1.0039	1.0049	1.0058	1.0068
700	1.0000	1.0000	1.0001	1.0005	1.0009	1.0018	1.0028	1.0037	1.0046	1.0056	1.0065
725	1.0000	1.0000	1.0001	1.0004	1.0009	1.0018	1.0026	1.0035	1.0044	1.0053	1.0062
750	1.0000	1.0000	1.0001	1.0004	1.0008	1.0017	1.0025	1.0034	1.0042	1.0051	1.0059
775	1.0000	1.0000	1.0001	1.0004	1.0008	1.0016	1.0024	1.0032	1.0040	1.0048	1.0057
800	1.0000	1.0000	1.0001	1.0004	1.0008	1.0015	1.0023	1.0031	1.0039	1.0046	1.0054

Table 9 Static Dielectric Constant ε – Continuation
Statische Dielektrizitätskonstante ε – Fortsetzung

t [°C]	p [MPa] 0.8	0.9	1	1.5	2	2.5	5	7.5	10	12.5	15
0	87.9328	87.9369	87.9411	87.9620	87.9829	88.0037	88.1078	88.2116	88.3150	88.4181	88.5209
5	85.9439	85.9480	85.9520	85.9723	85.9926	86.0129	86.1140	86.2148	86.3153	86.4155	86.5154
10	84.0021	84.0061	84.0100	84.0298	84.0496	84.0693	84.1679	84.2662	84.3642	84.4618	84.5591
15	82.1045	82.1083	82.1122	82.1315	82.1509	82.1702	82.2666	82.3626	82.4584	82.5538	82.6489
20	80.2491	80.2529	80.2567	80.2756	80.2946	80.3135	80.4080	80.5021	80.5960	80.6895	80.7827
25	78.4346	78.4383	78.4420	78.4607	78.4793	78.4979	78.5907	78.6832	78.7754	78.8673	78.9589
30	76.6599	76.6636	76.6672	76.6856	76.7039	76.7223	76.8137	76.9048	76.9956	77.0861	77.1762
35	74.9241	74.9277	74.9314	74.9495	74.9676	74.9857	75.0759	75.1659	75.2554	75.3447	75.4337
40	73.2263	73.2299	73.2335	73.2514	73.2693	73.2872	73.3765	73.4654	73.5540	73.6423	73.7302
45	71.5658	71.5693	71.5729	71.5907	71.6084	71.6261	71.7146	71.8027	71.8904	71.9778	72.0649
50	69.9417	69.9453	69.9488	69.9664	69.9840	70.0016	70.0894	70.1768	70.2638	70.3505	70.4369
60	66.8001	66.8036	66.8071	66.8245	66.8419	66.8593	66.9461	67.0326	67.1186	67.2042	67.2895
70	63.7957	63.7992	63.8027	63.8200	63.8373	63.8547	63.9409	64.0268	64.1123	64.1973	64.2820
80	60.9232	60.9266	60.9301	60.9474	60.9647	60.9820	61.0681	61.1538	61.2391	61.3239	61.4082
90	58.1769	58.1804	58.1839	58.2012	58.2186	58.2359	58.3222	58.4080	58.4934	58.5782	58.6626
100	55.5515	55.5550	55.5585	55.5760	55.5934	55.6108	55.6976	55.7838	55.8695	55.9547	56.0394
125	49.4805	49.4841	49.4877	49.5056	49.5235	49.5414	49.6306	49.7190	49.8068	49.8939	49.9804
150	44.0426	44.0464	44.0502	44.0690	44.0878	44.1066	44.1998	44.2922	44.3837	44.4744	44.5643
175	1.0345	39.1529	39.1570	39.1771	39.1972	39.2173	39.3170	39.4155	39.5128	39.6091	39.7043
200	1.0306	1.0347	1.0389	1.0610	34.7620	34.7841	34.8934	35.0011	35.1071	35.2117	35.3147
225	1.0274	1.0310	1.0347	1.0539	1.0747	1.0975	30.8394	30.9609	31.0800	31.1969	31.3117
250	1.0248	1.0280	1.0313	1.0482	1.0663	1.0855	27.0591	27.2024	27.3418	27.4778	27.6104
275	1.0225	1.0254	1.0284	1.0436	1.0595	1.0763	1.1788	23.6157	23.7895	23.9568	24.1183
300	1.0206	1.0232	1.0259	1.0396	1.0539	1.0688	1.1556	1.2763	20.2708	20.4988	20.7137
325	1.0189	1.0213	1.0238	1.0363	1.0492	1.0626	1.1382	1.2348	1.3705	16.8184	17.1654
350	1.0174	1.0197	1.0219	1.0333	1.0451	1.0573	1.1244	1.2060	1.3099	1.4534	1.6860
375	1.0161	1.0182	1.0203	1.0308	1.0416	1.0527	1.1131	1.1840	1.2696	1.3772	1.5207
400	1.0150	1.0169	1.0188	1.0285	1.0385	1.0487	1.1036	1.1664	1.2397	1.3271	1.4347
425	1.0140	1.0157	1.0175	1.0265	1.0358	1.0452	1.0954	1.1518	1.2160	1.2902	1.3774
450	1.0130	1.0147	1.0164	1.0248	1.0333	1.0421	1.0883	1.1394	1.1967	1.2612	1.3349
475	1.0122	1.0138	1.0153	1.0232	1.0312	1.0393	1.0820	1.1288	1.1804	1.2376	1.3016
500	1.0115	1.0129	1.0144	1.0217	1.0292	1.0368	1.0765	1.1196	1.1665	1.2179	1.2745
525	1.0108	1.0122	1.0135	1.0204	1.0274	1.0345	1.0716	1.1114	1.1545	1.2011	1.2518
550	1.0102	1.0115	1.0128	1.0193	1.0258	1.0325	1.0672	1.1042	1.1439	1.1866	1.2325
575	1.0096	1.0108	1.0121	1.0182	1.0244	1.0307	1.0632	1.0978	1.1345	1.1738	1.2158
600	1.0091	1.0103	1.0114	1.0172	1.0231	1.0290	1.0596	1.0919	1.1262	1.1626	1.2012
625	1.0086	1.0097	1.0108	1.0163	1.0219	1.0275	1.0563	1.0867	1.1187	1.1525	1.1882
650	1.0082	1.0092	1.0103	1.0155	1.0207	1.0260	1.0533	1.0819	1.1120	1.1435	1.1767
675	1.0078	1.0088	1.0098	1.0147	1.0197	1.0247	1.0506	1.0776	1.1058	1.1354	1.1664
700	1.0074	1.0084	1.0093	1.0140	1.0188	1.0235	1.0481	1.0736	1.1003	1.1281	1.1571
725	1.0071	1.0080	1.0089	1.0134	1.0179	1.0224	1.0457	1.0700	1.0952	1.1214	1.1486
750	1.0068	1.0076	1.0085	1.0128	1.0171	1.0214	1.0436	1.0666	1.0905	1.1152	1.1409
775	1.0065	1.0073	1.0081	1.0122	1.0163	1.0205	1.0416	1.0635	1.0862	1.1096	1.1339
800	1.0062	1.0070	1.0078	1.0117	1.0156	1.0196	1.0398	1.0606	1.0822	1.1044	1.1274

Table 9 Static Dielectric Constant ε – Continuation
Statische Dielektrizitätskonstante ε – Fortsetzung

t [°C]	p [MPa] 17.5	20	22.5	25	27.5	30	32.5	35	37.5	40	45
0	88.6234	88.7255	88.8273	88.9288	89.0300	89.1309	89.2315	89.3317	89.4317	89.5314	89.7298
5	86.6149	86.7142	86.8131	86.9117	87.0101	87.1081	87.2058	87.3032	87.4003	87.4971	87.6898
10	84.6561	84.7528	84.8493	84.9454	85.0411	85.1366	85.2318	85.3267	85.4213	85.5156	85.7034
15	82.7437	82.8382	82.9324	83.0263	83.1199	83.2132	83.3062	83.3990	83.4914	83.5835	83.7669
20	80.8756	80.9682	81.0605	81.1525	81.2442	81.3356	81.4267	81.5174	81.6079	81.6982	81.8777
25	79.0501	79.1411	79.2317	79.3221	79.4121	79.5018	79.5913	79.6804	79.7692	79.8578	80.0340
30	77.2661	77.3556	77.4448	77.5337	77.6223	77.7106	77.7986	77.8863	77.9737	78.0607	78.2340
35	75.5223	75.6106	75.6986	75.7863	75.8736	75.9607	76.0474	76.1338	76.2200	76.3058	76.4765
40	73.8178	73.9050	73.9920	74.0786	74.1649	74.2509	74.3365	74.4219	74.5069	74.5917	74.7602
45	72.1516	72.2380	72.3240	72.4097	72.4951	72.5802	72.6650	72.7494	72.8335	72.9173	73.0839
50	70.5228	70.6085	70.6938	70.7788	70.8634	70.9477	71.0317	71.1153	71.1986	71.2816	71.4466
60	67.3744	67.4590	67.5432	67.6270	67.7105	67.7936	67.8764	67.9588	68.0409	68.1226	68.2851
70	64.3662	64.4501	64.5336	64.6167	64.6994	64.7817	64.8637	64.9453	65.0265	65.1074	65.2681
80	61.4922	61.5757	61.6588	61.7415	61.8238	61.9057	61.9872	62.0683	62.1491	62.2294	62.3889
90	58.7466	58.8300	58.9131	58.9957	59.0779	59.1596	59.2409	59.3218	59.4023	59.4824	59.6413
100	56.1236	56.2073	56.2906	56.3733	56.4556	56.5374	56.6188	56.6997	56.7802	56.8603	57.0191
125	50.0663	50.1516	50.2364	50.3205	50.4041	50.4871	50.5695	50.6515	50.7329	50.8138	50.9740
150	44.6535	44.7418	44.8295	44.9164	45.0026	45.0881	45.1730	45.2572	45.3407	45.4237	45.5877
175	39.7984	39.8916	39.9839	40.0752	40.1656	40.2551	40.3438	40.4317	40.5187	40.6050	40.7753
200	35.4164	35.5168	35.6159	35.7137	35.8103	35.9058	36.0001	36.0934	36.1857	36.2769	36.4565
225	31.4245	31.5354	31.6445	31.7518	31.8575	31.9615	32.0641	32.1652	32.2649	32.3632	32.5561
250	27.7400	27.8666	27.9905	28.1118	28.2307	28.3472	28.4616	28.5739	28.6842	28.7927	29.0042
275	24.2744	24.4257	24.5724	24.7150	24.8537	24.9888	25.1205	25.2491	25.3747	25.4976	25.7357
300	20.9172	21.1109	21.2959	21.4732	21.6436	21.8077	21.9661	22.1193	22.2678	22.4118	22.6880
325	17.4770	17.7606	18.0217	18.2644	18.4919	18.7065	18.9098	19.1032	19.2880	19.4649	19.7985
350	13.3430	13.9564	14.4406	14.8500	15.2077	15.5273	15.8179	16.0859	16.3351	16.5687	16.9964
375	1.7316	2.1126	7.3679	10.2310	11.2131	11.8884	12.4194	12.8640	13.2500	13.5934	14.1887
400	1.5724	1.7589	2.0339	2.5035	3.5582	5.9302	7.7991	8.9146	9.7000	10.3126	11.2543
425	1.4820	1.6108	1.7744	1.9904	2.2899	2.7299	3.4028	4.3712	5.4980	6.5334	8.0894
450	1.4201	1.5199	1.6386	1.7823	1.9595	2.1820	2.4667	2.8348	3.3060	3.8823	5.2152
475	1.3737	1.4557	1.5496	1.6582	1.7850	1.9343	2.1114	2.3228	2.5753	2.8759	3.6278
500	1.3372	1.4069	1.4848	1.5725	1.6717	1.7842	1.9125	2.0592	2.2271	2.4190	2.8843
525	1.3072	1.3680	1.4347	1.5084	1.5900	1.6804	1.7810	1.8930	2.0178	2.1567	2.4819
550	1.2822	1.3360	1.3944	1.4580	1.5273	1.6029	1.6856	1.7761	1.8750	1.9831	2.2299
575	1.2608	1.3091	1.3610	1.4169	1.4771	1.5421	1.6123	1.6880	1.7698	1.8581	2.0557
600	1.2423	1.2860	1.3327	1.3826	1.4358	1.4928	1.5536	1.6187	1.6883	1.7627	1.9268
625	1.2260	1.2660	1.3084	1.3534	1.4011	1.4517	1.5054	1.5624	1.6229	1.6870	1.8269
650	1.2117	1.2485	1.2873	1.3282	1.3713	1.4168	1.4648	1.5155	1.5689	1.6252	1.7468
675	1.1989	1.2329	1.2687	1.3061	1.3455	1.3868	1.4302	1.4757	1.5234	1.5735	1.6809
700	1.1874	1.2190	1.2521	1.2867	1.3229	1.3607	1.4001	1.4414	1.4845	1.5296	1.6255
725	1.1770	1.2066	1.2373	1.2694	1.3028	1.3376	1.3738	1.4116	1.4508	1.4917	1.5783
750	1.1676	1.1953	1.2240	1.2539	1.2849	1.3171	1.3506	1.3853	1.4213	1.4586	1.5374
775	1.1590	1.1850	1.2120	1.2399	1.2688	1.2988	1.3298	1.3619	1.3951	1.4295	1.5017
800	1.1512	1.1757	1.2011	1.2273	1.2543	1.2823	1.3112	1.3410	1.3718	1.4036	1.4702

Table 9 Static Dielectric Constant ε – Continuation
Statische Dielektrizitätskonstante ε – Fortsetzung

t [°C]	p [MPa] 50	55	60	65	70	75	80	85	90	95	100
0	89.9270	90.1230	90.3179	90.5116	90.7042	90.8957	91.0861	91.2755	91.4639	91.6512	91.8376
5	87.8813	88.0717	88.2610	88.4491	88.6362	88.8221	89.0070	89.1909	89.3737	89.5556	89.7364
10	85.8900	86.0754	86.2597	86.4429	86.6250	86.8061	86.9861	87.1651	87.3430	87.5200	87.6960
15	83.9491	84.1302	84.3101	84.4890	84.6668	84.8435	85.0192	85.1938	85.3675	85.5401	85.7118
20	82.0561	82.2334	82.4095	82.5846	82.7585	82.9314	83.1033	83.2741	83.4439	83.6127	83.7805
25	80.2090	80.3829	80.5557	80.7274	80.8980	81.0675	81.2360	81.4034	81.5698	81.7352	81.8996
30	78.4062	78.5771	78.7470	78.9157	79.0833	79.2498	79.4153	79.5798	79.7432	79.9056	80.0670
35	76.6461	76.8145	76.9818	77.1479	77.3129	77.4768	77.6396	77.8014	77.9622	78.1219	78.2806
40	74.9276	75.0937	75.2587	75.4226	75.5853	75.7469	75.9074	76.0669	76.2252	76.3826	76.5390
45	73.2494	73.4136	73.5766	73.7385	73.8992	74.0588	74.2172	74.3746	74.5309	74.6862	74.8405
50	71.6104	71.7730	71.9343	72.0944	72.2534	72.4112	72.5678	72.7234	72.8779	73.0313	73.1837
60	68.4462	68.6061	68.7646	68.9220	69.0781	69.2330	69.3867	69.5393	69.6908	69.8411	69.9904
70	65.4274	65.5853	65.7419	65.8972	66.0512	66.2039	66.3554	66.5057	66.6549	66.8029	66.9499
80	62.5470	62.7036	62.8589	63.0127	63.1652	63.3164	63.4663	63.6149	63.7623	63.9086	64.0537
90	59.7987	59.9545	60.1089	60.2618	60.4133	60.5634	60.7121	60.8596	61.0058	61.1507	61.2945
100	57.1762	57.3318	57.4857	57.6381	57.7890	57.9385	58.0865	58.2332	58.3785	58.5226	58.6654
125	51.1323	51.2886	51.4432	51.5959	51.7469	51.8962	52.0439	52.1901	52.3347	52.4779	52.6198
150	45.7493	45.9087	46.0658	46.2209	46.3739	46.5249	46.6741	46.8215	46.9671	47.1111	47.2535
175	40.9426	41.1072	41.2691	41.4284	41.5853	41.7399	41.8922	42.0425	42.1906	42.3369	42.4813
200	36.6323	36.8047	36.9738	37.1397	37.3026	37.4627	37.6201	37.7748	37.9272	38.0773	38.2252
225	32.7441	32.9275	33.1068	33.2820	33.4535	33.6215	33.7860	33.9474	34.1059	34.2615	34.4145
250	29.2092	29.4081	29.6014	29.7895	29.9727	30.1514	30.3259	30.4963	30.6631	30.8263	30.9863
275	25.9643	26.1845	26.3970	26.6025	26.8016	26.9947	27.1822	27.3647	27.5424	27.7157	27.8849
300	22.9501	23.1999	23.4386	23.6675	23.8876	24.0997	24.3044	24.5024	24.6943	24.8805	25.0615
325	20.1091	20.4005	20.6755	20.9361	21.1842	21.4211	21.6480	21.8659	22.0757	22.2782	22.4740
350	17.3819	17.7344	18.0604	18.3645	18.6501	18.9196	19.1751	19.4183	19.6506	19.8732	20.0872
375	14.6975	15.1452	15.5470	15.9128	16.2496	16.5624	16.8550	17.1303	17.3907	17.6380	17.8737
400	11.9807	12.5797	13.0937	13.5465	13.9529	14.3228	14.6631	14.9788	15.2740	15.5514	15.8136
425	9.1766	10.0057	10.6791	11.2492	11.7458	12.1874	12.5862	12.9506	13.2869	13.5998	13.8926
450	6.4760	7.5122	8.3525	9.0505	9.6462	10.1664	10.6288	11.0460	11.4266	11.7773	12.1030
475	4.5285	5.4566	6.3155	7.0727	7.7334	8.3134	8.8284	9.2909	9.7108	10.0957	10.4513
500	3.4609	4.1203	4.8153	5.4964	6.1344	6.7189	7.2504	7.7340	8.1759	8.5823	8.9583
525	2.8757	3.3360	3.8480	4.3866	4.9291	5.4556	5.9555	6.4242	6.8613	7.2687	7.6491
550	2.5209	2.8575	3.2365	3.6487	4.0811	4.5211	4.9569	5.3803	5.7865	6.1733	6.5402
575	2.2835	2.5431	2.8340	3.1532	3.4948	3.8514	4.2152	4.5799	4.9401	5.2908	5.6301
600	2.1127	2.3215	2.5535	2.8075	3.0811	3.3705	3.6712	3.9783	4.2877	4.5960	4.8999
625	1.9832	2.1567	2.3477	2.5557	2.7797	3.0177	3.2671	3.5252	3.7887	4.0548	4.3211
650	1.8812	2.0289	2.1901	2.3647	2.5521	2.7513	2.9607	3.1788	3.4035	3.6328	3.8650
675	1.7984	1.9265	2.0653	2.2149	2.3747	2.5443	2.7227	2.9088	3.1015	3.2993	3.5012
700	1.7298	1.8425	1.9640	2.0941	2.2327	2.3793	2.5334	2.6942	2.8609	3.0327	3.2085
725	1.6717	1.7722	1.8798	1.9946	2.1164	2.2448	2.3797	2.5203	2.6661	2.8166	2.9709
750	1.6220	1.7124	1.8088	1.9111	2.0193	2.1332	2.2525	2.3768	2.5057	2.6387	2.7754
775	1.5788	1.6609	1.7479	1.8400	1.9371	2.0390	2.1455	2.2564	2.3714	2.4900	2.6122
800	1.5409	1.6159	1.6952	1.7788	1.8666	1.9585	2.0544	2.1541	2.2574	2.3641	2.4739

10 Refractive Index (Including Saturation State)

Brechungsindex (einschließlich des Sättigungszustandes)

For the single-phase region and for the saturation state, this table contains values of the

- Refractive index n.

The single-phase region covers temperatures from 0 °C to 500 °C and pressures from the triple-point pressure $p_t = 611.657$ Pa to 100 MPa. The refractive index was calculated from Eq. (4.11). For given pressure and temperature, the necessary density was determined from the industrial standard IAPWS-IF97. The line in the respective columns for an isobar indicates the transition from liquid to vapour state. Values of the refractive index for the saturated liquid (′) and saturated vapour (″) are listed at the end of this table. For the calculation of the saturation properties please refer to the introduction of Table 1, p. 103.

Diese Tafel enthält für das homogene Zustandsgebiet und für den Sättigungszustand Werte für den

- Brechungsindex n.

Das homogene Zustandsgebiet umfaßt den Temperaturbereich von 0 °C bis 500 °C bei Drücken vom Tripelpunktsdruck $p_t = 611{,}657$ Pa bis 100 MPa (Bereiche 1 bis 3 der IAPWS-IF97). Der Brechungsindex wurde aus Gl. (4.11) berechnet. Für vorgegebene Werte von Druck und Temperatur wurde die benötigte Dichte mit dem Industrie-Standard IAPWS-IF97 ermittelt. Die Linie in den jeweiligen Spalten für eine Isobare kennzeichnet den Übergang vom Flüssigkeits- zum Gaszustand. Werte für den Brechungsindex im Zustand der siedenden Flüssigkeit (′) und des gesättigten Dampfes (″) befinden sich am Ende dieser Tafel. Für die Ermittlung der Sättigungsgrößen gelten die in der Einführung zu Tafel 1, S. 103, gemachten Ausführungen.

Table 10 Refractive Index *n* (Including Saturation State)
Brechungsindex *n* (einschließlich des Sättigungszustandes)

t [°C]	p_t	p [MPa] 0.001	0.01	0.1	0.2	0.5	1	2.5	5	7.5	10
						λ = 0.2265 μm					
0	1.3945	1.3945	1.3945	1.3945	1.3945	1.3946	1.3947	1.3950	1.3955	1.3960	1.3965
25	1.0000	1.0000	1.3928	1.3928	1.3928	1.3929	1.3929	1.3932	1.3937	1.3941	1.3945
50	1.0000	1.0000	1.0000	1.3885	1.3886	1.3886	1.3887	1.3890	1.3894	1.3898	1.3903
75	1.0000	1.0000	1.0000	1.3827	1.3827	1.3827	1.3828	1.3831	1.3835	1.3840	1.3844
100	1.0000	1.0000	1.0000	1.0002	1.3755	1.3755	1.3756	1.3759	1.3764	1.3768	1.3773
125	1.0000	1.0000	1.0000	1.0002	1.0004	1.3672	1.3673	1.3676	1.3681	1.3686	1.3691
150	1.0000	1.0000	1.0000	1.0002	1.0004	1.3578	1.3579	1.3582	1.3588	1.3593	1.3599
175	1.0000	1.0000	1.0000	1.0002	1.0004	1.0009	1.3474	1.3478	1.3484	1.3490	1.3497
200	1.0000	1.0000	1.0000	1.0002	1.0003	1.0009	1.0018	1.3361	1.3368	1.3376	1.3383
225	1.0000	1.0000	1.0000	1.0002	1.0003	1.0008	1.0017	1.0046	1.3239	1.3248	1.3256
250	1.0000	1.0000	1.0000	1.0002	1.0003	1.0008	1.0016	1.0042	1.3091	1.3102	1.3113
275	1.0000	1.0000	1.0000	1.0001	1.0003	1.0007	1.0015	1.0039	1.0088	1.2932	1.2947
300	1.0000	1.0000	1.0000	1.0001	1.0003	1.0007	1.0014	1.0037	1.0080	1.0136	1.2745
325	1.0000	1.0000	1.0000	1.0001	1.0003	1.0007	1.0013	1.0035	1.0075	1.0122	1.0184
350	1.0000	1.0000	1.0000	1.0001	1.0003	1.0006	1.0013	1.0033	1.0070	1.0112	1.0162
375	1.0000	1.0000	1.0000	1.0001	1.0002	1.0006	1.0012	1.0032	1.0066	1.0104	1.0148
400	1.0000	1.0000	1.0000	1.0001	1.0002	1.0006	1.0012	1.0030	1.0063	1.0098	1.0137
425	1.0000	1.0000	1.0000	1.0001	1.0002	1.0006	1.0011	1.0029	1.0060	1.0093	1.0129
450	1.0000	1.0000	1.0000	1.0001	1.0002	1.0005	1.0011	1.0028	1.0057	1.0088	1.0121
475	1.0000	1.0000	1.0000	1.0001	1.0002	1.0005	1.0011	1.0027	1.0055	1.0084	1.0115
500	1.0000	1.0000	1.0000	1.0001	1.0002	1.0005	1.0010	1.0026	1.0052	1.0081	1.0110
						λ = 0.589 μm					
0	1.3343	1.3343	1.3343	1.3343	1.3344	1.3344	1.3345	1.3347	1.3352	1.3356	1.3360
25	1.0000	1.0000	1.3329	1.3329	1.3329	1.3329	1.3330	1.3332	1.3336	1.3339	1.3343
50	1.0000	1.0000	1.0000	1.3294	1.3294	1.3294	1.3295	1.3297	1.3301	1.3304	1.3308
75	1.0000	1.0000	1.0000	1.3245	1.3245	1.3246	1.3246	1.3249	1.3252	1.3256	1.3259
100	1.0000	1.0000	1.0000	1.0002	1.3186	1.3186	1.3187	1.3190	1.3193	1.3197	1.3201
125	1.0000	1.0000	1.0000	1.0002	1.0004	1.3118	1.3119	1.3121	1.3125	1.3129	1.3133
150	1.0000	1.0000	1.0000	1.0002	1.0003	1.3040	1.3041	1.3044	1.3048	1.3053	1.3058
175	1.0000	1.0000	1.0000	1.0002	1.0003	1.0008	1.2954	1.2957	1.2963	1.2968	1.2973
200	1.0000	1.0000	1.0000	1.0001	1.0003	1.0007	1.0015	1.2861	1.2867	1.2873	1.2879
225	1.0000	1.0000	1.0000	1.0001	1.0003	1.0007	1.0014	1.0039	1.2759	1.2767	1.2774
250	1.0000	1.0000	1.0000	1.0001	1.0003	1.0007	1.0014	1.0036	1.2636	1.2645	1.2655
275	1.0000	1.0000	1.0000	1.0001	1.0003	1.0006	1.0013	1.0034	1.0076	1.2504	1.2516
300	1.0000	1.0000	1.0000	1.0001	1.0002	1.0006	1.0012	1.0032	1.0069	1.0118	1.2347
325	1.0000	1.0000	1.0000	1.0001	1.0002	1.0006	1.0012	1.0030	1.0064	1.0106	1.0159
350	1.0000	1.0000	1.0000	1.0001	1.0002	1.0005	1.0011	1.0029	1.0060	1.0097	1.0140
375	1.0000	1.0000	1.0000	1.0001	1.0002	1.0005	1.0011	1.0027	1.0057	1.0090	1.0128
400	1.0000	1.0000	1.0000	1.0001	1.0002	1.0005	1.0010	1.0026	1.0054	1.0085	1.0119
425	1.0000	1.0000	1.0000	1.0001	1.0002	1.0005	1.0010	1.0025	1.0052	1.0080	1.0111
450	1.0000	1.0000	1.0000	1.0001	1.0002	1.0005	1.0009	1.0024	1.0049	1.0076	1.0105
475	1.0000	1.0000	1.0000	1.0001	1.0002	1.0005	1.0009	1.0023	1.0047	1.0073	1.0100
500	1.0000	1.0000	1.0000	1.0001	1.0002	1.0004	1.0009	1.0022	1.0045	1.0070	1.0095

Table 10 Refractive Index *n* (Including Saturation State) – Continuation
Brechungsindex *n* (einschließlich des Sättigungszustandes) – Fortsetzung

t [°C]	*p* [MPa] 15	20	25	30	40	50	60	70	80	90	100
						λ = 0.2265 μm					
0	1.3975	1.3985	1.3995	1.4004	1.4023	1.4041	1.4059	1.4077	1.4094	1.4111	1.4127
25	1.3954	1.3963	1.3972	1.3980	1.3997	1.4013	1.4030	1.4045	1.4061	1.4076	1.4091
50	1.3911	1.3919	1.3928	1.3936	1.3952	1.3968	1.3984	1.3999	1.4014	1.4029	1.4043
75	1.3853	1.3861	1.3869	1.3878	1.3894	1.3910	1.3926	1.3941	1.3956	1.3971	1.3985
100	1.3782	1.3791	1.3800	1.3808	1.3825	1.3842	1.3858	1.3874	1.3890	1.3905	1.3920
125	1.3701	1.3710	1.3720	1.3729	1.3747	1.3765	1.3783	1.3799	1.3816	1.3832	1.3847
150	1.3610	1.3620	1.3631	1.3641	1.3661	1.3680	1.3699	1.3717	1.3735	1.3752	1.3769
175	1.3509	1.3521	1.3533	1.3544	1.3566	1.3588	1.3608	1.3628	1.3647	1.3666	1.3684
200	1.3397	1.3411	1.3424	1.3438	1.3463	1.3487	1.3510	1.3532	1.3553	1.3574	1.3593
225	1.3273	1.3290	1.3306	1.3321	1.3350	1.3378	1.3404	1.3429	1.3452	1.3475	1.3497
250	1.3134	1.3154	1.3174	1.3192	1.3227	1.3259	1.3289	1.3318	1.3345	1.3371	1.3395
275	1.2975	1.3001	1.3025	1.3048	1.3091	1.3130	1.3166	1.3199	1.3230	1.3259	1.3287
300	1.2786	1.2822	1.2855	1.2885	1.2939	1.2988	1.3031	1.3070	1.3107	1.3141	1.3173
325	1.2541	1.2601	1.2650	1.2694	1.2767	1.2829	1.2883	1.2931	1.2975	1.3015	1.3051
350	1.0319	1.2285	1.2382	1.2455	1.2565	1.2650	1.2719	1.2779	1.2832	1.2879	1.2923
375	1.0262	1.0478	1.1909	1.2114	1.2315	1.2440	1.2535	1.2612	1.2677	1.2734	1.2786
400	1.0232	1.0367	1.0612	1.1334	1.1976	1.2188	1.2324	1.2426	1.2509	1.2579	1.2640
425	1.0212	1.0317	1.0463	1.0693	1.1473	1.1874	1.2079	1.2219	1.2325	1.2412	1.2486
450	1.0196	1.0286	1.0397	1.0543	1.1001	1.1501	1.1800	1.1990	1.2127	1.2235	1.2324
475	1.0184	1.0263	1.0355	1.0467	1.0771	1.1167	1.1510	1.1747	1.1918	1.2049	1.2155
500	1.0173	1.0245	1.0326	1.0418	1.0650	1.0946	1.1255	1.1512	1.1708	1.1860	1.1983
						λ = 0.589 μm					
0	1.3368	1.3376	1.3384	1.3391	1.3407	1.3422	1.3436	1.3450	1.3464	1.3478	1.3491
25	1.3350	1.3357	1.3364	1.3371	1.3385	1.3398	1.3411	1.3424	1.3437	1.3449	1.3461
50	1.3315	1.3321	1.3328	1.3335	1.3348	1.3361	1.3373	1.3386	1.3398	1.3410	1.3422
75	1.3266	1.3273	1.3280	1.3287	1.3300	1.3313	1.3326	1.3338	1.3351	1.3363	1.3374
100	1.3208	1.3216	1.3223	1.3230	1.3244	1.3257	1.3271	1.3283	1.3296	1.3308	1.3321
125	1.3141	1.3149	1.3157	1.3165	1.3180	1.3194	1.3208	1.3222	1.3235	1.3248	1.3261
150	1.3066	1.3075	1.3084	1.3092	1.3108	1.3124	1.3139	1.3154	1.3169	1.3183	1.3196
175	1.2983	1.2993	1.3002	1.3012	1.3030	1.3048	1.3065	1.3081	1.3097	1.3112	1.3127
200	1.2891	1.2902	1.2913	1.2924	1.2945	1.2964	1.2983	1.3001	1.3019	1.3036	1.3052
225	1.2788	1.2801	1.2815	1.2827	1.2851	1.2874	1.2896	1.2916	1.2936	1.2954	1.2972
250	1.2672	1.2689	1.2705	1.2720	1.2749	1.2776	1.2801	1.2824	1.2847	1.2868	1.2888
275	1.2539	1.2561	1.2581	1.2600	1.2636	1.2668	1.2698	1.2725	1.2751	1.2776	1.2798
300	1.2381	1.2411	1.2439	1.2464	1.2509	1.2549	1.2586	1.2618	1.2649	1.2677	1.2703
325	1.2176	1.2226	1.2267	1.2304	1.2365	1.2417	1.2462	1.2502	1.2538	1.2572	1.2602
350	1.0275	1.1959	1.2042	1.2103	1.2195	1.2266	1.2325	1.2375	1.2419	1.2459	1.2495
375	1.0227	1.0413	1.1641	1.1815	1.1985	1.2090	1.2170	1.2234	1.2289	1.2337	1.2380
400	1.0201	1.0317	1.0528	1.1150	1.1697	1.1877	1.1992	1.2078	1.2148	1.2207	1.2258
425	1.0183	1.0274	1.0400	1.0599	1.1268	1.1610	1.1785	1.1903	1.1993	1.2066	1.2128
450	1.0169	1.0247	1.0343	1.0469	1.0863	1.1292	1.1547	1.1709	1.1825	1.1916	1.1991
475	1.0159	1.0227	1.0307	1.0403	1.0665	1.1006	1.1299	1.1502	1.1647	1.1758	1.1848
500	1.0150	1.0211	1.0281	1.0361	1.0561	1.0816	1.1081	1.1301	1.1468	1.1598	1.1702

Table 10 Refractive Index *n* (Including Saturation State) Brechungsindex *n* (einschließlich des Sättigungszustandes)

t [°C]	p_t	p [MPa] 0.001	0.01	0.1	0.2	0.5	1	2.5	5	7.5	10
						λ = 0.70652 μm					
0	1.3313	1.3313	1.3313	1.3313	1.3314	1.3314	1.3315	1.3317	1.3321	1.3325	1.3330
25	1.0000	1.0000	1.3299	1.3299	1.3299	1.3300	1.3300	1.3302	1.3306	1.3310	1.3313
50	1.0000	1.0000	1.0000	1.3265	1.3265	1.3265	1.3266	1.3268	1.3271	1.3275	1.3278
75	1.0000	1.0000	1.0000	1.3217	1.3217	1.3217	1.3218	1.3220	1.3224	1.3227	1.3231
100	1.0000	1.0000	1.0000	1.0002	1.3158	1.3159	1.3160	1.3162	1.3166	1.3169	1.3173
125	1.0000	1.0000	1.0000	1.0002	1.0004	1.3091	1.3092	1.3094	1.3098	1.3102	1.3106
150	1.0000	1.0000	1.0000	1.0002	1.0003	1.3014	1.3015	1.3018	1.3022	1.3027	1.3031
175	1.0000	1.0000	1.0000	1.0002	1.0003	1.0008	1.2929	1.2932	1.2938	1.2943	1.2948
200	1.0000	1.0000	1.0000	1.0001	1.0003	1.0007	1.0015	1.2837	1.2843	1.2849	1.2855
225	1.0000	1.0000	1.0000	1.0001	1.0003	1.0007	1.0014	1.0039	1.2736	1.2744	1.2751
250	1.0000	1.0000	1.0000	1.0001	1.0003	1.0007	1.0013	1.0036	1.2615	1.2624	1.2633
275	1.0000	1.0000	1.0000	1.0001	1.0002	1.0006	1.0013	1.0034	1.0076	1.2484	1.2496
300	1.0000	1.0000	1.0000	1.0001	1.0002	1.0006	1.0012	1.0032	1.0069	1.0117	1.2328
325	1.0000	1.0000	1.0000	1.0001	1.0002	1.0006	1.0012	1.0030	1.0064	1.0105	1.0158
350	1.0000	1.0000	1.0000	1.0001	1.0002	1.0005	1.0011	1.0028	1.0060	1.0096	1.0139
375	1.0000	1.0000	1.0000	1.0001	1.0002	1.0005	1.0011	1.0027	1.0057	1.0090	1.0127
400	1.0000	1.0000	1.0000	1.0001	1.0002	1.0005	1.0010	1.0026	1.0054	1.0084	1.0118
425	1.0000	1.0000	1.0000	1.0001	1.0002	1.0005	1.0010	1.0025	1.0051	1.0080	1.0110
450	1.0000	1.0000	1.0000	1.0001	1.0002	1.0005	1.0009	1.0024	1.0049	1.0076	1.0104
475	1.0000	1.0000	1.0000	1.0001	1.0002	1.0004	1.0009	1.0023	1.0047	1.0072	1.0099
500	1.0000	1.0000	1.0000	1.0001	1.0002	1.0004	1.0009	1.0022	1.0045	1.0069	1.0094
						λ = 1.01398 μm					
0	1.3261	1.3261	1.3261	1.3261	1.3262	1.3262	1.3263	1.3265	1.3269	1.3273	1.3277
25	1.0000	1.0000	1.3248	1.3248	1.3248	1.3248	1.3249	1.3251	1.3255	1.3258	1.3262
50	1.0000	1.0000	1.0000	1.3215	1.3215	1.3215	1.3216	1.3218	1.3221	1.3225	1.3228
75	1.0000	1.0000	1.0000	1.3168	1.3168	1.3169	1.3169	1.3171	1.3175	1.3178	1.3182
100	1.0000	1.0000	1.0000	1.0002	1.3111	1.3112	1.3113	1.3115	1.3118	1.3122	1.3126
125	1.0000	1.0000	1.0000	1.0002	1.0003	1.3046	1.3047	1.3049	1.3053	1.3057	1.3061
150	1.0000	1.0000	1.0000	1.0002	1.0003	1.2971	1.2972	1.2975	1.2979	1.2983	1.2988
175	1.0000	1.0000	1.0000	1.0002	1.0003	1.0008	1.2888	1.2891	1.2896	1.2901	1.2906
200	1.0000	1.0000	1.0000	1.0001	1.0003	1.0007	1.0015	1.2798	1.2804	1.2809	1.2815
225	1.0000	1.0000	1.0000	1.0001	1.0003	1.0007	1.0014	1.0039	1.2699	1.2706	1.2713
250	1.0000	1.0000	1.0000	1.0001	1.0003	1.0007	1.0013	1.0036	1.2580	1.2589	1.2598
275	1.0000	1.0000	1.0000	1.0001	1.0002	1.0006	1.0013	1.0033	1.0075	1.2451	1.2463
300	1.0000	1.0000	1.0000	1.0001	1.0002	1.0006	1.0012	1.0031	1.0068	1.0116	1.2299
325	1.0000	1.0000	1.0000	1.0001	1.0002	1.0006	1.0011	1.0030	1.0063	1.0104	1.0156
350	1.0000	1.0000	1.0000	1.0001	1.0002	1.0005	1.0011	1.0028	1.0059	1.0095	1.0138
375	1.0000	1.0000	1.0000	1.0001	1.0002	1.0005	1.0010	1.0027	1.0056	1.0089	1.0126
400	1.0000	1.0000	1.0000	1.0001	1.0002	1.0005	1.0010	1.0026	1.0053	1.0083	1.0116
425	1.0000	1.0000	1.0000	1.0001	1.0002	1.0005	1.0010	1.0024	1.0051	1.0079	1.0109
450	1.0000	1.0000	1.0000	1.0001	1.0002	1.0005	1.0009	1.0024	1.0048	1.0075	1.0103
475	1.0000	1.0000	1.0000	1.0001	1.0002	1.0004	1.0009	1.0023	1.0046	1.0071	1.0098
500	1.0000	1.0000	1.0000	1.0001	1.0002	1.0004	1.0009	1.0022	1.0045	1.0068	1.0093

Table 10 Refractive Index *n* (Including Saturation State) – Continuation
Brechungsindex *n* (einschließlich des Sättigungszustandes) – Fortsetzung

t [°C]	*p* [MPa] 15	20	25	30	40	50	60	70	80	90	100
						λ = 0.70652 μm					
0	1.3337	1.3345	1.3353	1.3361	1.3376	1.3391	1.3405	1.3419	1.3433	1.3446	1.3459
25	1.3320	1.3327	1.3334	1.3341	1.3355	1.3368	1.3381	1.3393	1.3406	1.3418	1.3430
50	1.3285	1.3292	1.3299	1.3305	1.3318	1.3331	1.3343	1.3356	1.3368	1.3379	1.3391
75	1.3238	1.3244	1.3251	1.3258	1.3271	1.3284	1.3297	1.3309	1.3321	1.3333	1.3344
100	1.3180	1.3187	1.3195	1.3202	1.3215	1.3229	1.3242	1.3255	1.3267	1.3279	1.3291
125	1.3114	1.3122	1.3130	1.3137	1.3152	1.3166	1.3180	1.3194	1.3207	1.3220	1.3233
150	1.3040	1.3049	1.3057	1.3066	1.3082	1.3097	1.3112	1.3127	1.3141	1.3155	1.3169
175	1.2958	1.2968	1.2977	1.2986	1.3004	1.3022	1.3038	1.3055	1.3070	1.3085	1.3100
200	1.2867	1.2878	1.2889	1.2899	1.2920	1.2939	1.2958	1.2976	1.2993	1.3010	1.3026
225	1.2765	1.2778	1.2791	1.2804	1.2828	1.2850	1.2871	1.2892	1.2911	1.2930	1.2948
250	1.2650	1.2667	1.2683	1.2698	1.2726	1.2753	1.2778	1.2801	1.2823	1.2844	1.2864
275	1.2519	1.2540	1.2560	1.2579	1.2615	1.2646	1.2676	1.2703	1.2729	1.2753	1.2776
300	1.2362	1.2392	1.2419	1.2444	1.2489	1.2529	1.2565	1.2597	1.2627	1.2655	1.2682
325	1.2159	1.2208	1.2249	1.2285	1.2347	1.2398	1.2443	1.2482	1.2518	1.2551	1.2581
350	1.0273	1.1944	1.2026	1.2087	1.2178	1.2249	1.2307	1.2356	1.2400	1.2439	1.2475
375	1.0225	1.0410	1.1628	1.1801	1.1969	1.2074	1.2153	1.2217	1.2271	1.2319	1.2361
400	1.0199	1.0315	1.0524	1.1141	1.1684	1.1863	1.1977	1.2062	1.2131	1.2190	1.2240
425	1.0182	1.0272	1.0397	1.0594	1.1258	1.1598	1.1771	1.1888	1.1978	1.2050	1.2112
450	1.0168	1.0245	1.0341	1.0465	1.0857	1.1282	1.1536	1.1696	1.1811	1.1902	1.1976
475	1.0158	1.0225	1.0305	1.0400	1.0660	1.0999	1.1290	1.1491	1.1635	1.1745	1.1834
500	1.0149	1.0210	1.0279	1.0359	1.0557	1.0810	1.1074	1.1291	1.1457	1.1586	1.1690
						λ = 1.01398 μm					
0	1.3285	1.3293	1.3300	1.3308	1.3323	1.3337	1.3351	1.3365	1.3378	1.3391	1.3404
25	1.3269	1.3275	1.3282	1.3289	1.3302	1.3315	1.3328	1.3340	1.3353	1.3364	1.3376
50	1.3235	1.3241	1.3248	1.3254	1.3267	1.3280	1.3292	1.3304	1.3316	1.3327	1.3339
75	1.3189	1.3195	1.3202	1.3209	1.3222	1.3234	1.3247	1.3259	1.3270	1.3282	1.3293
100	1.3133	1.3140	1.3147	1.3154	1.3167	1.3181	1.3193	1.3206	1.3218	1.3230	1.3242
125	1.3069	1.3076	1.3084	1.3091	1.3106	1.3120	1.3134	1.3147	1.3160	1.3173	1.3185
150	1.2996	1.3005	1.3013	1.3021	1.3037	1.3053	1.3067	1.3082	1.3096	1.3109	1.3123
175	1.2916	1.2926	1.2935	1.2944	1.2962	1.2979	1.2995	1.3011	1.3026	1.3041	1.3056
200	1.2827	1.2838	1.2849	1.2859	1.2879	1.2899	1.2917	1.2935	1.2952	1.2968	1.2984
225	1.2727	1.2740	1.2753	1.2765	1.2789	1.2811	1.2832	1.2852	1.2871	1.2889	1.2907
250	1.2615	1.2631	1.2647	1.2662	1.2690	1.2716	1.2740	1.2763	1.2785	1.2806	1.2825
275	1.2486	1.2507	1.2527	1.2546	1.2580	1.2612	1.2641	1.2667	1.2693	1.2716	1.2739
300	1.2332	1.2362	1.2388	1.2413	1.2457	1.2496	1.2532	1.2564	1.2593	1.2621	1.2647
325	1.2132	1.2181	1.2221	1.2257	1.2317	1.2368	1.2412	1.2451	1.2486	1.2519	1.2549
350	1.0270	1.1920	1.2001	1.2061	1.2152	1.2221	1.2278	1.2327	1.2370	1.2409	1.2444
375	1.0222	1.0405	1.1609	1.1779	1.1946	1.2049	1.2127	1.2190	1.2244	1.2291	1.2332
400	1.0197	1.0311	1.0518	1.1128	1.1664	1.1841	1.1953	1.2037	1.2106	1.2163	1.2214
425	1.0180	1.0269	1.0393	1.0588	1.1244	1.1580	1.1751	1.1866	1.1955	1.2026	1.2087
450	1.0166	1.0242	1.0337	1.0460	1.0847	1.1268	1.1519	1.1677	1.1790	1.1880	1.1954
475	1.0156	1.0223	1.0301	1.0396	1.0653	1.0988	1.1275	1.1474	1.1617	1.1725	1.1814
500	1.0147	1.0208	1.0276	1.0355	1.0551	1.0802	1.1062	1.1277	1.1442	1.1569	1.1671

Table 10 Refractive Index *n* (Including Saturation State) – Continuation
Brechungsindex *n* (einschließlich des Sättigungszustandes) – Fortsetzung

	Saturation							
	λ = 0.2265 μm		λ = 0.589 μm		λ = 0.70652 μm		λ = 1.01398 μm	
t [°C]	*n*′	*n*″	*n*′	*n*″	*n*′	*n*″	*n*′	*n*″
0	1.3945	1.0000	1.3343	1.0000	1.3313	1.0000	1.3261	1.0000
0.01	1.3945	1.0000	1.3343	1.0000	1.3313	1.0000	1.3261	1.0000
10	1.3942	1.0000	1.3341	1.0000	1.3311	1.0000	1.3259	1.0000
20	1.3933	1.0000	1.3333	1.0000	1.3304	1.0000	1.3252	1.0000
30	1.3921	1.0000	1.3323	1.0000	1.3293	1.0000	1.3242	1.0000
40	1.3905	1.0000	1.3309	1.0000	1.3280	1.0000	1.3230	1.0000
50	1.3885	1.0000	1.3294	1.0000	1.3264	1.0000	1.3214	1.0000
60	1.3864	1.0000	1.3276	1.0000	1.3247	1.0000	1.3197	1.0000
70	1.3839	1.0001	1.3256	1.0001	1.3227	1.0001	1.3178	1.0001
80	1.3813	1.0001	1.3234	1.0001	1.3206	1.0001	1.3157	1.0001
90	1.3785	1.0002	1.3211	1.0001	1.3183	1.0001	1.3135	1.0001
100	1.3755	1.0002	1.3186	1.0002	1.3158	1.0002	1.3111	1.0002
110	1.3723	1.0003	1.3160	1.0003	1.3132	1.0003	1.3086	1.0003
120	1.3689	1.0004	1.3132	1.0004	1.3105	1.0004	1.3059	1.0003
130	1.3654	1.0005	1.3103	1.0005	1.3076	1.0005	1.3031	1.0005
140	1.3616	1.0007	1.3072	1.0006	1.3046	1.0006	1.3002	1.0006
150	1.3578	1.0009	1.3040	1.0008	1.3014	1.0008	1.2971	1.0008
160	1.3537	1.0012	1.3007	1.0010	1.2981	1.0010	1.2939	1.0010
170	1.3495	1.0015	1.2972	1.0013	1.2947	1.0013	1.2905	1.0013
180	1.3451	1.0019	1.2936	1.0016	1.2911	1.0016	1.2870	1.0016
190	1.3406	1.0023	1.2898	1.0020	1.2873	1.0020	1.2833	1.0020
200	1.3358	1.0029	1.2858	1.0025	1.2834	1.0025	1.2795	1.0024
210	1.3308	1.0035	1.2817	1.0030	1.2794	1.0030	1.2755	1.0030
220	1.3256	1.0042	1.2774	1.0037	1.2751	1.0036	1.2714	1.0036
230	1.3202	1.0051	1.2729	1.0044	1.2706	1.0044	1.2670	1.0043
240	1.3145	1.0061	1.2682	1.0053	1.2660	1.0053	1.2624	1.0052
250	1.3086	1.0073	1.2632	1.0063	1.2611	1.0063	1.2576	1.0062
260	1.3023	1.0087	1.2580	1.0075	1.2559	1.0074	1.2525	1.0073
270	1.2957	1.0103	1.2525	1.0089	1.2504	1.0088	1.2472	1.0087
280	1.2887	1.0121	1.2466	1.0105	1.2446	1.0104	1.2415	1.0103
290	1.2813	1.0143	1.2404	1.0124	1.2384	1.0123	1.2354	1.0121
300	1.2732	1.0169	1.2336	1.0146	1.2318	1.0145	1.2288	1.0143
310	1.2646	1.0199	1.2264	1.0172	1.2246	1.0171	1.2217	1.0169
320	1.2551	1.0236	1.2184	1.0204	1.2167	1.0203	1.2140	1.0200
330	1.2446	1.0282	1.2095	1.0243	1.2079	1.0242	1.2053	1.0239
340	1.2325	1.0339	1.1994	1.0293	1.1978	1.0291	1.1954	1.0288
350	1.2183	1.0417	1.1873	1.0360	1.1858	1.0357	1.1836	1.0353
360	1.1998	1.0529	1.1716	1.0457	1.1703	1.0454	1.1683	1.0448
370	1.1694	1.0746	1.1457	1.0644	1.1446	1.0639	1.1429	1.0632
373.946	1.1200		1.1035		1.1027		1.1015	